Vertebrate Life
TENTH EDITION

Vertebrate Life

TENTH EDITION

JOHN SIBBICK

F. Harvey Pough
Rochester Institute of Technology,
Emeritus

Christine M. Janis
Brown University, Emerita,
and University of Bristol

Chapter 26 "Primate Evolution and the Emergence of Humans"
by Sergi López-Torres, University of Toronto Scarborough and
Roman Kozlowski Institute of Paleobiology, Polish Academy of Sciences

SINAUER ASSOCIATES
NEW YORK   OXFORD
OXFORD UNIVERSITY PRESS

About the Cover

Artist's reconstruction of two of the earliest mammals, *Morganucodon* (left) and Kueh*neotherium* (right). Both of these Early Jurassic mammals were insectivores about the size of extant shrews. (Painting by John Sibbick, 2013, © Pamela Gill.)

Vertebrate Life, 10th Edition

Oxford University Press is a department of the University of Oxford. It furthers the University's objective of excellence in research, scholarship, and education by publishing worldwide. Oxford is a registered trade mark of Oxford University Press in the UK and certain other countries.

Published in the United States of America by Oxford University Press
198 Madison Avenue, New York, NY 10016, United States of America

Address editorial correspondence to:
Sinauer Associates
23 Plumtree Road
Sunderland, MA 01375 U.S.A.
publish@sinauer.com

Address orders, sales, license, permissions, and translation inquiries to:
Oxford University Press U.S.A.
2001 Evans Road
Cary, NC 27513 U.S.A.
Orders: 1-800-445-9714

Library of Congress Cataloging-in-Publication Data

Names: Pough, F. Harvey, author. | Janis, Christine M. (Christine Marie), 1950- author.
Title: Vertebrate life / F. Harvey Pough, Rochester Institute of Technology, Emeritus, Christine M. Janis, Brown University, Emerita, University of Bristol ; chapter 26 "Primate Evolution and the Emergence of Humans" by Sergi López-Torres, University of Toronto Scarborough, Roman Kozlowski Institute of Paleobiology, Polish Academy of Sciences
Description: Tenth edition. | New York : Oxford University Press, [2019] | Revised edition of: Vertebrate life / F. Harvey Pough, Christine M. Janis, John B. Heiser. 9th ed. Boston : Pearson, c2013. | Includes bibliographical references and index.
Identifiers: LCCN 2018005330 (print) | LCCN 2018005884 (ebook) | ISBN 9781605357218 (ebook) | ISBN 9781605356075 (casebound)
Subjects: LCSH: Vertebrates--Textbooks. | Vertebrates, Fossil--Textbooks.
Classification: LCC QL605 (ebook) | LCC QL605 .P68 2019 (print) | DDC 596--dc23
LC record available at https://lccn.loc.gov/2018005330

9 8 7 6 5

Printed in the United States of America

Brief Contents

Contents

Chapter 4

Living in Water 65

Chapter 5

Geography and Ecology of the Paleozoic Era 83

Chapter 6

Radiation and Diversification of Chondrichthyes 95

Chapter 7

Extant Chondrichthyans 103

Chapter 26

Primate Evolution and the Emergence of Humans 519

Preface

The sustainability of populations, and even the continued existence of some species of vertebrates, is becoming ever more problematic. In addition to overarching events that affect all living organisms—such as global climate change and acidification of the seas—each lineage of vertebrates faces threats that are intimately entwined with the biological characteristics of that lineage.

Thus, as we have prepared this tenth edition of *Vertebrate Life*, the study of vertebrates has taken on new urgency. Many areas of vertebrate biology have seen enormous advances since the previous edition:

- Phylogenies based on molecular data increasingly supplement, reinforce, and in some cases contradict phylogenies based on morphology. We have incorporated this information, including some of the cases where the two approaches have generated different hypotheses about the timing or sequence of evolutionary change.

- Our understanding of the genetic and epigenetic control of development has advanced greatly, and this evo-devo perspective provides a mechanistic understanding of evolutionary changes in phenotype.

- Newly discovered fossil sites, fossils, and even fossils that preserve soft tissue structures have added enormous detail (and often greater complexity) to our understanding of evolutionary lineages, especially early tetrapods, feathered nonavian dinosaurs, and humans.

This edition reflects these changes. What has not changed is the authors' view of vertebrates as complex and fascinating organisms that can best be understood by considering the interactions at multiple levels of biological organization that shape the biology of a species. This book presents vertebrates in a way that integrates all facets of their biology—from anatomy and physiology to ecology and behavior—in an evolutionary context

Authors

Two changes of authorship have occurred in this edition. Ill health has compelled John B. Heiser to withdraw from active participation. John has been an author since the first edition of *Vertebrate Life*, enriching the chapters on fishes with the experience gained from his field work in nearly every ocean. We thank him for his contributions, and wish him well.

We are delighted to have gained the assistance of Sergi López-Torres who authored the chapter on primates and the evolution of humans. Primate evolution is an extraordinarily active field of research, and Sergi brings the depth and breadth of knowledge of human evolution needed for an overview of this increasingly complex field.

Organization

The scope of vertebrate biology and of evolutionary time is vast, and encapsulating this multiplicity of themes and mountains of data into a book or a semester course is a continuing challenge. Following suggestions from users, the text has been extensively reorganized to improve the flow of information. Topics have been split or merged, and some have been deleted, resulting in presentation of information in more manageable segments. Each chapter includes a list of sources that will be useful to students, and all of the sources that we consulted in preparing this edition, as well as many sources from earlier editions, are available on the book's web page, oup.com/us/vertebratelife10e.

Sources

We have relied on these sources for the numbers of extant species and their common and scientific names.

- FishBase: http://www.fishbase.se/search.php
- AmphibiaWeb: https://amphibiaweb.org/
- The Reptile Database: http://www.reptile-database.org/
- IOC World Birds list: http://www.worldbirdnames.org/
- Avibase: http://avibase.bsc-eoc.org/checklist.jsp
- Mammal Species of the World: https://www.departments.bucknell.edu/biology/resources/msw3/
- ASM Mammal Diversity Database: https://mammaldiversity.org/
- IUCN Red List of threatened species: http://www.iucnredlist.org/

Acknowledgments

Authors are only the visible tip of the iceberg that is a textbook. Expert librarians are essential to any scholarly undertaking, and we are fortunate to have had the outstanding assistance of Adwoa Boateng and Morna Hilderbrand (Wallace Library, Rochester Institute of Technology) and Sue O'Dell (Hatch Science Library, Bowdoin College).

The figure legends cite the many artists and photographers who kindly allowed us to use their work, often provided additional information, and in some cases took new photos at our request.

We are grateful to the many colleagues who answered questions and suggested sources for data and photographs:

Albert Bennett
Lucille Betti-Nash
Daniel Blackburn
Caleb M. Brown
Grover Brown
Larry Buckley
Mark Chappell
Charles J. Cole
Alison Cree
Chris Crippen
Mark Dimmitt
Vladimir Dinets
Colleen Farmer
Vincent Fernandez
Luciano Fischer
Linda Ford

Margaret Fusari
Harry W. Greene
Katrina Halliday
James Hanken
Karsten E. Hartel
Gene Helfman
Peter E. Hillman
Andrew Holmes
Osamu Kishida
Harvey B. Lillywhite
Jessie Maisano
Patrick Moldowan
Andrew Moore
Kenneth Nagy
Darren Naish
Gavin Naylor

Stewart Nicol
Martin Nyffler
Kouki Okagawa
Todd Pierson
Theodore Pietsch
Christopher Raxworthy
Shawn M. Robinson
Caroline S. Rogers
Robert Rothman
Timothy B. Rowe
Michael J. Ryan
Sue Sabrowski
Allen Salzberg
Colin Sanders
Jay Savage
Kurt Schwenk

Wade C. Sherbrooke
Rick Shine
Matthew Simon
Megan Southern
Zachary Stahlschmidt
Hyla Sweet
Ryan Taylor
Frank Tiegler
Hugh Tyndale-Biscoe
Wayne van Devender
John K. VanDyk
David B. Wake
Cliff White
Mark Witton
Stephen Zozaya

We leaned especially heavily on many colleagues, and thank them for their generosity:

Robin M. Andrews, Virginia Tech
James Aparicio, Museo Nacional de Historia Natural de Bolivia
Paul M. Barratt, Natural History Museum, London
Robin M. D. Beck, University of Salford
David R. Begun, University of Toronto
William E. Bemis, Cornell University
Michael J. Benton, University of Bristol
Christopher Brochu, University of Iowa
Edmund D. Brodie, Jr., Utah State University
William S. Brown, Skidmore College
Emily A. Buchholtz, Wellesley College
Ann C. Burke, Wesleyan University
Kenneth Catania, Vanderbilt University
Jennifer A. Clack, University of Cambridge
René C. Clark, Tucson
Chris Crippen, Virginia Living Museum
Michael I. Coates, University of Chicago
A. W. "Fuzz" Crompton, Harvard University
Martha L. Crump, Utah State University
David Cundall, Lehigh University
John D. Damuth, University of California, Santa Barbara
Christine Dahlin, University of Pittsburgh at Johnstown
Dominique Didier, Millersville University

Kenneth Dodd, University of Florida
Phillip C. J. Donoghue, University of Bristol
William N. Eschmeyer, California Academy of Sciences
Robert E. Espinoza, California State University, Northridge
David E. Fastovsky, University of Rhode Island
Sharon Gilman, Coastal Carolina University
Pamela G. Gill, University of Bristol
Frank M. Greco, New York Aquarium
Gordon Grigg, The University of Queensland
David F. Gruber, City University of New York
Célio F. B. Haddad, Universidade Estadual Paulista "Júlio de Mesquita Filho"
Lindsay Hazely, The Southland Museum and Art Gallery
Axel Hernandez, Università di Corsica Pasquale Paoli
Thomas R. Holtz, University of Maryland
Carlos Jared, Instituto Butantan
Bruce C. Jayne, University of Cincinnati
Jeffrey Lang, University of North Dakota
Manuel S. Leal, University of Missouri
Agustín Martinelli, Museo Argentino de Ciencias Naturales "Bernardino Rivadavia"
Phillip Motta, University of South Florida
Sterling J. Nesbit, Virginia Tech

Dina L. Newman, Rochester Institute of Technology
Daniel Pinchiera-Donoso, University of Lincoln
P. David Polly, Indiana University
Malcolm S. Ramsay, University of Toronto
Mario Sacramento, Biologist & Photographer, Passos, Brazil
Mary Beth Saffo, National Museum of Natural History
Steven Salisbury, The University of Queensland
Colin Sanders, University of Alberta
Alan H. Savitzky, Utah State University
Nalani Schnell, Muséum National d'Histoire Naturelle, Paris
Mary T. Silcox, University of Toronto
Matthew G. Simon, Skidmore College
Janet M. Storey, Carleton University

Glenn Tattersall, Brock University
Christopher R. Tracy, University of California, Riverside
C. Richard Tracy, University of Nevada, Las Vegas
Jaime Troncoso-Palacios, Universidad de Chile
Peter Uetz, Virginia Commonwealth University
Laura Verrastro, Universidade Federal do Rio Grande do Sul
T. Bence Viola, University of Toronto
Richard Vogt, Instituto Nacional de Pesquisas da Amazônia
Christopher Walmsley, Monash University
Nicholas C. Wegner, NOAA Fisheries
Kentwood D. Wells, University of Connecticut
Lisa Whitenack, Allegheny College
L. Kate Wright, Rochester Institute of Technology

We are especially grateful to Sharon Gilman (Coastal Carolina University) who assembled and edited the active learning exercises that are available online.

And finally, we recognize that no textbook of this magnitude comes into being without the help of dedicated publishing professionals. Beth Wilbur of Pearson Higher Education was endlessly helpful as the editor of previous editions and wonderfully generous during the transition to Sinauer Associates. Words cannot express the gratitude and admiration we feel for the people we have worked with at Sinauer and Oxford University Press: Carol Wigg and Laura Green, whose skill, experience, and extraordinary patience accomplished the Herculean task of transferring the book to a new publisher in midstream; Elizabeth Pierson, whose editing identified inconsistencies and vastly improved style and diction; Michele Beckta, whose patience and knowledge of copyright law has kept us safe; Joan Gemme, whose design and layout artistry turned the many complicated pieces into an elegant and beautiful whole; Mark Siddall, who located obscure photographs and, when all else failed, photographed the needed material himself; Elizabeth Morales, who transformed our scribbled instructions into glorious and colorful finished art; Jason Dirks and Nathaniel Nolet, who organized the extensive instructors' resources; and of course, Andy Sinauer, whose energy, spirit, and commitment to excellence has always guided Sinauer Associates.

Media and Supplements

to accompany **Vertebrate Life**, TENTH EDITION

eBook

(ISBN 978-1-60535-721-8)

Vertebrate Life, Tenth Edition is available as an eBook, in several different formats, including RedShelf, VitalSource, and Chegg. All major mobile devices are supported.

For the Student

Companion Website
(oup.com/us/vertebratelife10e)

The following resources are available to students free of charge:

- **Active Learning Exercises**: Activities that engage students with topics discussed in the textbook.

- **Chapter References**: All of the sources that the authors consulted in preparing the Tenth Edition of *Vertebrate Life*, as well as many sources from earlier editions.

- **New References**: Important new papers published since the publication of the textbook, updated on a regular basis.

- **Video Links**: Links to interesting videos illustrating some of the organisms and processes related to each chapter, updated on a regular basis.

- **News Links**: Links to news items related to vertebrate biology, updated on a regular basis.

For the Instructor

Ancillary Resource Center
(oup-arc.com)

Includes a variety of resources to aid instructors in developing their courses and delivering their lectures:

- **Textbook Figures and Tables**: All the figures and tables from the textbook are provided in JPEG format, reformatted for optimal readability, with complex figures provided in both whole and split formats.

- **PowerPoint Resources**: A PowerPoint presentation for each chapter includes all of the chapter's figures and tables, with titles and captions.

- **Answers to Discussion Questions**: Sample answers are provided for all of the in-text discussion questions.

1

Evolution, Diversity, and Classification of Vertebrates

Iurii Konoval/123RF

Evolution is central to vertebrate biology because it provides a principle that organizes the diversity we see among living vertebrates and helps fit extinct forms into the context of extant (currently living) species. In this chapter, we present an overview of the environments, the participants, and the events that have shaped the evolution and biology of vertebrates.

1.1 The Vertebrate Story

Say the word "animal" and most people picture a vertebrate. Vertebrates are abundant and conspicuous parts of people's experience of the natural world. They are also remarkably diverse—the more than 67,000 extant species of vertebrates range in size from fishes weighing as little as 0.1 gram to whales weighing over 100,000 kilograms. Vertebrates live in virtually all of Earth's habitats. Bizarre fishes, some with mouths so large they can swallow prey bigger than their own bodies, live in the depths of the sea, sometimes luring prey with glowing lights. Some 15 kilometers above the fishes, migrating birds fly over the peaks of the Himalayas.

The behaviors of vertebrates are as diverse and complex as their body forms and habitats. Life as a vertebrate is energetically expensive, and vertebrates obtain the energy they need from the food they eat. Carnivores eat the flesh of other animals and show a wide range of methods of capturing prey. Some predators actively search the environment to find prey, whereas others remain stationary and wait for prey to come to them. Some carnivores pursue their prey at high speeds, and others pull prey into their mouths by suction. Many vertebrates swallow their prey intact, sometimes while it is alive and struggling, and other vertebrates have specific methods of dispatching prey. Venomous snakes, for example, inject complex mixtures of toxins, and cats (of all sizes, from house cats to tigers) kill their prey with a distinctive bite on the neck.

Herbivores eat plants. Plants cannot run away when an animal approaches, so they are easy to catch, but they are hard to chew and digest and frequently contain toxic compounds. Herbivorous vertebrates show an array of specializations to deal with the difficulties of eating plants. Elaborately sculptured teeth tear apart tough leaves and expose the surfaces of cells, but the cell walls of plants contain cellulose, which no vertebrate can digest directly. Herbivorous vertebrates rely on microorganisms living in their digestive tracts to digest cellulose and to detoxify the chemical substances that plants use to protect themselves.

Reproduction is a critical factor in the evolutionary success of an organism, and vertebrates show an astonishing range of behaviors associated with mating and reproduction. In general, males court females and females care for the young, but many species of vertebrates reverse those roles. At the time of birth or hatching, some vertebrates are entirely self-sufficient and never see their parents, whereas other vertebrates (including humans) have extended periods of obligatory parental care. Extensive parental care is found in seemingly unlikely groups of vertebrates—fishes that incubate eggs in their mouths, frogs that carry their tadpoles to water and then return to feed them, and birds that feed their nestlings a fluid called crop milk that is similar in composition to mammalian milk.

The diversity of living vertebrates is enormous, but the extant species are only a small proportion of the species of vertebrates that have existed. For each extant species, there may be hundreds of extinct species, and some of these have no counterparts among extant forms. For example, the dinosaurs that dominated Earth for 180 million years are so entirely different from extant animals that it is hard to reconstruct the lives they led. Even mammals were once more diverse than they are now. The Pleistocene saw giants of many kinds, such as ground sloths as big as modern rhinoceroses and raccoons as large as bears. The number of species of terrestrial vertebrates probably reached its maximum in the middle Miocene, between 14 and 12 million years ago, and has been declining since then.

Where and when the vertebrates originated, how they evolved, what they do, and how they work provide endless intriguing details. In preparing to tell this story, we first introduce some basic information, including what the different kinds of vertebrates are called, how they are classified, and what the world was like as their story unfolded.

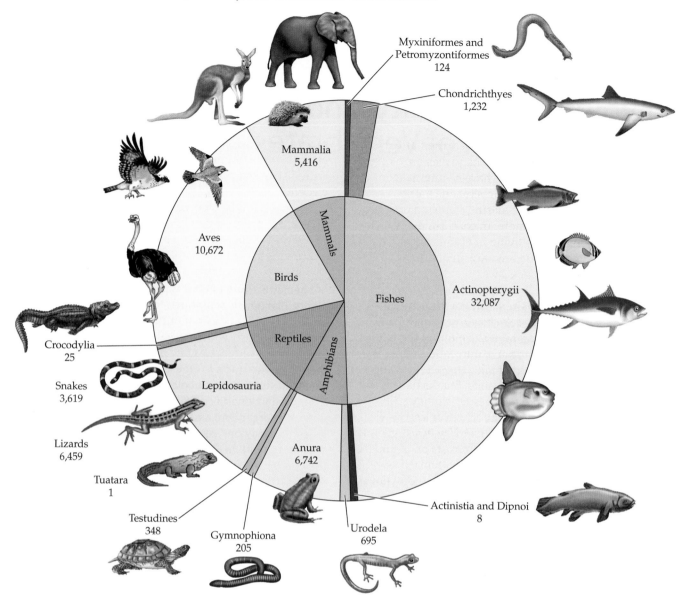

Figure 1.1 Diversity of extant vertebrates. Areas in the pie chart correspond to the approximate numbers of extant species in each group as of 2017; the numbers change as new species are described or existing species become extinct. Common names are in the center circle, with formal names for the groups shown in the outer circle. The two major lineages of extant vertebrates are Actinopterygii (ray-finned fishes; lavender) and Sarcopterygii (all other colors except gray), each of which includes more than 33,000 extant species. In its formal sense, Sarcopterygii includes the lineages Actinistia, Dipnoi, Urodela, Anura, Gymnophiona, Testudines, Lepidosauria, Crocodylia, Aves, and Mammalia (i.e, the lobe-finned fishes and all their descendants, including all amphibians, reptiles, birds, and mammals.)

Major extant groups of vertebrates

Two major groups of vertebrates are distinguished on the basis of an innovation in embryonic development: the appearance of three membranes formed by tissues that are generated by the embryo itself. The innermost of these membranes, the amnion, surrounds and cushions the embryo, and animals with this structure (the reptiles, birds, and mammals) are called **amniotes**. The split between non-amniotes and amniotes corresponds roughly to the division between aquatic and terrestrial vertebrates, although many amphibians and a few fishes lay non-amniotic eggs in nests on land. Among the amniotes, we can distinguish two major evolutionary lineages: the **sauropsids** (reptiles, including birds) and the **synapsids** (mammals). These lineages separated from each other in the mid-Carboniferous, before vertebrates had developed many of the characters we see in extant species. As a result, synapsids and sauropsids represent parallel but independent origins of basic characters such as lung ventilation, kidney function, insulation, and temperature regulation. **Figure 1.1** shows the major groups of vertebrates and the approximate number of extant species in each. Next we briefly describe these vertebrate groups.

Non-amniotes

The embryos of non-amniotes are enclosed and protected by membranes that are produced by the reproductive tract of the female. This is the condition seen among the invertebrate relatives of vertebrates, and it is retained in the non-amniotes (fishes and amphibians).

Hagfishes and lampreys: Myxiniformes and Petromyzontiformes Lampreys and hagfishes are elongate, limbless, scaleless, and slimy and have no internal bony tissues. They are scavengers and parasites and are specialized for those roles. Hagfishes are marine, living on the seabed at depths of 100 meters or more. In contrast, many species of lampreys are migratory, living in oceans and spawning in rivers. Hagfishes and lampreys are unique among extant vertebrates because they lack jaws.

Sharks, rays, and ratfishes: Chondrichthyes The name Chondrichthyes (Greek *chondros*, "gristle"; *ichthys*, "fish") refers to the cartilaginous skeletons of these fishes. Extant sharks and rays form a group called Neoselachii (Greek *neos*, "new"; *selachos*, "shark"), but the two kinds of fishes differ in body form and habits. Sharks have a reputation for ferocity that most species would have difficulty living up to. Some sharks are small (25 cm or less), while the largest species, the whale shark (*Rhincodon typus*), grows to 17 m long and is a filter feeder that subsists on plankton it strains from the water. Rays are mostly bottom feeders; they are dorsoventrally flattened and swim with undulations of their extremely broad pectoral fins.

The second group of chondrichthyans, the ratfishes or chimaeras, gets its name, Holocephali (Greek *holos*, "whole"; *kephale*, "head"), from the single gill cover that extends over all four gill openings. These are bizarre marine animals with long, slender tails. Some species have a bucktoothed face that looks rather like a rabbit. They forage on the seafloor and feed on hard-shelled prey, such as crustaceans and mollusks.

Bony fishes: Osteichthyes Osteichthyes (Greek *osteon*, "bone") are so diverse that any attempt to characterize them briefly is doomed to failure. Two broad categories can be recognized: the ray-finned fishes (Actinopterygii; Greek *aktis*, "ray"; *pteron*, "wing" or "fin") and the lobe-finned or fleshy-finned fishes (Sarcopterygii; Greek *sarco*, "flesh").

Actinopterygians have radiated extensively in both fresh and salt water, and more than 32,000 species have been named. Nearly 400 species have been described annually since 1997, and several thousand additional species may await discovery. A single project, the Census of Marine Life, is describing 150–200 previously unknown species of ray-finned fishes annually.

Actinopterygians can be divided into three groups: (1) the monophyletic and derived Neopterygii (almost all extant ray-finned fishes); (2) the more basal Chondrostei (sturgeons and paddlefish); and (3) Cladistia (polypterids, including bichirs and reedfish, which are swamp- and river-dwellers from Africa). Sturgeons are large fishes with protrusible, toothless mouths that suck food items from the bottom. Sturgeon eggs are the source of caviar, and many species have been driven close to extinction by overharvesting of females for their eggs. Paddlefishes (two species, one in the Mississippi system of North America and another nearly extinct species in China's Yangtze River) have a paddlelike snout with organs that locate prey by sensing electrical fields.

Neopterygii, the modern radiation of ray-finned fishes, can be divided into two lineages. One these—Holostei, the gars and bowfins—is a relict of an earlier radiation. These fishes have cylindrical bodies, thick scales, and jaws armed with sharp teeth. They seize prey in their mouths with a sudden gulp, and lack the specializations of the jaw apparatus that allow the bony fishes to use more complex feeding modes. The second neopterygian lineage, Teleostei, includes 95% of actinopterygians, embracing every imaginable combination of body form, ecology, and behavior. Most of the fishes that people are familiar with are teleosts, from sport fishes like trout and swordfishes, to food staples like tuna and salmon, to the "goldfish" and exotic tropical fishes in home aquaria. Modifications of the body and jaw apparatus have allowed many teleost species to be highly specialized in their swimming and feeding habits.

In one sense, only eight species of sarcopterygian fishes survive: the six species of lungfishes (Dipnoi) found in South America, Africa, and Australia; and two species of coelacanths (Actinistia), one from deep waters off the east coast of Africa and a second species discovered near Indonesia. These are the extant fishes most closely related to terrestrial vertebrates, however, and from an evolutionary standpoint the diversity of sarcopterygians includes all of their their terrestrial descendants—amphibians, turtles, lepidosaurs (the tuatara [*Sphenodon punctatus*], lizards, and snakes), crocodylians, birds, and mammals. From this perspective, bony fishes include two major evolutionary radiations—one in the water and the other on land—each containing more than 33,000 species.

Salamanders, frogs, and caecilians: Urodela, Anura, and Gymnophiona These three groups of vertebrates are popularly known as amphibians (Greek *amphis*, "double"; *bios*, "life") in recognition of their complex life histories, which often include an aquatic larval form (the larva of a salamander or caecilian and the tadpole of a frog) and a terrestrial adult. All amphibians have bare skins (i.e., lacking scales, hair, or feathers) that are important in the exchange of water, ions, and gases with their environment. Salamanders are elongate animals, mostly terrestrial, and usually with four legs; anurans (frogs) are short-bodied, with large heads and large hindlegs used for walking, jumping, and climbing; and caecilians are legless aquatic or burrowing animals.

Amniotes

A novel set of membranes associated with the embryo appeared during the evolution of vertebrates. These are called **fetal membranes** because they are derived from the embryo itself rather than from the reproductive tract of the mother. As mentioned at the start of this chapter, the amnion is the innermost of these membranes, and vertebrates with an amnion are called amniotes. In general, amniotes are more terrestrial than non-amniotes; but there are also secondarily aquatic species of amniotes (such as sea turtles and whales) as well as many species of salamanders and frogs that spend their entire lives on land despite being non-amniotes. However, many features distinguish non-amniotes (fishes and amphibians) from amniotes (mammals and reptiles, including birds), and we will use the terms to identify which of the two groups is being discussed.

By the Permian, amniotes were well established on land. They ranged in size from lizardlike animals a few centimeters long through cat- and dog-size species to the cow-size pareiasaurs. Some were herbivores; others were carnivores. In terms of their physiology, we can infer that they retained ancestral characters. They had scale-covered skins without an insulating layer of hair or feathers, a simple kidney that could not produce highly concentrated urine, simple lungs, and a heart in which the ventricle was not divided by a septum. Early in their evolutionary history, terrestrial vertebrates split into two lineages—synapsids (today represented by mammals) and sauropsids (modern reptiles, including birds).

Terrestrial life requires lungs to extract oxygen from air, a heart that can separate oxygen-rich arterial blood from oxygen-poor venous blood, kidneys that can eliminate waste products while retaining water, and insulation and behaviors to keep body temperature stable as the external temperature changes. These features evolved in both lineages, but because synapsids and sauropsids evolved terrestrial specializations independently, their lungs, hearts, kidneys, and body coverings are different.

Sauropsid amniotes   Extant sauropsids are the animals we call reptiles: turtles, the scaly reptiles (tuatara, lizards, and snakes), crocodylians, and birds. Extinct sauropsids include the forms that dominated the world during the Mesozoic—dinosaurs and pterosaurs (flying reptiles) on land and a variety of marine forms, including ichthyosaurs and plesiosaurs, in the oceans.

- **Turtles: Testudines** Turtles (Latin *testudo,* "turtle") are probably the most immediately recognizable of all vertebrates. The shell that encloses a turtle has no exact duplicate among other vertebrates, and the morphological modifications associated with the shell make turtles extremely peculiar animals. They are, for example, the only vertebrates with the shoulders (pectoral girdle) and hips (pelvic girdle) inside the ribs.

- **Tuatara, lizards, and snakes: Lepidosauria** These three kinds of vertebrates can be recognized by their scale-covered skin (Greek *lepisma,* "scale"; *sauros,* "lizard") as well as by characters of the skull. The tuatara, a stocky-bodied animal found only on some offshore islands of New Zealand, is the sole living remnant of an evolutionary lineage of animals called Rhynchocephalia, which was more diverse in the Mesozoic. In contrast, lizards and snakes (which are highly specialized lizards) are now at the peak of their diversity.

- **Alligators and crocodiles: Crocodylia** These impressive animals, which draw their name from the ancient Latin word for them (*crocodilus*) are in the same evolutionary lineage (Archosauria) as dinosaurs and birds. The extant crocodylians are semiaquatic predators with long snouts armed with numerous teeth. They range in size from the saltwater crocodile (*Crocodylus porosus*), which can grow to 6 m, to dwarf crocodiles and caimans that are 1.5 m long. Crocodylian skin contains many bones (osteoderms; Greek *osteon,* "bone"; *derma,* "skin") that lie beneath their scales and provide a kind of armor plating. Crocodylians are noted for the parental care they provide for their eggs and young.

- **Birds: Aves** Birds (Latin *avis,* "bird") are a lineage of dinosaurs that evolved flight in the Mesozoic. Feathers are characteristic of extant birds, and feathered wings are the structures that power a bird's flight. Discoveries of dinosaur fossils with feathers show that feathers evolved long before flight. This disparity between the time feathers are first seen and the time flight appeared illustrates an important principle: The function of a trait in an extant species is not necessarily the same as that trait's function when it first appeared. In other words, *current utility is not the same as evolutionary origin.* The original feathers were almost certainly structures used in courtship displays, and their modifications as airfoils, for streamlining, and as insulation in birds were secondary events.

Synapsid amniotes: Mammalia   The synapsid lineage contains the three kinds of extant mammals: the **monotremes** (prototheria; platypus and echidnas), **marsupials** (metatherians), and **placentals** (eutherians). Extinct synapsids include forms that diversified in the Paleozoic—pelycosaurs and therapsids—and the rodentlike multituberculates of the late Mesozoic.

The mammals (Latin *mamma,* "teat") originated in the Late Triassic. Extant mammals include about 5,400 species in three groups: placentals (by far the largest group), marsupials, and monotremes. The name "placentals" is misleading, because both placentals and marsupials have a placenta, a structure that transfers nutrients from the mother to the embryo and removes the waste products of the embryo's metabolism. Placentals have a long gestation period and have largely completed development at birth, whereas marsupials have a short gestation period and give birth to very immature young that continue their development attached to a nipple, often within a pouch on the mother's abdomen. Marsupials dominate the mammalian fauna only in Australia, although some 100 species are found in South and Central America, and one species, the Virginia opossum (*Didelphis virginiana*), has migrated from South America to North America. Kangaroos, koalas, and wombats are familiar Australian marsupials. Finally, the monotremes—platypus and echidnas—are unusual mammals whose young hatch from eggs. All mammals, including monotremes, feed their young with milk.

1.2 Classification of Vertebrates

The diversity of vertebrates (more than 67,000 extant species and perhaps 100 times that number of extinct species) makes organizing them into a coherent system of classification extraordinarily difficult. Yet classification has long been at the heart of evolutionary biology. Initially, classification of species was seen as a way of managing the diversity of organisms, much as an office filing system manages the paperwork of the office. Each species could be placed in a pigeonhole marked with its name; when all species were in their pigeonholes, the diversity of vertebrates would have been encompassed.

This approach to classification was satisfactory as long as species were regarded as static and immutable: once a species was placed in the filing system, it was there to stay. Acceptance of the fact that species evolve has made that kind of classification inadequate. Now biologists must express evolutionary relationships among species by incorporating evolutionary information in the system of classification. Ideally, a classification system should not only attach a label to each species but also encode the evolutionary relationships between that species and other species. Modern techniques of **systematics** (the evolutionary classification of organisms) have become methods for generating testable hypotheses about evolution.

Binominal nomenclature

Our system of naming species is pre-Darwinian. It traces back to methods established by the naturalists of the seventeenth and eighteenth centuries, especially those of Carl von Linné, the Swedish naturalist better known by his Latin pen name, Carolus Linnaeus. The Linnaean system identifies a species using two names—a genus name and a species name; hence, this convention is called **binomial nomenclature**. Species are arranged in hierarchical categories (family, order, and so on). These categories are called **taxa** (singular *taxon*), and the discipline of naming organisms is called **taxonomy** (Greek *taxo*, "to arrange"; *nomos*, "order").

The scientific naming of species became standardized when Linnaeus's monumental work *Systema Naturae* (*The System of Nature*) was published in sections between 1735 and 1758. Linnaeus attempted to give an identifying name to every known species of plant and animal. Familiar examples include *Homo sapiens* for human beings (Latin *homo*, "human" and *sapiens*, "wise"), *Passer domesticus* for the house sparrow (Latin *passer*, "sparrow" and *domesticus*, "belonging to the house"), and *Canis familiaris* for the domestic dog (Latin *canis*, "dog" and *familiaris*, "domestic").

Why use Latin? Latin was the early universal language of European scholars and scientists. It provided a uniform usage that scientists, regardless of their native language, continue to recognize worldwide. The same species may have different colloquial names, even in the same language. For example, *Felis concolor* (Latin, "uniformly colored cat") is known in various parts of North America as the cougar, puma, mountain lion, American panther, painter, and catamount. In Central and South America it is called león Colorado, león de montaña, pantera, onça vermelha, onça parda, yagua pytá, and suçuarana. But biologists of all nationalities recognize the name *Felis concolor* as referring to this specific kind of cat.

Phylogenetic systematics

All methods of classifying organisms, even Linnaean systems, are based on similarities shared by the included species, but some similarities are more significant than others. For example, nearly all terrestrial vertebrates have paired limbs, but only a few kinds of vertebrates have mammary glands. Consequently, knowing that two species have mammary glands tells you more about the closeness of their relationship than does knowing they have paired limbs. A way to assess the relative importance of different characteristics was developed in the mid-20th century by Willi Hennig, who introduced a method of determining evolutionary relationships called **phylogenetic systematics** (Greek *phylon*, "tribe"; *genesis*, "origin").

The groups of organisms recognized by phylogenetic systematics are called natural groups, and the members of these groups are linked by a nested series of characters that trace the evolutionary history of the lineage. Hennig's contribution was to insist that these groups can be identified only on the basis of homologous **derived characters**—that is, characters that have the same evolutionary origin (i.e., are homologous) and that differ from the ancestral condition (are derived). A derived character is called an **apomorphy** (Greek *apo*, "away from" and *morphe*, "form," which is interpreted as "away from the ancestral condition").

For example, the feet of terrestrial vertebrates have distinctive bones—the carpals, tarsals, and digits. This arrangement of foot bones is different from the ancestral pattern seen in lobe-finned fishes, and all lineages of terrestrial vertebrates either have that derived pattern of foot bones or had it at some stage in their evolution. Many groups of terrestrial vertebrates—horses, for example—have subsequently modified the foot, and some, such as snakes, have lost the limbs entirely. The significant point is that those evolutionary lineages include ancestral species that had the derived terrestrial pattern. Thus, the terrestrial pattern of foot bones is a **shared derived character** of terrestrial vertebrates. Shared derived characters are called **synapomorphies** (Greek *syn*, "together," so synapomorphy can be interpreted as "together away from the ancestral condition").

Of course, organisms also share ancestral characters—that is, characters that they have inherited unchanged from their ancestors. These are called **plesiomorphies** (Greek *plesios*, "near," in the sense of "similar to the ancestor"). The vertebral column of terrestrial vertebrates, for example, was inherited from lobe-finned fishes. Hennig called such shared ancestral characters **symplesiomorphies** (*sym*, like *syn*, is a Greek root that means "together"). A character can be either plesiomorphic or apomorphic, depending on the level at which the distinction is applied. A vertebral column is a symplesiomorphy of vertebrates, so it provides no information about evolutionary relationships of vertebrates to one another, but it is an apomorphy when vertebrates are compared with nonvertebrate chordates.

Applying phylogenetic criteria

The conceptual basis of phylogenetic systematics is straightforward, although applying the criteria to real organisms can become complicated. To illustrate phylogenetic classification, consider the examples presented in **Figure 1.2**. Each of the three hypothetical **phylogenies**—diagrams showing sequences of branching during evolution—illustrates a possible evolutionary relationship for the three

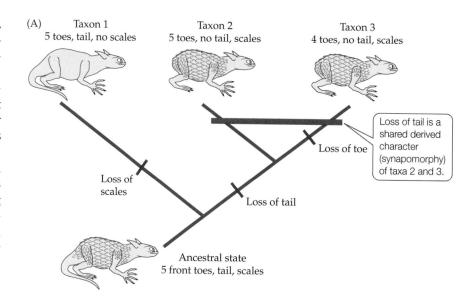

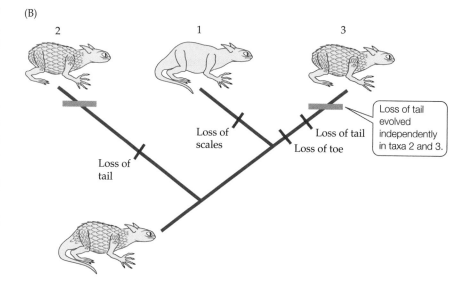

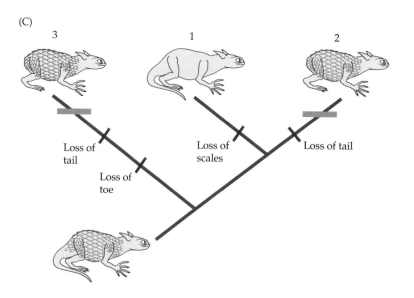

Figure 1.2 Phylogenies showing the possible evolutionary relationships of hypothetical taxa 1, 2, and 3. In this hypothetical example, the red bars identify changes (evolution) from the ancestral (plesiomorphic) character states of tail, skin, and number of toes on the front feet to three different derived (apomorphic) conditions. The green bar shows a shared derived character (a synapomorphy) of the lineage that includes taxa 2 and 3. Orange bars represent two separate origins of the same derived character state (loss of tail) that in these cases must be assumed to have occurred independently, since the apomorphy is not found in the most recent common ancestor of taxa 2 and 3. Phylogeny (A) requires a total of three changes from the ancestral condition to explain the distribution of characters in the extant taxa, whereas phylogenies (B) and (C) both require four changes. Because phylogeny (A) is more parsimonious (i.e., requires the smallest number of changes), it is considered to be the most likely evolutionary sequence.

taxa identified as 1, 2, and 3. To make the example more concrete, we consider three characters: the number of toes on the front feet, the skin covering (scales or no scales), and the presence or absence of a tail. For this example, in the ancestral character state there are five toes on the front feet, and in the derived state there are four toes. We'll say that the ancestral state is scaly skin, and the derived state is a lack of scales. As for the tail, it is present in the ancestral state and absent in the derived state.

How can we use the information in Figure 1.2 to decipher the evolutionary relationships of the three groups of animals? Notice that the derived number of four toes occurs only in taxon 3, and the derived tail condition (absent) is found in taxa 2 and 3. The **most parsimonious** phylogeny (i.e., the branching sequence requiring the fewest number of changes) is represented by Figure 1.2A. Only three changes are needed to produce the current distribution of character states: in the evolution of taxon 1, scales are lost; in the evolution of the lineage including taxa 2 and 3, the tail is lost; and in the evolution of taxon 3, a toe is lost.

The other two phylogenies shown in Figure 1.2 are possible, but they require tail loss to occur independently in taxa 2 and 3. Any change in a structure is an unlikely event, so the most plausible phylogeny is the one requiring the fewest changes. The second and third phylogenies each require four evolutionary changes, so they are less parsimonious than the first phylogeny, which requires only three changes.

A phylogeny (also called a phylogenetic tree or simply a tree) is a hypothesis about the evolutionary relationships of the groups included. Like any scientific hypothesis, it can be tested when new data become available. If it fails that test, it is falsified; that is, it is rejected, and a different hypothesis (a different branching sequence) is proposed. The process of testing hypotheses and replacing those that are falsified is a continuous one, and changes in the phylogenies in successive editions of this book show where new information has generated new hypotheses.

So far we have avoided a central issue of phylogenetic systematics: How do scientists know which character state is ancestral (plesiomorphic) and which is derived (apomorphic)? That is, how can we determine the direction (polarity) of evolutionary transformation of the characters? For that, we need additional information. Increasing the number of characters we are considering can help, but the preferred methods is to compare the characters we are interested in with those of an **outgroup**, a reference group that, although related to the **ingroup**—the organisms we are studying—is less closely related to any member of the ingroup than those members are to each other. For example, lobe-finned fishes are an appropriate outgroup for terrestrial vertebrates.

Morphology-based and molecular-based phylogenies

Initially, anatomical traits were the only characters that could be used to determine the relationships among organisms, but advances in molecular biology and genetics in the latter 20th century made it possible to include proteins and DNA in phylogenies. The American anthropologist Vincent Sarich pioneered the use of molecular characters in the 1960s, using immunological comparison of blood serum albumins to determine that chimpanzees are the apes most closely related to humans. In the 1970s, amino acid sequences of proteins and base sequences of mitochondrial and nuclear DNA were added to the repertoire of molecular phylogeny. At the same time, computer algorithms were devised that could sort and arrange large numbers of characters in a phylogeny. Any set of characters can produce multiple possible phylogenies, and these algorithms use various methods (e.g., parsimony, maximum likelihood, Bayesian inference) to identify the most plausible sequences of changes.

The late 20th century was also when personal computers came into general use, so that anybody could use these algorithms. Additionally, with the establishment of public-access repositories such as GenBank in 1982 and Morpho-Bank in 2005, vast quantities of genetic and morphological data became available on the internet. As a result, phylogenetic systematics has moved from human-based analyses of small numbers of characters to computer-based analyses of huge data sets.

Molecular characters rapidly came to be regarded as superior to morphological characters for creating phylogenies. The ease with which large quantities of molecular data can be obtained contributed to this perception, as did the assumption that molecular data would be free from some of the problems of interpretation associated with morphological data. More recently, molecular phylogenies have lost some of their luster. The assumptions that all types of molecular data are equal in creating a phylogeny and that more data produce better results are being questioned. Some characters are appropriate for looking at short-term changes and others at long-term processes; mixing the two can produce erroneous results.

Differences of that sort are easy to see in morphological data, but harder to see in molecular data. For example, coat color of mammals can be useful for distinguishing among different species within a genus, but it is not useful at higher taxonomic levels. Different patterns of striping distinguish different species of zebras, but stripes cannot be used as a character to ally zebras with tigers. Molecular characters are more difficult to assess in this fashion, because the functional significance of differences in base sequences or amino acids is rarely known—most molecular characters are black boxes. Furthermore, fossils seldom yield molecules, so molecular characters can rarely be used to determine the relationships of fossil organisms to one another or to extant species. This is a critical shortcoming. A morphology-based phylogeny can be drastically rearranged by the addition of a single fossil that alters the polarity of morphological characters, but molecular phylogenies cannot be tested in this manner.

Dating the time that lineages separated is another weakness of molecular phylogenies, because dates are based on assumptions about the rates at which mutations occur. In contrast, the age of a fossil can almost always be determined—provided that one has a fossil. The difficulty with fossils is the incompleteness of the fossil record. There are gaps, sometimes millions of years long, in the records of many taxa. For example, we have fossil lampreys from the Late Devonian (~360 million years ago, abbreviated Ma[1]), the Late Carboniferous (~300 Ma), and the Early Cretaceous (~145 Ma), but nothing between those dates.

Considering the difficulties with both molecular and morphological phylogenies, it's remarkable that they mostly agree about branching patterns. Disagreements usually center on dates of divergence. For example, molecular phylogenies indicate that the extant lineages of amphibians diverged in the Late Carboniferous (~315–300 Ma), whereas fossil data indicate divergence in the Late Permian (~260–255 Ma). However, molecular and morphological methods can agree; molecular evidence indicates that humans separated from their common ancestor with chimpanzees about 6.6 Ma, and this date fits well with the earliest fossil in the human lineage, *Ardipithecus* (5.8 Ma).

The best information comes from combining molecular and morphological data. Studies that include extant and extinct organisms often employ the technique of **molecular scaffolding**: the extant taxa are placed in their phylogenetic position by the relationships established by the molecular data, and then morphological data are used to integrate the fossil taxa with the extant taxa.

Using phylogenetic trees

Phylogenetic systematics is based on the assumption that organisms in a lineage share a common heritage, which accounts for their similarities. Because of that common heritage, we can use phylogenetic tress to ask questions about evolution. By examining the origin and significance of characters of extant animals, we can make inferences about the biology of extinct species. For example, the phylogenetic relationship of crocodylians, dinosaurs, and birds is shown in **Figure 1.3**.

We know that both crocodylians and birds display extensive parental care of their eggs and young. Some fossilized dinosaur nests contain remains of partly grown baby dinosaurs, suggesting that at least some dinosaurs cared for their young. Is that a plausible inference? Obviously, there is no direct way to determine what sort of parental care dinosaurs provided to their eggs and young. The intermediate lineages in the phylogenetic tree (pterosaurs and

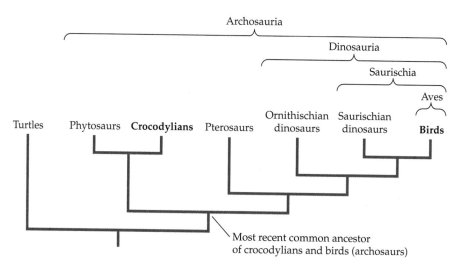

Figure 1.3   Using a phylogenetic tree to make inferences about behavior.   The tree shows the relationships of the Archosauria, the evolutionary lineage that includes today's crocodylians and birds. (Phytosaurs were crocodile-like animals that disappeared at the end of the Triassic; pterosaurs were flying reptiles of the Jurassic and Cretaceous.) Both extant archosaur groups (i.e., crocodylians and birds) display extensive parental care of eggs and young, but turtles do not. The most parsimonious explanation of this situation assumes that parental care evolved in the archosaur lineage after the separation from turtles, but before the separation of phytosaurs + crocodylians from pterosaurs + dinosaurs.

dinosaurs) are extinct, so we cannot observe their reproductive behavior. However, the phylogenetic tree in Figure 1.3 provides an indirect way to approach the question by examining the lineage that includes the closest living relatives of dinosaurs, crocodylians, and birds.

Crocodylians are the basal taxon in this tree (the lineage closest to the ancestral form), and birds are the most derived (farthest from the ancestral form). Together, crocodylians and birds form what is called an **extant phylogenetic bracket**. Both crocodylians and birds provide parental care. Looking at extant representatives of more distantly related lineages (outgroups), we see that parental care is not universal among fishes, amphibians, or turtles. The most parsimonious explanation of the occurrence of parental care in both crocodylians and birds is that it evolved in that lineage after the lineage separated from turtles and before crocodylians separated from dinosaurs + birds. We cannot prove that parental care did not evolve independently in crocodylians and in birds, but a single change to parental care is more likely than two separate changes. Thus, the most parsimonious hypothesis is that parental care is a derived character of the evolutionary lineage containing crocodylians + pterosaurs + dinosaurs + birds. That means we are probably correct when we interpret fossil evidence as showing that dinosaurs did indeed exhibit parental care.

Figure 1.3 also shows how phylogenetics has made talking about restricted groups of animals more complicated than it used to be. Suppose you wanted to refer to just the two lineages of animals that are popularly known as

---

[1]Ma is the abbreviation for mega-annums, "million years." It is used to mean both "million years" and, when referring to a specific date range, "million years ago." Analogously, ka = kilo-annums (thousand years, thousand years ago).

dinosaurs—ornithischians and saurischians. What could you call them? If you call them "dinosaurs," you're not being phylogenetically correct, because the Dinosauria lineage includes birds. So if you say "dinosaurs," you are including ornithischians + saurischians + birds, even though any 7-year-old would understand that you mean to restrict the conversation to the extinct Mesozoic animals.

In fact, there is no technically correct name in phylogenetic taxonomy for just those animals popularly known as dinosaurs. That's because phylogenetics recognizes only monophyletic lineages (see Section 1.3), and a monophyletic lineage includes an ancestral form and all of its descendants. As you can see in Figure 1.3, the most recent common ancestor of ornithischians + saurischians + birds lies at the intersection of the lineage of ornithischians with saurischians + birds, so Dinosauria is a monophyletic lineage. If birds are omitted, however, the lineage no longer includes all the descendants of the common ancestor. The lineage ornithischians + saurischians *minus* birds does not fit the definition of a monophyletic lineage and would be called **paraphyletic** (Greek *para*, "beside" or "near," meaning a taxon that includes the common ancestor and some, but not all, of its descendants).

Biologists who are interested in how organisms function often talk about paraphyletic groups. After all, dinosaurs (in the popular sense of the word) differ from birds in many ways. The only technically correct way of referring to the animals popularly known as dinosaurs is to call them "nonavian dinosaurs," and you will find that term and other examples of paraphyletic groups in this book. Sometimes even this construction does not work because there is no appropriate name for the part of the lineage you want to distinguish. In that situation, we will use quotation marks (e.g., "ostracoderms") to indicate that the group is paraphyletic.

Another important term is **sister group**, which refers to the monophyletic lineage most closely related to the monophyletic lineage being discussed. In Figure 1.3, for example, the lineage that includes crocodylians + phytosaurs is the sister group of the lineage that includes pterosaurs + ornithischians + saurischians + birds. Similarly, pterosaurs are the sister group of ornithischians + saurischians + birds, ornithischians are the sister group of saurischians + birds, and saurischians are the sister group of birds.

1.3   Crown and Stem Groups

Evolutionary lineages must have a single evolutionary origin; that is, they must be **monophyletic** (Greek *mono*, "one" or "single"), and they must also include all the descendants of the ancestor. The phylogenetic tree in **Figure 1.4** shows the hypothesis of the evolutionary relationships of the major extant groups of vertebrates that we will follow throughout this book. A series of dichotomous branches extends from the origin of vertebrates to the groups of extant vertebrates. Phylogenetic terminology assigns names to the lineages originating at each branch point.

This method of tracing ancestor–descendant relationships allows us to decipher evolutionary pathways that extend from fossils to extant groups, but a difficulty arises when we try to find names for groups that include fossils. The derived characters that define the extant groups of vertebrates did not all appear at the same time. On the contrary, derived characters appear in a stepwise fashion. The extant members of a group have all of the derived characters of that group because that is how we define the group today, but as we move backward through time to fossils that are ancestral to the extant species, we encounter forms that have a mosaic of ancestral and derived characters.

The farther back in time we go, the fewer derived characters the fossils have. What can we call the parts of a lineage that contain these fossils? They are not the same as the extant group because they lack some of the derived characters that define those groups, but the fossils in a lineage are more closely related to the animals in the extant group than they are to animals in other lineages. The solution to this problem lies in naming two types of groups: **crown groups** and **stem groups**. Crown groups have the character states of extant species. That is, members of the crown group have all the derived characters, but they don't have to be extant. An extinct species with all of the derived characters is included in a crown group. Stem groups are the extinct forms in the lineage that lack some of the derived characters. Basically, stem groups contain fossils with some derived characters, and crown groups contain extant species plus fossils that have all of the derived characters of the extant species. Stem groups are paraphyletic because they do not contain all of the descendants of the ancestor of the stem group.

1.4   Genetic Mechanisms of Evolutionary Change

"Descent with modification" is the phrase that Charles Darwin used to describe evolution. He drew his evidence from the animals and plants he encountered during his voyage aboard the HMS *Beagle* (1831–1836) and from his familiarity with selective breeding of domestic animals. Darwin emphasized the roles of natural selection and sexual selection as the mechanisms of evolution, although the basis of the traits he described was a mystery. Selective breeding of plants and domestic animals was practiced in the 19th century, but the phenotype of the offspring was thought to be a blend of the phenotypes of the parents. Gregor Mendel did not publish evidence of particulate inheritance (i.e., separate inheritance of specific genetic traits, such as yellow versus green and smooth versus wrinkled peas) until 1866. The rediscovery and extension of Mendel's work in the early 20th century firmly established genes as the basis of heritable traits, and the 1930s and 1940s saw

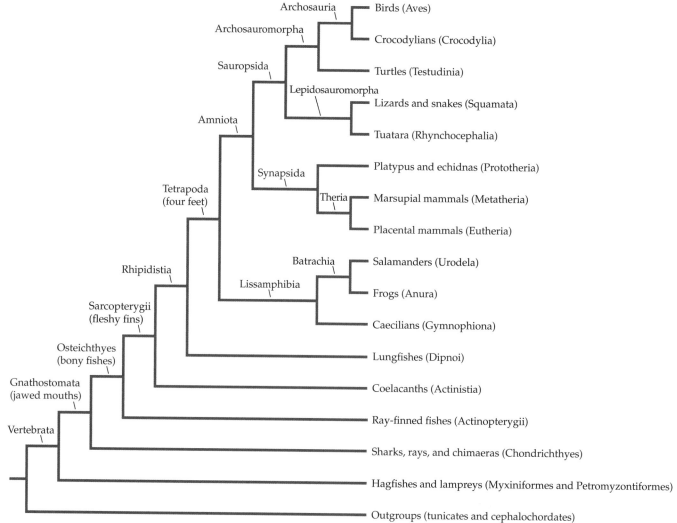

Figure 1.4 Phylogenetic relationships of extant vertebrates. This tree depicts the probable relationships among the major groups of extant vertebrates and is the phylogeny we follow in this book. Phylogenetic terminology assigns names to the lineages originating at each branch point. This process produces a nested series of groups, starting with the most inclusive. For example, Gnathostomata includes all vertebrate animals that have jaws. Because the lineages are nested, it is correct to say, for example, that mammals are both gnathostomes and osteichthyans.

a blending of genetics, natural selection, and population biology known as the **Modern Synthesis**.

Phenotypes and fitness

Most genes are polymorphic; that is, they have two or more alleles (forms of a gene that differ in base sequence). The phenotype of an organism is determined by its genotype—that is, by its combination of alleles—and selection acts on phenotypes via differential reproduction. In the Modern Synthesis, genetic mutations (new alleles) are considered to occur randomly, and heritable allelic variation produced by mutation is considered to be the raw material of evolution.

Although the phrase "survival of the fittest" conjures images of combat, evolutionary success is measured in

terms of reproductive success. Some phenotypes leave more descendants than others, and consequently the frequency of the alleles that produce those phenotypes increases from one generation to the next. **Darwinian fitness** is a shorthand term that refers to the genetic contribution of an individual to succeeding generations relative to the average contribution of all the individuals in that population. Positive selection is revealed by an increase in the frequency of a genetic trait in successive generations. For example, changes in allele frequencies in residents of the United Kingdom over the past 2,000 years have resulted in increased frequencies of alleles that allow adults to digest lactose (the sugar found in milk) and in alleles that produce blond hair and blue eyes. In addition, positive selection has increased the frequencies of genes associated with traits that are controlled by hundreds of genes: adult height, adult female hip size, and head circumference of infants. Other research has found evidence of negative selection acting against genetic variants associated with educational attainment.

Developmental regulatory genes

Natural selection is a series of compromises because it is possible only to tinker with what is present, not to redesign structures from scratch. The Modern Synthesis in combination with natural selection can explain the *survival* of the fittest, but it cannot account for the *arrival* of the fittest—that is, the origin of the phenotypic variation on which natural selection acts. For this, we need to consider the genetic mechanism of changes in embryonic development and the effects of these changes on phenotype. This is the field of evolutionary developmental biology ("evo-devo"), and it emphasizes interactions of developmental regulatory genes that are arranged in hierarchical networks.

Groups of cells can release molecular signals called paracrine growth factors that change the expression of genes in nearby cells, and these genes in turn produce other paracrine factors[2] that change gene expression in other neighboring cells, sometimes feeding back to influence genes in the cells that produced the first set of paracrine factors.

The genes encoding paracrine factors are grouped in families that include multiple structurally related forms of the factor, and genes from a relatively small number of families control a host of developmental processes. These genes occur in different parts of the body, affect diverse aspects of the phenotype, and can interact. For example, the bone morphogenetic protein (BMP) family of paracrine factors contains 23 members. Originally identified by their ability to induce bone formation, BMPs also regulate cell division, differentiation, migration, and apoptosis (programmed cell death). Sonic hedgehog (Shh), a member of the Hedgehog family of paracrine factors, participates in determining the left–right body axis, the proximal–distal axis of limbs, and the formation of feathers, among other processes. Gremlin, a member of the deadenylating nuclease (DAN) family, blocks the action of BMP and works with Shh to regulate limb growth.

The *Runx2* gene encodes a transcription factor that regulates expression of genes associated with the formation of bone, including expression of BMPs. Acting through BMPs, Runx2 protein stimulates development of bone by inducing the formation of osteoblasts (bone-forming cells) and by delaying the conversion of osteoblasts to mature bone cells (osteocytes) that no longer form new bone. Changes in the timing and extent of expression of Runx2 and BMPs have profound effects on the phenotype of a developing embryo because early formation of osteoblasts or delayed conversion to osteocytes allows more bone to develop.

Alleles of *Runx2* differ in the relative number of glutamine and alanine residues they contain, and alleles with higher glutamine/alanine ratios are expressed more strongly, resulting in more active synthesis of bone. The effect of

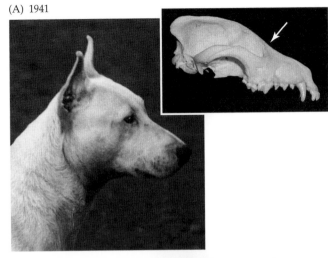

(A) 1941

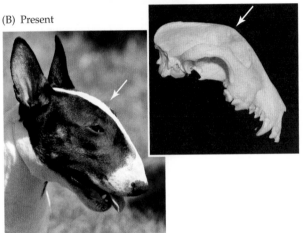

(B) Present

Figure 1.5 Rapid evolution of skull shape in bull terriers. Selective breeding has drastically changed the head shape of bull terriers. (A) In the early 20th century, the inflection point between the snout and the cranium (arrow) was at the level of the eyes, and the snout was horizontal. (B) In modern bull terriers, the inflection point has shifted behind the eyes, and the snout slopes downward. This change has resulted from selective breeding that favored alleles of *Runx2* with high glutamine/alanine ratios that cause strong expression of *Bmp*. Stronger expression of *Bmp* is associated with accelerated bone growth leading to the sharply angled snout. The *Runx2* gene from a bull terrier that died in 1931 has a glutamine/alanine ratio of 1.35, whereas the ratio for a modern bull terrier is 1.46. (Skull models from Fondon et al. 2004 © National Academy of Sciences, USA. Photographs: A, Mary Evans Picture Library/Alamy Stock Photo; B, Iurii Konoval/123RF.)

this variation in the glutamine/alanine ratio can be seen in changes in the heads of bull terriers that were produced by selective breeding between 1931 and 2004 (**Figure 1.5**). Human cleidocranial dysplasia (*kleis*, "clavicle"; *kranion*, "skull"; *dysplasia*, "abnormal form") is also the result of *Runx2* alleles that differ in glutamine/alanine ratio.

The effect of *Runx2* is not limited to selective breeding of domestic animals or to human deformities, however. Analysis of the glutamine/alanine ratio of *Runx2* alleles in 30 species of carnivores showed that the glutamine/alanine

---

[2]Paracrine factors are molecules that diffuse to neighboring cells and tissues, where they act through signaling cascades to activate transcription factors. Transcription factors are proteins that bind to DNA and control gene expression.

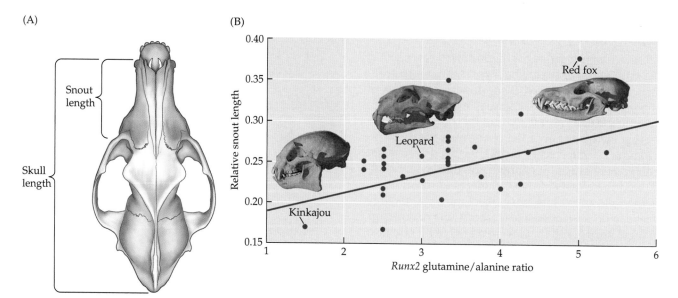

(A)

Snout
length

Skull
length

(B)

Kinkajou

Leopard

Red fox

Runx2 glutamine/alanine ratio

Figure 1.6 Changes in the snouts of carnivores. Increased glutamine/alanine ratios in *Runx2* among 30 species of Carnivora are associated with lengthening of the snout. (A) Relative snout length is expressed as the ratio of the snout length to the skull length. (B) A comparison of species shows that the relative snout length is proportional to the glutamine/alanine ratio of each species' *Runx2* gene. (B, data from Sears et al. 2007. Skulls from Digital Morphology; kinkajou by B. VanValkenburgh, 2006; leopard and fox by P. Owen, 2002.)

ratio of the *Runx2* alleles characteristic of each species correlates with the snout length of that species (**Figure 1.6**).

Three kinds of change in the expression of developmental genes can produce phenotypic variation that is subject to natural selection: the *time* during development that a gene is expressed, the *place* it is expressed, and *how strongly* it is expressed.

Heterochrony **Heterochrony** (Greek *heteros*, "different"; *chronos*, "time") refers to changes in the timing of gene expression during development. Heterochrony can involve the length of time during which a gene is expressed during development, as in the case of expression of the *Runx2* gene in bull terriers, or the time when one gene is expressed relative to expression of other genes. Heterochrony can occur at any stage of development, and can produce phenotypic changes in morphology, physiology, or behavior.

The body proportions of most vertebrates change substantially between infancy and maturity. For example, the head of a human infant is about 25% of its total body length, whereas the head of an adult is only about 13% of total body length. The cranium of an infant is large in proportion to its trunk, its snout is short, and its eyes are large in proportion to its head. The body proportions of infants explain why the rounded heads, big eyes, and short snouts of puppies and kittens are so appealing to humans. Infant fish also have proportionally larger heads and eyes than adult fish.

Because body proportions change during development, stopping development early produces an adult that retains body proportions that are characteristic of juveniles of its species, a phenomenon called **paedomorphosis** (Greek *pais*, "child"; *morph*, "form"). Paedomorphosis is a widespread form of phenotypic change among vertebrates, and the paedomorphic nature of the heads of birds compared with those of nonavian dinosaurs clearly reveals the role of heterochrony in the evolution of birds.

Once again, a comparison of the two extant groups of archosaurs—crocodylians and birds—provides evidence of developmental changes. Juvenile alligators may not be as cute as puppies, but they do have rounded heads, big eyes, and short snouts. Differential growth of the cranium, eyes, and snout during maturation leads to a very different morphology in adult alligators, which have flat heads, small eyes, and long snouts (**Figure 1.7A,B**). In contrast, the heads of juvenile and adult ostriches are much more alike (**Figure 1.7C,D**). Heterochrony—in this case early truncation of head development—has left adult birds with head proportions much like those of juveniles.

Heterotopy A change in the location of a gene's expression, called **heterotopy** ("different place"), can lead to dramatic phenotypic changes. During embryonic development, the fingers and toes of vertebrates are initially connected by a web of skin (remnants of this interdigital webbing are visible at the bases of your fingers and toes). Most of the webbing is lost before birth by apoptosis—except in web-footed aquatic vertebrates such as ducks, which retain interdigital webbing as adults. Embryonic ducks and chickens both have interdigital webbing, but chickens lose the webbing before hatching while ducks retain it. Apoptosis is initiated by *Bmp2*, *Bmp 4*, and *Bmp 7*, which are expressed in the webbing of both chickens and ducks. BMP-induced apoptosis removes the webbing between the toes of chickens, but the feet of ducks express Gremlin, an inhibitor of BMP that prevents apoptosis and leaves the webbing in place.

Figure 1.7 Paedomorphosis in the evolution of birds. The paedomorphic traits of extant birds can be understood by comparing alligators and ostriches. (A) A juvenile alligator has a short snout, rounded cranium, and eyes that are large relative to the rest of the head. (B) The head of an adult alligator has quite different proportions: the snout is long, the cranium is flat, and the eyes are small in proportion to the size of the head. (C) A juvenile ostrich also has a short snout, rounded cranium, and large eyes. (D) These proportions are little changed in the adult ostrich. (A © Gail Shumway/Getty Images; B, blickwinkel/Alamy Stock Photo; C, Penny Boyd/Alamy Stock Photo; D, PhotoKratky-Wildife/Nature/Alamy Stock Photo.)

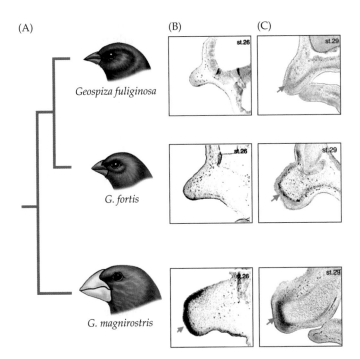

Figure 1.8 Heterometry, heterochrony, heterotypy, and beak shape. (A) Three species of ground-finches show different beak phenotypes, from the slim beak of *Geospiza fuliginosa* to the thick beak of *G. magnirostris*. (B) The dark color in the histological sections of developing beaks (indicated by arrow) shows the location and intensity of *Bmp4* expression at developmental stage 26. *Bmp4* is strongly expressed in the upper beak of *G. magnirostris*, is expressed less strongly in *G. fortis*, and is not expressed in *G. fuliginosa*. (C) At developmental stage 29, *Bmp4* expression is still intense in a large portion of the distal beak of *G. magnirostris* but is weaker and limited to a smaller area in the beaks of *G. fortis* and *G. fuliginosa*. (B,C from Abzhanov et al. 2004.)

The webbing between the digits in the wings of bats is also preserved by inhibition of BMP. In this case, both Gremlin and a second inhibitor of BMP, called fibroblast growth factor 8 (Fgf8), are expressed in the interdigital webbing, leaving the webbing intact to form the bat's wings. Additional examples of heterotopic changes include the reduction of limbs seen in many lizards and the complete absence of limbs in most extant snakes.

Heterometry Heterometry ("different measure") refers to a change in the *amount* of a gene product. A heterometric change in the production of Bmp4 is responsible for one of the classic examples in evolution, Darwin's finches. Found in the Galápagos and Cocos islands, this radiation of about 15 species, descended from a single ancestral species from South America, provides one of the best-studied examples of adaptive radiation and natural selection.

The ground-finches (*Geospiza*) forage for seeds on the ground, and the morphology of the beak is correlated with the kinds of seeds each species selects (**Figure 1.8**). Darwin's small ground-finch (*G. fuliginosa*) picks up small, soft seeds with its pointed beak, while the large ground-finch (*G. magnirostris*) uses its massive beak to crack hard-shelled seeds. The medium ground-finch (*G. fortis*) has a beak intermediate between those of the other two species and feeds on a broader range of seeds, but is less effective at opening both soft-shelled and hard-shelled seeds than either specialist.

The different beak phenotypes of these three species result from differences in expression of *Bmp4* in the upper beak. Early onset and a high level of expression of *Bmp4* produce the heavy beak of *Geospiza magnirostris*, whereas progressively later and weaker expression leads to the smaller beaks of *G. fortis* and *G. fuliginosa*.

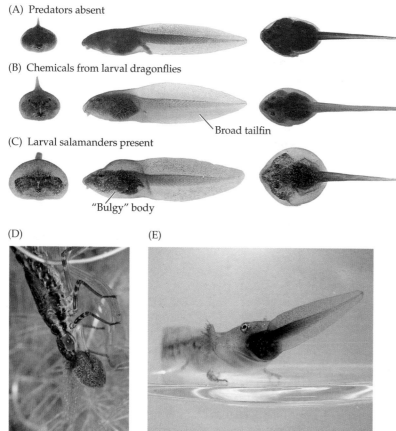

(A) Predators absent

(B) Chemicals from larval dragonflies

Broad tailfin

(C) Larval salamanders present

"Bulgy" body

(D)

(E)

Figure 1.9 Phenotypic plasticity of tadpoles. (A–C) Head-on, lateral, and dorsal views of tadpoles of the Hokaido brown frog (*Rana pirica*) show epigenetic changes produced by exposure to predators. (A) Tadpoles raised in the absence of predators have streamlined bodies with low tailfins. (B) Tadpoles exposed to chemicals released into the water by dragonfly larvae have streamlined bodies and broader tailfins. (C) Tadpoles living in the same aquaria as larval salamanders develop bulgy bodies and broad tailfins. (D) Dragonfly larvae stalk resting tadpoles, then seize and dismember them. In tadpoles, broad tails induced by exposure to chemicals released into the water by dragonfly larvae may increase acceleration, allowing tadpoles to evade attack. (E) Salamander larvae swallow tadpoles whole, and are thus gape-limited predators. A tadpole's bulgy body may make it impossible for a salamander larva to swallow the tadpole. (A–C, E courtesy of Osamura Kishida; D, Robert Henno/Alamy Stock Photo.)

Intragenerational versus transgenerational phenotypic modification

The changes in developmental regulatory genes we have described are transgenerational—that is, they are expressed in successive generations, and over evolutionary time they are responsible for the divergence of phylogenetic lineages.

A second class of developmental change is intragenerational. These changes affect only the generation that experiences them and are not passed on to the descendants of those individuals. Intragenerational developmental change results from modification of gene expression via **epigenetic** (Greek *epi*, "above") effects. Epigenetic mechanisms

modify the expression of genes without changing the DNA sequence of the gene.

A variety of factors, including diet, stress, temperature, and chemicals in the environment, can modify the behavior, physiology, or morphology of a developing organism, a phenomenon called **phenotypic plasticity**. For example, chemicals released into the water by predators of tadpoles lead to epigenetic changes in the tadpoles' body form and behavior that improve their chances of escaping predation (**Figure 1.9**). The phenotypic changes are limited to those individuals exposed to the stimulus; their offspring do not inherit the modified phenotypes.

1.5 Earth History and Vertebrate Evolution

Since their origin in the early Paleozoic, vertebrates have been evolving in a world that has changed enormously and repeatedly. These changes have affected vertebrate evolution both directly and indirectly. Understanding the sequence of changes in the positions of continents, and the significance of those positions regarding climates and interchange of faunas, is central to understanding the vertebrate story. These events will be described in Chapters 5, 13, and 23. The history of Earth has occupied three geological eons: the Archean, Proterozoic, and Phanerozoic. Only the Phanerozoic, which began about 541 million years ago, contains vertebrate life, and it is divided into three geological eras: Paleozoic (Greek *palaios*, "ancient"; *zoon*, "animal"), Mesozoic (Greek *mesos*, "middle"), and Cenozoic (Greek, *kainos*, "recent"). These eras are divided into periods, which can be further subdivided in a variety of ways, such as the epochs within the Cenozoic era.

Movement of landmasses—**continental drift**—has been a feature of Earth's history at least since the Proterozoic, and the course of vertebrate evolution has been shaped extensively by continental movements. The rate of continental drift remains low (movement of only about 1 millimeter a year) for tens of millennia. Then, as the strain begins to exceed the force holding a continent together, the rate of drift increases to 20 mm/year (about the rate at which fingernails grow) and a rift develops and widens in a geologically brief period. For example, the westward movement of North America relative to northern Africa remained at ~1 mm/year from 240 to 200 Ma, increased to ~12 mm/year from 200 to 190 Ma, increased again to ~20 mm/year from 190 to 180 Ma, then fell back to ~10 mm/year. A still greater acceleration, reaching 40 mm/year, accompanied the separation of southern Africa from South America between 128 and 120 Ma.

The continents still drift today. Indeed, Australia is moving north so rapidly that land-based latitudinal coordinates established in 1994 are now 1.5 m out of register with global coordinates determined by Global Positioning System (GPS) satellites. Because the movements are so complex, the sequence, the varied directions, and the precise timing of the changes are difficult to summarize. When the movements are viewed broadly, however, a simple pattern unfolds while vertebrates were diversifying: fragmentation during the Cambrian, coalescence by the Devonian, and a return to fragmentation by the Late Cretaceous.

Continents existed as separate entities more than 2 billion years ago. Some 300 million years ago, the separate continents combined to form a single landmass known as Pangaea, which was the birthplace of terrestrial vertebrates. Persisting and drifting northward as an entity, this huge continent began to break apart about 150 million years ago. Its separation occurred in two stages: first into Laurasia in the north and Gondwana in the south, and then into a series of units that have drifted and become the continents we know today.

The complex movements of the continents through time have had major effects on the evolution of vertebrates. Most obvious is the relationship between the location of landmasses and their climates. At the end of the Paleozoic, much of Pangaea was located on the Equator, and this situation persisted through the middle of the Mesozoic. Solar radiation is most intense at the Equator, and climates at the Equator are correspondingly warm. During the late Paleozoic and much of the Mesozoic, large areas of land enjoyed tropical conditions. Terrestrial vertebrates evolved and spread in these tropical regions. By the end of the Mesozoic much of Earth's landmass had moved out of equatorial regions, and by the middle of the Cenozoic most terrestrial climates in the higher latitudes of the Northern and Southern hemispheres were temperate instead of tropical.

A less obvious effect of the position of continents on terrestrial climates comes from changes in patterns of oceanic circulation. For example, the Arctic Ocean is now largely isolated from the other oceans and does not receive warm water via currents flowing from more equatorial regions. High latitudes are cold because they receive less solar radiation than do areas closer to the Equator, and the Arctic basin does not receive enough warm water to offset the lack of solar radiation. As a result, the Arctic Ocean has an extensive covering of ice, and cold climates extend well southward across the continents.

The Atlantic Meridional Overturning Circulation (AMOC) drives the Gulf Stream as it brings warm water north on the surface of the Atlantic and transports cold water south at greater depth, eventually to warm, rise to the surface, and move north again in the Gulf Stream. Heat transported by the AMOC is responsible for the relatively warm climates of northern North America and northern Europe. Sudden, drastic weakenings of the AMOC during the middle of the last ice age (~35,000 years ago) are associated with abrupt cooling events at intervals of about 1500 years. Icebergs released from the ice sheet covering Canada melted and diluted seawater in the North Atlantic, making it less dense so that it did not sink to form the southward flow of deep, cold water that drives the AMOC.

Another factor that influences climates is the relative levels of the continents and the seas. At some periods in Earth's history, most recently in the late Mesozoic and again in the first part of the Cenozoic, shallow seas flooded large parts of the continents. These **epicontinental seas** extended across the middle of North America and the middle of Eurasia during the Cretaceous and early Cenozoic, forming barriers between the eastern and western portions of those landmasses (see Chapters 13 and 23).

Water absorbs heat as air temperature rises, and then releases that heat as air temperature falls. Thus, areas of land near large bodies of water have maritime climates—they do not get very hot in summer or very cold in winter, and they are usually moist because water that evaporates from the sea falls as rain on the land. Continental climates, which characterize areas far from the sea, are usually dry with cold winters and hot summers. The draining of the epicontinental seas at the end of the Cretaceous probably contributed to the demise of the dinosaurs by making climates in the Northern Hemisphere more continental.

In addition to changing climates, continental drift has formed and broken land connections between the continents. Isolation of different lineages on different landmasses has produced dramatic examples of the independent evolution of similar types of organisms, such as the diversification of mammals in the mid-Cenozoic, a time when Earth's continents reached their greatest separation during the history of vertebrates.

Much of evolutionary history appears to depend on whether a particular lineage of animals was in the right place at the right time. This random element of evolution is assuming increasing prominence as more detailed information about the times of extinction of old groups and radiation of new groups suggests that competitive replacement of one group by another is not the usual mechanism of large-scale evolutionary change. The movements of continents and their effects on climates and the isolation or dispersal of animals are taking an increasingly central role in our understanding of vertebrate evolution.

On a continental scale, the advance and retreat of glaciers throughout the Pleistocene caused homogeneous habitats to split and merge repeatedly, isolating populations of widespread species and leading to the evolution of new species.

Summary

There are more than 67,000 extant species of vertebrates, ranging in size from 0.1 g to more than 100,000 kg.

Vertebrates occupy habitats extending from the depths of the sea to the tops of all but the highest mountains, and birds fly over those mountains.

Dietary specializations of vertebrates extend from carnivory (as predators, scavengers, and parasites) to herbivory, aided by specializations of the jaws, teeth, and digestive system.

Enormous as the diversity of extant vertebrates is, extinct species included forms so different from any living species that it is hard to reconstruct the lives they lived.

Vertebrates form two major categories—non-amniotes and amniotes—that differ in the origin of the membranes that surround the embryo.

The membranes that surround the embryos of non-amniotes are produced in the reproductive tract of the female, whereas those that surround the embryos of amniotes are outgrowths from the embryos themselves. The number of extant vertebrates is approximately evenly divided between non-amniotes and amniotes.

Fishes and amphibians are non-amniotes. Fishes are grouped in three evolutionary lineages:

- Jawless fishes: hagfishes and lampreys
- Cartilaginous fishes: sharks, rays, and chimaeras
- Bony fishes (Osteichthyes), which includes two radiations: ray-finned fishes (actinopterygians) and lobe-finned fishes (sarcopterygians). The sarcopterygian lineage includes all the descendants of the lobe-finned fishes, including amphibians, reptiles, birds, and mammals.

Amphibians include three lineages: salamanders, frogs, and caecilians.

Amniotes include two lineages, sauropsids and synapsids, which diverged in the mid-Carboniferous and represent independent paths to terrestrial life.

- Extant sauropsids include turtles; tuatara, lizards, and snakes; alligators and crocodiles; and birds.
- Extant synapsids include the three groups of mammals: monotremes, marsupials, and placentals.

Sauropsids, with more than 21,000 extant species, are much more diverse than synapsids (about 5,400 species).

Phylogenetic systematics groups organisms in evolutionary lineages on the basis of synapomorphies, homologous characters the organisms have in common and that differ from those of an ancestor.

Phylogenetic systematics produces branching diagrams of evolution (phylogenetic trees) showing the stepwise changes in characters. The new lineages formed at each branch point can be named, producing a nested series of named lineages of indeterminate length.

Parsimony is the basis for identifying the most likely sequence of evolutionary changes; any change is an unlikely event, so the phylogeny that requires the fewest changes to move from the ancestral distribution of characters to the current distribution is favored.

Branching sequences identified by phylogenies based on morphological or on molecular characters generally agree, and many phylogenies combine molecular and morphological characters. When molecular-based and morphology-based phylogenies disagree, the discrepancy usually lies in the estimated dates of divergence of lineages.

An extant phylogenetic bracket allows us to draw inferences about characters of the enclosed extinct lineages when direct evidence is lacking.

Crown groups are composed of the species that have all of the derived characters of the lineage; the members of a crown group can be extant or extinct. Stem groups include extinct taxa that lack some derived characters of other taxa in the lineage but are nonetheless more closely related to the taxa in the crown group than they are to taxa in other lineages.

Our understanding of evolutionary change has grown with increasing information about genetics and development.

Charles Darwin characterized evolution as "descent with modification" and invoked natural selection and sexual selection before Gregor Mendel described the particulate nature of genetic inheritance.

During the first part of the 20th century, the Modern Synthesis combined the perspectives of Darwinian selection and Mendelian genetics to explain the survival of the fittest as the result of differential reproduction of phenotypes. The Modern Synthesis does not explain the *origin* of the phenotypic variation on which selection acts—that is, the arrival of the fittest.

Summary (continued)

In the late 20th century, a blend of genetics and evolutionary developmental biology ("evo-devo") showed that genes acting early in embryonic development can produce profound phenotypic changes, providing the raw material for the action of selection.

Earth has changed dramatically during the half-billion years of vertebrate history.

Continents were scattered across the globe when vertebrates first appeared. They coalesced into one enormous continent, Pangaea, about 300 million years ago, and began to fragment again about 150 million years ago.

This pattern of fragmentation, coalescence, and fragmentation has isolated and renewed contact of major groups of vertebrates, producing the geographic distributions of lineages we see today.

Discussion Questions

1. Why doesn't phylogenetic systematics have a fixed number of hierarchical categories like Linnaean systematics?

2. What aspect of evolution does phylogenetic systematics represent more clearly than Linnaean systematics does?

3. Tetrapoda (see Figure 1.4) is a crown group; Tetrapodomorpha is the corresponding stem group. What organisms are included in Tetrapoda? In Tetrapodomorpha?

4. Look at Figure 1.4. Why can one human correctly say to another "you are a rhipidistian," but not "you are a lungfish?" How can you describe your relationship to lungfishes?

5. What inference can you draw from Figure 1.3 about parental care by pterosaurs?

6. For the sake of this question, suppose you have firm evidence that phytosaurs did not exhibit parental care. What would be the most parsimonious hypothesis about the appearance or disappearance of parental care in the archosaur lineage?

Additional Reading

Benton MJ, Donoghue PCJ, Asher RJ, Friedman M, Near TJ, Vinthe J. 2015. Constraints on the timescale of animal evolutionary history. *Palaeontologia Electronica* 18.1.1FC; 1-106; palaeo-electronica.org/content/fc-1.

Choi KY and 10 others. 2005. *Runx2* regulates FGF2-induced *Bmp2* expression during cranial bone development. *Developmental Dynamics* 233: 115–121.

De Queiroz A. 2014. *The Monkey's Voyage. How Improbable Journeys Shaped the History of Life.* Basic Books, New York.

Grant PR, Grant BR. 2002. Adaptive radiation of Darwin's finches. *American Scientist* 90: 130–140

Hand E. 2016. Crippled Atlantic currents triggered ice age climate change. http://www.sciencemag.org/news/2016/06/crippled-atlantic-currents-triggered-ice-age-climate-change.

Hennig W. 1966. *Phylogenetic Systematics.* University of Illinois Press, Urbana.

Hu Y and 10 others. 2017. Comparative genomics reveals convergent evolution between the bamboo-eating giant and red pandas. *Proceedings of the National Academy of Science* 114: 1081–1086.

Kishida O, Nishimura K. 2004. Bulgy tadpoles: Inducible defense morph. *Oecologia* 140: 414–421.

Pennisi E. 2001. Linnaeus's last stand? *Science* 291: 2304–2307.

Pennisi E. 2016. Tracking how humans evolve in real time. *Science* 352: 876–877.

Swiderski DL. 2001. Beyond reconstruction: Using phylogenies to test hypotheses about vertebrate evolution. *American Zoologist* 41: 485–607.

Varriale A. 2014. DNA methylation, epigenetics, and evolution in vertebrates: Facts and challenges. *International Journal of Evolutionary Biology* 2014, Article ID 475981.

2

What Is a Vertebrate?

Vertebrates are a diverse and fascinating group of animals. Because we are vertebrates ourselves, that statement may seem chauvinistic, but vertebrates are remarkable in comparison with most other animal groups. In this chapter, we discuss the relationship of vertebrates to other members of the animal kingdom, the structures that are characteristic of vertebrates, and the systems that make vertebrates functional animals. We need this understanding of the fundamentals of vertebrate anatomy, development, and physiology to appreciate the changes that have occurred during vertebrate evolution and to trace homologies between basal vertebrates and more derived ones.

Vertebrates are chordates, members of phylum **Chordata** (named for the notochord: Greek *notos*, "back"; *chorde*, "string"). Only the arthropods (Arthropoda, which includes the insects, crustaceans, and spiders) rival the vertebrates in diversity of forms and habitat. And it is only among mollusks (Mollusca, including snails, clams, and squids) that we find animals (such as octopuses and squids) that approach the large sizes of some vertebrates and that have a capacity for complex learning. **Urochordata** (Greek *oura*, "tail") (tunicates) and **Cephalochordata** (Greek *kephale*, "head") (cephalochordates) are grouped with vertebrates in Chordata. The relationship of chordates to other animal phyla is revealed by anatomical, biochemical, and embryonic characters.

2.1 Vertebrates in Relation to Other Animals

The "animal kingdom," more correctly called the Metazoa, is made up of more than 30 phyla, of which Chordata is one (**Figure 2.1**). All metazoans are multicellular heterotrophs (feeding on other organisms). They are also motile (capable of movement) for at least some part of their life cycle (many metazoans are motile as larvae but sessile as adults). Sponges and cnidarians are diploblastic, developing from two cell layers, the ectoderm and endoderm. All other phyla have three cell layers, including a mesoderm from which

the complex organ systems of the body cavity form. These derived phyla comprise the **Bilateria** (Latin, "two sides"); the name refers to the fact that at some point in their life cycle, whether as larvae or adults or both, bilaterians are distinguished by a body plan with two sides that are mirror images of each other.

Current molecular data place the origin of Bilateria late in the Proterozoic, probably between 688 and 596 Ma. Fossils of undisputed bilaterians are first seen at the very end of the Proterozoic (~553 Ma). Most bilaterian phyla first appear in the geological record during the early Cambrian (~541 Ma), and the term "Cambrian Explosion" has been used to describe this phenomenon. The apparent suddenness of this diversification of metazoans is probably related to two artifacts of the fossil record: (1) the frequency with which hard parts, such as shells, are preserved; and (2) a few fossil localities with exceptional preservation that record soft-bodied animals, which are usually absent from the fossil record.

There are two major divisions within Bilateria: **Protostomia** (Greek *pro*, a prefix meaning "earlier"; *stoma*, "mouth") and **Deuterostomia** (Greek *deuteros*, "second"). This division was originally based on embryonic features, including the development of the mouth and anus during **gastrulation**, another feature that defines all metazoans. During this period of early development, the undifferentiated cells of the very early embryo move into their distinctive cell layers and the body axes are defined.

The original embryonic gut opening is called the **blastopore** (Greek *poros*, "small opening"), and a second opening develops before gastrulation concludes. The classic view has been that the blastopore becomes the mouth of protostomes and the second opening becomes the anus, whereas in deuterostomes the second opening becomes the mouth and the blastopore becomes the anus. As with many ideas in science, new information has made the picture more complicated, and the mouths of protostomes and deuterostomes may actually be homologous. Nevertheless, protostomes and deuterostomes are firmly established as separate groups based on molecular differences.

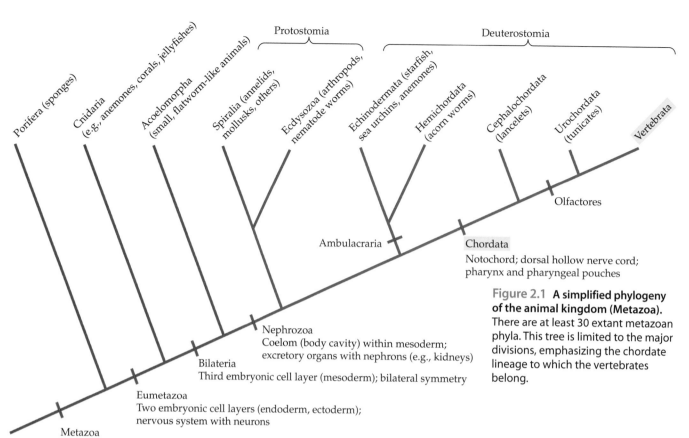

Figure 2.1 A simplified phylogeny of the animal kingdom (Metazoa). There are at least 30 extant metazoan phyla. This tree is limited to the major divisions, emphasizing the chordate lineage to which the vertebrates belong.

Protostomes comprise more than 90% of bilaterian phyla and are divided into two main groups: Ecdysozoa, including arthropods and nematodes; and Spiralia (sometimes called Lophotrochozoa), including mollusks and annelids (segmented worms, such as earthworms).

Deuterostomia is a much smaller grouping, but because it includes the chordates it is the division of interest here. Besides Chordata, there are two other deuterostome phyla: Echinodermata (starfishes, sea urchins, and similar animals) and Hemichordata (earthwormlike acorn worms and the fernlike pterobranchs). Both the fossil record and molecular data indicate that the common ancestor of extant deuterostomes was probably a small, marine, wormlike, free-swimming, filter-feeding animal.

Adult echinoderms lack a distinct head and have pentaradial (fivefold and circular) symmetry. Based on their appearance, it seems unlikely that they would be related to vertebrates, which superficially resemble other active animals, such as insects, in having distinct head and tail ends, jointed legs, and bilateral symmetry. However, this relationship is firmly established by both developmental and molecular information.

Hemichordates were long considered to be the sister group of chordates because both groups have pharyngeal slits, and hemichordates also have features of the pharynx that can be interpreted as the precursor to an endostyle (see Section 2.2). However, a mucus-producing, gill pore-bearing pharynx is now considered to be a primitive deuterostome feature, and this view is supported by a cluster of genes that are present in all deuterostomes and are expressed during pharyngeal development.

Although modern echinoderms lack pharyngeal slits, some extinct echinoderms appear to have had them. The diversity of extinct echinoderms was much greater than that of extant forms. Furthermore, ancestral echinoderm species were bilaterally symmetrical, meaning that the fivefold symmetry seen in adults of extant echinoderms is a derived character (echinoderm larvae are bilaterally symmetrical). Molecular characters currently unite echinoderms and hemichordates as **Ambulacraria** (see Figure 2.1). As we mentioned at the start of the chapter, lancelets and tunicates are placed with vertebrates in Chordata.

2.2 Characteristics of Chordates

Chordates are united by several shared derived features, which are seen in all members of the phylum at some point in their lives:

- A **notochord**, a dorsal (i.e., along the back) stiffening rod that gives phylum Chordata its name.
- A dorsal hollow nerve cord.

- A segmented, muscular postanal tail (i.e., a tail that extends beyond the gut region).
- A **pharynx** (throat region) with **pharyngeal slits**. Nonvertebrate chordates use the pharyngeal slits for filter feeding, and fishes use them for respiration (when they are properly called **gill slits**).
- An **endostyle**, a ciliated, glandular groove on the floor of the pharynx that develops as an outpocketing of the pharynx. In addition to secreting mucus that traps food particles during filter feeding, the endostyle takes up iodine and synthesizes thyroxine. Thus, the endostyle is homologous with the thyroid gland of vertebrates.

At least the first three of these features may have evolved in a larval stage that moved via a muscular tail rather than by ciliary action. The notochord serves not only as a stiffening rod but also as an attachment site for the segmented muscles that power swimming movements. In most vertebrates the notochord is a transient structure, replaced during embryonic development by the vertebral column and remaining in adults only as a portion of the intervertebral discs.

All chordates have a structure that pumps blood. In nonvertebrate chordates the pump is not a true heart, but simply an enlargement of the main ventral blood vessel, consisting of a single layer of contractile myocardial cells. Genes expressed in the development of this structure, however, are similar to those in the more complex hearts of vertebrates. Chordates also share some nervous system features, and the common ancestor probably had a brain similar to that of the cephalochordate lancelets.

Although chordates, like all bilaterians, are essentially bilaterally symmetrical animals, they have a left-to-right asymmetry within the body—for example, the heart is positioned on the left side while most of the liver is on the right side. (Rare individuals have a condition called *situs inversus*, in which the positions of the major body organs are reversed.)

Chordate origin and evolution

Compared with that of other metazoans, the internal organization of chordates appears to be "upside down." An earthworm, for example, has the nerve cord on the ventral (Latin *venter*, "belly") side and its heart and primary blood vessel on the dorsal (Latin *dorsum*, "back") side. In chordates, these positions are reversed. In the early 19th century, the French naturalist Étienne Geoffroy Saint-Hilaire proposed an "inversion hypothesis"—that the dorsal and ventral sides of chordates became inverted somewhere in their evolutionary history. This idea fell into disrepute for at least a century but has recently been revitalized by genetic and evolutionary developmental ("evo-devo") studies. During the course of development, chordates do indeed express genes on their dorsal side that are expressed on the ventral side in invertebrates. In addition, the position of the chordate mouth apparently moved. These changes seem to have happened at the evolutionary transition to chordates from other deuterostomes.

Although the notochord has long been considered a unique chordate feature, recent studies of gene expression indicate that it may be homologous with a ventral structure of annelids called the axochord. If this turns out to be so, it implies that some sort of notochord-like structure may be a basal bilaterian feature. Likewise, the endostyle may not be a unique chordate feature; genomic studies have identified a possible homolog in the epibranchial ridge of hemichordates.

Evidence from gene expression patterns during development shows that pharyngeal slits are homologous among all deuterostomes, to the extent that they represent endodermal outpocketings of the pharynx that perforate to the outside. Although pharyngeal slits are not found in extant echinoderms, there is evidence in the fossil record of their original presence, and they are present in their sister group, the hemichordates. Hemichordates have no mesodermal tissue associated with the slits, but they have associated collagenous supports formed from endodermal tissue that appear to be homologous with similar structures in cephalochordates.

All chordates have mesodermal tissue associated with the gill slits. (Gill supports are absent from tunicates, probably representing secondary loss, since tunicate gills are supported by the tunicate test—a semirigid outer covering, or "tunic," that gives these animals their name.) Vertebrates have complex pharyngeal arches between pharyngeal pouches, containing both mesoderm and supportive structures derived from neural crest cells (see Section 2.4). Although these supports are formed from a different developmental tissue than the supports in other deuterostomes, they are composed of a similar material: cartilage based on fibrillar collagen (although vertebrate cartilage is cellular, while that of other chordates is acellular).

Extant nonvertebrate chordates

The extant nonvertebrate chordates—lancelets (Cephalochordata) and tunicates (Urochordata)—are small marine animals. Other types of nonvertebrate chordates may have existed in the past, but such soft-bodied animals are rarely preserved as fossils.

Cephalochordates   The 25 species of lancelets, also known as amphioxus (Greek *amphis*, "double"; *oxys*, "sharp"), are small, superficially fishlike marine animals that are usually less than 5 cm long (**Figure 2.2**). Lancelets share with vertebrates some apparently derived anatomical features that are absent from tunicates. However, molecular analyses place lancelets as basal chordates, with tunicates as the sister group of vertebrates. Structures such as segmental muscles throughout the body and a tail fin are now considered to be basal chordate characters that have been lost in tunicates.

Lancelets employ a fishlike locomotion that is produced by **myomeres**—blocks of striated muscle fibers arranged along both sides of the body and separated by sheets of connective tissue (see Figure 2.10). Sequential contraction of

(A)

Anterior

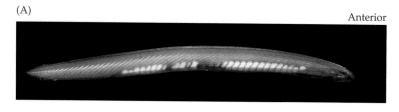

(B)

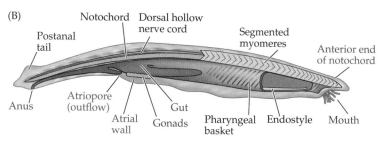

Postanal tail

Notochord

Dorsal hollow nerve cord

Segmented myomeres

Anterior end of notochord

Anus

Atriopore (outflow)

Atrial wall

Gut

Gonads

Pharyngeal basket

Endostyle

Mouth

Figure 2.2 Cephalochordates. (A) The lancelet *Branchiostoma lanceolatum* is a cephalochordate. Lancelets are widely distributed in marine waters of the continental shelves and are usually burrowing, sedentary animals as adults, although the adults of a few species retain an active, free-swimming behavior. (B) The mouth, gut, anus, and gonads (blue type) are ancestral bilaterian characters. Traditional characters of Chordata are in black type, and possible basal characters of chordates that have been lost in vertebrates are labeled in red. The cerebral vesicle (not visible in this view) is at the anterior end of the dorsal hollow nerve cord. (A, courtesy of Arthur Anker.)

(A)

Dorsal hollow nerve cord

Atrial siphon (outflow) and anus

Branchial siphon (inflow)

Pharynx

Endostyle

Intestine

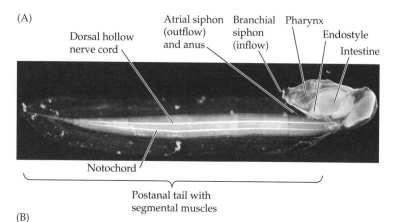

Notochord

Postanal tail with segmental muscles

(B)

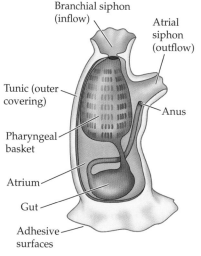

Branchial siphon (inflow)

Atrial siphon (outflow)

Tunic (outer covering)

Anus

Pharyngeal basket

Atrium

Gut

Adhesive surfaces

the myomeres bends the body from side to side, resulting in forward or backward propulsion. The myomeres of lancelets are simpler than those of vertebrates (V-shaped rather than W-shaped), and the left side is displaced by half a segment with respect to the right side.

The notochord acts as an incompressible elastic rod, extending the full length of the body and preventing the body from shortening when the myomeres contract. While the notochord of vertebrates ends midway through the head region, the notochord of lancelets extends from the tip of the snout to the end of the tail, projecting well beyond the region of the myomeres at both ends. This anterior elongation of the notochord apparently is a specialization that aids in burrowing.

The brain of lancelets, termed a **cerebral vesicle**, appears to be little more than a thickening at the front end of the spinal cord. But gene expression studies show that lancelets have all the genes that code for the vertebrate brain, with the exception of those directing formation of the front part of the vertebrate forebrain. These gene expression studies reinforce the growing understanding that differences in how, when, and where genes are expressed are as important as differences in what genes are present.

Urochordates Most adult tunicates are sessile animals, only a few centimeters long, that filter food particles from the water with a basketlike, perforated pharynx (**Figure 2.3**). Their thick cover (a "tunic" or "test") is made primarily from cellulose, a structural carbohydrate otherwise seen only in plants. Tunicates appear to have acquired the ability to synthesize cellulose by horizontal gene transfer from ancient

Figure 2.3 Urochordates. Tunicate larvae have a notochord, a dorsal hollow nerve cord, and a muscular postanal tail that moves in a fishlike swimming pattern. (A) Appendicularian tunicates are free-swimming throughout their lives. This species, *Oikopleura dioica*, is widely used in studies of development because the molecular basis of its development is similar to that of vertebrates. (B) The larvae of ascidian tunicates, such as *Ciona savignyi*, metamorphose into sedentary adults attached to the substrate. (C) Adults of these species have less obvious structural similarity to cephalochordates and vertebrates than their larval stages, but retain some modified chordate characters. (A, from Patel 2004, photo by Rüdiker Rudolf; B, Steve Lonnart/NOAA MBNMS.)

bacteria. Some species of tunicates, called larvaceans, retain the larval form throughout life.

Molecular analyses place tunicates and vertebrates as sister taxa, in a grouping called **Olfactores** (see Figure 2.1). The simple morphology of adult tunicates is considered to be highly derived rather than primitive, and tunicates have a greatly reduced genome in comparison with other chordates. This interpretation of tunicates implies that they have lost generalized chordate features such as segmentation, coeloms, and kidneys. Tunicates also share a few derived morphological features with vertebrates, such as a genetic distinction between the midbrain and hindbrain, embryonic cells that may be homologous with vertebrate neural crest cells (see Section 2.4), and a heart with a rhythmic pacemaker along with a cardiopharyngeal field during development.

2.3 What Distinguishes a Vertebrate?

Vertebrates take their name from vertebrae, the serially arranged bones that make up the spinal column, or backbone, of vertebrate animals. The vertebral column replaces (to a varying extent) the embryonic notochord. The earliest known vertebrates are found in Cambrian sediments. These fossils have segmented dorsal structures that have been interpreted as vertebrae.

Vertebrae form a **centrum** (plural *centra*) around the notochord during development and also encircle the nerve cord. Vertebrae with a centrum are found only among gnathostomes (vertebrates with jaws).

Although they are named for the vertebrae, vertebrates have additional unique characters, including the following:

- *Cranium.* All vertebrates have the uniquely derived feature of a cranium, or skull, which is a bony, cartilaginous, or fibrous structure surrounding the brain. The cranium, along with the vertebrae and gill

supports, comprises the endoskeleton, which would have been cartilaginous in basal vertebrates.

- *Head.* Vertebrates have a prominent head containing complex sense organs, including a nose, eyes, and ears. The vertebrate brain is tripartite, with genetically determined boundaries between the forebrain, midbrain, and hindbrain, and by a segmented hindbrain.

- *Pituitary.* Vertebrates have an anterior pituitary (adenohypophysis) that produces and regulates hormones. The anterior pituitary grows from the ectoderm lining the upper part of the mouth and joins the posterior part of the pituitary growing down from the brain.

- *Mineralized tissues.* Vertebrates deposit minerals (primarily calcium compounds) in tissues, creating rigid support structures (mineralized cartilage and bone) and surfaces with different degrees of resistance to abrasion (enamel, enameloid, dentine, and cementum).

Additional vertebrate features are described in **Table 2.1** and **Figure 2.4**, and are discussed in more detail in Sections 2.5 and 2.6.

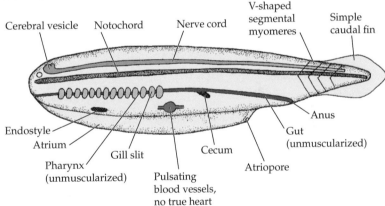

(A) **Amphioxus-like nonvertebrate chordate**

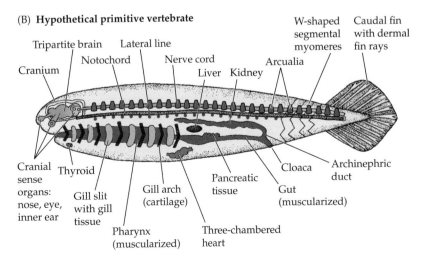

(B) **Hypothetical primitive vertebrate**

Figure 2.4 Generalized chordate structure compared with generalized vertebrate structure. (A) A generalized, lancelet-like nonvertebrate chordate. (B) A hypothetical ancestral vertebrate. The arcualia in the vertebrate are segmental structures flanking the nerve cord and are precursors of true vertebrae. The anus of the nonvertebrate chordate is now called the cloaca, because it is the exit not only for the gut but also for the urinary and reproductive systems.

Table 2.1	Comparison of features of nonvertebrate chordates and vertebrates
Generalized nonvertebrate chordates	**Vertebrates**[a]
Head and brain	
Notochord extends to anterior end of head (possibly a derived condition).	Head extends beyond the anterior tip of notochord.
Simple brain, no specialized sense organs.	Tripartite brain and multicellular sense organs.
Pharynx and respiration	
Gas exchange occurs across the body surface, gill arches are used for filter feeding.	Pharyngeal arches (= gill arches) are used primarily for respiration.
Pharynx is not muscularized, water is moved by ciliary action.	Muscular pharynx moves water by pumping.
Feeding and digestion	
Gut is not muscularized, food is moved by ciliary action.	Muscular contractions (peristalsis) move food.
Intracellular digestion: food particles are taken into cells lining the gut and digested within those cells.	Extracellular digestion: Food particles are digested in the gut and their breakdown products are absorbed.
Heart and circulation	
Open circulatory system, with few capillaries.	Closed circulatory system with an extensive capillary system.
No true heart; blood is moved by contracting regions of the vessels.	A three-chambered heart (sinus venosus, atrium, and ventricle) pumps blood.
No neural control of the heart.	Rate and force of cardiac contraction are under neural control.
No respiratory pigment or red blood cells; O_2 and CO_2 are transported in solution.	Red blood cells contain hemoglobin, which binds O_2 and CO_2 and aids in their transport and delivery to tissues.
Osmoregulation	
Body fluids have essentially the same osmolal concentration and ion composition as the external environment.	The kidneys and (in aquatic vertebrates) ion transporting cells in the skin and gills maintain internal osmolal concentrations and ion compositions that are different from the external environment.
Support and locomotion	
Cartilage is acellular.	Cartilage is cellular.
Notochord stiffens body.	Vertebral elements stiffen the body of fishes and support body weight of terrestrial vertebrates.
Myomeres are V-shaped.	Myomeres are W-shaped.
No fins (= limbs).	Fins (aquatic) or limbs (terrestrial) are present.

[a]Some exceptions exist among hagfishes and lampreys.

2.4 Vertebrate Embryonic Development

The development of a single fertilized egg into the differentiated cell types, tissues, and organs, and the organization of these elements into the functioning systems of a multicellular animal, is remarkable and complex. Scientists no longer adhere to the maxim "ontogeny recapitulates phylogeny"—the idea that an embryo passes through its ancestral evolutionary stages in the course of its development—proposed by the 19th-century embryologist Ernst Haeckel. Nevertheless, the study of embryonic development and its physical, functional, and evolutionary constraints and opportunities provides clues about the ancestral condition and about homologies between structures in different animal groups, and embryology provides important background information for many studies.

Development of the body

The bodies of all bilaterian animals are formed from three distinct tissue layers, called germ layers (Latin *germen*, "bud"). The fates of these germ layers have been largely conserved throughout vertebrate evolution.

- The outermost layer—the **ectoderm**—forms the superficial layers of skin (the epidermis), the linings of the most anterior and most posterior parts of the digestive tract, and the majority of the nervous system.
- The innermost layer—the **endoderm**—forms the rest of the digestive tract's lining, as well as the lining of glands associated with the gut, including the liver and pancreas. Endoderm also forms most respiratory surfaces of vertebrate gills and lungs, the taste buds, and the thyroid, parathyroid, and thymus glands.

- The middle layer—the **mesoderm**—forms everything else: muscles, skeleton (including the notochord), connective tissues, and the circulatory and urogenital systems (including the heart and kidneys).

In vertebrates, a split develops within the originally solid mesoderm layer, forming a **coelom**, or body cavity. The coelom is the cavity containing the internal organs: the peritoneal cavity around the viscera, and the pericardial cavity around the heart. In mammals, the pleural cavity around the lungs is a partition of the peritoneal cavity. These coelomic cavities are lined by thin sheets of mesoderm—the peritoneum (around the pleural or peritoneal cavity) and the pericardium (around the heart). The gut is suspended in the peritoneal cavity by sheets of peritoneum called mesenteries.

The vertebrate body develops in a segmented fashion from front to back, with each segment having an initial component of spinal nerves, major blood vessels, progenitors of bone and muscles, and other internal structures. This segmentation can be seen in the arrangement of human vertebrae and ribs. **Figure 2.5** shows how the developing mesoderm is divided into dorsal segmented portions that include the **somites** and nonsegmented lateral portions called the **lateral plate**:

- The derivatives of the somites are the axial skeleton, dermis, and striated muscles. The dorsally forming striated muscles and dermis later grow ventrally to encircle the body and are innervated by the somatic, or voluntary, nervous system.
- The derivatives of the lateral plate mesoderm are principally the viscera, smooth muscle lining the gut, and cardiac muscles of the heart, and are innervated by the visceral, or involuntary, nervous system.

This description applies to limbless vertebrates such as cyclostomes. When limbs are added, the situation becomes more complicated, as will be discussed in Chapter 3.

Development of the pharyngeal region

Cells of the neural crest (see below) form many of the structures in the anterior head region, including the cartilaginous arches supporting the gills. The muscles that move these arches to pump water are formed from mesoderm, but their tendons and supporting connective tissue (fascia) are formed from neural crest.

Figure 2.6 shows the stage in early embryonic development at which internal **pharyngeal pouches** make at least a fleeting appearance in all vertebrate embryos. **Pharyngeal arches** are delimited on the outside by grooves called the pharyngeal clefts and on the inside by the pharyngeal pouches. The pharyngeal arches are an astounding example of serial homology conserved across vertebrate lineages. Each arch is innervated by a particular cranial nerve and supplied with blood by a particular branch of the aortic arches, and the skeletal and muscular elements derived from the arch form homologous structures in the head and neck region (**Table 2.2**).

(A) Cross section

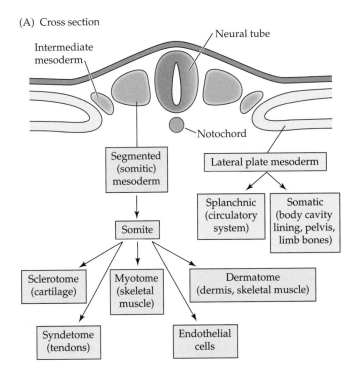

(B) Dorsal view

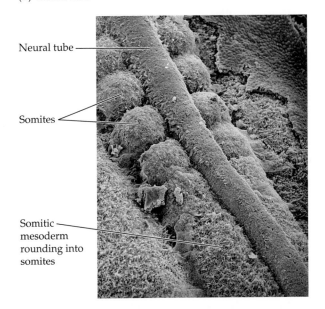

Figure 2.5 Major divisions of the mesoderm. (A) The segmented (somitic) mesoderm (also known as paraxial mesoderm) is located near the midline of the developing vertebrate embryo and gives rise to structures that are segmental (i.e., repeating in an anterior-to-posterior direction). The unsegmented lateral plate mesoderm, located laterally on the embryo, forms structures that are not segmentally compartmentalized. (Be careful to note the distinction between *somatic*, referring to the body, and *somitic*, referring to the somites.) Between these two types lies the intermediate mesoderm, or nephrotome, which gives rise to the kidneys and gonads. (B) When the surface ectoderm is peeled away, scanning electron microscopy reveals somites forming adjacent to the neural tube. Development proceeds from the anterior end of the embryo toward the posterior, where somites are just beginning to form. (A, after Gilbert and Barresi 2016; B, courtesy of Kathryn Tosney.)

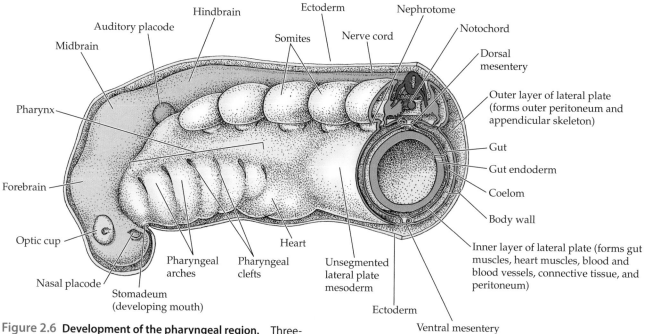

Figure 2.6 Development of the pharyngeal region. Three-dimensional view of a portion of a generalized vertebrate embryo at the developmental stage (called the pharyngula or phylotypic stage) when the developing pharyngeal pouches appear. The ectoderm is stripped off the left side, showing segmentation of the mesoderm in the trunk region and pharyngeal development. (After Kingsley 1912.)

Table 2.2	Fate of the pharyngeal arches and pouches					
Pouch	Cranial nerve supply	Aortic arch blood supply	Arch derivatives (fishes)	Arch derivatives (mammals)[a]	Pouch derivatives (bony fishes)	Pouch derivatives (mammals)
1 = Mandibular	V (trigeminal)	1st arch never develops as an independent arch	Upper and lower jaws and associated muscles	Incus and malleus in middle ear Muscles of mastication	Spiracle	Middle ear cavity and Eustachian tube
2 = Hyoid	VII (facial)	Stapedial artery	Hyomandibular and hyoid apparatus and associated muscles	Stapes (middle ear) Dorsal portions of hyoid, and stylohyoid muscle	Thymus gland	Nonmammalian amnotes: Mucus gland Mammals: Palatine tonsils
3 = 1st branchial	IX (glosso-pharyngeal)	Carotid artery	1st gill arch and associated muscles	Muscles of facial expression	Thymus gland	Thymus and inferior parathyroid gland
4 = 2nd branchial	X (vagus)	Systemic aorta	2nd gill arch and associated muscles	Thyroid cartilage Extrinsic laryngeal muscles	Thymus gland	Superior parathyroid gland
5 = 3rd branchial (absent in amniotes)	X (vagus)	Lost in adult tetrapods	3rd gill arch and associated muscles	—	Thymus Internal gill buds	—
6 = 4th branchial (rudimentary in amniotes)	X (vagus) (recurrent laryngeal branch in amniotes)	Pulmonary artery	4th gill arch and associated muscles	Arytenoid cartilages Intrinsic laryngeal muscles	Calcitonin-producing glands	Calcitonin-producing glands

[a] Only the major muscles are included

Table 2.3	Regions of the vertebrate brain	
Region	**Structure**	**Characteristics**
Forebrain	Telencephalon	Develops in association with the olfactory capsules and coordinates inputs from other sensory modalities. The area for olfaction appears to be a unique feature of vertebrates. The telencephalon becomes enlarged in derived vertebrate groups and is known as the cerebrum or cerebral hemispheres and is the area responsible for associative processing of information
	Diencephalon	The pineal organ, a median dorsal outgrowth, regulates responses to changes in photoperiod. In sauropsids the pineal organ responds to light that enters through an opening in the center of the skull (the pineal eye, which is present in tuatara and many extant lizards) or through the bone of the skull (other lizards and birds)
		The floor of the diencephalon (the hypothalamus) and the pituitary gland (a ventral outgrowth of the diencephalon) form the primary center for neural-hormonal coordination and integration
Midbrain	Mesencephalon	The tectum (dorsal portion) receives sensory information (auditory and touch). The tegmentum (ventral portion) is the pathway for incoming sensory information and outgoing responses to the forebrain
Hindbrain	Metencephalon	The cerebellum, a dorsal outgrowth, is present as a distinct structure only in gnathostomes, where it coordinates and regulates motor activities (whether reflexive, such as maintenance of posture, or directed, such as escape movements
	Myencephalon (medulla oblongata)	Controls automatic functions such as respiration and circulation

Development of the brain

Even at the very early embryonic stage shown in Figure 2.6, the three regions of the brain have started to differentiate (**Table 2.3**). The **forebrain** includes olfactory and visual sensory structures (seen in the figure developing from the optic cup and nasal placodes) and will become an integrative and associative area. The **midbrain** becomes the pathway for incoming information from organs of hearing and touch and for outgoing motor responses. The **hindbrain** controls involuntary motor activities, including respiration (breathing) and circulation (heartbeat).

Unique developmental features of vertebrates

Four phenomena built on the foundation of basal chordate genes may account for many of the differences between vertebrates and other chordates: duplication of the Hox gene complex, microRNAs, the appearance of a new embryonic tissue (the neural crest), and neural placodes.

Duplication of Hox genes **Hox genes** regulate expression of the hierarchical network of developmental genes described in Chapter 1. Hox genes occur in clusters that are arranged in the same linear sequence as the structures they control along the anterior-posterior axis of the body. The gene at the anterior end of a cluster is the first to be expressed, and is expressed in the head region. Expression of the following genes proceeds posteriorly, with the gene in the posteriormost position expressed at the tip of the abdomen.

Although all animals, with the possible exception of sponges, have Hox genes (they were first identified in fruit flies), vertebrates have more of these gene clusters than other groups. All invertebrates examined so far have a single cluster, while vertebrates have as many as 13 clusters.

This increase in the number of Hox gene clusters is the result of repeated duplication of the entire Hox gene complex during vertebrate evolution.

The first duplication event seems to have occurred at the start of vertebrate evolution, because lancelets and tunicates have a single Hox cluster, and cyclostomes probably had two clusters ancestrally (**Figure 2.7**). A second duplication event had taken place by the evolution of gnathostomes, because all jawed vertebrates have at least four Hox gene clusters. A third duplication event occurred in teleost fishes (derived bony fishes) and a fourth duplication occurred in salmonid fishes. Paralog[1] Hox genes have fine-tuned their functions, dividing control of a region that was originally controlled by a single gene among several paralog genes. These changes could allow variation in the timing and intensity of expression of downstream developmental genes, resulting in changes in the phenotype of an organism. Thus, doubling and redoubling of Hox gene clusters during vertebrate evolution is believed to have contributed to the structural complexity of vertebrates.

MicroRNAs MicroRNAs are noncoding RNA sequences about 22 bases long that have been added to the metazoan genome throughout evolutionary history. MicroRNAs regulate the synthesis of proteins by binding to complementary base sequences of messenger RNAs. Chordates are distinguished from other deuterostomes by the addition of two unique microRNAs. Another three are shared by vertebrates and tunicates, and all vertebrates possess an additional 41 unique microRNAs. Beyond these 41 shared nucleotide sequences, the different vertebrate lineages

---

[1] Paralogs arise when duplicated genes develop new functions (usually related to their original function).

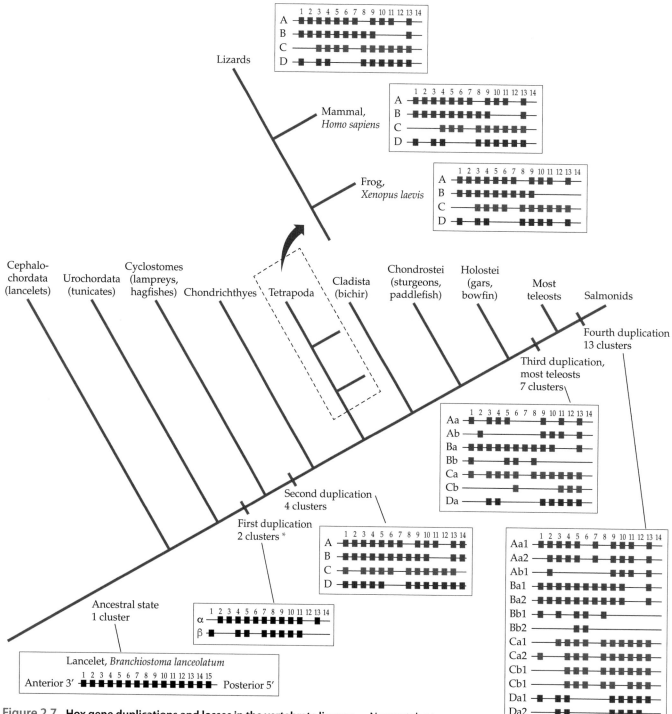

Figure 2.7 Hox gene duplications and losses in the vertebrate lineage.   Nonvertebrate chordates (cephalochordates and urochordates) have a single cluster of Hox genes. The first duplication corresponds with the origin of vertebrates (cyclostome ancestor) and resulted in two Hox gene clusters.* A second duplication occurred at the origin of gnathostomes, which ancestrally had four Hox clusters (*HoxA*, *HoxB*, *HoxC*, *HoxD*). Most teleosts have seven Hox clusters—the result of a third duplication followed by the loss of one cluster. A fourth duplication and another loss occurred in salmonids, which have 13 clusters. Lineage-specific losses have added variety. Among teleosts, for example, zebrafishes lost a duplicate *HoxD* cluster, whereas cichlids lost a duplicate *HoxC* cluster (see Kasahara 2007; Ravi et al. 2009; Heimberg and McGlinn 2012; Kadota et al. 2017). In addition, each lineage has lost specific Hox genes (indicated by missing boxes) within the clusters. (*Note that although the ancestral state for cyclostomes is believed to be two Hox clusters, both lampreys and hagfishes have lineage-specific Hox cluster duplications. As a result, extant lampreys have three or four Hox clusters, and hagfishes may have as many as seven.)

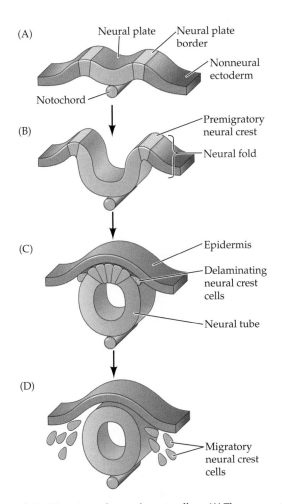

(A) Neural plate | Neural plate border
Nonneural ectoderm
Notochord

(B) Premigratory neural crest
Neural fold

(C) Epidermis
Delaminating neural crest cells
Neural tube

(D) Migratory neural crest cells

Figure 2.8 Migration of neural crest cells.   (A) The sequential steps in formation of neural crest cells begins with their specification at the border of the neural plate. (B) As the neural plate folds to form the neural tube, the neural crest cells are carried to the top of the fold. (C, D) As the neural tube closes, neural crest cells delaminate (separate from the epidermal sheet) and migrate out of the ectoderm. (After Gilbert and Barresi 2016.)

the aorta. Vertebrates more derived than cyclostomes have additional features formed from neural crest; these include the sympathetic nervous system and the dentine found in teeth and bony exoskeleton tissues. Neural crest tissue also interacts extensively with other tissues in forming these structures.

No neural crest-like cells are found in lancelets, but lancelet larvae do express a gene during the formation of the neural plate that appears to be homologous with a vertebrate gene known to be active in neural crest formation, and more definitively homologous genes are expressed in some tunicates. Cells resembling migratory neural crest cells have been identified in the larval stage of some tunicate species, where they differentiate into pigment cells. Thus, it appears that the neural crest of vertebrates had antecedents among nonvertebrate chordates.

Epidermal placodes   Another type of vertebrate embryonic tissue is somewhat similar to neural crest and forms placodes—thickenings of the embryonic epiderm that give rise to various sensory and cutaneous structures in different lineages, including teeth, hair, feathers, and mammary glands. Of special note are the cranial sensory placodes that give rise to the vertebrate ears, nose, and eyes. Nonvertebrate chordates have only epidermal sensory cells, but homologs to vertebrate sensory placodes may exist in tunicates as placodes that form the oral and atrial siphons, and even in lancelets, which may have a homolog to the placode that forms the anterior pituitary of vertebrates. Some placode cells migrate caudally to contribute, along with the neural crest cells, to the lateral line system of fishes and amphibians and to the cranial nerves that innervate it.

2.5 Basic Vertebrate Structures

At the whole-animal level, increased body size and increased activity levels distinguish vertebrates from nonvertebrate chordates. Early vertebrates generally had body lengths of 10 cm or more, which is about an order of magnitude larger than nonvertebrate chordates. Because of their relatively large size, vertebrates need specialized systems to carry out processes that are accomplished by diffusion or ciliary action in smaller animals.

Because vertebrates are more active than other chordates, they need organ systems that can carry out physiological processes at a greater rate. The transition from nonvertebrate chordate to vertebrate was probably related to the transition from filter feeding to a more actively predaceous mode of life, as shown by the features of the vertebrate head (largely derived from neural crest tissue) that enable suction feeding with a muscular pharynx, and a bigger brain and more complex sensory organs. A muscularized, chambered heart also relates to higher levels of activity and the need to transport oxygen in the circulatory system.

independently acquired yet more microRNAs of their own, and mammals in particular have a large number of novel microRNAs. The vertebrate microRNAs are involved in regulating the development of some derived structures, including the liver and kidneys.

Neural crest   The **neural crest** is seen only in vertebrate development, and it is perhaps the most important innovation in the origin of the vertebrate body plan. The neural crest can actually be considered as a fourth germ layer, in addition to ectoderm, mesoderm, and endoderm (**Figure 2.8**). The key features of neural crest cells are their migratory ability and their multipotency—their ability to differentiate into many different types of cells as the body develops. In fact, the neural crest gives rise to a greater number of cell types than does mesoderm.

Neural crest forms many uniquely vertebrate structures, including the adrenal glands, pigment cells in the skin, secretory cells of the gut, and the smooth muscle tissue lining

Here we provide a brief introduction to some aspects of vertebrate anatomical structure and function, contrasting the vertebrate condition with that of lancelets. Evolutionary changes and specializations of these structures and systems will be described in later chapters.

Adult tissue types

The body of a vertebrate includes four kinds of tissues with specific characteristics and functional properties:

1. **Epithelial tissues** consist of sheets of tightly connected cells and form the boundaries between the inside and outside of the body and between compartments within the body.

2. **Muscle tissues** contain the filamentous proteins actin and myosin, which together cause muscle cells to contract and exert force.

3. **Neural tissues** include neurons (cells that transmit information via electric and chemical signals) and glial cells (which form an insulative barrier between neurons and surrounding cells).

4. **Connective tissues** are diverse. They provide structural support, protection, and strength. Connective tissues include not only mineralized tissues such as bone and cartilage that form the skeleton (see the next section), but also adipose (fat) tissues, blood, and flexible tendons and ligaments.

Tissues combine to form larger units called organs, which usually contain most or all of the four basic tissue types. Organs in turn are combined to form systems, such as the circulatory, respiratory, and excretory systems.

Mineralized tissues

Some connective tissues are mineralized (Table 2.4). Mineralized tissues are composed of a complex matrix of collagenous fibers, cells that secrete a proteinaceous tissue matrix, and crystals of the calcium-containing compound hydroxyapatite. Hydroxyapatite crystals are aligned on a matrix of collagenous fibers in layers with alternating directions, much like the structure of plywood. This combination of cells, fibers, and minerals gives bone its complex latticework structure, which combines strength with relative lightness and helps prevent cracks from spreading. It is mineralized tissues that readily fossilize, and thus supply most of the information we have about extinct vertebrate groups; thus you will see a preponderance of discussion of these tissues throughout this book as we trace the evolutionary history of vertebrates.

Bone   There are three main types of bone in vertebrates. **Dermal bone** (sometimes called membrane bone) is formed in the skin and lacks a cartilaginous precursor. **Endochondral bone** is made up of osteocytes that form from a cartilaginous precursor deep within the body. Endochondral bone appears to be unique to bony fishes and tetrapods, in which it forms the internal skeleton (endoskeleton). **Perichondral bone** is like dermal bone in some respects and like endochondral bone in others. It is formed deep within the body, like endochondral bone, but forms in the perichondral membrane around cartilage or bone and thus has some characters of dermal bone. The genes expressed in the formation of the endoskeleton are absent from both lampreys and chondrichthyans (cartilaginous fishes), but the presence of bone in ostracoderms (extinct jawless fishes; see Chapter 3), which are less derived than any jawed vertebrate, indicates that bone is an ancestral character for gnathostomes.

Teeth   Odontodes were the toothlike components of the ancestral vertebrate dermal armor and are homologous with teeth. Recent studies indicate that odontodes were the primary elements, and that teeth evolved from cooption of the odontodes to become oral elements. Teeth are harder than bone and more resistant to wear because they are composed of dentine and enamel or enameloid, which are more mineralized than bone.

Teeth can form in either ectodermal or endodermal tissue (or a mixture of both), and can form with or without the presence of jaws. Ectoderm lines most of the vertebrate mouth, but endoderm lines the posterior mouth and the pharynx. The main determinate of tooth formation is the dentine-forming neural crest in the mesodermal layer. The enamel-forming ameloblasts are usually derived from the ectoderm, but they can also be derived from the endoderm.

The skeletomuscular system

The basic endoskeletal structural features of chordates are the notochord, acting as a dorsal stiffening rod running along the length of the body, and a gill skeleton keeping the gill slits open. The cranium surrounding the brain,

Table 2.4	Mineralized tissues of vertebrates	
Tissue	**Description**	**Mineral content**
Enamel, dentine	Found in teeth, in some fish scales, and in the mineralized exoskeleton of ancestral vertebrates	Enamel ~96% Dentine ~90%
Enameloid	Enamel-like tissue present in ancestral vertebrates; found today in the teeth and scales of extant cartilaginous fishes	~96%
Bone	Forms the internal skeleton of bony fishes and tetrapods, as well as certain external structures, such as antlers. *Dermal bone* forms in the skin without a cartilaginous precursor; *endochondral bone* forms within the body with a cartilaginous precursor; *perichondral bone* forms in the perichondral membrane around bones	~70%
Mineralized cartilage	Found in the vertebrae of chondrichthyans	~70%
Cementum	In some vertebrates (including mammals), fastens the teeth in the sockets	~45%

collagenous supports (radials) in the caudal fin, and vertebral elements were present in the earliest vertebrates. Later vertebrates added a dermal skeleton of external plates and scales, essentially forming an exoskeleton (**Figure 2.9**). We think of vertebrates as possessing only an endoskeleton, but most of our skull is dermal bone that forms a shell around our brain. The basic condition for vertebrates more derived than cyclostomes is a cartilaginous endoskeleton and a bony exoskeleton. The appendicular skeleton (bones of the limb skeleton and limb girdles) appeared in gnathostomes, although a pectoral (front) limb is seen in some derived ostracoderms.

The cranial skeleton   The skull, or cranium, of most extant vertebrates is formed from three components:

1. The **chondrocranium** (Greek *chondros*, "gristle"; New Latin *cranium*, "skull") formed anteriorly from neural crest tissue and posteriorly from regular mesoderm.

2. The **splanchnocranium** (Greek *splanchnon*, "viscera") forming the gill supports, formed from neural crest.

3. The **dermatocranium** (Greek *derma*, "skin") that forms in the skin as an outer cover—again formed anteriorly from neural crest tissue, and posteriorly from regular mesoderm.

A splanchnocranium of some sort is present in nonvertebrate chordates and hemichordates, although it is formed from endoderm rather than neural crest, and the splanchnocranium is probably a basal deuterostome feature. In vertebrates it becomes integrated with the chondrocranium. A dermatocranium was not a feature of the earliest vertebrates, and is absent in cyclostomes.

The splanchnocranial components of the vertebrate skeleton are known by a confusing variety of names:

visceral arches, gill arches, pharyngeal arches, and branchial arches. We will call these structures pharyngeal arches when we are discussing the embryonic elements of their development. In adult gnathostomes, the first two arches (the mandibular and hyoid arches) form the jaws and jaw supports. In adult fishes, arches 3–7 bear gill tissue, and we will call these gill arches. We will use "branchial" as a general term referring to these arches and the structures derived from them.

Cranial muscles   There are two main groups of striated muscles in the head of vertebrates: the extrinsic eye muscles and the muscles associated with the pharyngeal arches. The pharyngeal (branchiomeric) muscles of the mandibular arch form the muscles of mastication in all gnathostomes: in fishes, muscles of more posterior arches power gill pumping, while in tetrapods they have assumed other functions (see Table 2.2). A muscular pharynx is usually thought to be a unique vertebrate feature, but vertebrate branchiomeric muscles may be homologous with the muscles in the walls of the atrium in lancelets, and they share a common pattern of innervation.

Six muscles in each eye rotate the eyeball in gnathostomes; lampreys have seven muscles, but these muscles have apparently been lost in hagfishes along with the reduction of their eyes. In all vertebrates, the muscles moving the eyes derive from the premandibular mesoderm (formed from neural crest tissue in the region in front of the mouth).

Axial skeleton and musculature   The notochord is the original "backbone" of all chordates. It has a core of large, closely spaced cells packed with incompressible fluid-filled vacuoles wrapped in a complex fibrous sheath that is the site of attachment for segmental muscles and connective tissues.

In vertebrates, the notochord is supplemented by the vertebrae, and—in gnathostomes—additionally by ribs that articulate with the vertebrae. The notochord is lost in adult tetrapods, but portions remain as components of the intervertebral discs.

The axial musculature is composed of segmental myomeres that are complexly folded in three dimensions so that each one extends anteriorly and posteriorly

(A) Dermal skeleton (exoskeleton)

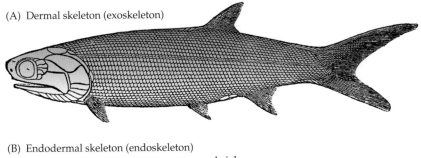

(B) Endodermal skeleton (endoskeleton)

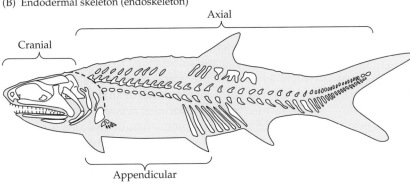

Figure 2.9 Dermal and endodermal components of the skeleton.   (A) The originally dermal bone exoskeleton. Unlike the scales of most extant fishes, these scales are bony. (B) The originally cartilaginous bone endoskeleton is formed by bone in extant bony fishes. The animal depicted is *Palaeoniscus macropomus*, a fish from the Permian. Like most palaeoniscids, it was completely covered by dermal bone, which formed rhomboidal scales on the body and plates on the head. After Lankester and Ridewood 1908 and Maisey 1996, © Patricia Wynne 1996.)

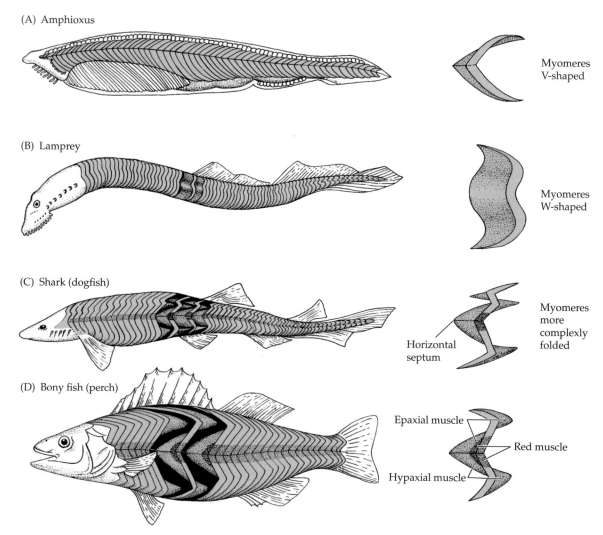

(A) Amphioxus

Myomeres
V-shaped

(B) Lamprey

Myomeres
W-shaped

(C) Shark (dogfish)

Horizontal
septum

Myomeres
more
complexly
folded

(D) Bony fish (perch)

Epaxial muscle

Red muscle

Hypaxial muscle

Figure 2.10 Chordate axial muscles are used for locomotion. Chordate characters include overlapping, sequential blocks of axial muscles called myomeres. When myomeres contract, they produce undulation of the body. (A) In the nonvertebrate chordate lancelet myomeres have a simple V shape. (B–D) The myomeres of vertebrates have a W shape. (B) Cyclostomes (hagfishes and lampreys) are jawless basal vertebrates with fairly simple myomeres. (C, D) The myomeres of jawed vertebrates are divided into epaxial (dorsal) and hypaxial (ventral) portions by a sheet of fibrous tissue called the horizontal septum. This division correlates with the evolution of paired fins, as we will discuss in Chapter 3. (After Bone and Marshall 1982.)

over several body segments (**Figure 2.10**). The segmental pattern of the axial muscles is clearly visible in fishes. It is easily seen in a piece of raw or cooked fish where the flesh flakes apart in zigzag blocks, each block representing a myomere. The pattern is less obvious in tetrapods, but the segmental pattern can be observed on the six-pack stomach of body builders, where each ridge represents a segment of the rectus abdominis muscle.

2.6 Basic Vertebrate Systems

An animal is more than the sum of its parts, and the next level of organization lies in integrated combinations of structures forming the systems that carry out basic functions. Here we describe the basic characteristics of several systems. Later chapters will describe their modifications in particular taxa.

The alimentary system

Feeding and digestion are accomplished by the alimentary system. Feeding includes getting food into the mouth, mechanical processing ("chewing" in the broad sense), and swallowing. Digestion includes the breakdown of complex compounds into small molecules that are absorbed across the wall of the gut and transported to the tissues.

Most vertebrates are particulate feeders—that is, they ingest food as bite-size pieces rather than as tiny particles, in contrast to the filter feeding of nonvertebrate chordates. Food is consumed through the mouth and moves through the gut by rhythmical muscular contractions (peristalsis). Digestion begins in the mouth with the breakdown of food by enzymes in saliva or venom. Digestion is continued in the stomach and small intestine by enzymes secreted by the liver and pancreas. (The pancreas also secretes the hormones insulin and glucagon, which are involved in the regulation of glucose metabolism and blood-sugar levels.) Absorption of food molecules (nutrients) occurs primarily in the small intestine. The large intestine, or colon, absorbs water and inorganic ions. Waste materials (feces) are excreted via the rectum, the terminal portion of the large intestine.

In the basal vertebrate condition, there is no stomach, no division of the intestine into small and large portions, and no distinct rectum. The intestine empties to the **cloaca**, which is the shared exit for the urinary, digestive, and (except in cyclostomes) reproductive systems in all vertebrates except therian mammals.

The cardiovascular system

Blood carries oxygen and nutrients through the vessels to the cells of the body, removes carbon dioxide and other metabolic waste products, and stabilizes the internal environment. Blood also carries hormones from their sites of release to their target tissues.

Blood is a fluid connective tissue composed of liquid plasma, red blood cells (erythrocytes) that contain the iron-rich protein hemoglobin that binds with oxygen and carbon dioxide, and several different types of white blood cells (leukocytes) that are part of the immune system. All vertebrates possess specialized cells (or cell-like structures) that promote clotting of blood.

Vertebrates have closed circulatory systems; that is, the arteries and veins are connected by capillaries. Arteries carry blood away from the heart, and veins return blood to the heart (**Figure 2.11**). Blood pressure is higher in the arterial system than in the venous system, and the walls of arteries have a layer of smooth muscle that is absent from veins.

The dorsal aorta is flanked by paired cardinal veins that return blood to the heart. Anterior cardinal veins (draining the head) and posterior cardinal veins (draining the body) unite on each side in a common cardinal vein that enters the atrium of the heart. In lungfishes and tetrapods, the posterior cardinal veins are functionally replaced by a single midline vessel, the posterior vena cava.

Blood vessels that lie between two capillary beds are called portal vessels. The hepatic portal vein, present in all vertebrates, lies between the capillary beds of the gut and the liver. Substances absorbed from the gut are transported directly to the liver, where toxins are rendered harmless and some nutrients are processed or removed for storage. Most vertebrates also have a renal portal vein between the veins returning from the tail and posterior trunk and the kidneys. The renal portal system is not well developed in jawless vertebrates and has been lost in mammals.

Interposed between the smallest arteries (arterioles) and the smallest veins (venules) are the capillaries, which are the sites of exchange between blood and tissues. Capillaries pass close to every cell and their walls are only one cell layer thick, so diffusion is rapid. Collectively, the capillaries provide an enormous surface area for the exchange of gases, nutrients, and waste products. Arteriovenous anastomoses connect some arterioles directly to venules, allowing blood to bypass a capillary bed, and normally only a fraction of the capillaries in a tissue have blood flowing through them.

The vertebrate heart develops as a muscular tube folded on itself and is constricted into three sequential chambers: the **sinus venosus**, the **atrium**, and the **ventricle** (a fourth chamber, the **conus arteriosus**, is added in gnathostomes). The heart wall is composed of a thick layer of cardiac muscle, the myocardium, with a thin outer epicardium and inner endocardium.

The sinus venosus is a thin-walled sac with few cardiac muscle fibers. Suction produced by muscular contraction draws blood anteriorly into the atrium, which has valves at each end that prevent backflow. The ventricle is

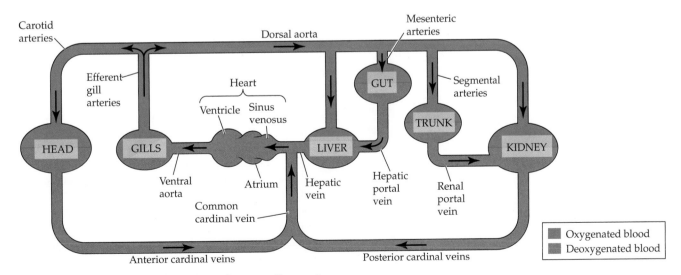

Figure 2.11 Diagrammatic plan of vertebrate cardiovascular circuit. All vessels are paired on the left and right sides of the body except for the midline ventral aorta and dorsal aorta.

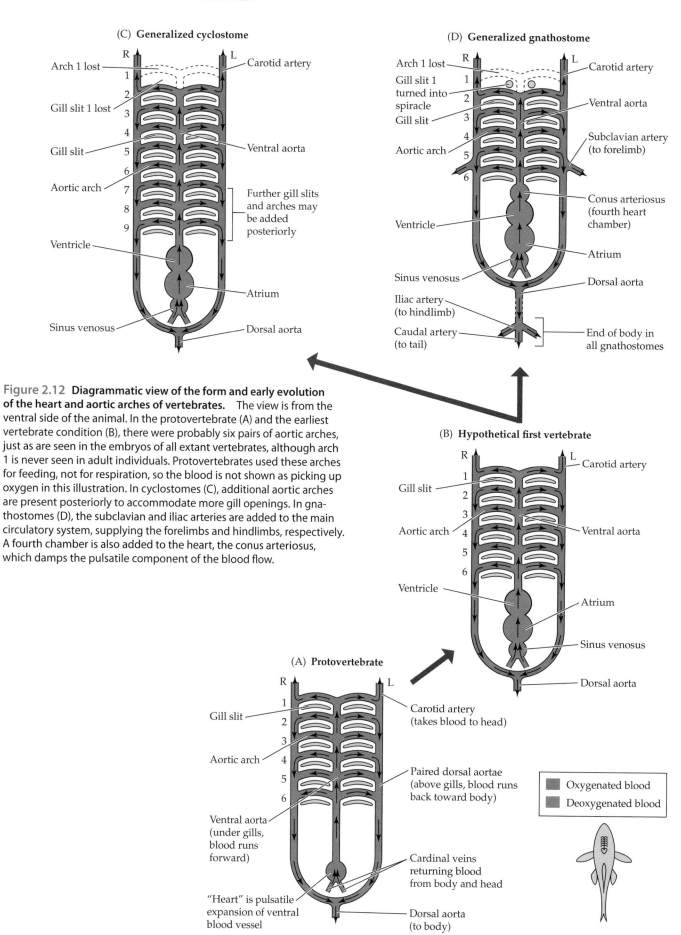

(C) Generalized cyclostome

Arch 1 lost
Gill slit 1 lost
Gill slit
Aortic arch
Ventricle
Sinus venosus

Carotid artery
Ventral aorta
Further gill slits and arches may be added posteriorly
Atrium
Dorsal aorta

(D) Generalized gnathostome

Arch 1 lost
Gill slit 1 turned into spiracle
Gill slit
Aortic arch
Ventricle
Sinus venosus
Iliac artery (to hindlimb)
Caudal artery (to tail)

Carotid artery
Ventral aorta
Subclavian artery (to forelimb)
Conus arteriosus (fourth heart chamber)
Atrium
Dorsal aorta
End of body in all gnathostomes

Figure 2.12 Diagrammatic view of the form and early evolution of the heart and aortic arches of vertebrates.   The view is from the ventral side of the animal. In the protovertebrate (A) and the earliest vertebrate condition (B), there were probably six pairs of aortic arches, just as are seen in the embryos of all extant vertebrates, although arch 1 is never seen in adult individuals. Protovertebrates used these arches for feeding, not for respiration, so the blood is not shown as picking up oxygen in this illustration. In cyclostomes (C), additional aortic arches are present posteriorly to accommodate more gill openings. In gnathostomes (D), the subclavian and iliac arteries are added to the main circulatory system, supplying the forelimbs and hindlimbs, respectively. A fourth chamber is also added to the heart, the conus arteriosus, which damps the pulsatile component of the blood flow.

(B) Hypothetical first vertebrate

Gill slit
Aortic arch
Ventricle

Carotid artery
Ventral aorta
Atrium
Sinus venosus
Dorsal aorta

(A) Protovertebrate

Gill slit
Aortic arch
Ventral aorta (under gills, blood runs forward)
"Heart" is pulsatile expansion of ventral blood vessel

Carotid artery (takes blood to head)
Paired dorsal aortae (above gills, blood runs back toward body)
Cardinal veins returning blood from body and head
Dorsal aorta (to body)

Oxygenated blood
Deoxygenated blood

thick-walled, and the muscular walls have an intrinsic pulsatile rhythm, which can be speeded up or slowed down by the nervous system. Contraction of the ventricle forces the blood into the ventral aorta. Therian mammals no longer have a distinct structure identifiable as the sinus venosus; rather, it is incorporated into the wall of the right atrium as the sinoatrial node, which controls the basic pulse of the heartbeat; the conus arteriosus is subsumed into the base of the aorta, and the atrium and ventricle are divided into left and right chambers.

The basic vertebrate circulatory plan consists of a heart that pumps blood into the single midline ventral aorta. Paired sets of aortic arches (originally six pairs) branch from the ventral aorta to supply the gills (**Figure 2.12**), where the blood is oxygenated and returns to the dorsal aorta. The dorsal aorta is paired above the gills, and the vessels from the most anterior arch run forward to the head as the carotid arteries. Behind the gill region, the two vessels unite into a single dorsal aorta that carries blood posteriorly.

The excretory and reproductive systems

Although the functions of the excretory and reproductive systems are entirely different, both systems form from the nephrotome, or intermediate mesoderm (see Figure 2.5), which forms the embryonic nephric ridge (**Figure 2.13**). The kidneys are segmental, whereas the gonads (ovaries in females, testes in males) are unsegmented. Both organs lie behind the peritoneum on the dorsal body wall (only in some therian mammals do testes descend into a scrotum).

Kidneys   The kidneys dispose of waste products, primarily nitrogenous waste from protein metabolism, and regulate the body's water and minerals—especially sodium, chloride, calcium, magnesium, potassium, bicarbonate, and phosphate. In tetrapods, the kidneys are responsible for almost all these functions; in fishes and amphibians, however, the gills and skin also play important roles (see Chapter 4).

The basic units of the kidney are microscopic structures called nephrons. Vertebrate kidneys work by ultrafiltration: high blood pressure forces water, ions, and small molecules through tiny gaps in the capillary walls. Nonvertebrate chordates lack true kidneys. Lancelets have excretory cells called solenocytes (possibly homologous with portions of the nephron) associated with the pharyngeal blood vessels that empty individually into the false body cavity (the atrium). The effluent is discharged to the outside via the atriopore.

Gonads   Although the gonads are derived from the mesoderm, the gametes (eggs and sperm) are formed in the endoderm and then migrate up through the dorsal mesentery to enter the gonads. The gonads are paired in gnathostomes but are single in cyclostomes; it is not clear which represents the ancestral vertebrate condition. The gonads also produce hormones, such as estrogen and testosterone.

In cyclostomes, probably representing the ancestral vertebrate condition, there is no special tube or duct for the passage of the gametes. Rather, the sperm or eggs erupt from the gonad and move through the coelom to pores that open to the base of the archinephric ducts. In gnathostomes, however, the gametes are always transported to the cloaca via specialized paired ducts (one for each gonad). In males, sperm are released directly into the archinephric ducts that drain the kidneys of non-amniotes and embryonic amniotes. In females, the egg is still released into the coelom but is then transported via the oviduct. The oviducts produce substances associated with the egg, such as the yolk or the shell. The oviducts can become enlarged and fused in various ways to form a single uterus or paired uteri in which eggs are stored and young develop.

Some vertebrates deposit eggs that develop outside the body (oviparity), while others retain the eggs within the mother's body until embryonic development is complete (viviparity). Shelled eggs must be fertilized in the oviduct before the shell and albumen are deposited. Many viviparous vertebrates and vertebrates that lay shelled eggs have some sort of intromittent organ—such as the pelvic claspers of sharks and the penis of amniotes—by which sperm are inserted into the female's reproductive tract.

The sense organs

We think of vertebrates as having five senses—taste, touch, sight, smell, and hearing—but this list does not reflect the ancestral condition, nor does it include all the senses of extant vertebrates. Many vertebrate lineages have complex, multicellular sense organs that are

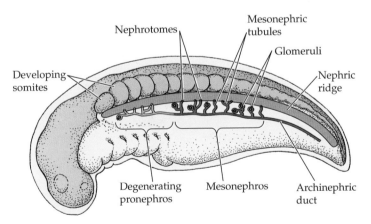

Nephrotomes

Mesonephric tubules

Glomeruli

Developing somites

Nephric ridge

Degenerating pronephros

Mesonephros

Archinephric duct

Figure 2.13 Kidney development in a generalized vertebrate embryo. In all vertebrate embryos, the kidney is composed of three portions: the pronephros, mesonephros, and metanephros. The pronephros is functional only in the embryos of extant vertebrates (possibly also in adult hagfishes). The kidney of adult fishes and amphibians includes the mesonephros and metanephros and is known as an opisthonephric kidney. The metanephric kidney seen in adult amniotes (the familiar compact, bean-shaped kidney) includes only the metanephros. The metanephros is drained by the ureter, derived from the basal portion of the archinephric duct. At the stage of development shown in this figure, the metanephros has barely started to form.

formed from epidermal placodes and tuned to the sensory worlds of those species.

Chemosensation: taste and smell   The senses of taste (gustation) and smell (olfaction) both involve the detection of dissolved molecules by specialized receptors. We think of these two senses as being interlinked; for example, our sense of taste is poorer if our sense of smell is blocked when we have a cold. However, the two senses are very different in their innervation. Smell is mediated by the somatic sensory system, sensing items at a distance, with the sensations received in the forebrain. Taste is mediated by the visceral sensory system, sensing items on direct contact, with the sensations received initially in the hindbrain.

Vision   All vertebrates have a similar type of eye, with a lens, an iris, muscles that move the eyes, and a retina that originates as an outgrowth of the brain. The retina contains three distinct layers: photoreceptors (light-sensitive cells), interneurons, and output neurons. The photoreceptors can be divided into cones (color-sensitive) and rods (primarily light-level-sensitive), although the distinction is not so clear in cyclostomes. Both cones and rods contain opsin photopigments. Color vision with four or five different opsins appears to be a primitive vertebrate feature.

Electroreception   The capacity to perceive electrical impulses generated by the muscles of other organisms is also a form of distance reception, but one that works only in water. Electroreception was probably an important feature of early vertebrates and is seen today primarily in fishes and monotreme mammals. Some extant fishes produce electrical discharges to sense their environment, communicate with other individuals, or deter predators.

Mechanoreception   Hair cells in the neuromast organs of the lateral line system of fishes and aquatic amphibians detect the movement of water around the body (see Chapter 4). Morphological and genomic information shows that these hair cells are homologous with the ectodermal sensory cells of nonvertebrate chordates.

Balance and orientation   Originally the structures of the inner ear—the **vestibular apparatus**—served to detect an animal's position in space, and these structures retain that function today in both aquatic and terrestrial vertebrates (**Figure 2.14**). The basic sensory cell of the vestibular apparatus is the hair cell, which detects the movement of fluid resulting from a change of position or the impact of sound waves. The vestibular apparatus (one on either side of the animal) is enclosed within the otic capsule of the skull, surrounded by a fluid called perilymph, and consists of a series of sacs and tubules containing a fluid called endolymph. The lower parts of the vestibular apparatus, the sacculus and utriculus, house sensory organs called maculae, which contain tiny crystals of calcium carbonate resting on hair cells. Sensations from the maculae tell the animal which way is up and detect

linear acceleration. The upper part of the vestibular apparatus contains the semicircular canals. Sensory areas located in swellings at the end of each canal (ampullae) detect angular acceleration through cristae, hair cells embedded in a jellylike substance, by monitoring the displacement of endolymph during motion. Extinct and extant jawless vertebrates have two semicircular canals (although hagfishes only have one), while all gnathostomes have three semicircular canals (the horizontal canal is the new addition).

We often fail to realize the importance of our own vestibular senses because we usually depend on vision to determine our position. We can sometimes be fooled, however, as when sitting in a stationary train or car and thinking that we are moving, only to realize from the lack of input from our vestibular system that it is the vehicle next to us that is moving.

Tetrapod hearing   The inner ear is also used for hearing (reception of sound waves) by tetrapods and by a few derived fishes. In tetrapods only, the inner ear contains the cochlea (the organ of hearing, also known as the lagena in non-mammalian tetrapods). The cochlea and vestibular apparatus together are known as the membranous labyrinth. Sound waves are transmitted to the cochlea, where they create waves of compression that pass through the endolymph. These waves stimulate the auditory sensory cells, which are variants of the basic hair cell.

Associated with hearing is the evolution of vocalization, important in communication. This is mainly a tetrapod feature, using expelled air through the larynx, but some bony fishes can produce sounds using the gas bladder. Investigation of the neural control of this function in fishes shows that it is homologous with that in tetrapods.

(A) Lamprey (2 semicircular canals)

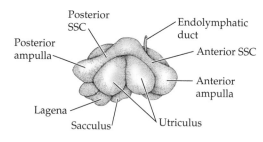

(B) Generalized gnathostome (shark) (3 semicircular canals)

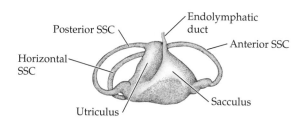

Figure 2.14 **Anatomy of the vestibular apparatus in fishes.** The lamprey (A) has two semicircular canals (SSC), whereas gnathostomes (represented by a shark) (B) have three semicircular canals. (After Kardong 2012.)

Summary

Vertebrates belong to the phylum Chordata, which along with the echinoderms and the hemichordates comprise the Deuterostomia.

Deuterostomes and protostomes are the subdivisions of the more complex animals with bilateral symmetry and bodies that develop from three layers of tissue (ectoderm, endoderm, and mesoderm)—Bilateria.

Non-chordate deuterostomes are all small marine animals, and are united by the presence of pharyngeal slits at some stage of their lives.

Chordates have unique characteristics.

The main characteristic of all chordates is the notochord, which gives the phylum its name. It is a dorsal stiffening rod to which muscles attach.

The notochord and other features (dorsal hollow nerve chord; segmented, muscular postanal tail; pharynx with pharyngeal slits; and endostyle) are seen at some point of development in all chordates. In most vertebrates, the notochord is a transient structure, replaced during embryonic development by the vertebral column.

In addition to vertebrates, Chordata contains two nonvertebrate groups of small, filter-feeding marine animals.

Cephalochordata contains the lancelets, also known as amphioxus.

Urochordata contains the tunicates, many of which are sessile as adults. Despite the fact that cephalochordates look more like vertebrates than do tunicates, molecular data show that tunicates and vertebrates are sister groups, united in the Olfactores.

Tunicates appear to have lost many ancestral chordate features and genes in their evolutionary history.

Vertebrates are defined by features in addition to the possession of vertebrae.

A major defining feature of vertebrates is the presence of a cranium, or skull, a bony, cartilaginous, or fibrous structure surrounding the brain.

Vertebrates have a prominent head end, containing a muscular pharynx (which pumps water over the gills in fishes), complex sense organs, and a tripartite brain. They also have a muscularized, chambered heart. These features reflect the fact that vertebrates are much larger than nonvertebrate chordates and have changed from a filter-feeding mode of life to a more active, predaceous one.

Vertebrate embryonic development is important for understanding the adult form.

The mesoderm in the body of the developing vertebrate is composed of a segmented dorsal portion (somites) and an unsegmented ventrolateral portion (lateral plate).

- Somites form the body's striated muscles, the dermis of the skin, and the axial skeleton (vertebral column and ribs).
- Lateral plate mesoderm forms principally the viscera, the smooth muscle lining of the gut, and the cardiac muscles of the heart. It also forms the limb bones, but remember that the original vertebrate plan lacked limbs.

The development of the pharynx, with the pharyngeal pouches, clefts, and arches, shows strong conservation throughout the vertebrates.

Each pharyngeal arch has its characteristic nerve and artery supply.

Several features of embryonic development have affected vertebrate evolution.

A doubling or quadrupling of the Hox genes increases control of gene expression during embryonic development.

A greater number of microRNAs than are present in other animals may contribute to the greater degree of complexity seen in vertebrates.

Neural crest, unique to vertebrates and considered a fourth germ layer, forms many structures, including most of the head region.

Placodes give rise to complex sense organs.

Vertebrates have distinctive tissue types with specific characteristics and functional properties.

The vertebrate body is composed of epithelial, muscle, neural, and connective tissues.

Mineralized connective tissues include bone, mineralized cartilage, dentine, enamel, enameloid, and cementum. The tissues have a structural basis in collagen, and mineral crystals are added to the collagen fibers. These crystals are made from a unique calcium compound, hydroxyapatite.

- Bone is found both in the skin (dermal bone) and in the endoskeleton, where it lines the cartilaginous bones (perichondral bone) or replaces the bone in development (endochondral bone). Endochondral bone is seen only in bony fishes and tetrapods. Some extinct jawless vertebrates (ostracoderms) were covered with dermal bone.

(Continued)

Summary (continued)

- Teeth are composed of dentine and enamel (or enameloid), which are more mineralized than bone and thus harder and more resistant. The odontodes on the dermal bone of some ostracoderms were also made of dentine and enamel and are basically homologous with teeth.

The vertebrate skeleton is a composite of exoskeletal and endoskeletal elements.

The basal vertebrate condition (above the level of cyclostomes) appears to be a cartilaginous endoskeleton and a bony exoskeleton. Humans retain some of this exoskeleton in the bones of our skull. The basic components of the endoskeleton are the cranial skeleton (in part), the axial skeleton (vertebral column and ribs) and, added in gnathostomes, the appendicular skeleton (limbs and limb girdles).

Many of the structures at the front of the head are formed from neural crest cells.

The axial skeleton was originally just the notochord in nonvertebrate chordates. Vertebrates added vertebral elements and, in gnathostomes, ribs.

The axial musculature is composed of complexly folded segmental myomeres, which are W-shaped in vertebrates rather than V-shaped as in cephalochordates.

Tissues combine to form organs, and organs combine to form systems.

Vertebrates are mostly particulate feeders, moving food down the gut via muscular contractions (peristalsis). Digestion is continued in the stomach and small intestine by enzymes secreted by the liver and pancreas. Primitive vertebrates lack the divisions of the gut seen in humans.

Blood carries food, respiratory gases, and hormones to the cells of the body. Blood is a fluid connective tissue composed of plasma, red blood cells (carrying the respiratory pigment hemoglobin), white blood cells (involved in the immune system), and specialized cells (or cell-like structures) that promote clotting.

Vertebrates have a closed circulatory system, with the arteries and veins connected by capillaries. In the basal condition (retained in fishes), deoxygenated blood returning from the body is pumped from the heart through the gills, and oxygenated blood is then supplied to the tissues via the dorsal aorta.

The vertebrate heart is basically a muscular tube that during development becomes folded and divided by valves into three chambers (cyclostomes) or four chambers (gnathostomes).

Vertebrate kidneys and gonads form from the nephrotome. The kidneys are composed of nephrons and dispose of waste products, although in fishes the gills and skin also play important roles in disposing of wastes.

Vertebrate kidneys function via ultrafiltration, and so require high blood pressure.

The gonads release gametes (originally formed in the endoderm), and in male gnathostomes they share a duct with the kidney. Females have a separate gonadal duct, the oviduct.

Vertebrates may lay eggs or retain fertilized eggs within the mother's body. In the latter case (and also with shelled eggs), internal fertilization is essential, and males may have a specialized intromittent organ.

Vertebrates have specialized sense organs.

Smell and taste are often considered to be closely linked, but they have different properties and are innervated in different ways.

Vision depends on the retina of the eye, which is actually an outgrowth of the brain containing light-sensitive cells, rods and cones.

Balance and orientation depend on the vestibular system in the inner ear.

Fishes have two senses that act in water: electroreception (the ability to perceive electrical stimuli generated by other animals; also present in monotreme mammals), and mechanoreception (the ability to detect vibrations in the water, via hair cells in the lateral line system; also present in aquatic amphibians).

Hearing—the reception of sound waves by the inner ear—is essentially a tetrapod feature.

Discussion Questions

1. Suppose new molecular data showed that cephalochordates and vertebrates are sister taxa. What difference would this make to our assumptions about the form of the original chordate animal? What additional features might this animal have possessed?

2. Why would evidence of sense organs in the head of a fossil nonvertebrate chordate suggest a close relationship with vertebrates; that is, which critical vertebrate feature would have to be present?

3. Vertebrates have been described as "dual animals," consisting of both segmented and unsegmented portions. How is this duality reflected in their embryonic development and the structure of the nervous system?

4. How could duplication of Hox genes lead to the structural complexity of vertebrates?

Additional Reading

Collin SP. 2009. Early evolution of vertebrate photoreception: Lessons from lampreys and lungfishes. *Integrative Zoology* 4: 87–98.

Diogo R and 7 others. 2015. A new heart for a new head in vertebrate cardiopharyngeal evolution. *Nature* 520: 466–473.

Donoghue PCJ, Purnell MA. 2009. The evolutionary emergence of vertebrates from among their spineless relatives. *Evolution Education and Outreach* 2: 204–212.

Garcia-Fernandez J, Benito-Gutierrez E. 2009. It's a long way from amphioxus: Descendants of the earliest chordate. *BioEssays* 31: 665–675.

Green SA, Simoes-Costa M, Bronner ME. 2015. Evolution of vertebrates as viewed from the crest. *Nature* 250: 474–482.

Heimberg A, McGlinn E. 2012. Building a robust anterior–posterior axis. *Current Genomics* 13: 278–288.

Hirasawa T, Kuratani S. 2015. Evolution of the vertebrate skeleton: Morphology, embryology, and development. *Zoological Letters* 1: 2.

Holland ND, Holland LZ, Holland PWH. 2015. Scenarios for the making of vertebrates. *Nature* 520: 450-455.

Janvier P. 2015. Facts and fancies about early fossil chordates and vertebrates. *Nature* 250: 483–489.

Kadota M and 6 others. 2017. CTCF bonding landscape in jawless fish with reference to Hox cluster evolution. *Scientific Reports* 7: 4957.

Kasahara M. 2007. The 2R hypothesis: An update. *Current Opinion in Immunology* 19: 547–552.

Patel NH. 2004. Time, space, and genomes. *Nature* 431: 28–29.

Ravi V, Lam K, Boon-Hull AT, Brenner S, Venkatesh B. 2009. Elephant shark (*Callorhunchus milii*) provides insights into the evolution of Hox gene clusters in gnathostomes. *Proceedings of the National Academy of Sciences USA*. 106: 16327–16332.

Satoh N. 2016. *Chordate Origins and Evolution: The Molecular Evolutionary Road to Vertebrates*. Elsevier, New York.

Shubin N. 2009. *Your Inner Fish: A Journey into the Billion-Year History of the Human Body*. Vintage Books, New York.

Venkatesh B and 32 others. 2014. Elephant shark genome provides unique insights into gnathostome evolution. *Nature* 505: 174–179.

Go to the Companion Website **oup.com/us/vertebratelife10e** for active-learning exercises, news links, references, and more.

Jawless Vertebrates and the Origin of Jawed Vertebrates

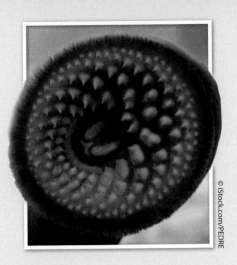

The earliest vertebrates represented an important advance over the nonvertebrate chordate filter feeders. The most conspicuous novel feature of these early, jawless vertebrates was an anatomically distinct head, containing a tripartite brain enclosed by a cartilaginous cranium (skull) and complex sense organs. These vertebrates used the newly acquired pharyngeal musculature that powered the gill skeleton to draw water into the mouth and over the gills, which were now used for respiration rather than for filter feeding. Early vertebrates were more active than the nonvertebrate chordates. Many of them also had external armor made of bone and other mineralized tissues. **Gnathostomes**—the jawed vertebrates, encompassing all extant vertebrate species other than lampreys and hagfishes—represent an increase in complexity in the vertebrate body plan for high levels of activity and predation. Jaws may have evolved originally as devices that improved the strength and effectiveness of gill ventilation, and later became modified for seizing and holding prey.

In this chapter, we trace the earliest steps in the radiation of vertebrates that began more than 500 million years ago. We discuss the biology of both the Paleozoic jawless vertebrates (ostracoderms) and the extant jawless forms (hagfishes and lampreys, the cyclostomes), as well as placoderms and acanthodians, two lineages of jawed fishes that did not survive the Paleozoic.

3.1 The Earliest Evidence of Vertebrates

Until recently, our most ancient evidence of vertebrates consisted of fragments of the dermal armor of the extinct jawless vertebrates colloquially known as **ostracoderms** (Greek *ostrakon*, "shell"; *derma*, "skin"). Bone fragments that can be assigned definitively to these vertebrates are known from the Ordovician, some 480 million years ago. This was about 80 million years before whole-body vertebrate fossils became abundant. The early radiation of vertebrates involved both jawed and jawless groups. Complete fossils of both ostracoderms and gnathostomes are known from the late Silurian (~425 Ma) from diverse fossil assemblages worldwide.

Discoveries of soft-bodied vertebrates from the Chengjiang Fauna of the early Cambrian of China have extended the vertebrate fossil record back another 100 Ma or so, to about 525 million years ago. The main taxon from this fauna is *Haikouichthys* (probably synonymous with *Myllokunmingia*), a small, fish-shaped animal about 3 cm long, known from hundreds of well-preserved individuals (**Figure 3.1A**). A slightly younger animal, *Metaspriggina*, is found in the middle Cambrian Burgess Shale. Evidence of a notochord and myomeres marks these animals as chordates, and the possible presence of a cranium and paired sensory structures (probably representing eyes) at the head end mark them as vertebrates, because these structures are formed from neural crest tissue and epidermal placodes (see Chapter 2). These animals also had additional vertebrate features: a dorsal fin and a ribbonlike ventral fin (but without the fin rays seen in other vertebrates); six to seven gill pouches with evidence of filamentous gills and a branchial skeleton; myomeres that were probably W-shaped (rather than V-shaped as in amphioxus); and segmental structures flanking the notochord that may have represented lampreylike vertebral rudiments. However, the early Cambrian vertebrates lack any evidence of bone or mineralized scales (as is also true of the cyclostomes, the extant jawless vertebrates; see Section 3.2). Additionally, there is evidence for serial segmental gonads, as in amphioxus (the gonads are nonsegmental in crown vertebrates). These Cambrian forms are generally considered to be stem vertebrates, because they lack features that would convincingly unite them with either cyclostomes or gnathostomes.

The earliest complete ostracoderm fossils are from the Late Ordovician of Bolivia, Australia, North America, and Arabia. These were torpedo-shaped jawless fishes, ranging from 12 to 35 cm in length (**Figure 3.1B**). The Ordovician was also a time of great radiation and diversification among marine invertebrates, in the wake of large-scale extinctions at the end of the Cambrian.

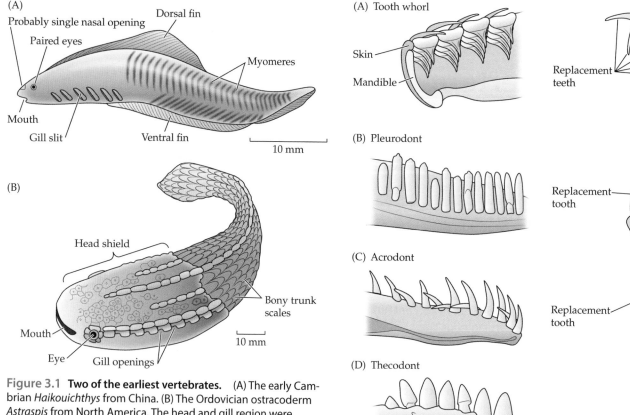

Figure 3.1 Two of the earliest vertebrates. (A) The early Cambrian *Haikouichthys* from China. (B) The Ordovician ostracoderm *Astraspis* from North America. The head and gill region were encased by many small, close-fitting, polygonal bony plates, and the body was covered by overlapping bony scales. The bony head shield shows the presence of sensory canals, special protection around the eye, and as many as eight gill openings on each side of the head. (B, after Elliott 1987.)

Figure 3.2 Gnathostome teeth. (A) Tooth whorl of a chondrichthyan. The teeth form within the skin, resulting in a tooth whorl that rests on the jawbone but is not actually embedded in it. (B–D) Bony fishes and tetrapods have teeth embedded in the dermal bones of the jaw. (B) Pleurodont, the basal condition: teeth set in a shelf on the inner side of the jawbone, seen in some bony fishes and in modern amphibians and some lizards. (C) Acrodont condition: teeth fused to the jawbone, seen in most bony fishes and in some reptiles (derived independently). (D) Thecodont condition: teeth set in sockets and held in place by peridontal ligaments, seen in archosaurian reptiles and mammals (derived independently). (A, from Hildebrand and Goslow 1998, B–D, after Smith 1960.)

The origin of bone and other mineralized tissues

Mineralized tissues composed of hydroxyapatite are a major feature that arose in the vertebrate lineage. Enamel (or enameloid) and dentine, which occur primarily in the teeth among extant vertebrates, are at least as old as bone and were originally found in intimate association with bone in the dermal armor of ostracoderms. However, mineralized tissues did not appear at the start of vertebrate evolution and are lacking in the extant jawless vertebrates.

The earliest mineralized tissues   The basic units of mineralized tissue appear to be odontodes, little toothlike elements formed in the skin, with similarities both to our own teeth and to the scales of sharks. They consist of projections of dentine with a base of bone, covered in some cases with an outer layer of enameloid.

The large scales, plates, and shields on the heads of many ostracoderms consisted of aggregates of odontodes and underlying bone. Note that these bony elements would not have been external to the skin like a turtle's shell. Rather, they were formed within the dermis of the skin and overlain by a layer of epidermis, as with our own skull bones.

The ancestral condition for vertebrate bone is to lack bone cells in the adult form; this type of acellular bone is known as aspidin. Except for its occurrence in osteostracans (derived ostracoderms related to gnathostomes), cellular bone is found only in gnathostomes.

The detailed structure of the tissues forming the bony armor suggests a function more complex than mere protection and has led to speculation about its original advantages. Suggestions include storage or regulation of minerals such as calcium and phosphorus (our own bones serve this function), and insulation around electroreceptive organs

(A)

(B)

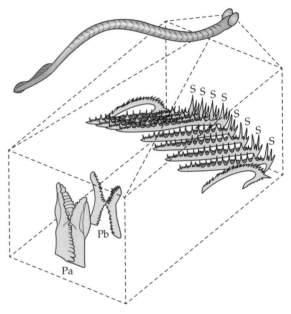

Figure 3.3 Traces of conodonts. (A) Conodont elements, assumed to be feeding structures, are abundant in some deposits. (B) Reconstructions of the conodont *Idiognathodus* and how its conodont elements might have been arranged. The elements were the only portions of these animals that fossilized, making such reconstructions speculative. (A, from Jarachowski and Munnecke 2016; B (above), © Bill Parsons; B (below), from Purnell 1994, adapted from Aldridge 1987.)

(like those of extant sharks) that enhanced detection of prey. These hypotheses are not mutually exclusive, and both functions may have been involved in the evolutionary origin of this exoskeleton.

The earliest teeth   Toothlike structures within the pharynx have been reported in some ostracoderms, but these may represent internal scales convergent with the teeth of gnathostomes. The placoderms, a basal jawed vertebrate group, had teeth, in the form of structures on tooth plates associated with the jaws—multiple cusps with a core of dentine and a cap of enameloid. The diversity of tooth types in extant gnathostomes is shown in **Figure 3.2**.

The mysterious conodonts

Microfossils known as conodont elements are widespread and abundant in marine deposits from the late Cambrian to the Late Triassic. Conodont elements are small (generally less than 1 mm long) spinelike or comblike structures composed of apatite, the mineralized calcium compound that is characteristic of vertebrate hard tissues (**Figure 3.3A**). Originally the elements were considered to be the partial remains of marine invertebrates, but the discovery of impressions of complete animals with features such as a notochord, myomeres, and large eyes has resulted in **conodonts** being categorized as vertebrates (**Figure 3.3B**). The conodont elements were found to be arranged within the pharynx in a complex apparatus. The elements were originally thought to have been formed from dentine and enamel, specifically vertebrate tissues, suggesting that conodonts were stem gnathostomes (but less derived than ostracoderms). However, it has recently been determined that

ancestral conodonts lacked mineralized tissues; so these conodont elements must have evolved convergently with the hard tissues seen in ostracoderms and gnathostomes. Conodonts are now considered to be stem vertebrates (more derived than other vertebrates known only from the Cambrian), or possibly basal gnathostomes or stem cyclostomes (**Figures 3.4** and **3.5**).

The environment of early vertebrate evolution

By the late Silurian, ostracoderms and early jawed fishes were abundant in both freshwater and marine environments. Under what conditions did the first vertebrates evolve? The vertebrate kidney is very good at excreting excess water while retaining biologically important molecules and ions. That is what a freshwater fish must do, because its body fluids are continuously being diluted by the osmotic inflow of water and it must excrete that water to regulate its internal concentration (see Chapter 4). Thus, the properties of the vertebrate kidney suggest that vertebrates evolved in fresh water.

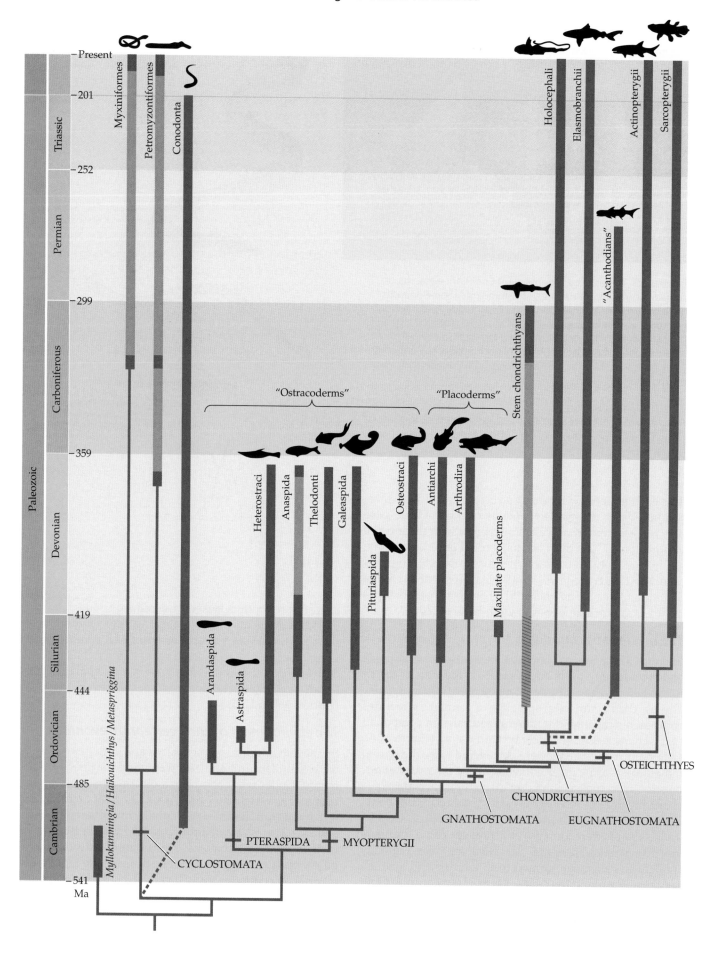

Figure 3.4 Phylogenetic relationships of early vertebrates.
This diagram depicts the probable relationships among the major groups of "fishes," including extant and extinct jawless vertebrates and the earliest jawed vertebrates. Only two of the four major lineages of placoderms are shown. Narrow lines show relationships only; they do not indicate times of divergence or the unrecorded presence of taxa in the fossil record. Lightly shaded bars indicate ranges of time when a taxon is believed to be present, because it is known from earlier and later times but is not recorded in the fossil record during this interval. Only the best-corroborated relationships are shown. The hatched bar in stem chondrichthyans lineage indicates possible occurrence based on limited evidence. Dashed lines indicate uncertainty about relationships; quotation marks indicate paraphyletic groups.

Despite the logic of that inference, however, a marine origin of vertebrates is now widely accepted. Osmoregulation is complex, and fishes use cells in the gills as well as the kidney to control their internal fluid concentrations. Probably the structure of the kidney is merely fortuitously suited to fresh water.

Two lines of evidence support the hypothesis of a marine origin of vertebrates:

1. The earliest vertebrate fossils are all from marine sediments.

2. All nonvertebrate chordates and deuterostome invertebrate phyla are exclusively marine, and they have body fluids with approximately the same osmolal concentration as their surroundings. Hagfishes also have concentrated body fluids, and these high body-fluid concentrations probably represent the original vertebrate condition.

Figure 3.5 Simplified phylogeny of early vertebrates. Only extant taxa and major extinct groups are shown. Quotation marks indicate paraphyletic groups. A dagger indicates an extinct group.

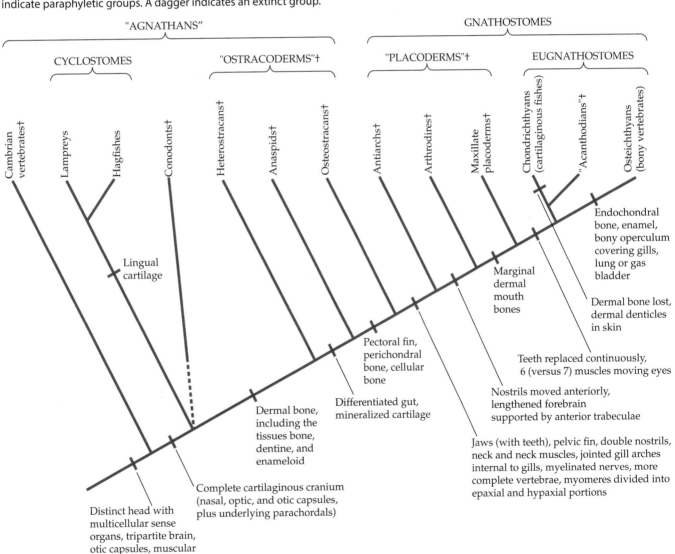

3.2   Cyclostomes: The Extant Jawless Fishes

The extant jawless vertebrates—hagfishes and lampreys, the cyclostomes—are often considered "primitive." Not only do they lack the gnathostome features of jaws and paired fins, but they also share some features of nonvertebrate chordates, such as the lack of specialized reproductive ducts or mineralized tissues. However, in other respects, such as their head anatomy, cyclostomes are highly derived in a fashion different from jawed vertebrates.

The relationship of hagfishes and lampreys to each other and to gnathostomes has been a subject of controversy. Most data (especially from molecular and genomic studies) now suggest that hagfishes and lampreys are sister taxa, and support the monophyletic grouping Cyclostomata for the extant jawless vertebrates.

Because cyclostomes appear to be more distantly related to gnathostomes than were the now-extinct armored ostracoderms of the Paleozoic (see Figures 3.4 and 3.5), we will consider cyclostomes before looking at the extinct jawless fishes in Section 3.3.

Characters of cyclostomes

Cyclostomes share several anatomical features. A single nostril is probably a basal vertebrate feature; pouched gills with branchial supports on the outside of the gill tissue are seen in ostracoderms and cyclostomes. Possibly unique cyclostome features include (1) a velum, a membranous flap in the pharynx that aids in pumping water over the gills; and (2) a large, muscular "tongue" (or tonguelike structure) that bears keratinous teeth and is supported by a prominent lingual cartilage plus accessory cartilages. The teeth are arranged in longitudinal (front to back) rows on cartilaginous supports. A complex series of muscles protrudes and retracts the tooth rows, and also moves them apart and together to effect a transverse bite. The cyclostome "tongue" is not homologous with the gnathostome tongue: we can tell this because the muscles are innervated by a different cranial nerve (V rather than XII).

Hagfishes and lampreys have rudimentary vertebral precursors called arcualia (see Figure 2.4B). Lampreys have segmental elements throughout the body above the notochord (equivalent to neural arches), and hagfishes have segmental elements below the notochord in the tail (equivalent to haemal arches); the common ancestor of cyclostomes and gnathostomes probably had both dorsal and ventral elements, and cyclostomes have simplified this condition. Cyclostomes also share a unique type of immune system that is different from the immunoglobulin-based system of the gnathostomes.

Although hagfishes and lampreys lack jaws, both have several head cartilages that cannot be homologized with head structures of gnathostomes. Both also have a single nasal sac and a

single nasal opening that is conjoined with the duct for the adenohypophysis (anterior pituitary). This structure is unlike the dual nasal sacs and openings of gnathostomes, and may represent a basal vertebrate feature because some ostracoderms apparently had a similar condition.

Fossil cyclostomes

The fossil record of cyclostomes is sparse because they are soft-bodied, and we know little of their evolutionary history. Lampreys are known from the Late Devonian (*Priscomyzon*, a short-bodied form, from South Africa); the late Carboniferous (*Hardistiella* from Montana, and *Mayomyzon* and *Pipiscius* from Mazon Creek, a Lagerstätt in Illinois); and the Early Cretaceous (*Mesomyzon* from southern China, the first known freshwater form). All of these fossil lampreys appear to have been specialized parasites, similar to the extant forms. All fossil hagfishes to date are from Mazon Creek: *Myxinikela* and *Myxineidus* are undisputed hagfishes, and *Gilpichthys* is a possible hagfish. Several other fossils have been considered as possible cyclostomes, the most dramatic of which is the "Tully Monster" (*Tullimonstrum*), also from Mazon Creek (**Figure 3.6**). Although some researchers question its vertebrate affinity, this animal provides a hint of a possibly vast diversity of extinct soft-bodied jawless vertebrates.

Extant hagfishes: Myxiniformes

There are about 75 species of hagfishes in two major genera, *Eptatretus* and *Myxine*. Adult hagfishes are elongated, scaleless, pinkish to purple in color, and about 0.5 m long (**Figure 3.7**). Hagfishes are entirely marine, with a nearly worldwide distribution except for the polar regions. They are primarily deep-sea, cold-water inhabitants and are the major vertebrate scavengers of the deep-sea floor, drawn in large numbers by their sense of smell to carcasses.

Structural characteristics   Hagfishes are unique in having a single opening in the front of the head connecting to the pharynx that is used for water intake for both olfaction and gill ventilation. The number of gill openings on each side of the body varies among species, from a single exhalent opening for all of the gill pouches, to up to 15 separate gill pouch openings. The pharynx (and gill slits) have been

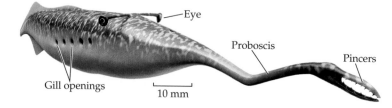

Figure 3.6 The "Tully Monster."   Seen here in a reconstruction, *Tullimonstrum gregarium* is a possible cyclostome from the Carboniferous. In addition to apparently having the vertebrate features of a notochord, cartilaginous arcualia, myomeres, gill pouches, and keratinous teeth like those of cyclostomes, it had eyes on lateral stalks and a long proboscis tipped with jawlike pincers that contained several rows of teeth. (From McCoy et al. 2016, courtesy of Sean McMahon/Yale University.)

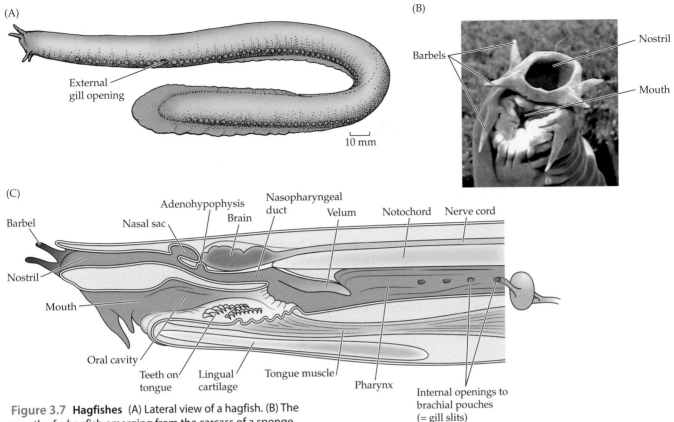

(A)

External
gill opening

10 mm

(B)

Barbels

Nostril

Mouth

(C)

Barbel

Nostril

Mouth

Oral cavity

Teeth on
tongue

Lingual
cartilage

Nasal sac

Adenohypophysis

Brain

Nasopharyngeal
duct

Velum

Notochord   Nerve cord

Tongue muscle

Pharynx

Internal openings to
brachial pouches
(= gill slits)

Figure 3.7 Hagfishes (A) Lateral view of a hagfish. (B) The mouth of a hagfish emerging from the carcass of a sponge. (C) Section through the head and pharynx of a hagfish. The single nasal opening at the front of the head connects to the nasopharyngeal duct that joins with the oral cavity at the level of the velum to form the pharynx. The single nasal sac and adenohypophyses are combined, lie underneath the forebrain, and open into the nasopharyngeal duct. Past the level of the branchial openings, the pharynx continues as the gut, and the heart lies behind the gills. (A,C, after Martini 1998; B, NOAA Okeanos Explorer Program/CC BY 2.0.)

moved posteriorly, probably in association with the hagfishes' feeding behavior of burying their head in their prey. The mouth is surrounded by three pairs of barbels (sensory tentacles) that are moved by muscles when a hagfish is searching for food, and hagfishes have a wider gape than that seen in many jawed fishes (see Figure 3.7B).

Hagfishes have the unique feature of **slime glands**, which open through the body wall via 90–200 pores and secrete enormous quantities of mucus and tightly coiled proteinaceous threads. The coiled threads straighten on contact with seawater and entrap the mucus close to the hagfish's body, apparently deterring predators. When danger has passed, the hagfish makes a knot in its body and scrapes off the mass of mucus.

Relationship to other vertebrates   Hagfishes had long been considered as basal to a group consisting of lampreys + gnathostomes because they appear to lack (or have only rudimentary versions of) features seen in other vertebrates. These features include small, skin-covered eyes without eye muscles to move them; the lack of

electroreception or a well-developed lateral line system; the presence of only one semicircular canal in the inner ear (all other vertebrates have at least two); a primitive kidney, with the persistence of the pronephros in the adult; a heart that does not receive autonomic innervation from the vagus nerve; and amphioxus-like accessory pumping areas in the circulatory system. An alternative interpretation is that these features have been simplified or lost in hagfishes because of their specialized lifestyle, an interpretation supported by the molecular and genetic data that links them with lampreys.

Osmoregulation and ion regulation   Hagfishes are unique among vertebrates in maintaining the nonvertebrate chordate condition of body fluids that have the same osmolal concentration as seawater. It is not clear whether this represents a secondary condition in hagfishes or evolutionary convergence between lampreys and gnathostomes. However, unlike nonvertebrate chordates, hagfishes *do* regulate the divalent ions calcium and magnesium and maintain an internal ion concentration different from that of seawater, and they also are capable of acid–base regulation. Hagfish slime may be important in the excretion of both monovalent (e.g., sodium and chloride, Na^+ and Cl^-) and divalent ions (e.g., calcium, Ca^{2+}).

Feeding   Hagfishes feed on dead or dying vertebrate prey, burying their heads in the prey's body (hagfishes have low metabolic rates and are able to tolerate the hypoxia that results from this feeding method). Hagfishes have two pairs of longitudinally arranged rows of keratinous teeth, sometimes referred to as tooth plates, on their tongue and their teeth are sharper than those of lampreys. The tooth plates can exert a force greater than that of some jawed fishes; however, bite speed is slower than in jawed fishes. Once attached to the prey, hagfishes tie a knot in their tail and pass it forward along their body until they are braced against their prey and can tear off the flesh in their pinching grasp. Hagfishes were long considered to be scavengers, but new information shows they can be active predators, using their slime to choke fishes by clogging their gills.

Reproduction   In most species, female hagfishes outnumber males by 100 to 1; the reason for this strange sex ratio is unknown. Some species are hermaphroditic. The yolky eggs, which are oval and more than 1 cm long, are encased in a tough, clear covering that is secured to the sea bottom by hooks, and the young appear to bypass the larval stage.

Lampreys: Petromyzontiformes

There are about 40 species of lampreys in two major genera, *Petromyzon* and *Lampetra*. Adults of the different species range in length from about 10 cm to up to 1 m (**Figure 3.8**). Lampreys are found primarily in temperate northern latitude regions, although a few species are known from southern temperate latitudes.

Structural characteristics   Lampreys have seven pairs of gill pouches that open to the outside just behind the head, and a round mouth located at the bottom of a large fleshy funnel (the oral hood). The development of this hood has displaced the single nasal opening to the top of the head. This opening is combined with the duct leading to the adenohypophysis and is known as a nasohypophyseal opening (NHO). A NHO was also seen in some ostracoderms. The eyes of lampreys are large and well developed (with good color vision), as is the **pineal body** (a light-sensitive structure homologous with the pineal gland of mammals), which lies under a pale spot just posterior to the NHO.

Respiration   Larval lampreys (like jawed fishes) employ **flow-through ventilation**: they draw water into the mouth and then pump it out over the gills. Adult lampreys, however, spend much of their time with their suckerlike mouths affixed to the bodies of other fishes, and during this time they cannot ventilate the gills in a flow-through fashion. Instead, they use a form of **tidal ventilation** by which water is both drawn in and expelled through the gill openings. The velum prevents water from flowing out of the pharyngeal tube into the mouth. The lampreys' mode of ventilation is not very efficient at oxygen extraction, but it is a necessary compromise given their specialized mode of feeding.

As in jawed fishes, chloride-transporting cells in the gills and well-developed kidneys regulate ions, water, and nitrogenous wastes, as well as overall concentration of body fluids, allowing lampreys to exist in a wide range of salinities.

Feeding   Lamprey larvae use their gills to filter feed, and produce mucus to entrap the food with an amphioxus-like endostyle (which turns into the thyroid gland in the adult). However, most adult lampreys are parasitic on other fishes, although some small freshwater species have nonfeeding adults. The parasitic species attach to the body of another vertebrate (usually a larger bony fish) by suction and rasp a shallow, seeping wound. Lampreys have a tongue with a single pair of longitudinal tooth rows and a single anterior transverse row of keratinous teeth. The inner surface of the oral hood is also studded with keratinous teeth. Together these structures allow tight attachment and rapid abrasion of the host's integument. An oral gland secretes an anticoagulant that prevents the victim's blood from clotting. Feeding is probably continuous as long as a lamprey is attached to its host.

The bulk of an adult lamprey's diet consists of body fluids of fishes. The digestive tract is straight and simple, as one would expect for an animal with a diet as rich and easily digested as blood and tissue fluids. Lampreys generally do not kill their hosts, but they do leave a weakened animal with an open wound. At sea, lampreys feed on several species of whales and porpoises in addition to fishes. Swimmers in the Great Lakes, after having been in the water long enough for their skin temperature to drop, have reported initial attempts by lampreys to attach to their bodies.

Reproduction   Nearly all lampreys are anadromous—that is, as adults they live in oceans or large lakes and ascend rivers and streams to breed. Some of the most specialized species live only in fresh water and do all of their feeding as larvae, with the adults acting solely as a reproductive stage in the life history of the species. Little is known of the habits of adult lampreys because they are generally observed only during reproductive activities or when captured attached to a host.

Female lampreys produce hundreds to thousands of eggs. The eggs are about 1 mm in diameter and are devoid of any specialized covering such as that found in hagfishes. Male and female lampreys construct a nest of rocks, creating a turbulent flow of water that oxygenates the eggs. The female attaches to one of the upstream rocks to lay her eggs, and the male wraps around her, fertilizing the eggs as they are extruded. Adult lampreys die after breeding once.

Lamprey larvae are radically different from their parents and were originally described as a distinct genus, *Ammocoetes* (see Figure 3.8D). Ammocetes are tiny (6–10 mm long) wormlike organisms with a large, fleshy oral hood and nonfunctional eyes hidden deep beneath the skin. Currents carry the ammocetes downstream to backwaters and quiet banks, where they burrow into the soft mud or sand and spend 3–7 years as sedentary filter feeders. Adult

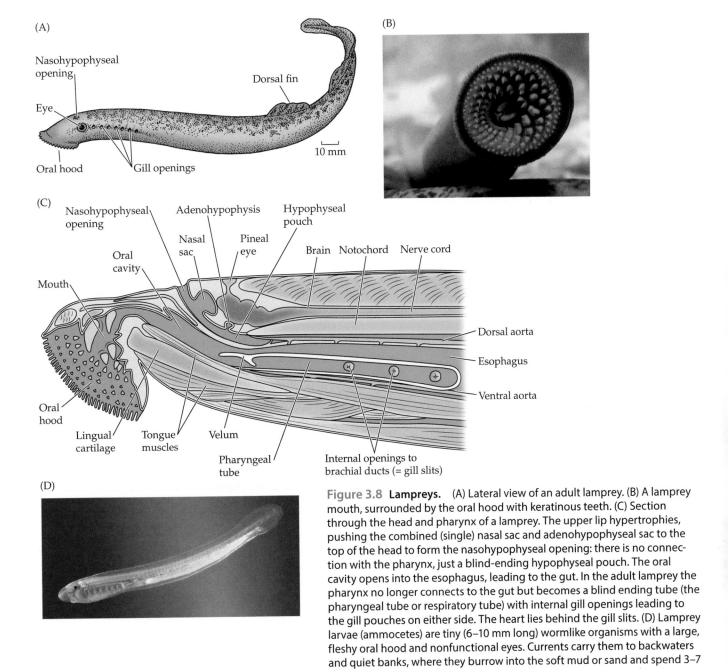

(A)

Nasohypophyseal opening

Dorsal fin

Eye

Oral hood

Gill openings

10 mm

(B)

(C)

Nasohypophyseal opening

Adenohypophysis

Hypophyseal pouch

Oral cavity

Nasal sac

Pineal eye

Brain Notochord Nerve cord

Mouth

Dorsal aorta

Esophagus

Ventral aorta

Oral hood

Lingual cartilage

Tongue muscles

Velum

Pharyngeal tube

Internal openings to brachial ducts (= gill slits)

(D)

Figure 3.8 Lampreys. (A) Lateral view of an adult lamprey. (B) A lamprey mouth, surrounded by the oral hood with keratinous teeth. (C) Section through the head and pharynx of a lamprey. The upper lip hypertrophies, pushing the combined (single) nasal sac and adenohypophyseal sac to the top of the head to form the nasohypophyseal opening: there is no connection with the pharynx, just a blind-ending hypophyseal pouch. The oral cavity opens into the esophagus, leading to the gut. In the adult lamprey the pharynx no longer connects to the gut but becomes a blind ending tube (the pharyngeal tube or respiratory tube) with internal gill openings leading to the gill pouches on either side. The heart lies behind the gill slits. (D) Lamprey larvae (ammocetes) are tiny (6–10 mm long) wormlike organisms with a large, fleshy oral hood and nonfunctional eyes. Currents carry them to backwaters and quiet banks, where they burrow into the soft mud or sand and spend 3–7 years as sedentary filter feeders. Adult life usually lasts no more than 2 years, and many lampreys return to spawn after 1 year. (B, © iStock.com/PEDRE; D, blickwinkel/Alamy Stock Photo.)

life usually lasts no more than 2 years, and many lampreys return to spawn after 1 year.

Cyclostomes and humans

In North America, humans and lampreys have been increasingly at odds for the past 100 years. The sea lamprey (*Petromyzon marinus*) seemed to have moved up the St. Lawrence River as far as Lake Ontario, but Niagara Falls prevented it from moving westward into the other Great Lakes. That situation changed with the construction and

subsequent expansions of the Welland Canal between 1829 and 1932. From the 1920s to the 1950s, lampreys expanded rapidly across the entire Great Lakes basin, reducing the populations of economically important fish species. Chemical lampricides as well as electrical barriers and mechanical weirs at the mouths of lamprey spawning streams have been used to bring Great Lakes lamprey populations down to their present levels.

We still know very little about hagfish ecology, or virtually any of the other information needed for good management of commercially exploited populations of hagfishes. Strangely enough, hagfishes do have an economic importance for humans. Almost all so-called eelskin leather products are made from hagfish skin. Worldwide demand for this leather has eradicated economically harvestable hagfish populations in Asian waters and in some sites along the West Coast of North America.

3.3 Ostracoderms: Extinct Jawless Fishes

Ostracoderms were the initial radiation of vertebrates and were the predominant forms in the Silurian. Ostracoderms were originally grouped with the extant jawless vertebrates (i.e., the cyclostomes) as Agnatha (Greek *a*, "without"; *gnathos*, "jaw"). The possession of true bone makes

(A) The heterostracan *Pteraspis* (Pteraspida)

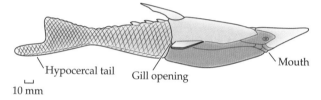

Hypocercal tail Gill opening Mouth

10 mm

(B) The anaspid *Pharyngolepis* (Cephalaspida)

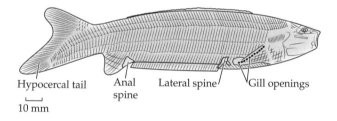

Hypocercal tail Anal spine Lateral spine Gill openings

10 mm

(C) The thelodont *Phlebolepis*

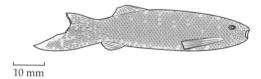

10 mm

(D) The osteostracan *Hemicyclaspis* (Cephalaspida)

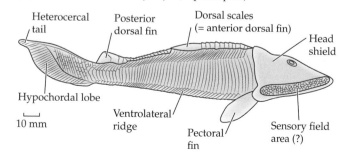

Heterocercal tail Posterior dorsal fin Dorsal scales (= anterior dorsal fin) Head shield

Hypochordal lobe

Ventrolateral ridge Pectoral fin Sensory field area (?)

10 mm

ostracoderms more derived than cyclostomes, but they are not a monophyletic evolutionary group, as some are clearly more closely related to gnathostomes than are others, making ostracoderms a paraphyletic assemblage of stem gnathostomes. Figure 3.5 shows how some of the characters seen in gnathostomes, but absent in cyclostomes, were acquired along the ostracoderm lineage.

Characters of ostracoderms

Most ostracoderms had an extensive dermal bone exoskeleton of large plates, but the exoskeleton sometimes took the form of smaller plates or scales (e.g., in anaspids), and some ostracoderms were relatively naked (e.g., thelodonts). Ostracoderms ranged in length from about 10 cm to more than 50 cm. Although they lacked jaws, many had movable mouth plates around a small, circular mouth. Most ostracoderms probably ate small, soft-bodied prey. As in extant jawless vertebrates, the notochord must have been the main axial support throughout adult life.

Ostracoderms usually had a midline dorsal fin, but only the most derived forms (osteostracans) had true pectoral fins, with an accompanying pectoral girdle and endoskeletal fin supports. Osteostracans also shared some other features with gnathostomes: perichondral bone, cellular bone, sclerotic rings of dermal bone around the eyes, an anal fin, and heterocercal (sometimes called epicercal) tailfin (i.e., with the upper lobe containing the vertebral column and larger than the lower lobe). Other ostracoderms had a hypocercal tailfin (lower lobe bigger than upper lobe). **Figure 3.9** depicts some representative ostracoderms.

Ostracoderm evolutionary patterns

Ostracoderms coexisted with early gnathostomes during the late Silurian and for most of the Devonian—a period of about 50 million years—so it is unlikely that the extinction of ostracoderms was related to the radiation of gnathostomes. Jawless and jawed vertebrates appear to represent two different basic types of animals that probably exploited different types of resources. The extinction of the ostracoderms in the Late Devonian occurred at the same time as mass extinctions among many marine invertebrate phyla, as well as the extinction of many of the jawed fishes, including all of the placoderms.

Figure 3.9 Ostracoderm diversity. (A) The heterostracan *Pteraspis* (late Silurian). Some heterostracans were fish-shaped, like *Pteraspis* (presumably pelagic), whereas others were dorsoventrally flattened (presumably bottom-living). All had a single exhalent gill opening. (B) The anaspid *Pharyngolepis* (late Silurian). Anaspids were covered by fine scales and had multiple gill openings and spines on the lateral and anal fins. (C) The thelodont *Phlebolepis* (late Silurian). Thelodonts had fine scales that barely covered the skin and were virtually naked. (D) The osteostracan *Hemicyclaspis* (Late Devonian). Osteostracans had multiple gill openings on the ventral surface of the head. (From Moy-Thomas and Miles 1971.)

3.4 The Basic Gnathostome Body Plan

Gnathostomes are derived, not only in having jaws and two sets of paired limbs (pectoral and pelvic), but in many other ways as well. Jaws and teeth allowed a variety of new feeding behaviors, including the ability to grasp objects firmly, and exploitation of a wide variety of food sources. The paired fins of jawed fishes allowed more sophisticated aquatic locomotion and maneuvering, providing lift, thrust, braking, and control of body orientation in three-dimensional space (see Figure 3.15).

Chondrichthyans have long been perceived as the more basal type of extant gnathostome, because they lack both the dermal and endochondral bone seen in osteichthyes. However, recent evidence shows that early osteichthyans exhibited more basal gnathostome features than chondrichthyans, which are highly derived in many respects, including in their loss of dermal bone.

Gnathostome biology

Gnathostomes are first known with certainty from the early Silurian, and by the Early Devonian they were taxonomically and morphologically diverse. Gnathostomes are characterized by many derived features, suggesting that jawed vertebrates represent a basic increase in level of activity and complexity from the jawless vertebrates, including improvements in locomotor and predatory abilities and in the sensory and circulatory systems (**Figure 3.10**). Interestingly, the form of the gnathostome lower jaw seems to have been most variable in the Silurian, when jawed fishes were not very common. By the Devonian, when the major radiation of jawed vertebrates began, the variety of jaw forms had stabilized.

You saw in Chapter 2 that all vertebrates show a duplication of the Hox gene complex from the single Hox complex in nonvertebrate chordates (see Figure 2.7). Gnathostomes show evidence of a second Hox duplication event. Gene duplication would have resulted in more extensive genetic control of development, perhaps necessary for building a more complex animal. However, it is difficult to know which of the derived features were unique to crown gnathostomes and which were acquired somewhere along the stem gnathostome lineage. Paired fins, for example, are usually considered to be a gnathostome character, but some osteostracans also had paired pectoral fins.

Gnathostomes have jointed gill arches and employ a "double-pump" mode of ventilation over the gills. In cyclostomes there is either an internal pump (the velum plus branchial arches in hagfishes and larval lampreys) or an

(A) Gnathostome body plan

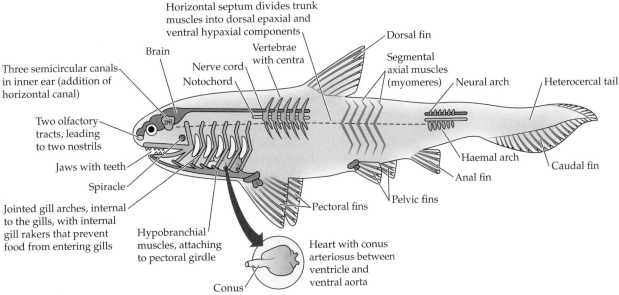

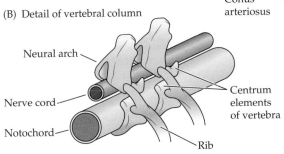

(B) Detail of vertebral column

Figure 3.10 Derived characters of gnathostomes. (A) A generalized gnathostome, showing derived features (compared with the jawless vertebrate condition). There is a clear distinction between the epaxial and hypaxial components of the trunk muscles, which are divided by a horizontal septum made of thin, fibrous tissue that runs the length of the animal. The lateral line system—containing the organs that sense vibrations in the surrounding water—lies in the plane of this septum, perhaps reflecting improved integration between locomotion and sensory feedback. The third (horizontal) semicircular canal in the inner ear may reflect an improved ability to navigate and orient in three dimensions. (B) Detail of the generalized gnathostome vertebral column, showing the central elements and rib attachments.

external pump (the "branchial basket" formed by the un-jointed gill arches and used in the tidal ventilation of adult lampreys). All gnathostomes use their mouth and jointed gill arches to generate an internal pump, but in bony fishes the operculum that covers the gill slits creates a second (parabranchial) pump, external to the gills. This powerful structure allows reduction of both the bony arches and the musculature of the branchial region. Cartilaginous fishes such as sharks have independently opening gill slits, but a septum extending from the gill bar beyond the gill tissue makes a small independent external parabranchial pump around each gill opening.

Extant gnathostomes have a vertebral column in which the segmental vertebrae are comprised of a centrum surrounding the notochord, in addition to neural arches throughout the vertebral column and haemal arches in the tail; these arches are homologous with the arcualia of cyclostomes. In tetrapods and some fishes, the centrum forms a complete bony ring attached to the neural and haemal arches, replacing the notochord completely in the adult. In other fishes the centrum is incomplete and is composed of several elements (central elements), and the notochord is retained. Ribs were another novel feature in gnathostomes. They articulate with the centrum and neural arch and lie in the connective tissue between the successive axial segmental muscles (myomeres) used in locomotion, providing anchorage for those muscles. Well-developed centra were not a feature of the earliest bony and cartilaginous fishes, and the earliest tetrapods also retained centra comprised of several portions (see Figure 3.10B). Both centra and ribs are unknown in the two extinct groups of jawed fishes—placoderms and acanthodians—although they may have been made of cartilage rather than bone, and so not preserved in the fossils.

What about soft tissues?

Hard tissues such as jaws and ribs can be observed in fossils, so we know that they are unique to gnathostomes. Although we cannot know for sure whether derived characters within the soft anatomy characterize gnathostomes alone or whether they appeared somewhere within the ostracoderm lineage, we can make some inferences:

- The insulating sheaths of myelin on the nerve fibers, which increase the speed of nerve impulses, are a derived character of gnathostomes. We can infer that myelinated nerves were present in placoderms but probably not in ostracoderms. The nerves supplying placoderm eye muscles, preserved as tracts within the skull, were much longer than those in ostracoderms—too long to have viable impulse transmission speeds without being insulated.

- Some characters of the nervous system that are seen only in gnathostomes among extant vertebrates were acquired by the earliest ostracoderms. For example, impressions on the inner surface of the dermal head shield reveal the presence of a cerebellum in the brain, but never show evidence of a horizontal semicircular canal.

- Gnathostomes have axial (trunk) muscles that are distinctly divided into epaxial (dorsal) and hypaxial (ventral) components.

- The gnathostome heart has an additional small chamber in front of the pumping ventricle, the conus arteriosus, which acts as an elastic reservoir that smooths the pulsatile nature of the blood flow produced by contractions of a more powerful heart.

- Gnathostomes have cellular bone and mineralized tissues in the endoskeleton. Cellular bone is seen in osteostracan and arandaspid ostracoderms. Mineralized cartilage is first seen in galeaspid ostracoderms, and perichondral bone is found lining the endoskeleton in osteostracans, but true endochondral bone is seen only in osteichthyans (chondrichthyans lack the requisite genes to produce endochondral bone).

- Cyclostomes lack the acid-secreting stomach that characterizes the digestive systems of elasmobranchs, bony fishes, and tetrapods, but fossil impressions of a possible stomach are seen in some derived ostracoderms (especially thelodonts and anaspids). Thus, the absence of an acid-secreting stomach in extant cyclostomes may be a derived character associated with their specialized diets and modes of feeding.

In addition, a particularly important gnathostome feature lies in the reproductive system, where the gonads have their own specialized ducts leading to the cloaca (see Chapter 2).

3.5   The Origin of Jaws

Our knowledge of the developmental origins of jaws has increased tremendously in the past decade, as scientists have learned more about gene expression and genetic homologies among jawed and jawless vertebrates. These studies have clarified our understanding of jaw evolution in some ways, but have made the issue more complex and mysterious in other ways.

In order to consider jaw evolution, we need to return to the pharyngeal pouches discussed in Section 2.4. **Figure 3.11A** shows the developmental domains of the head region characteristic of all vertebrates at the pharyngula stage. The apparent segmentation of the head suggests that the branchial arches and pouches once formed a complete series of gill arches and gill slits (**Figure 3.11B**). The original interpretation of early vertebrate head evolution was that cyclostomes and gnathostomes were divergent specializations from this generalized ancestral form.

Early hypotheses of jaw origins

Gnathostome jaws are known to be formed from the mandibular arch, with the hyoid arch forming the jaw supports

(A) Vertebrate embryo at pharyngula stage

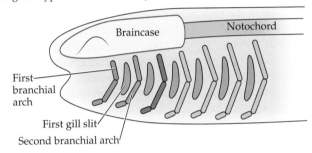

| Premandibular domain |
| Mandibular arch |
| Hyoid arch |
| 3rd branchial arch (first gill arch) |
| Branchial arches 4–7 |

Otic capsule
Nasal capsule
Eye
Hyomandibular pouch
Pharyngeal pouches (2–6)
Premandibular region
Mandibular arch
Hyoid arch
Post-hyoid branchial arches

| V_1 | V_{2+3} | VII | IX | X | X | X | X | Cranial nerve innervation |

Neural crest–derived cartilage
Branchiomeric mesoderm
Aortic arches
Hox gene expression
Gill tissue

(B) Original hypothesis of ancestral jawless condition

Braincase
Notochord
First branchial arch
First gill slit
Second branchial arch

(C) Gnathostome condition (shark)

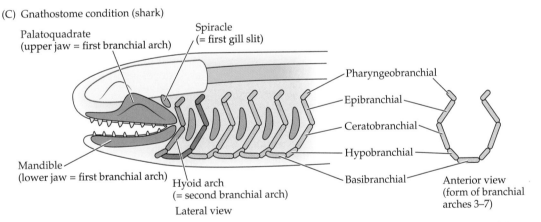

Palatoquadrate (upper jaw = first branchial arch)
Spiracle (= first gill slit)
Pharyngeobranchial
Epibranchial
Ceratobranchial
Hypobranchial
Mandible (lower jaw = first branchial arch)
Basibranchial
Hyoid arch (= second branchial arch)
Lateral view
Anterior view (form of branchial arches 3–7)

Figure 3.11 Evolution of the vertebrate jaw. (A) The pharyngula stage of embryonic development is seen in all vertebrates. There are six pharyngeal arches (the hyoid arch plus five post-hyoid branchial arches), associated with six pouches (which in fishes will develop into openings from the pharynx). The mesoderm associated with each arch has its own specified cranial nerve supply and blood supply from the aortic arches. The first pouch (the hyomandibular pouch) forms the spiracle of gnathostomes and the velar chamber of cyclostomes. The first true gill slit of gnathostomes comes from the second pouch, but in lampreys it is formed from the third pouch and in hagfishes from the fourth pouch. Each arch also contains a specified cartilage (branchial arch) and muscles (innervated by the designated cranial nerve): the branchial cartilage in the first arch is the mandibular arch, the branchial cartilage in the second arch is the hyoid arch, and the posterior branchial cartilages become the gill arches in the adult.

There is also a region in front of the mouth called the premandibular region. This contains neural crest–derived cartilage but has no associated muscles or blood vessels, although it does receive some sensory nerve supply from a portion of the trigeminal nerve (cranial nerve V) called V_1. (The portions of the trigeminal nerve supplying the mandibular arch are the sensory, V_2, and the motor, V_3.). (B) Original hypothesis of the ancestral (jawless) vertebrate condition, with the mandibular arch forming a functional gill arch and the first pouch forming a complete gill slit. (C) Condition in an adult shark. The mandibular arch forms the jaws, the hyoid arch forms the jaw supports, and the first pharyngeal pouch forms the spiracle. On the right is an anterior view of the form of the jointed gill arches in gnathostomes (branchial arches 3–7). The arches are jointed, meet along the midline, and lie internal to the gill tissue. (A, from Miyashita 2016.)

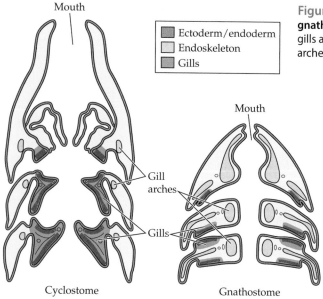

Figure 3.12 Gills are differently positioned in cyclostomes and gnathostomes. As these dorsally viewed horizontal sections show, the gills are positioned inside the gill arches in cyclostomes but outside the gill arches in gnathostomes. (From Gillis and Tidswell 2017 and Jarvik 1965.)

(**Figure 3.11C**). It was long thought that the mandibular arch was originally a gill-bearing branchial arch in jawless ancestors, and that the spiracle between the jaws and the hyoid supports was once a fully fledged gill slit. However, evolutionary-developmental (evo-devo) studies have revealed problems with this interpretation.

Despite their lack of jaws, cyclostome heads are highly complex. The mandibular arch is modified to form the lingual cartilage bars supporting the velum, and the premandibular mesoderm has been modified (independently in hagfishes and lampreys) to form an elaborate series of head cartilages in an expanded upper lip region (different in hagfishes and lampreys; see Figures 3.7C and 3.8C). Cyclostomes have unjointed gill arches that form a branchial basket around the outside of the gills, whereas the gill arches of gnathostomes lie internal to the gills (**Figure 3.12**).

Neither the fossil record nor embryological studies provide evidence that the mandibular arch ever bore gills, or that the first gill pouch formed a gill slit. On the contrary, developmental and genomic studies show that the first pouch and the mandibular arch are different from the rest of the series: they are not under the control of the Hox genes. Furthermore, the trigeminal nerve (cranial nerve V) that supplies this arch develops in a different fashion

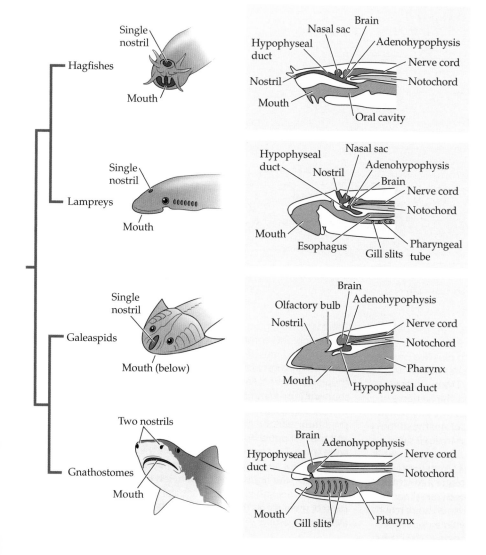

Figure 3.13 The number of nostrils and accompanying nasal sacs is an important distinction between jawed and jawless vertebrates. In cyclostomes (hagfishes and lampreys), the combination of the single nasal sac with the adenohypophysis (anterior pituitary) limits expansion of the forebrain and prevents development of an upper jaw. In gnathostomes (a shark is shown), the double nasal sac is separated from the adenohypophysis and the hypophyseal duct opens into the oral cavity. Derived stem gnathostomes such as galeaspids resembled the basal vertebrate condition in lacking jaws and an expanded forebrain, but the nasal sacs were paired (although the nasal opening was still single) and separated from the adenohypophysis, with the hypophyseal duct opening into the oral cavity as in gnathostomes—an anatomical change that enabled reorganization of the cranial anatomy in the gnathostome lineage. (From Gai et al. 2011.)

from the other arch-supplying cranial nerves, and also innervates the premandibular region (see Figure 3.11A). How, when, and why did the mandibular arch become incorporated into the branchial series in gnathostomes to form a jaw?

The importance of the nose

A crucial issue in our understanding of the origin of jaws traces back to the late 19th century, when Ernst Haeckel noted that the main difference between cyclostomes and gnathostomes is not jaws, but the presence of a single nostril (**monorhiny**) in cyclostomes and double nostrils (**diplorhiny**) in gnathostomes (**Figure 3.13**). This difference is associated with the position of the nasal (olfactory) sacs relative to the adenohypophysis: a single nasal sac combines with the opening for the adenophypophysis in cyclostomes, whereas gnathostomes have double nasal sacs separated from the adenohypophysis.

This anatomy also dictates the size and shape of the forebrain. In cyclostomes, the nasal capsules (the part of the chondrocranium overlying the forebrain and olfactory areas) lie very close to the brain, and the forebrain is short. In gnathostomes, the nasal capsules lie in a more anterior position; olfactory tracts (cranial nerve I) run between the nasal sacs and the forebrain; and the forebrain is longer, supported underneath by a pair of cartilaginous bars called anterior trabeculae. Note also the difference of the entry of the nasal sac in gnathostomes compared with cyclostomes: it is separate from the adenohypophysis rather than conjoined with it (see Figure 3.13). All of these differences between cyclostomes and gnathostomes are important in considering the origin of jaws.

Developmental studies of extant vertebrates

The cyclostome condition—with a single, midline nostril—precludes the development of an upper jaw during ontogeny. Cyclostomes may possess the basal vertebrate condition. We know that at least some ostracoderms had a single dorsal NHO, like that of lampreys, although heterostracans (basal ostracoderms) had paired nasal openings. Developmental studies show that the evolution of the jaw can take place only after the change from a single nasal sac to laterally displaced double sacs, and is effected by a heterotopic shift in tissue interactions and gene expression during head formation (**Figure 3.14**).

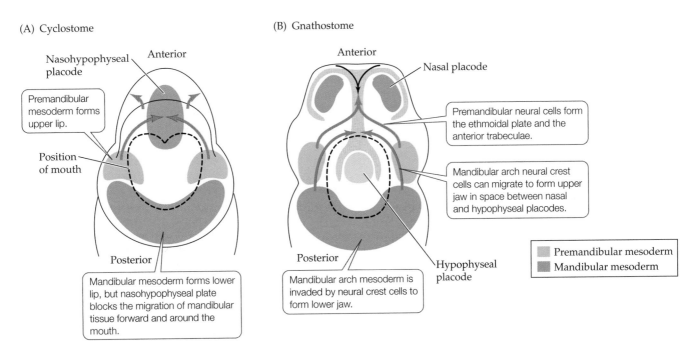

Figure 3.14 Development of the mandibular arch. Ventral views of developing cyclostome (A) and gnathostome (B) embryos at similar stages. Like other cranial sense organs, the olfactory organ is formed from a sensory epidermal placode (see Chapter 2). In early vertebrate development, the nasal (nose) and hypophyseal (adenohypophysis) placodes are combined in a single nasohypophyseal placode on the roof of the mouth. In the cyclostome, the nasal and hypophyseal placodes develop as a single unit. The premandibular mesoderm extends under this placode to form the upper lip. The midline position of the nasohypophyseal plate that develops from this placode in cyclostomes prevents the anterior migration of mandibular neural crest cells to form an upper jaw (palatoquadrate) in front of the mouth in the region of the upper lip. In gnathostomes, the nasal and hypophyseal placodes separate, the adenohypophysis grows up from the ectoderm of the roof of the mouth, and the nasal placode is paired and migrates to the front of the head. This leaves a space between the nasal and hypophyseal placodes into which neural crest cells from the mandibular arch can grow forward to form the upper jaw. In addition, the premandibular neural crest cells can grow forward to form the anterior trabeculae, the paired supports underlying the forebrain, and the midline cartilage (called the ethmoidal plate) that separates the nasal capsules. This is the only premandibular contribution to the head of gnathostomes, as opposed to the extensive contribution in cyclostomes. (From Kuratani 2004.)

Transitional anatomy in fossils

The chondrocranium of derived ostracoderms was lined with mineralized cartilage (among the galeaspids) or perichondral bone (among the osteostracans), preserving the internal head structures. A clue to the evolution of the gnathostome condition comes from some well-preserved specimens of the galeaspid *Shuyu* from the mid-Silurian of China. *Shuyu* had the gnathostome-like condition of double nasal sacs separated from the adenohypophysis opening in the roof of the mouth. However, the nasal sacs were still situated in a lamprey-like position between the eyes, and the forebrain was short (see Figure 3.13).

Thus, we have evidence of the occurrence of diplorhiny and the breakup of the limiting nasohypophyseal placode *before* the origin of jaws. More clues come from a basal gnathostome, the placoderm *Romundina*. This animal had jaws but retained an adult head anatomy very similar to that seen in the galeaspid *Shuyu*, with a considerable premandibular component to the head and a short forebrain. However, *Romundina* showed the more derived gnathostome feature of a complete separation of nasal sacs and adenohypophysis, with the hypophyseal duct opening into the roof of the mouth. Only in more derived placoderms did the nostrils move anteriorly and the cartilaginous anterior trabeculae support a lengthened forebrain.

The selective value of jaws

Although we now know *how* jaws evolved, at least in the developmental sense, we still know little about the evolutionary events that may have driven this profound shift in head development, and why jaws evolved from the mandibular arch. We do not know whether any ostracoderm had modified their mandibular arch into cyclostome-like lingual cartilages and velar bars. The presence of oral plates in some ostracoderms suggests the presence of premandibular cartilages (as seen in cyclostomes), and a toothed biting structure could have been formed from these. So why modify the mandibular cartilages into a biting apparatus?

The "ventilation hypothesis" of jaw formation may explain why the mandibular arch became co-opted into the pharyngeal series as a device for closing off the mouth, although it was not originally involved in gill ventilation. Closure of the mouth by jaws, combined with the interior positioning of the gill arches, would have been the first step in creating the double-pump system of gill ventilation seen in jawed fishes today, and would also have improved capacity for suction feeding. The double-pump system might have supplanted a cyclostome-like system of velar pumping, or might have been the first occurrence of such an improved means of gill ventilation in gnathostomes.

Another issue yet to be explained is how the gill arches shifted in their position from being external to the gills (which appears to be the case in both ostracoderms and cyclostomes) to being internal to the gills in gnathostomes (see Figure 3.12).

3.6 The Origin of Paired Appendages

The pectoral fins were apparently the first paired appendages to evolve, appearing in the osteostracan stem gnathostomes. Paired pelvic fins appear at the base of the jawed vertebrate radiation, in basal placoderms. These limbs and limb girdles are formed from the endochondral skeleton, although the pectoral fin has always had an intimate connection with the dermal head shield, and even the human pectoral girdle retains a dermal element (the clavicle; see Chapter 10). The pectoral and pelvic fins in extant chondrichthyans and basal bony fishes are tribasal, meaning that three main elements within the fin articulate (form joints) with the limb girdle in the body. Molecular studies back up anatomical ones in showing that this type of fin represents the ancestral gnathostome condition. Accompanying the paired fins in gnathostomes is the presence of a distinct neck region between the head and the pectoral fin. In fishes this area is filled with gills and their supporting structures and is relatively immobile.

The advantages of fins

Guiding a body in three-dimensional space is complicated. Fins act as hydrofoils, applying pressure to the surrounding water. Because water is practically incompressible, force applied by a fin in one direction against the water produces a thrust in the opposite direction. A tailfin increases the area of the tail, giving more thrust during propulsion, and allows the fin to exert the force needed for rapid acceleration. Rapid adjustments of the body position in the water may be especially important for active predatory fishes, including the early gnathostomes, and the unpaired fins in the midline of the body (dorsal and anal) control the tendency of a fish to roll (rotate around the body axis) or yaw (swing to the right or left; **Figure 3.15**). The paired fins (pectoral and pelvic) can control the pitch (vertical tilt) and act as brakes, and are occasionally specialized to provide thrust during swimming, as in the enlarged pectoral fins of skates and rays.

Fins have non-locomotor functions as well. Spiny fins are used in defense, and may become systems to inject poison when combined with glandular secretions. Colorful fins are used to send visual signals to potential mates, rivals, and predators.

Fin development and the lateral somitic frontier

Evo-devo studies have also greatly contributed to our understanding of fin evolution. In Chapter 2 we pointed out that limbs are an "add-on" in vertebrate evolution. In extant limbless vertebrates, the outer layers of the body are derived entirely from the somite (the upper segmented portion of the mesoderm) and the unsegmented lateral plate mesoderm (LPM) is confined to the lining of the coelom (somatic mesoderm) and viscera (splanchnic mesoderm). Limb development disrupts this dual-mesoderm system: appendages arise as outgrowths of the somatic portion of

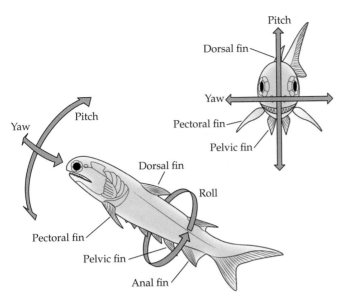

Figure 3.15 Fins stabilize a fish in three dimensions. Views of an early jawed fish (the acanthodian *Climatius*) from the side and front illustrate pitch, yaw, and roll and the fins that counteract these movements.

the LPM where it combines with the ectoderm to form the somatopleure. In lampreys, the somatopleure exists only in early development and is eliminated as the myotome expands ventrally, separating the ectoderm and the LPM. A

key feature of the gnathostome body plan appears to be the retention of the somatopleure and the contribution of LPM connective tissue to the structures arising from the body wall, including the paired limbs. The gnathostome trunk is composed of primaxial (somite only) and abaxial (mixed somite and LPM) domains. The dynamic boundary between these domains is termed the **lateral somitic frontier** (Figure 3.16). The frontier is not a simple dorsal–ventral divide of the body, as it is deflected by the ventrally extending ribs and intercostal muscles formed from the somite. Embryonic patterning genes appear to be regulated independently on either side of the frontier, indicating that the primaxial and abaxial domains are different developmental and evolutionary modules.

Although the paired fins of fishes are of abaxial origin, the midline fins (dorsal and tail) are of primaxial origin. The origin of most ostracoderm finlets was probably also primaxial. The gene expression program of the dorsal fins, which evolved early in vertebrate history, appears to be replicated in the paired fins of gnathostomes, even though they are of different developmental origin.

Origin of the neck region

The pectoral girdle is intimately involved with the branchial region in jawed fishes. Basal gnathostomes had a region between the pectoral girdle and the head with some important new muscles: the cucullaris and the hypobranchials

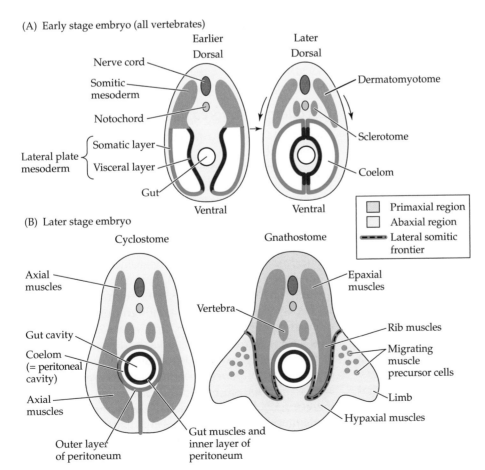

Figure 3.16 Development of gnathostome paired appendages. (A) The early development of cyclostome and gnathostome embryos is similar, but differences emerge at later stages (B). In both gnathostomes and lamprey, the axial skeleton is derived from the sclerotome portion of the somite and the dermatomyotome forms the dermis of the skin and the striated muscles. However, the gnathostome limbs and limb girdles (appendicular skeleton) are derived from a mixture of somitic myoblasts developing within lateral plate–derived connective tissue. The entire muscular body wall in the lamprey is primaxial (somite only), whereas the limbs and body wall of gnathostomes are abaxial (somite plus lateral plate), and there is a clear division of the axial muscles into epaxial (dorsal) and hypaxial (ventral) components. The boundary between primaxial and abaxial domains is the lateral somitic frontier. (After Shearman and Burke 2009, and Ann Burke, Pers. Comm.)

(Figure 3.17). These muscles are derived from the anterior trunk, but in their genetic development they are more similar to the head muscles than to the trunk muscles. Their connective tissue is formed from neural crest cells, and their innervation is derived, at least in part, from cranial nerves. The cucullaris and hypobranchials are involved in both gill ventilation and movement of the head. The cucullaris can aid in elevating the head, and the hypobranchials aid in depressing the jaw and hyoid apparatus, and also form the muscles of the tongue in tetrapods (jawed fish lack a fleshy tongue).

This fish anatomy is essentially retained in humans. The cucullaris of fishes is our trapezius, a broad muscle originating from the midline of the neck and trunk area and running down to the shoulder girdle, which we use to shrug our shoulders and turn our head. The muscles running down the front of our neck (e.g., the sternohyoid) are the fish hypobranchials and help open the mouth.

Cyclostomes lack a cucullaris, and while they have putative hypobranchials, these muscles are not involved in gill ventilation. A cucullaris was probably also absent in osteostracans, where the endochondral pectoral girdle was not separate from the dermal head shield, and was possibly also absent in basal placoderms. The pectoral girdle essentially forms a posterior border to the branchial area in jawed fishes, and developmental studies show that the clavicle marks the boundary between the head and trunk. Might the pectoral girdle originally have been developed in concert with improved gill ventilation, and only later become a structure supporting a pectoral fin? Interestingly, the pectoral fin shares some developmental similarities with the cartilaginous rays that extend from the branchial arches. More details are needed about the internal anatomy of osteostracans and placoderms to understand the origin of the crown gnathostome condition.

3.7 Extinct Paleozoic Jawed Fishes

The Phanerozoic has seen a succession of diversifications and extinctions of fishes and fishlike vertebrates. Conodonts flourished in the Ordovician, ostracoderms and acanthodians reached their greatest diversities in the late Silurian and Early Devonian, and placoderms were diverse in the mid Devonian. Osteichthyans and chondrichthyans also radiated in the Devonian and in the Carboniferous. Both lineages persisted with relatively low diversity until the Cretaceous, when they radiated into a wide variety of ecomorphs (Figure 3.18).

We have numerous fossils of the entire bodies of gnathostomes (rather than fragments such as teeth and scales)

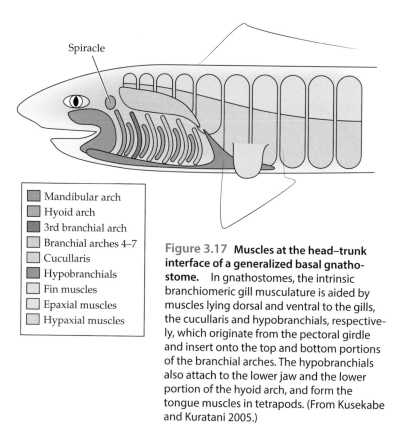

Spiracle

Mandibular arch
Hyoid arch
3rd branchial arch
Branchial arches 4–7
Cucullaris
Hypobranchials
Fin muscles
Epaxial muscles
Hypaxial muscles

Figure 3.17 Muscles at the head–trunk interface of a generalized basal gnathostome. In gnathostomes, the intrinsic branchiomeric gill musculature is aided by muscles lying dorsal and ventral to the gills, the cucullaris and hypobranchials, respectively, which originate from the pectoral girdle and insert onto the top and bottom portions of the branchial arches. The hypobranchials also attach to the lower jaw and the lower portion of the hyoid arch, and form the tongue muscles in tetrapods. (From Kusekabe and Kuratani 2005.)

from Devonian sediments. Gnathostomes can be divided into four distinct groups, two of them extinct—placoderms and acanthodians—and two of them extant—chondrichthyans (cartilaginous fishes) and osteichthyans (bony fishes and tetrapods). Before studying the extant groups of jawed fishes, we turn here to the placoderms and acanthodians to examine the variety of early gnathostomes.

Placoderms: Armored fishes

As their name (plax, "plate") implies, placoderms were covered with a thick, bony shield over the anterior third to half of their bodies. Placoderms retained some more primitive features seen in cyclostomes: seven (rather than six) muscles moving each eye and myomeres (trunk muscles) that were only weakly W-shaped and not distinctly separated into epaxial and hypaxial components. Placoderms are now generally accepted to be a paraphyletic grouping basal to other gnathostomes ("true" gnathostomes, or eugnathostomes), because some placoderm lineages share more features with eugnathostomes than others do (see Figure 3.5), although opposing views of monophyly exist.

A hallmark of placoderms was the division of the single bony head shield of ostracoderms into head and trunk portions, with an articulating joint between them, enabling the snout to be lifted independently of the body (Figure 3.19). Some exceptionally preserved fossils show elevator and depressor muscles attached to the head shield and spanning the joint between the head and trunk portions. Underneath this armor the endoskeleton was not

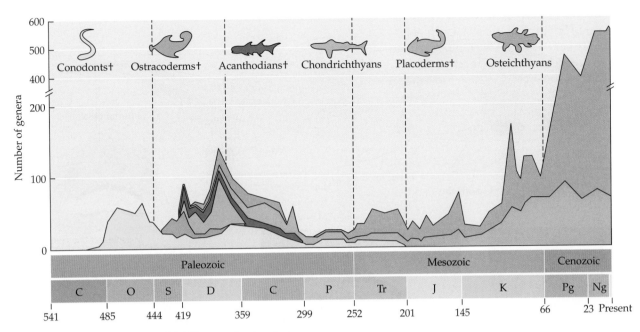

Figure 3.18 Diversity of fishes and fishlike vertebrates during the Phanerozoic. Six major lineages populated the seas during the Phanerozoic. Four of the lineages are now extinct (daggers). Dashed lines mark the five major extinction episodes. Ostracoderms and placoderms were badly hit by the Late Devonian extinctions, although they persisted until the end of the Devonian. Acanthodians persisted until the mid Permian. Conodonts were exterminated in the end Triassic extinction, although they were in decline throughout the second half of the period. (This figure shows lineages based on their significant presence in the fossil records; for more complete occurrence times, see Figure 3.4.) (From Friedman and Sallan 2012.)

too dissimilar from that of a chondrichthyan, although dermal elements were associated with both the endochondral pectoral girdle (as in all bony fishes) and the endochondral pelvic girdle. Dermal pelvic elements are not seen in any extant vertebrate, but there were dermal pelvic places in some basal ostheichthyans, and acanthodians had dermal pelvic spines. Placoderms also had a fusion of the anterior vertebrae of the spinal column into a structure called a synarcual, possibly to aid in the support of a heavy head. Synarcuals are also seen in some extant chondrichthyans.

History Placoderms are known from the early Silurian until the end of the Devonian. They were the most diverse vertebrates of their time in terms of number of taxa and also in morphological specializations. Many placoderms were flattened, bottom-dwelling forms, some even resembling extant skates and rays. Body sizes ranged from a few centimeters to 5–6 m, the size of a great white shark (*Carcharodon carcharias*). Placoderms were initially primarily marine, but a great many lineages became adapted to freshwater and estuarine habitats. Like the ostracoderms, placoderms suffered massive losses in the Late Devonian extinctions, but a few placoderm lineages continued until the end of the period.

Biology Placoderms have no modern analogues, and their extensive exoskeleton makes interpreting their lifestyle difficult. Many placoderms had dermal (gnathal) plates, found on the medial side of their cartilaginous endoskeletal jaws. It was originally thought that placoderms lacked teeth, but these gnathal ossifications were actually a composite of teeth. They were most obvious in the derived arthrodires, but they were also present in some basal placoderms. These teeth developed in succession, but unlike the teeth of other gnathostomes, they were not replaced; there was apparently no dental lamina, and the tooth and jaw integration was less well developed.

The dermal head shields of ostracoderms, placoderms, and osteichthyans must obviously be homologous at some level, even though the details of the bones are different. The dermal bones associated with the jaws of some placoderms (the "maxillate" placoderms) were not simple gnathal plates, but homologues of the dermal tooth-bearing marginal mouth bones of osteichthyans (retained in humans), although they lack actual teeth. These bones are intimately associated with the cartilaginous mandibular elements and also integrate with the dermal bones of the cheek.

These maxillate placoderms are known from the late Silurian of China, but despite their early appearance in the fossil record they are closest to crown gnathostomes in the vertebrate phylogeny (see Figures 3.4, 3.5). *Qilinyu* had a maxilla and a premaxilla in its upper jaw, and *Entelognathus* had these bones plus a dentary in the lower jaw. These dermal bones in placoderms bore palatal laminae (flanges of bone projecting from the medial side of the jaw) that were lost in crown gnathostomes.

A recent study of placoderms from the Late Devonian of Australia that preserve some soft-tissue structure has provided more information about these fishes, including information about their modes of reproduction. The

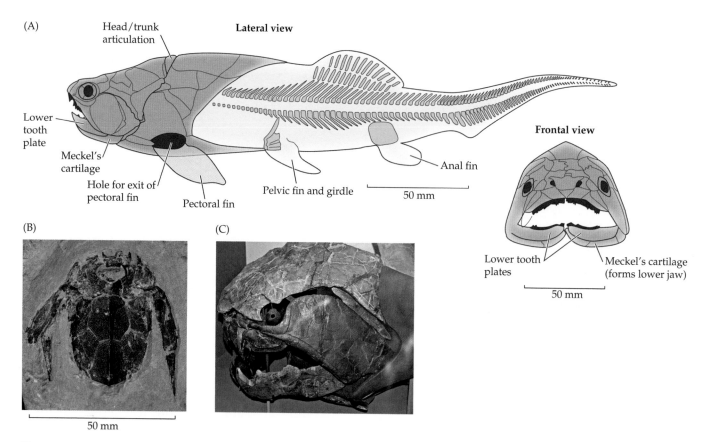

Figure 3.19 Extinct Paleozoic fishes: Placoderms. (A) Reconstruction of the arthrodire placoderm *Coccosteus* (Middle Devonian). More than half of the known placoderms were arthrodires (Greek, "jointed neck"), a predatory lineage most closely related to extant gnathostomes. (B) Fossil of the head and trunk shield of the antiarch placoderm *Bothriolepis* (Middle to Late Devonian) seen looking down from above. Antiarchs were basal placoderms that looked rather like extant armored catfishes. Their pectoral fins were also encased in the bony shield. (C) Fossilized head and trunk shield of the arthrodire *Dunkleosteus* (Late Devonian). This 8-m-long predator may be the best-known placoderm. Biomechanical studies show that it had an extraordinarily powerful bite and rapid gape expansion for engulfing prey. (A, after Moy-Thomas and Miles 1971; B, The Natural History Museum/Alamy Stock Photo; C, James St. John/CC BY 2.0).)

appropriately named *Materpiscis* (Latin *mater*, "mother"; *piscis*, "fish") had embryos and a structure interpreted as an umbilical cord preserved within the body cavity. This evidence of viviparity in at least some placoderms matches the observation that both basal and derived placoderms had pelvic claspers, resembling the pelvic claspers in male cartilaginous fishes that are used for internal fertilization (see Chapter 6). However, these claspers were not homologous with those of extant fishes; they were not situated on the pelvic fin, but were separate structures behind the pelvic fin. We can infer from the evidence for viviparity and pelvic appendages that placoderms, like extant chondrichthyans, had internal fertilization and probably had complex courtship behaviors.

Acanthodians

Acanthodians, or "spiny sharks," were small, generalized fishes, named for the stout spines (Greek *acantha*, "spine") anterior to their well-developed dorsal, anal, and paired fins (**Figure 3.20**) Acanthodians also sported additional pairs of ventrolateral fins (which did not connect to internal girdles). They lacked the extensive bony armor of placoderms but retained dermal head bones, and many had chondrichthyan-like denticles covering the body. The precise phylogenetic position of acanthodians has been a matter of dispute, although they are now considered to form a grouping basal to chondrichthyans (i.e., they are stem chondrichthyans).

History Acanthodians are known definitively from the early Silurian into the middle or early late Permian but had their major diversity in the Early Devonian. The earliest forms were marine, but by the Devonian they were predominantly a freshwater group. The acanthodids, the only group to survive into the Permian, were elongate, toothless, and had long gill rakers, indicating that they were probably plankton-eating filter feeders.

Biology Acanthodians had a basic fusiform fish shape with a heterocercal tailfin (i.e., with the upper lobe larger than the lower lobe; see Figure 3.15). Most acanthodians were no more than 20 cm long, although some species reached 2 m. Most acanthodians had a large head with a

(A)

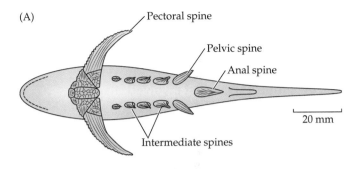

Pectoral spine

Pelvic spine

Anal spine

20 mm

Intermediate spines

(B)

Fin spine   Soft tissue of fin

10 mm

Figure 3.20 Extinct Paleozoic fishes: Acanthodians. (A) Ventral view of the acanthodian *Climatius* (Early Devonian), showing the multiple sets of ventrolateral paired fins (a lateral view of this animal can be seen in Figure 3.15). (B) Fossil of the acanthodian *Diplacanthus* (Late Devonian) lateral view. Here the outline of the soft portion of the fins can be seen behind the fin spines. (A, after Moy-Thomas and Miles 1971; B, from Woodward 1891.)

wide-gaping mouth. Some species were toothless, and in others the teeth formed a sharklike tooth whorl.

The surviving gnathostome groups

As we noted at the start of this section, two of the four distinct groups of gnathostomes are still extant: chondrichthyans and osteichthyans. We will discuss extant and extinct chondrichthyans in Chapters 6 and 7, and bony fish—the largest group of osteichthyans—in Chapters 8 and 9.

Summary

The earliest vertebrates were more complex and more active animals than the nonvertebrate chordates.

Novel vertebrate anatomical features included a tripartite brain enclosed by a cranium, and a muscular pharynx for using the gills for respiration rather than for filter feeding.

The gnathostomes—jawed vertebrates—represent an increase in complexity in the vertebrate body plan that reflects the vertebrate mode of life as an active predator.

Vertebrates are known from the start of the Paleozoic.

Soft-bodied vertebrates are known from the early Cambrian. The earliest known vertebrates with bone (ostracoderms) are from the Late Ordovician.

The earliest evidence of bone in vertebrates is in the odontodes of the dermal head shield of ostracoderms. Their bony tissues included bone (acellular), dentine, and enamaloid. The original functions of this "armor" may have been metabolic rather than protective.

Some ostracoderms had toothlike structures within the pharynx, but true teeth were not seen until the jawed vertebrates (gnathostomes).

Toothlike structures have been found in association with early soft-bodied vertebrates called conodonts, but conodont teeth do not contain true dentine, so are not homologous with the mineralized tissues in ostracoderms and gnathostomes.

The earliest environment of vertebrates was marine, although they rapidly moved into, and radiated in, freshwater habitats.

The extant jawless vertebrates are cyclostomes.

Hagfishes and lampreys—the cyclostomes—differ from gnathostomes in lacking paired fins as well as jaws. However, they have evolved complex head structures that diverge from the gnathostome condition, including a lingual cartilage that supports a muscular, tonguelike structure that is not homologous with the gnathostome tongue.

Cyclostomes lack mineralized tissues but have keratinous teeth and/or tooth plates on their tongue. They also differ from gnathostomes in having a single nasal opening, and their vertebrae consist only of rudiments called arcualia. Their gill arches (supporting gill tissue and muscles) are unjointed and lie external to the gill tissue (ostracoderms were apparently similar).

Hagfishes are entirely marine, and are more primitive than other vertebrates in many ways, including their less specialized sense organs and primitive kidney, and especially in the fact that their body fluids have the same osmolal

(Continued)

Summary (*continued*)

concentration as seawater. Hagfishes were once considered to be the sister taxon to other vertebrates, but molecular studies have indicated that they are related to lampreys.

Hagfishes usually feed on dead or dying fishes, although they can be active predators. A unique feature of hagfishes is the production of slime via slime glands running the length of the body. Hagfish slime deters predators and may aid in ion regulation.

Lampreys live in both salt water and fresh water. Larval lampreys are filter feeders, but adult lampreys are parasites on other vertebrates (usually larger bony fishes), sucking blood by affixing themselves to prey via their oral hood. They obtain oxygen via tidal ventilation in and out of the gill slits.

Both hagfishes and lampreys have had negative interactions with humans. Hagfish skin is used to make "eelskin" leather products, and demand for these products has eradicated some populations. Lampreys have colonized the North American Great Lakes and reduce stocks of economically important fishes.

The earliest radiation of vertebrates was of the armored jawless ostracoderms.

Many ostracoderms were encased in dermal bone, but some had fine scales or were virtually naked. They represent a paraphyletic assemblage of stem gnathostomes: some ostracoderms had more features in common with gnathostomes than others did, such as pectoral fins.

Ostracoderms were diverse and successful during the Silurian and Devonian. During most of this time they lived alongside a diversity of jawed fishes. There is no evidence that ostracoderms were outcompeted by gnathostomes; rather, they disappeared in the Late Devonian extinctions that also claimed several gnathostome lineages, as well as marine invertebrates.

Gnathostomes have many derived features in addition to jaws and two sets of paired limbs.

Gnathostomes have undergone a further duplication of the Hox genes. They have jointed gill arches situated internal to the gill tissue, with a double-pump mode of gill ventilation; myelinated nerves; a cerebellum in the brain (also seen in ostracoderms); a third (horizontal) semicircular canal; a conus arteriosus in the heart; cellular bone and mineralized internal tissues (shared with some derived ostracoderms); and specialized ducts connecting the gonads to the cloaca. All of these features reflect a step up in levels of activity and complexity from the agnathan (jawless) condition.

Chondrichthyans have long been perceived as primitive jawed vertebrates, but they are actually derived in many ways, including in their loss of dermal bone. In many

features, osteichthyans appear to be closer to the original gnathostome body plan.

The evolution of jaws first involved changes to the nose.

Jaws develop from the mandibular arch, which is divergently specialized in cyclostomes into the lingual and velar cartilages. In contrast to what was proposed in the original hypothesis about jaw origins, the mandibular cartilage was not originally part of a branchial series— that is, it was never a gill-bearing structure.

Developmental studies show that an upper jaw cannot form in cyclostomes because of the central position of the single, dorsally placed nostril that combines with the duct for the adenohypophysis (anterior pituitary).

The transition from the cyclostome condition of monorhiny (single nostril) to the gnathostome condition of diplorhiny (double nostril) can be seen in the fossil record. Derived ostracoderms (galeaspids) showed the first stage of this process, primitive gnathostomes (basal placoderms) showed the second stage, and derived placoderms showed the third stage, in which the forebrain was also lengthened.

Jaws may have evolved in the context of the mandibular arch being "co-opted" by the branchial arches posterior to it (i.e., the gill arches). A mouth that could close would have increased the force of the water pumped over the gills, and the internal jointed gill arches appear to be part of this system of improved gill ventilation.

Paired fins required changes in the development of the body.

Paired pectoral fins were first seen in derived ostracoderms (osteostracans), and paired pelvic fins in basal placoderms. Paired fins are important for maneuvering in three-dimensional space, braking, and in some species, propulsion. Fins also have non-locomotory functions, such as defense and communication.

The paired fins have an internal skeleton of cartilage or bone and attach to limb girdles in the body. The pectoral girdle was initially involved with elements of the dermal bone of the head shield, a condition retained in humans (with the clavicle) but lost in cartilaginous fishes, which have lost all dermal bone.

Paired fins are an "add-on" to the basal vertebrate body plan. In cyclostomes, all of the body musculature is derived from the somite, the segmental portion of the mesoderm, but in gnathostomes the somatic layer of the lateral plate becomes involved in forming the limbs and the hypaxial (ventral) musculature. In gnathostomes, the lateral somitic frontier is the dividing line in axial development between structures derived from the somite

Summary (continued)

(primaxial) and structures derived from a mixture of somite and lateral plate (abaxial).

Gnathostomes have a distinct neck area between the head and pectoral girdle, with neck muscles (cucullaris and hypo-branchials) that originate from the pectoral girdle and are involved in movements of the head and gill arches. These neck muscles are derived from the trunk, but in developmental terms are more akin to the muscles of the head.

There were four major lineages of jawed fishes in the Paleozoic.

The cartilaginous fishes (Chondrichthyes) and the bony vertebrates (Osteichthyes) that are alive today date back to the Silurian, but the two other lineages—placoderms and acanthodians—did not survive the Paleozoic. Both of these lineages represent paraphyletic groupings. Placoderms are stem gnathostomes, and acanthodians are stem chondrichthyans.

Placoderms are known from the early Silurian until the end of the Devonian, and they were the dominant fishes of the Devonian in terms of both taxonomic and morphological diversity. Like the ostracoderms, placoderms were covered in dermal armor, but they had a distinct hinge between their head and trunk shields. Some placoderms (presumed males) had pelvic claspers, most likely for internal fertilization, like those of chondrichthyans but not homologous to them. There is also evidence of viviparity in placoderms.

Placoderms had gnathal plates attached to their endoskeletal jaws, and were once thought to lack teeth. However, it is now apparent that in some placoderms, including basal ones, these plates were formed from a composite of teeth. Some highly derived placoderms also had dermal marginal mouth bones homologous with the tooth-bearing mouth bones seen in osteichthyans (including humans).

Acanthodians are known from the early Silurian through the middle Permian, with their major diversity in the Early Devonian. They retained dermal head bones but had sharklike dermal denticles on their bodies and were characterized by spines in front of the dorsal, anal, and paired fins, and multiple pairs of ventrolateral fins in addition to the pectoral and pelvic fins.

Discussion Questions

1. Did jawed vertebrates outcompete the jawless ones? Does the fossil record provide any evidence?

2. How did the realization that conodonts were vertebrates change our ideas about the pattern of vertebrate evolution?

3. Cyclostomes are often thought of as "primitive" because they lack jaws. In what ways could the heads of cyclostomes be considered to be derived in their own right?

4. Ventilating the gills by taking water in through the gill openings, which are also used for water ejection, is not very efficient. Why, then, do adult lampreys ventilate their gills this way?

5. We usually think of jaws as structures whose original function was biting. What might have been a different original use of jaws, and what is the evidence for this?

6. Recent fossil evidence shows a placoderm with a developing embryo inside its body. In the absence of such evidence, how might we have been able to speculate that placoderms were viviparous?

Additional Reading

Brazeau MD, Friedman M. 2015. The origin and early phylogenetic history of jawed vertebrates. *Nature* 520: 490–497.

Clark AJ, Summers AP. 2007. Morphology and kinematics of feeding in hagfish: possible functional advantages of jaws. *Journal of Experimental Biology* 210: 3897–3909.

Donoghue PCJ, Keating JN. 2014. Early vertebrate evolution. *Palaeontology* 57: 879–893.

Friedman M, Sallan LC. 2012. Five hundred million years of extinction and recovery: a Phanerozoic survey of large-scale diversity patterns in fishes. *Palaeontology* 55: 707–742.

Gillis JA, Dahn RD, Shubin NH. 2009. Shared developmental mechanisms pattern the vertebrate gill arch and paired fin skeletons. *Proceedings of the National Academy of Sciences USA* 106: 5720–5724.

Janvier P. 2013. Led by the nose. *Nature* 493: 169–170.

Janvier P. 2015. Facts and fancies about early fossil chordates and vertebrates. *Nature* 520: 483–489.

Long JA. 2011. *The Rise of Fishes*, 2nd ed. Johns Hopkins University Press, Baltimore, MD.

McCoy VE and 15 others. 2016. The "Tully monster" is a vertebrate. *Nature* 532: 496–499.

Ota KG, Oisi Y, Fujimoto S, Shigeru K. 2014. The origin of developmental mechanisms underlying vertebral elements: implications from hagfish evo-devo. *Zoology* 117: 77–80.

Venkatesh B and 32 others. 2014. Elephant shark genome provides unique insights into gnathostome evolution. *Nature* 505: 174–179.

Living in Water

Although life evolved in water and the earliest vertebrates were aquatic, the physical properties of water create some difficulties for animals. To live successfully in open water, a vertebrate must adjust its buoyancy to remain at a selected depth and force its way through a dense medium to pursue prey or to escape its own predators. Heat flows rapidly between an animal and the water around it, and it is difficult for an aquatic vertebrate to maintain a body temperature that is different from the water temperature. This phenomenon was dramatically illustrated when the *Titanic* sank; in the frigid water of the North Atlantic, most of the victims died from hypothermia rather than by drowning.

Ions and water molecules move readily between the external environment and an animal's internal body fluids, so maintaining a stable internal environment can be difficult. On the plus side, ammonia is extremely soluble in water, so disposal of nitrogenous waste products is easier in aquatic environments than on land. The concentration of oxygen in water is lower than it is in air, however, and the density of water imposes limits on the kinds of gas-exchange structures that can be effective.

In this chapter, we consider how aquatic vertebrates—fishes and amphibians—have responded to some of the challenges of living in water. Fishes, especially the bony fishes, have diversified into an enormous array of sizes and ways of life. Aquatic amphibians include some highly specialized aquatic forms (such as the European olm, an eyeless white salamander that lives in deep caves of the Adriatic region, and a frog that lives in tidal areas where it experiences twice-daily changes in salinity), but the major diversification of amphibians has occurred on land.

4.1 The Aquatic Environment

Water covers 73% of Earth's surface. Most of this water is held in the ocean basins, which are populated everywhere by vertebrates. Freshwater lakes and rivers hold a negligible amount of the water—about 0.01%. This is much less than the amount of water tied up in the atmosphere, ice, and groundwater, but freshwater habitats are exceedingly rich biologically, and nearly 40% of all bony fishes live in fresh water.

Water and air are both fluids at biologically relevant temperatures and pressures, but they have different physical properties that make them drastically different environments in which to live (**Table 4.1**). In air, for example, gravity is an important force acting on an animal, but fluid resistance to movement (i.e., air resistance) is trivial for all but the fastest birds. In water, however, the opposite relationship holds. Gravity is negligible, but fluid resistance to movement is a major factor with which aquatic vertebrates must contend, and most fishes are streamlined. Although each major clade of aquatic vertebrates has dealt with environmental challenges in somewhat different ways, the basic specializations needed by all aquatic vertebrates are the same.

Obtaining oxygen from water: Gills

Most aquatic vertebrates have gills, which are specialized structures where oxygen and carbon dioxide are exchanged, although amphibians use cutaneous respiration in addition to, or even instead of, gills. Teleosts are derived ray-finned fishes (Actinopterygii), the group that includes the majority of species of extant freshwater and marine fishes. The gills of teleosts are enclosed in pharyngeal pockets called the opercular cavities (**Figure 4.1**).

Because water is dense and viscous, it takes more energy to move water than air. Unlike the tidal (in–out) flow that air-breathing animals use to ventilate their lungs, the flow of water across a fish's gills is unidirectional—in through the mouth and out past the gills. Flaps just inside the mouth and flaps at the margins of the gill covers (**opercula**; singular *operculum*) of bony fishes act as valves to prevent backflow. The respiratory surfaces of the gills are delicate projections from the lateral side of each gill arch. Two columns of gill filaments extend from each gill arch. The tips of the filaments from adjacent arches meet when the filaments are extended. As water leaves the buccopharyngeal cavity, it passes over the filaments. Gas exchange takes place at

(A) Lateral view of head

(B) Detail of gill filaments

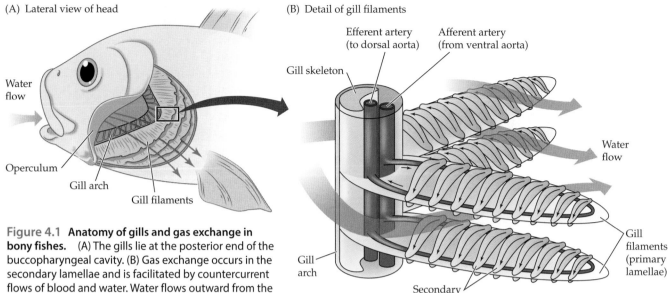

Water flow

Operculum

Gill arch

Gill filaments

Gill skeleton

Efferent artery (to dorsal aorta)

Afferent artery (from ventral aorta)

Water flow

Gill arch

Secondary lamellae

Gill filaments (primary lamellae)

Figure 4.1 **Anatomy of gills and gas exchange in bony fishes.** (A) The gills lie at the posterior end of the buccopharyngeal cavity. (B) Gas exchange occurs in the secondary lamellae and is facilitated by countercurrent flows of blood and water. Water flows outward from the buccopharyngeal cavity past the secondary lamellae to the exterior, while deoxygenated blood flows in the opposite direction through the secondary lamellae. (A from Townsend 2012.)

the numerous microscopic projections from the filaments called **secondary lamellae**.

The pumping action of the mouth and pharyngeal region, called **buccopharyngeal pumping**, creates a positive pressure across the gills so that the respiratory current is only slightly interrupted during each pumping cycle. Some filter-feeding fishes and many pelagic fishes—including mackerel, certain sharks, tunas, and swordfishes—have reduced or even lost the ability to pump water across the gills. These fishes create a respiratory current by swimming with their mouths open, a method known as **ram ventilation**, and they must swim continuously. Many other fishes rely on buccopharyngeal pumping when they are at rest and switch to ram ventilation when they are swimming.

The arrangement of blood vessels in the gills maximizes oxygen exchange. Each gill filament has two arteries: an afferent vessel running from the gill arch to the filament tip, and an efferent vessel returning blood to the arch. Each

secondary lamella is a blood space connecting the afferent and efferent vessels (**Figure 4.2**). The direction of blood flow through the lamellae is opposite to the direction of water flow across the gill. This arrangement, known as **countercurrent exchange**, assures that as much oxygen as possible diffuses into the blood. Pelagic fishes (such as tunas) that sustain high levels of activity for long periods have skeletal tissue reinforcing the gill filaments, large gill exchange areas, and a high oxygen-carrying capacity per milliliter of blood compared with sluggish bottom-dwelling fishes (such as toadfishes and flat fishes).

Obtaining oxygen from air: Lungs and other respiratory structures

Although the vast majority of fishes depend on gills to extract dissolved oxygen from water, fishes that live in water with low oxygen levels cannot obtain enough oxygen via gills alone. These fishes supplement the oxygen they get from their gills with additional oxygen obtained from the air via lungs or accessory air respiratory structures.

Table 4.1	Physical properties of fresh water and air at 20°C		
Property[a]	**Fresh water**	**Air**	**Comparison**
Density	1 kg/L	0.0012 kg/L	Water is about 833 times as dense as air.
Dynamic viscosity	1.002 mPa/s	0.0186 mPa/s	Water is about 55 times as viscous as air.
Heat capacity	4.18 kJ/L/°K	0.0012 kJ/L/°K	It takes ~3,500 times more heat to raise a unit volume of water 1°K.
Heat conductivity	0.6 W/m/°K	0.024 W/m/°K	Heat moves through water about 25 times faster than through air.
Oxygen content	6 mL/L	209 mL/L	A unit volume of air has ~35 times as much oxygen as a unit volume of water.
Oxygen diffusion rate	0.000021 cm²/s	0.176 cm²/s	Oxygen diffuses nearly 8,500 times faster in air than in water.
Velocity of sound	1,481 m/s	343 m/s	Sounds travels 4.3 times faster in water than in air.
Refractive index	1.33	1.00	The refractive index of water is nearly the same as that of the cornea of the eye.

[a]Most of these properties change with temperature and atmospheric pressure, and some are affected by the presence of solutes as well.

(A) Gill filament

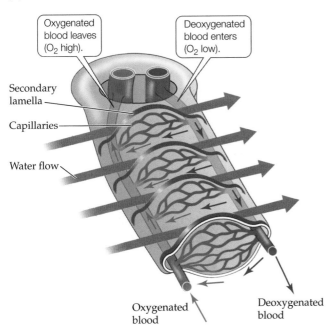

(B) With countercurrent flow

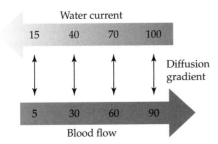

(C) Without countercurrent flow

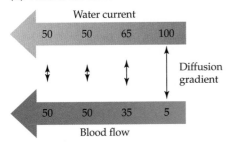

Figure 4.2 Countercurrent exchange in the gills of bony fishes. (A) The direction of water flow across the gill is opposite to the flow of blood through the secondary lamellae. Blood cells are separated from oxygen-rich water only by the thin epithelial cells of the capillary wall, as shown in this cross section of a secondary lamella. (B) Countercurrent flow maintains a difference in oxygen concentration (a diffusion gradient) between blood and water for the full length of a lamella and results in a high oxygen concentration in the blood leaving the gills. (C) If water and blood flowed in the same direction, the difference in oxygen concentration and the diffusion gradient would be high initially, but would drop to zero as the concentration of oxygen equalized. No further exchange of oxygen would occur, and the blood leaving the gills would have a low oxygen concentration. (A from Townsend 2012.)

The accessory structures used to take up oxygen from air include enlarged lips that are extended just above the water surface and a variety of internal structures into which air is gulped. The anabantid fishes of tropical Asia (including the bettas and gouramis seen in pet stores) have a vascularized chamber in the rear of the head, called a labyrinth. Air is sucked into the mouth and transferred to the labyrinth, where gas exchange takes place. Many of these fishes are facultative air breathers; that is, they switch oxygen uptake from their gills to accessory respiratory structures when the level of oxygen in the water becomes low. Others, like the electric eel (*Electrophorus electricus*) of South America and some of the snakeheads, are obligatory air breathers. The gills alone cannot meet the respiratory needs of these fishes, even if the water is saturated with oxygen, and they drown if they cannot reach the surface to breathe air.

We think of lungs as being the respiratory structures used by terrestrial vertebrates, as indeed they are, but lungs first appeared in fishes and preceded the evolution of tetrapods by millions of years. Lungs develop embryonically as outpocketings (evaginations) of the pharyngeal region of the digestive tract, originating from its ventral or dorsal surface. The lungs of bichirs (a group of air-breathing fishes from Africa), lungfishes, and tetrapods originate from the ventral surface of the gut, whereas the lungs of gars (a group of primitive bony fishes) and the lungs of the derived bony fishes (the teleosts) originate embryonically from the dorsal surface.

Lungs used for gas exchange need a large surface area, which is provided by ridges or pockets in the wall. This structure is known as an alveolar lung, and it is found in gars, lungfishes, and tetrapods. Increasing the volume of the lung by adding a second lobe is another way to increase the surface area, and the lungs of lungfishes and tetrapods consist of two symmetrical lobes. Bichirs have non-alveolar lungs with two lobes, but one lobe is much smaller than the other; gars have single-lobed alveolar lungs.

Adjusting buoyancy

Holding a bubble of air inside the body changes the buoyancy of an aquatic vertebrate, and bichirs and teleost fishes use lungs and gas bladders to regulate their position in the water. Air-breathing aquatic vertebrates (whales, dolphins, seals, and penguins, for example) can adjust their buoyancy by altering the volume of air in their lungs when they dive.

Bony fishes   Many bony fishes are neutrally buoyant (i.e., they have the same density as water). These fishes do not have to swim to maintain their vertical position in the water column. The only movement they make when at rest is backpedaling with the pectoral fins to counteract the forward thrust produced by water as it is ejected from the gills and a gentle undulation of the tailfin to keep them level in the water. Fishes capable of hovering in the water like this usually have a well-developed gas bladder.

(A)

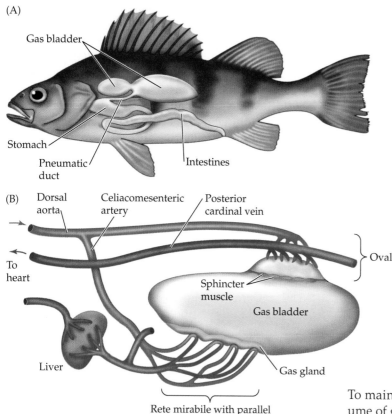

(B)

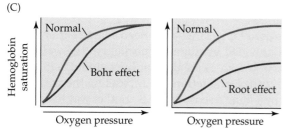

(C)

Figure 4.3 Gas bladder of bony fishes. (A) The gas bladder lies in the coelomic cavity just beneath the vertebral column. Guanine crystals in the wall of the gas bladder make it impermeable to oxygen and give it a silvery appearance. The drawing is of a physostomous fish, in which the gas bladder retains its ancestral connection to the gut via the pneumatic duct. Physostomous fishes gulp air into the gas bladder and burp air out to adjust the bladder's volume. (B) The gas bladder of physoclistous fishes has lost the connection to the gut; oxygen is added via the gas gland and removed via the oval. Removing oxygen to decrease the volume of the gas bladder is straightforward: sphincter muscles allow oxygen from the gas bladder to enter the oval, where the oxygen binds to hemoglobin and enters the systemic circulation. The blood supply to the oval comes from the dorsal aorta; that blood is already at 75–80% of full oxygenation so most of the oxygen released via the oval travels in solution in the blood plasma. (C) The Bohr effect (left) is a reduction in the ability of hemoglobin to bind oxygen in the presence of acid and is a general property of the hemoglobins of vertebrates. The Root effect (right), which is a property of the blood of bony fishes, is a reduction in the maximum amount of oxygen that hemoglobin can bind. Acidification of blood in the gas gland causes hemoglobin to release oxygen that enters the gas bladder.

The gas bladder is located between the peritoneal cavity and the vertebral column (see Figure 4.3A), occupying about 5% of the body volume of marine teleosts and 7% of the volume of freshwater teleosts. The difference in volume corresponds to the difference in density of salt water and fresh water—salt water is denser, so a smaller bladder is sufficient. The gas bladder, which has walls composed of interwoven collagen fibers without blood vessels, is virtually impermeable to gas.

Neutral buoyancy produced by a gas bladder works as long as a fish remains at one depth, but if a fish swims vertically up or down, the hydrostatic pressure that the surrounding water exerts on the bladder changes, which in turn changes the volume of the bladder. For example, when a fish swims deeper, the additional pressure of the water column above it compresses the gas in its bladder, making the bladder smaller and reducing the buoyancy of the fish. When the fish swims toward the surface, water pressure decreases, the gas bladder expands, and the fish becomes more buoyant. To maintain neutral buoyancy, a fish must adjust the volume of gas in its bladder as it changes depth.

A bony fish regulates the volume of its gas bladder by secreting gas into the bladder to counteract the increased external water pressure when it swims down and removing gas when it swims up. Primitive teleosts such as eels, herrings, anchovies, salmon, and minnows retain a connection—the pneumatic duct—between the gut and gas bladder (**Figure 4.3A**). These fishes are called **physostomous** (Greek *physeter*, "bellows"; *stoma*, "mouth"); goldfishes are a familiar example of this group. Because they have a connection between the gut and the gas bladder, they can gulp air at the surface to fill the bladder and burp gas out to reduce its volume.

Adult teleosts from more derived clades lack a pneumatic duct, a condition called **physoclistous** (Greek *cleistos*, "closed"). Physoclistous teleosts increase the volume of the gas bladder by secreting gas from the blood into the bladder and decrease the volume by absorbing gas from the bladder and releasing it at the gills. Both physostomes and physoclists have a gas gland located in the anterior ventral floor of the gas bladder (**Figure 4.3B**). Underlying the gas gland is an area with many capillaries arranged to give countercurrent flow of blood entering and leaving the area. This structure, known as a **rete mirabile** (Latin, "wonderful net"; plural *retia mirabilia*), moves gas (especially oxygen) from the blood to the gas bladder. It is remarkably effective at extracting oxygen from the blood and releasing it into the gas bladder, even when the pressure of oxygen in the bladder is many times higher than its pressure in blood. Gas secretion occurs in many deep-sea fishes despite the hundreds of atmospheres of gas pressure within the bladder.

The gas gland secretes oxygen by releasing lactic acid and carbon dioxide, which acidify the blood in the rete mirabile. Acidification causes hemoglobin to release oxygen into solution (the Bohr and Root effects; **Figure 4.3C**). Because of the anatomy of the rete mirabile, which folds back on itself in a countercurrent multiplier arrangement, oxygen released from the hemoglobin accumulates and is retained within the rete until its pressure exceeds the oxygen pressure in the gas bladder. At this point, oxygen diffuses into the bladder, increasing its volume. The maximum multiplication of gas pressure that can be achieved is proportional to the length of the capillaries of the rete mirabile, and deep-sea fishes have very long retia. A large Root effect is characteristic only of the blood of ray-finned fishes, and it is essential for the function of the gas gland.

Physoclistous fishes have no connection between the gas bladder and the gut, so they cannot burp to release excess gas from the bladder. Instead, physoclists open a sphincter muscle to allow gas to enter a region called the oval that is located in the posterior dorsal region of the bladder. The oval contains a capillary bed, and the high internal pressure of oxygen in the gas bladder causes it to diffuse into the blood when the oval sphincter is opened.

Cartilaginous fishes   Sharks, rays, and chimaeras do not have a gas bladder. Instead, many of these fishes use the liver to create neutral buoyancy. The liver of a shark has a high oil content and a mass density of only 0.95 g/mL (which is lighter than water), and can account for as much as 25% of the total body mass. Not surprisingly, bottom-dwelling sharks, such as the nurse shark (*Ginglymostoma cirratum*), have livers with fewer and smaller oil vacuoles in their cells, and these sharks are negatively buoyant.

Nitrogen-containing compounds in the blood of cartilaginous fishes also contribute to their buoyancy. Urea and trimethylamine oxide in the blood and muscle tissue provide positive buoyancy because they are less dense than an equal volume of water. Chloride ions are also lighter than water and provide positive buoyancy, whereas sodium ions and protein molecules are denser than water and are negatively buoyant. Overall these solutes provide positive buoyancy.

Deep-sea fishes   Many deep-sea fishes have deposits of lightweight oil or fat in the gas bladder, and others have reduced or lost the gas bladder entirely and have lipids distributed throughout the body. These lipids provide static lift, like the oil in shark livers. Because a smaller volume of the bladder contains gas, the amount of secretion required for a given vertical descent is less. Nevertheless, a long rete mirabile is needed to secrete oxygen at high pressures, and the gas gland in deep-sea fishes is very large. Fishes that migrate over large vertical distances depend more on lipids such as wax esters than on gas for buoyancy, whereas their close relatives that do not undertake such extensive vertical movements depend more on gas.

Water has properties that influence the behaviors of fishes and other aquatic vertebrates. Light is absorbed by water molecules and scattered by suspended particles. Objects more than 100 m distant are invisible even in the clearest water, whereas distance vision is virtually unlimited in clear air. Fishes supplement vision with other senses, some of which can operate only in water. The most important of these aquatic senses is mechanical and consists of detecting water movement via the lateral line system. Small currents of water can stimulate the sensory organs of the lateral line because water is dense and viscous. Electrical sensitivity is another sensory mode that depends on the properties of water and does not operate in air. In this case it is the electrical conductivity of water that is the key.

Vision

Vertebrates generally have well-developed eyes, but the way an image is focused on the retina is different in terrestrial and aquatic animals. Air has an index of refraction of 1.00, and light rays bend as they pass through a boundary between air and a medium with a different refractive index. The amount of bending is proportional to the difference in indices of refraction. Water has a refractive index of 1.33, and the bending of light as it passes between air and water causes underwater objects to appear closer to an observer in air than they really are.

The corneas of the eyes of terrestrial and aquatic vertebrates have an index of refraction of about 1.37, so light is bent as it passes through the air-cornea interface. As a result, the cornea of a terrestrial vertebrate plays a substantial role in focusing an image on the retina. This relationship does not hold in water, however, because the refractive index of the cornea is too close to that of water for the cornea to have much effect in bending light. That is why your vision is blurred underwater; a diving mask restores the air-cornea interface and allows you to see clearly.

The lens plays the major role in focusing light on the retina of an aquatic vertebrate, and fishes have spherical lenses with high refractive indices. The entire lens is moved toward or away from the retina to focus images of objects at different distances from the fish. Terrestrial vertebrates have flatter lenses, and muscles in the eye change the shape of the lens to focus images. Aquatic mammals such as whales and porpoises have spherical lenses like those of fishes.

Hearing

Sound travels about four times faster in water than in air, and for any given frequency the wavelength in water is about four times longer than in air. A frequency of 34.3 kHz (thousand cycles per second) has a wavelength of 1 cm in air, but achieving that wavelength in water requires a frequency of 148.4 kHz. That relationship is significant for echolocation, because only signals with wavelengths shorter than the diameter of an object are reflected. Thus, toothed whales must emit higher frequency echolocation calls than bats to detect objects of the same size.

Sound generally travels farther in the open sea than in terrestrial environments for two reasons:

1. Sound rarely has an unobstructed path on land; solid objects reflect sound waves, and vegetation absorbs sound energy. The open sea, in contrast, provides an unobstructed path for sound waves.

2. In air, sound energy obeys the inverse square law—that is, as sound propagates from a source, the energy spreads over larger and larger areas. Doubling the distance from the source reduces the energy by a factor of four. In the open ocean, sound waves can reflect off thermoclines (interfaces between water masses with different temperatures) and remain within a sound channel rather than spreading outward. As a result, sound energy can travel many kilometers in the open ocean, allowing whales to communicate over distances of hundreds of kilometers and potentially allowing all of the whales in an ocean basin to be in direct or indirect contact with each other.

Chemosensation: Taste and smell

Fishes have taste-bud organs in the mouth and around the head and anterior fins. In addition, olfactory organs on the snout detect soluble substances. Sharks and salmon can detect odors at concentrations of less than 1 part per billion. Sharks, and perhaps bony fishes, compare the time of arrival of an odor stimulus on the left and right sides of the head to locate the source of the odor. Homeward-migrating salmon are directed to their stream of origin from astonishing distances by a chemical signature from the home stream that was permanently imprinted when they were juveniles. Plugging the nasal olfactory organs of salmon destroys their ability to home.

Detecting water displacement

Mechanical receptors detect touch, sound, pressure, and motion. Like all vertebrates, fishes have an internal ear (the labyrinth organ, not to be confused with the organ of the same name that assists in respiration in anabantid fishes) that detects changes in speed and direction of motion. Fishes also have gravity detectors at the base of the semicircular canals that allow them to distinguish up from down.

In fishes and aquatic amphibians, clusters of sense cells called **hair cells** form **neuromast organs** that are dispersed over the surface of the head and body. In jawed fishes, neuromast organs are often located in a series of canals on the head, and one or more canals pass along the sides of the body onto the tail. This surface receptor system of fishes and aquatic amphibians is referred to as the **lateral line system** (Figure 4.4).

Lateral line systems are found only in aquatic vertebrates because air is not dense enough to stimulate the neuromast

organs. Amphibian larvae have lateral line systems, and permanently aquatic species of amphibians retain lateral lines throughout their lives. Terrestrial species of amphibians lose their lateral lines when they metamorphose into adults, and terrestrial vertebrates that have secondarily returned to the water, such as whales and porpoises, do not have lateral line systems.

Neuromasts of the lateral line system are distributed in two configurations: within tubular canals, or exposed in epidermal depressions. Many kinds of fishes have both arrangements. Hair cells have a **kinocilium** placed asymmetrically in a cluster of microvilli called **stereocilia**. Hair cells are arranged in pairs with the kinocilia positioned on opposite sides of adjacent cells. A neuromast contains many such hair-cell pairs. Each neuromast has two afferent nerves: one transmits impulses from hair cells with kinocilia in one orientation, and the other carries impulses from cells with kinocilia positions reversed by 180 degrees. This arrangement enables a fish to determine the direction of displacement of the kinocilia.

All kinocilia and microvilli are embedded in a gelatinous secretion, the **cupula** (Latin, "small tub"). Displacement of the cupula causes the kinocilia to bend. The resultant deformation either excites or inhibits the neuromast's nerve discharge. Each hair-cell pair therefore signals the direction of cupula displacement. The excitatory output of each pair has a maximum sensitivity to displacement along the line joining the kinocilia, and falling off in other directions. The net effect of cupula displacement is to increase the firing rate in one afferent nerve and to decrease it in the other nerve. These changes in lateral line nerve firing rates thus inform a fish of the direction of water currents on different surfaces of its body.

Several surface-feeding fishes and the African clawed frog (*Xenopus laevis*) provide vivid examples of how the lateral line organs act under natural conditions. These animals find insects on the water surface by detecting surface waves created by their prey's movements. Each neuromast group on the head of a killifish provides information about surface waves coming from a different direction (**Figure 4.5**). The groups of neuromasts have overlapping stimulus fields, allowing the fish to determine the precise location of the insect. Removing a neuromast group from one side of the head disturbs the directional response to stimuli, showing that a fish combines information from groups on both sides of the head to interpret water movements.

The large numbers of neuromasts on the heads of some fishes might be important for sensing vortex trails in the wakes of adjacent fishes in a school. Many of the fishes that form extremely dense schools (such as herrings) lack lateral line organs along the flanks and retain canal organs only on the head. These well-developed cephalic canal organs concentrate sensitivity to water motion in the head region, where it is needed to sense the turbulence into which the fish is swimming, and the reduction of flank lateral line elements would reduce noise from turbulence beside the fish.

(A)

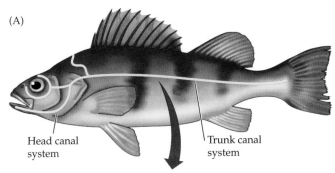

Head canal
system

Trunk canal
system

Figure 4.4 **Lateral line systems.**   (A) The hair
cells of most bony fishes are contained in lateral
line canals that ramify in the dermal bones of
the head and beneath the scales of the trunk. (B)
Pores open to the canals from the exterior, and
movement of water around the fish is transmitted
through the canals to sensory neuromast cells. (C)
Each hair cell has a kinocilium placed asymmetrically in a cluster of stereocilia. The kinocilia are embedded in a gelatinous material called the cupula. As
water flows in the canals, it displaces the cupulae,
bending the kinocilia. (D) Bending the kinocilia
excites or inhibits the neuromast's nerve discharge.
Hair cells are arranged in pairs, with the kinocilia
positioned on the opposite sides of adjacent cells.
Bending a kinocilium in one direction increases the
cell's firing rate, and bending in the other direction
inhibits firing. Each hair-cell pair therefore signals the
direction of water movement. (C,D after Flock 1967.)

(B)

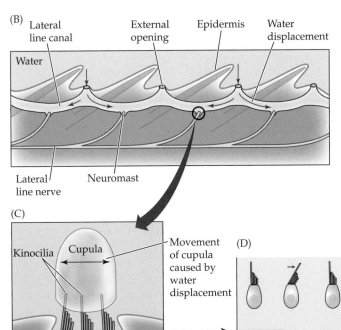

Lateral
line canal

External
opening

Epidermis

Water
displacement

Water

Lateral
line nerve

Neuromast

(C)

Kinocilia

Cupula

Movement
of cupula
caused by
water
displacement

Recording
sites

(D)

Hyperpolarization

Hair cell receptor
potential

Depolarization

Excitation

Static
discharge

Inhibition

Nerve
impulse
discharge

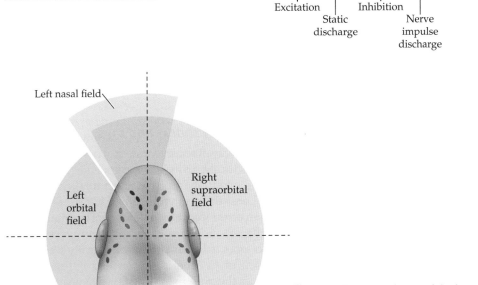

Left nasal field

Left
orbital
field

Right
supraorbital
field

Figure 4.5 **Fields of view of the lateral line canal organs.**
The wedge-shaped areas indicate the fields of view for each
group of canal organs on the head of the killifish *Aplocheilus
lineatus*. The sensory fields overlap on opposite sides as well as
on the same side of the body, allowing the lateral line system to
localize the source of a water movement. (After Schwartz 1974.)

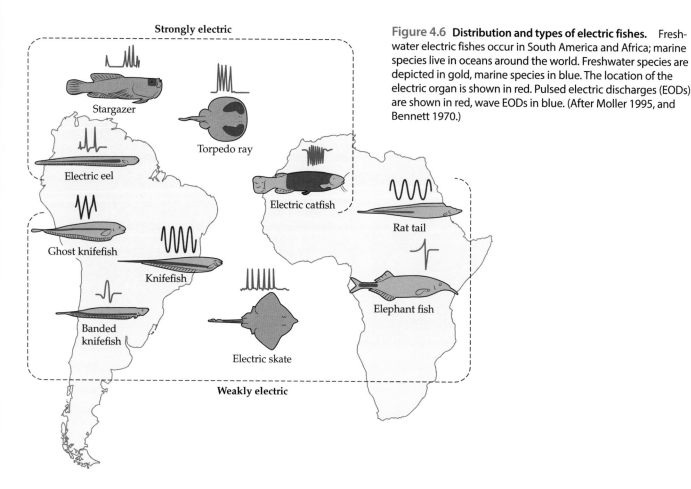

Figure 4.6 Distribution and types of electric fishes. Freshwater electric fishes occur in South America and Africa; marine species live in oceans around the world. Freshwater species are depicted in gold, marine species in blue. The location of the electric organ is shown in red. Pulsed electric discharges (EODs) are shown in red, wave EODs in blue. (After Moller 1995, and Bennett 1970.)

Electrical discharge

Unlike air, water conducts electricity, and about 350 fish species use modified muscle or nerve cells to produce action potentials large enough to create an electric field in the water surrounding them. In most species, the electric organ is located in the tail, but in some species it extends for most of the length of the body (**Figure 4.6**). The body of the fish creates a dipole during electric organ discharge (EOD), with the positive pole near the head and the negative pole near the tail. The EOD of some species is a continuous, nearly sinusoidal wave, whereas other species produce individual pulses. Most electric fishes are only weakly electric, with EODs that are less than 1 volt in amplitude. Weakly electric species use EODs for orientation, detection of prey and predators, species identification, courtship, and social interactions. Dominant individuals of many species produce low-frequency EODs, whereas subordinate individuals emit signals at higher frequencies.

A few electric fishes produce EODs strong enough to stun prey and deter predators. The electric organ of the electric eel runs nearly the full length of the body and sums the output of thousands of specialized muscle cells to produce EODs of up to 600 V. Despite the amplitude of this EOD, the shock administered by an electric eel depends on the location of the animal receiving the shock in relation to the positive and negative ends of the eel's dipole. Michael Faraday, the famed English pioneer in the study of electricity, reported in 1839 that he felt only a weak shock when he placed one or both hands at the same location on the body of an electric eel, but that the shock was very strong when he moved his hands close to the poles (**Figure 4.7**).

Electric fishes are found in tropical fresh waters of Africa and South America (see Chapter 9). Few marine forms can generate electrical discharges. Among marine cartilaginous fishes, only the torpedo ray (*Torpedo*), the ray genus *Narcine*, and some skates are electric; and among marine teleosts, only the stargazers (Uranoscopidae) produce electric discharges (see Figure 4.6).

Electroreception by sharks and rays

The high conductivity of seawater makes it possible for sharks to detect the electrical activity that accompanies muscle contractions of their prey. Sharks have structures known as the **ampullae of Lorenzini** on their heads (**Figure 4.8A,B**), and rays have them on the pectoral fins as well. The ampullae are sensitive electroreceptors. The canal connecting the receptor to the surface pore is filled with an electrically conductive gel, and the wall of the canal is nonconductive (**Figure 4.8C**). Because the canal runs for some distance beneath the epidermis, the sensory cell can detect a difference in electric potential between the tissue in which it lies (which reflects the adjacent epidermis and environment) and the distant pore opening. Thus, it can detect electric fields, which are changes in electric potential in space.

(A)

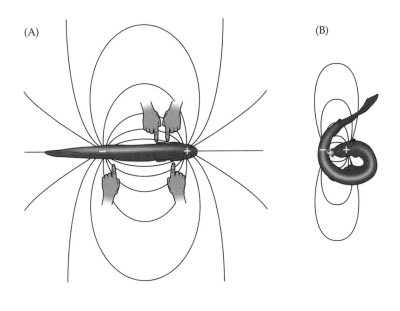

(B)

Figure 4.7 **Intensity of the shock delivered by an electric eel.** The eel can generate electric potentials up to 600 V. The intensity of the shock depends on the position of the receiver relative to the eel's positive and negative poles. (A) Black lines show the distribution of electric potential in the water around an electric eel. Touching the eel produces only a small shock when both hands are in nearly the same place along the eel's body, but moving the hands closer to the two poles increases the shock. (B) After an eel seizes a fish, it bends its body, placing the prey between its positive and negative poles and thus increasing the voltage. Repeated electrical discharges overstimulate the prey's muscles, leading to muscle fatigue that renders the prey unable to escape. (C) Images from a video show an electric eel attacking a partially submerged predator by leaping from the water and pressing its body against the predator. (Here the "predator" is a model of an alligator head equipped with LED lights that respond to the eel's electrical discharges.) By leaping higher from the water, the eel increases the voltage the predator receives, as shown by the greater intensity of the lights in the rightmost panel. (A,B from Catania 2015; C from Catania 2016, courtesy of Kenneth C. Catania.)

(C)

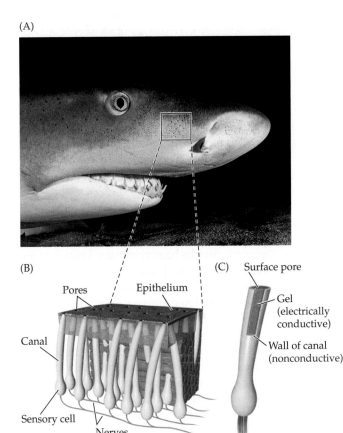

(A)

Electroreceptors of sharks respond to minute changes in the electric field surrounding an animal. They act like voltmeters, measuring differences in electric potentials across the body surface. Ampullary organs are remarkably sensitive, with thresholds lower than 0.01 microvolt per centimeter, a level of detection achieved by only the best voltmeters.

Sharks use their electrical sensitivity to detect prey. All muscle activity generates electric potential: motor nerve cells produce extremely brief changes in electric potential, and muscular contraction generates changes of longer duration. In addition, a steady potential issues from an aquatic organism as a result of the chemical imbalance between an organism and its surroundings. A shark can locate and attack a hidden fish using these differences in electric potential (**Figure 4.9**).

(B)

Pores   Epithelium

Canal

Sensory cell

Nerves

(C) Surface pore

Gel (electrically conductive)

Wall of canal (nonconductive)

Figure 4.8 **Ampullae of Lorenzini.** Ampullae of Lorenzini are clustered in fields on the heads of sharks. The pores marking the openings of the ampullae are clearly visible (A) on the head of this lemon shark (*Negaprion brevirostris*) and (B) in the close-up drawing below. (C) A single ampullary organ consists of a sensory cell connected to the surface pore by a canal filled with a gel that conducts electricity. (A, WalterFrame/Alamy Stock Photo; B,C from Josberger et al. 2016, CC BY-NC 4.0.)

(A)

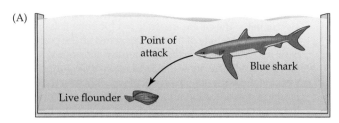

(B)

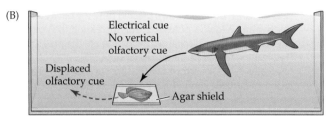

(C)

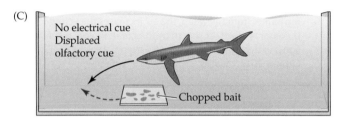

(D)

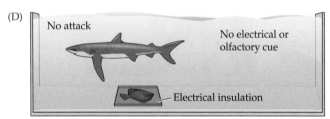

(E)

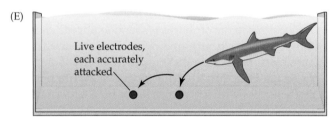

Figure 4.9 Electrolocation capacity of sharks. In a classic series of experiments, Adrianus Kalmijn tested the ability of blue sharks (*Prionace glauca*) to locate hidden prey. (A) Flounders normally rest on the bottom, covered with a layer of sand. The shark had no difficulty locating the flounder, but it could have been using electroreception, olfaction, or both sensory modalities. (B) When the flounder was covered by an agar shield, which allowed electrical impulses to pass but blocked olfactory cues, the shark directed its attack at the flounder, showing that it was responding to the electrical activity of the flounder's muscles as the flounder ventilated its gills. (C) When the flounder was replaced by pieces of fish that emitted an odor but no electrical activity, the shark swam past the bait without detecting it, but responded when it reached the odor plume. (D) When the flounder was covered with electrical insulation (blocking the electric signal) and with agar (blocking the odor), the shark was unable to detect it. (E) The shark attacked electrodes emitting an electric signal duplicating the activity of a flounder's respiration. (After Kalmijn 1971.)

lack a kinocilium and fire when the environment is positive relative to the cell, and nerve impulses are sent to the lateral portions of the brain rather than to the midline.

4.3 The Internal Environment of Vertebrates

Most vertebrates are 70–80% water, and the chemical reactions that release energy and synthesize new chemical compounds take place in an aqueous environment. The body fluids of vertebrates contain a complex mixture of ions and other solutes. Some ions are cofactors that control the rates of metabolic processes; others are involved in the regulation of pH, the stability of cell membranes, or the electrical activity of nerves. Metabolic substrates and products must diffuse from sites of synthesis to the sites of utilization. Almost everything that happens in the body tissues of vertebrates involves water, and maintaining the concentrations of water and solutes within narrow limits is a vital activity.

Water sounds like an ideal place to live for an animal that itself consists mostly of water, but in some ways an aquatic environment can be too much of a good thing. Freshwater vertebrates—especially fishes and amphibians—face the threat of being flooded with water that flows into them from their environment, and saltwater vertebrates must prevent the water in their bodies from being sucked out into the sea. Terrestrial organisms must conserve water by controlling evaporative and urinary loss, and access to water for rehydration is a critical factor in habitat selection for many terrestrial animals.

Temperature, too, is a critical factor for living organisms because chemical reactions are temperature-sensitive. In general, the rates of chemical reactions increase as temperature increases, but not all reactions have the same sensitivity to temperature. A metabolic pathway is a series of chemical reactions in which the product of one reaction

Other electroreceptive vertebrates Electrogenesis and electroreception are not restricted to a single group of aquatic vertebrates, and monotremes (platypus and echidnas, early offshoots off the main mammalian lineages that still lay eggs) use electroreception to detect prey. Electrosensitivity was probably an early feature of vertebrate evolution. The brain of the lamprey responds to electric fields, and it seems likely that the earliest vertebrates had electroreceptive capacity. All fishlike vertebrates of lineages that evolved before the neopterygians have electroreceptor cells. These cells, which have a prominent kinocilium, fire when the environment around the kinocilium is negative relative to the cell. Their impulses pass to the midline region of the posterior third of the brain.

Electrosensitivity was apparently lost by basal neopterygians, and teleosts have at least two separate new evolutions of electroreceptors. Electrosensitivity of teleosts is distinct from that of other vertebrates: teleost electroreceptors

is the substrate for the next, yet each of these reactions may have a different sensitivity to temperature, so a change in temperature can mean that too much or too little substrate is produced to sustain the next reaction in the series. To complicate the process of regulation of substrates and products even more, the chemical reactions take place in a cellular milieu that itself is changed by temperature because the viscosity of plasma membranes is temperature-sensitive. Clearly, the smooth functioning of metabolic pathways is greatly simplified if an organism can limit the range of temperatures its tissues experience.

Water temperature is more stable than air temperature because the heat capacity of water is 3,500 times that of air. Thus, the stability of water temperature simplifies the task of maintaining a constant body temperature, as long as the body temperature the animal needs to maintain is the same as the temperature of the water around it. An aquatic animal has a hard time maintaining a body temperature different from water temperature, however, because water conducts heat 25 time faster than air does. Heat flows out of the body if an animal is warmer than the surrounding water and into the body if the animal is cooler than the water.

4.4 Exchange of Water and Ions

An organism can be described as a leaky bag of dirty water. That is not an elegant description, but it accurately identifies the two important characteristics of a living animal: it contains organic and inorganic substances dissolved in water, and this fluid is enclosed by a permeable body surface. Exchange of matter and energy with the environment is essential to the survival of the organism, and much of that exchange is regulated by the body surface. Water molecules and ions pass through the skin quite freely, whereas larger molecules move less readily. The significance of this differential permeability is particularly conspicuous in the case of aquatic vertebrates, but it applies to terrestrial vertebrates as well. Vertebrates use both active and passive exchange to regulate their internal concentrations in the face of varying external conditions.

Nitrogen excretion

Carbohydrates and fats are composed of carbon, hydrogen, and oxygen, and the waste products from their metabolism are carbon dioxide and water molecules that are easily voided. Proteins and nucleic acids are another matter, for they contain nitrogen. When protein is metabolized, the nitrogen is enzymatically reduced to ammonia through a process called deamination. Ammonia is very soluble in water and diffuses readily, but it is also extremely toxic. Rapid excretion of ammonia is therefore crucial.

Differences in how nitrogenous wastes are excreted are partly a matter of the toxicity of different compounds, the cost of their production, and the availability of water. Most vertebrates eliminate nitrogen as a mixture of ammonia, urea, or uric acid. Excretion of nitrogenous wastes primarily as ammonia is called **ammonotely**, excretion primarily as urea is **ureotely**, and excretion primarily as uric acid is **uricotely**.

- *Ammonotely:* Bony fishes are primarily ammonotelic and excrete ammonia through the skin and gills as well as in urine. Because ammonia is produced by deamination of proteins, no metabolic energy is needed to produce it.

- *Ureotely:* Mammals are primarily ureotelic, although they excrete some nitrogenous wastes as ammonia and uric acid. Normal values for humans, for example, are 82% urea and 2% each ammonia and uric acid. The remaining 14% of nitrogenous waste is composed of other nitrogen-containing compounds, primarily amino acids and creatine. Urea is synthesized from ammonia in a cellular enzymatic process called the **urea cycle**. Urea synthesis requires more energy than does ammonia production, but urea is less toxic than ammonia. Because urea is not very toxic, it can be concentrated in urine, thus conserving water.

- *Uricotely:* Reptiles, including birds, are primarily uricotelic, but here again all three of the major nitrogenous compounds are present. The pathway for synthesis of uric acid is complex and requires more energy than synthesis of urea. The advantage of uric acid lies in its low solubility: it precipitates from the urine and is excreted as a semisolid paste. The water that was released when the uric acid precipitated is reabsorbed, so uricotely is an excellent method of excreting nitrogenous wastes while conserving water. Some species of reptiles change the proportions of the three compounds depending on the water balance of the animal, excreting more ammonia and urea when water is plentiful and shifting toward uric acid when it is necessary to conserve water.

The vertebrate kidney

An organism can tolerate only a narrow range of concentrations of its body fluids and must eliminate waste products before they reach harmful levels. The molecules of ammonia that result from the breakdown of protein are especially important because they are toxic. Vertebrates have evolved superb capacities for controlling water balance and excreting wastes, and the kidney plays a crucial role in these processes.

The adult vertebrate kidney consists of hundreds to millions of tubular **nephrons**, each of which produces urine. The primary function of a nephron is removing excess water, salts, waste metabolites, and foreign substances from the blood. In this process, the blood is first filtered through the **glomerulus**, a structure unique to vertebrates (**Figure 4.10**). Each glomerulus is composed of a leaky arterial capillary tuft encapsulated within a sievelike filter. Arterial blood pressure forces fluid into the nephron to form an

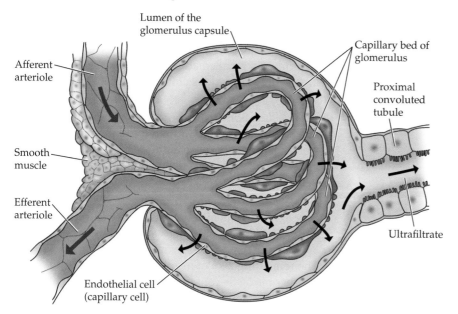

Lumen of the glomerulus capsule

Afferent arteriole

Capillary bed of glomerulus

Proximal convoluted tubule

Smooth muscle

Efferent arteriole

Endothelial cell (capillary cell)

Ultrafiltrate

Figure 4.10 A typical mammalian glomerulus. Blood pressure forces fluid through the walls of the capillary into the lumen of the glomerular capsule, filtering out large molecules such as proteins while allowing ions and small molecules to pass. The blood flow to each glomerulus is regulated by smooth muscles that can close off the afferent and efferent arterioles to adjust the glomerular filtration rate of the kidney as a whole. The ultrafiltrate passes from the glomerular capsule into the proximal convoluted tubule, where the process of adding and removing specific substances begins.

ultrafiltrate, composed of blood minus blood cells and larger molecules. The ultrafiltrate is then processed to return water and essential metabolites (e.g., glucose, amino acids) to the general circulation. The fluid that remains after this processing is urine.

Regulation of ions and body fluids

The salt concentrations in the body fluids of many marine invertebrates are similar to those in seawater, as are those of hagfishes (**Table 4.2**). It is likely that the first vertebrates also had ion levels similar to those in seawater. In contrast, salt levels are lower in the blood of all other vertebrates.

In the context of body fluids, a **solute** is a small molecule that is dissolved in water or blood plasma. Salt ions, urea, and some small carbohydrate molecules are the solutes primarily involved in the regulation of body fluid concentrations. The presence of solutes lowers the potential activity of water. Water moves from areas of high potential to areas of lower potential; therefore, water flows from a dilute solution (one with a high water potential) to a more concentrated solution (with a lower water potential).[1] This process is called **osmosis**.

Seawater has a solute concentration of approximately 1,050 millimoles per kilogram (mmol/kg) of water. Most marine invertebrates and hagfishes have body fluids that are in osmotic equilibrium with seawater; that is, they are **isosmolal** to seawater. Body fluid concentrations of marine teleosts and lampreys are between 300 and 350 mmol/kg. Therefore, water flows outward from their blood to the sea (i.e., from a region of high water potential to a region of lower water potential). Cartilaginous fishes retain urea and other nitrogen-containing compounds, raising the osmolality of their blood to slightly above that of seawater so

water flows from the sea into their bodies. These osmolal differences are specified by the terms **hyposmolal** (lower solute concentrations than the surrounding water, as seen in marine teleosts and lampreys) and **hyperosmolal** (higher solute concentrations than the surrounding water, as seen in coelacanths and cartilaginous fishes).

Monovalent ions such as sodium (Na^+) and chloride (Cl^-) can diffuse through the surface membranes of an animal, so the water and salt balance of an aquatic vertebrate in seawater is constantly threatened by outflow of water and inflow of ions, and in fresh water by inflow of water and outflow of ions.

Most fishes are **stenohaline** (Greek *stenos*, "narrow"; *hals*, "salt"), meaning they inhabit either fresh water or seawater and tolerate only modest changes in salinity. Because these fishes remain in one environment, the magnitude and direction of the osmotic gradient to which they are exposed are stable. Some fishes, however, move between fresh water and seawater and tolerate large changes in salinity. These fishes are called **euryhaline** (Greek *eurys*, "wide"); water and salt gradients are reversed in euryhaline species as they move from one medium to the other.

4.5 Vertebrates in Different Environments

Vertebrates live in the sea, in estuaries and tidal rivers, in fresh water, and on land. In aquatic habitats, water and ions diffuse in or out of the body. On land, water evaporates from the skin and respiratory passages. Nonetheless, with only a few exceptions, vertebrates in all habitats maintain stable and similar internal water and ion concentrations.

Marine vertebrates

Vertebrates evolved in a marine environment, but the internal osmotic and ion concentrations of most marine vertebrates are quite different from seawater. Seawater is

---

[1] The relationship of water potential to osmolality can be counterintuitive. Remember that more water means a higher water potential, and when solutes are present they occupy some of the volume of the solution, so the presence of *more* solutes (higher osmolality) means there is *less* water (lower water potential).

Table 4.2	Osmolality of vertebrate blood				
Environment	**Osmolality**[a] **(mmol/kg)**	**Na$^+$**	**Cl$^-$**	**Urea**	**TMAO**[b]
Seawater	1,050	475	550	—	—
Fresh water	<10	5	5	—	—
Type of animal					
MARINE VERTEBRATES					
Hagfishes	1,035	485	515	—	—
Teleosts	350	150	135	—	—
Coelacanths	930	200	190	380	120
Dogfish (shark)	1,080	255	240	440	70
FRESHWATER VERTEBRATES					
Teleosts	300	140	125	—	—
Lungfish	240	100	45	—	—
Aquatic amphibians	200	100	90	—	—
EURYHALINE VERTEBRATES					
European flounder					
in seawater	365	195	165	—	—
in fresh water	305	157	115	—	—
Bull shark					
in seawater	1,070	290	295	370	45
in fresh water	640	210	200	190	20
Crab-eating frog adult					
in 80% seawater	830	250	225	350	—
in fresh water	290	125	100	40	—
Crab-eating frog tadpole					
in seawater	525	265	255	—	—
in fresh water	270	125	100	—	—
TERRESTRIAL VERTEBRATES					
Terrestrial amphibians	250	125	90	—	—
Birds	350	160	140	—	—
Mammals	340	160	110	—	—

[a] Concentrations are expressed in millimoles of solute per kilogram of water and have been rounded to the nearest 5 units.

[b] TMAO, trimethylamine oxide.

more concentrated than the body fluids of vertebrates, so there is a net outflow of water by osmosis and a net inward diffusion of ions. Hagfishes have few problems with ion balance because they regulate only divalent ions and reduce osmotic water movement by being nearly isosmolal to seawater. However, most other aquatic vertebrates maintain internal osmotic concentrations and ion concentrations that are quite different from the water that surrounds them.

Coelacanths and cartilaginous fishes   Coelacanths and cartilaginous fishes minimize osmotic flow by maintaining the internal concentration of the body fluid close to that of seawater. These animals retain nitrogen-containing compounds, primarily urea and trimethylamine oxide (TMAO) to produce internal concentrations that are usually slightly hyperosmolal to seawater (see Table 4.2). As a result,

they gain water by osmotic diffusion across the gills and do not need to drink seawater.

This influx of water allows hyperosmotic fishes to have large kidney glomeruli that produce high filtration rates and therefore rapid elimination of metabolic waste products from the blood. Urea is very soluble and diffuses through most biological membranes, and that could be a problem for fishes that retain urea to remain hyperosmotic to their surroundings. The gills of cartilaginous fishes are nearly impermeable to urea, however, and the kidney tubules actively reabsorb it. With internal ion concentrations that are low relative to that of seawater, cartilaginous fishes experience ion influxes across the gills, as do marine teleosts. Unlike the gills of marine teleosts, those of cartilaginous fishes have low ion permeabilities (less than 1% those of teleosts).

(A) Freshwater teleost

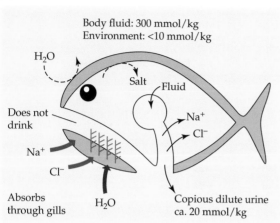

(B) Marine teleost

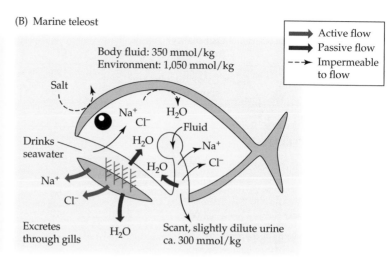

Figure 4.11 Water and salt regulation by freshwater and marine teleosts. (A) The body fluids of freshwater teleosts are more concentrated than the water surrounding them; consequently, they gain water by osmosis and lose sodium and chloride by diffusion. They do not drink water; they actively absorb sodium and chloride through the gills; and they have kidneys with large glomeruli that produce a large volume of dilute urine. (B) Marine teleosts have body fluids that are less concentrated than the water they live in; consequently, they lose water by osmosis and gain sodium and chloride by diffusion. Marine teleosts drink seawater and actively excrete sodium and chloride through the gills. Their kidneys have small glomeruli that produce small volumes of concentrated urine.

Cartilaginous fishes generally do not have highly developed salt-excreting cells in the gills. Rather, they achieve ion balance by secreting from the rectal gland a fluid that is approximately isosmolal to body fluids and seawater but contains higher concentrations of sodium and chloride ions than do the body fluids.

Teleosts The body surface of fishes has low permeability to water and to ions. However, a fish cannot entirely prevent osmotic exchange. Gills are permeable to oxygen and carbon dioxide, and also to water. As a result, most water and ion movements take place across the gill surfaces (**Figure 4.11**). The kidney glomeruli of marine teleosts are small and the glomerular filtration rate is low (see Figure 4.12). Little urine is formed, so the amount of water lost by this route is reduced. Marine teleosts lack a water-impermeable distal convoluted tubule. The urine is always hyposmolal to blood, but the volume of urine leaving the nephron is small. To compensate for osmotic dehydration, marine teleosts do something unusual: they drink seawater. Sodium and chloride ions are actively absorbed across the lining of the gut, and water flows by osmosis into the blood.

Estimates of seawater consumption vary, but many species drink in excess of 25% of their body weight each day and absorb 80% of this ingested water. Drinking seawater to compensate for osmotic water loss, however, increases the influx of sodium and chloride ions. To compensate for this salt load, chloride cells in the gills actively pump chloride ions outward against a large concentration gradient.

Active transport of ions is metabolically expensive—the energy cost of osmoregulation may account for more than one-fourth of the daily energy expenditure of some fishes.

Freshwater vertebrates: Teleosts and amphibians

Several mechanisms are involved in the salt and water regulation of vertebrates that live in fresh water. Water is gained by osmosis, and ions are lost by diffusion. A freshwater teleost does not drink water because osmotic water movement is already providing more water than the fish needs—drinking would only increase the amount of water it had to excrete via the kidneys. To compensate for the influx of water, the kidney of a freshwater fish or amphibian produces a large volume of urine. Salts are actively reabsorbed to reduce salt loss. Indeed, urine processing in a freshwater teleost provides a simple model of vertebrate kidney function.

The large glomeruli of freshwater teleosts produce a copious flow of urine, but the glomerular ultrafiltrate is isosmolal to the blood and contains essential blood salts (**Figure 4.12**). To conserve salt, ions are reabsorbed across the **proximal** and **distal convoluted tubules**. Reabsorption is an active process that consumes metabolic energy.

Because the distal convoluted tubule is impermeable to water, water remains in the tubule and the urine becomes less concentrated as ions are removed from it. Ultimately, the urine becomes hyposmolal to the blood. In this way, the water that was absorbed across the gills is removed and ions are conserved. Nonetheless, some ions are lost in the urine in addition to those lost by diffusion across the gills. Salts from food compensate for some of this loss, and teleosts have ionocytes in the gills that take up chloride ions from the water. The chloride ions are moved by active transport against a concentration gradient, and this process requires energy. Sodium ions also enter the gills, passively following the chloride ions.

Freshwater amphibians face the same osmotic problems as freshwater fishes. The entire body surface of amphibians is involved in the active uptake of ions from the water. Like freshwater fishes, aquatic amphibians do not drink because the osmotic influx of water more than meets their needs.

(A) Freshwater teleost

Plasma: 300 mmol/kg

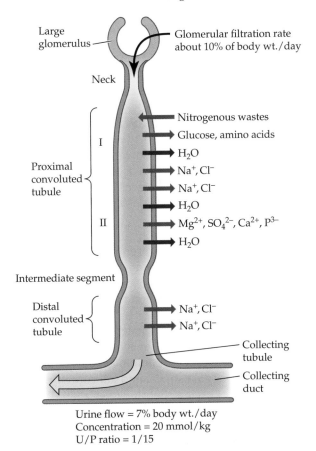

Urine flow = 7% body wt./day
Concentration = 20 mmol/kg
U/P ratio = 1/15

(B) Marine teleost

Plasma: 350 mmol/kg

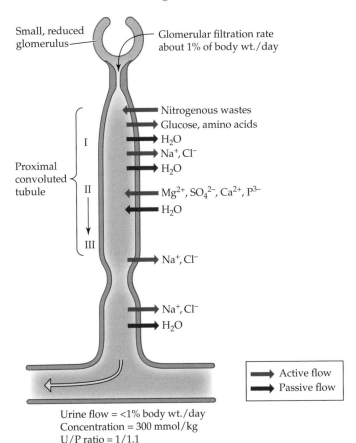

Urine flow = <1% body wt./day
Concentration = 300 mmol/kg
U/P ratio = 1/1.1

Figure 4.12 Kidney structure and function of marine and fresh-water teleosts. Freshwater teleosts must rid themselves of water that enters their bodies by osmosis and so produce large volumes of dilute urine. Marine teleosts lose water by osmosis, and produce small volumes of more concentrated urine. Arrows pointing into the lumen of the kidney tubules show movement of substances into the forming urine, and arrows pointing outward show movement from the urine back into the body fluids. Red arrows show active movements (i.e., those that involve a transport system); blue arrows show passive flow. The glomerular filtration rate, the rate at which ultrafiltrate is formed, is expressed as percentage of body weight per day. U/P is the urine/plasma concentration ratio. Freshwater teleosts have high glomerular filtration rates and low U/P ratios; marine teleosts have low glomerular filtration rates and high U/P ratios. Two segments [I and II] of the proximal convoluted tubule are recognized in both freshwater and marine teleosts. Segment III of the proximal convoluted tubule of marine fishes is sometimes equated with the distal convoluted tubule of freshwater fishes. Substances a fish needs to conserve (primarily glucose and amino acids) are actively removed from the ultrafiltrate in the proximal convoluted tubule, and nitrogenous waste products (ammonia and urea) are actively added to the forming urine. Freshwater fishes actively recover divalent ions (magnesium, sulfate, calcium, and phosphorus) from the forming urine, whereas marine fishes actively excrete those ions into the urine. Na^+ and Cl^- are also removed from the forming urine. That makes sense for freshwater fishes because they are trying to conserve those two ions, but it is surprising for marine fishes because they are battling an influx of excess Na^+ and Cl^- from seawater. The explanation of that paradox is that movement of Na^+ and Cl^- is needed to produce a passive flow of water back into the body. Freshwater fishes continue to reabsorb Na^+ and Cl^- in the collecting tubule, but the walls of their CTs are not permeable to water so there is no passive uptake at this stage. In contrast, the collecting tubules of marine fishes are permeable to water, allowing further recovery of water at this point. The net effect of the differences in glomerular filtration rate, inward and outward movements, and permeabilities allows freshwater fishes to produce copious amounts of dilute urine that rids them of excess water and allows marine fishes to produce scanty amounts of more concentrated urine that conserves water. Finally, the urine passes into the collecting duct, and from there to the outside of the body via the archinephric duct (not shown).

Also like fishes, aquatic amphibians lose ions by diffusion. The skin of these amphibians contains cells that actively take up ions from the surrounding water. Acidity inhibits this active transport of ions in both amphibians and fishes, and inability to maintain internal ion concentrations is one of the causes of death of these animals in habitats acidified by acid rain or drainage of water from mine wastes.

Euryhaline vertebrates

Many species of teleosts are euryhaline, moving between fresh water and the sea as part of their life history. **Anadromous** species, such as the European flounder (*Platichthys flesus*) shown in Table 4.2, breed in fresh water, migrate to the sea as juveniles, and as adults return to fresh water to spawn. Catadromous species, such as American and European eels (*Anguilla rostrate* and *A. anguilla*), follow the opposite pattern: they breed in the sea, migrate to fresh water as juveniles, and return to the sea as adults. Some teleosts live in estuaries, where the salt concentration of the water rises and falls with the tides. In all these situations, teleosts maintain stable osmotic and concentrations despite the changes in their environment.

About 50 species of cartilaginous fishes are euryhaline. The bull shark (*Carcharhinus leucas*) readily enters rivers and may venture thousands of kilometers from the sea. This species has been recorded from Indiana (in the Ohio River) and from Illinois (in the Mississippi River), and a landlocked population lives in Lake Nicaragua. In seawater, bull sharks retain high levels of urea and TMAO, but in fresh water their blood levels decline (see Table 4.2).

Although most amphibians are found in fresh water or terrestrial habitats, nearly 150 species occur in saline coastal or inland habitats. The Southeast Asian crab-eating frog (*Fejervarya* [formerly *Rana*] *cancrivora*) inhabits intertidal mudflats and is exposed to 80% seawater at each high tide. During seawater exposure, the frog allows its blood ion concentrations to rise and thus reduces the ionic gradient. In addition, ammonia is removed from proteins and rapidly converted to urea, which is released into the blood. The blood urea concentration rises from 40 to 350 mmol/kg, and the frogs become hyperosmolal to the surrounding water. In this sense, *F. cancrivora* acts like a shark and absorbs water osmotically. Frog skin is permeable to urea, however,

so the urea that these frogs synthesize is rapidly lost. To compensate for this loss, the activity of their urea-synthesizing enzymes is very high.

Tadpoles of the crab-eating frog face a more difficult situation because, unlike adult frogs, they cannot move out of the water to wait on land until the tide ebbs; they are trapped in pools on the mudflats and are exposed to high salt concentrations when water evaporates from the pools. Like most tadpoles, they lack urea-synthesizing enzymes until late in their development. Thus, *F. cancrivora* tadpoles use a different method of osmoregulation: they have salt-excreting cells in their gills and pump ions outward even as the ions diffuse inward. This pumping action does not match the rate at which ions diffuse inward, however, and as a result the osmolality of the tadpoles' blood doubles during low tide.

Terrestrial vertebrates

Living in air instead of water presents a different set of challenges to the internal fluid and ion concentrations of vertebrates, and the two lineages of terrestrial vertebrates (mammals and reptiles) have responded in different ways as we will discuss in detail in Chapter 13.

Terrestrial amphibians are ureoteles when they are on land, although when they are in water they can excrete some nitrogenous waste as ammonia, as freshwater amphibians do. The internal osmotic and ion concentrations of terrestrial amphibians are slightly higher than those of aquatic species, but the difference is small compared to the variation between marine and freshwater chondrichthyans.

Summary

Primary aquatic vertebrates (fishes and amphibians) exchange oxygen and carbon dioxide via gills and/or the skin.

Water is both dense and viscous, and in bony fishes the flow of water is unidirectional, rather than tidal (in–out). Buccopharyngeal pumping draws water into the mouth and expels it across the gills. A countercurrent flow of water and blood in the gills maximizes transfer of oxygen and carbon dioxide.

Some fishes supplement gill respiration with air-breathing, using a variety of structures that range from enlarged lips to lungs.

Most fishes can adjust their buoyancy to remain at a selected depth without needing to swim up or down. Bony fishes change the volume of the gas bladder by adding or releasing oxygen, whereas sharks have large, oil-filled livers that are lighter than water.

The sensory world of water differs from that of air.

Vision is an important sensory modality for behavior and predation, but vision in water rarely extends beyond a few tens of meters, and aquatic vertebrates supplement vision with other senses.

Sound travels about four times faster in water than in air, and in the open sea sound can travel for hundreds of kilometers. These properties make passive listening and echolocation effective sensory modalities in water. The wavelength of a sound frequency is four times longer in water than in air, so echolocation of small objects requires emitting higher frequency calls in water than in air.

Fishes have taste-bud organs in the mouth and around the head and anterior fins, and chemosensation can be exquisitely sensitive, detecting odors at concentrations below 1 part per billion.

The density of water allows detection of small water movements, and fishes and aquatic amphibians have mechanical sensors (neuromasts) distributed over the body in the lateral line system.

Hair cells detect water movement, and overlapping sensory fields allow the direction of a stimulus to be identified.

Summary (*continued*)

Water conducts electricity, and about 350 species of fishes use modified muscles or nerve cells to create an electric field in the water around their bodies. Electric fishes can use their electric organ discharge (EOD) to navigate, detect predators, and communicate with other individuals. A few species, notably the electric eel, can generate electric potentials large enough to be effective in predation and defense.

Electrosensitivity was probably widespread among early vertebrates and has been retained in elasmobranchs, which use the electrical discharges produced by muscular activity to locate hidden prey. Electrosensitivity was apparently lost during the evolution of neopterygians, but has re-evolved at least twice. Monotremes (echidnas and the platypus) use electroreception to locate prey.

Water makes up 70–80% of the mass of vertebrates, and cellular and tissue processes take place in aqueous solutions.

Physiological stability depends on regulating the concentration of solutes in body fluids, and the ease with which water and many ions to move in and out of aquatic organisms makes regulation a challenge.

Temperatures of aquatic environments are more stable than those of terrestrial environments because the heat capacity of water (the amount of heat energy needed to change its temperature) is greater than that of air.

It is hard for an aquatic vertebrate to maintain a body temperature different than water temperature because heat conduction (the speed with which heat energy is transferred) is higher in water than in air.

Disposal of nitrogenous waste products is easier in aquatic environments than on land.

Ammonia, urea, and uric acid are the major forms in which vertebrates excrete nitrogenous wastes. Many vertebrates can produce all three compounds, balancing energy costs, toxicity, and conservation of water.

Ammonia is the product of deamination, so it requires no additional energy for synthesis, but it is highly toxic and must be excreted quickly. Ammonia is soluble in water, and ammonotely (excretion of ammonia as the primary nitrogenous waste) is characteristic of fishes.

Urea is less toxic than ammonia, but its synthesis requires energy. Urea is more soluble than ammonia, so it can be excreted at high concentration to conserve water. Ureotely is characteristic of mammals and some terrestrial amphibians.

Uric acid requires more energy for synthesis, but it has low solubility in water and precipitates from solution,

conserving water. Uricotely is characteristic of reptiles, including birds.

The vertebrate kidney regulates salt and water balance.

Nephrons are the site of urine production and processing.

- In the glomerulus of the nephron, arterial blood pressure forms an ultrafiltrate of blood plasma by forcing water and small molecules out through the leaky capillaries of the glomerulus.
- The tubules of the nephron modify the ultrafiltrate, reabsorbing essential molecules and adjusting the volume to retain or excrete water.

Ions and water move down concentration gradients.

Ions move from areas of high activity (concentration) to areas of low activity. The internal concentrations of sodium and chloride of bony fishes and amphibians lie between those of fresh water and seawater.

- Ions diffuse *from* a freshwater fish *into* the surrounding water. Freshwater bony fishes have cells in the gills that take up chloride ions from the water and actively transport it inward, and sodium ions follow passively. In freshwater amphibians, the skin actively transports chloride ions inward, and sodium ions follow passively.
- Ions diffuse *from* seawater *into* a fish, and additional ions enter in the water the fish drinks. Ion-transporting cells in the gills excrete the ions.

Water moves from areas of high water potential (i.e., dilute solutions) to areas of low water potential (i.e., concentrated solutions). Bony fishes and amphibians have internal water concentrations between those of fresh water and seawater.

- Water diffuses *from* fresh water *into* a fish. Freshwater fishes do not drink, and their kidneys produce large volumes of dilute urine to excrete the excess water.
- Water diffuses *from* a fish *into* seawater. Marine fishes drink seawater, and their kidneys produce small quantities of concentrated urine to conserve water.

Some marine fishes have internal osmolal concentrations similar to that of seawater.

Hagfishes minimize water and ion regulation by having internal ion and water concentrations that are close to those of seawater.

Coelacanths and cartilaginous fishes retain nitrogen-containing compounds (primarily urea and trimethylamine oxide), raising their internal concentrations slightly above those of seawater. Thus, water flows inward.

(Continued)

Summary *(continued)*

Cartilaginous fishes excrete excess sodium and chloride via the rectal gland.

Some sharks and rays live in fresh water; these species have reduced levels of urea and trimethylamine oxide.

Although most aquatic amphibians inhabit fresh water, nearly 150 species occur in saline habitats. The crab-eating frog lives in tidal areas and is exposed to 80% seawater at every high tide. As the tide rises, the frog synthesizes urea, raising its internal concentration enough to produce an inward flow of water. Enzymes for urea synthesis are not expressed in tadpoles; instead, tadpoles maintain hyposmolal internal concentrations by using active transport to pump ions outward.

Discussion Questions

1. What is the difference between the effects of the density and the viscosity of water from the perspective of an aquatic vertebrate?

2. When you walk past the tanks in an aquarium store, you see both physostomous and physoclistous fishes. Goldfishes, for example, are physostomous, and cichlids are physoclistous. Conduct a thought experiment comparing the responses of the two kinds of fishes to changes in air pressure. (Don't actually conduct this experiment, because it would be painful for the fish.) Imagine that you put a goldfish and a cichlid into a jar half filled with water and then screw on a lid to make an airtight seal. Protruding from the lid is a tube to which you can attach a vacuum pump and a valve that allows you to close off the tube.

   a. Pump some air out of the jar. What effect will that have on the air pressure inside the jar? How will each of the fishes respond to that change in pressure? How are those responses related to being physostomous or physoclistous?

   b. Now open the valve and allow air to enter the jar, returning the pressure quickly to its starting level. How will each of the fishes respond to this change in pressure? How are these responses related to being physostomous or physoclistous?

3. When deep-sea fishes are pulled quickly to the surface, they often emerge with their gas bladder protruding from their mouth.

   a. Why does this happen, and what does it tell you about whether these deep-sea fishes are physoclistous or physostomous?

   b. Why could you have figured out whether deep-sea fishes are physostomous or physoclistous simply by considering how air enters the gas bladder of a fish?

4. The text points out that the gas bladders of deep-sea fishes that migrate over large vertical distances are nearly filled with lipids, leaving space for only a small volume of gas, whereas deep-sea fishes that do not make large vertical migrations have more gas and less lipid in their gas bladders. What is the functional significance of this difference?

5. If you look at pictures of frogs, you will see that a few species have conspicuous lateral lines. What does the presence of lateral lines tell you about the ecology of those species?

Additional Reading

Archer R. 2016. Stunning suppers with an electric attack. *Journal of Experimental Biology* 219: 612.

Bleckmann H, Zelick R. 2009. Lateral line system of fish. *Integrative Zoology* 4: 13–25.

Carlson BA. 2016. Animal behavior: Electric eels amp up for an easy meal. *Current Biology* 25: R1070–1072.

Catania KC. 2016. Leaping eels electrify threats, supporting Humboldt's account of a battle with horses. *Proceedings of the National Academy of Sciences USA* 113: 6979–6984.

Evans DH, Piermarini PM, Choe KP. 2005. The multifunctional fish gill: Dominant site of gas exchange, osmoregulation, acid–base regulation, and excretion of nitrogenous waste. *Physiological Reviews* 85: 97–177.

Feulner PGD, Plath M, Engelmann J, , Kirschbaum F, Tiedemann R. 2009. Electrifying love: Electric fish use species-specific discharge for mate recognition. *Biology Letters* 5: 225–228.

Gardiner JM, Atema J. 2010. The function of bilateral odor time differences in olfactory orientation of sharks. *Current Biology* 20: 1187–1191.

Helfman GS, Collette BB, Facey DE, Bowen BW. 2009. *The Diversity of Fishes: Biology, Evolution, and Ecology*, 2nd Ed. Wiley-Blackwell, Chichester, UK.

Kimber JA, Sims DW, Bellamy PH, Gill AB. 2011. The ability of a benthic elasmobranch to discriminate between biological and artificial electric fields. *Marine Biology* 158: 1–8.

Moller P. 1995. *Electric Fishes: History and Behavior*. Chapman and Hall, London.

Nelson ME. 2011. Electric fish. *Current Biology* 21: R528–R529.

Go to the Companion Website **oup.com/us/vertebratelife10e** for active-learning exercises, news links, references, and more.

CHAPTER

5

Geography and Ecology of the Paleozoic Era

Vertebrates are known from the portion of Earth's history known as the **Phanerozoic** (Greek *phaneros*, "visible"; *zoon*, "an animal") eon. The Phanerozoic began 541 million years ago and comprises the Paleozoic (Greek *palaios*, "ancient"), Mesozoic (Greek *mesos*, "middle"), and Cenozoic (Greek *kainos*, "recent") eras. Each era contains several periods, and each period contains several subdivisions called epochs (see the inside front cover). The current portion of time, the Holocene epoch (sometimes referred to as the Recent), lies within the Cenozoic era. At least 99% of all fossil species are from the Phanerozoic, although the oldest known fossils are about 3.4 billion (3,400 million) years old.

The time before the Phanerozoic is informally referred to as the **Precambrian**, and it represents the vast majority of Earth's 4.6-billion-year history (**Figure 5.1A**). The Precambrian has two formal divisions, the Archean (Greek *archaios*, "ancient"), commencing with the cooling of Earth's crust and the formation of the continents about 4 billion years ago; and the Proterozoic (Greek *pro-*, a prefix meaning "earlier" or "before"), which began about 2.5 billion years ago. The lifeless period between the formation of Earth about 4.6 billion years ago and the start of the Archean 4 billion years ago is informally known as the Hadean, and no rocks are preserved from this time. "Hades" is the classical term for Hell, reflecting the notion that the Earth of this time would have been pretty hellish.

The Hadean and Archean worlds would have looked rather like today's South Pacific, with small volcanic islands separated by immense tracts of ocean. Life in the Archean appears to have consisted of microbial mats (**Figure 5.1B**); organisms more complex than unicellular prokaryotes are not known before the Proterozoic. The start of the Proterozoic (~2,500 million years ago) was marked by the appearance of the large continental blocks seen in today's world.

Substantial levels of atmospheric oxygen were not present until about 2.2 billion years ago—some 300 hundred million years into the Proterozoic. This rise was a result of the evolution of photosynthesis among unicellular cyanobacteria and the ensuing release of free oxygen (O_2, a by-product of photosynthesis). Atmospheric oxygen did not approach present-day levels until late in the Proterozoic, about 700 Ma. The evolution of eukaryotic organisms, which depend on oxygen for respiration, commenced about 1.8 billion years ago. There is some evidence of the first animals (sponges) appearing about 635 Ma, following the second rise in levels of atmospheric oxygen. However, metazoans (animals; see Chapter 2) did not start their major radiation until near the end of the Proterozoic, some 550 Ma.

The first of the three Proterozoic eras was the Paleozoic, beginning with the Cambrian period about 541 Ma and extending through the end of the Permian, about 252 Ma (**Figure 5.2**). The Paleozoic world was very different from today's world. The continents were in different places, climates were different, and initially most life was aquatic—there was little structurally complex life on land. By the Early Devonian, however, terrestrial environments supported a substantial diversity of plants and invertebrates, setting the stage for the first terrestrial vertebrates (the earliest tetrapods), which appeared in the Late Devonian.

5.1 Shifting Continents and Changing Climates

Our present-day pattern of global climate, including such features as ice at the poles and the directions of major winds and water currents, results in large part from the positions of the continents. Continental positions have varied over Earth's history, and the global climate has varied with them. Earth today is generally colder and drier than in many past times (**Figure 5.3**). Another significant difference is that the continents are now widely separated from one another, and the main continental landmass is in the Northern Hemisphere—conditions that did not exist during most of vertebrate evolution.

Our understanding of Earth's dynamic nature and variable climate over geological time is fairly recent. The notion of mobile continents, or **continental drift**, was established in

(A)

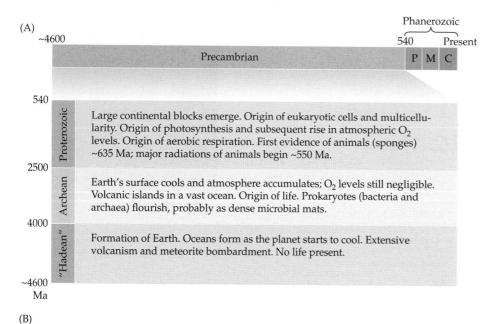

Figure 5.1 The Precambrian accounts for most of Earth's geologic history. (A) Of the 4.6 billion (4,600 million) years since the formation of Earth, the proliferation of multicellular and structurally complex animals occurred only over the Phanerozoic, which is divided into the Paleozoic, Mesozoic, and Cenozoic eras. The more than 4 billion years prior to the Phanerozoic is referred to as the Precambrian. (B) Life during most of the Precambrian consisted of unicellular prokaryotes (bacteria and archaea) that could withstand the environment. Much of this life may have taken the form of dense microbial mats, as can be seen even today in certain harsh environments such as the volcanic hot springs of Yellowstone National Park in the U.S. (B, James St. John/CC BY 2.0.)

Within the timeline:

Proterozoic (540–2500): Large continental blocks emerge. Origin of eukaryotic cells and multicellularity. Origin of photosynthesis and subsequent rise in atmospheric O_2 levels. Origin of aerobic respiration. First evidence of animals (sponges) ~635 Ma; major radiations of animals begin ~550 Ma.

Archean (2500–4000): Earth's surface cools and atmosphere accumulates; O_2 levels still negligible. Volcanic islands in a vast ocean. Origin of life. Prokaryotes (bacteria and archaea) flourish, probably as dense microbial mats.

"Hadean" (4000–~4600): Formation of Earth. Oceans form as the planet starts to cool. Extensive volcanism and meteorite bombardment. No life present.

(B)

the late 1960s as the theory of **plate tectonics**. Oceanographic research has demonstrated the spreading of the seafloor as a plausible mechanism for movements of the tectonic plates that underlie the continents. These movements are responsible for the sequence of fragmentation, coalescence, and re-fragmentation of the continents that has occurred during Earth's history. Plants and animals were carried along as continents slowly drifted, collided, and separated.

As continents collided, terrestrial floras and faunas that had evolved in isolation mixed, and populations of marine organisms were separated. A recent (in geological terms) example of this phenomenon is the joining of North and South America about 2.8 Ma. The faunas and floras of the two continents mingled, which is why we now have opossums (of South American origin) as far north as southern Canada and why some species of deer (of North American origin) are found in Argentina. In contrast, the marine organisms originally found in the sea between North and South America were separated by the rise of the Central American isthmus, and the populations on the Atlantic and Pacific sides of the land bridge have become increasingly different from each other with the passage of time.

The positions of continents affect the flow of ocean currents, and because ocean currents transport enormous quantities of heat, changes in their flow affect climates worldwide. For example, the breakup and northward movements of the continents during the late Mesozoic and Cenozoic led first to the isolation of Antarctica and the formation of the Antarctic ice cap (~45 Ma) and eventually to the isolation of the Arctic Ocean, with the formation of Arctic ice some 5 Ma. The presence of this ice cap influences global climate conditions in a variety of ways, and Earth today is colder and drier than it was earlier in the Cenozoic.

5.2 Continental Geography of the Paleozoic

The world of the early Paleozoic contained at least six major continental blocks (**Figure 5.4**). **Laurentia** included most of present-day North America plus Greenland and Scotland. **Gondwana** (sometimes called Gondwanaland) included most of what is now the Southern Hemisphere— the continents of South America, Africa, Antarctica, and Australia—as well as southern portions of Asia and Europe. Four smaller blocks contained other parts of what is now the Northern Hemisphere.

In the late Cambrian (~500 Ma), Gondwana and Laurentia straddled the Equator, and the north–south orientations of the modern continents within Gondwana were different than they are today—Africa and South America appear to be upside down (see Figure 5.4A). By the late Silurian (~420 Ma), the eastern portion of Gondwana was over the South Pole, and Africa and South America were in orientations similar to those they occupy today, but were entirely south of the Equator (see Figure 5.4B).

From the Devonian through the Permian, the continents were drifting closer together (**Figure 5.5**). The continental blocks that correspond to parts of modern North America,

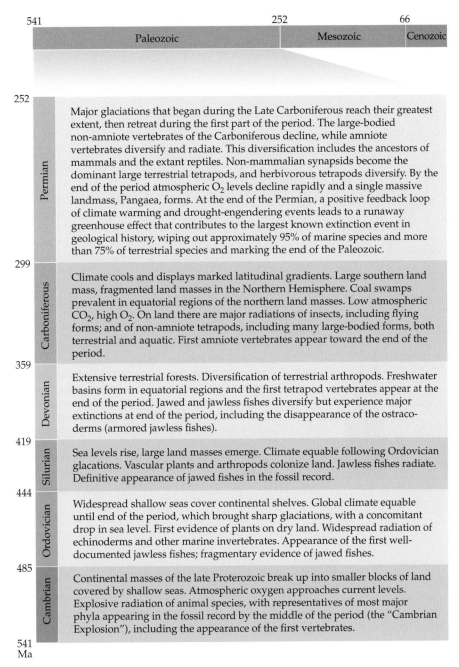

541   252   66

| Paleozoic | Mesozoic | Cenozoic |

Permian (252–299)

Major glaciations that began during the Late Carboniferous reach their greatest extent, then retreat during the first part of the period. The large-bodied non-amniote vertebrates of the Carboniferous decline, while amniote vertebrates diversify and radiate. This diversification includes the ancestors of mammals and the extant reptiles. Non-mammalian synapsids become the dominant large terrestrial tetrapods, and herbivorous tetrapods diversify. By the end of the period atmospheric O_2 levels decline rapidly and a single massive landmass, Pangaea, forms. At the end of the Permian, a positive feedback loop of climate warming and drought-engendering events leads to a runaway greenhouse effect that contributes to the largest known extinction event in geological history, wiping out approximately 95% of marine species and more than 75% of terrestrial species and marking the end of the Paleozoic.

Carboniferous (299–359)

Climate cools and displays marked latitudinal gradients. Large southern land mass, fragmented land masses in the Northern Hemisphere. Coal swamps prevalent in equatorial regions of the northern land masses. Low atmospheric CO_2, high O_2. On land there are major radiations of insects, including flying forms; and of non-amniote tetrapods, including many large-bodied forms, both terrestrial and aquatic. First amniote vertebrates appear toward the end of the period.

Devonian (359–419)

Extensive terrestrial forests. Diversification of terrestrial arthropods. Freshwater basins form in equatorial regions and the first tetrapod vertebrates appear at the end of the period. Jawed and jawless fishes diversify but experience major extinctions at end of the period, including the disappearance of the ostracoderms (armored jawless fishes).

Silurian (419–444)

Sea levels rise, large land masses emerge. Climate equable following Ordovician glaciations. Vascular plants and arthropods colonize land. Jawless fishes radiate. Definitive appearance of jawed fishes in the fossil record.

Ordovician (444–485)

Widespread shallow seas cover continental shelves. Global climate equable until end of the period, which brought sharp glaciations, with a concomitant drop in sea level. First evidence of plants on dry land. Widespread radiation of echinoderms and other marine invertebrates. Appearance of the first well-documented jawless fishes; fragmentary evidence of jawed fishes.

Cambrian (485–541)

Continental masses of the late Proterozoic break up into smaller blocks of land covered by shallow seas. Atmospheric oxygen approaches current levels. Explosive radiation of animal species, with representatives of most major phyla appearing in the fossil record by the middle of the period (the "Cambrian Explosion"), including the appearance of the first vertebrates.

541 Ma

Figure 5.2 Key events during the Paleozoic. The Cambrian, the first period of the Paleozoic era, saw an "explosion" of evolutionary radiations. By the end of the Cambrian, representatives of all extant major animal taxa, including the vertebrates, are found in the fossil record. The Paleozoic ended with a massive extinction event that eliminated many taxa, especially marine species.

or "supercontinent," **Pangaea** (sometimes spelled Pangea). At its maximum extent, the landmass of Pangaea covered 36% of Earth's surface, compared with 31% for the present arrangement of continents. Pangaea persisted for 160 Ma, from the mid-Carboniferous to the mid-Jurassic.

5.3 Paleozoic Climates

During the early Paleozoic, sea levels were at or near an all-time high for the Phanerozoic. Atmospheric carbon dioxide levels were also apparently very high—at least 10 times today's level—while oxygen levels were low, probably two thirds or less of today's level of 21%. The high levels of carbon dioxide created a greenhouse effect, with Earth experiencing hot and dry terrestrial climates during the Cambrian and much of the Ordovician. However, a major glaciation in the Late Ordovician combined with falling levels of atmospheric carbon dioxide, resulted in cooler and moister conditions (see Figure 5.3). These changes made the land more hospitable for the first terrestrial plants and set the scene for the development of terrestrial ecosystems during the Silurian.

A relatively equable climate continued throughout the Silurian to the Middle Devonian. Glaciation recurred in the Late Devonian, with ice sheets covering much of Gondwana from the mid-Carboniferous until the Middle Permian. The waxing and waning of glaciers created oscillations in sea level, which resulted in the cyclic formation of coal deposits from buried plant material (the word "carboniferous" means "coal-bearing"), especially in eastern North America and western Europe (see Figure 5.5B). The climate over Pangaea was fairly uniform in the Early Carboniferous (also known as the Mississippian), but glaciation led to significant regional differences in flora in the Late Carboniferous and Early Permian. Most vertebrates were found in equatorial regions during this time. Glaciers regressed in the Middle Permian, and an essentially ice-free world persisted until the mid-Cenozoic.

While the terrestrial flora of the early Paleozoic had a significant effect on the biosphere, the spread of derived land plants in the Devonian, such as trees with deep root

Greenland, and western Europe had come into proximity along the Equator. With the later addition of Siberia, these blocks formed a northern supercontinent known as **Laurasia**. Most of Gondwana was in the far south, overlying the South Pole. The Tethys Sea, which separated Gondwana from Laurasia, did not close completely until the Late Carboniferous (also known as the Pennsylvanian), when Africa moved northward to collide with the east coast of North America (see Figure 5.5B). By the Permian, most of the continental surface was united in a single continent,

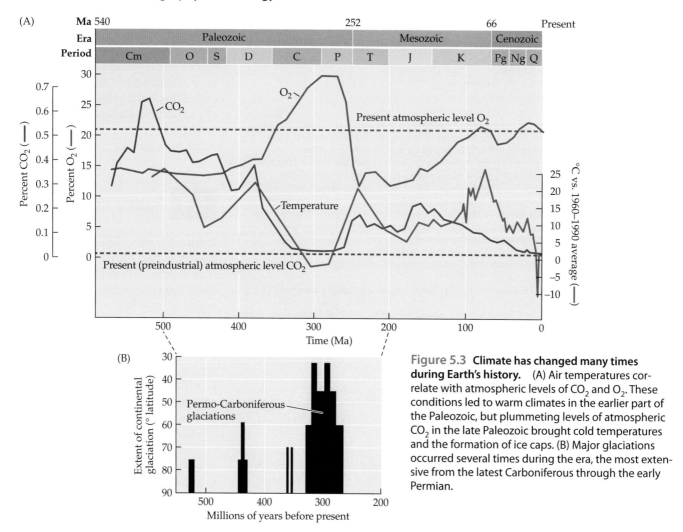

Figure 5.3 Climate has changed many times during Earth's history. (A) Air temperatures correlate with atmospheric levels of CO_2 and O_2. These conditions led to warm climates in the earlier part of the Paleozoic, but plummeting levels of atmospheric CO_2 in the late Paleozoic brought cold temperatures and the formation of ice caps. (B) Major glaciations occurred several times during the era, the most extensive from the latest Carboniferous through the early Permian.

systems and their association with symbiotic fungi, resulted in modification of terrestrial soils, with profound effects on the Earth's atmosphere and climate. Soils formed as the underlying rocks were broken down by penetration of plant roots, and the roots' organic secretions plus decomposition of dead plant material created acids that dissolved minerals in the rocks. This process caused an increased flux of calcium to the oceans, with subsequent deposition of marine carbonates that formed from the combination of this calcium with atmospheric carbon dioxide. Removal of atmospheric carbon dioxide led to climatic cooling by the Late Devonian.

Atmospheric levels of carbon dioxide reached an extreme low during the Late Carboniferous and Early Permian, resembling the levels of today. (Even with the recent anthropogenic increase in atmospheric carbon dioxide, concentrations are much lower now than they were for most of the Phanerozoic.) The reverse greenhouse effect of reduced atmospheric carbon dioxide (see Figure 5.3) was probably the cause of the extensive Permo-Carboniferous glaciations. In contrast, atmospheric oxygen levels were high during this time, reaching present-day levels by the start of the Carboniferous and peaking at about 30% in the Early Permian before declining later in the period.

5.4   Paleozoic Ecosystems

The most dramatic radiation of animal life commenced during the early Cambrian, with the so-called Cambrian Explosion of marine animals with hard (and thus preservable) parts. We also have a vision of the soft-bodied animals of this era from sites of exceptional preservation such as Canada's Burgess Shale. Most Cambrian animals were invertebrates, although the first chordates, including soft-bodied vertebrates and conodonts, are known from the late-early and middle Cambrian. Many lineages of Cambrian invertebrates were extinct by the end of the period, although animal life rebounded in the Ordovician, and many of the groups that would dominate the ecosystems of the rest of the Paleozoic appeared and radiated at this time.

Aquatic life

All Ordovician vertebrates were marine, and they consisted mainly of jawless vertebrates such as conodonts and basal ostracoderms. Phosphatic microremains from the latest Cambrian and Early Ordovician possibly represent the earliest of the armored ostracoderm vertebrates, but

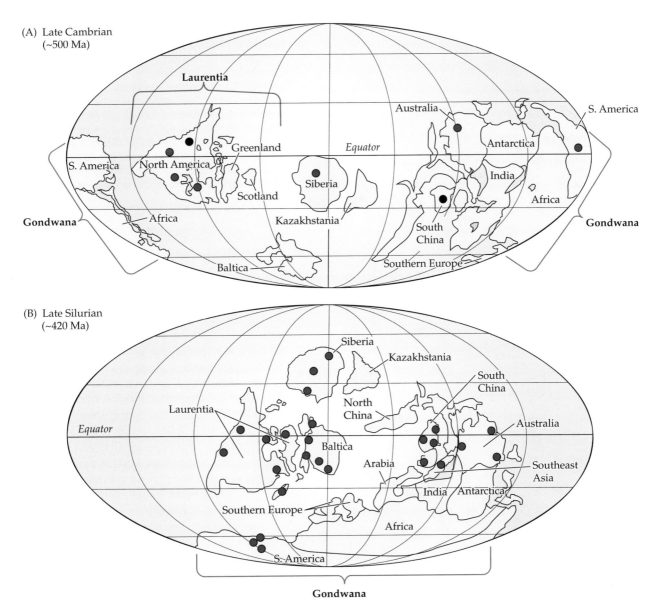

Figure 5.4 Location of continental blocks in the early Paleozoic.
(A) Late Cambrian. The two black dots indicate locations of fossil vertebrates dating from the Cambrian; red dots mark the locations of Ordovician vertebrate fossils. Baltica consisted of northeastern Europe plus part of western Asia. (B) Late Silurian. Baltica has now combined with western Europe and parts of northeastern North America. Red dots indicate fossil vertebrate locations. (After Scotese and McKerrow 1990.)

more definitive remains are not known until the Middle Ordovician. Fragmentary scales of possible basal cartilaginous fishes are known from the Middle Ordovician. Ostracoderms and basal jawed fishes (acanthodians and placoderms) diversified in the Silurian, and basal bony fishes appeared toward the end of that period.

The Devonian is popularly known as the "Age of Fishes." Ostracoderms remained prominent in the Early Devonian, and placoderms were the most diverse of the Devonian vertebrates, but both were extinct by the end of the period. Cartilaginous and bony fishes commenced their major radiations in the Devonian, with both freshwater and marine

forms. Acanthodians persisted into the Permian, and conodonts until the end of the Triassic.

Cartilaginous fishes were especially prominent in the Carboniferous, and basal members of extant groups appeared at the end of the Paleozoic. The prominent Paleozoic ray-finned bony fishes (actinopterygians) were the heavily armored basal forms; more modern kinds of ray-finned bony fishes appeared at the end of the Carboniferous. The lobe-finned bony fishes (sarcopterygians) were diverse in the Devonian, including a radiation of the tetrapodomorph fishes, but by the Middle Permian lungfishes and coelacanths were the only remaining sarcopterygians.

Tetrapods were initially a mainly aquatic radiation in the Late Devonian, and in the Carboniferous and Permian there were still many aquatic non-amniote tetrapods in fresh waters. The only Paleozoic amniotes to return to the seas were the mesosaurs, small aquatic sauropsids of the Permian.

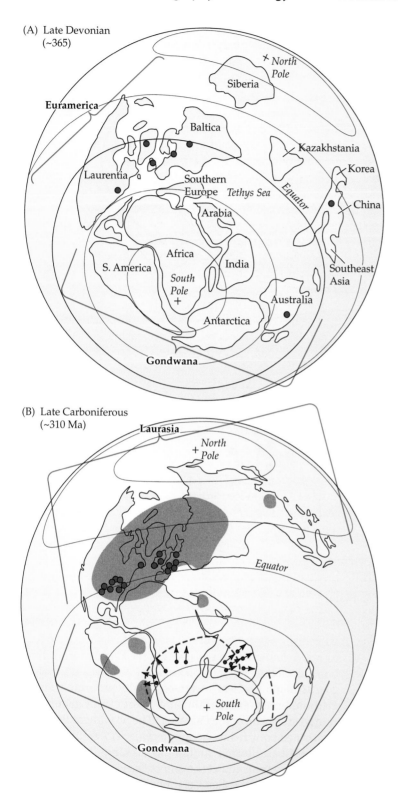

(A) Late Devonian (~365)

(B) Late Carboniferous (~310 Ma)

Figure 5.5 Location of continental blocks later in the Paleozoic. Red dots indicate locations of fossil tetrapods. (A) Late Devonian. Laurentia and Baltica (i.e., the continental areas including present-day North America, Greenland, and northern Europe) lie on the Equator. An arm of the Tethys Sea extends westward between Gondwana and the northern continents. (B) Late Carboniferous. This map illustrates an early stage of the supercontinent Pangaea. The location and extent of continental glaciation are shown by the dashed lines and arrows radiating outward from the South Pole. Gray shading shows the extent of the forests that formed today's coal beds.

whole-body fossils from the early Silurian, but a low-growing terrestrial flora (revealed by spores of primitive moss-like plants) and arthropods (represented by fossil trackways) were evident by the Middle Ordovician, and may date back as far as the Cambrian. The Silurian landscape would have looked bleak by our standards—mostly barren, with a few kinds of low-growing vegetation limited to moist areas.

Land plants (also known as **embryophytes**) represent a single terrestrial invasion from a particular group of green algae, the charophytes. Charophytic algae are an ancient lineage, dating to at least a billion years ago; extant species all live in fresh water, and land plants may have had their origin in fresh water. Land plants comprise two main groups: the low-lying **bryophytes**, represented today by mosses, liverworts, and hornworts, and the **tracheophytes**, or vascular plants. Bryophytes, which form the stem group to the vascular plants, lack internal channels for the conduction of food and water through the tissues and are thus low-growing. Bryophytes probably formed a sparse terrestrial flora prior to the invasion of vascular plants: definitive fossils of bryophytes are not known until the Devonian, but they are less likely to fossilize than the larger and tougher vascular plants.

Vascular plants, which do have internal channels for the conduction of water, are first known from the Silurian and are able to grow taller than bryophytes. Early vascular plants included plants that reproduce by spores (ferns, horsetails, lycopods, and some extinct groups) and plants that reproduce via seeds (**gymnosperms** [conifers and cycads] and seed ferns; the latter were extinct by the end of the Mesozoic). **Angiosperms**, the flowering seed plants that are predominant today, did not appear until the Mesozoic.

Terrestrial floral ecosystems

Photosynthesizing bacteria (cyanobacteria, once called "blue-green algae") have probably existed in wet terrestrial habitats since the Archean, and terrestrial algae, lichens, and fungi probably date from the late Proterozoic. Complex terrestrial ecosystems, with evidence of associated plants, fungi, and animals (arthropods), are known from

Silurian ecosystems consisted of a minimal terrestrial food web of primary producers (small plants), decomposers (fungi), secondary consumers (fungus-eating arthropods such as millipedes, springtails, and mites), and predators (larger arthropods such as centipedes, spiders, and scorpions). Terrestrial ecosystems increased in complexity through the Early and Middle Devonian, with a greater diversity of plant species and the appearance of more-derived vascular plants, some now containing the supportive material lignin. Some fossils of Devonian plants show damage from arthropods, but most Devonian arthropods were probably detritivores, consuming dead plant material and fungi. This process would have speeded recycling of nutrients and enhanced the formation of soils.

By the Middle Devonian, vascular plants attained heights of about 2 m, and the canopy they created would have made the microclimate on the ground cooler and wetter than in open areas—a step toward the variation in terrestrial microclimates that both plants and animals exploit today. Treelike forms evolved independently among several ancient plant lineages, and stratified forest communities consisted of plants of different heights. However, these plants were not closely related to modern trees and were different in structure. Devonian trees had narrow trunks and would not have provided enough woody tissue to make furniture. Plants with large leaves first appeared in the Late Devonian, and at this time forests of the progymnosperm *Archaeopteris* (an early seed plant with a trunk up to 1 m in diameter and a height of at least 10 m) developed and spread.

The vegetation of the Early Carboniferous was still dominated by spore-bearing vascular plants. Many species of clubmosses (lycopods) reached heights of several meters, and a few of these plants survive today, mainly as small ground plants (although large tree ferns still occur in some regions). The predominant seed-bearing plants of the Early Carboniferous were the seed ferns. Carboniferous terrestrial vegetation would have looked superficially as it does today, although the actual types of plants present were completely different.

Swamps were abundant across much of the Earth during the Devonian and for most of the Carboniferous, but a fundamental transition in global vegetation types occurred near the end of the Carboniferous (~305 Ma), coincident with a major reduction in atmospheric CO_2 to an all-time low. At least across the paleocontinent Euramerica (the combination of the North American continent with Greenland and Northern Europe), complex forests indicative of at least seasonally dry conditions replaced the earlier types of rainforest biomes that had formed the coal deposits. These forests had an increased diversity of vines and ground cover; they were dominated by tree ferns and contained the first conifers and cycads. This vegetational change had a profound effect on the fauna: the first herbivorous tetrapods appeared at this time, and herbivorous insects underwent a major radiation.

By the Permian, forests typical of drier environments characterized terrestrial habitats worldwide. These forests were dominated by conifers and contained many deciduous plants, in contrast to the earlier evergreen forests, although swamplike forests persisted in China. By the Late Permian, highly productive forests in Antarctica reflected a return to an ice-free Earth.

Terrestrial faunal ecosystems

Three different arthropod lineages colonized the land independently: myriapods (millipedes and centipedes) and arachnids (spiders, scorpions, and mites) appeared in the late Silurian, and hexapods (springtails and primitive flightless insects) appeared in the Early Devonian. Terrestrial mollusks (slugs and snails) and annelids (earthworms) are not known until the Carboniferous.

Terrestrial invertebrates diversified during the Carboniferous. Flying insects are known from the late Early Carboniferous, and fossils of damaged leaves show that herbivorous insects had diversified by the end of the Early Carboniferous. Higher levels of atmospheric oxygen at this time would have made it easier for flight to evolve, in part because flight requires high levels of activity and thus large quantities of oxygen. Giant dragonflies (one species had a wingspan of 63 cm) flew through the air, and predatory arthropleurids (extinct millipede-like arthropods) were nearly 2 m long. New arthropod types appearing in the Permian included hemipterans (bugs), beetles (coleopterans), and forms resembling but not closely related to mosquitoes.

By the Early Carboniferous (~355 Ma), a fauna of terrestrial tetrapods had been established that consisted a diversity of taxa the size of extant amphibians, in contrast to the large (1–2 m) Devonian aquatic forms. Both terrestrial and semi-aquatic tetrapods diversified during the Carboniferous. The first definitive fossils of amniotes appeared early in the Late Carboniferous (~312 Ma), and from this time amniotes could be divided into two major lineages—one leading to mammals (synapsids, the most diverse lineage at that time) and the other to reptiles, including birds (sauropsids). However, fragmentary fossils from the Early Carboniferous provide hints that amniotes may have been present from the start of the period, a date of divergence from modern amphibians (~355 Ma) that agrees with the molecular data.

The first herbivorous tetrapods appeared near the end of the Late Carboniferous (~305 Ma), and with them came the capacity to directly exploit the primary production of plants (**Figure 5.6**). Fossil herbivores can be identified from their blunt teeth and their capacious rib cages, housing the large guts required to digest vegetation. (Multicellular animals are unable to digest cellulose, the structural material of plants, and must rely on symbiotic microorganisms in the gut to ferment the plant material.) The diversity of herbivores—mainly large-bodied synapsids and "parareptiles" (an extinct group of basal sauropsids)—reached a peak in the Late Permian. Following the end-Permian extinctions, new types of herbivorous tetrapods emerged in the Triassic.

The Late Carboniferous and Early Permian terrestrial faunas showed less regional differentiation than terrestrial faunas of earlier times, probably reflecting greater connectivity between communities following the demise of the rainforests. However, species diversity decreased over the Carboniferous/Permian boundary before slowly rebounding in the Permian. During the Early Permian, the numbers of non-amniote tetrapods declined, while the numbers of

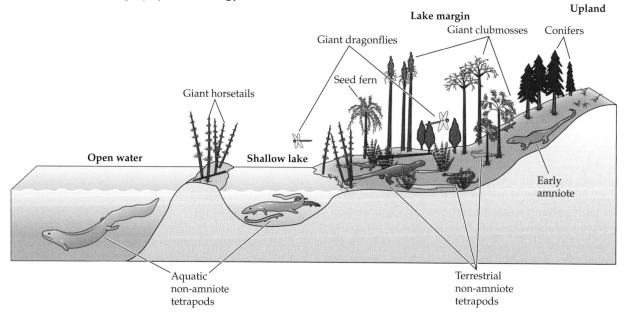

Figure 5.6 Terrestrial and aquatic ecosystems in the Carboniferous. A reconstruction of a Late Carboniferous lake in Europe and its surroundings. The Carboniferous was characterized by a diversity of semiaquatic early tetrapods, but toward the end of the period there was a dramatic extinction of many types of semiaquatic tetrapods and an increase in the diversity of terrestrial forms, including both amniotes and some derived non-amniotes, such as the herbivorous *Diadectes*. The plants and animals are not shown to scale. The largest animals would have been 1–1.5 m long. (After Benton 2014.)

amniotes increased. Several amniote lineages gave rise to small, insectivorous predators that were rather like modern lizards. Larger synapsids (up to 1.5 m long), like the well-known "sailback" *Dimetrodon* (see Figure 24.4), were the principal terrestrial predators. By the Late Permian, terrestrial communities had begun to assume the pattern we know today, with a multitude of herbivorous vertebrates supporting a smaller variety of carnivorous forms, although the kinds of plants and animals in those ecosystems were still almost entirely different from those of today.

5.5 Paleozoic Extinctions

Major extinctions of marine invertebrates occurred at the end of both the Cambrian and Ordovician; the record of vertebrates from that time is too poor, however, for us to know whether these extinction events affected them as well. The next two major extinctions occurred during the Late Devonian. The first of these, the Kellwasser event (~374 Ma), affected mainly marine invertebrates; the armored ostracoderm fishes (see Chapter 3) had been declining through the Middle Devonian and had disappeared before this event. The second extinction, the Hangenberg event, occurred at the end of the Devonian (~359 Ma) and affected both marine and freshwater vertebrates. An estimated 44% of jawed vertebrate lineages became extinct at this time, including all of the

placoderms, many of the families of lobe-finned fishes, and most of the initial radiation of tetrapods. The end-Devonian extinction resulted in the restructuring of marine vertebrate communities, and the rise to dominance of cartilaginous fishes and ray-finned bony fishes.

A significant extinction event (although not a mass extinction like those at the end-Devonian and end-Permian) occurred in the mid-Permian, about 260 Ma. Fifteen families of tetrapods became extinct, with a 75% loss of genetic richness. This extinction claimed many non-amniote tetrapods and synapsids. The mid-Permian extinction may have been related to climate changes associated with the end of the Permo–Carboniferous glaciations, including a decrease in the level of atmospheric oxygen and an increase in the level of carbon dioxide (see Figure 5.3). These changes would have resulted in hypoxic stress for the tetrapods that had evolved and radiated under conditions of oxygen abundance during the Early Permian. Furthermore, lower atmospheric oxygen levels may have limited the maximum elevation at which vertebrates could live. Restricting vertebrates to lower elevations would have reduced habitat diversity available to them and limited their ability to migrate.

The greatest extinction of the Paleozoic—indeed, the greatest extinction of all time— occurred at the end of the Permian, 252 Ma. Approximately 57% of marine invertebrate families (including the trilobites) and 95% of all marine species, including 12 families of fishes, disappeared at the end of the Permian. Among the tetrapods, 4 of 11 non-amniote families and 17 of 32 amniote families disappeared, with especially heavy losses among the remaining synapsids; only the cynodonts (the group that gave rise to mammals) and the tusked dicynodonts survived.

Hypotheses abound for the reason for the massive end-Permian extinction, including the possibility of an asteroid impact. But the extinction took place in several pulses, and appears to have extended over about 180,000 years. A more likely cause is massive volcanic eruptions in Siberia that

(A)

(B)

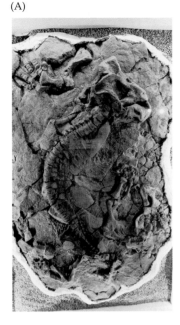

15 cm

Figure 5.7 Fossil vertebrates show evidence of end-Permian drought conditions. (A) Skeleton of a *Lystrosaurus* (a dicynodont therapsid) from the latest Permian of the Karoo basin of South Africa. (B) Artist's reconstruction of this specimen showing the carcass prior to its preservation in a lake bed. This animal died and was mummified during a drought, and the mummy was subsequently buried in lake sediments. Mummification preserved the skeleton intact and with the bones articulated in the position assumed soon after death. Fossilized mummified skin has been found on many of the skeletons preserved in the Karoo at about the Permo–Triassic boundary. (From Smith and Botha-Brink 2014.)

were coincident with the Permo–Triassic boundary and were concentrated in a time period of less than a million years. Enough molten lava was released to cover an area about half the size of the United States. Carbon isotope changes in the boundary sediments indicate that massive amounts of carbon dioxide and/or methane were released, probably from the interaction of volcanic magma with surface rocks rich in organic materials (e.g., oil shales).

This rapid change in atmospheric composition would have resulted in profound and rapid global warming and ocean acidification, and the ozone layer may have been destroyed. Such warming would have had a direct effect not only on organisms but also on oceanic circulation, resulting in stagnation and low oxygen levels, with profound effects on marine organisms. Charcoal and soot-bearing sediments are evidence of extensive wildfires on land, which would have resulted in habitat destruction and terrestrial ecosystem collapse. The fossil record of end-Permian therapsids, with many specimens mummified before preservation, suggests that they succumbed during droughts (**Figure 5.7**).

The Permo–Triassic boundary was marked by a runaway greenhouse effect, with a positive feedback loop of events causing increasing global warming that disrupted the normal global environmental mechanisms for hundreds of thousands of years and resulted in the extinction of almost all life on Earth. Recovery of terrestrial ecosystems did not occur until at least 5 million years into the Triassic.

Summary

Earth's history is divided into the Precambrian and the Phanerozoic.

No life was present during the first part of the Precambrian, informally known as the Hadean. The Archean eon supported only bacterial life forms. Eukaryotic organisms did not appear until the second eon, the Proterozoic.

The Phanerozoic represents only the last 540 million years of Earth's 4-billion-year history, yet it is from this time that most multicellular life is known.

The Phanerozoic is divided into three eras, the Paleozoic, Mesozoic, and Cenozoic. We live in the Holocene epoch of the Quaternary period of the Cenozoic era.

Earth's climate is determined by the positions of the continents.

The continental blocks move over time, and their positions affect how heat is transferred via oceanic circulation, and whether there is ice at the poles. Earth today is colder and drier than it has been in most of the Phanerozoic.

During the Paleozoic the continents were initially widespread, but by the end of the period they came together, forming the supercontinent Pangaea.

The world of the Paleozoic was very different from today's world.

There was initially only sparse life on land. The evolution of land plants in the Silurian paved the way for terrestrial faunas, providing a terrestrial habitat with soils. Through a complex chain of biochemical reactions involving many elements of the environment, these soils reduced the atmospheric levels of carbon dioxide, resulting in the cooling of the planet.

(Continued)

Summary (continued)

Terrestrial ecosystems of the Devonian and Carboniferous were largely swamp-based. During the Carboniferous, oscillations in sea level resulted in the formation of coal beds from buried vegetation.

Paleozoic climates were influenced by glaciations.

A major glaciation in the Late Ordovician cooled Earth, resulting in suitable conditions for the invasion of terrestrial life in the Silurian.

A second major glaciation occurred from the mid-Carboniferous to the mid-Permian, accompanied by low levels of atmospheric carbon dioxide and high levels of atmospheric oxygen.

The later Permian saw a return to a "hothouse" ice-free world; the poles were free of ice until the mid-Cenozoic and there were no major glaciations until the late Cenozoic ice ages, which we are still in progress.

Aquatic vertebrates diversified in the Paleozoic.

The jawless conodonts, known almost entirely from microscopic teeth, appeared in the mid-Cambrian and survived until the end of the Paleozoic.

The jawless armored ostracoderms were the original marine vertebrates in the Ordovician, but they were extinct by the end of the Devonian.

Various jawed fishes diversified in the Silurian and Devonian. Placoderms were the most diverse of the Devonian fishes, but they were extinct by the end the period.

Cartilaginous and bony fishes started to radiate in the Devonian, but many lineages of lobe-finned fishes were extinct by the Middle Permian.

The earliest terrestrial land cover was not formed by plants.

Early Paleozoic terrestrial ecosystems were composed of bacteria, algae, lichens, and fungi. Land plants—a single lineage related to the charophytes, a group of green algae—did not gain a foothold until the Silurian.

Early terrestrial floras consisted of vascular and nonvascular plants.

The nonvascular plants, or bryophytes, are represented today by mosses, liverworts, and hornworts.

The vascular plants, or tracheophytes, have internal channels for the conduction of food and water and are able to grow taller than nonvascular plants.

In the Devonian and Carboniferous, vascular plants were represented by some groups that are now extinct (seed ferns) or uncommon (lycopods, clubmosses).

Conifers and cycads (gymnosperms) appeared near the end of the Carboniferous, but the predominant plants of today—the flowering seed plants, or angiosperms—were unknown during the Paleozoic; they did not appear until the Mesozoic.

A profound shift in terrestrial vegetation occurred near the end of the Carboniferous.

During the Devonian and throughout most of the Carboniferous, the predominant global vegetation was the swampy rainforest that formed today's coal beds. Near the end of the Carboniferous this ecosystem collapsed in most of the world and was replaced by more complex forests characteristic of seasonally dry environments.

The radiation of herbivorous tetrapods and herbivorous insects dates from this shift in vegetation near the end of the Carboniferous. Permian forests were dominated by conifers and contained many deciduous plants.

Invertebrates appeared on land later than the plants.

Three different groups of arthropods colonized the land: myriapods (centipedes and millipedes) and arachnids (spiders, scorpions, and mites) in the late Silurian and springtails and primitive flightless insects in the Early Devonian.

Flying insects were not known until the late in the Early Carboniferous. Slugs, snails, and earthworms also first appeared in the Carboniferous.

Tetrapods were the last major group to invade the land.

Late Devonian tetrapods were predominantly aquatic, but terrestrial tetrapod communities were established early in the Carboniferous.

The non-amniote tetrapods (terrestrial and semiaquatic) remained prominent through the Carboniferous, but in the Early Permian they began to decline and amniotes became more common.

The earliest amniotes are known from the first part of the Late Carboniferous, and from their first appearance amniotes fall into two lineages: the synapsids (mammals) and sauropsids (reptiles, including birds).

By the Late Permian, vertebrate communities were similar in structure to those of today, but with different players: the herbivores were mainly synapsids and parareptiles, the insectivores were small eureptiles, and the principal predators were large synapsids.

Summary *(continued)*

There were numerous extinctions during the Paleozoic, with the largest extinction coming at the end of the era.

The Ordovician extinction affected mainly marine invertebrates. Vertebrates (both fishes and tetrapods) were badly hit by the extinction at the end of the Devonian.

The extinction at the end of the Permian was the biggest in Earth's history; volcanic eruptions resulted in a runaway greenhouse effect, with global warming and ocean acidification adversely affecting all life on Earth. The end-Permian extinction terminated the dominance of synapsids as the large terrestrial vertebrates, a role they would not play again until the evolution of large mammals in the Cenozoic.

Discussion Questions

1. We think of the presence of ice caps at the poles as the "normal" condition for Earth, but in fact this situation is less common over geological history than are ice-free polar regions. How does continental movement relate to this?

2. What other aspects of Earth's history might make the world colder, with ice at the poles, and when was this a particular issue for tetrapods?

3. What effect did the Late Carboniferous/Early Permian climate changes have on global vegetation, and how did this affect vertebrate evolution?

4. We think of amniotes as being more derived types of tetrapods than non-amniotes. Is there evidence that the replacement of terrestrial faunas dominated by non-amniotes with faunas dominated by amniotes was the result of competition?

5. What profound change happened to the community structure of tetrapods at the end of the Carboniferous that led to a change for the entire terrestrial vertebrate community?

6. We know that invertebrates colonized the land before vertebrates did. What is a possible reason for this? Did things have to happen this way?

Additional Reading

Benton MJ. 2010. The origins of modern biodiversity on land. *Philosophical Transactions of the Royal Society B* 365: 3667–3679.

Delwiche CF, Cooper ED. 2015. The evolutionary origin of a terrestrial flora. *Current Biology* 25: R899–R910.

Dunne EM and 6 others. 2018. Diversity change during the rise of tetrapods and the impact of the "Carboniferous rainforest collapse." *Proceedings of the Royal Society B* 285: 20172730.

Kenrick P, Wellman CH, Scheider H, Edgecombe GD. 2012. A timeline for terrestrialization: consequences for the carbon cycle during the Palaeozoic. *Philosophical Transactions of the Royal Society B* 367: 519–536.

Lozano-Fernandez J and 9 others. 2016 A molecular palaeobiological exploration of arthropod terrestrialization. *Philosophical Transactions of the Royal Society B* 371: 20150133.

Montañez IP and 7 others. 2016. Climate, p_{CO_2}, and terrestrial carbon cycle linkages during the late Paleozoic glacial-interglacial cycles. *Nature Geosciences* 9: 824–828.

Morris JL and 10 others. 2015. Investigating Devonian trees as geo-engineers of past climates: Linking palaeosols to palaeobotany and experimental geobiology. *Palaeontology* 58: 787–801.

Morris, JL and 9 others. 2018. The timescale of early landplant evolution. Proceedings of the National Academy of Sciences USA. In press.

Pearson MR, Benson RBJ, Upchurch P, Fröbisch J, Kammerer CF. 2013. Reconstructing the diversity of early terrestrial herbivorous tetrapods. *Palaeogeography, Palaeoclimatology, Palaeoecology* 372: 42–49.

Reisz RR, Fröbisch J. 2014. The oldest caseid synapsid from the Late Pennsylvanian of Kansas, and the evolution of herbivory in terrestrial vertebrates. *PLoS One* 9(4): e94518.

Schachat SR and 5 others. 2018. Phanerozoic p_{O_2} and the early evolution of terrestrial animals. *Proceedings of the Royal Society B* 285: 20172631.

Smith RMH, Botha-Brink J. 2014. Anatomy of a mass extinction: Sedimentological and taphonomic evidence for drought-induced die-offs at the Permo-Triassic boundary in the main Karoo Basin, South Africa. *Palaeogeography, Palaeoclimatology, Palaeoecology* 396: 99–118.

Go to the Companion Website **oup.com/us/vertebratelife10e** for active-learning exercises, news links, references, and more.

CHAPTER

6

Radiation and Diversification of Chondrichthyes

The appearance of jaws and internally supported paired appendages (described in Chapter 3) were the basis for a new radiation of vertebrates with diverse predatory and locomotor specializations. The extant cartilaginous fishes—sharks, rays, and chimaeras—are the descendants of one clade of this radiation. Chondrichthyans are often thought of as primitive fishes, but the more we come to understand them, the more apparent it becomes that they are highly derived in their own fashion, even if they lack many of the derived features of osteichthyans. Most modern chondrichthyans are marine, although there are a handful of freshwater sharks and rays.

6.1 Chondrichthyes: The Cartilaginous Fishes

Chondrichthyans make a definite first appearance in the fossil record in the Early Devonian, although isolated scales and possible teeth appear in Ordovician sediments, and undoubted chondrichthyan scales are known from the early Silurian. The oldest articulated fossil (i.e., a fossil with the bones connected) of a chondrichthyan is *Doliodus problematicus* from the Early Devonian of Canada. Although sharklike in appearance, *Doliodus* was not a neoselachian; that group (which contains the extant sharks) is unknown until the Mesozoic.

Chondrichthyans can be divided into two groups (**Figures 6.1** and **6.2**), although the fossil lineages make this subdivision a little more complex.

- Elasmobranchii (Greek *elasma*, "plate"; *branchia*, "gills") have multiple gill openings on each side of the head. This group includes the extant Neoselachii (Greek *neos*, "new"; New Latin *selachos*, "cartilaginous fish"): sharks (mostly torpedo-shaped forms with five to seven gill openings on each side of the head) and rays (including skates) (dorsoventrally flattened forms with five pairs of gill openings on the ventral surface

of the head). Neoselachians are also characterized by an independently mobile upper jaw (**hyostyly**; see Chapter 7 for a discussion of chondrichthyan jaws). The sharklike form was probably the ancestral chondrichthyan condition.

- Holocephali (Greek *holos*, "whole, entire"; *kephale*, "head") is named for the undivided appearance of the head that results from having a single external gill opening. Individual gill slits are covered by an operculum formed by soft tissue that is supported on the hyoid arch by an opercular cartilage. The common names of extant holocephalans—chimaeras, rabbitfishes, ratfishes, and ghost sharks—reflect their bizarre forms. They have a fishlike body and a long, flexible tail. Some species have a head that resembles a caricature of a rabbit, with big eyes and broad tooth plates that look a bit like a rabbit's teeth.

In contrast to the mobile upper jaw of extant elasmobranchs, the upper jaw of chimaeras and their fossil relatives is fused with the skull (**holostyly**, a condition convergent with that of humans). In addition, the dorsal elements of the hyoid arch are separate (pharygeohyal and epihyal, equivalent to the pharyngeobranchial and epibranchial elements of the gill arches) and are not fused together into a hyomandibula, as in other extant gnathostomes.

Distinctive characters of chondrichthyans

Chondrichthyans are defined by two unique characters:

- A cartilaginous skeleton of individual small calcium-containing units with a distinctive microstructure called tesserate prismatic calcifications.

- The presence of pelvic claspers in males, which are used for internal fertilization. (You saw in Chapter 3 that some placoderms had pelvic claspers, but these were not homologous with the pelvic claspers of chondrichthyans.)

A cartilaginous endoskeleton is the ancestral condition in vertebrates, seen today in the hagfishes and lampreys. Some

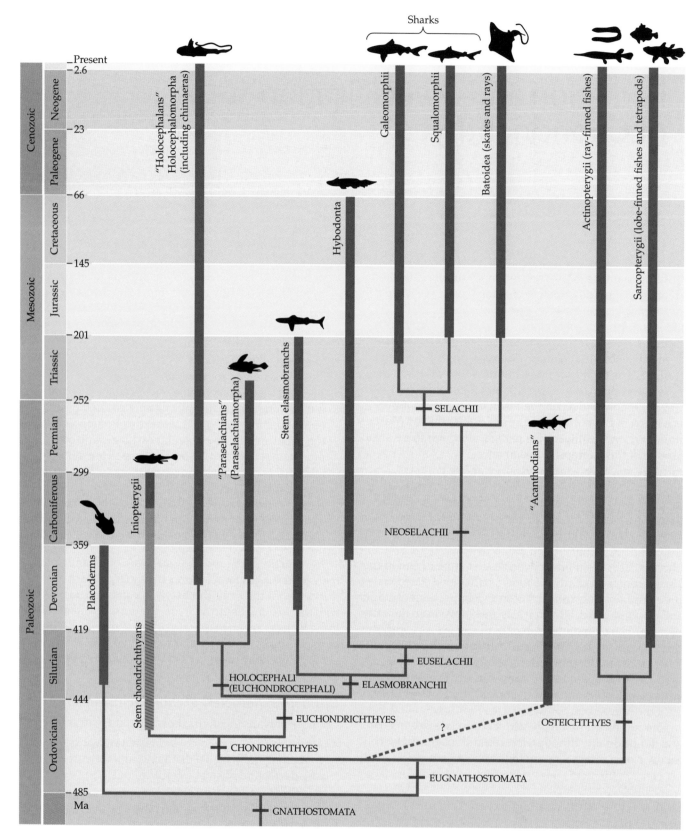

Figure 6.1 Phylogenetic relationships of cartilaginous fishes. Tree showing the general diversification of jawed vertebrates, focusing on the best-corroborated probable relationships among the major groups of chondrichthyans. The paraselachians, shown here as a distinct lineage, were in fact probably paraphyletic with respect to the holocephalans. The narrow lines show interrelationships only; they do not indicate times of divergence or the unrecorded presence of taxa in the fossil record. Blue bars indicate the group's presence, and the light blue bar represents the possible presence of the group. The dashed line indicates current controversies about relationships. The hatched bar indicates possible occurrence based on limited evidence.

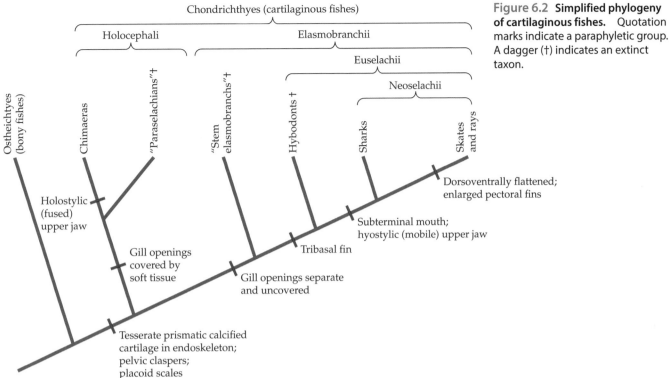

Figure 6.2 **Simplified phylogeny of cartilaginous fishes.** Quotation marks indicate a paraphyletic group. A dagger (†) indicates an extinct taxon.

chondrichthyans mineralize portions of the skeleton, and may have perichondral bone surrounding the mineralized cartilage, but only osteichthyans (including tetrapods) have an endoskeleton composed entirely of bone, both perichondral and endochondral. Cartilage is not necessarily weaker than bone; calcified cartilage can be made extremely strong when there are multiple layers and internal supports, as in the jaws of rays that crunch mollusks.

All vertebrates more derived than the extant agnathans have a bony exoskeleton of some sort, formed from dermal bone. You saw this external covering in the ostracoderms and placoderms in Chapter 3. Indeed, the bones just underneath the skin on the top of our skull are a remnant of this exoskeleton (see Chapter 2). Although chondrichthyans have evidently lost the extensive plates of dermal bones seen in other jawed fishes, they retain an exoskeleton of scales composed of dentine, enameloid, and even traces of bone, and also have these tissues in their teeth. These unique scales are called **placoid scales**, also known as **dermal denticles**, and they develop in the same way as teeth—that is, from a single dental papilla. Many chondrichthyans, especially chimaeras, also have complex labial cartilages associated with their jaws.

The nature of the cartilaginous skeleton means that, unlike bone, it tends to fall apart after death; thus the extinct chondrichthyans are known largely from fossil teeth. Fortunately, chondrichthyans have a lot of teeth. Their teeth are not embedded in the jaw like those of extant bony fishes. Rather, they usually form in a **tooth whorl**, by which the teeth are continually replaced as they wear down and eventually fall off (**Figure 6.3**). Each tooth

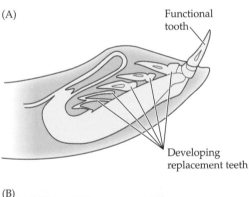

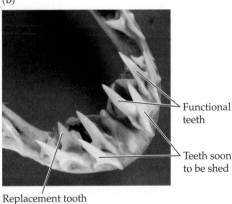

Figure 6.3 **Tooth replacement by chondrichthyans.** (A) Cross section of the jaw of an extant shark, showing a single functional tooth backed by a band of replacement teeth in various stages of development. Together, the functional and developing teeth form a tooth whorl. (B) The replacement teeth are visible behind the functional teeth in this skeletal shark jaw. (B, courtesy of Lisa Whitenack.)

on the functional edge of the jaw is but one member of a tooth whorl, attached to a ligamentous band (the dental lamina) that courses down the inside of the jaw cartilage deep below the fleshy lining of the mouth. A series of developing teeth is aligned in each whorl in a row directly behind the functional tooth.

Tooth replacement in extant sharks can be rapid. Young sharks replace each lower-jaw tooth as often as every 8.2 days and each upper-jaw tooth every 7.8 days, although the rate of replacement varies considerably with species, age, and the general health of the shark, as well as with environmental factors such as water temperature.

Tooth whorls are also seen in acanthodians, which are now considered to be the sister group to the chondrichthyans. They are also found in some basal osteichthyans (as are rooted teeth). Thus, tooth whorls are probably a primitive feature of crown gnathostomes (i.e., all gnathostomes excluding the now-extinct placoderms). The grinding tooth plates of chimaeras, which are not replaced, are a secondary modification of the original dental condition. Many of the extinct related forms, such as paraselachians, also had tooth whorls, as did some of the earlier holocephalans.

Extant chondrichthyans also have distinctive features of their internal anatomy and physiology, notably a large, lipid-filled liver that renders them neutrally buoyant (i.e., relatively stationary in the water, neither sinking nor floating upwards). In addition, chondrichthyans retain nitrogenous compounds that raise their internal osmotic concentrations, and ketones are the primary metabolic substrates of their skeletal and cardiac muscles.

6.2 Evolutionary Diversification of Chondrichthyes

The classification of extant chondrichthyans into Elasmobranchii and Holocephali reflects a division that dates back almost to the start of chondrichthyan history. Both the fossil record and molecular dating techniques indicate that this division had occurred by the start of the Devonian. Elasmobranchii contains the extant Neoselachii plus extinct relatives. Holocephali includes chimaeras and related extinct forms. When fossils are considered, Holocephali can be seen to be part of a larger group that also includes paraselachians (extinct sharklike forms, paraphyletic with respect to holocephalans). This larger grouping is also known as the Euchondrocephali.

Paleozoic chondrichthyan radiations

Chondrichthyans probably exhibited their greatest diversity during the Paleozoic, both in terms of number of species and ecomorphological types. Most species were less than 50 cm long, and body forms included almost every shape known among extant fishes, from dorsoventrally flattened through fusiform to ribbonlike.

Paleozoic stem chondrichthyans Fragmentary remains of chondrichthyans are known from the Middle Ordovician, but definitive forms are not known until the Devonian. The Ordovician and Silurian remains consist of scales that share derived features with scales of crown chondrichthyans. The stem chondrichthyan iniopterygians are known from the Carboniferous. These were smallish (20–60 cm) and rather chimaera-like in form, with large, dorsally positioned pectoral fins. Iniopterygians were originally thought to be related to holocephalans, but various features (including the primitive nature of the braincase) now place them outside the crown chondrichthyans.

Paleozoic elasmobranchs The first elasmobranchs appeared in the Early Devonian. Many were generally shark-like in form, although they lacked several derived characters of modern sharks. A conspicuous difference between the stem elasmobranchs and modern sharks is the position of the mouth; stem elasmobranchs had a terminal mouth located at the front of the head, rather than a subterminal mouth beneath a protruding snout (or rostrum) as in modern sharks. They also lacked the independently mobile upper jaw of extant sharks and rays.

Cladoselache is a representative example of a stem elasmobranch (**Figure 6.4A**). It was about 2 m long when fully grown, with large fins, a large mouth, and five separate external gill openings on each side of the head. Like many other stem elasmobranchs, *Cladoselache* was probably a predator that swam in a sinuous manner, engulfing its prey whole or slashing them with daggerlike teeth.

Xenacanthans were a lineage of mainly freshwater bottom-dwellers. These stem elasmobranchs had robust fins and heavily calcified skeletons that may have decreased their buoyancy. Some had elongated, eel-like bodies (**Figure 6.4B**). Xenacanthans appeared in the Devonian and survived until the end of the Triassic (possibly surviving into the Jurassic), at which time they died out without leaving direct descendants.

Stethacanthid stem elasmobranchs from the Early Carboniferous showed remarkable sexual dimorphism. The five species of stethacanthids that have been described ranged in length from about 30 cm to nearly 3 m. The first dorsal fin and spine of males in this family were elaborated into a structure that was probably used in courtship. In two genera of stethacanthids, *Orestiacanthus* and *Stethacanthus*, this structure (the spine-brush complex) projected more or less upward and ended in a blunt surface that was covered with spines (**Figure 6.4C**). Males of two other genera of stethacanthids, *Damocles* and *Falcatus*, had a swordlike nuchal spine that projected forward parallel to the top of the head (**Figure 6.4D**). One of the fossils of *Falcatus* may be a pair that died in a precopulatory courtship position, with the female grasping the male's dorsal spine in her jaws. These spines are odd-looking structures, but they are not unique; males of some extant chimaeras have a modification of the nuchal spine called a cephalic clasper that is used during courtship.

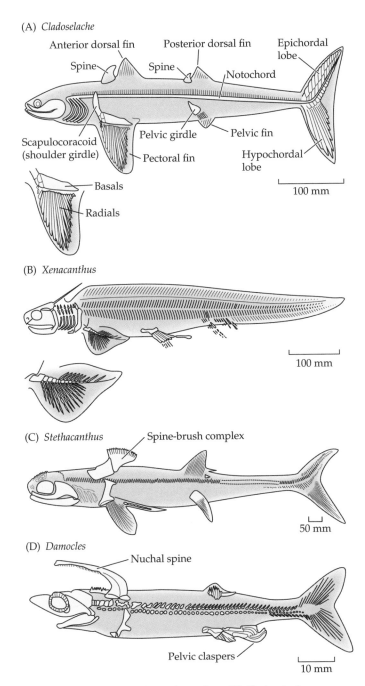

Figure 6.4 Paleozoic elasmobranchs.   (A) *Cladoselache*.
(B) *Xenacanthus*. (C) A male *Stethacanthus*, showing the first dorsal
fin and spine elaborated into a brushlike structure atop the head.
(D) Male *Damocles serratus*, a 15-cm-long, sharklike form from
the Late Carboniferous, showing the sexually dimorphic nuchal
spine and pelvic claspers (females had neither structure). (A,B after
Moy-Thomas and Miles 1971; C after Benton 2005, adapted from
Zanerl 2004; D after Lund 1986.)

Paleozoic holocephali   Both holocephalans and paraselachians first appeared in the Middle Devonian, although the
extant holocephalan order Chimaeriformes did not appear
until the Late Carboniferous. The paraselachians (six orders)
and the other orders of holocephalans were mainly Paleozoic,
although a few forms to survived into the Triassic.

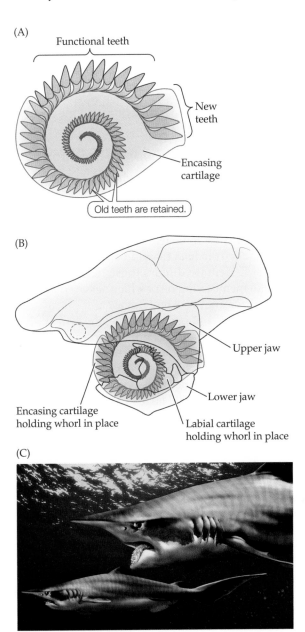

Figure 6.5 Tooth whorl of *Helicoprion*.   (A) Lateral view
of the symphysial (middle of the lower jaw) tooth whorl of
the paraselachian *Helicoprion*, showing the chamber into
which the lifelong production of teeth spiraled. (B) The
tooth whorl was attached to the lower jaw by encasing cartilage. (C) Reconstruction of *Helicoprion*. (A,B after Ramsay
et al. 2015; C © Jamie Chirinos/Science Source.)

As their name suggests (Greek *para*, "beside" or "near"),
paraselachians looked like sharks; they had sharklike teeth,
and their upper jaw was not fused to the cranium as in
extant chimaeras. One of these forms was *Helicoprion*, a Triassic survivor, known for its bizarre, spiraling tooth whorl
(**Figure 6.5**). There have been many weird and wonderful
reconstructions of what *Helicoprion* would have looked like
in life, but a detailed biomechanical analysis shows that this
tooth whorl was contained entirely within the mouth. *Helicoprion* probably fed on soft-bodied prey, using the tooth
whorl rather like a circular saw, serially trapping, piercing,

and cutting its prey as they advanced into its mouth.

Another Paleozoic group was the menaspoids, heavily armored fishes with plates of dermal bone on their heads, almost like placoderms, and numerous paired spines projecting from the side of their mouths.

The Mesozoic chondrichthyan radiation

Members of the modern elasmobranch radiation, Neoselachii, first appeared in the Late Triassic (although molecular data would place the split between sharks and rays as Permian). Chimaeras were present as holdovers from the Paleozoic. These modern radiations will be considered in Chapter 7. Here we discuss those radiations of the stem members (none of which survived into the Cenozoic) of the extant lineages.

Elasmobranchs Xenacanths, freshwater eel-like stem elasmobranchs, survived until the end of the Triassic (possibly into the Jurassic), but the Mesozoic elasmobranch radiations consisted primarily of more derived forms, with changes in their feeding and locomotor systems. These were the hybodont elasmobranches, which form the stem lineage to the modern sharks and rays. Hybodont "sharks" and extant elasmobranchs are grouped together in Euselachii.

Hybodonts (Hybodontiformes) first appeared in the Late Devonian and flourished until the end of the Cretaceous in both marine and freshwater environments. *Hybodus* is a well-known genus of the Late Triassic through the Cretaceous (**Figure 6.6**). Complete, 2-m-long skeletons have been found, although many hybodonts were much smaller than this.

Hybodonts were distinguished by a heterodont dentition (different tooth shapes in different regions of the jaw). The anterior teeth had sharp cusps and may have been used for piercing, holding, and slashing softer foods. The posterior teeth were stout, blunt versions of the anterior teeth and may have crushed hard-bodied prey, such as crabs.

The hybodonts also showed advances in the structure of the pectoral and pelvic fins that made them more mobile than the broad-based fins of earlier elasmobranchs. Both pairs of fins were supported on narrow stalks formed by three narrow, platelike basal cartilages that replaced the long series of basals seen in earlier elasmobranch. The narrow base allowed the fin to be rotated to different angles as the hybodont swam up or down, allowing greater control of locomotion. This narrow-based fin had shortened radials, with a large extent of the fin now composed of flexible fin rays (**ceratotrichia**).

Along with changes in the paired fins, the caudal fin assumed new functions, and an anal fin was now present. Reduction of the lower (hypochordal) lobe altered the caudal fin shape so that the upper lobe was distinctively larger than the lower lobe. This tailfin arrangement is known as

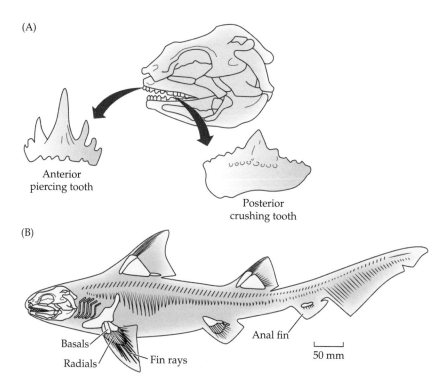

Figure 6.6 The Late Cretaceous elasmobranch *Hybodus*. (A) Dentition consisted of pointed teeth at the front of the mouth and blunt teeth in the rear. (B) Hybodonts looked very much like modern sharks except that the mouth was terminal. (After Moy-Thomas and Miles 1971.)

epicercal or heterocercal (Greek *heteros*, "different"; *kerkos*, "tail"). The value of the euselachian heterocercal tail lies in its flexibility and the control of shape made possible by the intrinsic musculature. When it is undulated from side to side, the fin twists so that the flexible lower lobe trails behind the stiff upper one. This distribution of force produces forward and upward thrust that can lift a fish from a resting position or counteract its tendency to sink as it swims horizontally.

Other morphological changes in euselachians included the appearance of a complete set of haemal arches (arches below the centrum) in the caudal (tail portion) vertebral column that protected the arteries and veins running below the notochord; well-developed ribs; and narrow, more pointed dorsal-fin spines at the leading edge of the dorsal fins. These spines were ornamented with ridges and grooves and studded with barbs on the posterior surface, suggesting that they were used in defense.

Despite the variety and success of hybodonts during the Mesozoic, they became increasingly rare and disappeared at the end of the Mesozoic.

Holocephali A few lineages of paraselachians, such as *Helicoprion*, survived into the Early Triassic. Most of the lineages of holocephalans were extinct by the end of the Paleozoic. The survivors were mainly members of Chimaeriformes, with the modern families diverging in the Late Jurassic to Early Cretaceous. Another holocephalan lineage that persisted into the Mesozoic was the myriacanthids, some of which had small armored plates on their heads (modern chimaeras have none). Myriacanthids persisted only until the end of the Jurassic.

Paleozoic and Mesozoic chondrichthyan paleobiology

Devonian chondrichthyans shared their environment with a diversity of other fishes, including early bony fishes, plus various more ancient lineages of fishes: ostracoderms, placoderms, and acanthodians. During this time chondrichthyans were probably the most numerous and diverse types of fishes, although they were not yet playing the role of large marine predator that was the preserve of the arthrodire placoderms such as *Dunkleosteus*. The eel-like xenacanths may have been the apex predators in the freshwater realm, although they would have had competition from the bony fishes of the lineage that would give rise to tetrapods (tetrapodomorph fishes; see Chapter 7).

By the Carboniferous, placoderms were extinct, and tetrapodomorph fishes were rare. Chondrichthyans diversified into this empty ecomorph space in both marine and freshwater environments, becoming even more diverse and numerous than in the Devonian. A rich array of fishes is known from the late Early Carboniferous Bear Gulch Limestone in Montana, which represents a shallow tropical marine bay. Here 60% of the fishes present were chondrichthyans. About one-third of these were holocephalans, one-fourth were elasmobranchs (some euselachians, but mostly more basal forms such as stethacanthids and a few iniopterygians), and the rest were paraselachians. These fishes represented nine different ecological types, including midwater predators and bottom-dwelling mollusk crunchers.

A similarly diverse array of fishes is known from the middle Late Carboniferous, from the Minto Formation in New Brunswick, Canada. Here the deposits appear to represent a span of environments from marine through brackish to freshwater, favoring species that could tolerate a variety of water salinities (euryhaline species). Chondrichthyans, mostly xenacanth elasmobranchs, were found throughout the different salinity environments, making up 65% of the fauna. Also present were acanthodians, bony fishes, and tetrapods.

Most of these mid-Carboniferous chondrichthyan lineages continued into the later Carboniferous and Permian, with a few stragglers making it into the Early Triassic, but aquatic faunas became increasingly dominated by bony fishes in the late Paleozoic. Neoselachians started off as nearshore predators in the Mesozoic and expanded into the open oceans by the mid-Cretaceous. This shift was probably related to the radiation of neopterygian bony fishes, which would have provided predatory neoselachians with a prey base. The hybodont radiation, prominent throughout most of the Mesozoic, was brought to an end in the Late Cretaceous, probably by competition from neoselachians.

Summary

Chondrichthyans are not primitive fishes, but are highly derived in their own fashion.

The first definitive chondrichthyan fossils are from the Early Devonian, although chondrichthyan-like scales are known from the early Silurian, and possibly the Ordovician.

The two main branches of chondrichthyans reflect a division that had occurred by the start of the Devonian.

- Elasmobranchii includes the extant neoselachians (sharks and rays) and extinct relatives. Sharks have 5–7 pairs of gill openings on the side of the head, whereas rays have 5 pairs of gill openings on the ventral surface of the head. Both sharks and rays have a mobile upper jaw.
- Holocephali includes the extant chimaeras and related extinct forms. Chimaeras have a single gill opening covered by soft tissue and an immobile upper jaw fused with the skull. Paraselachians, a mainly Paleozoic paraphyletic grouping of sharklike forms, were included within Holocephali.

Chondrichthyans have distinctive features of their skeleton, scales, and teeth.

Chondrichthyans share two unique anatomical features: a cartilaginous skeleton that, if calcified, is mineralized in a unique fashion (differently from bone); and pelvic claspers in males, used for internal fertilization.

Chondrichthyans also have distinctive scales over their body—placoid scales, composed of enameloid, dentine, and traces of bone at their base. Their teeth are contained in a tooth whorl, in which new teeth are continually developing. Chimaeras do not have tooth whorls, and instead have grinding tooth plates that are not replaced.

Extant chondrichthyans have some distinctive features of their internal anatomy and physiology, including a large, lipid-filled liver and the retention of nitrogenous compounds in their tissues for osmotic regulation in the marine environment.

(Continued)

Summary *(continued)*

Chondrichthyans had an extensive and diverse fossil history.

Paleozoic Elasmobranchii were diverse in form and function and were found in both marine and freshwater environments. They looked rather like modern sharks, but their mouth was terminal rather than ventral, and their upper jaw was not highly mobile. The stethacanthid stem elasmobranches were sexually dimorphic, with males displaying elaborate courtship spines on their head.

Paleozoic Holocephali are first known from the Middle Devonian. The extant order Chimaeriformes appeared in the Late Carboniferous. The sharklike paraselachians were an almost entirely Paleozoic radiation.

Mesozoic Elasmobranchii mainly comprised the hybodont "sharks," which together with the Neoselachii form the Euselachii. These elasmobranchs had many skeletal modifications associated with advances in feeding and locomotion.

Mesozoic Holocephali mainly comprised the ancestors of the modern chimaeras. *Helicoprion*, a paraselachian that survived into the Early Triassic, had a complicated tooth whorl that it probably used like a circular saw to consume its prey.

Paleozoic chondrichthyans were highly diverse and successful.

Chondrichthyans made up the majority of Devonian fish faunas. The continuation of placoderms probably inhibited the evolution of large marine predatory chondrichthyans, but the eel-like xenacanths may have been the apex predators in freshwater environments.

Chondrichthyans were also highly diverse in the Carboniferous, but in the later Carboniferous and into the Permian their numbers declined, while the diversity of bony fishes increased.

Mesozoic chondrichthyans were moderately diverse and successful.

The prime radiation of neoselachians in the Cretaceous paralleled the increasing diversity of neopterygian bony fishes and may have brought about the extinction of hybodonts.

Discussion Questions

1. *Chondrichthyes* means "cartilaginous fishes." Biologists used to assume that bone did not evolve until the Osteichthyes, the bony fishes. What clues do we have that shark ancestors, at some level, had bone? How can we infer this from (a) evidence from the fossil record and (b) evidence from comparative anatomy?

2. Why do we call Paleozoic chondrichthyans such as *Cladoselache* "sharks," and why isn't that terminology strictly correct?

Additional Reading

Friedman M, Sallan LC. 2012. Five hundred million years of extinction and recovery: A Phanerozoic survey of large-scale diversity patterns in fishes. *Paleontology* 55: 707–742.

Gogáin A and 9 others. 2016. Fish and tetrapod communities across a marine to brackish salinity gradient in the Pennsylvanian (early Moscovian) Minto formation of New Brunswick, Canada, and their palaeoecological and palaeogeographical implications. *Paleontology* 59: 689–724.

Heinicke MP, Naylor GJP, Hedges SB. 2009. Cartilaginous fishes (Chondrichthyes). *In* Hedges SB, Kumar S (eds.) *The Timetree of Life*, pp. 320–327. Oxford University Press, Oxford.

Lund R, Greenfest-Allen E, Grogan ED. 2012. Habitat and diversity of the Bear Gulch fish: Life in a 318-million-year-old marine Mississippian bay. *Palaeogeography, Palaeoclimatology, Palaeoecology* 342–343: 1–16.

Maisey JG. 2012. What is an elasmobranch? The impact of palaeontology in understanding elasmobranch phylogeny and evolution. *Journal of Fish Biology* 80: 918–951.

Ramsay JB and 6 others. 2015. Eating with a saw for a jaw: Functional morphology of the jaws and tooth-whorl in *Helicoprion davisii*. *Journal of Morphology* 276: 47–64.

Smith MP, Donoghue PCJ, Sansom IJ. 2002. The spatial and temporal diversification of Early Palaeozoic vertebrates. *Geological Society, London, Special Publications* 194: 69–83.

Tapanila L and 6 others. 2013. Jaws for a spiral-tooth whorl: CT images reveal novel adaptation and phylogeny in fossil *Helicoprion*. *Biology Letters* 9: 20130057.

CHAPTER
7

Extant Chondrichthyans

© iStock.com/Matt Potenski

As you saw in Chapter 6, Chondrichthyes contains two main lineages: Elasmobranchii and Holocephali. The first representatives of the extant radiation of elasmobranchs, Neoselachii, appeared at least as early as the Triassic. By the Jurassic, species of modern appearance had evolved. Most of the extant families date back to the Cretaceous, and a surprising number of Jurassic and Cretaceous genera are still extant. The characters distinguishing neoselachian clades are subtle, and the interrelationships among the extant lineages are not yet fully understood. Molecular studies unite the sharks (Selachii) and the skates and rays (Batoidea) as Elasmobranchii (**Figure 7.1**). Chimaeras (Holocephali) are the extant sister group of Elasmobranchii. The first chimaeras are known from the Late Carboniferous, although close relatives of extant forms were not apparent until the Cretaceous.

Although most species of chondrichthyans are marine, more than 50 species of elasmobranchs occur in estuarine or freshwater habitats. Estuarine species include sawfishes, several species of sharks, one species of skate, and several species of stingrays. Other species of stingrays are found only in fresh water.

7.1 Morphology of Extant Chondrichthyans

Chondrichthyans are often portrayed as primitive, but the extant sharks, skates, rays, and chimaeras are highly derived fishes that differ from the Paleozoic forms described in the previous chapter and from the more familiar bony fishes.

Skeleton

Although chondrichthyans are called "cartilaginous fishes," their skeletons contain substantial mineralized tissue. Although the mineral—hydroxyapatite—is the same as in bone, the way this mineral is crystalized is unlike the internal structure of bone and is unique to the chondrichthyans (see Chapter 6). The vertebral centra of extant sharks are densely calcified. Between centra, spherical remnants of the notochord fit into depressions on the faces of adjacent vertebrae. Because the notochord is flexible, the axial skeleton can flex from side to side between the rigid centra.

Jaws

The primitive chondrichthyan condition (probably the primitive gnathostome condition), as seen in the primitive chondrichthyan *Debeerius ellefseni* (Holocephali), was an **autodiastylic** jaw articulation (Greek *auto*, "self"; *dia*, "across"; *stylos*, "pillar"). Projections from the upper jaw provided a firm attachment to the cranium and prevented the upper jaw from being moved in relation to the cranium (**Figure 7.2**). The hyoid arch was unmodified and was not involved in jaw suspension. These connections were loosened in the jaw articulation of ancestral elasmobranchs, which was **amphistylic** (Greek *amphis*, "double"): the anterior end of the upper jaw was attached directly to the cranium, and the posterior end was supported by the hyomandibula (the modified upper part of the hyoid arch). Further changes in the connections between the upper jaw and the cranium have occurred in extant chondrichthyans.

Most extant elasmobranchs have a derived type of **hyostylic** jaw suspension (Greek *hyo*, from *hyoeides*, "shaped like the letter upsilon [Y]," referring to the shape of the hyoid arch). The upper jaw articulates with the cranium via the hyomandibula, and the other connections of the upper jaw to the cranium are flexible ligaments. This articulation allows the upper jaw of sharks to be protracted and retracted during feeding (see Figure 7.2C). Batoids have developed an extreme hyostylic jaw suspension that allows the upper jaw to drop below the fish's ventral surface (see Figure 7.2D). Chimaeras have moved in the opposite direction, developing a **holostylic** jaw articulation (Greek *holos*, "whole"); their upper jaw is fused to the cranium, and the lower jaw pivots like a nutcracker to crush prey (see Figure 7.2E).

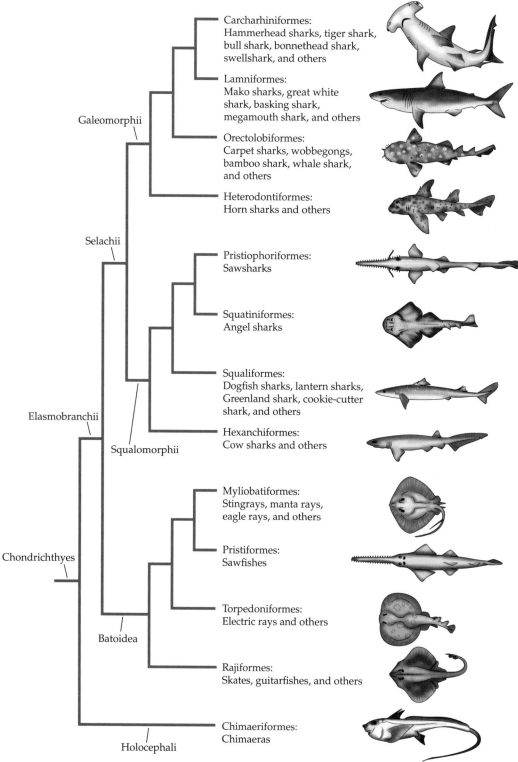

Figure 7.1 Phylogenetic relationships of extant chondrichthyans. Species described in the text are identified. (Tree adapted from Vélez-Zuazo and Agnarsson 2011; Aschliman et al. 2012.)

Skin

The placoid scales of early elasmobranchs were often found in clusters, and individual scales often fused to form larger scales as an animal grew larger. Extant species have small placoid scales with a single cusp and single pulp cavity, much like their teeth, and add additional scales as they grow. These scales reduce turbulence in the flow of water next to the body surface and increase the efficiency of swimming.

Extant sharks have a strong, helically wound layer of collagen fibers firmly attached to thick septa of collagen that run between muscle segments to attach on the vertebral column. This elastic jacket with anchors to the axial

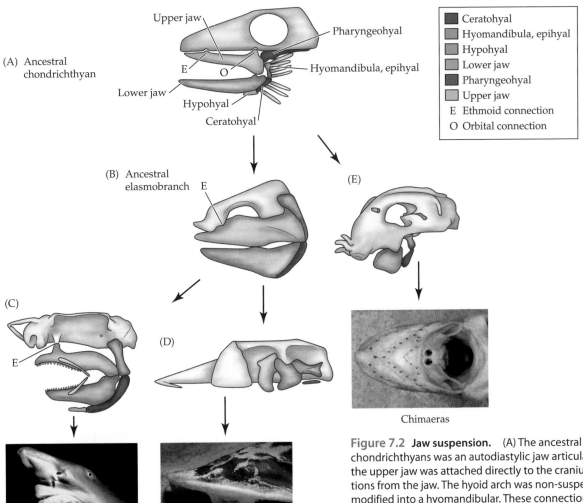

(A) Ancestral chondrichthyan

Upper jaw

Pharyngeohyal

Hyomandibula, epihyal

Lower jaw

Hypohyal

Ceratohyal

E

O

■	Ceratohyal
■	Hyomandibula, epihyal
■	Hypohyal
■	Lower jaw
■	Pharyngeohyal
■	Upper jaw
E	Ethmoid connection
O	Orbital connection

(B) Ancestral elasmobranch   E

(E)

(C)

E

(D)

Chimaeras

Sharks

Skates and rays

Figure 7.2 Jaw suspension.   (A) The ancestral condition for chondrichthyans was an autodiastylic jaw articulation, meaning the upper jaw was attached directly to the cranium by two projections from the jaw. The hyoid arch was non-suspensory and not modified into a hyomandibular. These connections prevented the upper jaw from being protracted or retracted during feeding. (B) Ancestral sharks had an amphistylic jaw articulation, in which the upper jaw was attached to the cranium anteriorly and articulated posteriorly with the hyomandibula (the modified upper part of the hyoid arch). (C) Extant sharks have a hyostylic jaw articulation in which the upper jaw is hinged to the skull posteriorly by the hyomandibula, and its anterior attachment is a flexible ligament. The partially open mouth of a sand tiger shark (*Carcharias taurus*) shows the independence of the jaws from the chondrocranium. (D) The hyostylic jaw articulation of skates and rays is still more flexible. The jaws of these bottom-dwelling chondrichthyans extend downward to capture prey on the seafloor, as seen in the photo. (E) Chimaeras have a holostylic articulation in which the upper jaw is fused with the cranium and the lower jaw pivots upward as the mouth closes. Their blunt teeth are used to crush hard-shelled prey such as crustaceans and mollusks. (Art: A, B, D, E, from Wilga 2002; C, from Goodrich 1909. Photos: C, © Jeff Rotman/Naturepl.com; D, Hubert Yann/Alamy Stock Photo; E, © Edward Farrell.)

skeleton appears to store and release energy during the side-to-side oscillations of swimming. Muscles pulling on the skin probably contribute more to swimming than do muscles pulling on the axial skeleton.

Bioluminescence and biofluorescence   The skin of some species of sharks and rays emits light (bioluminescence) or absorbs light of one wavelength and re-emits it at a different wavelength (biofluorescence). Two families of small, deepwater squaliform sharks have specialized luminous cells (**photophores**). Light production is believed to be controlled by hormones. Photophores located on the ventral surface may camouflage fish by counterillumination—that is, by emitting light matching the intensity of down-welling light from the water surface, thereby concealing the silhouette of the fish from predators looking upward. Photophores on the sides of the body or on a dorsal spine may be used for intraspecific recognition.

Biofluorescence has evolved at least three times in chondrichthyans: once in rays and twice in squaliform sharks.

Blue is the predominant color in the undersea environment, because blue light penetrates deeper than other wavelengths. Biofluorescent sharks and rays absorb blue light and transform it to green light, revealing dramatic patterns (**Figure 7.3**). These patterns, which are both species-specific and sexually dimorphic, probably play a role in intraspecific communication.

Figure 7.3 Biofluorescence in a shark. Below certain depths, the skin of the swellshark (*Cephaloscyllium ventriosum*) absorbs the penetrating blue light and re-emits it as fluorescent green light (above). This vivid pattern is not visible in the broader spectrum light that penetrates shallower depths (below) and may be involved in intraspecific communication, including courtship and mating. (Photos © David F. Gruber.)

7.2 Sharks (Selachii)

Throughout their evolutionary history, sharks have been consummate carnivores. In the third radiation of the elasmobranchs in the mid-Mesozoic, derived locomotor, trophic, sensory, and behavioral characteristics produced predatory sharks that dominate the top levels of marine food webs today. Sharks show enormous diversity in size (**Figure 7.4**). The largest extant forms attain lengths of at least 12–15 m and may grow to 17 m, whereas a few miniature forms that inhabit deep seas off the continental shelves are less than 25 cm long.

- Galeomorphii: Galeomorph sharks include many of the sharks often featured on television—the large carnivores such as great white shark (*Carcharodon carcharias*) and the hammerhead sharks (*Sphyrna*). The enormous whale shark (*Rhincodon typus*), which grows to a length of 17 m, and the slightly smaller basking shark (*Cetorhinus maximus*; up to 15 m m) are also galeomorphs. The largest sharks—the whale shark, basking shark, and megamouth shark (*Megachasma pelagios*; up to 7 m)—are filter feeders. Whale and basking sharks feed at or near the surface, whereas the megamouth shark forages at depths down to 600 m. Not all galeomorphs are large, however; mako sharks (*Isurus*) are 3–4 m long and horn sharks (*Heterodontus*) reach lengths of only 1–1.5 m.

- Squalomorphii: Squalomorph sharks are generally smaller than galeomorphs. Among the largest species are cow sharks (*Hexacanthus*; up to 5 m) and angel sharks (*Squatina*; 1.5–2 m), and the spiny dogfish (*Squalus acanthias*; 1.5 m). This lineage includes species that live in cold, deep water, such as the green lantern-shark (*Etmopterus virens*; 23 cm) and the cookie-cutter shark (*Isistius brasiliensis*; 0.5 m). Even these deep-water sharks are limited to depths of 3,000 m or less,

however; there are no abyssal chondrichthyans to rival the bony fishes that can live at depths exceeding 8,000 m.

The diverse environments inhabited by sharks are reflected in the variety of their body forms (**Figure 7.5**).

Sensory systems and prey detection

The sensory systems of extant sharks are refined and diverse, and include specialized characteristics of the visual and chemosensory systems as well as exquisitely sensitive mechano- and electroreceptive systems.

Chemoreception Sharks have been described as "swimming noses," so acute is their sense of smell. Experiments have shown that some sharks respond to chemicals in concentrations as low as 1 part per 10 billion, turning in the direction of the nostril that first receives an odor stimulus to locate its source. Not all sharks are swimming noses, however. The size of the olfactory bulbs in the brain probably corresponds to chemoreceptive sensitivity. Coastal and pelagic (open-ocean) species, especially species that engage in migrations, have large olfactory bulbs. Species that live at depths where light is dim also have large olfactory bulbs, but most sharks that occur on coral reefs have small olfactory bulbs.

Mechanoreception The lateral line system detects the turbulent wake of a swimming fish and vibrations such as those produced by a struggling fish.

Vision Many sharks feed at dusk or at depths in the sea where little sunlight penetrates, and their vision at low light intensities is especially well developed. This sensitivity is due to a rod-rich retina and cells with numerous platelike crystals of guanine that are located just behind the retina in the eye's choroid layer. Collectively called the **tapetum lucidum**, the shiny crystals of guanine in these cells act like mirrors to reflect light back through the retina and increase the chance that photons will be absorbed. (A tapetum lucidum is found in many nocturnal animals and accounts for the eyeshine of animals seen in the headlights of a car.)

Electroreception In Chapter 4 we described the extraordinary sensitivity of the neuromast organs and ampullae of Lorenzini. The ampullae of Lorenzini, which are composed of soft tissues, do not normally fossilize, but their concentration on the rostrum of extant sharks and the near-universal appearance of a projecting rostrum as a derived character in the extant radiation of sharks make a strong case for the role of electrosensitivity in the success of the extant forms. The strange heads of hammerhead sharks of the genus *Sphyrna* may spread the ampullae of Lorenzini over a larger area than does the standard shark rostrum. This arrangement might increase the sensitivity to electrical impulses from buried prey and to minute geomagnetic gradients used for navigation.

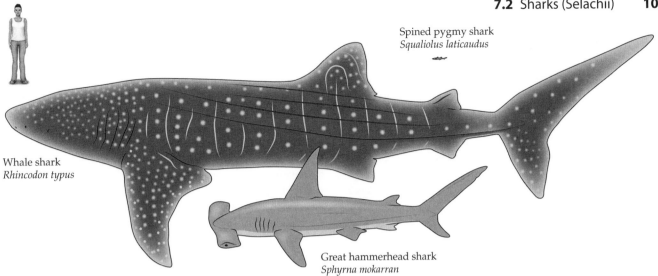

Spined pygmy shark
Squaliolus laticaudus

Whale shark
Rhincodon typus

Great hammerhead shark
Sphyrna mokarran

Figure 7.4 Size range of sharks.   Sharks range in size from tiny pygmy and dwarf sharks less than 25 cm to the huge (but filter-feeding) whale shark. Three species are shown here at the same scale, along with a human for comparison. (From Townsend 2012.)

(A)

(B)

(C)

(D)

Figure 7.5 Representative extant sharks.   (A,B) Galeomorph sharks. (A) The great white shark (*Carcharodon carcharias*) reaches 5–6 m in length. Compare the rostrum that extends forward of the mouth, as in most extant sharks, to the terminal position of the mouth in *Hybodus* in Figure 6.6. (B) The bonnethead shark (*Sphyrna tiburo*; ~4 m) is a species of hammerhead shark. The peculiar shape of the head may increase the directional sensitivity of the mechanoreceptors in the lateral lines and of the electroreceptors in the ampullae of Lorenzini. Bonnetheads are the only sharks known to be omnivorous; they consume and digest seagrass. (C,D) Squalomorph sharks. (C) The rostrum of the longnose sawshark (*Pristiophorus cirratus*; ~1.25 m) makes up more than one-fourth of the shark's total length. The sides of the rostrum are studded with teeth, and a pair of barbels (presumed to be sensory) are located near its midpoint. (D) Angel sharks are raylike in morphology and ecology. The Japanese angel shark (*Squatina japonica*; ~1.5 m) lies buried during the day with only its eyes protruding, and swims in the water column at night. (A, iStock.com/cdascher; B, Felix Manuel Cobos Sánchez/CC BY-SA 2.0; C, Stephen Frink Collection/Alamy Stock Photo; D, Ryo Sato/CC BY-SA 2.0.)

Integration of sensory information for predation

Sharks use their sensory modalities to locate, identify, and attack prey (**Figure 7.6**). The sequence in which the senses are employed varies with sea conditions. Olfaction is often the first sense that alerts a shark to potential prey, especially when the prey is wounded or otherwise releasing body fluids. Because of its exquisite sensitivity, a shark can use smell as a long-distance sense. Next, mechanoreception is added to olfaction as the shark tracks the wake of its prey.

Once a shark is close to the stimulus, vision takes over as the primary mode of prey detection. If the prey is easily recognized, a shark may proceed directly to an attack. Unfamiliar prey is treated differently, as studies aimed at developing shark deterrents have discovered. A circling shark may suddenly turn and rush toward unknown prey. Instead of opening its jaws to attack, however, the shark bumps and scrapes the surface of the object with its rostrum. Opinions differ on whether this is an attempt to determine texture through mechanoreception, make a quick electrosensory appraisal, or use the rough placoid scales to abrade the surface and release fresh olfactory cues. In the last moments before contact, some sharks draw an opaque eyelid (the nictitating membrane) across each eye to protect it. At this point, it appears that sharks shift entirely to electroreception to track prey.

Ecology of sharks

Large predatory sharks, such as the great white shark, tiger shark (*Galeocerdo cuvier*), and great hammerhead (*Sphyrna mokarran*), are **apex predators**, meaning that they are at the top of their food web. They eat species smaller than themselves, influencing the population dynamics of species at lower trophic levels. These apex predators are pelagic, moving hundreds or thousands of kilometers across oceans. In contrast, most coastal and reef-dwelling sharks are mesopredators, a sharing a feeding niche with large bony fishes, such as groupers. Some bottom-dwelling sharks, such as the horn shark (*Heterodontus francisci*) and bonnethead shark (*Sphyrna tiburo*), are durophagous (Latin *durus*, "hard"; Greek *phago*, "eat"), feeding primarily on hard-shelled prey such as crabs and sea urchins, whereas carpet sharks are bottom-dwelling ambush predators that capture octopuses and bony fishes.

Although many species of sharks appear to spend most of their lives in a restricted area, others move hundreds of kilometers along a coast, and some pelagic species travel enormous distances. Great white sharks travel thousands

(A)

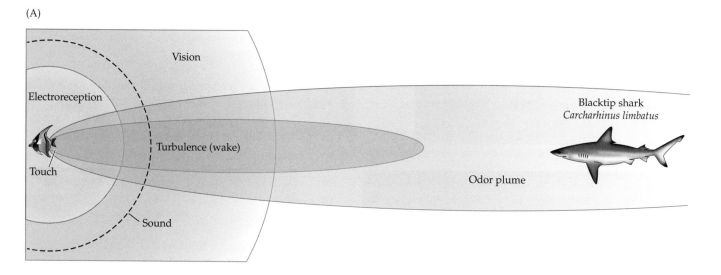

(B)

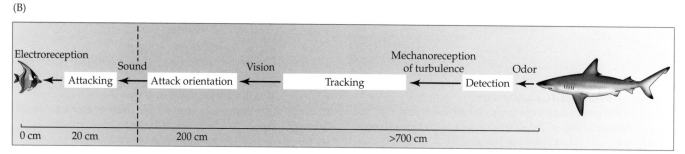

Figure 7.6 Integration of sensory information by sharks.
(A) A potential prey generates a variety of signals that can be detected by a shark downstream of it. (B) During the day, a blacktip shark (*Carcharhinus limbatus*) detects the presence of prey using olfaction and tracks its odor plume and wake. When it sees the prey, it switches to vision to orient and attack. At close range, the shark's snout blocks its view of the prey, and the attack is adjusted using turbulence detected by the lateral line system. Finally, the shark switches to electroreception and touch to capture the prey. (From Gardiner et al. 2014.)

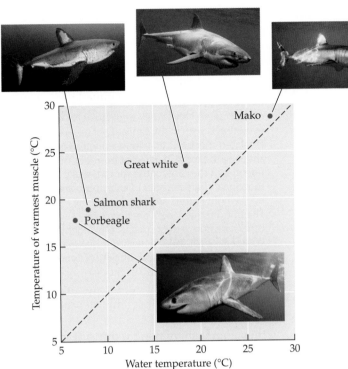

Figure 7.7 Muscle temperatures of lamnid sharks are warmer than water temperatures. The muscles of large lamnid sharks are as much as 10°C warmer than the water in which they are swimming. Species that occur in polar waters, such as the porbeagle (*Lamna nasus*) and salmon shark (*Lamna ditropis*), can maintain greater temperature differentials than species that occur in warmer water, such as the great white (*Carcharodon carcharias*) and mako sharks (*Isurus oxyrinchus* and *I. paucus*). (Graph from Carey et al. 1985. Photos: porbeagle and salmon shark, Doug Perrine/Alamy Stock Photo; great white, © iStock.com/Whitepointer; mako, © Alessandro De Maddalena/Shutterstock.)

of kilometers across ocean basins, returning annually or biennially to favored feeding and breeding areas. Bull sharks move from the ocean to fresh water and back and have been found in the Mississippi River as far north as Illinois.

Heterothermy

Regional heterothermy—having different temperatures in different parts of the body—is characteristic of the mackerel sharks (Lamnidae). The five species of lamnids include the shortfin mako shark (*Isurus oxyrinchus*) and great white shark (**Figure 7.7**). All are large marine species, reaching lengths of 6 m or more, with pointed snouts and cylindrical bodies. They are fast-swimming predators with a worldwide distribution, preying on bony fishes, other sharks, sea turtles, marine mammals, and (occasionally) humans.

Countercurrent heat exchange in an extensive system of retia captures heat produced by muscular contractions as a lamnid shark swims, raising its muscle temperatures above water temperature, a physiological process known as endothermy (Greek *endon*, "within"; *thermos*, "heat"; see Chapter 9). A second retia network carries cold blood from the gills through capillaries that traverse a venous sinus filled with blood warmed by the swimming muscles, keeping the eye temperature above water temperatures. As a result, the shark's eyes and the muscles it uses for swimming are warmer than other parts of its body, a condition called endothermal heterothermy. Retaining heat in the muscles may increase power output, and hence swimming speed allowing lamnid sharks to penetrate into deep, cold waters and to range far north and south into cold seas. Keeping the eyes warm probably increases their ability to detect prey in dim light.

Feeding

Chondrichthyans employ three basic mechanisms of prey capture, often in combination. Biting consists of opening the mouth and closing it on large prey, usually tearing out a chunk of flesh and leaving the horrendous wounds featured in sensational news reports of shark bites. Engulfing prey with an open mouth ("ramming") and using suction to draw prey into the mouth are two feeding methods that are effective with smaller prey items.

Extant sharks have a wide variety of tooth shapes, ranging from sharp, narrow points to broad, flat surfaces, and many sharks have teeth of different shapes in different parts of their jaws. The horn shark gets its generic name, *Heterodontus* (Greek *heteros*, "different"; *odous*, "tooth"), from this characteristic; it has sharp teeth at the front of the mouth and broad, flat, molarlike teeth at the rear. The shortfin mako has narrow, pointed teeth at the anterior end of the jaw and broader, more bladelike teeth farther back in the tooth row, while the knifetooth dogfish shark (*Scymnodon ringens*) has pointed teeth in the upper jaw and broader, bladelike teeth in the lower jaw. The variety of tooth shapes is probably related to differences in the functions and durability of differently shaped teeth; very sharp teeth cut well but dull quickly, whereas less sharp teeth are not as good at cutting but dull more slowly.

The shape of a shark's teeth can sometimes tell us something about their function (tearing, cutting, or crushing) and, by extension, what kind of prey that shark eats. A biomechanical analysis of the functional characteristics of shark teeth by Lisa Whitenack and her colleagues found that the shapes of sharks' teeth show no clear relationship to their phylogeny; examples of parallel or convergent evolution are widespread (**Figure 7.8**). In addition, most of the teeth tested were able to carry out at least two of the three functions, and almost half of the teeth tested could do all three of them.

The way that shark teeth are fastened to the jaw appears to be important in allowing a single tooth type to perform multiple functions. Teeth can implant only in dermal bone, and sharks have lost dermal bone. Thus, sharks' teeth are not set rigidly in bony sockets or fused to the jaw. Instead, collagen holds them in place within

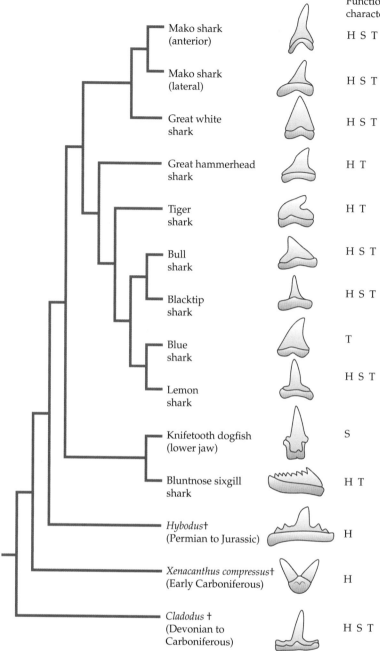

Functional characteristics		H	Penetrates hard-shelled prey
		S	Penetrates soft-bodied prey
		T	Tears flesh

Mako shark (anterior) — H S T
Mako shark (lateral) — H S T
Great white shark — H S T
Great hammerhead shark — H T
Tiger shark — H T
Bull shark — H S T
Blacktip shark — H S T
Blue shark — T
Lemon shark — H S T
Knifetooth dogfish (lower jaw) — S
Bluntnose sixgill shark — H T
Hybodus† (Permian to Jurassic) — H
Xenacanthus compressus† (Early Carboniferous) — H
Cladodus † (Devonian to Carboniferous) — H S T

Figure 7.8 Shapes and functional characteristics of the teeth of extant and fossil sharks. The shapes of shark teeth vary widely, and parallel or convergent evolution is seen in the similarity of teeth of distantly related species. Even within a single species, the shape of the teeth in different positions may be different, as is the case for the shortfin mako shark. The functional characteristics of each tooth are shown on the right (letters are explained in the key above). The bottom three teeth are from extinct species, the others are from extant species. (Phylogeny based on Whitenack et al. 2011; drawings from Whitenack and Motta 2010, © The Linnean Society of London.)

they approach, making it difficult for the prey to detect the shark.

Sharks kill mammalian prey such as seals and sea lions by exsanguination—that is, by bleeding them to death. A shark holds a seal in its jaws until it is dead, but sharks seize and release sea lions repeatedly until they die from blood loss, probably because sea lions, unlike seals, have powerful front flippers that can be used effectively in defense. These behavioral observations support a hypothesis that emerged from the biomechanical analysis of shark teeth: perhaps the main selective factor that acts on shark dentition is not efficiency, but rather the severity of damage that a bite inflicts.

Although sharks are formidable predators, they do not turn down an easy meal. Raine Island off the coast of Queensland, Australia is an important nesting area for green sea turtles (*Chelonia mydas*), and tiger sharks assemble near the island during the nesting season, scavenging dead turtles and preying on weakened individuals.

Reproduction

Much of the success of the extant neoselachians may be attributed to their sophisticated breeding mechanisms. Internal fertilization is universal, as was probably true for all chondrichthyans. Males have two pelvic claspers, one associated with each pelvic fin. The claspers have a solid skeletal structure that may increase their effectiveness. Each clasper is associated with a muscular subcutaneous sac extending anteriorly beneath the skin of the male's pelvic fins. These siphon sacs have a secretory lining and are filled with seawater by the pumping of the male's pelvic fins before copulation. The male uses one clasper during copulation, depending on how he is holding

the tooth whorl. This arrangement provides flexibility that allows limited movement and rotation of each tooth as force is applied during a bite. In addition, each tooth acts in concert with adjacent teeth because the bases of teeth overlap and forces exerted on one tooth may be transmitted to adjacent teeth.

Hunting behavior Rare observations of sharks feeding under natural conditions indicate that these fishes are versatile and effective predators. Great white sharks swim back and forth parallel to the shoreline of islands with seal and sea lion breeding sites, attempting to intercept the adults as they come and go from the rookeries. On sunny days, great white sharks place the sun behind them as

the female. During copulation (**Figure 7.9**), the clasper is bent at 90° to the long axis of the male's body, so the dorsal groove on the clasper lies directly under the cloacal papilla from which sperm are emitted. This flexed clasper is inserted into the female's cloaca and locked there by an assortment of barbs, hooks, and spines near the clasper's tip. Simultaneously, the siphon sacs contract and sperm from the genital tract are ejaculated into the clasper groove. Seminal fluid from the clasper's siphon sac washes sperm down the groove of the clasper into the female's cloaca, from which point the sperm swim up the female's reproductive tract.

Oviparity, viviparity, and fetal nutrition With the evolution of internal fertilization, sharks adopted a reproductive strategy favoring the production of a small number of offspring that are retained, protected, and nourished for varying periods within the female's body. A female can invest energy in nourishing the embryo in the form of an egg yolk (as in birds) or deliver energy to the developing embryo from the reproductive tract of the female.

Oviparity—the reproductive mode in which a mother deposits eggs that develop outside her body—is the ancestral mode of reproduction of chondrichthyans. However, **viviparity**—the reproductive mode in which a mother gives birth to a fully developed baby—is universal in some lineages. **Lecithotrophy** (Greek *lekithos*, "yolk of an egg"; *trophos,* "one who feeds") refers to the situation in which yolk supplies the nourishment for the embryo, whereas **matrotrophy** (Latin *mater*, "mother") means the reproductive tract of the mother supplies the energy. Lecithotrophy and matrotrophy are at the ends of a continuum, and many sharks (and other vertebrates) use a combination of lecithotrophy and matrotrophy, with the embryo receiving some nourishment from a yolk and some from the reproductive tract of the mother.

The embryo of an oviparous species that develops outside the body of its mother must be lecithotrophic. Most oviparous elasmobranchs (skates, rays, and chimaeras as well as sharks) produce large eggs with large yolks (the size of a chicken yolk or larger). A specialized structure at the anterior end of the oviduct, the nidimental gland, secretes a proteinaceous case around the fertilized egg (**Figure 7.10**). The embryo obtains nutrition exclusively from the yolk during development, which lasts from several weeks to 15 months. Movements of the embryo produce a flow of water through openings in the case that flushes out organic wastes and brings in dissolved oxygen. The young are miniature replicas of the adults when they hatch and seem to live much as they do when mature.

Prolonged retention of the fertilized eggs in the reproductive tract was a significant step in the evolution of shark reproduction. The eggs hatch within the oviducts, and the young may spend as long in their mother after hatching as they did within the shell, eventually emerging as miniatures of the adults. Viviparous sharks employ a mix of lecithotrophy and matrotrophy. Nonplacental viviparity

(A)

(B)

Figure 7.9 Copulation by the whitespotted bamboo shark (*Chiloscyllium plagiosum*). (A) The siphon sacs beneath the male's pelvic fins, visible here as the bulge at the ventral area of the male (arrow), have been inflated with seawater, and the male has grasped the female's right pectoral fin. (B) The right clasper of the male is bent at a 90° angle to his body and is inserted into the female's cloaca (arrow). The male's siphon sacs have nearly been emptied. (Photographs © Tony Wu/Naturepl.com.)

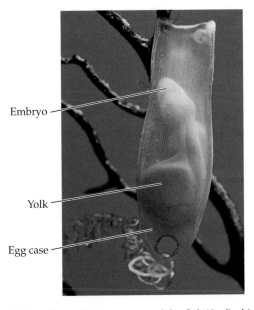

Figure 7.10 Embryo of a lesser spotted dogfish (*Scyliorhinus canicula*) developing in an egg case. Protuberances on the egg cases become tangled with vegetation or wedged into protected sites on the substrate. (Sander van der Wel/CC BY-SA 3.0.)

describes a range of situations—embryos may obtain all their energy from the yolk, or they may initially depend on the yolk and switch to maternal energy when the yolk has been exhausted. Maternal energy can be provided by absorption of uterine fluid, by cannibalizing siblings, or by eating unfertilized eggs that the mother continues to produce. Placental viviparity means the mother nourishes the embryos via a vascular placenta.

7.3 Skates and Rays (Batoidea)

Batoids, the skates and rays, are dorsoventrally compressed. This body form places the mouth and gill slits on the ventral surface of the head and the eyes and the spiracle on the dorsal surface. When an animal is resting on the bottom, water is drawn in through the spiracle and exhausted through the gill slits.

Batoids have enlarged pectoral fins that are attached to the side of the head and used for locomotion. The anterior-most basal elements fuse with the chondrocranium in front of the eye and with one another in front of the rest of the head. Skates and rays swim by undulating these massively enlarged pectoral fins. The placoid scales so characteristic of the integument of a shark are absent from large areas of the body and pectoral fins of skates and rays. Differences between skates and rays are not conspicuous:

- In general, skates have a circular body outline and rays are kite-shaped, although there are many exceptions to this generalization.

- Most skates have a thick tail that lacks spines; most rays have a whiplike tail, sometimes with spines that can inject venom.

- Skates and rays differ most in their reproductive modes: skates are oviparous, depositing eggs in stiff cases, whereas rays are viviparous, with varying degrees of matrotrophy.

Morphology

The body forms of batoids are more diverse than those of sharks (**Figure 7.11**). The characteristic suite of specializations in batoids relates to their bottom-dwelling, durophagous habit. The teeth are almost all flat-crowned plates that form a pavement-like dentition that crushes hard-bodied prey. The mouth is often extremely protrusible, providing powerful suction used to dislodge shelled invertebrates from the substrate. Skates and rays have undergone considerable diversification in the adaptations of their jaws and teeth for feeding, and many similar-appearing morphologies in extant forms have evolved independently.

Ecology

Many skates and rays are ambush hunters, resting on the seafloor and covering themselves with a thin layer of sand.

(A)

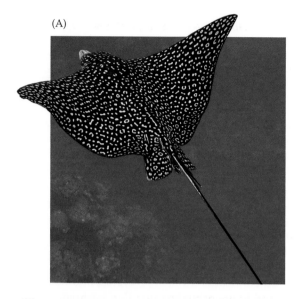

(B)

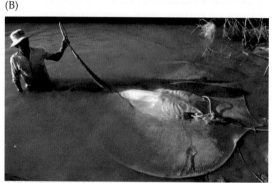

(C)

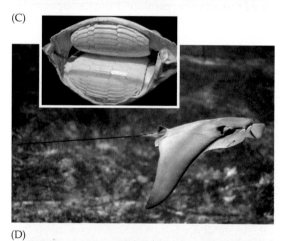

(D)

◀ Figure 7.11 **Representative extant skates and rays.** (A) Spotted eagle ray (*Aetobatus narinari*; ~3 m). Eagle rays are wide-ranging predators that feed primarily on mollusks and crustaceans. (B) The freshwater whipray (*Himantura* [*chaophraya*] *polylepis*), found in rivers and estuaries from the Mekong basin in China to northern Australia, reaches a length of 2.5 m and a weight of 600 kg. (C) The cownose stingray (*Rhinoptera bonasus*; ~1 m) has broad, flat teeth (inset) that crush the shells of mollusks and crustaceans. The tails of stingrays have one or more barbed spines that inject venom. When a ray whips its spine against a predator, the barb penetrates the cutaneous covering that contains the venom-producing tissue, and venom is carried into the wound by the barb. (D) Common sawfish (*Pristis pristis*; ~7.5 m). Sawfishes (Pristidae) have toothed rostra similar to those of sawsharks (Pristiophoridae) but lacking barbels (see Figure 7.5C). (A, © iStock.com/Matt Potenski; B, courtesy of Zeb Hogan; C, Citron/CC BY-SA 3.0; C, inset James St. John/CC BY 2.0; D, Simon Fraser University/CC BY 2.0.)

They spend hours partially buried and nearly invisible except for their prominent eyes, surveying their surroundings and lunging at a prey item that approaches too closely. During the day, electric rays are ambush feeders: they surge upward from the sand, enclose the prey in their pectoral fins, and emit an electrical discharge that stuns the prey. At night, rays hover in the water column 1–2 m above the bottom and drop on a fish that passes beneath them, cupping it in their pectoral fins while they stun it.

The largest rays, like the largest sharks, are plankton strainers. Manta rays are up to 6 m wide. These highly specialized rays swim through the open sea with winglike motions of the pectoral fins, filtering plankton from the water as they go.

Courtship and reproduction

Male skates and rays hold the female's pectoral fin during mating. The dentition of many benthic rays is sexually dimorphic; males have sharp teeth in the front of the jaw that they use to hold a female before and during copulation. Males of the Atlantic stingray (*Dasyatis sabina*) have blunt teeth like those of females for most of the year, but during the breeding season, males grow sharp-cusped teeth.

Skates are oviparous, laying eggs enclosed in horny shells popularly called "mermaid's purses." Fetal nutrition is entirely lecithotrophic, and embryonic development can extend over many months. Rays, in contrast, have nonplacental viviparity. In the early stages of development, the egg yolk nourishes the embryos, and when the yolk is exhausted the embryos consume a nutrient secretion produced by the walls of the oviduct.

7.4 Chimaeras (Holocephali)

The extant species of chimaeras, none much longer than 1 m in length, have a soft anatomy more similar to that of sharks and rays than to any other extant fishes (**Figure 7.12**). Chimaeras have a long body that may end in a whiplike tail, and they swim with lateral undulations of the body that throw the long tail into sinusoidal waves and with fluttering movements of the large, mobile pectoral fins.

(A)

(B)

(C)

Figure 7.12 **Representative extant chimaeras.** (A) Close-up of the head of a species of *Chimaera* shows how this group got its common name, rabbitfishes. (B) A longnose chimaera, (*Rhinochimaera*). Most longnose chimaeras are deepwater species, living at depths of 2,000 m or more. (C) The head of a plownose chimaera (*Callorhinchus milii*), showing the distinctive rostrum that gives this group its name. (A, NOAA/CC BY-SA 2.0; B, NOAA OKEANOS Explorer Program, 2013 Northeast U.S. Canyons Expedition/CC BY 2.0; C, © Doug Perrine/Naturepl.com.)ww

Shortnose chimera

Longnose chimera

Plownose chimera

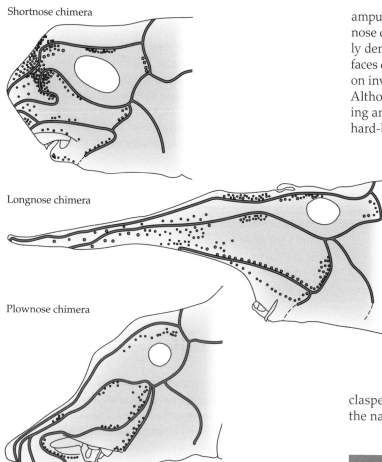

Figure 7.13 Specialized sensory receptors in chimaeras. The heads and rostra of chimaeras are richly supplied with mechano-receptive lateral line canals (shown here as brown lines) and electroreceptive ampullary organs (brown dots) that can detect even buried prey in the lightless depths that chimaeras inhabit. (From Didier 1995.)

Most species of chimaeras live at depths of 500 m or more, and many have geographic ranges that include entire ocean basins. Male and female chimaeras may spend most of the year in separate areas and come together only during an annual inshore spawning migration. Some chimaeras have the same life-history characteristics that make sharks vulnerable to overexploitation (see Section 7.5)—they take 10–12 years to reach maturity, reproduce once a year, and produce only a few young. All extant chimaeras are oviparous, but fossil evidence suggests that some extinct species were viviparous.

Chimaeras have a suite of unique features. They have four gill openings covered by a soft tissue flap, so that only one opening is visible to the outside; hence the name Holocephali ("complete head"). Chimaeras have no spiracle (the branchial skeleton is beneath the cranium rather than behind it, as in sharks), and their teeth have been modified into tooth plates.

Plownose and longnose chimaeras have rostra that are densely studded with lateral line mechanoreceptors and ampullary electroreceptors, as are the blunt snouts of short-nose chimaeras (**Figure 7.13**). These receptors are especially dense around the mouth, which is relatively small and faces downward. Chimaeras prefer soft substrates, feeding on invertebrates and small fishes that live on the seafloor. Although they consume soft-bodied organisms, including anemones and jellyfishes, their tooth plates can crush hard-bodied prey, such as crabs. Like other fishes that have crushing tooth plates, chimaeras have a derived type of jaw suspension (holostyly), where the upper jaw is completely fused to the cranium (see Figure 7.2E).

Some species of chimaeras have a poison gland that is associated with a mobile dorsal spine. The spine can be erected when the chimaera is attacked, and a predator that is stabbed in the mouth as it tries to swallow a chimaera might well decide to release the fish and seek less noxious prey. Males of some species of chimaeras have a spine-encrusted cephalic clasper that is used in courtship. Oddly, although the clasper is on the top of the head, it is controlled by muscles that also move the lower jaw. During copulation, the cephalic clasper is believed to fit into a corresponding hollow on the nape of the female's neck.

7.5 Declining Shark Populations: An Ecological Crisis

Sharks have been successful since their appearance 400–350 million years ago, but the life-history characteristics that have served them so well are not well suited for a world dominated by humans. Sharks grow slowly, mature late, and have only a few young at a time (**Figure 7.14**). The Greenland shark (*Somniosus microcephalus*) is an extreme example, living for at least 400 years (and perhaps longer) and requiring more than 150 years to reach sexual maturity. Females of many viviparous species reproduce only every other year because they must accumulate enough energy to nourish their embryos.

The worldwide decline in the number of sharks has been dramatic. The 2017.3 version of the IUCN Red List classified 28% of shark species as Critically Endangered, Endangered, or Vulnerable, and some groups are far worse off than that. About 70% of large predatory species of sharks fall into one of the three IUCN categories of risk, and some species have been particularly hard-hit; hammerhead shark populations have declined by anywhere from 76% to more than 99% in the Atlantic, Pacific, and Mediterranean ocean basins.

Fishing is the primary cause of most shark population declines. People eat sharks, although they often don't realize it. In England, spiny dogfish is served as fish and

	Greenland shark *Somniosus microcephalus*	Great white shark *Carcharodon carcharias*	Scalloped hammerhead *Syphyrna lewini*	Spiny dogfish *Squalus acanthius*
Age to maturity (years)	~150	♂ 9–10, ♀12–14	♂ 4–10, ♀4–15	♂ 6–14, ♀ 10–12
Size at maturity	~4 m	♂ 3.5–4.1 m, ♀ 4.0–4.3 m	♂ 1.4–2.8 m, ♀ 1.5–3.0 m	♂ 65 cm, ♀ 85 cm
Lifespan (years)	400	36	35	75
Reproductive mode	Viviparous, nonplacental (oophagy?)	Viviparous, nonplacental (oophagy)	Viviparous, placental	Viviparous, nonplacental (yolk sac)
Litter size	10 pups	10–14 pups	15–23 pups	2–14 pups
Reproductive frequency	Biennial?	Biennial?	(?)	Biennial
Gestation period (months)	(?)	>12	9–10	18–24

Figure 7.14 Life-history characteristics of sharks. The life-history characteristics of sharks (late maturation, small litters, and biennial reproduction) make them vulnerable to overexploitation. (Modified from Klimley 1999.)

chips, and shark steaks are marketed as an alternative to swordfish in many American supermarkets. Shark fins are harvested to make shark fin soup. Because the fins are the most valuable part of the shark, "finners" cut the fins from live sharks and throw the rest of the animal, often still alive, back into the sea. Sharks are also caught as bycatch in fisheries that are targeting other fishes, and most of these sharks die before they are heaved back into the sea.

The ecological effects of declining shark populations extend far beyond the sharks themselves. Apex predators often exert top-down control on populations of predators at lower trophic levels, stabilizing the structures of the communities in which they occur. As populations of apex-predator sharks have fallen over the past 30 years, populations of the elasmobranchs that the great sharks consume (skates, rays, and smaller sharks) have increased. Without top-down control, these mesopredators could increase to the point at which they decimate populations of their prey (**Figure 7.15**).

Developing management and conservation programs for marine species that move across national boundaries and spend large parts of their lives at sea where they are beyond any nation's jurisdiction is extremely difficult. International concern among shark biologists and fisheries managers, and in some cases the public at large, has led to proposals of policies to conserve sharks. Most policies fall into one of two general categories:

- Target-based policies focus on sustainable exploitation of individual species. This approach has shown promise for small, fast-growing species of sharks that occur in the waters of developed nations that have the fisheries management infrastructure necessary to plan and implement regulations. The Canadian spiny dogfish fishery is the first shark fishery to obtain marine stewardship council certification.

- Limit-based policies ban all exploitation of sharks. Fin bans, for example, prohibit sale or purchase of shark fins from any species of shark. Although that phrase is pleasing, the reality is less happy. Fin bans allow sharks to be caught and sold as long as the fins are not sold separately from the rest of the shark. Protected areas where all fishing is prohibited and shark sanctuaries where only sharks are protected are approaches to a limit-based policy. Although regulations are often poorly enforced, protected areas have been effective in rebuilding some populations of bony fishes.

The life-history characteristics of sharks work against them, however. Recruitment of new individuals to a population is slow, and even a century of protection may be insufficient to restore populations of the large species of sharks. There is one small but brighter note in the generally dark prospects for sharks. Viewing sharks and rays in the wild has become a tourist attraction in several parts of the world, and may give living sharks an economic value that rivals that of dead sharks.

(A)

(B)

Figure 7.15 Proposed mechanism of top-down control by apex-predator sharks Predator–prey relationships are shown by arrows. (A) Predation by large sharks controls the population sizes of mesopredators (smaller sharks, large predatory bony fishes such as groupers and barracudas, and mollusk-eating skates and rays). (B) When apex predators are removed, populations of mesopredators increase and exert more pressure on their prey species, causing those populations to decline. Habitat disturbance may increase (e.g., from increased numbers of rays making feeding pits in their search for buried mollusks), and ecosystem services (such as the cleaning services provided by wrasses) may be lost, leading to increased parasite loads.

Summary

The extant radiation of chondrichthyans includes three lineages that differ in body form and ecology.

Elasmobranchs (Elasmobranchii) include the sister groups sharks (Selachii) and skates and rays (Batoidea). Chimaeras (Holocephali) are the sister group of elasmobranchs.

Although most elasmobranchs are marine, more than 50 species occur in estuarine or freshwater habitats. All chimaeras are marine.

Despite the name "cartilaginous fishes," the skeletons of chondrichthyans are stiffened by mineralized tissue, and the vertebral centra of sharks are densely calcified.

The jaws of extant chondrichthyans are specialized for different feeding modes.

The upper jaw of extant chondrichthyans articulates with the chondrocranium via the hyomandibular. The upper jaw of sharks can be protracted and retracted as the mouth opens and closes. The upper jaw of skates and rays can drop below the ventral surface of the body.

The upper jaw of chimaeras is fused to the chondrocranium, providing a solid surface against which to crush the shells of mollusks and crustaceans.

The placoid scales of extant elasmobranchs reduce turbulence in the flow of water across the body surface. As an elasmobranch grows, it adds additional rows of scales.

The skin of some deepwater sharks contains photophores that emit light (bioluminescence). Luminescent ventral surfaces probably match down-welling light from above, concealing the shark from predators looking upward, whereas photophores on the sides of the body or on a dorsal spine are probably used for species and sex recognition.

Other sharks and rays are biofluorescent; their skin absorbs blue light and emits green light. Fluorescence makes these species conspicuous to conspecifics and is probably used for species and sex recognition.

Derived locomotor, sensory, trophic, and behavioral specializations make the extant sharks consummate carnivores.

Galeomorph sharks are cylindrical, with a projecting rostrum containing lateral line mechano- and electroreceptors. Great white sharks, tiger sharks, makos, hammerheads, and the bull shark are midwater predators that eat other sharks, bony fishes, and mammals. The largest galeomorphs—the basking shark, whale shark, and megamouth shark—are filter feeders.

Squalomorph sharks are smaller than galeomorphs, and more diverse in body form and ecology. Many are bottom-dwellers that include crustaceans in their diets, and a few live as deep as 3,000 m.

Sharks employ chemoreception (olfaction), vision, mechanoreception, and electroreception (frequently in that sequence) to detect, track, identify, and attack prey.

Large sharks move hundreds or thousands of kilometers in a year, some crossing entire ocean basins and returning to favored feeding or breeding sites annually.

Species in one family of galeomorph sharks, Lamnidae, have developed regional heterothermy. Retia capture and retain heat produced by muscular activity, maintaining muscle temperatures and eye temperatures above water temperature.

The teeth of sharks vary from being broad and relatively flat to pointed with serrated cutting edges; some species have differently shaped teeth in different parts of the jaws. There is no clear relationship between tooth shape and phylogeny, indicating that specializations have evolved independently in multiple lineages.

Tearing, cutting, and crushing are important components of many sharks' predatory behavior. Prey items that have the potential to harm a shark are repeatedly seized and released until they die from blood loss.

Internal fertilization is universal among chondricthyans.

Male sharks insert one of their paired claspers into the cloaca of a female to transfer sperm.

Oviparity is the ancestral form of reproduction for chondrichthyans, but viviparity has evolved independently in several lineages.

Oviparous species are lecithotrophic, meaning the energy for embryonic development is provided entirely by the egg yolk.

Most viviparous species exhibit a combination of lecithotrophy and matrotrophy, with the embryo receiving some nourishment from the yolk and some from the reproductive tract of the mother. In nonplacental viviparous species, the additional energy may come from uterine secretions, from cannibalizing fetal siblings, or from eating unfertilized eggs that the female produces while she is pregnant. Some viviparous species are placental, transferring nutrients to embryos via a vascularized placenta.

Skates and rays are dorsoventrally flattened. The mouth and gill slits are on the ventral surface, and large pectoral fins extend from the head to the pelvic region.

Few external differences distinguish skates and rays. In general, skates have rounded bodies and rays are

(Continued)

Summary *(continued)*

kite-shaped. Most skates have a thick tail without spines; rays have a whiplike tail, and some species have one or more spines on the tail that can inject a venom-carrying barb into an attacker.

Skates are oviparous. Rays have various forms of nonplacental viviparity.

Most batoids are bottom-dwellers that feed on crustaceans and mollusks, which they crush with flat-crowned teeth. Their protrusible mouths can suck even buried prey from the seafloor. Manta rays are pelagic filter feeders.

Chimaeras are deepwater chondrichthyans.

Most chimaeras live at depths of 500 m or more, and many species have geographic ranges than extend across entire ocean basins. Males and females may spend most of the year in different parts of the species' range, coming together only during the breeding season.

Modifications of the rostrum give chimaeras a bizarre appearance and are the basis for the common names of the three lineages: shortnose chimaeras (also known as rabbitfishes), longnose chimaeras, and plownose

chimaeras. The rostra have dense concentrations of electro- and mechanoreceptors.

Chimaeras have a derived jaw suspension allows them to crush hard-bodied prey. They feed on bottom-dwelling fishes, crustaceans, and mollusks.

Declining shark populations are a reason for concern.

The life-history characteristics of apex-predator sharks make them vulnerable to overfishing. Sharks are slow to mature, produce few young in a reproductive season, and may breed only every other year.

Overall, the IUCN classifies 24% of the species of chondrichthyans as Critically Endangered, Endangered, or Vulnerable. More than 70% of the species of large predatory sharks are in one of those three categories, and populations of some apex-predator sharks have decreased as much as 99% in the past three decades.

Because of the importance of large sharks as apex predators, reductions in their populations can reduce predation on mesopredators (in the next trophic level down), with potentially disruptive effects for marine ecosystems.

Discussion Questions

1. The peculiar heads of hammerhead sharks are called "cephalofoils," a term that reflects an interpretation of the head as a hydrodynamic structure that helps direct the shark's head upward while it is swimming or to make the shark more maneuverable. An alternative hypothesis is that the head increases the shark's chemosensitivity or electrosensitivity. Proponents of the two interpretations have used anatomical, behavioral, and evolutionary information to support their views and to discredit the alternative hypothesis.

Hydrodynamic function

- The cross section of the cephalofoil is shaped like a wing, and the angle of its trailing edge can be altered in a manner reminiscent of changing the angle of the flaps on the trailing surface of an airplane's wing to control the amount of lift.

- The sizes of the two anterior planing surfaces, the cephalofoil and the pectoral fins, are inversely related. That is, species with wide cephalofoils have relatively small pectoral fins, and vice versa. As a result of that relationship, the total surface area of those surfaces is fairly constant.

- Hammerheads are among the least buoyant species of sharks.

Sensory function

- Hammerheads swing the head from side to side when they are searching for prey buried in the sediment. This is the same way a person searches for buried objects with a metal detector.

- The ampullae of Lorenzini are distributed over the entire ventral surface of the hammer.

Additional observations

- The least-derived extant species of hammerhead, the winghead shark (*Eusphyra blochii*), has the broadest hammer; its width is nearly 50% of the total length of the shark.

- The size of the hammer in more derived species of hammerheads does not change unidirectionally according to current phylogenetic interpretations of the lineage (that is, there appears to be no tendency for the size of the hammer to have increased or decreased during the evolution of the extant hammerheads).

- The nostrils of the winghead shark are near the midline of the hammer, but they lie toward the ends of the hammer in more derived species.

- The nostrils of the scalloped hammerhead (*Sphyrna lewini*), a well-studied species, lie near the ends of the hammer and collect water via a prenarial groove that probably increases the volume of water flowing across the olfactory epithelium.

- The surface area of the olfactory epithelium of the scalloped hammerhead is no larger than that of non-hammerhead sharks.

It appears that observations have taken us as far as they can and that it's time to try a different approach to testing hypotheses about the functional significance of the hammerhead morphology.

    a. What experimental tests can you propose to evaluate the two hypotheses? (Don't worry too much about how you would carry out a manipulation. If a test involves putting sharks into a flow tank, for example, assume that you have access to a tank and to sharks of the appropriate species and sizes.)

    b. Is there another possibility that you miss if you consider the hydrodynamic and sensory functions to be alternative hypotheses?

2. John Musick and Julia Ellis have proposed that the ancestral reproductive mode for elasmobranchs was "yolk-sac viviparity." That is, embryos were retained throughout their development in the oviducts of the female and emerged as miniatures of the adults (i.e., viviparity), but nutrition was provided by yolk that was deposited at the time the egg was formed, not from the mother during development (i.e., lecithotrophy, not matrotrophy). They suggest that oviparity (depositing eggs that develop outside the body of the mother) was associated with the evolution of small body size because it increased the fecundity of small species of elasmobranchs. What is their reasoning? That is, why would oviparity provide greater fecundity than viviparity for small species of elasmobranchs? What other factors might make one mode superior to the other?

3. A hyostylic jaw suspension in chondrichthyans allows protraction of the upper jaw independent of the lower jaw. Evolutionary biologists have assumed that the ability to protract the upper jaw is advantageous for feeding—in other words, that it is an adaptive derived character. An alternative hypothesis is that a hyostylic jaw suspension is a neutral ancestral character (i.e., one that is neither advantageous nor disadvantageous). These two hypotheses (i.e., hyostylic jaw is advantageous vs. hyostylic jaw is neutral) generate different predictions about the phylogenetic distribution of hyostyly. What are those predictions, and which one is supported by the phylogenetic distribution of hyostyly?

Additional Reading

Claes JM and 6 others. 2014. Photon hunting in the twilight zone: Visual Features of mesopelagic bioluminescent sharks. *PLoS ONE* 9(8): e104213. doi:10.1371/journal.pone.0104213.

Corn KA, Farina SC, Brash J, Summers AP. 2016. Modelling tooth–prey interactions in sharks: the importance of dynamic testing. *Royal Society Open Science* 3: 160141.

Dulvy NK, Reynolds JD. 1997. Evolutionary transitions among egg-laying, live-bearing and maternal inputs in sharks and rays. *Proceedings of the Royal Society B* 264: 1309–1315.

Gardiner JM, Atema J, Hueter RE, Motta PJ. 2014. Multisensory integration and behavioral plasticity in sharks from different ecological niches. *PLoS ONE* 9(4): e93036. doi:10.1371/journal.pone.0093036.

Gruber DF and 9 others. 2016. Biofluorescence in catsharks (Scyliorhinidae): Fundamental description and relevance for elasmobranch visual ecology. *Scientific Reports* 6: 24751.

Jaiteh VF and 6 others. 2016. Higher abundance of marine predators and changes in fishers' behavior following spatial protection within the world's biggest shark fishery. *Frontiers in Marine Science* 3: 43.

Lucifora LO, de Carvalho MR, Kyne PM, White WT. 2016. Freshwater sharks and rays. *Current Biology* 25: R965–R979.

Shiffman DS, Hammerschlag N. 2016. Shark conservation and management policy: a review and primer for non-specialists. *Animal Conservation* 19: 401–412.

Smith WL, Stern JH, Girard MG, Davies MP. 2016. Evolution of venomous cartilaginous and ray-finned fishes. *Integrative and Comparative Biology* 56: 950–961.

Sparks JS and 6 others. 2014. The covert world of fish biofluorescence: a phylogenetically widespread and phenotypically variable phenomenon. *PLoS ONE* 9(1): e83259. doi:10.1371/journal.pone.0083259.

Whitenack LB, Simkins DC Jr., Motta PJ. 2011. Biology meets engineering: the structural mechanics of fossil and extant shark teeth. *Journal of Morphology* 272: 169–179.

Go to the Companion Website **oup.com/us/vertebratelife10e** for active-learning exercises, news links, references, and more.

CHAPTER
8

Radiation and Diversity of Osteichthyes

Courtesy of Brian Choo

By the end of the Silurian, agnathous fishes had diversified and cartilaginous gnathostomes had begun to radiate. The stage was set for the appearance of the largest extant group of vertebrates, the bony fishes. The first fossils of bony fishes are known from the late Silurian. The osteichthyan radiation was in full bloom by the Middle Devonian, with two major groups diverging: ray-finned fishes (Actinopterygii) and lobe-finned fishes and tetrapods (Sarcopterygii). Recall from Chapter 1 the Greek derivations of these names: Osteichthyes, *osteon*, "bone"; *ichthys*, "fish"; Actinopterygii, *aktis*, "ray"; *pteron*, "wing, fin"; and Sarcopterygii, *sarkos*, "flesh." Osteichthyes technically includes aquatic bony fishes (i.e., "fishes," more than 32,000 species) and terrestrial bony fishes (tetrapods, more than 34,000 species).

Specialization of feeding mechanisms is a key feature when describing the evolution of the major vertebrate groups. Increasing mobility among the bones of the skull and jaws allowed the ray-finned fishes to exploit a wide range of prey types and predatory modes. Specializations in locomotion, habitat, behavior, and life histories accompanied the specializations of feeding mechanisms.

8.1 The Origin of Bony Fishes

Bony fishes, as their name suggests, possess bone, but as you have seen in earlier chapters, dermal bone (or its remnants, such as placoid scales) is seen at some location in all other types of jawed fishes and also in the jawless ostracoderms. The novel feature in osteichthyans was an ossified endoskeleton with endochondral bone in addition to the dermal and perichondral bone seen in basal jawed fishes such as placoderms. Osteichthyans have several additional important features that distinguish them, and many of these features are still present in humans:

- A unique pattern of dermal bones surrounds the jaws and braincase, although these bones around the mouth are also seen in some placoderms.

- All osteichthyans, even stem members, have teeth embedded in their dermal marginal mouth bones, and they shed their teeth by basal resorption rather than via a tooth whorl, as seen in chondrichthyans and acanthodians (although some basal osteichthyans retained a tooth whorl at the front of the jaw).

- All osteichthyans possess true enamel (epidermal origin) as the outermost layer on their scales (where it is termed *ganoine* in ray-finned fishes), rather than the external layer of enameloid (dermal origin) seen in other vertebrates.

- The dermal bones extend into the roof of the mouth to cover the palate, and many osteichthyans have teeth on the palate.

- Some of the posterior dermal head bones (known as the supracleithral series) extend posteriorly to attach to the pectoral girdle; dermal bones also form the operculum, which covers the gills and aids in gill ventilation.

- A fanlike series of dermal bones, called **branchiostegal rays**, forms the floor of the gill chamber and aids in the rapid expansion of the mouth for suction feeding and respiration.

- The fin rays are formed from jointed bony elements and are called **lepidotrichia**.

- A gas-containing structure, derived from the embryonic gut tube and used for buoyancy (gas bladder), for breathing (lung), or for both functions, is an ancestral condition of osteichthyans, although it has been lost in some bottom-dwelling lineages, such as darters and flatfishes. This phylogenetic pattern (lung present in basal actinopterygians) suggests that breathing was the original function. Molecular genomic and developmental studies show that gas bladders and lungs are probably homologous in all osteichthyans.

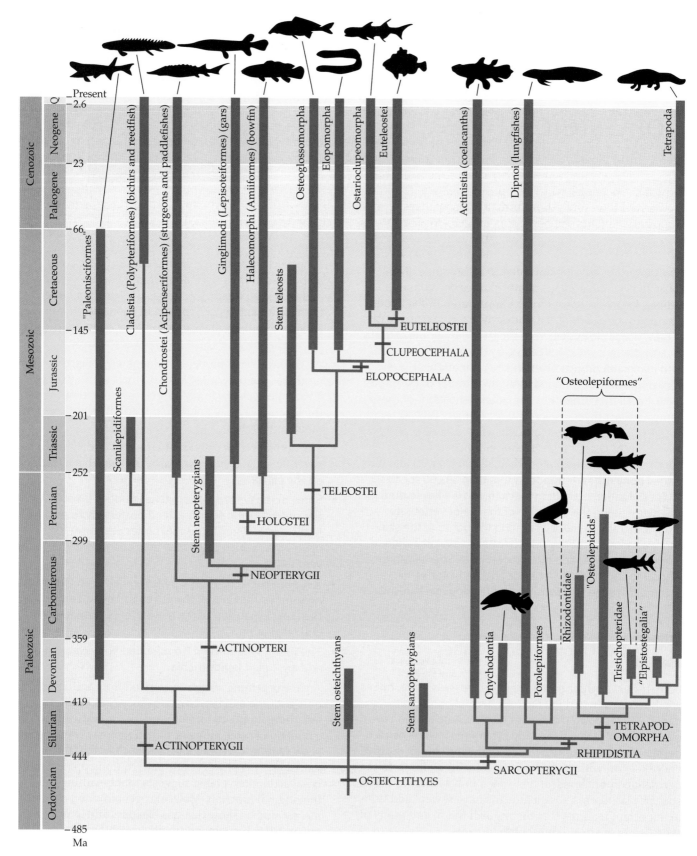

Figure 8.1 Phylogenetic relationships of osteichthyans. This tree depicts the probable relationships among the major groups of bony fishes. Only well-supported relationships are shown. The narrow lines show interrelationships only; they do not indicate times of divergence or the unrecorded presence of taxa in the fossil record. Bars denote the taxon's fossil record. Quotation marks indicate paraphyletic groups.

Earliest osteichthyans and the major groups of bony fishes

Osteichthyes includes two major groups: ray-finned fishes (Actinopterygii) and lobe-finned fishes (Sarcopterygii). Actinopterygii accounts for the largest number of aquatic fishes, and all terrestrial vertebrates are Sarcopterygii, so each group includes more than 32,000 extant species.

The fossil record   Fragmentary remains of bony fishes (scales and tooth-containing jaws) are known from the late Silurian, representing Lophosteiformes (the genera *Andreolepis* and *Lophosteus*), which comprised stem osteichthyans less derived than any extant form. Representatives of this group survived into the beginning of the Devonian. More recently, complete skeletons of fishes determined to be stem sarcopterygians (*Guiyu oneiros* and *Sparalepis tingi*) have been discovered in late Silurian sediments (~ 425 Ma). These were small fishes, around 30 cm in length.

Bony fishes radiated in the Devonian, but placoderms were the most diverse Devonian fishes in terms of number of species and types of body forms. Acanthodians were also prominent, and sarcopterygians were the most diverse bony fishes during the Devonian. Only with the extinctions at the end of the Devonian did actinopterygians and

chondrichthyans became the predominant aquatic vertebrates, a condition that persists to this day. **Figures 8.1** and **8.2** show how bony fishes are interrelated.

Fins   As their names suggest, actinopterygians and sarcopterygians differ in the structure of their fins, although the fins of early actinopterygians had lobate bases like those of sarcopterygians. The ancestral condition for the ostheichthyan fin (**Figure 8.3**) is rather like that of chondrichthyans—a row of basals that articulate with the endochondral limb girdles (scapulocoracoid for the pectoral girdle, ischiopubic plate for the pelvic girdle); then a row of radials, which articulate with the basals; and finally, fin rays that branch out from the radials and support the web of the fin. The basals and radials are endochondral bone, while the fin rays are dermal bone. The fin rays of osteichthyans (lepidotrichia) are segmented and made of dermal bone, while the ceratotrichia of chondrichthyans are unsegmented and made of a keratinlike protein.

The ancestral form of the osteichthyan fin is seen in some extant basal actinopterygians, such as the Polypteriformes (bichirs and the reedfish, *Erpetoichthyes calabaricus*) and Acipenseriformes (sturgeons and paddlefishes). The tendency in actinopterygian evolution has been to reduce the

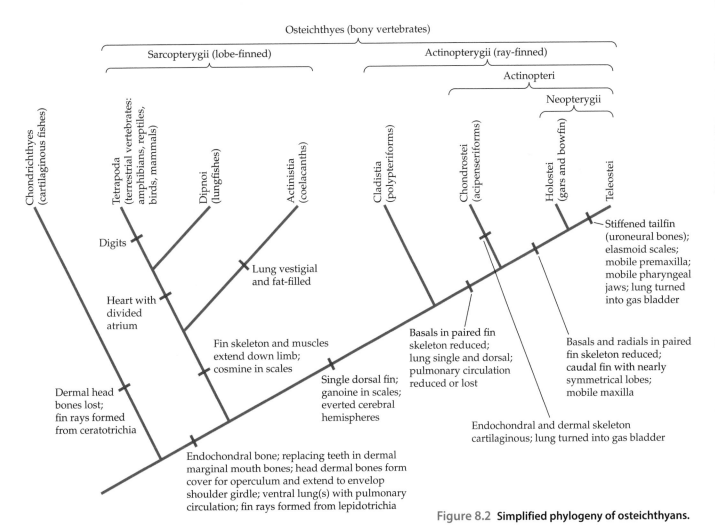

Figure 8.2 Simplified phylogeny of osteichthyans.

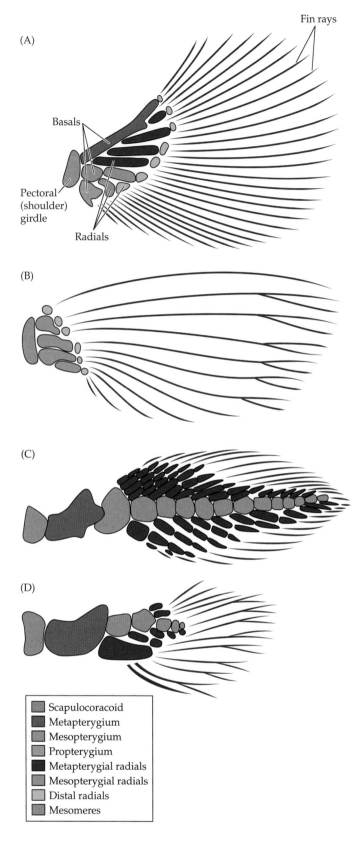

(A)

Fin rays

Basals

Pectoral (shoulder) girdle

Radials

(B)

(C)

(D)

Scapulocoracoid
Metapterygium
Mesopterygium
Propterygium
Metapterygial radials
Mesopterygial radials
Distal radials
Mesomeres

Figure 8.3 Fin structure of osteichthyans. All drawings show a left pectoral fin with the animal facing left and the preaxial (leading) edge of the fin downward. (A) Hypothetical ancestral type of osteichthyan fin, with similarities to those seen today in actinopterygians such as *Polyodon*. (B) Derived type of actinopterygian fin, typical of teleosts. (C) Derived type of sarcopterygian fin, resembling that seen in the extant lungfish *Neoceratodus*. All but the earliest sarcopterygians have a monobasal fin that retains only the metapterygium out of the original basals (the other basals being the propterygium and mesopterygium). The metapterygial bony axis extends down into the fin as bony elements called mesomeres, with radials branching off to one or both sides, and with the fin muscles following the bones to form the fleshy lobe of the fin. (D) In tetrapodomorph sarcopterygian fishes (the precursors of tetrapods), the fin skeleton was asymmetrical. The radials branched mainly from the preaxial edge of the skeleton so that the mesomeres and radials formed a "one bone, two bones" pattern, foreshadowing the pattern of tetrapod limbs.

make a single ray. This allows muscular action at the fin base to curve the fin, resisting fluid loading and drag.

It has long been assumed that the symmetrical fin skeleton seen in the extant lungfishes and coelacanths is the ancestral sarcopterygian condition, and that the alternating pattern of tetrapodomorphs is derived. However, recent skeletal material of a primitive Devonian coelacanth, *Shoshonia arctopteryx* (see Figure 8.3D), shows an asymmetrical pattern, suggesting that asymmetry may be the ancestral sarcopterygian condition.

Scales Although most extant bony fishes (teleosts) have reduced, thin (**elasmoid**) scales, early forms had thick scales, which differed between the two groups: actinopterygian scales had a thick layer of enamel (ganoine), while sarcopterygian scales in lineages more derived than coelacanths and relatives (but not seen in any extant form) had a thick layer of a dentine-like material (cosmine).

Skulls and brains Other ways in which actinopterygians and sarcopterygians differ include their feeding mechanisms and skull anatomy. Although basal actinopterygians, such as polypterids, retain a complete dermal skull roof and have no mobility of their upper jaw, more derived actinopterygians have reduced the number of dermal bones in their skull and have obtained greater mobility of the marginal mouth bones of the upper jaw, the **maxillae** and **premaxillae**. Sarcopterygians have not reduced the dermal bone of the skull, and if they have jaw mobility, it is derived from movement within the skull itself, between the anterior and posterior portions of the chondrocranium, a condition retained in the extant coelacanths (*Latimeria*).

Many sarcopterygians (including humans) develop large brains, but they retain the ancestral form of brain development seen in other vertebrates, whereby the cerebral hemispheres fold inward on themselves during growth (that is, they become inverted). In contrast, actinopterygians have brains that develop by eversion, folding the cerebral hemispheres outward.

endoskeletal fin elements. A fin made almost entirely of fin rays can be collapsed against the body, creating a streamlined profile. Actinopterygian fin rays are also unique in having a bilaminar structure—two halves combining to

8.2 Evolution of Actinopterygii

Although fragmentary fossils of late Silurian actinopterygians exist, complete fossil skeletons are not known before the late Middle Devonian. The earliest actinopterygians had unique features of the skull, and teeth that were capped by enamel (more derived actinopterygians have replaced this with an enameloid tissue called acrodin). The presence of a single dorsal fin was a characteristic feature of early actinopterygians; early sarcopterygians had two dorsal fins.

Early actinopterygians were mostly small (5–25 cm, although some were longer than 1 m) with a single dorsal fin and a caudal fin that was strongly heterocercal (upper lobe larger than lower lobe), forked, and had little webbing. The paired fins retained the ancestral condition of multiple basals (see Figure 8.3A).

Basal actinopterygians

Actinopterygians were rare in the Devonian but are known from both marine and freshwater environments. They started to diversify in the Carboniferous and apparentlywere little affected by the end-Permian extinctions, becoming the dominant fishes in the Triassic.

Paleonisciformes A diversity of basal fishes known as Palaeonisciformes form a paraphyletic grouping on the actinopterygian stem. Palaeonisciforms are primarily known from the Devonian and Carboniferous and were the only ray-finned fishes known until the Late Carboniferous, although they are present in the fossil record through the Mesozoic. Many palaeonisciforms were armored with thick bony scales; one example is the Middle Devonian *Cheirolepis* (**Figure 8.4**), one of the most basal of all the actinopterygians. Within the crown group Actinopterygia, the basal branch is Cladistia, and all other ray-finned fishes are grouped in Actinopteri (see Figure 8.1). Although basal ray-finned fishes originally included both freshwater and marine forms, all extant non-teleost fishes live in fresh water.

Cladistia Cladistians (Greek *klados*, "branch"; *histion*, "sail") include the extant Polypteriformes (= polypterids)—bichirs and reedfish, elongate, heavily armored fishes from Africa (see Chapter 9). Polypterids retain heavy scales with a layer of ganoine, among other primitive features, and they retain true paired ventral lungs with a pulmonary circulation. All of the more derived actinopterygians have a single dorsal lung or gas bladder and have reduced or lost the pulmonary circulation. Polypterids are not known in the fossil record until the mid-Cretaceous, but a related fossil group, the Scanilepidiformes, date back to the Triassic and have been recognized as stem polypterids. Molecular clock data indicate a Miocene origin for crown polypterids.

Chondrostei Chondrosteans (Greek *chondros*, "gristle"; *osteon*, bone) secondarily lack endochondral bone and have lost much of the original dermal skeleton; in fact, they are unique among vertebrates in having dermal head bones that are formed of cartilage. They are also the only gnathostomes to have lost the vertebral centra. They resemble teleosts in having an upper jaw that is not fused to the cranium and in having transformed their lung into a nonrespiratory gas bladder, although these features evolved independently of the teleost condition. Stem chondrosteans are first known from the Late Permian. The extant forms (Acipenseriformes, the sturgeons and paddlefishes; see Chapter 9) are first known from the Late Jurassic.

Neopterygii

Neopterygians (Greek *neos*, "new") include Lepisosteiformes (gars), Amiiformes (the bowfin, *Amia calva*), and Teleostei. Neopterygians were the predominant fishes of the Mesozoic and have remained so in the Cenozoic. As their name suggests, they are characterized by a novel type of fin. In the caudal fin, the upper and lower lobes are nearly symmetrical. In the paired fins, the fin membranes are supported by fewer bony rays: the basal elements of the fin skeleton are reduced to a small metapterygium, or are

(A)

(B)

Figure 8.4 Basal non-teleost fishes. *Moythomasia* (A) and *Cheirolepis* (B) were stem actinopterygians of the Devonian. Their cylindrical bodies and long, tooth-filled jaws indicate that they were midwater predators that fed on other fish. (A, © Esben Horn 10tons.dk; B, © By Smokeybjb (own work) CC BY-SA 3.0.)

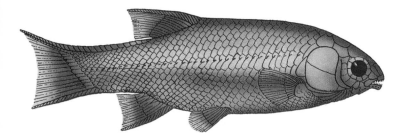

Figure 8.5 A stem neopterygian. *Acentrophorus* of the Permian was an early member of the late Paleozoic neopterygian radiation. It was about 10–15 cm long and probably fed on small crustaceans that it captured one by one. (After Gill 1923.)

lost all together in the more derived teleosts (see Figure 8.3B). These morphological changes probably increased the flexibility of the fin.

The dermal scales of neopterygians have also been reduced compared with those of their ancestors. The changes in fins and armor may have been complementary—more mobile fins mean more versatile locomotion, and a greater ability to avoid predators may have permitted a reduction in heavy armor.

Neopterygians are first known from the Permian, and possibly from as early as the Late Carboniferous. Non-teleost neopterygians (Lepisosteiformes and Amiiformes) are grouped as Holostei. Morphological data were originally considered to group the bowfin more closely to teleosts than to gars, and Holostei was long considered to be paraphyletic. However, molecular data have united the bowfin and gars as sister taxa.

A diversity of stem neopterygians lived in the Paleozoic and Mesozoic (**Figure 8.5**). One of these, *Leedsichthys* of Jurassic England, was the largest bony fish ever known, with a maximum length of 16 m. It appears to have been a filter feeder like the extant whale shark (*Rhincodon typus*).

Teleostei Fossils of teleosteans (Greek *teleos*, "entire") are first known from the Late Triassic, and conservative evaluation of molecular data and the fossil record places the origin of teleosts in the Late Permian. By the Jurassic, teleosts were well established in both marine and freshwater habitats, and most of the more than 400 families of extant teleosts had evolved by the start of the Cenozoic.

Most extant fishes are teleosts, and Teleostei contains more species than any other lineage of vertebrates. The entire genome was duplicated in the common ancestor of all extant teleosts.[1] This teleost-specific whole genome duplication (TSWGD) doubled the number of chromosomes and the DNA content of teleosts. Several different methods of inferring the ancestral diploid number of chromosomes (before TSWGD) give estimates of 24–26, leading to a post-duplication diploid complement of 48–52

chromosomes. More than half of all teleosts that have been examined have 48–50 chromosomes. The TSWGD and subsequent gene duplication is believed to have contributed to teleosts' potential for evolutionary adaptation and innovation. The number of sets of Hox genes would have been doubled during the TSWGD, and one set has subsequently been lost in many lineages, leaving most, but not all, teleosts with seven sets of Hox genes.

Teleosts share many characters of caudal and cranial structure, including highly protrusable jaws. Many less-derived teleosts use the gas bladder for breathing as well as for buoyancy control. A posterior location of the pelvic fins is another ancestral feature retained by some teleosts. In many derived forms, the pelvic fins have moved to an anterior position where they are attached to the base of the pectoral girdle, and some teleosts have lost the pelvic fins altogether.

Teleosts continued and expanded the changes in fins and jaws that contributed to the diversity of more-basal neopterygians. Teleosts have radiated into more than 32,000 species occupying almost every ecological niche imaginable for an aquatic vertebrate, and including some kinds of fishes that spend substantial amounts of time on land.

Evolution of jaw protrusion

Early actinopterygians had tooth-bearing marginal mouth bones of the upper jaw, and the premaxilla and maxilla were firmly attached to the cranium, as they are in humans and other sarcopterygians. The lower jaw snapped closed in a scissors action by contraction of the adductor mandibulae muscle, moving the jaw rapidly but with little crushing force. The close-knit dermal bones of the cheeks left no space for a larger adductor mandibulae muscle. This condition is retained in the extant polypterids, but the success of later neopterygians results largely from changes in the structure of the jaws that allowed them to develop functions ranging from the application of crushing force to a precise pincerlike action.

Neopterygians have a mobile upper jaw, and protrusion of the jaws allows effective suction feeding. Non-teleost neopterygians have jaws with a short maxilla that is free at its posterior end from the other bones of the cheek. The dermal bones of the cheek region are reduced, allowing the nearly vertically oriented hyomandibula to swing out laterally when the mouth opens. This action rapidly increases the volume of the **orobranchial chamber** (the mouth and gill region) and produces a powerful suction that draws prey into the mouth.

In teleosts, the bones of the opercular series are connected to the mandible so that expansion of the orobranchial chamber aids in opening the mouth. The dermal cheekbones are further reduced, providing space for larger jaw muscles. The anterior articulated end of the maxilla has developed a ball-and-socket joint with the chondrocranium. Because of its ligamentous connection to the mandible, the free posterior end of the maxilla rotates forward as the

---

[1] An additional whole-genome duplication occurred in salmon, and an independent duplication occurred in the paddlefish lineage.

(A)

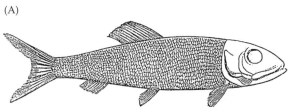

(B)

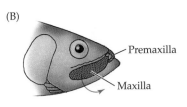

Figure 8.6 An early teleost. (A) *Leptolepis* lived in freshwater and marine habitats from the Late Triassic to the Early Cretaceous. (B) It had a derived jaw structure with enlarged mobile maxillae that rotated to form a nearly circular mouth when the jaws were fully opened. Membranes of skin closed the gaps behind the bony elements, increasing the effectiveness of prey capture. (A, from Gunther 1880, courtesy of Marine Image Bank, U. of Washington.)

mouth opens. This rotation points the maxilla's marginal teeth forward and helps grasp prey. The folds of skin covering the maxilla change the shape of the gape from a semicircle to a circular opening (**Figure 8.6**). These changes increase the suction produced during opening of the mouth. The result is greater directionality of suction and elimination of a possible side-door escape route for prey.

In more-derived teleosts, the premaxilla is free of the maxilla and contributes to the jaw protrusion. The fossil record shows that both the average and the maximum amount of jaw protrusion have increased in teleosts over their evolutionary history, with modern teleosts having greater protrusion than earlier teleosts (**Figure 8.7**). The initial increase appears to have been related to a shift in predatory dynamics following the end-Cretaceous[2] extinctions.

In addition to rapid and forceful suction, many teleosts have great mobility in the skeletal elements that rim the mouth. This mobility allows the grasping margins of the jaws to be extended forward from the head, often at remarkable speed. The functional result, called **protrusible jaws**, has evolved three or four times in different teleost clades, as shown by differences in the details of jaw anatomy. (Sturgeons and some sharks also generate suction but with completely different mechanisms of jaw protrusion.)

Pharyngeal jaws

Early neopterygian fishes had numerous dermal tooth plates in the pharynx. These plates were aligned with, but not fused to, both dorsal and ventral skeletal elements of the gill arches. Increasingly, these tooth plates became fused to one another and to a few gill arch elements above and below the esophagus. These consolidated tooth plates and gill arches were originally not very mobile, a condition retained today in the bowfin and gars, where the jaws are used primarily to hold and manipulate prey in preparation

for swallowing it whole. In teleosts, however, these gill arches attained great mobility and are now termed **pharyngeal jaws**. This second set of jaws has been modified convergently among different teleost lineages.

In the most derived teleosts, the muscles associated with the branchial skeletal elements supporting the pharyngeal jaws have undergone radical evolution, resulting in a variety of powerful movements of the pharyngeal jaw

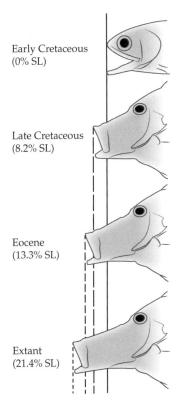

Early Cretaceous
(0% SL)

Late Cretaceous
(8.2% SL)

Eocene
(13.3% SL)

Extant
(21.4% SL)

Figure 8.7 Evolution of jaw protrusion in teleosts. The ability of teleosts to protrude their jaws has increased progressively from the Early Cretaceous to the present. Protrusion is measured as a percentage of the standard length (SL) of a fish (the distance from the tip of the snout to the end of the last caudal vertebra). Extant teleosts can protrude their jaws by more than 20% of their standard length, allowing them to engage in forceful and precise suction feeding. (From Bellwood et al. 2015.)

---

[2] We refer to the transition from the Cretaceous to the Paleogene as the "end-Cretaceous." Other notations used in the literature to identify this time period are K–Pg (Cretaceous–Paleogene) and, less frequently now, K–T (Cretaceous–Tertiary). Traditionally, the Cenozoic was divided into the Tertiary and Quaternary, but most authorities now divide the Cenozoic into three periods: the Paleogene, Neogene, and Quaternary.

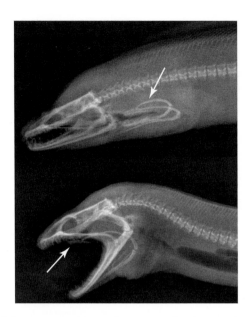

Figure 8.8 Pharyngeal jaws of a moray eel.   A moray first seizes and holds prey with the teeth in its maxillae and mandibles, then advances the pharyngeal jaws (arrows) into its mouth to grasp the prey. The mandibles and maxillae release their hold on the prey, which is then dragged into the throat by the pharyngeal jaws. (Courtesy of Rita Mehta, UC Santa Cruz.)

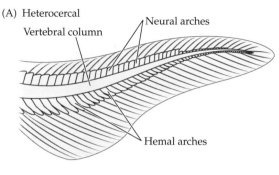

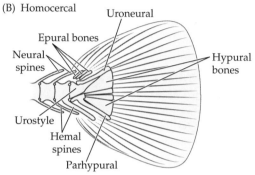

Figure 8.9 Tail structures of actinopterygian fishes.   (A) Sturgeons and gars have a heterocercal tail in which the vertebral column extends into the upper lobe and the fin rays attach to the neural and hemal arches. (B) Teleosts have a homocercal tail and the fin rays attach to epural, uroneural, and hypural bones. The urostyle marks the end of the tail vertebrae.

tooth plates. Not only are the movements of these second jaws completely unrelated to the movements and functions of the primary jaws, but in a variety of derived teleosts the upper and lower tooth plates of the pharyngeal jaws move independently of each other. Some moray eels can extend their pharyngeal jaws out of their throat and into their oral cavity to grasp struggling prey and pull it back into the throat and esophagus (**Figure 8.8**). With so many separate systems to work with, it is little wonder that some of the most extensive adaptive radiations among teleosts have been in fishes endowed with protrusible primary jaws and specialized mobile pharyngeal jaws.

Specializations of fins

The specializations of the paired fins of neopterygians are carried to an extreme in teleosts, where the basals have been lost, the radials attach directly to the limb girdles, and the visible external part of the fin consists of only the fin rays (see Figure 8.3B).

The tails of sturgeons and gars are **heterocercal** (Greek *heteros*, "different"; *kerkos*, "tail") in which the vertebral column extends into the upper lobe and fin rays attach to the neural and hemal spines (**Figure 8.9A**). Modified posterior neural and hemal spines (epurals, uroneurals, hypurals, and parhypurals) that support the fin rays are a derived character of teleosts (**Figure 8.9B**). The caudal fin of teleosts is symmetrical, a condition referred to as **homocercal**. The posterior margin is extremely flexible, bending left or right to serve as a rudder, or passing a sine wave upward to hold the fish level in the water.

In conjunction with a gas bladder that adjusts buoyancy, a homocercal tail allows a teleost to swim horizontally without using its paired fins for lift, as sharks must. In burst and sprint swimming, the tail produces a symmetrical force, but during steady speed swimming, intrinsic muscles in the tail may produce an asymmetrical action that increases maneuverability without requiring the use of the lateral fins. Relieved of responsibility for controlling lift, the paired fins of teleosts are flexible, mobile, and diverse in shape, size, and position. They have become specialized for activities that include food gathering, courtship, sound production, walking, and gliding over the surface of the water, as well as turning and braking during swimming.

8.3 Evolution of Sarcopterygii

Early sarcopterygians were cylindrical and 20–70 cm long. They had two dorsal fins, an distinct upper lobe (the epichordal lobe) on the heterocercal caudal fin, and paired fins that were fleshy, scaled, and had a bony central axis (see Figure 8.3C,D).

Three major lineages of sarcopterygians are known from the Early Devonian (Actinistia, Dipnoi, and Tetrapodomorpha; **Figure 8.10**), and all have members existing today,

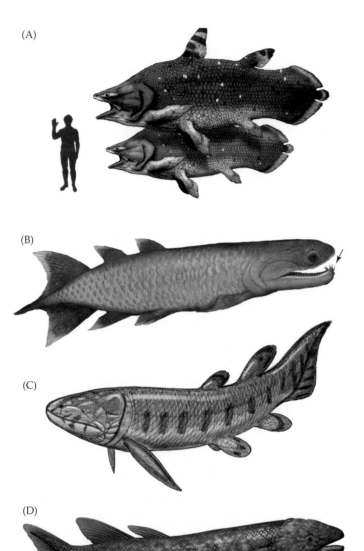

Figure 8.10 Early sarcopterygian fishes.   (A) The Cretaceous coelacanth *Mawsonia gigantea* reached lengths in excess of 4 m but otherwise was much like the two species of extant coelacanths (genus *Latimeria*). (B) The Devonian predator *Qingmenodus yui* was an onychodont, related to coelacanths, and about 20 cm long. Note the tooth whorl at the front of the jaw (arrow) that was seen in some Devonian bony fishes. (C) *Porolepis* was marine. The oldest known porolepiform, it was about 1.5 m long. (D) *Dipterus*, a Devonian lungfish, was about 12 cm long. (A, artwork by Joschua Knüppe; B, courtesy of Dr. Brian Choo; C, © Dmitry Bogdanov CC BY-SA 4.0; D, © Corey A. Ford/123RF.)

The interrelationships of these lineages have been the subject of considerable debate, but genomic data confirm coelacanths as the basal forms, with lungfishes and tetrapodomorphs being sister taxa (grouped together as Rhipidistia). Extant lungfishes are highly specialized, which led earlier workers to group the more superficially similar coelacanths and tetrapodomorph fishes (previously known as rhipidistians) into Crossopterygii (tassel-finned fishes), a name that is now obsolete.

Actinistia

Actinistians are first known from the Early Devonian. Coelacanths (Greek *koilos*, "hollow"; *akantha*, "spine") have two derived characters: a first dorsal fin that is supported by a plate of bone but lacks an internal lobe, and a symmetrical three-lobed tail with a central fleshy lobe that ends in a fringe of rays. Coelacanths lack a maxilla and branchiostegal rays, and they have a unique pattern of bones in their paired fins.

Extant coelacanths retain the ancestral fusiform fish shape, but during the Carboniferous there was an explosive radiation resulting in a greater diversity of body forms, including some deep-bodied species with reduced fin lobes that were originally mistaken for actinopterygians. Some early coelacanths lived in shallow fresh waters, but the fossil remains of coelacanths from the Mesozoic are largely marine, as is the extant genus *Latimeria* (although unlike *Latimeria*, early coelacanths probably were not deepwater bottom-dwellers). There are four extinct families of coelacanths and one extant family, Latimeriidae, which can be traced back to the Triassic, although coelacanths are unknown from the fossil record after the end of the Cretaceous. The lung of extant coelacanths is vestigial, although evidence exists for larger lungs in some fossil forms, surrounded by calcified plates that perhaps aided ventilation. In extant coelacanths the lung is filled with fat and serves as a buoyancy organ.

Related to coelacanths were the onychodonts, large, mostly marine predatory fishes known from the Devonian of China, Germany, and Australia. They were characterized by a spiky tooth whorl at the front of the lower jaw (see Figure 8.10B).

Dipnoi

Dipnoans (Greek *dis*, a prefix meaning "two"; *pneuma*, "breathing—i.e., lungs") are first known from the Early Devonian. Lungfishes are distinguished by the derived features of the loss of the tooth-bearing dermal bones—the premaxilla, maxilla, and **dentary** (the last being the anteriormost tooth-bearing bone of the lower jaw of vertebrates)—and the fusion of the palatoquadrate to the cranium (autostyly). Teeth are scattered over the dermal bones of the palate and fused into tooth ridges (tooth plates) along the lateral palatal margins to form a dentition capable of crushing hard prey. This **durophagous** apparatus has persisted throughout lungfish evolution. Extant lungfishes

although tetrapodomorphs persist as the predominantly terrestrial tetrapods, not as fishes. Sarcopterygian fishes were most diverse in the Paleozoic, primarily in the Devonian and Carboniferous. Only a few are known as fossils from the Mesozoic and Cenozoic.

(A)

(B)

Figure 8.11 Devonian tetrapodomorph fishes. (A) The tristichopterid *Eusthenopteron* had a cylindrical body, short snout, and four unpaired fins in addition to the paired pectoral and pelvic appendages. (B) The elpistostegalian *Panderichthys* had a dorsoventrally flattened body with a long, broad snout and eyes on top of the head. The spiracular chamber was also enlarged compared with that of *Eusthenopteron*, indicating air breathing via the spiracle. The dorsal and anal fins had been lost, and the caudal fin had been reduced in size. In *Eusthenopteron*, the ribs were short and probably extended dorsally. *Panderichthys* had larger ribs than *Eusthenopteron*, and they projected laterally and ventrally. (Nobu Tamura/CC BY-SA 3.0.)

have the largest-known genome among vertebrates and, as their name suggests, retain paired lungs.

The earliest lungfishes were marine, although the later forms and all extant forms live in fresh water. The group was diverse during the Paleozoic, with a dozen or so families; by the Mesozoic, most species belonged to the lineages that include the extant forms. Extant lungfishes (further discussed in Chapter 9) comprise three families, each containing a single genus: the Neoceratodontidae (in the order Ceratodontiformes) and the Lepidosirenidae and Protopteridae (in the order Lepidosireniformes). Both of these orders can be traced back to the Devonian.

By the Carboniferous, lungfishes had evolved a body form quite distinct from that of the other Osteichthyes. The dorsal, caudal, and anal fins are fused into a single continuous fin that extends around the entire posterior third of the body. In addition, the caudal fin has changed from heterocercal to symmetrical, and the mosaic of small dermal bones seen in the earliest dipnoan skulls has changed to a pattern of several large elements. Most of this transformation can be explained as a result of paedomorphosis (Greek *pais*, "child"; *morphe*, "form"), the retention of juvenile characters in the adult.

A Devonian lineage called Porolepiformes, known from Canada, Australia, and northern Europe, is related to dipnoans (see Figure 8.10C). Porolepiforms were once classified with tetrapodomorph fishes, mainly on the basis of generalized characters. However, the fin structure of porolepiforms shows that they are more closely related to lungfishes.

Tetrapodomorpha

Tetrapodomorph fishes (Greek *tetra*, "four"; *pous* "foot") were found all over the world, in marine and freshwater environments. They are known mainly from the late Middle Devonian to early Late Devonian. Tetrapodomorph fishes became extinct in the Early Permian, but terrestrial vertebrates (tetrapods), first seen in the Late Devonian, are their descendants.

Tetrapodomorph fishes can be distinguished from other Devonian fishes by their fin skeleton (see Figure 8.3C) and by the presence of **choanae** (internal nostrils). Several families of tetrapodomorph fishes are often grouped as "osteolepiforms," a paraphyletic assemblage that includes the following:

Rhizodonts, the most basal group of tetrapodomorph fishes, were Devonian to Carboniferous freshwater predatory fishes with long, recurved teeth. They were the largest bony fish predators of their time, some reaching 7 m.

Osteolepidids—a paraphyletic assemblage of cylindrical-bodied, large-headed fishes with thick scales—were generalists and probably shallow-water predators.

Tristichopterids had a worldwide distribution during the Devonian. They resembled early tetrapods in having paired crescentic vertebral centra and teeth with labyrinthine infoldings of enamel. Tristichopterids had lost the cosmine (dentine) and enamel layers typical of the scales of other sarcopterygians, reducing their scaly covering to just bone. *Eusthenopteron*, the best-known member of this family, had some microscopic patterns of bone growth similar to that seen in tetrapods.

Elpistostegalians are a paraphyletic assemblage of tetrapodomorph fishes more closely related to tetrapods than the "osteolepiforms" (see Figure 8.1). They had eyes on the top of their head (like crocodylians) and probably lived in shallow water. Tetrapod-like features associated with this lifestyle included the loss of the dorsal and anal fins, a greatly reduced caudal fin, a dorsoventrally flattened body, and a flat head with a long snout (**Figure 8.11**). These fishes also shared with early tetrapods a derived form of the humerus (upper arm bone), indicating powerful forelimbs that would have been capable of propping the front end of the animal out of the water, and many details of the skull, including the anatomy of the ear region. The most derived elpistostegalian, *Tiktaalik* from the early Late Devonian of the Canadian Arctic, will be discussed in Chapter 10 in the context of the origin of tetrapods.

Summary

Osteichthyans, the bony fishes, have several unique features.

Derived characters of bony fishes include a skeleton made of endochondral bone and the pattern of dermal bones in the head and shoulder region, including the tooth-bearing marginal mouth bones.

A lung was probably also a basal feature of the group (transformed into a gas bladder in most extant forms).

Osteichthyans comprise two groups, actinopterygians (ray-finned fishes) and sarcopterygians (lobed-finned fishes and tetrapods).

Bony fishes first appeared in the Late Silurian but did not become the dominant fishes until the end of the Devonian.

The ancestral osteichthyan fin had a moderately fleshy base, which is retained in some extant basal actinopterygians. Most extant actinopterygians, however, have reduced the skeletal elements so that only the dermal fin rays remain.

All but the earliest sarcopterygians have monobasal fins, in which the supporting elements close to the body are reduced to one. The branching axis of fin elements is symmetrical in extant sarcopterygians but was asymmetrical in tetrapodomorph fishes—the "one bone, two bones" pattern of our own limbs. This may be the basal sarcopterygian condition.

Ray-finned fishes have reduced the dermal elements of the skull and evolved a protrusable upper jaw that aids in suction feeding. Lobe-finned fishes developed movement between the anterior and posterior portions of the chondrocranium, a condition retained in the extant coelacanths (*Latimeria*).

Lobe-finned fishes retain the original type of vertebrate brain development, whereby the cerebral hemispheres fold inward on themselves, while ray-finned fishes have evolved "inside-out" development of the cerebral hemispheres.

The original scales in both groups of osteichthyns were multilayered, although most extant forms have thin, elasmoid scales. The scales of the basal actinopterygians had a thick layer of ganoine, formed from enamel, while the scales of the basal sarcopterygians had a thick layer made of cosmine, which is was formed from dentine.

There are several groups of basal ray-finned fishes

The heavily armored Paleonisciformes, known from the Paleozoic and Mesozoic, are a paraphyletic grouping of basal actinopterygians.

The most basal extant branch is Cladistia, the extant heavily armored polypterid fishes, but fossils are not known until the Late Cretaceous.

The most derived of the basal actinopterygians are the Chondrostei, first known from the latest Permian, which survive today as the Acipenseriformes (sturgeons and paddlefishes). The skeleton is secondarily cartilaginous, and the lung has been converted into a gas bladder.

Neopterygii are the most derived ray-finned fishes.

Neopterygii were the predominant fishes of the Mesozoic and Cenozoic. They have derived types of fins with a reduced internal skeleton, protrusable upper jaws, and reduced scale covering. Neopterygians are divided into Holostei (including Lepisosteiformes [gars] and Amiiformes [the bowfin]) and Teleostei (the great majority of extant fishes).

Teleosts are characterized by extremely protrusable jaws and often by mobile pharyngeal jaws. A duplication of the entire genome occurred at the base of the teleost lineage. Teleosts have thin, proteinaceous (elasmoid) scales and a homocercal caudal fin with the dorsal side supported by bones called uroneurals.

Derived teleosts have a gas bladder that is a buoyancy device that is not used for gas exchange. The pelvic fins have moved forward to lie beneath the pectoral fins.

Today teleosts are found in both freshwater and marine environments, whereas all of the extant non-teleost fishes live in fresh water.

Sarcopterygian fishes include two extant groups, coelacanths and lungfishes, both of which made their appearance in the Devonian.

Coelacanths lack a maxilla and have a distinctive three-lobed tail, and extant forms have a vestigial fat-filled lung. Coelacanths are unknown from the fossil record after the Cretaceous but persist today as the genus *Latimeria*.

Some early coelacanths were freshwater fishes, but later ones (including the extant genus) were marine. Related forms were the Devonian onychodonts.

Lungfishes are more closely related to tetrapomorphs than are coelacanths. Lungfishes retain paired lungs, but they have lost the tooth-bearing dermal bones around the mouth and instead have crushing tooth plates.

There were about a dozen families of lungfishes in the Paleozoic and Mesozoic, but only three survive today. Some Paleozoic lungfishes were marine, but later forms and all of the extant genera are freshwater fishes.

(Continued)

Summary (*continued*)

Tetrapodomorph fishes, an extinct group of sarcopterygians, gave rise to tetrapods.

Tetrapodomorph fishes are known mainly from the Devonian but persisted into the Early Permian and were found in a diversity of environments. They gave rise to tetrapods in the Late Devonian.

Tetrapodomorph fishes were distinguished by an alternating pattern of bones in the fins and by the presence of choanae (internal nostrils).

Different types of basal tetrapodomorph fishes included rhizodontids, the paraphyletic osteolepidids, and tristichopterids. Tristichopterids shared some derived features with early tetrapods, including teeth with labyrinthine infoldings of enamel.

Elpistostegalians are the tetrapodomorph fishes most closely related to tetrapods.

Elpistostegalians resembled tetrapods in having a flattened body, a flattened and long-snouted head with dorsally placed eyes, a reduced caudal fin, and no dorsal or anal fins.

Elpistostegalians probably lived in shallow water, where their powerful forelimbs allowed them to raise their heads above the surface.

Elpistostegalians form a paraphyletic stem to the earliest tetrapods, which first appeared in the Late Devonian. *Tiktaalik*, from the early Late Devonian of the Canadian Arctic, is the elpistostegalian most closely related to the tetrapods.

Discussion Questions

1. Today, lungfishes are considered to be more closely related to tetrapods than the coelacanths are, but in earlier times coelacanths were thought to be the more closely related forms. How have changes in our ways of thinking about phylogenetic relationships changed our opinion in this case?

2. What do the pharyngeal jaws of bony fishes do? How is this different from the actions of the anterior jaws (the premaxilla and maxilla)?

Additional Reading

Bellwood, DR and 5 others. 2015. The rise of jaw protrusion in spiny-rayed fishes closes the gap on elusive prey. *Current Biology* 25: 2696–2700.

Clack JA. 2012. *Gaining Ground: The Origin and Evolution of Tetrapods*, 2nd ed. Indiana University Press, Bloomington, IN.

Friedman M. 2015. The early evolution of ray-finned fishes. *Palaeontology* 58: 213–228.

Giles S, Xu G-H, Near T, Friedman M. 2017. Early members of "living fossil" lineage imply later origin of modern ray-finned fishes. Nature 549: 265–268.

Glasauer SMK, Neuhauss SCF. 2014. Whole-genome duplication in teleost fishes and its evolutionary consequences. *Molecular Genetics and Genomics* 289: 1045–1060.

Dial KP, Shubin N, Brainerd EL (ed.). 2015. *Great Transformations in Vertebrate Evolution*. The University of Chicago Press, Chicago, IL.

Long JA. 2011. *The Rise of Fishes: 500 Million Years of Evolution*, 2nd ed. The Johns Hopkins University Press, Baltimore, MD.

Lu J, Giles S, Friedman M, den Blaauwen JL, Zhu M. 2016. The oldest actinopterygian highlights the cryptic history of the hyperdiverse ray-finned fishes. *Current Biology* 26: 1602–1608.

Lu J and 6 others. 2016. A Devonian predatory fish provides insights into the early evolution of modern sarcopterygians. *Science Advances* 2(e1600154). doi:10.1126/sciadv.1600154.

Mehta RS, Wainwright PC. 2008. Raptorial jaws in the throat help moray eels swallow large prey. *Nature* 449: 79–82.

Tatsumi N and 6 others. 2016. Molecular developmental mechanism in polypterid fish provides insight into the origin of vertebrate lungs. *Scientific Reports* 6(30580): doi:10.1038/srep30580.

Zhu M and 5 others. 2009. The oldest articulated osteichthyan reveals mosaic gnathostome characters. *Nature* 458: 469–474.

CHAPTER
9

Extant Bony Fishes

Courtesy SEFSC Pascagoula Laboratory

Bony fishes are the success story of vertebrate evolution. Each of the two major lineages of bony fishes—actinopterygians (ray-finned fishes) and sarcopterygians (lobe-finned fishes)—contains more than 33,000 extant species. Actinopterygians are the most diverse group of aquatic vertebrates, and all terrestrial vertebrates are sarcopterygians.

9.1 Actinopterygians: Ray-Finned Fishes

In this chapter, we discuss the aquatic bony fishes, concentrating primarily on actinopterygians. Following a brief overview of the major lineages of ray-finned fishes, shown in **Figure 9.1**, we describe some examples of the diversity of ray-finned fishes.

Non-teleosts

As we explained in Chapter 8, current opinion recognizes Polypteriformes (bichirs and the reedfish *Erpetoichthyes calabaricus*) as the sister group of the other two lineages of actinopterygians: Acipenseriformes (sturgeons and paddlefishes) + Neopterygii (Holostei—gars and the bowfin *Amia calva*) and Teleostei, which includes almost all extant fishes). Although stem actinopterygians were largely replaced by neopterygians during the early Mesozoic, a few specialized forms of these early ray-finned fishes have survived (**Figure 9.2**). Polypteriforms and acipenseriforms have retained many ancestral characters, including basals in the fins and a spiral intestinal valve.

Polypteriformes: Bichirs and the reedfish Polypterids are the sole extant representatives of Cladistia, and thus the sister group of Actinopteri, which includes all the other ray-finned fishes. Polypterids are found in stagnant freshwater habitats and use their lungs to supplement gill respiration. Bichirs are heavy-bodied ambush predators that feed primarily on other fishes. The reedfish is slimmer and more eel-like than bichirs; it moves sinuously through thick vegetation and feeds primarily on snails and invertebrates.

Acipenseriformes: Sturgeons and paddlefishes Acipenseriforms are the extant members of Chondrostei, the sister group of Neopterygii (which contains the remaining ray-finned fishes). Sturgeons are large (1–6 m), active, benthic (bottom-dwelling) fishes found in the Northern Hemisphere. The beluga sturgeon (*Huso huso*) of eastern Europe and Asia is the world's largest freshwater fish, growing to more than 8 m and a weight of 1,300 kg. Some sturgeons live in fresh water, while others are marine forms that ascend to fresh water to breed. Most sturgeons have five rows of enlarged armorlike scales (called scutes) along the body that represent the remnants of ganoid scales. The protrusible jaws of sturgeons make them effective suction feeders, and they slurp small crustaceans and insect larvae from the substrate.

The paddlefishes are closely related to the sturgeons (although the lineages have been separate since the Late Cretaceous) but have a greater reduction of dermal ossification. Their most outstanding feature is a greatly elongated and flattened rostrum which is richly innervated with ampullary organs that detect minute electric fields. Only two species of paddlefishes remain, one in North America and the other in China. The Chinese paddlefish (*Psephurus gladius*) occurs in the Changjiang (Yangtze) River valley of China. It has a protrusible mouth and feeds on small fishes and crustaceans that it sucks from the river bottom. It may be extinct; the last adult specimens were recorded in 2002, and an intensive search of the upper Changjiang River from 2006 to 2008 failed to find any individuals. The Mississippi paddlefish (*Polyodon spathula*) occurs in the Mississippi River drainage, from western New York to central Montana and from Canada to Louisiana. Unlike the bottom-feeding Chinese paddlefish, the American paddlefish is a filter feeder that captures copepods in its gaping mouth as it swims.

Lepisosteiformes: Gars Also called garpikes, gars grow to lengths of 1–4 m. They live in stagnant freshwater habitats in North and Central America and on Cuba and use a vascularized gas bladder as a lung. Gars feed on other fishes, which are taken unaware when a seemingly lethargic

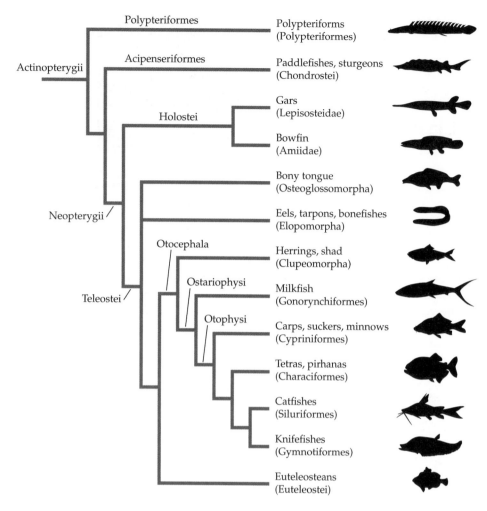

Figure 9.1 Phylogenetic relationships of extant actinopterygian fishes. The phylogeny of extant actinopterygians is complex and still in a state of flux. This is a widely accepted classification, but you will encounter other arrangements in the literature. (Courtesy of William E. Bemis.)

phylogenetic analysis reveals that this genome duplication played a relatively small role in creating the current diversity. About 88% of the diversity of extant teleosts has resulted from the radiations of Ostariophysi and Euteleostei (especially Perciformes) that occurred about 100 Ma, and genome duplication did not play a role in these events.

Jaw mobility has been a major factor in the radiation of teleosts; the premaxillae and maxillae are not attached to the skull and are protruded when the mouth is opened. In addition, teleosts have homocercal tails in which the fin rays are attached to a fan-shaped array of hypural bones. Four teleost clades—Osteoglossomorpha, Elopomorpha, Otocephala, and Euteleostei—are generally recognized.

and well-camouflaged gar eases alongside them and, with a sideways flip of its body, grasps its prey with its needlelike teeth. The interlocking multilayered ganoid scales of gars are similar to those of many Paleozoic and Mesozoic actinopterygians, and alligators are the only natural predators able to cope with the thick armor of an adult gar.

Amiiformes: The bowfin The single species of bowfin occurs in weed-filled lakes and slow-flowing rivers from Quebec to Texas. It is a very different kind of predator than gars. The bowfin is 0.5–1 m long and eats almost any organism smaller than itself, using suction to draw prey into its mouth. The bowfin's scales are comparatively thin and made up of a single layer of mineralized tissue, as in teleost fishes; however, the bowfin's asymmetrical caudal fin is very similar to the heterocercal caudal fin of many Paleozoic and Mesozoic actinopterygians.

Teleosts

Teleost fishes are an impressive success story—more than 95% of extant ray-finned fishes are teleosts (Greek *teleos*, "entire"; *osteon*, bone). Teleosts had a period of rapid diversification early in their history, and it is tempting to suggest that this diversification was triggered by a duplication of the entire genome that occurred about 200 Ma. However,

Osteoglossomorpha The extant osteoglossomorphs (Greek *glossa*, "tongue"; hence "bony tongue") are considered to be the most primitive extant teleosts. Arawana (genus *Osteoglossum*; **Figure 9.3A**) are predators from the Amazon basin, familiar to tropical fish enthusiasts. The pirarucu (*Arapaima gigas*), an even larger Amazonian predator, may be the largest freshwater teleost. Before intense fishing reduced their populations, pirarucus were known to reach lengths of at least 3 m and perhaps as much as 4.5 m. *Gnathonemus petersii* (**Figure 9.3B,C**), one of the African elephant-trunk fishes (Mormyridae), is a small bottom feeder that uses weak electric discharges to communicate with conspecifics and to detect objects in its environment. As dissimilar as the various osteoglossomorphs seem, they are united by unique bony characters of the mouth (including teeth on the lower portions of the hyoid arch that resemble a bony tongue and give the group its name) and by the mechanics of their jaws.

Elopomorpha Elopomorphs (Greek *elopos*, "a marine fish") had appeared by the Late Jurassic. A specialized leptocephalus (Greek *leptos*, "small"; *kephale*, "head") larva is a unique character of elopomorphs. These larvae spend a long time drifting at the ocean surface and are widely dispersed by currents.

Figure 9.2 **Extant non-teleost actinopterygian fishes.**
(A) Polypteriformes: Saddled bichir (*Polypterus endlicheri*).
(B,C) Acipenseriformes: Shortnose sturgeon (*Acipenser brevirostrum*; B) and Mississippi paddlefish (*Polyodon spathula*; C).
(D,E) Holosteans: Longnose gar (Lepisosteiformes: *Lepisosteus osseus*; D) and bowfin (Amiiformes; *Amia calva*; E). The ocellus, or eyespot, on the bowfin's tail confuses predators and causes them to attack the wrong end of the fish. (A, Koen van Uitert; B,D, Chris Crippen; C, Doug Canfield/USFWS; E, Ryan Hagarty/USFWS.)

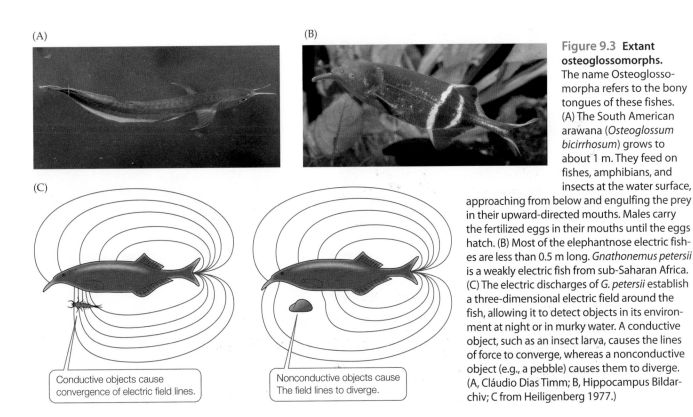

Figure 9.3 **Extant osteoglossomorphs.** The name Osteoglossomorpha refers to the bony tongues of these fishes. (A) The South American arawana (*Osteoglossum bicirrhosum*) grows to about 1 m. They feed on fishes, amphibians, and insects at the water surface, approaching from below and engulfing the prey in their upward-directed mouths. Males carry the fertilized eggs in their mouths until the eggs hatch. (B) Most of the elephantnose electric fishes are less than 0.5 m long. *Gnathonemus petersii* is a weakly electric fish from sub-Saharan Africa. (C) The electric discharges of *G. petersii* establish a three-dimensional electric field around the fish, allowing it to detect objects in its environment at night or in murky water. A conductive object, such as an insect larva, causes the lines of force to converge, whereas a nonconductive object (e.g., a pebble) causes them to diverge. (A, Cláudio Dias Timm; B, Hippocampus Bildarchiv; C from Heiligenberg 1977.)

Conductive objects cause convergence of electric field lines.

Nonconductive objects cause The field lines to diverge.

(A)

(B)

Figure 9.4 **Extant elopomorphs.** Although adult eels (*Anguilla*; A) and tarpons (*Megalops*; B) could scarcely look more different, they are united by having leptocephalus larvae. (A, Sean Landsman; B, Robert Patzner.)

Most elopomorphs are eel-like marine fishes, although some species live in fresh water and some, such as tarpons and bonefishes, are popular game fishes (**Figure 9.4**). Like other eels, the common American eel (*Anguilla rostrata*) has an unusual life history. After growing to sexual maturity (which takes as long as 25 years) in rivers, lakes, and ponds, the **catadromous** (migrating from fresh water to seawater to breed) eels enter the ocean and migrate to the Sargasso Sea, an area in the central North Atlantic between the Azores and the West Indies. Here they are thought to spawn and die, presumably at great depth. The eggs and newly hatched leptocephalus larvae float to the surface and drift in the

currents. Larval life continues until the larvae reach continental margins, where they transform into miniature eels and ascend rivers to feed and mature.

The European eel (*Anguilla anguilla*) spawns in the same region of the Atlantic as its American kin, but may choose a somewhat more northeasterly part of the Sargasso Sea and perhaps a different (but also deep) depth at which to spawn. Its larvae remain in the clockwise currents of the North Atlantic (principally the Gulf Stream) and ride to northern Europe as well as North Africa, the Mediterranean, and even the Black Sea before entering river mouths to migrate upstream and mature.

In contrast to eels, the two other major lineages of elopomorphs—bonefishes and tarpons—are prized by sport fishers. Both are fast-swimming, predatory fishes that inhabit shallow coastal waters in the tropics and subtropics. Bonefishes reach maximum lengths of about 40 cm, and female tarpons (which are larger than the males) can be as long as 2.5 m.

Otocephala   This group includes two large lineages, Clupeomorpha (herrings, shads, sardines, and anchovies; **Figure 9.5A**) and Ostariophysi (catfishes, milkfish, carps, suckers, minnows, tetras, and pirhanas; **Figure 9.5B–E**). Ostariophysi also includes the highly derived electric knifefishes of Central and South America (**Figure 9.5F**).

Clupeomorphs are silvery, mostly marine fishes that feed on plankton. They are of great commercial importance, and are a critical part of the marine food web. Several species are **anadromous**, migrating from seawater to freshwater to spawn. During the colonial period, springtime migrations of American shad (*Alosa sapidissima*) from the North Atlantic into rivers on the east coast of North America involved millions of individuals. Those enormous shad runs have been greatly depleted by dams and pollution.

◀ **Figure 9.5   Otocephalans include more than 10,000 extant species.** (A) Sardines (*Sardina pilchardus*), herrings, and other fish that form dense schools lack lateral lines on the flanks and retain canal organs only on the head. (B) The black bullhead (*Ictalurus melas*) is a catfish. Catfish (Siluriformes) include about 35 families. Commonalities include scaleless skins, sensory barbels, and a spine at the front of the dorsal and pectoral fins. In some species, the spine is covered by toxin-producing cells. (C) The blue shiner minnow (*Cyprinella caerulea*). Although the term "minnow" is used informally to describe any small, silvery fish, taxonomically "minnow" is the common name for cyprinid fishes. (D) The red-bellied piranha (*Pygocentrus nattereri*) is a characin. Found throughout the Amazon basin, this is one of the carnivorous species that give piranhas their fearsome (but largely unwarranted) reputation. (E) Some species of piranhas are herbivorous. The pacu (*Colossoma macropomum*) swims deep into flooded forests to feed on fruit and is an important seed disperser for Amazonian trees. (F) The electric eel (*Electrophorus electricus*), found in the Amazon and Orinoco basins of South America, is not a true eel but a knifefish (gymnotid), a member of the most derived group of ostariophysans. Growing to a length of 2.5 m, *E. electricus* uses powerful electrical discharges to immobilize prey and for defense (see Section 4.2). (A, Kouki Okagawa; A inset, Alessandro Duci; B, Chris Crippen; C, U.S. Geological Survey photo; D,E, Frank Greco; F, Brian Gratwicke.)

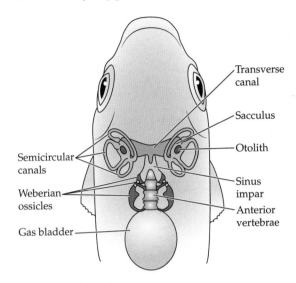

Figure 9.6   The Weberian apparatus.   The sound-detection system of otophysan fishes includes the gas bladder and anterior vertebrae. The Weberian ossicles are a series of small bones—modified vertebral elements—that transmit sound vibrations from the gas bladder to the inner ear.

The presence of an alarm substance (a pheromone) in the skin is a synapomorphy of Ostariophysi. A chemical signal ("Schreckstoff") is released into the water when the skin is damaged and produces a fright reaction in nearby conspecifics and other ostariophysan species, causing nearby fishes to rush for cover or to form a tighter school.

Ostariophysans are predominant in the world's fresh waters, representing about 80% of freshwater species and 25–30% of all extant fishes. Many ostariophysans have protrusible jaws and pharyngeal teeth in the throat that process a range of food items. Many species have fin spines or armor for protection, and the skin typically contains glands that produce substances used in olfactory communication.

The **Weberian apparatus**, consisting of small bones that connect the gas bladder with the inner ear, is a synapomorphy of a lineage of ostariophysans called Otophysi (**Figure 9.6**). Using the gas bladder as an amplifier and the chain of bones as conductors, the Weberian apparatus greatly enhances the hearing sensitivity of these fishes. Ostariophysans are more sensitive to sounds and have a broader frequency range of detection than other fishes.

Euteleostei   The vast majority of extant teleosts belong to the clade Euteleostei (Greek *eu*, "true"), which evolved sometime before the Late Cretaceous. With many thousands of species, it is impossible to give more than a brief overview here.

Stem euteleostians are probably represented today by the generalized Salmoniformes, although this interpretation is a matter of dispute. The esocid and salmonid fishes include important commercial and game fishes (**Figure 9.7A,B**). The esocids (pickerels, pikes, muskellunges, and their relatives) are at the base of the radiation of more derived teleosts. They are ambush predators that eat other fishes, relying on cryptic colors and patterns to conceal them until a prey fish has

(A)

(B)

(C)

(D)

Figure 9.7 Euteleosts include more than 17,500 extant species. This vast lineage includes both primitive and highly derived species. Salmoniformes: pikes, pickerels, the muskellunge (A, *Esox masquinongy*), salmon, and trout (B, brown trout, *Salmo trutta*) are among the least derived euteleosts and have relatively primitive jaw protrusion. Perciformes are more derived euteleosts. (C,D) The largemouth bass (*Micropterus salmoides*; C), preys on another derived euteleost, the pumpkinseed sunfish (*Lepomis gibbosus*; D). Bass are open-water predators that prowl the margins of aquatic weed beds, whereas sunfishes stay close to aquatic plants and flee into them when a bass approaches. The broad head, large and downturned mouth, and ringed eyes of the largemouth bass are characteristics of predatory fishes that elicit a fright response in other fishes. (A, B, C inset, D, Chris Crippen; C, Ryan Hagerty, USFWS.)

approached close enough to be captured by a sudden lunge. Salmonids include the anadromous salmon, which usually spend their adult lives at sea and make epic journeys to inland waters to breed, as well as the closely related trouts, which usually live entirely in fresh water. Perciformes (the perchlike fishes) includes more than 10,000 species and is the second recent explosive diversification of teleosts (**Figure 9.7C,D**).

The body form of a euteleost reflects its habitat and habits. Species that live in open water and engage in linear pursuits of prey are fusiform (torpedo-shaped), whereas species that frequently change direction as they thread their way through complex habitats, such as beds of aquatic plants or coral reefs, are often laterally flattened. Species that gather food from the water surface have an upward-facing mouth, and bottom feeders have a mouth that faces downward.

Many euteleosts have body forms and ways of life that stand out even among the remarkable diversity of teleosts. Examples include:

- Flying fishes glide through the air to escape predators, using their expanded pectoral fins as airfoils (**Figure 9.8A**); some species have expanded pelvic fins as well. To initiate flight, a fish swims rapidly toward the water surface with its fins furled and leaps though the surface

at a shallow angle, spreading its fins and beating the water with its tail to gain enough speed to glide. As the glide slows and the fish descends toward the water, it may beat the water with its tail to prolong the glide.

- Billfishes (marlins, swordfishes, and sailfishes) are among the largest and fastest pelagic predators. The bill is formed by an extension of the premaxillae, laterally flattened in swordfishes and rounded in marlins and sailfishes. The bill is used as a club to stun prey and sometimes as a lance to impale large fishes. Sailfishes engage in cooperative hunting. Groups of 6–40 sailfish have been observed attacking schools of sardines, clubbing or slashing them with their bills (**Figure 9.8B**). Relatively few sardines are captured outright in these attacks, but large numbers are injured and are pursued and captured.

- Eight species of archerfishes that occur in brackish and freshwater habitats from India to the Philippines, Australia, and Polynesia have a unique method of capturing terrestrial prey: they spit a jet of water to knock insects off plants overhanging the water (**Figure 9.8C**). An archerfish can strike insects as much as 2 m away, despite the difficulty created by the refraction of light as it passes through the water surface, which makes the prey appear to be in a different position from its actual location. Archerfishes have a remarkable ability to discriminate patterns, even recognizing different human faces.

- Adult flatfishes (dabs, flounders, halibuts, plaice, soles, and turbots) rest with one side of the body (either right or left) on the seafloor (**Figure 9.8D**). They begin life

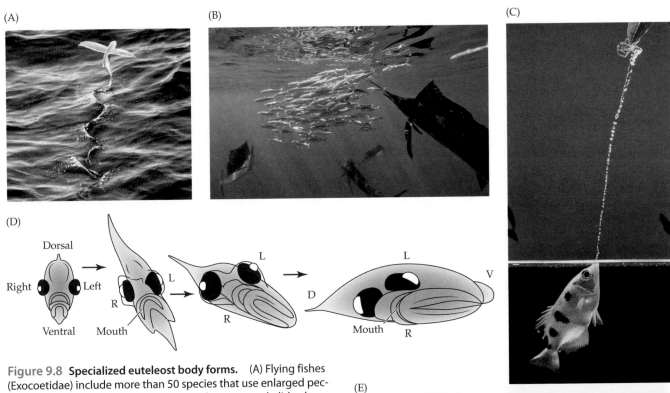

Figure 9.8 Specialized euteleost body forms. (A) Flying fishes (Exocoetidae) include more than 50 species that use enlarged pectoral fins as airfoils, allowing them leave the water and glide above the surface for significant distances. This behavior is defensive, a means for these fishes to escape their predators. (B) An individual Atlantic sailfish (*Istiophorus albicans*) usually attacks prey from either the right or left side, and capture success is higher for attacks from the favored side. Hunting groups of sailfish typically include approximately equal numbers of fish that attack from right and left, making it difficult for the prey (here, sardines) to determine the direction of the next attack. (C) With the opercula closed, compression of the buccal cavity forces a jet of water out of the mouth of an archerfish (*Toxotes jaculatrix*). Archerfish can hit an insect on an overhanging plant several body lengths away and can accurately predict where the falling prey will hit the water, enabling it to reach the prey before another fish steals it. (D) Stages in the metamorphosis of a Japanese flounder (*Paralicthys olivaceous*). Flatfishes such as flounder begin life as normal-looking bilateral pelagic larvae, but one eye migrates across the top of the head to the opposite side of the body. (E) Adult flatfishes rest on their side on the seafloor, and both eyes are on the same side of the body (either the left, as seen here, or the right). (A, courtesy of NOAA; B, © wildestanimal/Shutterstock.com; C, Avalon/Bruce Coleman Inc./Alamy Stock Photo; D, after Okada et al. 2003; E, Christian Gloor/CC BY 2.0.)

as normal-looking bilateral pelagic larvae, but during larval development one eye migrates across the top of the head (**Figure 9.8E**). Both eyes of adult flatfishes are on the same side of the head, although the jaws retain their original orientation.

9.2 Swimming

Perhaps the single most recognizable characteristic of fishes is their mode of locomotion. Fish swimming is immediately recognizable, aesthetically pleasing, and—when first considered—rather mysterious, at least compared with the locomotion of most land animals. Fish swim with anterior-to-posterior sequential contractions of the muscle segments along one side of the body and simultaneous relaxation of the corresponding muscle segments on the opposite side. Thus, a portion of the body momentarily bends, the bend moves posteriorly, and a fish oscillates from side to side. These lateral undulations are most visible in elongated fishes such as lampreys and eels (**Figure 9.9**).

Most fishes power swimming with muscles in the posterior region of the trunk and caudal peduncle and use the paired fins for steering and stopping, but a few fishes rely on fin movements for power. A variety of swimming movements have been described (**Figure 9.10**):

- *Anguilliform.* This movement is typical of highly flexible fishes capable of bending into more than half a sinusoidal wavelength. It is named for the type of locomotion seen in true eels, Anguilliformes.

- *Carangiform.* Undulations are limited mostly to the caudal region, with the body bending into less than

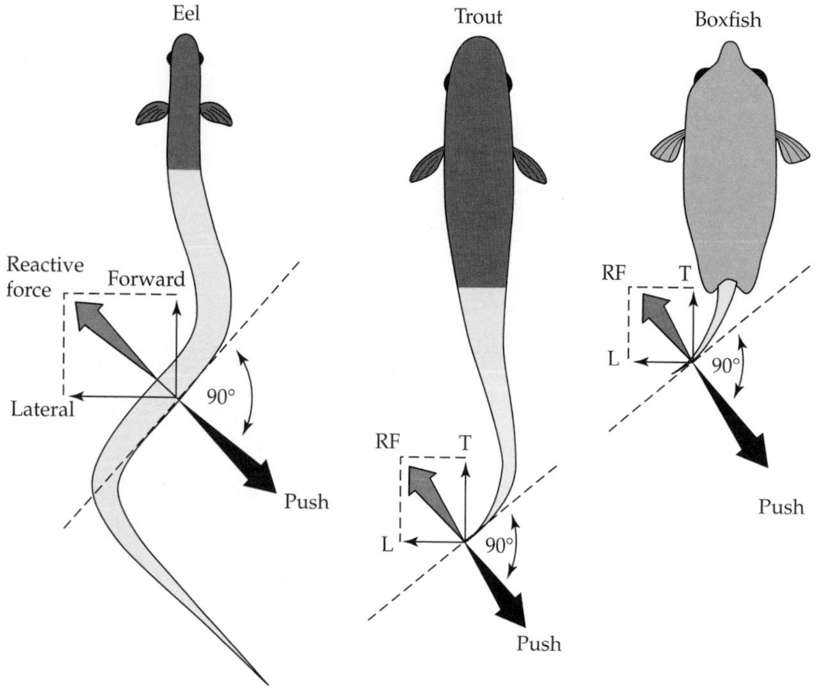

Eel

Trout

Boxfish

Reactive force

Forward

Lateral

90°

Push

RF   T

L

90°

Push

RF   T

L

90°

Push

Figure 9.9 Some basic movements of swimming fish. In general, fish swim forward by pushing backward on the water. For every active force, there is an opposite reactive force (Newton's third law of motion). Undulations of the body or fins produce an active force directed backward and also lateral forces. A vector diagram of the forces exerted on the water shows how lateral undulations can propel a fish forward. The lateral component of one undulation is canceled by that of the next (oppositely directed) undulation, so the fish swims in a straight line. The forward component of each undulation is in the same direction and thus is additive, so the fish moves forward through the water.

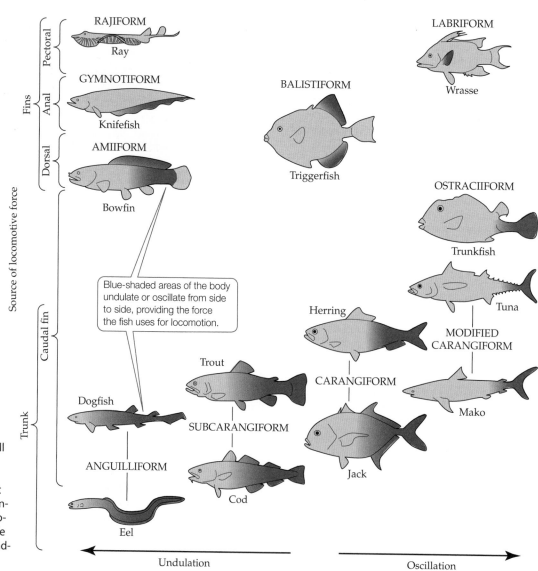

RAJIFORM
Ray

GYMNOTIFORM
Knifefish

AMIIFORM
Bowfin

LABRIFORM
Wrasse

BALISTIFORM
Triggerfish

OSTRACIIFORM
Trunkfish

Tuna

MODIFIED CARANGIFORM

Mako

Herring

CARANGIFORM

Jack

Blue-shaded areas of the body undulate or oscillate from side to side, providing the force the fish uses for locomotion.

Trout

SUBCARANGIFORM

Dogfish

Cod

ANGUILLIFORM

Eel

Undulation

Oscillation

Pectoral

Anal

Dorsal

Fins

Source of locomotive force

Caudal fin

Trunk

Figure 9.10 Types of movements. Words in all capitals (e.g., LABRIFORM) describe the major types of locomotion; they do not imply phylogenetic relationships, and the types of locomotion they describe grade into one another. (After Lindsey 1978.)

half a wavelength. The movement is named for *Caranx*, a genus of fast-swimming marine fishes known as jacks.

- *Ostraciiform.* The body is inflexible, and undulation is limited to the caudal fin. The movement is named for the boxfishes, trunkfishes, and cowfishes (Ostraciidae) whose fused scales form a rigid box around the body, preventing undulations.

- *Labriform.* Most of the force used for locomotion is provided by the pectoral fins, and the caudal fin plays a relatively small role. This movement is characteristic of wrasses and other fishes that maneuver through complex three-dimensional habitats, such as coral reefs.

- *Rajiform, amiiform, gymnotiform, and balistiform.* Sine waves are passed along the elongated pectoral fins (rajiform), the dorsal fin (amiiform), the anal fin (gymnotiform), or both the dorsal and anal fins (balistiform). Usually, several complete waves can be observed along the fin, and fine adjustments can be made in the direction of motion, including swimming backward. Because only the fins are moving, these types of locomotion can be stealthy and maneuvering can be precise.

Although these forms of locomotion were named for the groups of fishes that exemplify them, many other types of fishes use these swimming modes. For example, hagfishes, lampreys, most sharks, sturgeons, arawanas, many catfishes, and countless elongated spiny-rayed fishes are not true eels, yet they use an anguilliform swimming mode.

Minimizing drag

A swimming fish experiences two forms of **drag** (backward force opposed to forward motion): viscous (or frictional) drag, which is caused by friction between the fish's body and the water, and inertial drag, which is caused by pressure differences created by the fish's displacement of water. The two types of drag respond differently to speed and body form:

- *Viscous drag* is relatively constant over a range of speeds and is sensitive to the surface area of the body and the smoothness of the surface. A thin body has high viscous drag because it has a large surface area relative to its muscle mass; a scaleless skin (such as an eel's) has low viscous drag because it is smooth.

- *Inertial drag* increases as speed increases and is sensitive to the shape and cross-sectional area of the body. A thick body induces high inertial drag because it displaces a large volume of water as it moves forward. Streamlined (teardrop) shapes produce minimum inertial drag when their maximum width is about one-fourth their length and is situated about one-third body length from the leading tip.

The fastest swimmers, such as mackerels and tunas, use modified carangiform swimming, undulating only the caudal peduncle and caudal fin. These fishes have a caudal peduncle that is flattened from top to bottom and edged by sharp scales, so it creates minimum inertial drag as it sweeps from side to side. Swordfishes, which are reputed to be the fastest fishes, release a drag-reducing oil from pores on the head.

The caudal fin creates turbulent vortexes (whirlpools of swirling water) in a fish's wake that are part of the inertial drag. The total drag created by the caudal fin depends on its shape. When the aspect ratio of the fin (the dorsal-to-ventral height divided by the anterior-to-posterior length) is high, the amount of thrust produced relative to drag is high. The stiff, sickle-shaped fin of mackerels, tunas, swordfishes, and certain sharks results in a high aspect ratio and efficient forward motion. Even the cross section of the forks of these caudal fins assumes a streamlined teardrop shape, further reducing drag. Many species with these specializations swim continuously.

The caudal fins of trouts, minnows, and perches are not stiff and seldom have high aspect ratios. These fishes bend the body more than carangiform swimmers, in a swimming mode called subcarangiform (see Figure 9.10). Subcarangiform swimmers spread or compress the caudal fin to modify thrust and stiffen or relax portions of the fin to produce vertical movements of the posterior part of the body. The caudal peduncle of subcarangiform swimmers is laterally compressed and deep from top to bottom, and it contributes a substantial part of the total force of propulsion. These fishes often swim in bursts, usually accelerating rapidly from a standstill.

Steering, stopping, and staying in place

Fishes use all of their fins for maneuvering through their three-dimensional world. Observing fishes in an aquarium or in an online video will illustrate the behaviors described here.

A fish steers by extending the fins on one side of its body to create drag, just as a rower allows an oar to drag in the water to turn a boat. Any combination of fins can be used to steer—extending just a pectoral fin produces a gradual turn, whereas extending additional fins increases the drag and sharpens the turn. A fish also uses its fins to stop. The pectoral and pelvic fins rotate outward, and the posterior portions of the dorsal, anal, and caudal fins are flexed sideward to create drag.

Watching a fish stay in place in the water is fascinating. Teleosts use the gas bladder to make themselves neutrally buoyant, so you would think a fish could float in the water just as you can, with no movement of its appendages. A moment's observation will disprove that hypothesis, however—a motionless fish is actually moving several fins continuously. The pectoral fins maintain a sculling motion, and careful observation will show that the power stroke is directed forward. Looking a bit farther forward, you will see that the opercula are opening and closing as the fish breathes. Each time a fish exhales, water jets backward from beneath the opercula, creating a thrust that would drive the fish forward. The backward sculling movements of the pectoral fins counteract the forward thrust of water leaving the gill chambers.

Acute observation will reveal that a sine wave travels down the posterior margin of the caudal fin and sometimes the margins of the dorsal and anal fins as well. A downward sine wave creates an upward thrust, which should rotate the posterior end of the fish upward, yet the fish does not somersault—it remains horizontal in the water. The position of the gas bladder explains this observation. The volume of gas in the gas bladder can be adjusted to make the fish neutrally buoyant, but the gas bladder is located anterior to the center of gravity of the fish. As a result, a fish is tail-heavy and it would float tail-down without the upward thrust produced by the sine wave in its caudal fin.

9.3   Actinopterygian Reproduction

Actinopterygians show more diversity in their reproductive biology than any other vertebrate taxon.

- Most species lay eggs, which may be buried in the substrate, deposited in nests, or released to float in the water, and a few species lay eggs out of water, but viviparity has evolved in several lineages and nutrition of the embryos ranges from lecithotrophy to matrotrophy.

- Parental care ranges from none at all, through attending eggs in nests and protecting the young after they hatch, to carrying eggs in the mouth during development and allowing the young to flee back into the parent's mouth when danger threatens.

- Even sex determination shows diversity—the sex of most species of teleosts is genetically determined and fixed for life, but some species change sex partway through their lives, others are hermaphroditic, and a few species consist entirely of females.

The diverse modes of reproduction seen among fishes mean there are exceptions to every generalization. The following discussion focuses on the typical characteristics of the groups of fishes described.

Oviparity

The vast majority of ray-finned fishes are oviparous, and freshwater and marine species show contrasting specializations of oviparous reproduction (**Figure 9.11**).

Freshwater habitats   Teleosts that inhabit fresh water generally produce and care for a relatively small number of large, yolk-rich **demersal** (buried in gravel, placed in a nest, or attached to the surface of a rock or plant) eggs. Attachment is important in flowing water that might carry a floating egg away from habitats suitable for development. Because the eggs remain in one place, parental care is possible; males often guard the nests and sometimes the young.

The eggs of freshwater teleosts hatch into young that often have large yolk sacs containing a reserve of yolk that supports their growth for some time after hatching. When the yolk reserve has been consumed, the young fishes (called fry) generally resemble their parents.

Marine habitats   Teleosts that live in the sea generally release large numbers of small, buoyant, transparent eggs into the water. These eggs are left to develop and hatch while drifting in the open sea. The larvae are also small and usually have little yolk reserve remaining after they hatch. They begin feeding on microplankton soon after hatching.

Marine larvae are often specialized for life in the oceanic plankton, feeding and growing while adrift at sea for weeks or months, depending on the species. The larvae are generally very different in appearance from their parents, and many larvae have been described for which the adult forms are unknown. The larvae eventually settle into the juvenile or adult habitats appropriate for their species. Although the larvae drift with currents, increasing evidence suggests that many larvae actively choose their settling habitats, and groups of larvae can remain together until they settle.

The strategy of producing planktonic eggs and larvae that are exposed to a prolonged and risky pelagic existence appears to be wasteful of gametes. Nevertheless, complex life cycles of this sort are the principal mode of reproduction among marine fishes. Pelagic spawning may offer advantages:

- Predators on eggs may be abundant in the parental habitat but scarce in the pelagic realm.

- Microplankton (bacteria, algae, protozoans, rotifers, and minute crustaceans) are abundant where sufficient nutrients reach sunlit waters. The same winds and currents that transport larvae transport these food items, and both larvae and their food aggregate in convergence zones.

- Producing floating, current-borne eggs and larvae increases the chances of colonizing all patches of appropriate adult habitat in a large area. A widely dispersed species is not vulnerable to local environmental changes that could extinguish a species with a restricted geographic distribution. Perhaps the predominance of pelagic spawning species in the marine environment reflects millions of years of extinctions of species with reproductive behaviors that did not disperse their young as effectively.

Terrestrial habitats   Anyone who has lived near the California shore has probably heard of grunion runs. California grunions (*Leuresthes tenuis*) are small marine fishes that lay their eggs at the top of the tide mark on beaches from March through August. Tides reach their maximum for about two nights on each side of the full moon and the new moon, and this is when grunions spawn, riding a wave ashore. Females squirm tail-first into the wet sand as the wave recedes. One or more males cluster around a

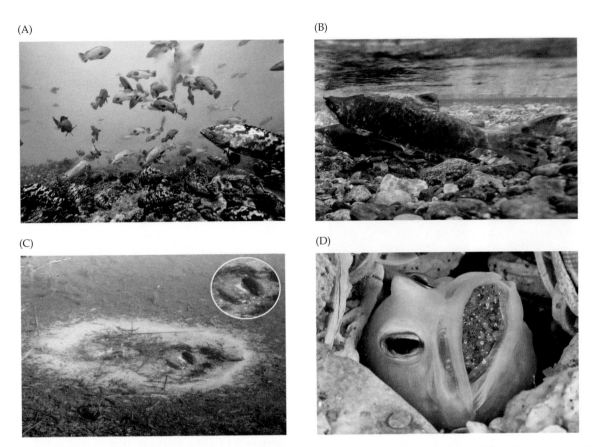

Figure 9.11 Spawning and parental care by oviparous ray-finned fishes. Reproductive modes of fishes extend from broadcasting eggs and sperm to extensive parental care. (A) Many marine fishes release eggs and sperm into the sea, where fertilized eggs hatch into larvae that drift with currents. Here, a spawning group of camouflage groupers (*Epinephelus polyphekadion*) are releasing a cloud of eggs and sperm. (B) Female sockeye salmon (*Oncorhynchus nerka*) move streambed pebbles to prepare a nest while a dominant male attends her. The female deposits eggs that are fertilized by the dominant male, then moves to the upstream edge of the nest and digs a new nest, covering the eggs in in the original nest at the same time. Adult sockeye salmon die after spawning. (C) A "nest ring." Male bluegill sunfish (*Lepomis macrochirus*) clear debris from a wide area to create a nest site. The male remains in attendance, defending the eggs against predators, which include other male bluegills, like the one shown in the inset. (D) Males in several families brood eggs in their mouths, as seen in this yellow-head jawfish (*Opistognathus aurifrons*). Individuals of this species are monogamous; the members of a pair live in adjacent burrows, sometimes trading burrows or occupying the same burrow. (A, Paul Mckenzie/wildencounters.net; B, Ryan Hagerty, USFWS; C, Scott Costello; D, Kevin Bryant.)

female and release sperm as she releases eggs, and then the fishes ride a receding wave back into the sea. Grunion eggs require about 10 days to hatch, and during this period of lower tides the nests are above the reach of waves. At the next high tide cycle—about 2 weeks after the eggs were laid—the grunion fry hatch and are swept back into the sea.

Even more un-fishlike are the reproductive behaviors of at least one species of mudskipper (*Periophthalmodon schlosseri*) and the rockhopper blenny (*Andamia tetradactyla*). Mudskippers are marine fishes of the Indo-Pacific region that live on mudflats, where they construct burrows. During high tide the mudskippers hide in their flooded burrows, and at low tide they emerge to forage on the mud. The burrows contain an upward bend that creates a chamber that is filled with air, even when the burrow entrance is covered by the tide, and the adhesive eggs are deposited on the roof of this chamber, well above the water and exposed only to air. Other mudskipper species also deposit eggs in their burrows, and it is likely that some of those eggs also develop in air.

Rockhopper blennies forage during low tide on wave-splashed mats of algae on rocky coastlines and shelter during high tide in crevices in the rocks. Females deposit eggs in the crevices, and males attend the nests, which are entirely out of the water except during the height of the tide or when they are washed by waves during storms.

Viviparity

The widespread occurrence of viviparity among familiar aquarium fishes in Poeciliidae (guppies, mollies, platys, swordtails, and related species) gives the impression that it is a common mode of reproduction for teleosts, but that is not true. Although viviparity is believed to have evolved in at least 12 lineages of teleosts, only 3% of teleosts are viviparous, and more than one-third of viviparous teleosts are poeciliids. This scarcity of viviparous lineages of teleosts is a bit surprising considering the diversity that teleosts display in other aspects of their reproductive biology.

(A)

Brood pouch

(B)

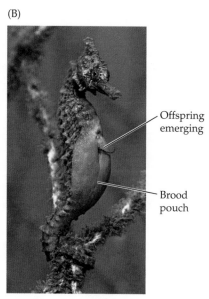

Offspring emerging

Brood pouch

Figure 9.12 Male seahorses brood the eggs. (A) Male and female seahorses mating. Egg transfer and fertilization occur during a seconds-long period when the female presses her abdomen against the male, aligning the opening of her oviducts with the opening of the brood pouch, which will close immediately after the eggs have been transferred. (B) The eggs develop within the brood pouch for 2–4 weeks. Here a male White's seahorse (*Hippocampus whitei*) is seen expelling a juvenile, which emerges tail-first (A, varrqnuht on Visualhunt.com/CC BY-SA; B, © Alex Mustard/Minden Pictures.)

Syngnathidae (seahorses, sea dragons, and pipefishes) provides an especially interesting example of male gestation. Males of all the species in this family carry the fertilized eggs until they hatch, but the details vary. Some species carry the eggs externally, adhering to the male's abdomen or his tail; in other species the eggs are partially enclosed in a pouch that remains open to the external environment. In the most derived species, including the yellow seahorse (*Hippocampus kuda*) of the Indo-Pacific region, the eggs are held in a closed pouch on the male's abdomen. The male and female mate by facing each other and pressing their abdomens together (**Figure 9.12A**). During this stage, which is very brief, the eggs are apparently released and pass across the opening of the sperm duct while sperm are released, and the eggs and sperm are carried through the opening of a male's brood pouch, which then closes tightly. The eggs develop within the pouch, and after a gestation period (2–4 weeks for most species), the male expels the young using muscular contractions (**Figure 9.12B**). Unfertilized eggs and some of the embryos die during development, and the brooding male reabsorbs nutrients from these tissues.

9.4 The Sex Lives of Teleosts

As mammals, humans are accustomed to thinking that the sex of an individual is determined genetically at the moment of conception, with female mammals carrying the XX genotype and males being XY. If we were birds, we would be equally comfortable with the assumption that females are always ZW and males are always ZZ. However, only about 10% of the fish species that have been studied have heteromorphic sex chromosomes. Some species are XX XY or ZZ ZW, while others have more complicated systems, such as X1 X2 Y, XY1 Y2, or ZW1 W2. Even closely related

species often have different sex chromosome systems, and the genes that actually determine sex vary. Furthermore, the environment plays a role in sex determination of some species.

Further complicating their sex lives, some fish species begin life as one sex and change to the other after they are adults (**Figure 9.13**). The principle that underlies these differences in life history is Darwinian fitness—that is, what allocation of time and energy to reproduction yields the highest representation of the alleles of an individual in succeeding generations relative to the alleles of other individuals in the population?

Protandry

Some teleost species exhibit **protandry** (Greek *prot*, "first"; *andros*, "male"), meaning individuals start life as males and change to females. Anemone fishes (about 30 species in the genus *Amphiprion*) are protandrous. Many people are familiar with anemone fishes from the popular animated film *Finding Nemo*. Nemo and his family are clown anemone fish (*Amphiprion ocellaris*), a species that lives in pairs or small groups that defend a territory within the tentacles of a sea anemone. The largest individual is always female and the second-largest is male, and both of these individuals are reproductively mature. Smaller individuals, if they are present, remain sexually immature. If the female dies, the male changes sex and the largest immature individual in the group becomes male. (The movie errs in this detail; when Nemo's mother died, his father should have become a female.)

The "size advantage hypothesis" proposes that the benefit of protandry lies in the relationship between body size and sperm and egg production. Large females produce more eggs than small ones, but even a small male can produce all the sperm needed to fertilize the eggs of a large female. Because of this relationship, a clown anemone fish can maximize its lifetime reproductive success by starting

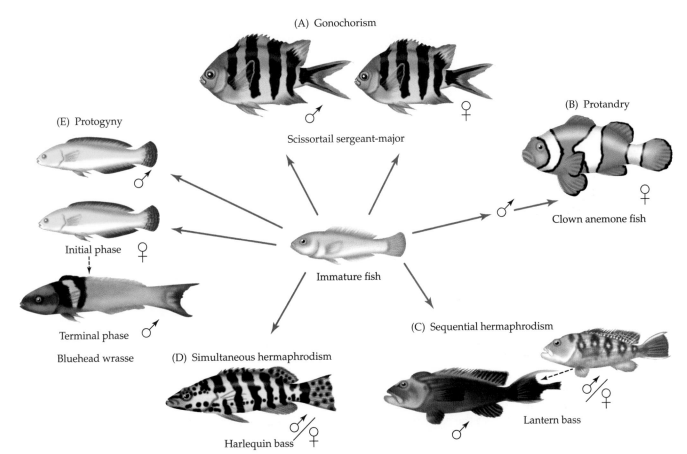

Figure 9.13 Sex determination in teleosts. Teleosts display an enormous variety of sex-determination mechanisms and life-history patterns. (A) In gonochorism, sex does not change during the lifetime of an individual. About 88% of teleosts are gonochoristic, illustrated here by the scissortail sergeant-major (*Abudefduf sexfasciatus*). (B) Protandrous fishes, illustrated by the clown anemone fish (*Amphiprion ocellaris*), begin life as males and change to females. (C) Some species of sea basses, including the lantern bass (*Serranus baldwini*), start life as simultaneous hermaphrodites and change to males when they are large enough to defend a harem of smaller hermaphrodites. (D) Fishes that are simultaneous hermaphrodites, such as the harlequin bass (*Serranus tigrinus*), alternate male and female roles during spawning. (E) Some protogynous fishes, such as the bluehead wrasse (*Thalassoma bifasciatum*), begin life as either males or females and grow into an initial phase that is identified by a pattern that is the same for both sexes. Males and females can reproduce in the initial phase, but the harem-based social system of this species greatly limits the reproductive success of initial-phase males.

life as a male and fertilizing the eggs of a large female, and then changing to a female when it has grown to a large size.

Protogyny

Protogyny (Greek *gyne*, "female") is the reverse of protandry. In protogyny, juveniles mature initially as females and subsequently change sex to become males. Protogyny is characteristic of species such as the bluehead wrasse (*Thalassoma bifasciatum*), in which territorial males defend mating territories. In this situation, large body size is advantageous to a male. Both male and female bluehead wrasses start adult life in a mostly yellow color pattern called the initial phase. Some initial-phase individuals (both males and females) subsequently change into terminal-phase males, which have a blue head, a black saddle, and a green posterior region.

In small populations of bluehead wrasse, spawning takes place over coral heads that protrude above the general height of the reef, and a terminal-phase male defends a territory centered on a coral head, mating with females

that come to the coral head to spawn. Initial-phase males also breed, but instead of defending a breeding site they swim in groups of a dozen or more individuals and try to intercept females. When a female spawns, all of the initial-phase males in the group release sperm. The mating success of initial-phase males is much lower than that of terminal-phase males. An initial-phase male has only one or two mating opportunities a day, and even then his sperm must compete with sperm from all the other males in his group. In contrast, a terminal-phase male mates 40–100 times a day without sperm competition from other males.

Clearly, the potential reproductive success of a terminal-phase male is greater than that of even the largest female because the male mates so often. In this situation it is advantageous for a large individual to be a male, and when a terminal-phase male disappears from its territory the largest female in the group changes to a male. The transition is astonishingly rapid. A field study found that the largest female began to display male behavior within minutes of the removal of the terminal-phase male from its territory, and individuals that had spawned as females one day mated as males on the following day.

Hermaphroditism

Simultaneous hermaphroditism means that an individual has functional ovaries and functional testes at the same time and can mate either as a male or as a female. This

life-history pattern is uncommon, but it does occur in about 20 families of teleosts, including in some sea basses (*Hypoplectrus, Serranus*), which have three mating patterns:

* Egg traders (or parcelers) alternate male and female roles during a single mating session. The member of the pair that is acting as a female releases a portion of its eggs (the parcel that gives this mating pattern its name),

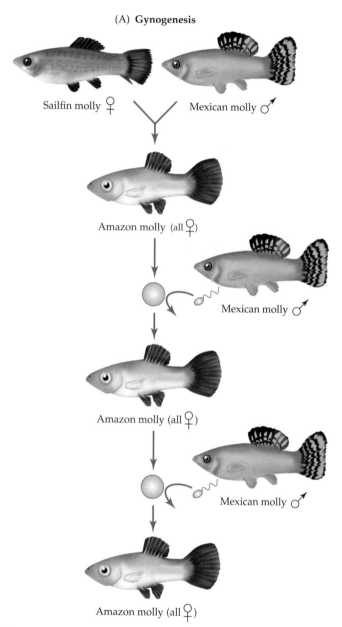

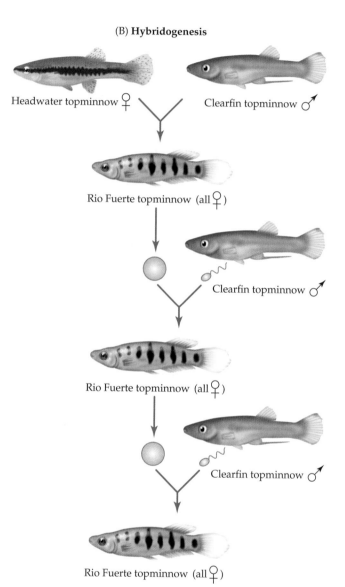

Figure 9.14 Gynogenesis and hybridogensis are forms of asexual reproduction. (A) In gynogenesis, a diploid egg must be activated by a sperm from a male of a bisexual species before it begins embryonic development. The DNA from the sperm never enters the egg, which develops into a clone of its mother. The Amazon molly (*Poecilia formosa*) is an all-female gynogenetic species that originated from hybridization between two bisexual species, the sailfin molly (*P. latipinna*) and the Mexican molly (*P. mexicana*). Amazon mollies produce diploid eggs that are activated by sperm from a male Mexican molly. (B) In hybridogensis, male DNA enters the egg but is eliminated at each generation. The Rio

Fuerte topminnow (*Poeciliopsis monacha-lucida*) is an all-female hybridogenetic species resulting from interbreeding of a female headwater topminnow (*P. monacha*) and a male clearfin topminnow (*P. lucida*). Elimination of the male's genome during meiosis produces a haploid egg carrying only the maternal DNA. This egg is fertilized by sperm from a male of a bisexual species, and in this case the DNA from the sperm *does* enter the egg, producing a diploid female with a combined genome. The DNA from the male is eliminated at meiosis, again producing an egg with only the DNA of the female parental species, and that egg is fertilized by sperm from a male of a bisexual species.

and the individual that is acting as the male releases sperm. Then the two individuals switch roles and continue alternating through the remainder of the spawning session, which may last for more than an hour.

- Reciprocating monogamous pairs alternate roles between spawning sessions; that is, the individual that releases eggs in one spawning session releases sperm in the next spawning session, which may occur some days later.

- Harem polygynous species combine an initial period of simultaneous hermaphroditism with sex change later in life. They live in harems consisting of as many as seven small hermaphrodites and a single large individual. These large fishes have lost their ovarian tissue, becoming functional males and mating with the hermaphrodites, which act as females.

Although the hermaphrodites in the harem could mate with each other, they almost never do so. The largest hermaphrodites have more testicular tissue than small ones, but it is not known whether these individuals, called "transitional hermaphrodites," will leave the harem to establish their own harems in new territories, or if they remain in the transitional phase until the male in their harem dies.

Only one species of self-fertilizing hermaphrodite has been documented among fishes, *Kryptolebias* (formerly *Rivulus*) *marmoratus*. This small fish (7.5 cm) lives in shallow brackish waters from Florida through the Caribbean. About 95% of the individuals are born as self-fertilizing hermaphrodites, which produce homozygous clones of the parent when they reproduce, and the remaining 5% are males—females have never been described in this species. In areas with temperatures below 20°C, about 60% of the hermaphrodites have transformed to secondary males after 3–4 years, but in warmer areas where the water temperature is 25°C or higher, the fish remain hermaphrodites throughout their lives.

All-female species

Although **parthenogenesis** (Greek *parthenos*, "virgin"; *genesis*, "origin")—reproduction by females without fertilization by males—is rare among fishes, four families of freshwater fishes include species that consist entirely of females. These species arose by hybridization between two bisexual species and are able to reproduce without any genetic contribution by a male. Two types of parthenogenesis are seen in fishes: gynogenesis and hybridogensis.

Gynogenesis In **gynogenesis** a diploid egg must be activated by a sperm from a male of a bisexual species, but the DNA from the sperm never enters the egg (**Figure 9.14A**). The gynogenetic Amazon molly (*Poecilia formosa*) is a unisexual hybrid of the bisexual sailfin molly (*P. latipinna*) and the bisexual Mexican molly (*P. mexicana*). Meiosis in Amazon mollies omits the final reduction division, so it produces diploid eggs. These eggs must be activated by sperm from a male Mexican molly in order to begin embryonic development, but the genetic material from the male never enters the egg, which develops as a clone of its mother.

Hybridogenesis In **hybridogenesis**, male DNA enters the egg but is eliminated at each generation (**Figure 9.14B**). The Rio Fuerte topminnow (*Poeciliopsis monacha-lucida*) is a unisexual hybrid that originated from the insemination of a female headwater topminnow (*P. monacha*) by a male clearfin topminnow (*P. lucida*). All Rio Fuerte topminnows are diploid females that carry a haploid genome from each of the parental species. During meiosis the paternal (clearfin) genome is eliminated, producing eggs with only the maternal (headwater) genome. Female Rio Fuerte topminnows mate with male clearfin topminnows, and in this case the male's DNA does enter the egg, producing a new generation of diploid female Rio Fuerte topminnows. These females again eliminate the clearfin genome at meiosis. Thus, only the genetic contribution of the maternal (headwater) species persists from generation to generation; the chromosomes from the paternal (clearfin) species are lost and replaced at each generation.

9.5 Teleosts in Different Environments

Given the great numbers of both species and individuals of extant fishes, especially the teleosts, it is little wonder that they inhabit an enormous diversity of watery environments, from the abyssal depths of oceans to high-elevation streams and lakes. Examining two vastly different habitats, the deep sea and coral reefs, allows us to better understand the amazing adaptability of teleost fishes. These two habitats differ in many ways, the most important of which are light and food.

Deep-sea fishes

As a place to live, the deep sea presents two major problems:

- There's no light. Even in the clearest water, light does not penetrate deeper than 1,000 m, so the depths of the sea are perpetually dark.

- There's very little food. Photosynthetic organisms are largely confined to the upper 200 m of the sea (the epipelagic zone; **Figure 9.15**). The food chain of the ocean depths is largely dependent on falling detritus, which can range in size from microscopic remains of plankton to carcasses of large fishes and whales. The plankton biomass can reach 500 mg per cubic meter of water at the surface, but by the time it has fallen to 10,000 m only 0.5 mg of plankton per cubic meter of water remains.

Despite these challenges, a distinctive and bizarre array of fishes lives in the mesopelagic and bathypelagic zones. Because there is not much food, deep-sea fishes are small (the average length is about 5 cm) and their populations are sparse. Both small size and sparse populations present problems: small individuals are vulnerable when they venture into regions with larger predators; and it is difficult to locate a mate when it's pitch dark and there is only one potential mate in 100,000 cubic meters of water. Many of the behavioral, anatomical, and life-history characteristics of deep-sea fishes are related to these two challenges.

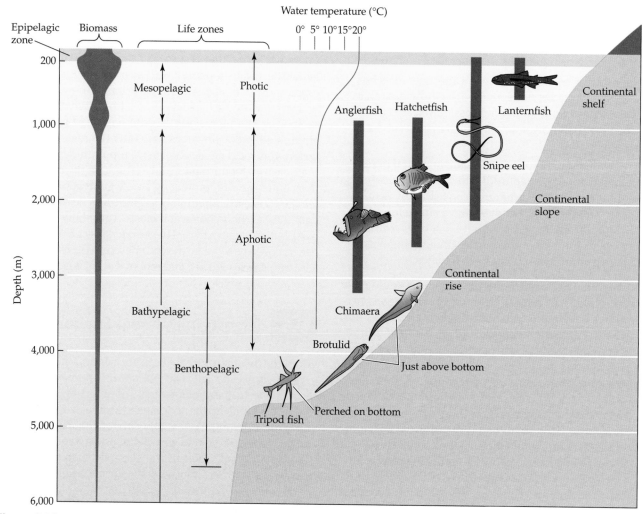

Figure 9.15 Life zones of the deep sea.   Most photosynthesis is limited to the upper 200 m, known as the epipelagic zone, and most of the animal biomass (indicated by the width of the green "Biomass" bar) is above 1,000 m. Water temperature remains constant below 1,000 m. Many bathypelagic fishes move up to the mesopelagic zone at night to feed where prey is more abundant and return to deeper water during the day. The ranges of vertical movement for these fish are indicated with bars. Benthopelagic fishes are deepwater bottom-dwellers and do not undertake vertical movements.

Mesopelagic fishes   One way of coping with the scarcity of food in the deep sea is to move to shallower water to feed, and many fishes and invertebrates that live in the mesopelagic zone do this. Mesopelagic fishes have large, upward-directed eyes that detect prey silhouetted by light from above (**Figure 9.16**). They move toward the surface at dusk to enter areas with more food and descend again near dawn. This vertical migration has both benefits and costs. By rising at dusk, mesopelagic fishes enter a region of higher productivity, where food is more concentrated so they can gather more energy. But the shallower water is also warmer, so their metabolic rates increase and they use energy faster. The daytime descent into cooler waters lowers their metabolism and conserves energy.

Bathypelagic fishes   The costs in energy and time to migrate several thousand meters from the bathypelagic zone to the surface outweigh the energy that would be gained from invading the rich surface waters. Instead of migrating, bathypelagic fishes are specialized for life in the depths.

• *Light organs.* Many deep-sea fishes and invertebrates are emblazoned with startling patterns formed by **photophores**, organs that emit blue light, and distinctive bioluminescent patterns characterize the males and females of many bathypelagic fishes. Tiny, light-producing photophores are arranged on their bodies in species-specific and even sex-specific patterns. The light is produced by symbiotic species of *Photobacterium* and groups of bacteria related to *Vibrio*. Some photophores, such as those in the barbels of dragonfishes, probably attract prey (**Figure 9.17A,B**). Others may act as signals to potential mates in the darkness of the deep sea. These sources of light are dim and intermittent, however, and the sensitivity of the vertebrate visual system has been pushed to its extreme by bathypelagic fishes.

(A)

(B)

White light

Blue light

Figure 9.16 Mesopelagic fishes. Light is dim in the mesopelagic zone (200–1,000 m) and fishes that live in this region have large eyes. (A) The silver John Dory (*Zenopsis conchifer*) descends to 600 m. Fishes with reflective silvery skins are hard to see in the open ocean. (B) The greeneye (*Chlorophthalmus acutifrons*) also has large eyes, as seen in surface-level light (left). This species reaches depths of 950 m, where sunlight is very dim; at these depths the fish is fluorescent, absorbing blue light end emitting it as green light (right). (A, SEFSC Pascagoula Laboratory, Collection of Brandi Noble, NOAA/NMFS/SEFSC; B, Edie Widder/NOAA Ocean Explorer.)

(A)

Lure

(B)

Lure

(C)

Eyes

(D)

Pelvic fins   Caudal fins

Pectoral fins with sensory elements

Figure 9.17 Bathypelagic fishes: Life without sunlight. Many bathypelagic fishes have biofluorescent organs that attract prey and identify the species. (A) The light organ of *Stomias affinis*, a barbled dragonfish, is at the end of a barbel (a whiskerlike sensory organ near the mouth) on the lower jaw. (B) *Photostomias guernei*, another species of barbled dragonfish, has a lighted lure on a barbel as well as light organs on the sides of the head. (C) *Ipnops murrayi* has no common name; the generic name can be translated as "sleepy eye." Instead of a compound eye with a lens that forms an image, *Ipnops* has an extended naked retina. This modification increases the sensitivity of the eye at the expense of directionality; the eyes of *Ipnops* can determine only that a light source is to the left or right of the fish. (D) The tripod fish (*Bathypterois grallator*) is benthic. It rests on its pelvic and caudal fins (top), while the long, flexible rays of its pectoral fins (bottom) are directed forward and sense the presence of its prey both by direct contact and by chemoreception. (A and inset, SEFSC Pascagoula Laboratory, collection of Brandi Noble, NOAA/NMFS/SEFSC; B, NOAA/Ocean Explorer; C, NOAA Okeanos Explorer Program, Gulf of Mexico 2012 Expedition; D, NOAA Okeanos Explorer Program, INDEX-SATAL 2010, NOAA/OER.)

(A)

(B)

(C)

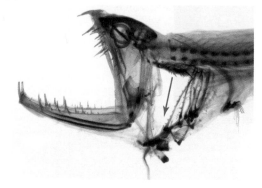

Figure 9.18 Bathypelagic fishes: All jaws and stomachs.
Prey are few and far between in the depths, and these fishes must be able to eat nearly anything they encounter. (A) Fearsome arrays of piercing teeth are characteristic of dragonfish such as *Chauliodus sloani*. (B) Like many bathypelagic fishes, the pelican eel (*Eurypharynx pelecanoides*) has a large mouth and expandable stomach and can swallow prey larger than itself. (C) Unlike most actinopterygian fishes, the barbled dragonfish (*Grammatostomias dentatus*) can bend its neck to increase the jaws' gape. A gap in the vertebral column (arrow) allows the notochord to act as a flexible joint. (A, Professor Francesco Costa, CC BY-SA 3.0; B, NOAA Okeanos Explorer Program, Index-SATAL 2010, NOAA/OER; C, courtesy of Nalani Schnell.)

- Vision is of little use in the perpetual dark of the deep-sea floor. Some deep-sea crustaceans and a few fishes have "naked retinas" that increase sensitivity (**Figure 9.17C**), others rely on tactile and chemosensation to detect prey (**Figure 9.17D**).

- *Large mouths and stomachs.* If a fish rarely encounters potential prey, it is important to have (1) a mouth large

Light organ   Parasitic male

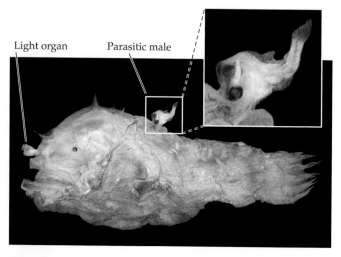

Figure 9.19 Parasitic males.   The low population densities of bathypelagic fishes make encounters between males and females chancy. The males of many species of ceratioid anglerfishes are obligate parasites. This 46 mm female *Photocorynus spiniceps* has a 6.2 mm male attached to her back. (Photos by Theodore W. Pietsch.)

enough to engulf nearly anything it does meet and (2) a gut that can expand to accommodate nearly any meal. The jaws and teeth of deep-sea fishes are often enormous in proportion to the rest of the body. Many bathypelagic fishes can be described as a large mouth accompanied by a stomach (**Figure 9.18A–C**).

The life history of ceratioid anglerfishes dramatizes how selection adapts a vertebrate to its habitat. The adults typically spend their lives in lightless regions below 1,000 m. Fertilized eggs, however, rise to the surface, where they hatch into larvae. The larvae remain mostly in the upper 30 m, where they grow, and later descend to the lightless region. Descent is accompanied by metamorphic changes that differentiate females and males. During metamorphosis, young females descend to great depths, where they feed and grow slowly, reaching maturity after several years.

Female anglerfishes feed throughout their lives, whereas males feed only during the larval stage. Males have a different future—reproduction, literally by lifelong matrimony. During metamorphosis, males cease eating and begin an extended period of swimming. The olfactory organs of males enlarge at metamorphosis, and the eyes continue to grow. These changes suggest that adolescent males spend a brief free-swimming period finding a female. The journey is precarious, for males must search vast, dark regions for a single female while running a gauntlet of other deep-sea predators. The sex ratio of young adults is unbalanced—often more than 30 males for every female.

Having found a female, a male does not want to lose her. He ensures a permanent union by attaching himself as a parasite to the female, biting into her flesh and attaching himself so firmly that their circulatory systems fuse (**Figure 9.19**). Preparation for this encounter begins during metamorphosis

(A) Midday

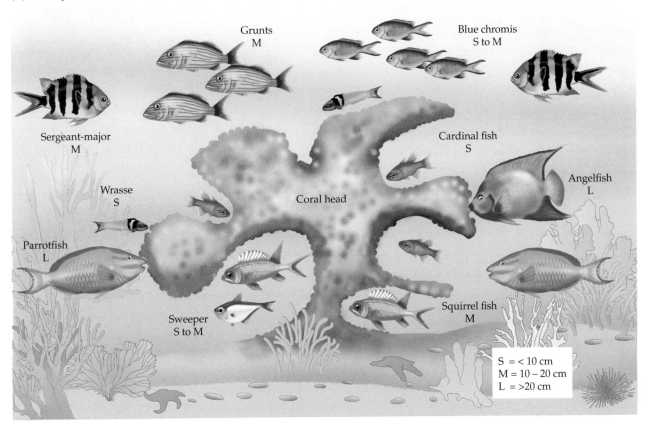

(B) Midnight

Figure 9.20 Day and night on a Caribbean coral reef. The activity and diversity of coral reef fishes change as diurnal species (A) retreat into the reef and nocturnal species (B) emerge.

when the male's teeth degenerate and strong toothlike bones develop at the tips of the jaws. A male remains attached to the female for life, and in this parasitic state, he grows and his testes mature. Monogamy prevails in this pairing, for females usually have only one attached male.

Coral reef fishes

The assemblages of vertebrates associated with coral reefs are the most diverse on Earth. These vertebrates are almost all teleosts—more than 600 species can be found on a single Indo-Pacific reef. Each of these species occupies a unique ecological niche, and the niches available to species on a coral reef are defined by the time of day a species is active, where it hides when it is inactive, what it eats, and how it captures its prey.

Ancestral actinopterygians had relatively unspecialized jaws and preyed on invertebrates. Many reef invertebrates became nocturnal to avoid these predators, limiting their activity to night and remaining concealed during the day. In response to the nocturnal activity of their prey, early acanthopterygians evolved large eyes that are effective at low light intensities. To this day, their descendants—squirrel fishes and cardinal fishes—disperse over the reef at night to feed. They use the irregular contours of the reef to conceal their approach and rely on a large protrusible mouth and suction to capture prey.

The evolution of jaws specialized to take food items hidden in the complex reef surface was a major advance. Some species rely on suction, whereas others use a forceps action of their protrusible jaws to extract small invertebrates from their daytime hiding places or to snip off coral polyps, bits of sponges, and other exposed reef organisms. This mode of predation demands high visual acuity that can be achieved only in the bright light of day. These selection pressures produced fishes capable of maneuvering through a complex three-dimensional habitat in search of food.

A coral reef has day and night shifts. At dusk the colorful diurnal fishes seek nighttime refuges in the reef, while the nocturnal fishes leave their hiding places and replace the diurnal fishes in the water column (**Figure 9.20**). The timing of the shift is controlled by light intensity, and the precision with which each species leaves or enters the protective cover of the reef day after day indicates that this is a strongly selected behavioral and ecological characteristic. Space, time, and the food resources available on a reef are partitioned through this activity pattern, but the growing exposure of coastal habitats to artificial light at night increases predatory behavior and has the potential to alter predator–prey relationships.

9.6   Heterothermal Fishes

The intimate contact between water and the gills of fishes that is required for gas exchange makes it difficult for a fish to maintain a body temperature that differs from the water temperature. As blood passes through the gills, it

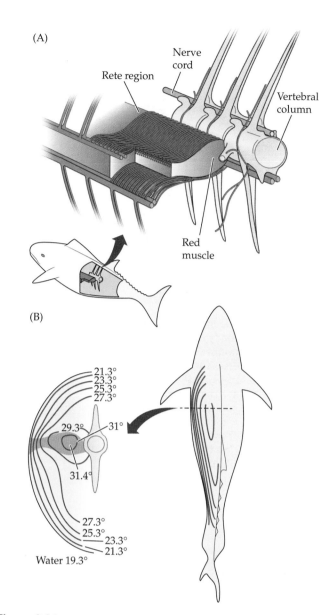

Figure 9.21 Regional heterothermy of bluefin tunas results from the location of red muscle.   (A) The red muscle and retia (rete regions, which are elaborate networks of blood vessels) of a bluefin tuna (*Thunnus*) are located deep within the body, adjacent to the vertebral column. (B) Muscular activity warms the red muscle, and the retia maintain a 12°C temperature gradient between muscle (31.4°C) and water (19.3°C). (After Carey and Teal 1966.)

comes into temperature equilibrium with the water. Heat produced by metabolic activity is lost via the gills, and most fishes have body temperatures that are the same as the water temperature. Nonetheless, a few species of fishes are able to keep parts of their bodies warm. In addition to the regionally heterothermal sharks described in Chapter 7, regional heterothermy has evolved at least four times among bony fishes: three times in Scombroidei, a group that includes mackerels, tunas, and billfishes, and in the opah (*Lampris guttatus*), a mesopelagic species that lives at depths of 100–500 m.

Warm muscles

Tunas have no specialized thermogenic (heat-producing) tissues. Instead, an arrangement of retia (blood vessel

(A)

(B)

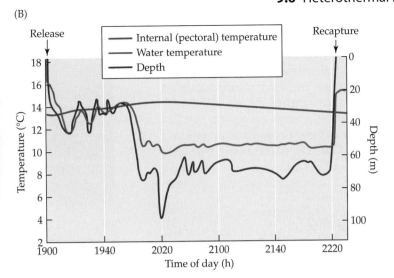

Figure 9.22 **The opah is the only fish known to keep its entire body warm.** (A) An adult opah (*Lampris guttatus*) aboard a NOAA scientific research vessel. The large eye is characteristic of mesopelagic fishes that forage at depths where light is dim (see Figure 9.16). Most of the heat production occurs in the muscles that move the large pectoral fins. (B) The core body temperature of a 39.2-kg opah changed by less than 1°C as the fish swam from the surface to a depth of 100 m and back to the surface, passing through a 6°C gradient of water temperature. (A, photograph by Adriana Gonzalez Pestana; B from Wegner et al. 2015.)

networks) in the brain, muscles, and viscera retains heat produced by myoglobin-rich swimming muscles (red muscles) and by digestion. The swimming muscles are located deep in the body (**Figure 9.21**), and the muscles of some species remain close to 30°C in water temperatures that range from 7°C to 23°C.

The opah keeps most of its body warm and maintains a stable body temperature as it moves from the surface layer of warm mixed water to deeper, colder water and back to the surface (**Figure 9.22**). Heat production is centered in the muscles that move the pectoral fins. These muscles are insulated by a layer of fat nearly 1 cm deep.

Unlike heterothermal fishes, the opah has countercurrent retia inside the gill arches that isolate the respiratory surfaces from the rest of the body, thus minimizing the loss of heat across the gills that limits warming in regionally heterothermal fishes. As a result, the opah's heart is warm, even when it dives into cold water. In contrast, the regional heterothermy of scombroids does not include the heart. Weaker cardiac contractions in cold water are believed to limit the ability of scombroids to make deep, prolonged excursions into cold, deep water (**Figure 9.23**).

Hot eyes

The billfishes warm only the brain and eyes. The superior rectus eye muscle is thermogenic, having changed its function from contraction to heat production. Mitochondria

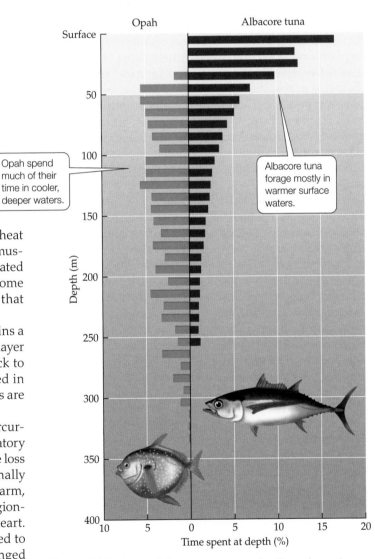

Figure 9.23 **The opah forages at greater depths and remains in deep water longer than the albacore tuna.** During the day, an opah spends most of its time at depths of 100–500 m, which is below the warm, mixed-surface layer of water. The albacore tuna (*Thunnus alalunga*), however, rarely goes deeper than 200 m and returns frequently to the warmer surface layer. (After Wegner et al. 2015; tuna data from Childers et al. 2011.)

occupy more than 60% of the cell volume, and changes in cell structure and biochemistry result in the release of heat by the calcium-cycling mechanism that is usually associated with contraction of muscles. The butterfly mackerel (*Gasterochisma melampus*) also warms the eye and brain, but in this species it is the lateral rectus eye muscle that has evolved thermogenic capacity and the same structural and biochemical characteristics as the superior rectus muscle of billfishes.

9.7 Sarcopterygians: Lobe-Finned Fishes

Two lineages of sarcopterygian fishes that first appeared in the Devonian have survived as aquatic forms to the present day: dipnoans (lungfishes) and actinistians (coelacanths). A third lineage, the Paleozoic tetrapodomorph fishes, contains the ancestors of tetrapods. Although they were abundant in the Devonian, the number of sarcopterygian fishes dwindled in the late Paleozoic and Early Mesozoic. Today only four non-tetrapod genera remain: the freshwater dipnoans or lungfishes (*Neoceratodus* in Australia, *Lepidosiren* in South America, and *Protopterus* in Africa) and the marine actinistian *Latimeria* (coelacanths) in waters 100–300 m deep off East Africa and central Indonesia (**Figure 9.24**).

Actinistians: Coelacanths

The coelacanth fossil record terminates in the Cretaceous, and until about 70 years ago these fishes were thought to be extinct. Then, in 1938 a coelacanth was caught off the coast of East Africa and named *Latimeria chalumnae* in honor of Marjorie Courtenay-Latimer, the South African naturalist who recognized the unusual nature of the fish. *Latimeria chalumnae* is large (specimens range from 75 cm to more than 2 m in length and weigh up to 90 kg) and is steel blue-gray in color with irregular pinkish white spots. It is a deepwater, bottom-dwelling fish, which may explain its absence from the Cenozoic fossil record, as such habitats rarely preserve.

Since the original discovery of *L. chalumnae*, more than 150 specimens have been caught in the Comoro Archipelago or in nearby Madagascar or Mozambique. Coelacanths are hooked near the bottom, usually in 260–300 m of water about 1.5 km offshore. In 1998 a second species of coelacanth, *L. menadoensis*, was discovered in Indonesian waters. DNA data show that it separated from the East African species about 6 Ma. Coelacanths are nocturnal, hiding in underwater caves during the day and venturing out at night to dine on small fishes, squids, and octopuses. Unlike extant lungfishes, coelacanths do not use their paired fins as props or to walk across the bottom, but when they swim the pectoral and pelvic appendages are moved in diagonal pairs, just as tetrapods move their limbs on land.

Coelacanths have several rather surprising features. Rather than having a ventral lung that can be used for

(A)

(B)

(C)

Figure 9.24 Extant sarcopterygian fishes. (A) *Latimeria chalumnae*, found in the West Indian Ocean, is the better known of the two extant coelacanths. (B) The Australian lungfish (*Neoceratodus forsteri*) is the only lungfish that retains clearly lobed fins (other lungfishes have threadlike fins). It is a facultative air breather that can survive without surfacing and shows no parental care. (C) The West African lungfish (*Protopterus annectans*) is an obligatory air breather and drowns if it cannot reach the water surface. Males attend their eggs, which are deposited in deep burrows. (A, © Peter Scoones/Science Source; B, Helen Sanders; C, Hippocampus Bildarchiv.)

respiration, as in lungfishes and basal actinopterygians, *Latimeria* have a vestigial lung that is filled with fat, which is less dense than water. Thus, the lung is probably a hydrostatic organ, akin to the lipid-filled livers of sharks. Although it has no function in gas exchange, this organ retains a pulmonary circulation like that seen in the primitive extant ray-finned fishes. *Latimeria* also have a unique rostral organ, a large cavity in the midline of the snout with three gel-filled tubes that open to the surface through a series of six pores. This organ is almost certainly an electroreceptor.

Latimeria resemble chondrichthyans in retaining urea to maintain a blood osmolality close to that of seawater. This osmoregulatory physiology may be an ancestral gnathostome condition. Like chondrichthyans, *Latimeria* have a rectal gland that excretes salt. *Latimeria* are viviparous; internal fertilization must occur, but how copulation is achieved is unknown because males have no specialized intromittent organ.

Coelacanths are the only extant taxa to retain the intracranial hinge between the front and back portions of the braincase. Biomechanical modeling shows that this joint, which is moved by the axial muscles, probably enables suction feeding and allows generation of high bite forces. Coelacanths are often called "living fossils," but this is a misnomer. Extant coelacanths are morphologically distinct from their extinct relatives, as well as several times bigger than most of them. However, studies of the coelacanth genome show that its protein-coding genes are evolving more slowly than those of tetrapods.

Dipnoans: Lungfishes

Extant lungfishes share several features of soft anatomy with tetrapods, in addition to the lung (which is probably a primitive bony fish feature), and molecular data also show that lungfishes are more closely related to tetrapods than are coelacanths. Most notably, lungfishes have a heart with a divided atrium and a pulmonary vein that feeds back into the right atrium. (The pulmonary vein of basal actinopterygians such as *Polypterus* feeds back into the sinus venosus, and the atrium is undivided.) Lungfishes also resemble tetrapods in having choanae, or internal nostrils, but this feature is now thought to have arisen convergently.

Extant lungfishes are found in the Southern Hemisphere. The Australian, or Queensland, lungfish (*Neoceratodus forsteri*) is similar in morphology to Paleozoic and Mesozoic dipnoans, retaining lobelike fins and heavy scales. The other genera, the African *Protopterus* and the South American *Lepidosiren*, have reduced scales, long eel-like bodies, and threadlike fins with a reduced internal skeleton. Despite their reduced fins, African lungfishes can actually walk and bound underwater, using their pelvic fins to lift themselves off the substrate. *Neoceratodus* has well-developed gills, and although it uses its lung to obtain additional oxygen, it is not dependent on air-breathing. The other lungfishes have reduced gills and die if denied access to air. Chemical senses seem important to lungfishes, and their mouths contain numerous taste buds.

The Australian lungfish attains a length of 1.5 m and a reported weight of 45 kg. It inhabits permanent bodies of water, large lakes and rivers, and does not estivate. Australian lungfishes have a complex courtship that may include male territoriality, and they are selective about the vegetation on which they lay their adhesive eggs; however, no parental care has been observed after spawning. The jelly-coated eggs, 3 mm in diameter, hatch in 3–4 weeks, but the young are elusive and nothing is known of their juvenile life.

Surprisingly little is known about the South American lungfish (*Lepidosiren paradoxa*). Males develop vascularized extensions on their pelvic fins during the breeding season that probably deliver oxygen from a male's blood to the eggs and young that he is guarding in a nest cavity.

The four species of *Protopterus*, the African lungfishes, are better known. They live in areas that flood during the wet season and bake during the dry season—habitats that are not available to actinopterygians except by immigration during floods. When the floodwaters recede, a lungfish digs a vertical burrow in the mud with an enlarged chamber at the end, and becomes increasingly lethargic as drying proceeds. Eventually the water in the burrow dries up, and the lungfish enters the final stages of estivation—folded into a U shape with its tail over its eyes. Heavy secretions of mucus that the fish has produced since entering the burrow dry and form a protective envelope around its body. Only an opening at its mouth remains, to permit breathing. Estivation is an ancient trait of dipnoans. Fossil burrows containing lungfish tooth plates have been found in Carboniferous and Permian deposits in North America and Europe.

9.8 Pollution, Overfishing, and Fish Farming

Fishes confront a host of problems, and many species that were once common have been brought to the verge of extinction. Like all extant organisms, fishes suffer from threats that stem directly or indirectly from human activities, including climate change, habitat loss, pollution, and disease. Some of these issues create particular problems for fishes.

Freshwater fishes

Nearly 40% of the world's fish species live in fresh water, and all of them are threatened by the alteration and pollution of lakes, rivers, and streams. Draining, damming, canalization, and diversion of rivers create habitats that no longer sustain indigenous fishes. In addition to the loss and physical degradation of fish habitat, fresh waters in much of the world are polluted by silt and toxic chemicals of human origin.

Lakes and rivers are vulnerable to pollution. Some pollutants, such as oil, are rapidly toxic, whereas heavy metals (e.g., arsenic, cadmium, cobalt, copper, lead, manganese, nickel, uranium, and zinc) accumulate in body tissues and weaken fishes, rendering them susceptible to other stresses, such as high temperatures or low oxygen levels. Heavy-metal-containing drainage from coal and mineral mines in North America has a greater impact on fishes than agriculture, runoff from roads, or urbanization.

Like other vertebrates, including humans, fishes are susceptible to endocrine-disrupting chemicals, including compounds in pesticides and herbicides, bisphenol A from plastics, and estrogens from birth control pills excreted in human urine. Feminization of male fishes is a common effect of endocrine-disrupting chemicals, which also affect social and parental behavior and survival on a transgenerational basis.

Nearly 800 species of native freshwater fishes once lived in the United States, but 61 of those species are extinct, and almost 40% of the remaining species are imperiled. The situation is no better in Europe, where 12 species are extinct and 38% of the remaining freshwater fish species are threatened with extinction. Assessments of this sort for Asian countries are only now being initiated.

Marine fishes

Commercial overfishing is a sadly familiar problem. Many of the world's richest fisheries are on the verge of collapse. Georges Bank, which lies between Cape Cod, Massachusetts, and Cape Sable Island, Nova Scotia, provides an example of what overfishing can do. For years, conservation organizations called for a reduction in catches of Atlantic cod (*Gadus morhua*), yellowtail flounder (*Pleuronectes ferruginea*), and haddock (*Melanogrammus aeglefinus*) on Georges Bank, but their concerns were not heeded.

Not surprisingly, the populations of many commercial fish species crashed dramatically in the 1990s. By October 1994, the situation in northeast North America was so bad that a government and industry group, the New England Fishery Management Council, directed its staff to devise measures that would reduce the catch of those species essentially to zero. Nearly 25 years later, a few encouraging signs indicate that these draconian measures have allowed some species to increase. Populations of cod and haddock are on the rebound, although the average size of the fishes is less than half what it was in the pre-crash years, and full recovery, if it is possible, will require decades.

These few gains, however, must be balanced against losses. In 2011, the European Fisheries Commission determined that the fisheries policy that the European Union had been following since 1983 has been a failure and called for far more stringent regulations beginning in 2013.

Reproductive biology and overfishing Cod, flounders, and haddock mature relatively fast (in as little as 2 years) and produce vast numbers of eggs—up to 11 million eggs per year for a large female cod. Early maturity and high fecundity are characteristics that give a species high reproductive potential. Species of this sort are relatively resistant to overharvesting, and they are able to rebuild populations when excessive harvesting is stopped.

Maturing slowly and producing a small number of eggs relative to adult body size are characteristics that make a species vulnerable to overfishing, and large predatory fishes, such as tunas and billfishes, fall into this category. On a unit weight basis, the egg production of bluefin tunas is less than 3% that of cod, and bluefins take more than twice as long as cod to reach reproductive size. To make the situation even worse, tuna flesh is considered a delicacy and bluefin tunas fetch high market prices, which rise even higher as the supply falls.

A recent analysis by Bruce Collette of the National Marine Fisheries Service and colleagues from similar organizations around the world found that eight of the ten large predatory fish species that have generation times of 6 years or longer meet the IUCN criteria for being Vulnerable, Endangered, or Critically Endangered. The study concluded that high commercial value and long life create double jeopardy for these species.

Fish farming Overfishing is obviously directly attributable to humans. A different kind of threat to marine fishes results from another human activity, fish farming. As populations of commercially important, overfished species have plummeted, the role of fish farming has increased. Many fish farms are huge, netted enclosures that are anchored in rivers or in coastal waters (**Figure 9.25A**). Young fishes from a hatchery are placed in the enclosures and fed until they have grown to harvestable size. Although fish farming is promoted as an ecologically sound alternative to catching wild fishes, it creates both short-term and long-term problems.

In the short term, concentrating thousands of fishes in a small volume of water and feeding them artificial food create a tremendous accumulation of feces and uneaten food that promotes blooms of bacteria and algae. In addition, the fishes crowded into pens are susceptible to disease and parasites that can be transferred to wild fishes. Sea lice are tiny crustaceans that parasitize fishes, and farmed salmon are heavily infected with them (**Figure 9.25B**). When wild salmon swim past the farm pens, they become infected with sea lice. A recent study found that a single fish farm in British Columbia increased the infection of juvenile wild salmon by 73-fold, and the effect extended for 30 km along the migration paths the wild fishes followed.

An analysis of 35 years of wild salmon populations compared 7 rivers that flow into channels containing fish farms with 64 rivers in which wild salmon were not exposed to fish farms. Until 2001, when sea lice appeared in the fish farms, there were no differences in the year-to-year population changes in salmon in the two groups of rivers. From 2001 onward, the wild salmon populations in rivers that were exposed to fish farms shrank every year, whereas the populations in rivers that were not exposed to fish farms remained constant. The wild populations exposed to fish farms are shrinking so fast that, if the current trend continues, they will be 99% gone in four fish generations. Furthermore, a drug used to control sea lice is toxic to other crustaceans, and may do more harm than good.

The rationale for fish farming is that the availability of farmed fishes will reduce the need to catch wild fishes. This reasoning is only partly correct, because farmed salmon and sea trouts are raised on fish meal; about 5 kg of meal are needed to produce 1 kg of farmed salmon. The meal is

(A)

(B)

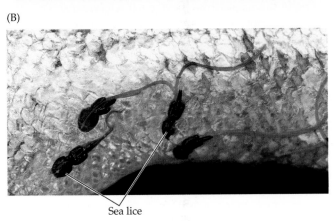

Sea lice

Figure 9.25 Fish farms create problems for wild fishes.
(A) A salmon farm in the Broughton Archipelago of British Columbia comprises a series of large pens in a waterway that is used by wild salmon. Pathogens and parasites flourish and can freely move out of the pens to infect wild fishes. (B) At the right of this photograph, blood flecks can be seen on the skin of an escaped juvenile farmed salmon, the result of infestation with sea lice (parasitic copepods, which are tiny crustaceans). (A, Chris Cheadle/Alamy Stock Photo; B, Robin Lund/Alamy Stock Photo.)

produced by catching small fishes and invertebrates, thereby depleting populations of the food base needed by wild fishes. In the long run, taking increasing amounts of forage fishes to feed farmed fishes will be disastrous for marine ecosystems.

The conspicuous problems associated with fish farming, such as pollution, disease, parasites, and depletion of prey species, are not the only concerns. Recent work has focused attention on the interaction of evolutionary processes with ecological and commercial elements of fishery biology. For example, the reproductive physiologies of farmed and wild salmon are different. One of the general features of the life history of animals is a trade-off between the number of eggs produced and the size of each egg—many small eggs versus fewer large eggs.

The large eggs laid by wild salmon produce large hatchlings that can survive the risks that wild hatchlings face. In contrast, hatchlings of farm-raised salmon are protected from most of the hazards that confront wild hatchlings. As a result, farmed salmon have shifted toward producing many small eggs. In just four generations, the egg size of farmed salmon can decrease by 25%.

That change in egg size would not be a problem if farmed fishes remained in farms, but not all of them do—escapes are a regular occurrence and sometimes involve large numbers of fishes. Storms have allowed millions of farmed salmon to escape from their pens and enter the wild salmon population, bringing with them genotypes that produce small eggs. The influence of this sudden genetic input on wild populations of salmon is unknown, but it is potentially harmful to the wild genotype that has been shaped by millennia of natural selection.

Summary

Osteichthyes (in a phylogenetic sense) includes most of the extant species of vertebrates.

The two lineages of bony fishes include a similar number of species, but they have diversified in different environments.

Actinopterygians (ray-finned fishes) are the most diverse aquatic vertebrates, having radiated in freshwater and marine habitats from mountain lakes to the abyssal depths of the ocean.

Extant sarcopterygians (lobe-finned fishes) are nearly all terrestrial and are as diverse on land as their sister group is in water.

Understanding the phylogeny of the more than 33,000 species of ray-finned fishes is an ongoing endeavor.

Most extant actinopterygians are teleosts, but a few specialized forms of non-teleost radiations have survived.

Polypteriformes (bichirs and the reedfish) is the sister group of the other two lineages of actinopterygians, Acipenseriformes (sturgeons and paddlefishes) and Neopterygians. Neopterygii consists of Holostei (gars and the bowfin) and Teleostei.

(Continued)

Summary (continued)

More than 95% of extant ray-finned fishes are teleosts.

Duplication of the entire genome in the ancestor of teleosts played a role in their diversification, but the more recent diversifications of two lineages, Ostariophysi and Perciformes, account for about 88% of teleost diversity. Genome duplication was not a factor in those two radiations.

Osteoglossomorphs ("bony tongues") are considered to be the most primitive teleosts. The arawanas and pirarucu are large fishes from the Amazon basin. Mormyrids, found in sub-Saharan Africa, use discharges from an electric organ to sense the environment and to communicate with conspecifics.

Elopomorphs include the true eels as well as tarpons and bonefishes. The character that unites the elopomorphs is their leptocephalus ("small head") larva.

Otocephalans include two large lineages: clupeomorphs (herrings and other fishes that use their gill rakers to strain zooplankton from the water) and ostariophysans. Ostariophysans have two important derived characters: the Weberian apparatus, which uses a chain of bones to transmit vibrations from the gas bladder to the inner ear, endowing ostariophysans with more sensitive hearing than other teleosts; and an alarm substance (pheromone) in the skin that produces a fright reaction in nearby conspecifics or other ostariophysan species.

Euteleosts are the largest lineage of teleost fishes, and the clade has diversified in freshwater and marine habitats worldwide. Many euteleosts fit the image of a generic fish, but the group includes species with remarkable morphological and behavioral specializations. Billfishes (marlins, swordfishes, and sailfishes) use a greatly elongated premaxilla to club or impale prey, and sailfishes improve their hunting success by cooperating to attack schooling prey. Archerfishes shoot insects from overhanging vegetation by spitting a stream of water. Flatfishes begin life as bilaterally symmetrical larvae, but as they develop one of their eyes migrates across the top of the head; as adults, they lie on their sides on the seafloor.

Swimming is the most distinctive thing that fishes do, and fishes have evolved many specialized modes of swimming.

Fish swim with undulations that press alternately left and right; the lateral thrust components cancel each other out, and the backward thrust forces the fish forward.

Most fishes use the posterior part of the body and the tail to generate thrust, but some species use the fins.

Swimming modes are classified according to the portion of a fish's body that exerts thrust and are named for groups of fishes that use that mode. This convention can be misleading, because fishes outside the clade on which a name is based may also use the mode. For example, the anguilliform (eel-like) swimming mode is characteristic of many elongated species in addition to eels.

Fish use their fins to steer and stop, much as a rower uses oars to steer and stop a boat. Even hovering motionless in the water requires fin movements, to prevent water jetting from the gills from moving a fish forward and to compensate for the asymmetrical location of the gas bladder.

Actinopterygians show more diversity in reproductive modes than any other lineage of vertebrates.

Oviparity is the ancestral and the most common mode of reproduction among extant fishes. Males of many species construct nests and attend the eggs and sometimes even the young after they hatch.

Many marine fishes spawn in the water column, with pairs or groups of males and females releasing eggs and sperm simultaneously. Fertilized eggs hatch into larvae that are dispersed by the current and can travel long distances before metamorphosing.

Sex-determination mechanisms in fishes range from heteromorphic sex chromosomes to environmental effects, with many intermediates. Even when sex determination is genetic, the sex-determining genes can differ in related species.

Some fishes change sex during their lives, starting either as male or as female and switching sex at some point. Other species are hermaphrodites and function as males *and* females, sometimes alternating on a regular basis. Some species of fishes consist only of females and reproduce without incorporating genetic material from males.

Teleosts have colonized aquatic habitats from the depths of the sea to high elevations, revealing remarkable abilities to adapt to different environments.

Deep-sea fishes face two major challenges: darkness and limited food. Light does not penetrate below 1,000 m, so depths below this are permanently dark. Without light, there is no photosynthesis, and deep-sea communities depend on organic detritus that falls from closer to the surface.

Many fishes that live at moderate depths (mesopelagic fishes) perform daily vertical migrations, rising toward the surface at night to a region where more food is available and sinking deeper during the day.

Bathypelagic fishes live at depths too great to make vertical migration effective; instead, many have enormous mouths and stomachs that allow them to eat fishes larger than themselves, or have lures that attract prey.

Population densities of deep-sea fishes are low, and locating a mate can be challenging. Male anglerfishes take no chances—when a male encounters a female, he attaches to her as a parasite, fusing his circulatory system to hers.

Coral reefs and support dense, multispecies communities of fishes that are specialized for different modes of feeding. Not even darkness stops fish activity on a reef; the diurnal species of fish retreat to sheltered hiding places while the nocturnal species emerge.

The intimate contact between blood and water in the gills that is necessary for gas exchange makes it difficult for fishes to maintain body temperatures different from that of the surrounding water, but the ability to maintain parts of the body at a temperature higher than the surrounding water has developed in some teleosts.

Although there are more than 33,000 extant terrestrial species of sarcopterygians, only 8 aquatic species survive: 6 species of lungfishes (three genera) and 2 species of coelacanths (one genus).

Although actinistians were thought to have been extinct since the Cretaceous, two extant species of coelacanths have been discovered within the past 100 years—one off the east coast of Africa and the other in the Indonesian Archipelago. Both species are deepwater and nocturnal, hiding during the day and foraging at night. Although they do not use their limbs to walk on the seafloor, they do move them in diagonal pairs (like tetrapods) when they swim.

Coelacanths exhibit several unusual features: a vestigial lung filled with fat, used as a buoyancy device; a rostral organ that is almost certainly electrosensitive; and retention of the basal sarcopterygian intracranial hinge,

which enables both suction feeding and the generation of high bite forces. Coelacanths resemble chondrichthyans in retaining urea for osmoregulation in the marine environment, and like many chondrichthyans, they are viviparous.

Three genera of lungfishes (dipnoans) are found in rivers and lakes of the Southern Hemisphere, and one genus of coelacanth (actinistians) is found in the Indian Ocean.

Lungfishes are more closely related to tetrapods (the remaining sarcopterygians) than coelacanths are, and they share with tetrapods several features of soft anatomy, including a heart with a divided atrium. Lungfishes also have choanae (internal nostrils), but this may be a convergent feature.

The Australian lungfish (*Neoceratodus forsteri*) is most like the ancestral lungfish condition, retaining heavy scales, robust fins, and a substantial gill system. The South American lungfish (*Lepidosiren paradoxa*) and the African lungfishes (*Protopterus*) have a reduced scale cover, eel-like bodies with ribbonlike fins, and reduced gills. While all lungfishes breathe air, *Lepidosiren* and *Protopterus* will die if denied access to air. African lungfishes estivate in burrows on land during the dry season, a behavior also recorded in fossil lungfishes.

Fish confront many anthropogenic threats.

Habitat modification and loss are major threats to freshwater fishes, as rivers are dammed or canalized and lakes and marshes are drained. Pollutants flushed into surface waters can be directly toxic (e.g., oil spills) or have cumulative deleterious effects (many heavy metals).

Overfishing is a major threat to marine fishes, and impending collapse of historically rich fisheries has resulted in draconian limits on capture quotas. Signs of recovery of some fish stocks are appearing, but full recovery will require decades, and other fisheries have sunk even deeper into decline.

Fish farming, touted as the solution to overfishing of wild fish, has been at best a mixed blessing. Crowded into netted pens, farmed fish are vulnerable to diseases and parasites that spread to wild fishes, and production of food for farmed fishes is consuming increasing quantities of the small fishes and crustaceans on which wild fishes feed. In addition, huge numbers of farmed fishes escape during storms, interbreeding with wild fishes and contaminating the gene pools of wild populations.

Discussion Questions

1. How do actinopterygian fishes and sarcopterygian fishes use their pectoral fins? Evaluate the structure of the pectoral fins in those two lineages in the context of the differences in how they are used.

2. Explain briefly why modifications of the jaws of early neopterygians (see Chapter 8) represented an important advance in feeding, and how further modifications seen in the jaws of teleosts allow still more specialization.

3. In general, freshwater teleosts lay adhesive eggs that remain in one place until they hatch, whereas marine teleosts release free-floating eggs that are carried away by water currents. (There are exceptions to that generalization in both freshwater and marine forms.) What can you infer about differences in the biology of newly hatched freshwater and marine teleosts on the basis of that difference in their reproduction?

4. Teleosts exhibit a bewildering variety of patterns of sex change, the most common of which are protogyny and protandry. Describe the evolutionary principle that underlies those two patterns.

5. Cuckolding males are a behavioral variant that occurs in some species of fishes in which territorial males build and defend nests. While a female releases eggs and the territorial male releases sperm at the nest, a cuckolder male sneaks into the nest and releases sperm that fertilize some of the eggs. The bluegill sunfish (*Lepomis macrochirus*) provides a classic example of this mating tactic:

- Territorial males do not reproduce until they are 7 years old.

- Cuckolder males can begin to fertilize eggs in their second year.

- Females begin to reproduce in their fourth year.

   a. What are the advantages and disadvantages of the two male tactics?

   b. In their second and third years, small cuckolder males hide in vegetation to escape notice by the territorial male, and emerge only to make a rapid dash through the nest to release sperm. Starting in the fourth year, however, cuckolder males mimic females. What advantage do they gain from female mimicry, and why do they wait until they are 4 years old to initiate this behavior?

Additional Reading

Bhandari RK and 8 others. 2015. Effects of the environmental estrogenic contaminants bisphenol A and 17 α-ethinyl estradiol on sexual development and adult behaviors in aquatic wildlife species. *General and Comparative Endocrinology* 214: 195–219.

Block BA, Finnerty JR, Stewart AFR, Kidd J. 1993. Evolution of endothermy in fish: Mapping physiological traits on a molecular phylogeny. *Science* 260: 210–214.

Burnette MF, Ashley-Ross MA. 2015. One shot, one kill: The forces delivered by archer fish shots to distant targets. *Zoology* 118: 302–311.

Casane D, Laurente P. 2013. Why coelacanths are not "living fossils." *Bioessays* 35: 332–339.

Collette BB and 32 others. 2011. High value and long life: Double jeopardy for tunas and billfishes. *Science* 333: 291–292.

De Busserolles F, Marshall NJ. 2017. Seeing in the deep-sea: Visual adaptations in lanternfishes. *Philosophical Transactions of the Royal Society B* 372: 1717.

Helfman GS, Collette BB, Facey DE, Bowen BW. 2009. *The Diversity of Fishes*, 2nd ed. Wiley-Blackwell, Hoboken, NJ.

Liu H, Todd EV, Lokman PM, Lamm MS, Godwin JR, Gemmell NJ. 2016. Sexual plasticity: A fishy tale. *Molecular Reproduction and Development* 84: 171–194..

Nelson ME. 2011. Electric fish. *Current Biology* 21: R528–R529.

Tacon AGJ, Metian M. 2009. Fishing for feed or fishing for food: Increasing global competition for small pelagic forage fish. *Ambio* 38: 294–302.

Wainwright PC, Longo SJ. 2017. Functional innovations and the conquest of the oceans by acanthomorph fishes. *Current Biology* 27: R550–sR557.

Wegner NC, Snodgrass OE, Dewar H, Hyde JR. 2015. Whole-body endothermy in a mesopelagic fish, the opah, *Lampris guttatus*. *Science* 348: 786–789.

Go to the Companion Website **oup.com/us/vertebratelife10e** for active-learning exercises, news links, references, and more.

10

Origin and Radiation of Tetrapods

The earliest tetrapods of the Late Devonian were aquatic animals, and many of the anatomical changes that originally evolved in the water were later useful for life on land. Tetrapods underwent a rapid radiation in the Carboniferous: many lineages were probably amphibious, but some became secondarily fully aquatic, while others became increasingly specialized for terrestrial life.

By the Mississippian (Early Carboniferous), tetrapods had split into two lineages: (1) batrachomorphs, comprising modern amphibians—the lissamphibians—and their fossil relatives; and (2) reptiliomorphs, comprising modern amniotes (reptiles, birds, and mammals) and their fossil relatives. The greatest diversity of the stem members of these groups (that is, those taxa outside the grouping of the extant lineages) was from the Pennsylvanian (Late Carboniferous) to Early Permian; most were extinct by the end of the Permian. By the start of the Cenozoic, the only remaining non-amniote tetrapods were the extant amphibians: frogs, salamanders, and caecilians.

Amniotes have been the most abundant tetrapods since the late Paleozoic. Amniote diversification shows an initial early split into two major lineages—one (the synapsid lineage) leading to mammals, and the other (the sauropsid lineage) leading to reptiles, including birds (**Figures 10.1 and 10.2**). Amniotes have radiated into many terrestrial adaptive zones previously occupied by non-amniotes, and they developed novel feeding and locomotor specializations.

10.1 Tetrapod Origins

Fossils of Devonian tetrapods and their fish relatives are mostly found at high latitudes. Today places such as Greenland and the Baltic states have polar or cold temperate climates, but in the Devonian these land masses lay on the paleoequator (that is, the position of the Earth's equator at that point in geological time), and thus the first tetrapods evolved in warm tropical environments.

Tetrapodomorph fishes

The evolution and diversification of tetrapodomorph lobe-finned fishes (sarcopterygians) on the tetrapod stem lineage were described in Chapter 8. Elpistostegalians (derived tetrapodomorph fishes such as *Panderichthys* and *Elpistostege*) were large—up to 2 m long—with long snouts and large teeth. Their lobed fins may have been useful in slow, careful stalking in dense vegetation on the bottom of a lagoon, and the development of digits, wrists, and ankles would have enhanced maneuverability.

The most derived of the tetrapodomorph fishes was *Tiktaalik*. (Pronounced with the accent on the second syllable, the name means "a large freshwater fish seen in the shallows" in the local Inuktitut language.) Skulls and incomplete skeletons of *Tiktaalik* have been found on Ellesmere Island in the Canadian Arctic, in deposits dating from the early Late Devonian, approximately 375 Ma. This age places *Tiktaalik* approximately 22 million years before the earliest evidence of tetrapod fossils, but 2–3 million years later than its relative *Panderichthyes.* The deposits in which *Tiktaalik* remains were found formed in a shallow river floodplain.

These fossils bridge a morphological gap between the previously known most-derived sarcopterygian fishes and the first tetrapods (**Figure 10.3**), although the retention of fin rays means that *Tiktaalik* was definitely on the "fish" side of this transition. But despite their fishlike elements, the derived tetrapodomorph fishes had most of the skeletal features that we find in ourselves (**Figure 10.4**).

Tiktaalik displayed a mosaic of fish- and tetrapod-like characters (see Figure 10.3A). The most tetrapod-like features (some of which were approached by *Panderichthys*) were the loss of the bony operculum, the large pelvic girdle of similar size as the pectoral girdle, the robust overlapping ribs, the forelimb with wrist bones (at least two were homologous with the bones of tetrapods), fingerlike bones distal to the wrist, flexion in the forelimb at areas corresponding to our wrist and elbow, and evidence of strong flexor muscles running between the humerus and the shoulder girdle. (Fingerlike bones were not unique to *Tiktaalik*, however; other tetrapodomorph fishes such as *Panderichthys* and the

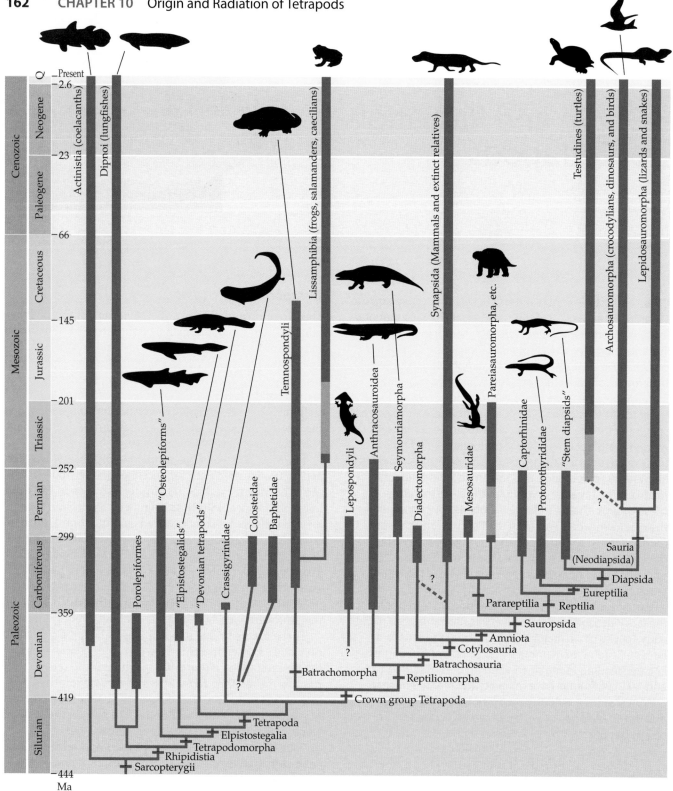

Figure 10.1 Phylogenetic relationships of the major groups of sarcopterygian fishes and early tetrapods. Only well-corroborated relationships are shown; alternative relationships of lissamphibians and lepospondyls are discussed in the text. Note that some authors restrict the term "tetrapod" to the crown group, which encompasses only extant taxa and those extinct taxa that fall within this taxonomic bracket. The relationship of turtles to other Eureptilia has been a matter of contention; on morphological grounds, turtles are grouped with the pararaptiles, but molecular phylogenies establish turtles as the sister taxon to Archosauromorpha. (The extension of turtles back to the Late Permian reflects the possible inclusion of *Eunotosaurus* as a stem turtle.) Quotation marks indicate paraphyletic groups. Narrow blue lines show interrelationships only; they do not indicate times of divergence or the unrecorded presence of taxa in the fossil record. Dark bars indicate presence in the fossil records, and light bars indicate ranges of time when the taxon is believed to be present because it is known from earlier times, but is not recorded in the fossil record during this interval.

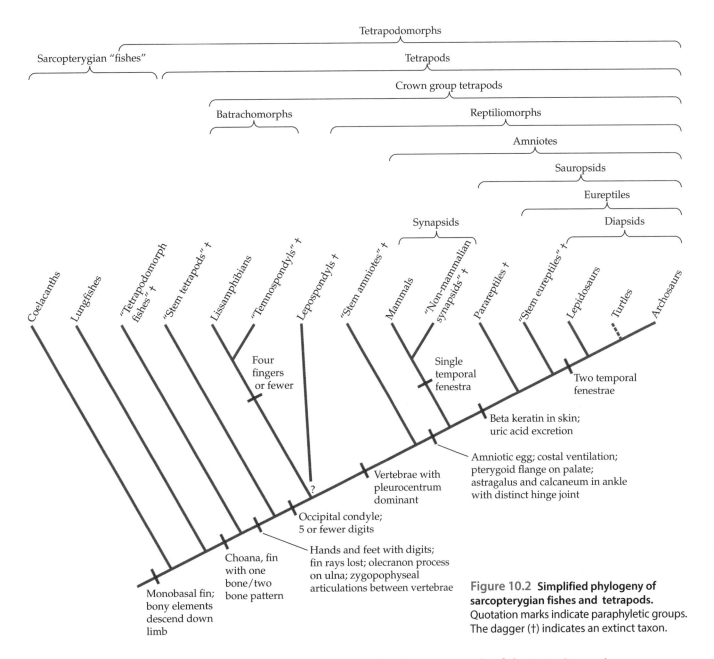

Figure 10.2 Simplified phylogeny of sarcopterygian fishes and tetrapods. Quotation marks indicate paraphyletic groups. The dagger (†) indicates an extinct taxon.

rhizodontid *Sauripterus* were also apparently "experimenting" at this time with the development of fingerlike bones at the end of the fin.) *Tiktaalik*, unlike any other fish, was able to lift its chest off the substrate and move its head relative to its body, while resisting torsion through the trunk.

The loss of the operculum, along with some of the bones linking the skull to the shoulder girdle, resulted in *Tiktaalik* having the tetrapod character of a distinct neck. This change in anatomy may have been related to lateral snapping at prey on land or in the water, where independent movement of the head relative to the body would be advantageous (see Figure 10.7A). *Tiktaalik* remained fishlike in its retention of fin rays, lack of contact between the pelvic girdle and backbone, lack of an ischium in the pelvis, and short ulna lacking a definitive olecranon process (the portion of the elbow to which the triceps tendon attaches in tetrapods).

Earliest tetrapods of the Late Devonian

The earliest known tetrapod fossils are from near the end of the Late Devonian, about 360 Ma. *Acanthostega* and *Ichthyostega* (see Figure 10.3B, C) were discovered in East Greenland in sediments from a river basin. Both animals were primarily aquatic rather than terrestrial. These two iconic tetrapods were discovered in the 1930s, and only in the past four decades have other Devonian tetrapods been described. Devonian tetrapods ranged from about 0.5 to 1.2 m in length, and differed enough from each other to show that by the end of the Devonian tetrapods occupied a diversity of adaptive zones.

Both *Acanthostega* and *Ichthyostega* had fore- and hindlimbs more capable of weight support and propulsive "push-off" than *Tiktaalik* had. Their hindlimbs had definitive contact with the vertebral column (the sacrum), and there were distinct articulations between the vertebrae

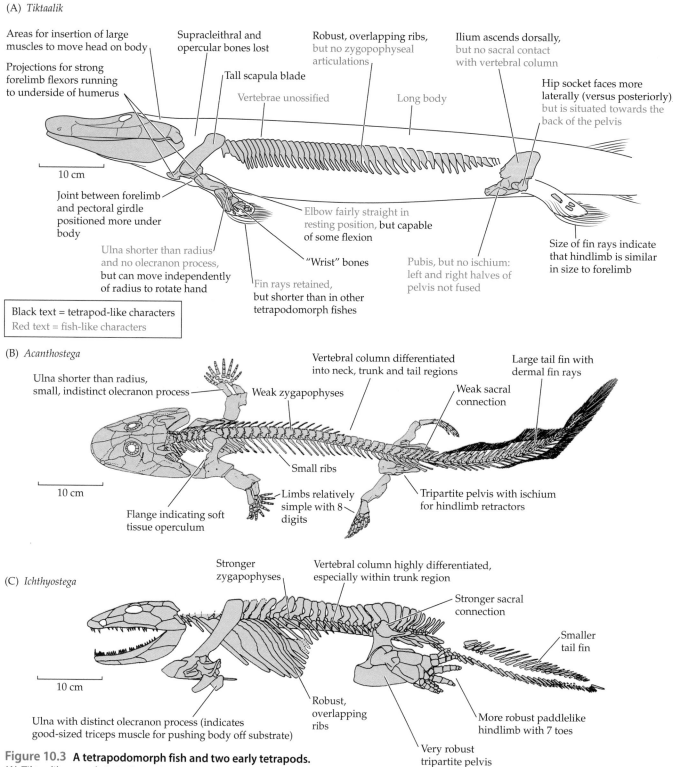

(A) *Tiktaalik*

Areas for insertion of large muscles to move head on body

Projections for strong forelimb flexors running to underside of humerus

Supracleithral and opercular bones lost

Tall scapula blade

Vertebrae unossified

Robust, overlapping ribs, but no zygopophyseal articulations

Long body

Ilium ascends dorsally, but no sacral contact with vertebral column

Hip socket faces more laterally (versus posteriorly) but is situated towards the back of the pelvis

Joint between forelimb and pectoral girdle positioned more under body

Ulna shorter than radius and no olecranon process, but can move independently of radius to rotate hand

Elbow fairly straight in resting position, but capable of some flexion

"Wrist" bones

Fin rays retained, but shorter than in other tetrapodomorph fishes

Pubis, but no ischium: left and right halves of pelvis not fused

Size of fin rays indicate that hindlimb is similar in size to forelimb

Black text = tetrapod-like characters
Red text = fish-like characters

(B) *Acanthostega*

Ulna shorter than radius, small, indistinct olecranon process

Weak zygapophyses

Vertebral column differentiated into neck, trunk and tail regions

Large tail fin with dermal fin rays

Weak sacral connection

Small ribs

Flange indicating soft tissue operculum

Limbs relatively simple with 8 digits

Tripartite pelvis with ischium for hindlimb retractors

(C) *Ichthyostega*

Stronger zygapophyses

Vertebral column highly differentiated, especially within trunk region

Stronger sacral connection

Smaller tail fin

Ulna with distinct olecranon process (indicates good-sized triceps muscle for pushing body off substrate)

Robust, overlapping ribs

Very robust tripartite pelvis

More robust paddlelike hindlimb with 7 toes

Figure 10.3 A tetrapodomorph fish and two early tetrapods. (A) *Tiktaalik*, an early Late Devonian tetrapodomorph fish. (B, C) Two Late Devonian aquatic stem tetrapods, *Acanthostega* (B) and *Ichthyostega* (C). (A, from Shubin et al. 2013 and Daeschler et al. 2006; B, from Coates 1996; C, from Ahlberg et al. 2005.)

(zygapophyses), reflecting resistance of the vertebral column to gravity on land. *Acanthostega* had a flange on the dermal portion of the shoulder girdle (technically the "postbranchial lamina of the cleithrum," which supports the posterior wall of the opercular chamber in bony fishes), indicating that it had a soft-tissue operculum covering the gills. It was not until more derived tetrapods that the head had a distinct articulation with the neck vertebrae, via the occipital condyle on the skull.

Ichthyostega was the more specialized of the two. Its differentiated vertebral column, robust ribs, and strong

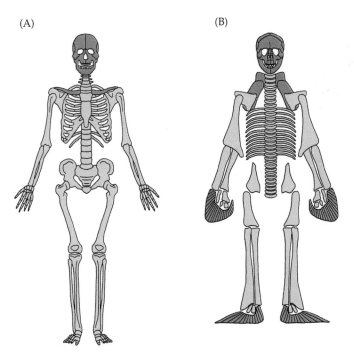

(A)　　　　　　(B)

Figure 10.4 Humans are derived fishes. The skeleton of a human (A) compared with the skeleton of a derived tetrapodomorph fish (B), with the fish skeleton scaled to the size of the human's and oriented in a humanlike pose (and with a humanlike skull). Red shading indicates dermal bone and, in fishes, the dermal fin rays. (From Long 2011 © John A. Long.)

forelimbs indicate that *Ichthyostega* was capable of terrestrial locomotion, but its hindlimbs were paddlelike (more specialized than in *Acanthostega*), and its joint anatomy shows that it would have been unable to rotate its hindlimb to place its feet under its body. Its ear region was apparently specialized for underwater hearing.

The limbs of Devonian tetrapods bore more than the five digits of post-Devonian and extant tetrapods. These polydactyl hands and feet looked more like flippers for maneuvering in shallow water than like appendages for walking on land. Devonian tetrapods also retained portions of the gill skeleton, with grooves on the surface of a portion of the segmented gill arches (the ceratobranchials) showing the presence of blood vessels supplying functional internal gills (different from the external gills seen today in larval amphibians).

Although the presence of gills in Devonian tetrapods was originally surprising, subsequent discoveries showed that gills were also present in many of the more derived Paleozoic tetrapods, especially those related to extant amphibians (temnospondyls). An extant phylogenetic bracket (lungfishes and extant tetrapods) supports the inference that these animals had lungs, even though lungs are not preserved in fossils (and, as we saw in Chapter 8, lungs are a basal feature of bony fishes in general).

How much ability early tetrapods had to engage in terrestrial activity is unclear. The skull sutures in *Acanthostega* have been interpreted as being similar to those of tetrapods feeding on land and biting food rather than using fishlike

suction, but the biting could have occurred underwater. There is evidence that *Acanthostega* had a juvenile stage of several years, during which the limb bones were unossified, implying that these juveniles were entirely aquatic. The skeleton of *Ichthyostega*, discussed below, shows a front end specialized for terrestrial locomotion, with substantial shoulders, and a hind end specialized for swimming, with paddlelike hindfeet. The Devonian tetrapod *Tulerpeton* from Russia had a skeleton that looked more like those of later tetrapods and may have engaged in salamander-like walking and walking-trot gaits (see Chapter 11).

For many years, tetrapods were presumed to have had freshwater origins, in part because modern amphibians do not tolerate salt water. But many of the tetrapodomorph fossils, as well as fossils of some of the Devonian tetrapods, are found in marine or estuarine deposits, and the original environment of the fish–tetrapod transition remains unclear. At a minimum, early tetrapods appear to have been saltwater-tolerant.

Apparent tetrapod footprints have been found in the Middle Devonian of Poland in a marine lagoon, predating the earliest known tetrapods by about 15 million years. The identity of that track maker has been questioned, but uncontested tetrapod tracks are known from Ireland only a few million years later, and there are also Late Devonian tracks from Scotland and Australia. These fossilized footprints hint at an earlier diversity of tetrapods, and possibly the existence of more terrestrial forms, than the ones currently known.

10.2 Moving onto Land

Tetrapod characters such as lungs and digits were advantageous for animals that were still living in water, perhaps making occasional forays onto land (**Figure 10.5**). But how did the transition from the aqueous environment to the terrestrial one actually happen?

The classic story of tetrapod evolution proposed that the Devonian was a time of seasonal droughts and that fishes living in lakes and ponds used limbs to crawl to another pond when their pond dried up. In other words, this hypothesis proposed that limbs originally aided movement across land. It now appears that the limbs of early tetrapods improved their locomotion in water, and only later became advantageous on land. What advantages might these animals have gained from terrestrial activity? Hypotheses (which are not mutually exclusive) include searching for food, dispersal of juveniles, laying eggs in moist environments, and basking in the sun to elevate body temperature.

Terrestrial and walking fishes today

Many fishes today walk rather than swim. Walking catfishes pivot their body over their pectoral spines to move over land, and mudskippers use a "crutching" form of locomotion on land, moving both front fins together and propping

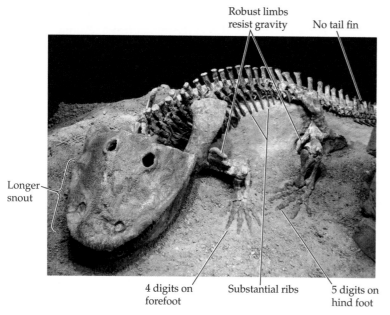

Robust limbs
resist gravity   No tail fin

Longer
snout

4 digits on   Substantial ribs   5 digits on
forefoot   hind foot

Figure 10.5 *Eryops*, **an Early Permian batrachomorph tetrapod.** Although probably semiaquatic, *Eryops* had more terrestrial characters than other post-Devonian tetrapods. The tail fin was lost, the limbs and limb girdles were robust, and the snout had lengthened. Post-Devonian tetrapods had five or fewer digits on their hands and feet. (Tim Evanson/CC BY-SA 2.0.)

up their bodies with anteriorly placed strutlike hindfins. *Cryptotora*, a blind cave fish, can climb up waterfalls using a diagonal type of walking-trot gait and has evolved a tetrapod-like connection between its pelvic girdle and vertebral column. African lungfishes (*Protopterus*) walk and bound underwater, propelling themselves with their ribbonlike pelvic fins, and can locomote on land by planting their head on the surface and pivoting their body around it.

All of these examples of extant walking fishes provide clues as to how terrestrial locomotion might have evolved in basal tetrapods. When bichirs (*Polypterus*), which are basal actinopterygians, were forced to move onto land for prolonged periods, the fish planted their pectoral fins progressively closer to the body and their locomotor performance improved. Bone is remodeled by stress, and forced terrestrial locomotion caused changes in the shape of some of the pectoral girdle bones (the clavicle and cleithrum) of bichirs, making their bones more similar to those of basal tetrapods. Although such changes would not be heritable, selection acts on developmental plasticity, leading to future advantageous phenotypic changes.

How are fins made into limbs?

Whereas actinopterygians show a diversity of complex patterns in their endochondral fin bone anatomy, all sarcopterygians share a common pattern: (1) a basal stylopod (corresponding to the human humerus and femur), (2) a more distal zeugopod (human ulna and radius, or tibia and fibula), and (3) a series of distal radials from which the fin rays branch. Fin rays are made of dermal bone, formed in

the skin, and unlike the rest of the limb, they have their origin in neural crest cells. But how did tetrapods come to lose their dermally derived fin rays and develop hands and feet—that is, the autopod (the most distal portion of the limb)?

Developmental studies show that fins and limbs are homologous structures, coded for by the same set of Hox genes in development (the *HoxA/D* cluster). The tetrapod autopod is not a truly novel structure, as originally thought, but has genetic homologies with structures in fish fins; tetrapod limbs express some of the same genes during development as do fish fin rays (**Figure 10.6A**).

Teleosts have a highly reduced fin skeleton, but paddlefishes (*Polyodon*) retain a robust endochondral fin structure. Studies comparing paddlefish with mice (a representative tetrapod) show two phases of expression of the Hox genes in limb formation in both groups: an early phase that results in formation of the proximal part of the limb, and a later phase that forms the distal radials of the fish and the autopod of the tetrapod.

In the later phase of the fish development, *Hox11* and *Hox13* are expressed in an overlapping fashion, whereas in tetrapods these genes split their domains of expression, with *Hox11* becoming restricted to the zeugopod and *Hox13* to the autopod. This does not mean that fish radials are strictly homologous with tetrapod digits, but merely that these structures share a common pattern of gene expression.

Thus, these studies show that hands and feet emerge as a property of differential expression of genes present in fishes, rather than requiring completely new genetic information for their evolution, although we still do not know what makes the tetrapod autopod such a distinctive structure. The modern tetrapod expression of Hox genes may have been incomplete in the Devonian forms: experiments with mice show that certain disruptions in Hox gene expression results in polydactyl limbs.

Figure 10.6B shows the sequence of digit development in an extant tetrapod and a Late Devonian tetrapod. In all extant tetrapods the formation of the digits commences with digit 4 (the ring finger on our hands) and proceeds in an arc across the base of the limb toward the thumb, with the fifth digit (the "pinky") being added in the opposite direction. The branching of digits from this arc ends with digit 1 (the thumb) or sooner in animals that have lost digit 1. If the process of developmental branching continues, however, a polydactylous condition results with additional digits beyond the thumb (see the *Acanthostega* forelimb in Figure 10.6B). This situation sometimes occurs as an abnormal condition in extant vertebrates, and it was apparently the normal condition in Late Devonian tetrapods: their additional digits are clearly beyond the thumb. In addition, when digits are lost in more derived tetrapods (whether limbless lizards, or mammals and birds specialized for more rapid locomotion), the loss happens

(A) Genes controlling forelimb development in fish and tetrapods

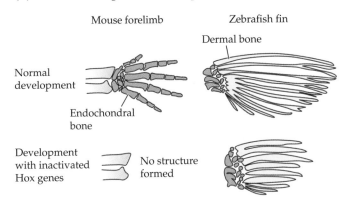

(B) Sequence of embryonic formation of digits

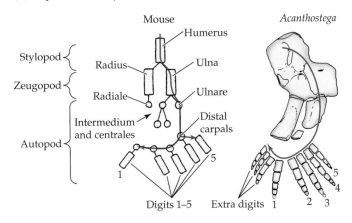

Figure 10.6 Forelimb development in fishes and tetrapods.
(A) The evolution of autopods (hands and feet) in tetrapods may have resulted from the hijacking of genes shaping the fin rays of fishes. The *Hox13* genes (*Hoxa13* and *Hoxd13*) are known to be essential for autopod formation in tetrapods. If these genes are inactivated in mice (left), the autopod fails to develop. However, fishes (right) that have the same genes inactivated develop fins with shorter fin rays (dermal bone) and an increased number of distal endochondral bones. Thus, although fin rays and tetrapod digits are formed from different tissues, they share a common population of developing (mesenchymal) cells, and a common genetic pattern of regulation. (B) Sequence of embryonic formation of digits in mice (left) and the Late Devonian tetrapod *Acanthostega* (right). Red arrows show the developmental order of digit formation, starting with digit 4 and moving towards digit 1. The *Acanthostega* forelimb shows the axis of digit development and the polydactylous condition of eight digits. The carpals were unossified in this early tetrapod. (A, from Saxena and Cooper 2017; B left, from Ahlberg and Milner 1994; B right from Clack and Coates 1990.)

in the reverse pattern. Digit 1 is invariably the first to be lost, followed by digit 5 and then 2; digits 3 and/or 4 tend to be retained.

Body support and locomotion

A fish in water is neutrally buoyant, and its notochord counteracts the compressive force produced by the lateral undulations of swimming. On land, gravity becomes a downward force acting on the body, and torsional (twisting) forces are experienced along the trunk during locomotion (**Figure 10.7A**). The dorsoventrally flattened body of the elpistostegalian fishes and early tetrapods amplified torsional forces on land. In fishes, the axial muscles are used only for propulsion, but in tetrapods the hypaxial muscles have been modified to resist torsion and stabilize the trunk. Enlargement of the endochondral limb girdles and ribs also aids trunk stability, and the beginnings of this change can be seen in the derived tetrapodomorph fishes.

The limbs of fishes project horizontally, even in tetrapodomorph fishes with joints functionally equivalent to elbows, so the limbs extend out from the sides of the body. The upper arm (humerus) and upper leg (femur) of early

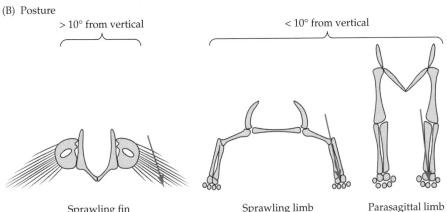

Figure 10.7 Forces exerted on terrestrial vertebrates. (A) When a derived fish or a primitive tetrapod turns its head to the left to snap at prey on land, the right limb must support a resultant force created by the inertial force and the pull of gravity. (B) The transition from fin to limb and from the sprawling stance of basal tetrapods to the upright stance of some derived tetrapods affects the orientation of gravitational force (red arrows) and hence the weight-bearing structure of the skeleton. (A from Hohn-Schulz et al. 2013; B from Kawano and Blob 2013.)

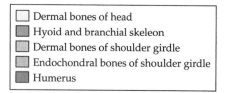

☐ Dermal bones of head
■ Hyoid and branchial skeleton
■ Dermal bones of shoulder girdle
■ Endochondral bones of shoulder girdle
■ Humerus

tetrapods also extended more or less horizontally—as do those of extant salamanders and lizards—and the elbows and knees were bent to allow the lower limbs and feet to support the body. This arrangement provides a stable base of support but exerts torsional forces on the elbow and knee joints.

The pectoral fins of *Tiktaalik* were more ventrally positioned than those of other fishes, making it better able to place the limbs under the body. In addition, there is bony evidence for strong flexor muscles (pectoralis) running from the shoulder girdle to the underside of the humerus which would have counteracted torsion at the shoulder joint. The humerus and femur of derived mammals and the femur of non-avian dinosaurs and birds have assumed a vertical orientation, reducing the torsional forces on the joints seen in the limb posture of less derived tetrapods (**Figure 10.7B**).

Tetrapod muscles are used for support of the body on land as well as for locomotion and are bolstered by internal ligaments, which makes tetrapod flesh (e.g., a steak) tougher to cut and to consume than the flesh of fishes. Likewise, tetrapod ribs (except those in extant amphibians) are much more robust than the ribs of fishes.

How did the earliest tetrapods move on land? It was originally assumed that they would have had a salamander-like walk or walking-trot, but salamanders are highly derived as well as much smaller than Devonian vertebrates, so they are probably not good models for the locomotion of early vertebrates. Computed tomography (CT) scans of *Ichthyostega* show that the limbs appear to have had a limited range of motion, moving mainly backward and downward, with little ability to rotate the limb to face the palm forwards or backwards. *Ichthyostega* probably walked on land by crutching like a mudskipper, dragging the belly and hind end. *Ichthyostega*'s backbone would have flexed in a dorsoventral plane during this locomotion, and ossified midline elements, perhaps contained within a cartilaginous sternum, would have supported the rib cage.

Lung ventilation and dermal bone

Water pressure helps air-breathing fishes empty their lungs, but on land tetrapods do not have a water column to help them exhale. Tetrapods contract the transverse abdominius layer of the hypaxial muscles (which are associated with the ribs in amniotes) to force air out of the lungs. Derived tetrapodomorph fishes and early tetrapods were probably bimodal breathers, perhaps using their spiracles for inhaling air at the water surface (**Figure 10.8**), and using gills for aquatic gas exchange. In the water, gills also remain important for ion exchange and nitrogen excretion—functions carried out by the kidneys in extant tetrapods.

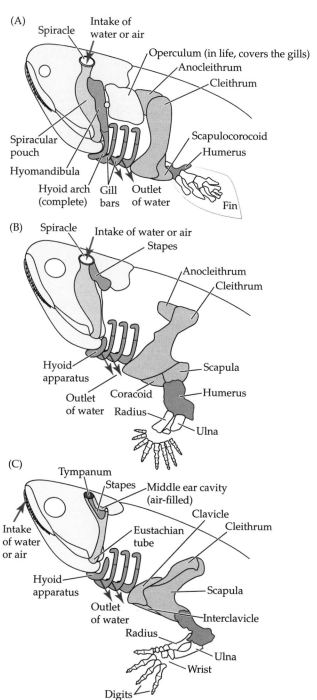

Figure 10.8 Evolution of the spiracular cavity and the middle ear in tetrapods. (A) The tetrapodomorph fish *Eusthenopteron*. The spiracle probably functioned for water intake (as it does in some extant elasmobranchs) or for air intake (as in the basal actinopterygian *Polypterus*). Air would pass into the lungs and be expelled via the spiracle or the mouth. (B) The stem tetrapod *Acanthostega*. With the loss of the operculum, the hyomandibula (now technically the stapes) detached from the hyoid arch and may have controlled passage of air through the spiracle. (C) A generalized temnospondyl, showing a froglike middle ear. In more derived tetrapods, the dorsal portion of the spiracular pouch has become the middle ear cavity, with the stapes (the old hyomandibula) conducting sound to the oval window on the otic capsule. The ventral portion of the spiracular pouch has become the Eustachian tube, connecting the middle ear to the pharynx. (After Schoch 2014.)

Obtaining oxygen on land is easy, as oxygen is plentiful in the air. Getting rid of carbon dioxide, however, would have posed difficulties for early tetrapods. Carbon dioxide is soluble in water and easily lost from the gills and skin, but it is harder to lose to the air. Buccal pumping is sufficient for extant amphibians for both oxygen gain and carbon dioxide loss because they are small and their skins are permeable to gases. Many early tetrapods, especially temnospondyls, were semiaquatic and retained internal gills. Among temnospondyls, only the small terrestrial forms (dissorophids) lost their gills; these species would have been small enough for cutaneous gas exchange, as seen in extant amphibians.

Most early tetrapods were considerably larger than modern amphibians, and large body size reduces the efficiency of cutaneous carbon dioxide loss, because larger animals have relatively less surface area for their volume. Furthermore, early tetrapods retained the dermal scales of their fish ancestors, although they had lost the enamel and dentine layers, and their skin was probably not very permeable to gases, unlike the thin skin of extant amphibians.

When carbon dioxide is retained in the blood it forms carbonic acid, a process called respiratory acidosis. If the earliest tetrapods were like extant lungfishes and amphibians, they used buccal pumping to inflate their lungs, a function for which their large, flat heads were well suited. But buccal pumping is not a particularly effective means of lung ventilation because the lungs cannot be filled with a single breath, and so lung ventilation cannot be frequent and rapid enough for the expulsion of carbon dioxide. Evidence for costal ventilation—muscular expansion of the rib cage to draw in air—is not known until amniotes or their near relatives among non-amniotes. Consequently, respiratory acidosis might have been a problem for early terrestrial tetrapods.

How could an animal like *Eryops* (see Figure 10.5) have coped with respiratory acidosis? Many of these animals had an armor of ornamented dermal bones over their body, and this bone provides a clue. Dermal bone is used by extant turtles and crocodylians to buffer acidosis when they are underwater, exchanging protons from carbonic acid with calcium and magnesium ions from bone. The dermal bone of early tetrapods might have had a buffering capacity while the animals were on land. These layers of bone were reduced or lost in amniotes and their near relatives when they were better able to get rid of excess carbon dioxide via the lungs.

10.3 Radiation and Diversity of Non-Amniote Tetrapods

For more than 200 million years, from the Late Devonian to the Early Cretaceous, non-amniote tetrapods radiated into a great variety of terrestrial and aquatic forms.

It is misleading to think of Paleozoic tetrapods as being like modern amphibians, even though they may have had amphibious habits, and the term "amphibian" (or, more formally, "lissamphibian") is now reserved for the extant non-amniote tetrapods—frogs, salamanders, and caecilians. Many of the Paleozoic tetrapods were much larger than any extant amphibians and would have been somewhat crocodile-like in appearance, and many of them were actually more closely related to amniotes than to modern amphibians. **Table 10.1** lists the major groups of Paleozoic and Mesozoic non-amniote tetrapods.

Current consensus recognizes two main groups of large non-amniote Paleozoic tetrapods: the stem lissamphibians (basal batrachomorphs) and the stem amniotes (basal reptiliomorphs), as shown in Figures 10.1 and 10.2. Some Paleozoic non-amniote tetrapods do not fall into either group and are stem tetrapods. Lepospondyls—small forms, some of which were similar to extant amphibians—are usually considered to be a monophyletic grouping. Their phylogenetic relationships are cause for debate: some researchers place them as the sister taxon to amniotes, while others consider that they contain the ancestry of at least some lissamphibians.

Stem lissamphibians include temnospondyls; stem amniotes include anthracosauroids, seymouriamorphs, and diadectomorphs. Stem lissamphibians were in general more aquatic than stem amniotes (although the anthracosauroid embolomeres were a specialized aquatic lineage of reptiliomorphs). Stem lissamphibians were characterized by flat, nonkinetic skulls and a reduction on the hand to four fingers, whereas stem amniotes had domed skulls that retained some kinetic ability and had a five-fingered hand. **Figures 10.9, 10.10,** and **10.11** illustrate a variety of basal tetrapods.

Evolutionary patterns Until recently, there was little evidence of any tetrapods from the Mississippian (Early Carboniferous), a period of time colloquially known as "Romer's Gap." However, extensive explorations have now unearthed a diverse assemblage of tetrapods from this time at localities in Nova Scotia and the northern British Isles. These new taxa show that the Devonian extinctions did not have as dramatic an effect on tetrapod evolution as previously thought. Some of these taxa appear to be stem tetrapods, a few are probably relics of the Devonian fauna, and others may be stem lissamphibians. The last category potentially pushes the amniote–lissamphibian divergence back to about 355 Ma.

Non-amniote tetrapods reached their peak diversity in the Pennsylvanian (Late Carboniferous) and Early Permian, when they included fully aquatic, semiaquatic, and terrestrial forms. Most lineages were extinct by the Middle Permian, although some specialized aquatic lineages of temnospondyls were moderately common in the Triassic, and in Australia one survived into the Early Cretaceous. Although all extant lissamphibian groups have their origins in the Mesozoic, the generic diversity of non-amniote tetrapods did not return to Permian levels until the mid-Cenozoic.

Table 10.1	Paleozoic and Mesozoic non-amniote tetrapods described in this chapter

Stem tetrapods

Late Devonian

Acanthostega and *Ichthyostega* (Figure 10.3B,C) were discovered in Greenland in the 1930s, although *Acanthostega* was not fully described until the 1980s and 1990s. These forms are known from multiple complete skeletons.

Tulerpeton from Russia is more derived than *Acanthostega* and *Ichthyostega*. The robust skeleton suggests a degree of terrestrial adaptation, and both the forefeet and hind feet had 6 digits.

Late Mississippian

Crassigyrinus (Figure 10.9A) was an aquatic moray eel-like form from Scotland and North America.

Pederpes is more derived than earlier taxa. It had 5 definitive digits on the forefoot and a possible small 6th digit (the hind foot is unknown) and a high, narrow skull.

Batrachomorpha (lissamphibians and stem lissamphibians)

Temnospondyli: From the late Mississippian to the Early Cretaceous worldwide, this diverse and long-lived group had a worldwide distribution. Temnospondyls had large heads with akinetic skulls and palatal vacuities and had only 4 fingers on the forefeet. The group included both semiaquatic eryopoids such as *Eryops* (Figures 10.5 and 10.9C) and fully aquatic stereospondyls, such as *Trematosaurus* (Figure 10.9E). Some trematosaurids evolved the elongated snout characteristic of specialized fish-eaters and are the only known possibly marine non-amniote tetrapods. Other stereospondyls were the capitosaurids (e.g., *Cyclotosaurus*, Figure 10.9F) and plagiosaurids (e.g., *Gerrothorax*, Figure 10.9G). Dissorophoid temnospondyls, such as *Amphibamus* and *Gerobatrachus* (Figure 10.9D), were small terrestrial forms and may lie in the ancestry of extant amphibians.

Reptiliomorpha (amniotes and stem amniotes)

Uncertain phylogenetic position: *Westlothiana* (Figure 10.9B) and *Casineria* are small, lizardlike forms from the early Mississippian of Scotland.

Anthracosauroidea: Known from the late Mississippian to the earliest Triassic of North America and Europe; this diverse, long-lived group is second only to the temnospondyls in diversity and geographic distribution. Anthracosauroids had deeper skulls than temnospondyls, with prominent tabular horns. They retained cranial kinesis. This group included semiaquatic forms, especially the embolomeres (e.g., *Pteroplax*; Figure 10.10A) and gephyrostegids, small terrestrial forms from the Pennsylvanian (e.g., *Gephyrostegus*; Figure 10.10B).

Seymouriamorpha: Early Permian forms from North America were large and fully terrestrial (e.g., *Seymouria*; Figure 10.10C). Later Permian forms from Europe and China were semiaquatic.

Diadectomorpha: Known from the Pennsylvanian to the Early Permian of North America and Europe, these large, fully terrestrial forms are now considered to be the sister group of amniotes (or possibly even were within the amniote crown group as the sister group to synapsids). Diadectids like *Diasparactus* (Figure 10.10D) had laterally expanded cheek teeth like those of extant herbivorous lizards. Limnoscelids and tseajaiids had sharp, pointed teeth and were probably carnivorous.

Lepospondyli (crown tetrapods of uncertain affinity)

Microsaurs: Known from the Mississippian to the Early Permian of North America and Europe; probably not a monophyletic group. Many were terrestrial and rather lizardlike, with deep skulls and elongated bodies, but had only 4 toes (*Pantylus*; Figure 10.11A). Others had external gills as adults and were probably aquatic.

Nectrideans: Known from the Pennsylvanian to the Early Permian of North America, Europe and North Africa. Aquatic, with long tails and small, poorly ossified limbs. Some, such as *Diplocaulus* (Figure 10.11B), had broad, flattened skulls with tabular bones elongated into horns. These horns, which would have been covered by a flap of skin extending back to the shoulder, may have acted like a hydrofoil in locomotion as well as providing additional area for gas exchange.

Aïstopodans: Known from the Mississippian to the Early Permian of North America and Europe. Elongated bodies (with up to 200 trunk vertebrae) lacking both limbs and limb girdles (e.g., *Ophiderpeton*; Figure 10.11C). They may have been aquatic or burrowed through leaf litter. Derived forms had snakelike skulls that may have allowed them to swallow large prey items.

Ecological and adaptive trends   One of the most striking aspects of early tetrapods is the number and diversity of forms that returned to a fully aquatic mode of life. Among extant amphibians, some salamanders and frogs are fully aquatic as adults, and this was also true of a diversity of Paleozoic forms.

A key event in the radiation of early tetrapods may have been the great diversification of insects in the Pennsylvanian, probably in response to the increasing quantity and diversity of terrestrial vegetation. Now the terrestrial food supply could support a diverse fauna of terrestrial vertebrates. Carnivores (including piscivores and invertebrate-eaters) dominated the initial radiation of non-amniotes. With the exception of diadectids, no Paleozoic non-amniote tetrapod appears to have been herbivorous, and neither is any extant adult amphibian.

Some late Mississippian taxa appear to have been quite terrestrial. Included in this group are the stem tetrapod *Pederpes*, which may have retained a sixth finger on the hand; the long-bodied, limbless lepospondyl *Lethiscus*; and possible early amniotes (or close relatives), the small, lizard-size (and possibly ecologically lizardlike) *Westlothiana* (see Figure 10.9B) and *Casineria*.

Temnospondyls were the only group of non-amniote tetrapods (aside from lissamphibians) to survive the Paleozoic, and almost all of the temnospondyls that persisted into the Mesozoic were large, flattened, fully aquatic predators. The dissorophid temnospondyls were small terrestrial forms, known from the Pennsylvanian to the Early Triassic, and this group probably contains the ancestry of modern amphibians. Lepospondyls (**Figure 10.11**) were probably the forms most similar to present-day amphibians. Various lepospondyl lineages independently acquired an elongated body, together with the reduction or loss of the limbs, as seen today in caecilians and some salamanders.

In contrast to stem lissamphibians, stem amniotes appear to have been predominantly terrestrial as adults, and many were once mistaken for early reptiles (especially *Seymouria* and diadectids such as *Diasparactus*; see Figure 10.10C, D). Terrestriality also evolved convergently among other early tetrapods, predominantly in the dissorophids and the microsaur lepospondyls. These animals independently acquired skeletal features associated with terrestriality, such as longer and more slender (but highly ossified) limb bones.

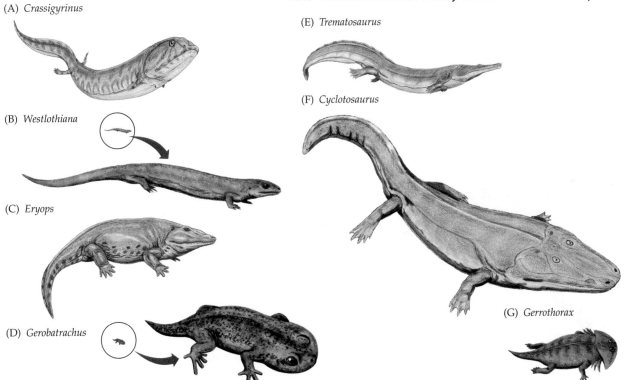

(A) *Crassigyrinus*

(B) *Westlothiana*

(C) *Eryops*

(D) *Gerobatrachus*

(E) *Trematosaurus*

(F) *Cyclotosaurus*

(G) *Gerrothorax*

Figure 10.9 **Stem tetrapods and temnospondyls.** (A–D) Paleozoic taxa. (A) *Crassigyrinus*, an aquatic stem tetrapod. (B) *Westlothiana*, a small, lizardlike animal of uncertain phylogenetic position, possibly related to amniotes. (C) *Eryops*, a semiterrestrial eryopoid temnospondyl. (D) *Gerobatrachus*, a terrestrial dissorophoid temnospondyl, close to the ancestry of modern amphibians. (E–G) Mesozoic taxa. Triassic tetrapods were almost all large, secondarily aquatic forms. (E) *Trematosaurus*, an aquatic (possibly marine) trematosaurid temnospondyl. (F) *Cyclotosaurus*, an aquatic capitosaurid temnospondyl. (G) *Gerrothorax*, an aquatic plagiosaurid temnospondyl. *Eryops* was about 2 m long; other animals are drawn approximately to scale, with insets indicating the relative size of the small animals. (A,C,E,F Dmitry Bogdanov/CC BY-SA 3.0; B,D,G Nobu Tamura/CC BY-SA 3.0.)

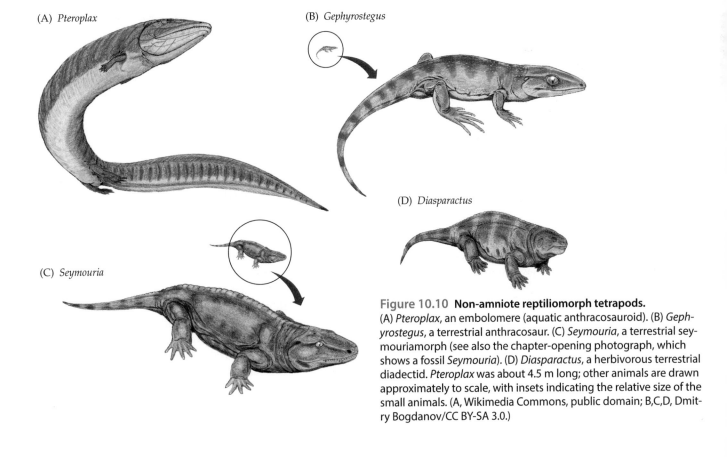

(A) *Pteroplax*

(B) *Gephyrostegus*

(C) *Seymouria*

(D) *Diasparactus*

Figure 10.10 **Non-amniote reptiliomorph tetrapods.** (A) *Pteroplax*, an embolomere (aquatic anthracosauroid). (B) *Gephyrostegus*, a terrestrial anthracosaur. (C) *Seymouria*, a terrestrial seymouriamorph (see also the chapter-opening photograph, which shows a fossil *Seymouria*). (D) *Diasparactus*, a herbivorous terrestrial diadectid. *Pteroplax* was about 4.5 m long; other animals are drawn approximately to scale, with insets indicating the relative size of the small animals. (A, Wikimedia Commons, public domain; B,C,D, Dmitry Bogdanov/CC BY-SA 3.0.)

(A) *Pantylus*

(B) *Diplocaulus*

(C) *Ophiderpeton*

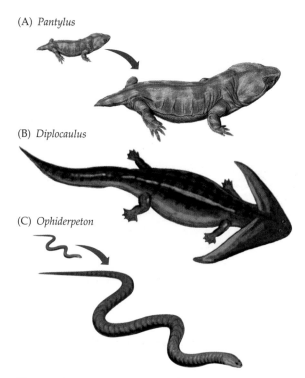

Figure 10.11 Lepospondyls. (A) *Pantylus*, a terrestrial micro-saur. (B) *Diplocaulus*, an aquatic nectridian. (C) *Ophiderpeton*, an aquatic (or possibly terrestrial burrower) aïstopod. *Diplocaulus* was about 1 m long; other animals are drawn approximately to scale, with insets indicating the relative size of the small animals. (A, Dmitry Bogdanov/CC BY 3.0; B,C, Nobu Tamura/CC BY-SA 3.0.)

Origins of extant amphibians The origins of the extant amphibians (Lissamphibia) are a matter for debate. Some workers propose that lissamphibians are a monophyletic group derived from within the dissorophoid temnospondyls (as shown in Figure 10.1), and the terrestrial dissorophoid *Gerobatrachus* (see Figure 10.9D) has been proposed as a possible ancestor of Lissamphibia. Other researchers propose that frogs and salamanders are derived from temnospondyls while caecilians are derived from microsaur lepospondyls, or even that all lissamphibians are derived from lepospondyls.

10.4 Amniotes

Amniotes include most of the tetrapods alive today. Their name refers to the amniotic egg, which is one of the most obvious features distinguishing extant amniotes from extant amphibians. **Table 10.2** details the major groups of Paleozoic amniotes, and **Figure 10.12** illustrates some examples.

Amniotes were well established by the mid-Carboniferous, although they were a minor part of the fauna. Their first major radiation was in the Permian. By the Late Carboniferous (Pennsylvanian) the three major groups of amniotes had appeared:

Table 10.2	**Major groups of Paleozoic amniotes described in this chapter**

Synapsida

Synapsids are known from the Pennsylvanian (Late Carboniferous) to the present, with a worldwide distribution. Non-mammalian synapsids are discussed in Chapter 24. Early synapsids (e.g., the pelycosaur *Haptodus*) were somewhat larger than early eureptiles, and their larger heads and teeth suggest that they preyed on other tetrapods. Small synapsids appeared in the Pennsylvanian and diversified into larger forms in the Permian.

Sauropsida: Parareptilia

Mesosauridae: Known from Early Permian freshwater deposits in South Africa and South America, these are the first secondarily aquatic amniotes. Aquatic features include large, probably webbed, hindfeet, a laterally flattened tail, and heavily ossified ribs that may have acted as ballast in diving. The long jaws and slender teeth may have been used to strain small crustaceans from the surrounding water (*Mesosaurus*; Figure 10.12B).

Procolophonidae: Known from the Late Permian to the Late Triassic, procolophonids had a worldwide distribution that included Antarctica. Derived forms had laterally expanded peglike teeth that were apparently specialized for crushing or grinding, indicative of herbivory (*Procolophon*; Figure 10.12C).

Pareiasauria: Pareiasaurs were the dominant terrestrial herbivores of the later Permian, found in Europe, Asia and Africa. The largest forms, such as *Bradysaurus* (Figure 10.12D), were nearly 2.5 m long. Their teeth were laterally compressed and leaf-shaped, like the teeth of extant herbivorous lizards.

Sauropsida: Eureptilia

Protorothyrididae: Known from the Pennsylvanian to the Early Permian of North America and Europe, these small, relatively short-legged, lizardlike forms were probably insectivorous (*Hylonomus*; Figure 10.12A).

Captorhinidae: Known worldwide (except for Australia and Antarctica) from the Pennsylvanian to the Late Permian, these lizardlike forms had more robust skulls and flatter teeth than protorothyridids and early diapsids and may have had an omnivorous diet that required crushing (*Captorhinus*).

Araeoscelidia: Known from the Pennsylvanian to the Early Permian of North America and Europe, these small stem diapsids had shorter bodies and longer legs than protorothyridids. They were probably insectivorous (*Petrolacosaurus*).

Synapsida, which gave rise to mammals. Carboniferous synapsids were small insectivores, but in the Permian synapsids diverged into larger carnivores and herbivores (see Chapter 24).

Parareptilia included secondarily aquatic forms (mesosaurs) and large herbivores (pareiasaurs). Parareptilia was mainly a Late Permian and Early Triassic radiation, with only a single lineage extending until the end of the Triassic.

Eureptilia, which gave rise to extant reptiles, including birds. Eureptilians were mainly small insectivores in the Paleozoic. Parareptilia and Eureptilia comprise Sauropsida.

On the basis of their anatomies and the depositional environments of the sites where they have been found, early amniotes are considered to have been a primarily terrestrial radiation. However, the microanatomy of bones suggests that at least one Early Permian amniote, *Captorhinus*, was amphibious.

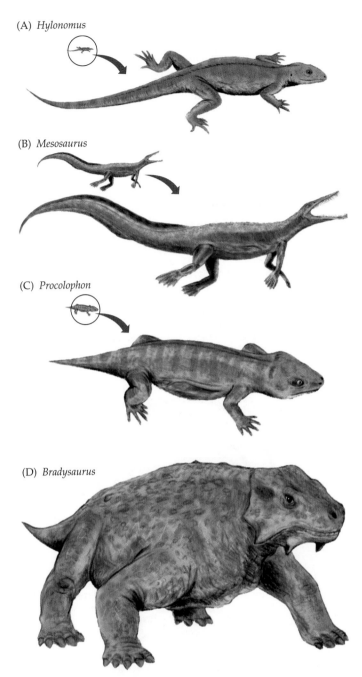

(A) *Hylonomus*

(B) *Mesosaurus*

(C) *Procolophon*

(D) *Bradysaurus*

Figure 10.12 Diversity of Paleozoic amniotes. Early amniotes varied in length from a few centimeters to a few meters, and their ecological roles were equally diverse. (A) *Hylonomus*, an insectivore. (B) *Mesosaurus*, a piscivorous mesosaur. (C) *Procolophon*, an omnivorous procolophonid. (D) *Bradysaurus*, a herbivorous pareiasaur. *Bradysaurus* was about 2 m long; other animals are drawn approximately to scale, with insets indicating the relative size of the small animals. (Nobu Tamura/CC BY-SA 3.0.)

The only potential non-amniote herbivores were diadectids from the Early Permian (see Figure 10.10D). Diadectids had flat, crushing teeth like those of some extant herbivorous lizards. Carboniferous amniotes, such as *Hylonomus* (see Figure 10.12A), appear to have been insectivores.

From the start of the Permian onward, terrestrial habitats were dominated by a series of radiations of amniote tetrapods that now included herbivores such as pareiasaurs (see Figure 10.12D) and the synapsid caseids and edaphosaurs discussed in Chapter 24. The short, deep jaws of herbivorous tetrapods greatly increased the diversity of jaw types seen among Palaeozoic tetrapods. Other new body types seen among Permian amniotes included the bipedal eureptile *Eudibamus* and the arboreal synapsid *Suminia*.

Derived features of amniotes

Traditionally, extant amniotes have been distinguished by the amniotic egg (sometimes called the cleidoic egg; Greek *cleistos*, "closed") and a keratinized skin. The embryonic membranes that contribute to the placenta of therian mammals (marsupials and placentals) are homologous with the extraembryonic membranes in the eggs of other amniotes. But amniotes have more going for them than their eggs; there are many other ways in which amniotes represent a more derived kind of tetrapod than either the extant amphibians or the Paleozoic non-amniote tetrapods.

Skin elaborations Amniotes have a great variety of skin elaborations—scales, hair, feathers, nails, beaks, and horns—all formed from keratin. Scales, hair, and feathers are homologous; all of them develop from a common epidermal thickening—the anatomical placode—and all express a similar suite of developmental signaling molecules. Thus, hair and feathers are not independently derived structures, nor did they have to evolve by anatomical transformation of scales; indeed, there is no evidence of such transitional structures in the fossil record. Rather, a change in gene expression in development caused the anatomical placode (a shared ancestral character of amnioles) to produce different types of epidermal structures from the same primordium.

Feeding Although all tetrapods (unlike fishes) have a muscular tongue with extrinsic muscles, amniotes also have intrinsic (i.e., internal) tongue muscles. In addition, basal amniotes had a complex pattern of teeth on the palate (the roof of the mouth), and the tongue and teeth probably worked in concert to form a bolus of food in the mouth for swallowing. The palatal teeth were reduced or lost in various lineages of more derived amniotes, probably in relation to improved feeding mechanisms. The palatal teeth are absent in most extant amniotes but are present in the tuatara (*Sphenodon punctatus*) and in snakes. They were lost in synapsids by the level of cynodonts.

Lung ventilation and necks Amniotes have costal (rib) ventilation of the lungs, in contrast to the buccal pumping of amphibians. Amniotes also have stout, ossified ribs that connect via cartilaginous extensions to a sternum. Extant lissamphibians have short ribs, and although many extinct non-amniotes had substantial ribs, there is little evidence of them connecting to a sternum. Costal ventilation allows an animal to have a longer neck because movement of the ribs (unlike buccal pumping) can produce a pressure differential large enough to move air down a long, thin tube, such as the trachea. In addition, some of the muscles involved in buccal pumping insert into the shoulder girdle, and this arrangement may limit the development of a long neck in non-amniote tetrapods.

A longer neck separating the head from the thorax also provides space for elaboration of the nerves that supply the forelimb. These nerves leave the spinal cord in the neck and join together in a complex called the brachial plexus. The brachial plexus in extant amphibians is simple: only two nerves are involved, whereas the plexus of amniotes has at least five nerves, increasing control of the forelimbs and their ability for manipulation.

Extant amphibians have very short necks, and they have not moved the heart posteriorly from its embryonic position at the base of the branchial area in the head. Amniotes, with their longer necks, have displaced the heart caudally, into the thorax.

Locomotion   Amniotes are in general more reliant on limb propulsion than are non-amniotes, which rely on side-to-side flexion of the trunk. This is reflected in the structure of the ankle joint (discussed in Chapter 14). In early tetrapods, the ankle joint lacked a distinct plane of motion, but the ankles of amniotes, including some stem amniotes, have a distinct hinge joint between the proximal and distal rows of tarsal bones that makes the foot a better propulsive lever, and the proximal tarsals have been consolidated into two distinct bones, the astragalus and the calcaneum. Frogs independently evolved a similar condition.

Reproduction   An amniotic egg requires internal fertilization. Most male amniotes have an intromittent organ (the penis), which is probably homologous in all amniote lineages (a penis also evolved independently the caecilian lissamphibians). Unlike the genetic sex determination of mammals and most birds, ancestral amniotes probably had environmentally determined sex determination, as seen in many nonavian reptiles today.

The amniotic egg

The amniotic egg is a remarkable example of biological complexity. A flexible, leathery eggshell is probably the ancestral amniote condition, and it persists in many lizards, snakes, and turtles as well as in monotreme mammals. In some other lizards and turtles and in crocodylians and birds, the shell is rigid because of the inclusion of calcium deposits. Equally important are the extraembryonic membranes in the shells of amniotes.

The shell is the first line of defense against mechanical damage, while pores in the shell permit the movement of water vapor, oxygen, and carbon dioxide. (Amniotic eggs must be deposited in air.) Albumen (egg white) gives further protection against mechanical damage and provides a reservoir of water and protein. The large yolk is the energy supply for the developing embryo.

While all vertebrates have a **yolk sac** membrane enclosing the yolk, amniotes have three additional extraembryonic

membranes: the **chorion**, **amnion**, and **allantois** (Figure 10.13). The allantois appears to have evolved as a storage place for nitrogenous wastes produced by the metabolism of the embryo, and the urinary bladder of the adult grows out from its base. The allantois also serves as a respiratory organ during later development because it is vascularized and can transport oxygen from the surface of the egg to the embryo and carbon dioxide from the embryo to the surface. The allantois is left behind when the embryo emerges,

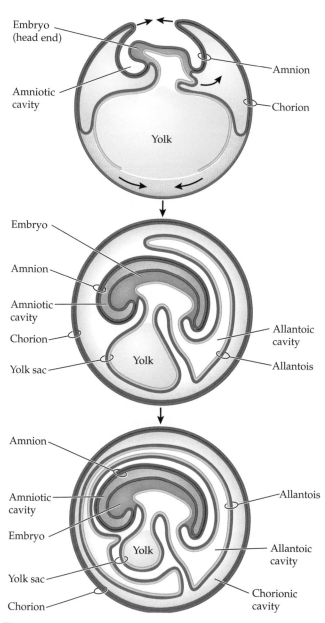

Figure 10.13 Distinctive features of the amniotic egg. The chorion and amnion develop from outgrowths of the body wall at the edges of the developing embryo and spread outward and around the embryo until they meet. At their junction, the chorionic and amniotic membranes merge and leave an outer membrane (the chorion), which surrounds the entire contents of the egg, and an inner membrane (the amnion), which surrounds just the embryo. The allantoic membrane develops as an outgrowth of the hindgut posterior to the yolk sac and lies within the chorion.

so the nitrogenous wastes stored in it do not have to be reprocessed.

The embryo in an amniotic egg bypasses the larval stage typical of amphibian embryos and does not form external gills at any stage in development. All traces of the lateral line are also lost. This loss of the larval form relates to one significant disadvantage of the amniotic egg—it can no longer be laid in water because, lacking gills, the embryo would drown. Marine amniotes must either come ashore to lay eggs, as sea turtles and penguins do, or be viviparous, as most sea snakes and all cetaceans (dolphins, porpoises, and whales) and sirenians (mantees and the dugong) are.

Origin of the amniotic egg What advantages does an amniotic egg provide? Many species of extant amphibians and fishes lay non-amniotic eggs that develop quite successfully on land, although these eggs are smaller than those of amniotes and more vulnerable to environmental conditions. The greater availability of oxygen on land is an important benefit, but both amniotic and non-amniotic eggs must be laid in relatively moist conditions to avoid desiccation. Both types of eggs are usually buried in the soil or deposited under objects such as rocks and logs. (Birds and some nonavian dinosaurs are exceptions to these generalizations.)

Various explanations for the development of the amniotic egg have been proposed. For example, the extraembryonic membranes may facilitate gas exchange, and the shell may provide mechanical support on land; together these features would allow the evolution of a large egg that produced a large hatchling that in turn grew into a large adult because egg size is related to adult size. These are speculations, however, and there is no convincing explanation of the origin of the amniotic egg, although its significance in the subsequent evolution of amniotes is clear.

How can scientists tell if an extinct tetrapod laid an amniotic egg when features such as extraembryonic membranes are not preserved in the fossil record? Scientists can estimate the latest point of origin of the amniotic egg from the tetrapod phylogeny. Because the extraembryonic membranes of extant amniotes are homologous, all tetrapods included within the radiation of extant forms must have inherited an amniotic egg from their common ancestor (see Figure 10.1). A more difficult question is whether any stem amniotes might have laid an amniotic egg. We know that this was not the case for seymouriamorphs because there are fossils of larval seymouriamorphs with external gills and lateral lines. Diadectomorphs share with amniotes possession of keratinous claws on the tips of their digits, and diadectomorphs may be the sister group of the synapsids, which would make them crown amniotes.

Patterns of amniote temporal fenestration

Amniotes have traditionally been classified according to the number of holes in their head—that is, on the basis of **temporal fenestration** (Latin *fenestra*, "a window"). The

major configurations that give names to different lineages of amniotes are **anapsid** (Greek *an*, a prefix meaning "without"; *apsis*, "a loop"—i.e., an opening in the side of the skull, and thus no bony arch), seen in early amniotes and in turtles; **synapsid** (single arch; Greek *syn*, a prefix meaning "together"), seen in mammals and their ancestors; and **diapsid** (double arch; Greek *dis*, a prefix meaning "two"), seen in birds and other reptiles. The term "arch" refers to the temporal bars lying below and between the holes. **Figure 10.14** shows the patterns in the different lineages. We can infer that temporal fenestrations arose independently in the synapsid and diapsid lineages because basal sauropsids lack fenestrae.

This classification system was devised more than 100 years ago, and although molecular phylogenies largely uphold these relationships as they were originally proposed, the nature of the temporal fenestrae is more variable and more plastic than originally thought. For example, the anapsid skulls of turtles have probably secondarily roofed over the fenestrae from an originally diapsid condition, and the apparently primitive diapsid condition of the tuatara (*Sphenodon*) is a secondary redevelopment of the lower temporal bar.

Even though the skull of extant mammals is highly modified from the ancestral synapsid condition, you can still feel these skull features in yourself. If you put your hands on either side of your eyes, you can feel your cheekbone (the zygomatic arch)—that is the temporal bar that lies below your synapsid skull opening. Then, if you clench your jaws, you can feel the muscles bulging above the arch. The muscles are passing through the temporal opening, running from their origin on the top of your skull to their insertion on your lower jaw. The temporal opening allows the muscles that run from the lower jaw to the outside of the skull roof to bulge when they contract, and the margins of the openings initially provided an origin for the muscles.

The evolution of the temporal fenestrae may be related to changes in the complexity and orientation of the jaw-closing (adductor) muscles. The large, flat skulls of non-amniotes, which may be important for their buccal-pumping mode of respiration, did not permit a change in the orientation of the jaw muscles from the basic fish condition (**Figure 10.15A**). With the evolution of costal ventilation, amniotes evolved smaller, more domed skulls, allowing differentiation of the jaw adductors into the adductor mandibularis and the pterygoideus (**Figure 10.15B**). The pterygoideus originates from the pterygoid flange on the base of the skull, which is a characteristic feature of amniotes.

The advantage of this change in musculature would have been a change in feeding abilities. Fishes and non-amniote tetrapods close their jaws with a single snap (inertial feeding), whereas amniotes can snap the jaws closed and then apply pressure with the teeth when the jaws are closed (static pressure feeding). This difference may have allowed amniotes to develop more complex types of oral food processing, such as the ability of herbivores to nip off

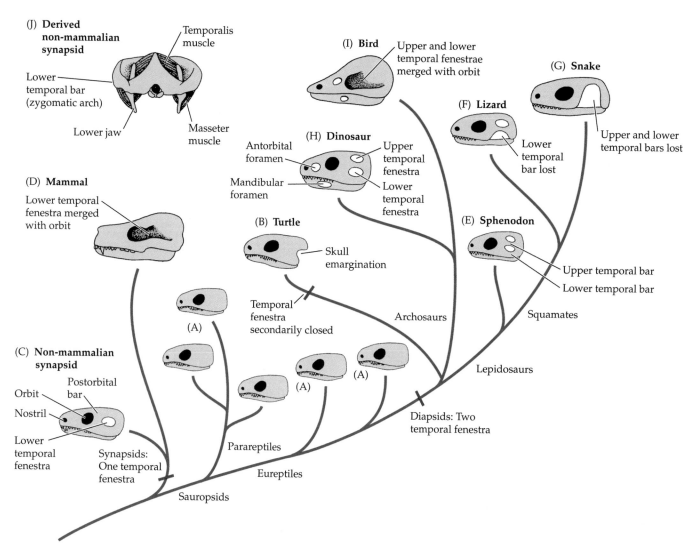

Figure 10.14 Patterns of amniote skull fenestration.
(A) Ancestral anapsid condition, as seen in parareptiles and eureptiles. (B) Modified anapsid condition (derived from diapsid condition), with emargination of the posterior portion of the skull, as seen in turtles. (C) Ancestral synapsid condition, with lower temporal fenestra only. (D) Derived mammalian synapsid condition, in which the orbit has become merged with the temporal opening and dermal bone has grown down from the skull roof to surround the braincase. (E) Ancestral diapsid condition. Both upper and lower temporal fenestrae are present (the presence of the lower temporal bar in *Sphenodon* is a reversal from a lizardlike condition). (F) Lizardlike condition, typical of most squamates, where the lower temporal bar has been lost. (G) Snake condition, with the upper temporal bar lost in addition to the lower bar.
(H) Ancestral archosaur diapsid condition, seen in most dinosaurs, with the addition of an antorbital foramen and, in some taxa, a mandibular foramen. (I) Derived avian archosaur condition, convergent with the condition in mammals. The orbit has become merged with the temporal openings, and the braincase is enclosed in dermal bone. (J) Posterior view through the skull of a derived non-mammalian synapsid (a cynodont) showing how the temporal fenestra allows muscles to insert on the outside of the skull roof. The temporalis and masseter muscles are divisions of the original amniote adductor muscle complex.

vegetation with their front teeth, or to crush tough food by applying pressure with their cheek teeth.

Reorientation of the adductor mandibularis in the more domed skull of basal amniotes may be the underlying explanation for skull fenestration in derived amniotes. The development of even a small hole in the skull roof (in an area where three bones fail to completely meet during development) would allow the muscle to develop an area of origin on the outside of the skull (see Figure 10.15B, cross-sectional view) rather than being confined to the interior of the skull (see Figure 10.15A, cross-sectional views). Subsequent enlargement of the hole could accommodate a larger muscle, allowing a more powerful bite. Perhaps differences in muscle actions, relating to different feeding modes, led to the differences in temporal fenestration in the synapsid and diapsid lineages.

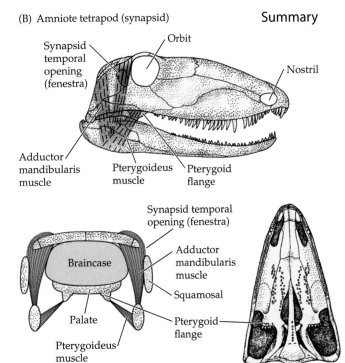

(A) Non-amniote tetrapod

(B) Amniote tetrapod (synapsid)

Figure 10.15 Skull fenestration and jaw muscles in early tetrapods. The skulls show the differences in temporal fenestrations and jaw muscles in a non-amniote tetrapod (A) and an amniote tetrapod (B). The cross-sectional views are at the level of the skull just behind the orbit. Note that in the non-amniote tetrapod the skull is flat, and a single adductor muscle mass originates from the braincase and the underside of the dermal skull roof, as in a fish. If a portion of this muscle were oriented more vertically, it would be too short to allow the jaw to open, because muscles can be stretched by only one-third of their resting length. In the amniote, the skull is tall enough for a jaw muscle to originate from the top of the head. The original muscle mass is now differentiated into the adductor mandibularis (part or all of which may exit through the temporal fenestra to insert on the top of the skull roof) and the pterygoideus (which now originates from the pterygoid flange on the dermal palate). (Some parts of this figure are based on Kemp 1982 and Romer and Price 1940.)

Summary

The earliest tetrapods are known from the Late Devonian. Their immediate fish ancestors are known from slightly earlier in the period.

The tetrapodomorph fish *Tiktaalik* had a mosaic of fish- and tetrapod-like characters, making it an intermediate form between fishes and tetrapods. Although it would have been able to move its head on its body and push itself up from the substrate, like a tetrapod, it retained fin rays, and so is classified as a fish.

The best-known Devonian tetrapods are *Acanthostega* and *Ichthyostega*. Both were primarily aquatic, although they may have made forays onto land. They retained gills and had limbs that were polydactylous and probably more suitable for swimming than for walking on land.

Tetrapods were originally thought to have freshwater origins, but new evidence suggests that they may have originated in marine or brackish environments. Footprints hint at tetrapods existing up to 15 million years before the first known fossils.

Many of the tetrapod features that were useful on land were originally advantageous for life in the water.

There are many reasons why a fish might benefit from terrestrial activity, and indeed many fishes do so today.

Many extant fishes use their limbs to walk on land or underwater. These fish all provide clues as to how the earliest tetrapods might have made their initial terrestrial forays.

Fins and limbs are not as different as previously thought.

No fish possesses limbs with digits, and it was once thought that these were new inventions in tetrapods. Evolutionary developmental studies have shown that the same genes code for limb development in fishes and tetrapods, with the major difference lying in the timing of gene expression in the distal part of the developing limb.

(Continued)

Summary *(continued)*

There are more challenges to locomotion on land than simply overcoming gravity.

Compared with fishes, tetrapods have larger limbs and limb girdles—features that resist the compressive and torsional forces encountered by bodies on land. Early tetrapods had elbows that could flex to push the body off the ground and strong muscles that could lift the body and brace the forelimbs. More derived tetrapods also have limbs that point downward rather than out to the side.

Studies of *Ichthyostega* show that the earliest tetrapods did not walk like modern ones. *Icththyostega* had a limited range of limb motion and probably dragged itself over the ground by "crutching" (moving both forelimbs together), as extant seals do. The anatomy of *Tiktaalik* was intermediate between that of other fishes and the earliest tetrapods.

For early tetrapods, eliminating carbon dioxide on land may have been a bigger challenge than obtaining oxygen.

The earliest tetrapods would have inherited lungs from their fish ancestors, and many early tetrapods were at least semi-aquatic and retained their gills. They may have inhaled air via the spiracle, as do the basal actinopterygians *Polypterus* today.

Carbon dioxide is easily lost over the gills in water but is harder to lose in air. The buccal pumping mode of lung ventilation of the early tetrapods (seen in modern lungfishes and amphibians) is not sufficient for carbon dioxide loss. Amniotes evolved costal ventilation (using their ribs), enabling all of the carbon dioxide to be lost via the lungs.

Extant amphibians lose carbon dioxide over their thin skins, but this would have not been an option for the thicker-skinned, large-bodied early tetrapods. The dermal bony armor of the more terrestrial early tetrapods might have counteracted acidosis resulting from carbon dioxide build up in the body.

There were two major lineages of non-amniote Paleozoic tetrapods.

Batrachomorphs included the ancestry of modern amphibians, and reptiliomorphs included the ancestry of modern amniotes. Lepospondyls, small forms that somewhat resembled extant amphibians, may be related to one of these lineages or may form an independent minor lineage.

Some Paleozoic tetrapods were stem forms, basal to the main lineages.

Many of the Paleozoic tetrapods were much larger than any extant amphibians and were probably rather crocodilelike in appearance.

Non-amniote tetrapods reached their peak diversity in the Late Carboniferous and Early Permian and included aquatic, semiaquatic, and terrestrial forms. Most lineages were extinct by the late Paleozoic, but some specialized aquatic forms survived into the Mesozoic.

The diversification of insects in the Carboniferous, following the development of a land flora, may have been critical for the initial radiation of tetrapods. Herbivorous tetrapods are not known until the Permian.

The first amniotes coexisted with non-amniote tetrapods.

The three major lineages of amniotes—the synapsids (ancestors of mammals), eureptiles (ancestors of modern reptiles, including birds), and parareptiles (extinct by the end of the Triassic)—were all established by the Late Carboniferous, but their major radiations took place in the Permian.

There's a lot more to amniotes than the amniotic egg.

There are many ways in which amniotes are more derived than other tetrapods.

Amniotes have a great diversity of skin elaborations that are all formed from keratin: scales, hair, feathers, nails, beaks, and horns.

Costal ventilation allows amniotes to have longer necks, because movement of the ribs (unlike buccal pumping) can produce a pressure differential large enough to move air down the trachea. A longer neck also permits elaboration of the nerves supplying the forelimbs, making possible the evolution of fine manipulation.

Amniotes are more dependent on limbs for locomotion than are most other tetrapods, and have ankles with a distinct hinge joint.

The amniotic egg requires internal fertilization, and most male amniotes have a penis for this function. This structure is probably homologous among the different amniote lineages.

The amniotic egg is remarkably complex.

While all vertebrates have a yolk sac enclosing the yolk, amniotes have three additional extraembryonic membranes: the chorion, amnion, and allantois.

The shells of amniotic eggs allow them to be larger than non-amniotic eggs and thereby produce larger hatchlings.

Summary (continued)

Amniotes lack a larval stage, and amniote embryos lack gills and will drown if the egg is laid in water. Some stem amniotes (seymouriamorphs) are known to have had aquatic larvae and thus could not have had a amniotic egg.

Amniote lineages are characterized by their pattern of temporal fenestration.

The three main patterns of temporal fenestration are anapsid (no opening), synapsid (one opening), and diapsid (two openings). The basal condition is for no openings, and the synapsid and diapsid conditions evolved independently of each other.

Temporal fenestration provided an early basis for dividing amniotes into major lineages, and this classification is supported by molecular phylogenies, although the patterns of temporal fenestration appear to be more plastic than originally thought.

Differentiation of the jaw adductor muscles of amniotes follows a change in the shape of the skull when buccal pumping was replaced by costal ventilation. This muscle differentiation allows amniotes to apply pressure with closed jaws in addition to snapping the jaws shut. This diversification of jaw muscles opened the door to the evolution of new feeding behaviors.

Discussion Questions

1. What information about the Devonian tetrapods leads to the conclusion that tetrapods evolved in the water rather than on land?

2. What is the definition (in phylogenetic terms) of a tetrapodomorph fish?

3. What features of the elpistostegalian fishes lead us to infer that they were shallow-water forms?

4. What is the main new piece of information about early tetrapods that contradicts the old "drying pond" hypothesis of the origins of terrestriality?

5. How can we be sure that fossil amniotes, such as the captorhinomorphs, laid amniotic eggs? How can we be sure that at least most of the non-amniote reptiliomorphs (below the level of diadectomorphs) did not?

6. How can we determine whether an animal was herbivorous? Propose a hypothesis to explain why herbivory is limited to amniotes and their closest relatives.

Additional Reading

Bever GS, Lyson TR, Field DJ, Bhullar B-AS. 2016. The amniote temporal roof and the diapsid origin of the turtle skull. *Zoology* 6: 471–473.

Brennan PLR. 2016. Evolution: One penis after all. *Current Biology* 26: R22–R40.

Clack JA. 2012. *Gaining Ground: The Origin and Evolution of Tetrapods*, 2nd ed. Indiana University Press, Bloomington, IN.

Daeschler EB, Shubin NH, Jenkins FA Jr. 2006. A Devonian tetrapod-like fish and the evolution of the tetrapod body plan. *Nature* 440:757-763.

Janis CM, Keller JC. 2001. Modes of ventilation in early tetrapods: Costal aspiration as a key feature of amniotes. *Acta Palaeontologica Polonica* 46: 137–170.

Janis CM, Devlin K, Warren DE, Witzmann F. 2012. Dermal bone in early tetrapods: A palaeophysiological hypothesis of adaptation for terrestrial acidosis. *Proceedings of the Royal Society B*, 279: 3035–3048.

Martin KL, Carter AL. 2013. Brave new propapgules: Terrestrial embryos in anamniotic eggs. *Integrative and Comparative Biology* 53: 223–247.

Musser JM, Wagner GP, Prum RO. 2015. Nuclear β-catenin localization supports homology of feathers, avian scutate scales, and alligator scales in early development. *Evolution & Development* 17: 185–194.

Nakamura T, Gehrke AR, Lemberg J, Szymaszek J, Shubin NH. 2016. Digits and fin rays share common developmental histories. *Nature* 537: 225–228.

Pierce SE, Hutchinson JR, Clack JA. 2013. Historical perspectives on the evolution of tetrapodomorph movement. *Integrative and Comparative Biology* 53: 209–223.

Schoch RR. 2014. *Amphibian Evolution: The Life of Early Land Vertebrates*. Wiley Blackwell, West Sussex, UK.

Go to the Companion Website **oup.com/us/vertebratelife10e** for active-learning exercises, news links, references, and more.

Extant Amphibians

© Twan Leenders

The extant amphibians, Lissamphibia (Greek *lissos*, "smooth"), are tetrapods with moist, scaleless skins. The group includes three distinct lineages: Anura (frogs), Urodela (salamanders), and Gymnophiona (caecilians). Common names of amphibians can be confusing: "Frogs, toads, and treefrogs" and "salamanders and newts" are widely used, and there are any number of regional variations, such as "waterdogs" and "spring lizards." We use "frogs" as the general term for all anurans; "toads" are members of the family Bufonidae, and thus are frogs. Similarly, all urodeles are salamanders, and "newts" are aquatic species in the family Salamandridae, so all newts are salamanders. In contrast to the names "toads" and "newts," which have a phylogenetic basis, "treefrog" is applied to arboreal species of frogs in many families.

Although the three lineages of extant amphibians have very different body forms, they are identified as a monophyletic evolutionary lineage by several shared derived characters—many of which play important roles in the functional biology of amphibians.

11.1 Diversity of Lissamphibians

Most amphibians have four well-developed limbs, although a few salamanders and all caecilians are limbless. Frogs lack tails (hence the name Anura, Greek *an*, a prefix meaning "without"; *oura*, "tail"), whereas salamanders have long tails. The tails of caecilians are short, as are those of other kinds of elongated, burrowing animals.

At first glance, the three lineages of amphibians appear to be very different kinds of animals. Frogs have long hind limbs and short, stiff bodies that don't bend when they walk; salamanders have forelimbs and hind limbs of equal size and move with lateral undulations; and caecilians are limbless and burrow with alternating extensions and contractions known as concertina locomotion. These obvious differences are all related to locomotor specializations, however, and closer examination shows that amphibians have many shared derived characters (**Table 11.1**).

Perhaps the most important derived character of extant amphibians is a moist, permeable skin. Many of the Paleozoic non-amniote tetrapods had bony dermal armor; a permeable, unadorned skin is a derived character shared by amphibians.

All extant adult amphibians are carnivorous, and relatively little morphological specialization is associated with different dietary habits within each group. Amphibians eat almost anything they can catch and swallow. The tongue of aquatic forms is broad, flat, and relatively immobile, but some terrestrial amphibians can protrude the tongue from the mouth to capture prey.

The size of the head is an important determinant of the maximum size of prey that can be taken, and salamander species that live in the same habitat frequently have markedly different head sizes, suggesting that this character may reduce competition for food. Frogs in the Neotropical genera *Lepidobatrachus* and *Ceratophrys*, which feed largely on other frogs, have such large heads that they are practically walking mouths (**Figure 11.1**).

A function has recently been discovered for the unique green rod cells in the retinas of amphibians. Viewed as a curiosity since they were first described 1877, these cells are now known to allow frogs to distinguish colors, even at extremely low light levels. This visual sensitivity suggests that the conspicuous colors and markings found on the flanks of many frogs (such as the red-eyed leaf frog, *Agalychnis callidryas*; see Figure 11.5E) function in species recognition at night.

Salamanders

Salamanders have the most generalized body form and locomotion of the extant amphibians. Salamanders are elongated, and all but a few completely aquatic species have four functional limbs (**Figure 11.2**). Their walking-trot gait is probably similar to that employed by the earliest tetrapods. It combines the lateral bending characteristic of fish locomotion with leg movements.

The genome size (total amount of DNA in a haploid cell) of salamanders sets them apart from frogs and caecilians, and indeed from nearly all other vertebrates. The genome

(A)

Figure 11.1 **Some frogs prey on other frogs**   The horned frogs (*Ceratophrys*) of South America are terrestrial sit-and-wait predators that prey on other frogs. (A) Blending with fallen leaves on the forest floor, a horned frog (*C. cornuta*) waits motionless until another frog approaches. At this point, the horned frog begins to twitch one of its toes to lure the prey close enough to seize. (B) This micro-CT image shows a ventral view of a Bell's horned frog (*C. ornata*) that had swallowed a frog nearly as large as itself. The horned frog is green, the frog it ate is red. The prey's head presses against the left body wall of the horned frog and its hind leg extends through the horned frog's esophagus into its buccal cavity. (A, Bernard Dupont, CC BY-SA 2.0; B, Thomas Kleinteich.)

(B)

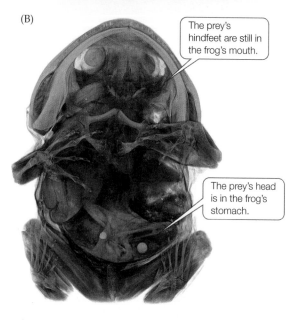

The prey's hindfeet are still in the frog's mouth.

The prey's head is in the frog's stomach.

other amphibians. These large cells require morphological and developmental adjustments, especially for structures in which form and function are intimately intertwined, such as kidney tubules and brain cells.

The ten families of salamanders with their approximately 695 species are almost entirely limited to the Northern Hemisphere; their southernmost occurrence is in northern South America (**Table 11.2**). North and Central America have the greatest diversity of species—more salamander species are found in Tennessee than in all of Europe and Asia combined. Paedomorphosis is widespread among salamanders, and several families of aquatic salamanders consist solely of paedomorphic forms. Paedomorphs retain larval characters, including larval tooth and bone patterns, the absence of eyelids, a functional lateral line system, and (in some cases) external gills.

The largest extant salamanders are the Japanese and Chinese giant salamanders (*Andrias*), which can reach lengths of more than 1 m. The related North American hellbender (*Cryptobranchus alleganiensis*) grows to 60 cm. All are members of Cryptobranchidae and are paedomorphic and permanently aquatic. As their name indicates (Greek *kryptos*, "hidden"; *branchia*, "gills"), they do not retain external gills, although they do have other larval characters. Congo eels (three species of aquatic salamanders in the genus *Amphiuma*) are another group of aquatic salamanders

size for most vertebrates ranges from 1 to 7 picograms (pg), but the mean value for salamanders is 35 pg, with a maximum of 120 pg. Expansion of the genome originated at the base of crown group salamanders. Genome size and cell size are correlated, and salamanders have larger cells than

Table 11.1	**Shared Derived Characters of Lissamphibians**
Moist, permeable skin and substantial cutaneous gas exchange: All amphibians have mucus glands that keep the skin moist. A considerable part of an amphibian's exchange of oxygen and carbon dioxide with the environment takes place through the skin. All amphibians also have granular (poison) glands in the skin.	**Papilla amphibiorum:** This special sensory area in the sacculus of the inner ear is present in all amphibians and is sensitive to sound frequencies below 1,000 hertz (Hz; cycles per second). A second sensory area, the papilla basilaris, detects frequencies above 1,000 Hz.
Pedicellate, bicuspid teeth: Nearly all extant amphibians have teeth in which the crown and base (pedicel) are composed of dentine and are separated by a narrow zone of uncalcified dentine or fibrous connective tissue. The teeth have two cusps, one on the lingual (inner) side and the other on the labial (outer) side. A few temnospondyls (the lineage from which lissamphibians are derived) had pedicellate teeth.	**Green rods:** In addition to the red rods that are found in all vertebrates and are sensitive to red light, salamanders and frogs have green rods, a distinct type of retinal cell that has maximum sensitivity to blue light. Caecilians apparently lack green rods; however, the eyes of caecilians are extremely reduced, and the green rods may have been lost.
Operculum-columella complex: Most amphibians have two bones that are involved in transmitting sounds to the inner ear. The columella, which is derived from the hyoid arch, is present in salamanders, caecilians, and most frogs. The operculum (not homologous to the operculum of fishes) is a bony or cartilaginous structure that attaches to the ear capsule and is connected to the shoulder by the opercular muscle. The operculum transmits ground vibrations to the inner ear.	**Structure of the levator bulbi muscle:** This muscle, present in salamanders and frogs and in modified form in caecilians, is a thin sheet in the floor of the eye socket that is innervated by the fifth cranial nerve. It causes the eyes to bulge outward, thereby enlarging the buccal cavity.

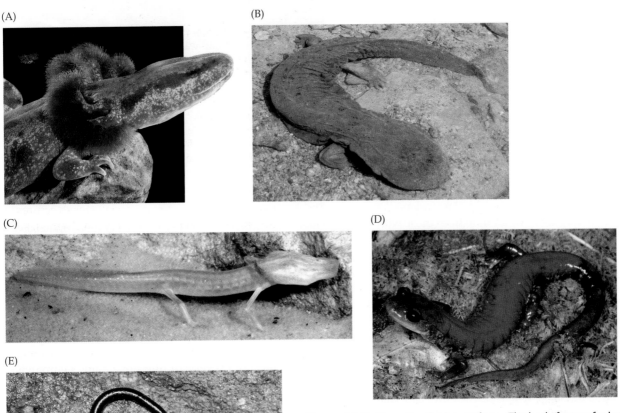

Figure 11.2 Diversity of salamanders The body forms of salamanders reflect differences in their life histories and habitats. (A) Some aquatic salamanders retain gills as adults, as in the North American mudpuppy (*Necturus maculosus*). (B) Other aquatic salamanders, such as the North American hellbender (*Cryptobranchus alleganiensis*), have folds of skin that are used for gas exchange. (C) Specialized cave-dwelling salamanders, such as the Texas blind salamanders (*Eurycea rathbuni*) are white and lack eyes. (D) Many terrestrial salamanders have sturdy legs, like the Yonahlossee salamander (*Plethodon yonahlossee*). (E) Some small terrestrial salamanders, like this Central American species of *Oedipina*, have small legs and use sinusoidal locomotion to move rapidly. (A, B, D, courtesy of Todd Pierson; C, E, courtesy of Wayne Van Devender.)

Table 11.2 Families of Extant Salamanders[a]

Plethodontidae: Aquatic or terrestrial salamanders, some with aquatic larvae, others with direct development (458 species in North and Central America, northern South America, Mediterranean Europe, the Korean peninsula; up to 30 cm)

Amphiumidae: Elongated aquatic salamanders lacking gills (3 species in North America; up to 1 m)

Rhyacotritonidae: Semiaquatic salamanders with aquatic larvae (4 species in the Pacific Northwest of North America; up to 10 cm)

Proteidae: Paedomorphic aquatic salamanders with external gills (5 species of *Necturus* in North America, 1 species of *Proteus* in Europe; up to 30 cm)

Dicamptodontidae: Terrestrial salamanders with facultative metamorphosis and a single paedomorphic aquatic species (4 species in the northwestern U.S., barely extending into Canada; up to 35 cm)

Ambystomatidae: Small to large terrestrial salamanders with aquatic larvae (32 species in North America; up to 30 cm)

Salamandridae: Terrestrial and aquatic salamanders, most with aquatic larvae; some species of *Salamandra* are viviparous (115 species in Europe, Asia, North America, and extreme northwestern Africa; up to 20 cm)

Sirenidae: Elongated aquatic salamanders with external gills, lacking the pelvic girdle and hindlimbs, and probably with external fertilization (4 species in the southeastern U.S.; up to 75 cm)

Hynobiidae: Terrestrial or aquatic salamanders with external fertilization of eggs and aquatic larvae (66 species in Asia; up to 30 cm)

Cryptobranchidae: Very large paedomorphic aquatic salamanders with external fertilization of eggs (1 species in North America, 2 species in Asia; up to 1.8 m)

[a]Families are arranged in order from the least derived at the bottom to most derived at the top. Identified in text: Cryptobranchidae, Plethodontidae, Hynobiidae, Sirenidae, Proteidae.

without external gills. Congo eels live in the lower Mississippi Valley and coastal plain of the United States. They have well-developed lungs and can estivate in the mud of dried ponds for up to 2 years.

Mudpuppies (*Necturus*) are another group of large aquatic salamanders, but mudpuppies do retain external gills. They occur in lakes and streams in eastern North America. Sirens are also aquatic salamanders with external gills.

Several lineages of salamanders have adapted to life in caves. The constant temperature and moisture of caves make them good salamander habitats, and cave-dwelling invertebrates supply food. The brook salamanders (*Eurycea*) include species that form a continuum from those with fully metamorphosed adults inhabiting the twilight zone near cave mouths to fully paedomorphic forms in the depths of caves or sinkholes. The Texas blind salamander (*Eurycea rathbuni*) is a highly specialized cave dweller—blind, white, and with external gills, extremely long legs, and a flattened snout used to probe underneath pebbles for food. The Texas blind salamander is a plethodontid, and the European olm (*Proteus anguinus*), a proteid, has converged on the same body form.

Terrestrial salamanders such as the North American mole salamanders (*Ambystoma*) and the European salamanders (*Salamandra*) have aquatic larvae that lose their gills at metamorphosis. The most fully terrestrial salamanders, the lungless plethodontids (such as the northern slimy salamander [*Plethodon glutinosus*]), include species in which the young hatch from eggs as miniatures of the adult and there is no aquatic larval stage. Larval salamanders and adults of aquatic species use suction feeding to capture prey, whereas terrestrial adults seize prey with the jaws, using the tongue to manipulate it in the mouth. Species of salamanders that return to ponds to breed, such as the European and North American newts, switch feeding methods to match their current habitat.

Projectile tongues of plethodontid salamanders

Lungs seem an unlikely organ for a terrestrial vertebrate to abandon, but among salamanders the evolutionary loss of lungs has been a successful tactic. Plethodontidae is characterized by the absence of lungs and contains more species and has a wider geographic distribution than any other salamander lineage. Furthermore, many plethodontids have evolved specializations of the hyobranchial apparatus that allow them to project the tongue a considerable distance from the mouth to capture prey. This ability has not evolved in salamanders that have lungs, probably because the hyobranchial apparatus in these forms is an essential part of the respiratory system.

Bolitoglossine (Greek *bole*, "dart"; *glossa*, "tongue") plethodontids can project the tongue a distance equivalent to their head plus trunk length and can pick off moving prey (**Figure 11.3**). This ability requires fine visual discrimination of distance and direction, and the eyes of bolitoglossines are placed more frontally on the head than are the eyes of less specialized plethodontids, giving

Figure 11.3 A European bolitoglossine salamander, *Hydromantes* This species captures prey on the sticky tip of its tongue, which can be projected from the mouth. (From Deban et al. 1997, photo by Stephen M. Deban.)

bolitoglossines binocular vision. Furthermore, the eyes of bolitoglossines have a large number of nerves that travel to the ipsilateral (same side) visual centers of the brain as well as the strong contralateral (opposite side) visual projection that is typical of vertebrates. Because of this neuroanatomy, bolitoglossines have a complete dual projection of the binocular visual fields to both hemispheres of the brain, and can estimate the distance to prey precisely and rapidly.

Anurans

Whereas salamanders have a limited number of species and a restricted geographic distribution, anurans include more than 50 families containing more than 6,700 species and occur on every continent except Antarctica (**Table 11.3**). Specialization of the body for jumping is the most conspicuous skeletal feature of anurans (**Figure 11.4**).

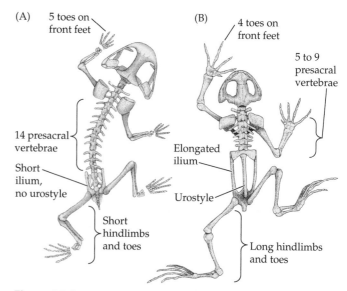

Figure 11.4 *Triadobatrachus* and a derived anuran The Triassic fossil *Triadobatrachus* (A) is considered the sister group of the modern anuran (B). The derived characters of anurans visible in this comparison are shortening of the body, elongation of the ilia, and fusion of the posterior vertebrae to form a urostyle.

The hindlimbs and muscles of anurans form a lever system that can catapult the animal into the air. Numerous morphological specializations are associated with this type of locomotion:

- The hindlegs are elongated, and the tibia and fibula are fused.

- A powerful pelvis strongly fastened to the vertebral column is clearly necessary, as is stiffening of the vertebral column.

- The ilium is elongated and reaches far anteriorly, and the posterior vertebrae are fused into a solid rod, the **urostyle**. The pelvis and urostyle render the posterior half of the trunk rigid.

- The vertebral column is short, with only five to nine presacral vertebrae, and these are strongly braced by articulations between adjacent vertebrae (zygapophyses) that restrict lateral bending.

- The strong forelimbs and flexible pectoral girdle absorb the impact of landing.

- The eyes are large and are placed well forward on the head, providing binocular vision.

The anuran body form probably evolved from a more salamander-like starting point. Both jumping and swimming have been suggested as the mode of locomotion that made the change advantageous. Salamanders and caecilians swim as fishes do—by passing a sine wave down the body. Anurans have inflexible bodies and swim with simultaneous thrusts of the hindlegs.

Some paleontologists have proposed that the anuran body form evolved because of the advantages of using the hindlimbs for swimming. An alternative hypothesis traces the anuran body form to the advantage gained by an animal that could rest near the edge of a body of water and escape aquatic or terrestrial predators with a rapid leap followed by locomotion on either land or water.

The diversity of anurans exceeds the number of common names that can be used to distinguish various ecological specialties (**Figure 11.5**). Animals called frogs usually have long legs and move by jumping, and this body form is found in many lineages. Semiaquatic frogs are moderately streamlined and have webbed hindfeet. Stout-bodied terrestrial anurans that make short hops instead of long leaps are often called toads. They usually have blunt heads, heavy bodies, relatively short legs, and little webbing between the toes. True toads (Bufonidae) have this body form, and very similar body forms are found in other families, including the spadefoot toads of North America (*Spea, Scaphiopus*) and the horned frogs of South America (*Ceratophrys*).

Arboreal frogs usually have large heads and eyes and often slim waists and long legs. Arboreal frogs in many different families move by quadrupedal walking

---

Table 11.3 Major Families of Extant Anurans[a]

Pixicephalidae: Terrestrial, semiaquatic, and arboreal frogs; most have aquatic tadpoles (80 species in sub-Saharan Africa; up to 25 cm)

Ranidae: Aquatic and terrestrial frogs; most have aquatic tadpoles (387 species worldwide except for extreme southern South America, South Africa, Madagascar, and most of Australia; up to 30 cm)

Mantellidae: The major group of frogs on Madagascar; most are terrestrial, but the family includes arboreal, aquatic, and fossorial species; species in the genus *Mantella* resemble Neotropical dendrobatids in being brightly colored terrestrial frogs with alkaloid toxins in the skin (212 species in Madagascar and the Comoros Islands; up to 10 cm)

Rhacophoridae: Asian treefrogs that have converged on the body form of hylids and have large heads, large eyes, and enlarged toe pads (405 species in sub-Saharan Africa, southern Asia, Japan, and the Philippines; up to 12 cm)

Microhylidae: Terrestrial and arboreal frogs; many have aquatic larvae, but some species have nonfeeding tadpoles and others lay direct-developing eggs on land (599 species in North, Central, and South America, sub-Saharan Africa, India, and Korea to northern Australia; up to 10 cm)

Bufonidae: Mainly short-legged terrestrial frogs; most have aquatic larvae, but some species of *Nectophrynoides* are viviparous (596 species worldwide except for Australia, Madagascar, and the Oceanic islands; up to 25 cm)

Ceratophryidae: Large, short-legged terrestrial frogs with enormous mouths that consume small vertebrates as well as insects; tadpoles of *Ceratophrys* and *Lepidobatrachus* are primarily carnivorous (12 species in South America; up to 15 cm)

Dendrobatidae: Small terrestrial frogs, many brightly colored and toxic; terrestrial eggs hatch into tadpoles that are transported to water by an adult (306 species in Central America and northern South America; up to 6 cm)

Hylidae: Most species are treefrogs, but a few species are aquatic or terrestrial (964 species in North, Central, and South America, the West Indies, Europe, Asia, and the Australo-Papuan region; up to 15 cm)

Leptodactylidae: Semiaquatic and terrestrial frogs with aquatic larvae (209 species in southern North America, Central America, northern South America, and the West Indies; up to 25 cm)

Centrolenidae: Small arboreal frogs commonly called glass frogs because the internal organs are visible through the skin of the ventral surface of the body; oviparous with aquatic larvae; most centrolenids deposit eggs on leaves or rocks above the water (154 species in Central America and northern South America; up to 3 cm)

Eleutherodactylidae: Small arboreal frogs; internal fertilization may be widespread, and many species have direct development; one species, *Eleutherodactylus jasperi* in Puerto Rico, retains the eggs in the oviducts until the young have passed through metamorphosis and emerge as froglets; this species has not been seen since 1981 and is believed to be extinct (213 species in Central America, northern South America, and the Greater and Lesser Antilles; up to 9 cm)

Scaphiopodidae: Terrestrial frogs with aquatic larvae; they use keratinized structures on the hindfeet to burrow backward into soil (7 species from southern Canada to northern Mexico; up to 8 cm)

Pipidae: Specialized aquatic frogs; most species of *Xenopus* are polyploids, with some having up to 12 sets of chromosomes; *Xenopus, Hymenochirus, Pseudohymenochirus,* and some species of *Pipa* have aquatic larvae; other species of *Pipa* have direct development (41 species in South America [*Pipa*] and sub-Saharan Africa; up to 15 cm)

Ascaphidae: Streambank frogs, and the only anurans with an intromittent organ (2 species in the mountains of northwestern North America; up to 50 mm)

[a]More than 50 families of anurans have been named, with a total of 6,714 species. Only some of the families are included here and they are arranged in order from the least derived at the bottom to most derived at the top.

Figure 11.5 **Diversity of anurans** Body shape and limb length of anurans reflect specializations for different habitats and different methods of locomotion. (A) Aquatic frogs, like the American bullfrog (*Rana catesbeiana*), have sturdy hind legs and extensive webbing between the toes. (B) Specialized aquatic frogs, such as the African clawed frog (*Xenopus laevis*), retain lateral lines as adults, whereas terrestrial and semiaquatic species lose their lateral lines during metamorphosis. (C) Terrestrial leapers, like *Craugastor xucanebi* from the highlands of Guatemala, have long legs and toes without webbing. (D) Terrestrial walkers and hoppers, such as Budgett's frog (*Lepidobatrachus laevis*), have short limbs and toes without webbing. (E) Arboreal frogs, like the red-eyed leaf frog (*Agalychnis callidryas*), have long limbs and toe pads. (F) Frogs that burrow into the ground head first, like the eastern narrow-mouthed toad (*Gastrophyrne carolinensis*), have short limbs and pointed snouts. (A, © Dee Browning/ShutterStock; B, R. D. Bartlett; C, F, courtesy of Todd Pierson; D, © Twan Leenders; E, © iStock.com/NajaShots.)

and climbing as much as by leaping. Toe pads have evolved independently in several frog lineages and show substantial convergence in structure. Many arboreal species of Hylidae and Rhacophoridae have enlarged toe pads and are called treefrogs. Expanded toe pads are not limited to arboreal frogs, however; some terrestrial species that move across fallen leaves on the forest floor also have toe pads.

The surface of the toe pads consists of an epidermal layer of polygonal plates separated by deep channels (**Figure 11.6**). Mucus glands distributed over the pads secrete a viscous solution of long-chain, high-molecular-weight polymers in water. Arboreal frogs use a mechanism known as wet adhesion to stick to smooth surfaces. (This is the same mechanism by which a wet scrap of paper sticks to glass.) The watery mucus secreted by the glands on the toe pads forms a layer of fluid between the pad and the surface and establishes a meniscus at the interface between air and fluid at the edges of the toes. As long as no air bubble enters the fluid layer, a combination of surface tension and viscosity holds the toe pad and surface together.

Frogs can adhere to vertical surfaces and even to the undersides of leaves. Cuban treefrogs (*Osteopilus*

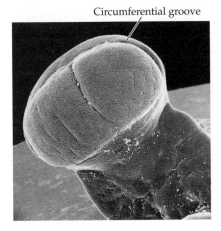

Circumferential groove

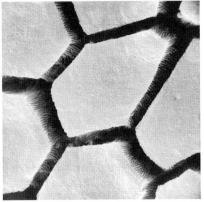

Figure 11.6 Toe pad of a hylid frog (A) The pad lies on the ventral surfaces of the toe tips and is defined by a circumferential groove. (B) The surface of the pad is formed by polygonal plates of epithelial cells separated by deep channels. (Courtesy of Sharon B. Emerson, University of Utah.)

septentrionalis) can cling to a sheet of smooth plastic as it is rotated past the vertical; the frogs do not begin to slip until the rotation reaches an average of 151°—61° past vertical. Adhesion and detachment of the pads alternate as a frog walks across a leaf. As it moves forward, its pads are peeled loose, starting at the rear, as the toes are lifted.

Treefrogs are not able to rest facing downward, because in that orientation the frog's weight causes the toe pads to peel off the surface. Frogs invariably orient their bodies facing upward or across a slope, and they rotate their feet if necessary to keep the toes pointed upward. When a frog must descend a vertical surface, it backs down. This orientation keeps the toes facing upward. During backward locomotion, toes are peeled loose from the tip backward by a pair of tendons that insert on the dorsal surface of the terminal bone of the toe.

Several aspects of the natural history of anurans appear to be related to their different modes of locomotion. In particular, short-legged species that move by hopping are frequently wide-ranging predators that cover large areas as they search for food. This behavior makes them conspicuous to their own predators, and their short legs prevent them from fleeing rapidly enough to escape. Many of these anurans have potent defensive chemicals that are released from glands in the skin when they are attacked.

Species of frogs that move by jumping, in contrast to those that hop, are usually sedentary predators that wait in ambush for prey to pass their hiding places. These species are usually cryptically colored, and they often lack chemical defenses. If they are discovered, they rely on a series of rapid leaps to escape. Anurans that forage widely encounter different kinds of prey from those that wait in one spot, and differences in dietary habits may be associated with differences in locomotor mode.

Aquatic anurans use suction to engulf food in water, but most species of semiaquatic and terrestrial anurans have viscoelastic tongues that can be flipped out to trap prey and carry it back to the mouth. Most terrestrial anurans use a catapult-like mechanism to project the tongue (**Figure 11.7**). The tongue flips over as it leaves the mouth, and the portion that makes contact with the prey has a complex microstructure that assists sticky saliva in fastening to the prey

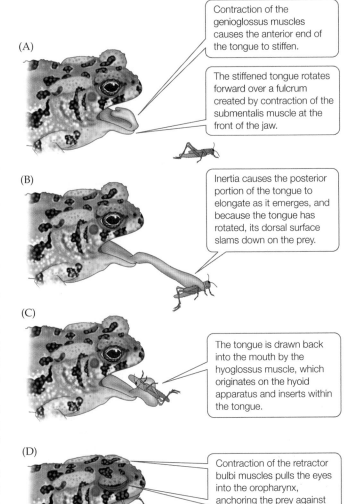

(A) Contraction of the genioglossus muscles causes the anterior end of the tongue to stiffen.

The stiffened tongue rotates forward over a fulcrum created by contraction of the submentalis muscle at the front of the jaw.

(B) Inertia causes the posterior portion of the tongue to elongate as it emerges, and because the tongue has rotated, its dorsal surface slams down on the prey.

(C) The tongue is drawn back into the mouth by the hyoglossus muscle, which originates on the hyoid apparatus and inserts within the tongue.

(D) Contraction of the retractor bulbi muscles pulls the eyes into the oropharynx, anchoring the prey against the tongue and pushing it toward the esophagus.

Figure 11.7 Terrestrial anurans project their tongues to capture prey The tongue is attached at the front of the lower jaw and pivots around the stiffened submentalis muscle as it is flipped out. The tip of the tongue (the portion that is at the rear of the mouth when the tongue is retracted) has glands that excrete sticky mucus that adheres to the prey as the tongue is retracted into the mouth. Retraction of the eyes assists in forcing the prey into the throat.

Table 11.4   Families of Extant Caecilians[a]

Dermophiidae: Subterranean caecilians; viviparous (14 species in Central America, northern South America, tropical west Africa, and Kenya; up to 60 cm)

Siphonopidae: Subterranean caecilians; oviparous with direct development; larvae of two species are known to be dermatophagous (26 species in northern and central South America; up to 0.5 m)

Indotyphlidae: Subterranean caecilians; both oviparous and viviparous species, some with aquatic larvae (22 species in tropical Africa, Seychelles Islands, and India; up to 20 cm)

Typhlonectidae: Aquatic and semiaquatic caecilians; viviparous; embryos scrape lipid-rich cells from the mother's oviduct (14 species in Central America and northern South America; up to 1 m)

Caeciliidae: Subterranean caecilians; oviparous and possibly also viviparous species; no aquatic larval stage (42 species in Central and South America; up to 1.5 m)

Chikilidae: Subterranean caecilians; oviparous with direct development (4 species in northeastern India; up to 25 cm)

Herpelidae: Subterranean caecilians; oviparous with direct development; hatchlings of at least 1 species are dermatophagous (9 species in tropical Africa; up to 35 cm)

Scolecomorphidae: Subterranean caecilians; both oviparous and viviparous species; in this family, the eye is attached to the tentacle and moves out from the skull when the tentacle is extended (6 species in tropical Africa; up to 45 cm)

Ichthyophiidae: Subterranean caecilians; oviparous with aquatic larvae (57 species in the Philippines, India, Thailand, southern China, mainland Malaysia, Sumatra, and Borneo; up to 50 cm)

Rhinatrematidae: Terrestrial caecilians, possibly surface dwellers beneath leaf litter; believed to have aquatic larvae (11 species in northern South America; up to 30 cm)

[a]Families are arranged in order from the least derived at the bottom to most derived at the top.

Figure 11.8 Caecilians are elongated, limbless, and nearly eyeless   This terrestrial Asian species of *Ichthyophis* is oviparous. A female deposits eggs in a burrow in moist soil and remains with the clutch, rotating the eggs frequently. It is likely that antimicrobial peptides from the female's skin protect the eggs, because unattended clutches quickly become infected by fungi. (Courtesy of Lee Grismer.)

muscles, the retractor bulbi, has become the retractor muscle for the tentacle; the levator bulbi moves the tentacle sheath; and the Harderian gland (which moistens the eye in other tetrapods) lubricates the channel of the tentacle of caecilians. The eyes of caecilians in the African group Scolecomorphidae are attached to the sides of tentacles. When the tentacle is protruded, the eye is carried along with it, moving out of the tentacular aperture beyond the roofing bones of the skull.

The caecilian tentacle is probably a sensory organ that allows chemical substances to be transported from the animal's surroundings to the **vomeronasal organ** (an olfactory organ in the roof of the mouth of tetrapods). Terrestrial caecilians feed on small or elongated prey—termites, earthworms, and larval and adult insects—and the tentacle may allow them to detect the presence of underground prey. Aquatic caecilians capture insect larvae, fishes, and tadpoles.

11.2   Life Histories of Amphibians

Of all the characters of amphibians, none is more remarkable than their range of reproductive modes. Most species of frogs and salamanders lay eggs. The eggs may be deposited in water or on land, and they may hatch into aquatic larvae or develop directly into miniatures of the terrestrial adults. The adults of some species of frogs carry eggs attached to the surface of their bodies. Others carry their eggs in pockets in the skin of the back or flanks, in the vocal sacs, or even in the stomach. In still other species, females retain the eggs in the oviducts and give birth to metamorphosed young. Many frogs and salamanders provide no parental care to their eggs or young, but in many other species a parent remains with the eggs and transports tadpoles from the nest to water. The adults of some species of frogs remain with their tadpoles, and in a few species the mother even feeds her tadpoles.

Salamanders and frogs have two characteristics that make their population ecology hard to study. First, fluctuation in size appears to be a normal feature of amphibian populations. Many species of amphibians lay hundreds of eggs, and the vast majority of these eggs never reach maturity. In a good year, however, the survival rate can be unusually high and a large number of individuals may be

and pulling it back into the mouth. The tongues of horned frogs (*Ceratophrys*) have an adhesive strength greater than the frogs' body weight. When the prey is in the mouth, retraction of the eyes assists in swallowing.

Caecilians

The third group of extant amphibians, Gymnophiona (Greek *gymnos*, "naked" [i.e., scaleless]; *ophis*, "snake"), is the least known and does not even have an English common name (**Table 11.4**). These are the caecilians (Latin *caecilia*, "slow worm"), ten families with 205 species of legless, burrowing or aquatic amphibians that occur in tropical habitats around the world (**Figure 11.8**). The eyes of caecilians are greatly reduced and covered by skin or even by bone. Some species lack eyes entirely, but the retinas of other species have the layered organization that is typical of vertebrates, and these species appear to be able to detect light. Conspicuous folds in the skin called **annuli** encircle the bodies of caecilians. The annuli overlie vertebrae and mark the position of the ribs.

Many species of caecilians have dermal scales in pockets in the annuli; scales are not known in the other groups of extant amphibians. A second unique feature of caecilians is a pair of protrusible tentacles, one on each side of the snout between the eye and the nostril. Some structures that are associated with the eyes of other vertebrates have become associated with the tentacles of caecilians. One of the eye

Figure 11.9　A spermatophore　Male salamanders deposit spermatophores that contain a capsule of sperm supported on a gelatinous base. The spermatophore shown here is from an axolotl (*Ambystoma mexicanum*). The sperm is enclosed in the cream-colored structure; the clear base anchors the spermatophore to the substrate. (Courtesy of Michael Guglielmelli © A Lot'l Axolotls.)

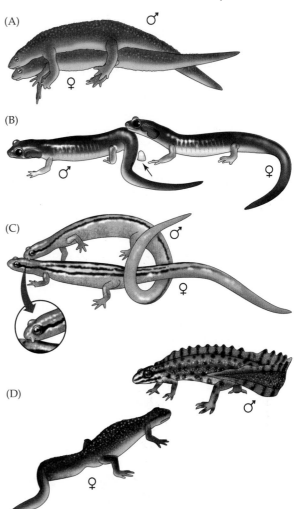

Figure 11.10　Transfer of pheromones by male salamanders during courtship　(A) A male rough-skinned newt (*Taricha granulosa*) rubbing his chin on a female's snout. (B) A female Jordan's salamander (*Plethodon jordani*) following a male in the tail-walk behavior rests her chin on the base of the male's tail as the pair walks forward. The male has just deposited a spermatophore (arrow). (C) A male two-lined salamander (*Eurycea bislineata*) using enlarged teeth to scrape secretions from his mental gland onto the top of the female's head. (D) A male smooth newt (*Lissotriton vulgaris*), using his tail to waft pheromones toward the female.

added to the population. Conversely, in a year of drought, the entire reproductive output of a population may die and no individuals will be added to the population. Thus, year-to-year variation in recruitment creates annual fluctuations in population size that can obscure long-term trends.

Second, many species of salamanders and frogs live in metapopulations in which individual animals move among local populations that are often centered on breeding sites. In the shifting existence of a metapopulation, breeding populations may disappear from some sites while a healthy metapopulation continues to exist and breed at other sites. A limited study might conclude that a species was vanishing, whereas a broader analysis would show that the total population of the species had not changed.

Salamanders

Most groups of salamanders have internal fertilization, but Cryptobranchidae, Hynobiidae, and probably Sirenidae retain external fertilization. Internal fertilization in salamanders is accomplished by transferring a packet of sperm (the **spermatophore**) from the male to the female (**Figure 11.9**). The form of the spermatophore differs among species of salamanders, but all consist of a sperm cap on a gelatinous base.

Females of the hynobiid salamander *Ranodon sibiricus* deposit egg sacs on top of a spermatophore. Males of the Asian salamandrid genus *Euproctus* deposit a spermatophore on the body of a female and then, holding her with their tail or jaws, use their feet to insert the spermatophore into her cloaca. In derived species of salamanders, the male deposits a spermatophore on the substrate, and the female picks up the cap with her cloaca. The sperm are released as the cap dissolves, and fertilization occurs in the oviducts.

A group of all-female salamanders related to *Ambystoma jeffersonianum* is characterized by kleptogensis (Greek *kleptes*, "a thief"; *genesis*, "origin"), a reproductive mode that is unique among vertebrates. Females produce eggs that must be activated by sperm from a male of a bisexual species, and males of five different species can be sperm donors. Furthermore, a female can gather spermatophores from different species of males and incorporate variable amounts

of DNA from those males into her offspring. The result is an enormously complex genome, varying from triploid to pentaploid, in which mtDNA from the mother is the only consistent component.

Courtship　Courtship patterns are important for species recognition, and they show great interspecific variation among salamanders. Pheromones probably contribute to species recognition and may stimulate endocrine activity that increases the receptivity of females.

Pheromone delivery by most salamanders that breed on land involves physical contact between a male and female, during which the male applies secretions of specialized courtship glands (hedonic glands) to the nostrils or body of the female (**Figure 11.10**). Several modes of pheromone delivery have been described:

- Males of many plethodontids have a large gland beneath the chin (the mental gland), and secretions of the gland are applied to the nostrils of the female with a slapping motion.
- The anterior teeth of males of many species of *Desmognathus* and *Eurycea* (both Plethodontidae) hypertrophy during the breeding season. Males of these species spread secretions from the mental gland onto the female's skin and then abrade the skin with their enlarged teeth, inoculating the female with the pheromone.
- Males of two small species of *Desmognathus* use specialized mandibular teeth to bite and stimulate the female.
- Male salamandrids rub the female's snout with hedonic glands located on their cheeks, chin, or cloaca, or vibrate their tails to waft pheromones toward a female.

The males of many species of newts (aquatic salamandrids) perform elaborate courtship displays. Two trends are illustrated by the degree of sexual dimorphism exhibited by European newts: an increase in diversity of the sexual displays performed by the male and an increase in the importance of positive feedback from the female (**Figure 11.11**). The behaviors seen in the alpine newt (*Ichthyosaura alpestris;*

Figure 11.11A) may represent the ancestral condition. This species shows little sexual dimorphism, and a male's display consists only of fanning (a display in which the tail is folded back against the flank nearest the female and the tail tip is vibrated rapidly). The male's behavior is nearly independent of response by the female—a male may perform his entire courtship sequence and deposit a spermatophore without any response from the female he is courting.

A group of large newts, including the great crested newt (*Triturus cristatus;* Figure 11.11B) and banded newt (*Ommatotriton vittatus;* Figure 11.11C), is highly sexually dimorphic. Males of these species defend display sites. Their displays are relatively static and lack rapid fanning movements of the tail. A male of these species does not deposit a spermatophore unless the female he is courting touches his tail with her snout.

A group of small-bodied newts, including the smooth newt (*Lissotriton vulgaris;* Figure 11.11D) and Bosca's newt (*L. boscai;* Figure 11.11E), shows less sexual dimorphism than the large species, and males have a more diverse array of behaviors, including a nearly static lateral display, whipping the tail violently against the female's body, and fanning with the tail tip. Response by the female is an essential component of courtship for these species. A male will not progress

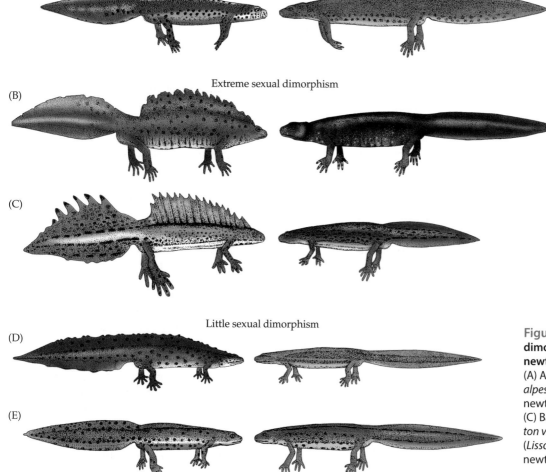

Males   Females

(A)

Extreme sexual dimorphism

(B)

(C)

Little sexual dimorphism

(D)

(E)

Figure 11.11 Sexual dimorphism of European newts (Salamanadridae) (A) Alpine newt (*Ichthyosaura alpestris*). (B) Great crested newt (*Triturus cristatus*). (C) Banded newt (*Ommatotriton vittatus*). (D) Smooth newt (*Lissotriton vulgaris*). (E) Bosca's newt (*L. boscai*).

from a static display to the next phase of courtship unless the female approaches him repeatedly, and he will not deposit a spermatophore until the female touches his tail.

The ancestral display might have included a single component, and females mated with the males that performed it most vigorously. That kind of selection by females would have produced a population of males that all display vigorously, and males that added novel components to their courtship might have been more attractive to females than their less innovative rivals.

Eggs and larvae   In most cases, salamanders that breed in water lay their eggs in water. The eggs may be laid singly or in a mass of transparent gelatinous material. The eggs hatch into gilled aquatic larvae that, except in permanently aquatic species, transform into terrestrial adults. Some families—for example, Plethodontidae—include species that have dispensed in part or entirely with an aquatic larval stage. The northern dusky salamander (*Desmognathus fuscus*) lays its eggs beneath a rock or log near water, and the female remains with them until after they have hatched. The larvae have small gills at hatching and may either take up an aquatic existence or move directly to terrestrial life. The red-backed salamander (*Plethodon cinereus*) lays its eggs in a hollow space in a rotten log or beneath a rock. The embryos have gills, but these are reabsorbed before hatching, so the hatchlings are miniatures of the adults.

Parental care is usually provided by the female, as would be expected when fertilization is internal and the male is not present when the eggs are deposited. Fertilization is external in cryptobranchids, hynobiids, and probably sirenids, and the male is the attending parent in cryptobranchids and sirenids (parental care has not been demonstrated in hynobiids).

Viviparity   Only a few species of salamanders, in the genera *Salamandra* and *Lyciasalamandra*, are viviparous. The European alpine salamander (*Salamandra atra*) gives birth to one or two fully developed young, each about one-third the adult body length, after a gestation period that lasts from 2 to 4 years. Initially the clutch contains 20–30 eggs, but only 1 or 2 of these eggs are fertilized and develop into embryos. When the energy in their yolk sacs is exhausted, these embryos consume the unfertilized eggs; when that source of energy is gone, the embryos scrape the reproductive tract of the female with specialized teeth.

Females in some populations of the European fire salamander (*Salamandra salamandra*) produce 20 or more small larvae, each about one-twentieth the length of an adult. The embryos probably get all the energy needed for growth and development from egg yolk. The larvae are released in water and have an aquatic stage that lasts about 3 months. In other populations of this species, the eggs are retained in the oviducts, and when all of the unfertilized eggs have been consumed, some of the embryos cannibalize other embryos. The surviving embryos pass through metamorphosis in the oviducts of the female.

Paedomorphosis   Paedomorphosis is the rule in Cryptobranchidae and Proteidae, and it characterizes most cave dwellers. Paedomorphosis appears in populations of some other kinds of salamanders in regions where the terrestrial habitat is inhospitable for adults. It also appears as a variant in the life history of species of salamanders that usually metamorphose, and paedomorphosis can be a short-term response to changing conditions in aquatic or terrestrial habitats.

The life history of the small-mouthed mole salamander (*Ambystoma talpoideum*) illustrates the flexibility of paedomorphosis. This species breeds in the autumn and winter, and some larvae metamorphose during the following summer to become terrestrial juveniles. These animals become sexually mature by autumn and return to the ponds to breed when they are about a year old. Ponds in South Carolina also contain paedomorphic larvae that remain in the ponds throughout the summer and mature and breed in the winter. Some of these paedomorphs metamorphose after breeding, whereas others do not metamorphose and remain in the ponds as permanently paedomorphic adults.

Anurans

Anurans are the most familiar amphibians, largely because of the vocalizations associated with their reproductive behavior. It is not even necessary to be in a rural area to hear them. In springtime, a weed-choked drainage ditch beside a highway or a trash-filled marsh at the edge of a parking lot is likely to attract a few toads or treefrogs that have not yet succumbed to human usurpation of their habitat.

The mating systems of anurans can be divided roughly into **explosive breeding**, in which the breeding season is very short (sometimes only a few days), and **prolonged breeding**, with breeding seasons that may last for several months. Explosive breeders include many species of toads and other anurans that breed in temporary aquatic habitats, such as vernal ponds or pools that form after rainstorms in the desert. Because these water bodies do not last very long, breeding congregations of anurans usually form as soon as the site is available. Males and females arrive at the breeding sites nearly simultaneously and often in large numbers. The numbers of males and females present are approximately equal because the entire population breeds in a short time. Time is the main constraint on how many females a male is able to court, and mating success is usually similar for all the males in a chorus.

In species with prolonged breeding seasons, the males usually arrive at the breeding sites first. Males of some species, such as green frogs (*Rana clamitans*), establish territories in which they spend several months, defending the spot against the approach of other males. The males of other species move between daytime retreats and nocturnal calling sites on a daily basis. Females come to the breeding site to breed and leave when they have finished. Only a few females arrive every day, and the number of males at the breeding site is greater than the number of females every night. Mating success may be skewed, with many males not mating at all and a few males mating several times.

Males of anuran species with prolonged breeding seasons compete to attract females, usually by vocalizing. The characteristics of a male frog's vocalization (pitch, length, and repetition rate) may provide information that a female can use to evaluate his quality as a potential mate. This is an active area of study in anuran behavior.

Vocalizations   Anuran calls are diverse; they vary from species to species, and most species have two or three different sorts of calls used in different situations. The most familiar calls are the ones usually referred to as mating calls, although a less specific term such as **advertisement calls** is preferable. These calls range from the high-pitched *peep* of a spring peeper (*Pseudacris crucifer*) to the nasal *waaah* of a spadefoot toad or the bass *jug-o-rum* of an American bullfrog (*Rana catesbeiana*). The characteristics of a call identify the species and sex of the calling individual. Many anuran species are territorial, and males of at least one species, the American bullfrog, can distinguish neighbors from intruders by their calls.

An advertisement call is a conservative evolutionary character, and among related species there is often considerable similarity in advertisement calls. Superimposed on the basic similarity are the effects of morphological factors, such as body size, as well as ecological factors that stem from characteristics of the habitat. Most toads have an advertisement call that consists of a train of repeated pulses, and the pitch of the call varies with body size, extending downward from 5,200 Hz for the oak toad (*Anaxyrus quercicus*), which is only 2–3 cm long; to 1,800 Hz for the American toad (*A. americanus*), about 6 cm long; to 600 Hz for the Sonoran Desert toad (*Incilius alvarius*), nearly 20 cm long. The Borneo tree hole frog (*Metaphrynella sundana*) calls from cavities in trees, and males adjust the frequency of their calls to match the resonant frequency of the hole from which they are calling, thereby increasing the amplitude of the call so it carries farther and can be heard by more potential mates.

Female frogs are responsive to the advertisement calls of males of their species for a brief period when their eggs are ready to be laid. The hormones associated with ovulation are thought to sensitize specific cells in the auditory pathway that respond to the species-specific characteristics of the male's call. Mixed choruses of anurans are common in the mating season; a dozen species may breed simultaneously in one pond. A female's response to her own species' mating call is a mechanism for species recognition in that situation.

Anthropogenic noise pollution interferes with the vocalizations of anurans (as it does also for birdsong). Frogs near roads alter the characteristics of their advertisement calls, and playback of recordings of road noise induces the same changes in vocalization. Playback of traffic noise to male European treefrogs (*Hyla arborea*) led to increased levels of corticosterone (a hormone indicating stress), suppressed their immune systems, and reduced the intensity of the color of their normally dark red vocal sacs. In choice experiments, female treefrogs prefer males with bright red vocal sacs; thus, traffic noise is likely to have an impact on sexual selection in this species.

Costs and benefits of vocalization   The vocalizations of male frogs are costly in two senses. The amount of energy that goes into call production can be very large, and the variations in calling pattern that accompany social interactions among males in a breeding chorus can increase the energy cost per call. Another cost of vocalization for a male frog is the risk of predation. A critical function of vocalization is to permit a female frog to locate a male, but female frogs are not the only animals that can use vocalizations as a cue to find male frogs; predators also find that calling males are easy to locate.

The túngara frog (*Engystomops pustulosus*) is a small terrestrial species that lives in Central America (**Figure 11.12**). It breeds in small pools, and breeding assemblies range from a single male to several hundred males. The advertisement call of a male túngara frog is a strange noise, a whine that sounds as if it would be more at home in a video-game arcade than in the tropical night. The whine starts at a frequency of 1,000 Hz and sweeps downward to 500 Hz in about 400 ms.

The whine may be produced by itself, or it may be followed by one or several "chucks." When a male túngara frog is calling alone in a pond, it usually gives only the whine portion of the call; however, as additional males join a chorus, more and more of the frogs produce calls that include chucks. Male túngara frogs calling in a breeding pond added chucks to their calls when they heard playbacks of calls of other males. That observation suggested that it was the calls of other males that stimulated frogs to make their own calls more complex by adding chucks to the end of the whine.

What advantage would a male frog in a chorus gain from using a whine-chuck call instead of a whine? When female frogs were released individually in the center of an arena where two speakers gave them a choice of a call without chucks and one with chucks, 14 of the 15 frogs tested moved toward the speaker broadcasting the whine-chuck call.

If female frogs are attracted to whine-chuck calls in preference to whine calls, why do male frogs give whine-chuck calls only when other males are present? Male frogs that give whine-chuck calls are more vulnerable to predators than frogs that give only whine calls. Túngara frogs in breeding choruses are preyed on by frog-eating bats, and the bats locate the frogs by homing on their vocalizations. In a series of playback experiments, bats were given a choice of speakers playing the calls of túngara frogs. In five experiments at different sites, the bats approached speakers broadcasting whine-chuck calls twice as frequently as those playing simple whines (168 approaches vs. 81). Thus, female frogs are not alone in finding whine-chuck calls more attractive than simple whines—frog-eating bats also locate male frogs that are producing chucks more readily than males that emit only whines.

(A)

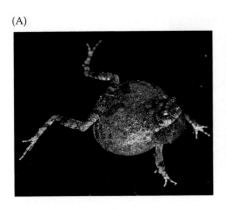

(B)

Figure 11.12 Male túngara frogs change the complexity of their calls As the frog calls, air is forced from the lungs (A) into the vocal sac (B). (C) A sonogram is a graphic representation of a sound: time is shown on the horizontal axis and frequency on the vertical axis. The calls increase in complexity from top (a whine only) to bottom (a whine followed by three chucks). (A, B, courtesy of Kentwood D. Wells; C, from Ryan 1992.)

(C)

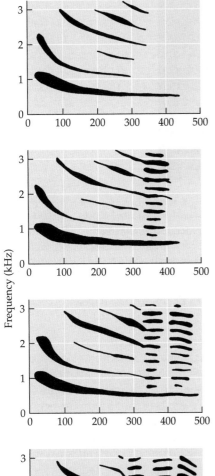

large. Under those conditions, it is apparently advantageous for a male túngara frog to give simple whines. However, as chorus size increases, competition with other males also increases while the risk of predation falls. In that situation, the advantage of giving a complex call apparently outweighs the risks.

Modes of reproduction The variety of reproductive modes of amphibians exceeds that of every other lineage of vertebrates except for bony fishes (**Figure 11.13**). External fertilization is ancestral for anurans and is retained in many lineages: the male uses his forelegs to clasp the female in the pectoral region (axillary amplexus) or pelvic region (inguinal amplexus). Amplexus may be maintained for several hours or even days before the female lays eggs.

Internal fertilization has been demonstrated in a few species of anurans and may be more common than we realize. Male tailed frogs (*Ascaphus*) of the Pacific Northwest have an extension of the cloaca (the "tail" that gives them their name) that is used to introduce sperm into the cloaca of the female. The Puerto Rican coquí (*Eleutherodactylus coqui*) lacks an intromittent organ but has internal fertilization, nonetheless; internal fertilization may be widespread among frogs that lay eggs on land. Fertilization must also be internal for the few species of viviparous anurans.

Many arboreal frogs lay their eggs on the leaves of trees overhanging water (see Figure 11.13E). The eggs undergo embryonic development beyond the reach of aquatic egg predators, and when the tadpoles hatch they drop into the water and take up an aquatic existence. Other species, such as the túngara frog, achieve the same result by constructing foam nests that float on the water surface. The female simultaneously releases mucus and a few eggs, and the male uses his hind legs to whip the mucus into foam that suspends the eggs (see Figure 11.13C). when the tadpoles hatch, they drop through the foam into the water.

Although these methods reduce egg mortality, the tadpoles are subjected to predation and competition. Some anurans avoid both problems by finding or constructing breeding sites free from competitors and predators. Some frogs, for example, lay their eggs in the water that accumulates in bromeliads—epiphytic tropical plants that grow in trees and are morphologically specialized to collect rainwater. A large bromeliad may hold several liters of water, and the frogs pass through egg and larval stages in that protected microhabitat.

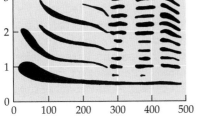

When a male frog shifts from a simple whine to a whine-chuck call, it increases its chances of attracting a female, but it simultaneously increases its risk of attracting a predator. In small choruses, the competition from other males for females is relatively small, and the risk of predation is relatively

(A)

(B)

(C)

(D)

(E)

Figure 11.13 Reproductive modes of anurans (A) Wood frogs (*Rana sylvatica*) retain the ancestral mode of depositing large numbers of small eggs in a pond and providing no parental care. (B) Male blacksmith frogs (*Hypsiboas faber*) excavate a pool in on a streambank and call to attract a female. The eggs are deposited in the predator-free pool, and the male remains to guard the clutch. (C) A male túngara frog (*Engystomops pustulosus*) takes the eggs and egg jelly with his hind legs as the female releases them and beats the jelly into a foam nest. Embedded in the foam, the eggs float on the surface of the water. When the tadpoles hatch, they release an enzyme that dissolves the foam and allows them to drop into the water. (D) Male and female Brazilian chocolate-foot treefrogs (*Phasmahyla cochranae*) in amplexus; the slightly smaller male is on the female's back and grasps her pelvic region. (E) Many tropical treefrogs deposit eggs on the surface of a leaf overhanging water. *Phasmahyla cochranae* folds the leaf so that the eggs are concealed during their development. (A, courtesy of Kentwood D. Wells; B, E, from Crump 2015, photos by Celio Haddad; C, courtesy of Ryan Taylor; D, courtesy of Mario Sacramento.)

Many tropical frogs lay eggs on land near water. The eggs or tadpoles may be released from the nest sites when pond levels rise after a rainstorm. Other frogs construct pools in the mud banks beside streams. These volcano-shaped structures are filled with water by rain or seepage and provide a favorable environment for the eggs and tadpoles. Some frogs have eliminated the tadpole stage entirely. These frogs lay large eggs on land that develop directly into little frogs. This reproductive mode, called direct development, is characteristic of about 20% of anuran species.

Parental care Adults of many frog species guard their eggs. Frogs that lay their eggs over water remain with them, either sitting beside the eggs or resting on top of them. Most of the terrestrial frogs that lay direct-developing eggs remain with the eggs and will attack an animal that approaches the nest. Removing the guarding frog frequently results in the eggs being eaten by predators or desiccating and dying before hatching.

Some frogs, including the Neotropical dart-poison frogs and the Bornean guardian frog, deposit their eggs on the ground, and one of the parents remains with the eggs until they hatch into tadpoles (**Figure 11.14A,B**). The tadpoles adhere to the adult and are transported to water. Females of the Panamanian frog *Colostethus inguinalis* carry their tadpoles for more than a week, and the tadpoles increase in size during this period. Small amounts of plant material have been found in the stomachs of the largest of these tadpoles, suggesting they had begun to feed while they were still being transported by their mother. Females of the strawberry poison frog (*Oophaga pumilio*) release their tadpoles in small pools of water in the leaf axils of plants and then return at intervals to the pools to deposit unfertilized eggs that the tadpoles eat (**Figure 11.14C**).

Other anurans, instead of remaining with their eggs, carry the eggs with them (**Figure 11.14D**). The male European midwife toad (*Alytes obstetricans*) gathers the egg strings about his hindlegs as the female lays them. He carries them with him until they are ready to hatch, at which time he releases

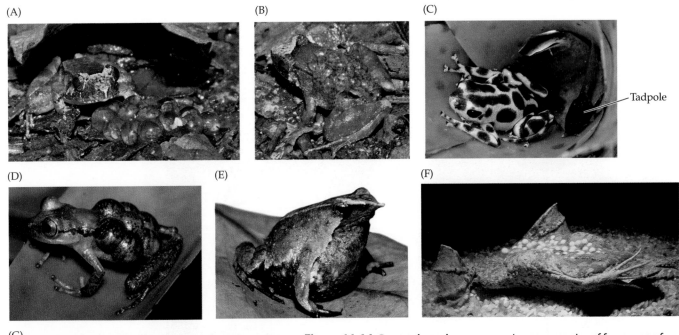

(A)

(B)

(C)

Tadpole

(D)

(E)

(F)

(G)

Figure 11.14 Parental care by anurans In many species of frogs, one of the parents remains with the eggs, and some species extend care to the tadpoles. (A) A male Bornean guardian frog (*Limnonectes palavanensis*) guarding its clutch. (B) When the tadpoles hatch, the male transports them on his back to a stream. (C) Both parents care for the eggs and tadpoles of strawberry poison frogs (*Oophaga pumilio*). The male guards the clutch, and when the eggs hatch the female transports tadpoles and deposits them one at a time in water-filled leaf axils of plants. She returns to each tadpole several times during the 6- to 8-week period of development to deposit trophic eggs for the tadpoles, which are obligate egg-eaters. (D) Females of some species of treefrogs carry the eggs on their backs, either adhering the surface of the skin or, as in this Brazilian species of *Fritziana*, within a pouch of skin. The young of some species emerge as tadpoles, whereas other species complete development in the pouch. (E) The male Darwin's frog (*Rhinoderma darwinii*) picks up the eggs as the female deposits them and transfers them to his vocal sacs, where the eggs develop through metamorphosis. (F) A female Surinam toad (*Pipa pipa*) carries eggs on her back. The eggs of this species complete development and hatch as tiny frogs, whereas other species of *Pipa* hatch as tadpoles. (G) Male African bullfrogs (*Pyxicephalus adspersus*) guard the eggs, and then accompany the tadpoles, eventually digging a channel to allow the tadpoles to reach deeper water. (A, B, courtesy of Johana Goyes; C, © Mark Moffett/Minden Pictures; D, from Crump 2015, photo by Celio Hadddad; E, from Crump 2015, courtesy of Martha L. Crump; F, Harvey Pough; G, Nature Picture Library/Alamy Stock Photo.)

the tadpoles into water. The male of the terrestrial Darwin's frog (*Rhinoderma darwinii*) of Chile carries the eggs in his vocal sacs, which extend back to the pelvic region (**Figure 11.14E**). The embryos pass through metamorphosis in the vocal sacs and emerge as fully developed froglets.

Male frogs are not alone in caring for eggs. Females of a group of Neotropical treefrogs known as marsupial frogs carry their eggs on their back in an open, oval depression, a closed pouch, or individual pockets. The eggs develop into miniature frogs before they leave their mother's back. Transfer of nutrients from the mother to the embryos has been demonstrated for a Peruvian species, *Gastrotheca excubitor*.

Females of the completely aquatic Surinam toad (*Pipa pipa*) also carry their eggs on their back (**Figure 11.14F**). In the breeding season, the skin of a female's back thickens and softens. During egg laying, the male and female in amplexus swim in vertical loops in the water. On the upward part of the loop, the female is above the male and releases

a few eggs, which fall onto his ventral surface. He fertilizes them and, on the downward loop, presses them against the female's back. The eggs sink into the female's soft skin, and a cover forms over each one, enclosing it in a small capsule.

Some species of frogs care for their tadpoles. Male African bullfrogs (*Pyxicephalus adspersus*) guard their eggs and then continue to guard the tadpoles after they hatch (**Figure 11.14G**). Tadpoles of several large species in the Neotropical genus *Leptodactylus* follow their mother around the pond.

Viviparity Species in the African bufonid genus *Nectophrynoides* show a spectrum of reproductive modes. One species deposits eggs that are fertilized externally and hatch into aquatic tadpoles, two species produce young that are nourished by yolk, and other species have embryos that feed on secretions from the walls of the oviduct. The golden coquí (*Eleutherodactylus jasperi*) of Puerto Rico also gives birth to metamorphosed young, but in this case the energy and

Figure 11.15 The body forms and mouth structures of tadpoles reflect differences in habitat and diet (A) The tadpole of the Kwangshien spadefoot toad (*Xenophrys minor*) is a surface feeder. The mouthparts unfold into a platter at the water surface, and water and floating particles are drawn into the mouth. (B) The tadpole of the red-legged frog (*Rana aurora*) scrapes food from rocks and other submerged objects. (C) The tadpole of the red-eyed leaf frog (*Agalychnis callidryas*) is a midwater suspension feeder that filters particles of food from the water column. This species shows the large fins and protruding eyes that are typical of midwater tadpoles. It maintains its position in the water column with rapid undulations of the end of its tail, which is thin and nearly transparent.

(D) The torrent-dwelling tadpole of the Australasian frog *Litoria nannotis* adheres to rocks in swiftly moving water with a suckerlike mouth while scraping algae and bacteria from the rocks. Its low fins and powerful tail are characteristic of tadpoles living in swift water. (E) The tadpole of the Pe-ret' toad (*Otophryne robusta*) buries itself in sand on the floor of a stream with the tip of its elongated spiracle projecting into the current. The Bernoulli effect draws water into the tadpole's mouth and out the spiracle, creating a passive filter-feeding system. The tadpole's needlelike teeth, which were originally interpreted as indicating a carnivorous diet, prevent sand grains from entering the oral cavity.

nutrients come from the yolk of the egg. (The golden coquí has not been seen since 1981 and is presumed to be extinct.)

The ecology of tadpoles

Although many species of frogs have evolved reproductive modes that bypass an aquatic larval stage, a life history that includes a tadpole has certain advantages. A tadpole is a completely different animal from an adult anuran, both morphologically and ecologically, and tadpoles and frogs are specialized for different habitats and feeding modes (**Figure 11.15**).

Most tadpoles filter food particles from a stream of water, whereas all adult anurans are carnivores that catch prey items individually. Because of these differences, tadpoles can exploit resources that are not available to adult anurans. This advantage may be a factor that has led many species of frogs to retain the ancestral life-history pattern in which an aquatic larva matures into a terrestrial adult. Many aquatic habitats experience annual flushes of primary production when nutrients washed into a pool by rain or melting snow stimulate the rapid growth of algae. The energy and nutrients in this algal bloom are transient resources that are available for a brief time to the organisms able to exploit them.

Tadpoles are eating machines, and feeding and ventilation of the gills are related activities. As the stream of water passes through the branchial basket, small food particles are trapped in mucus secreted by epithelial cells. The mucus, along with the particles, is moved from the gill filters to ciliary grooves on the margins of the roof of the pharynx, and then transported posteriorly to the esophagus.

Larvae of many species of frogs and salamanders are able to distinguish siblings from nonsiblings, and they associate preferentially with siblings. They probably recognize siblings by scent. Tadpoles of some species can distinguish full siblings (both parents the same) from maternal half-siblings (only the mother the same), and they can distinguish maternal half-siblings from paternal half-siblings.

Individual tadpoles of species that are normally herbivorous can develop into a cannibalistic morph with a sharp, keratinized beak that allows them to bite off bits of flesh. Eating the freshwater shrimp that occur in some breeding ponds causes tadpoles of the Mexican spadefoot toad (*Spea multiplicata*) to transform into the cannibal morph (**Figure 11.16**). In addition to eating the shrimp, the cannibals prey on other tadpoles, attacking unrelated individuals more often than their own kin.

The feeding mechanisms that make tadpoles such effective collectors of food particles suspended in the water allow them to grow rapidly, but that growth contains the seeds of its own termination. As tadpoles grow bigger, they become less effective at gathering food because of the changing relationship between the size of food-gathering surfaces and the size of their bodies. The branchial surfaces that trap food particles are two-dimensional. Consequently, the food-collecting apparatus of a tadpole increases in size

Figure 11.16 Tadpoles of *Spea multiplicata* express alternative phenotypes
Some individuals develop as typical omnivorous anuran larvae, with small heads, small jaw muscles, smooth beaks, and long, coiled intestines. In some populations, other individuals develop as carnivorous morphs, with large heads, large jaw muscles, notched and serrated beaks, and a short, relatively uncoiled gut. Carnivorous morphs eat omnivorous morphs, although they avoid eating their kin. (Copyright © Wild Horizon/Getty Images.)

exhausted, the fetuses emerge from their egg membranes, uncurl, and align themselves lengthwise in the oviducts. The fetuses apparently bite the walls of the oviduct, stimulating secretion and stripping some epithelial cells and muscle fibers that they swallow with the uterine milk. Small fetuses are regularly spaced along the oviducts. Large fetuses have their heads spaced at intervals, and the body of one fetus overlaps the head of the next. This spacing probably gives all the fetuses access to the secretory areas on the walls of the oviducts.

Dermatophagy (Greek *derma*, "skin"; *phago*, "to eat") is a remarkable method of feeding the young employed by *Boulengerula taitana*, a herpelid from Kenya, and *Siphonops annulatus* and *Microcaecilia dermatophaga*, siphonopids from South America. Females give birth to young that are in a relatively undeveloped stage, and the young remain with their mother in subterranean nests (**Figure 11.17A**). The cells

approximately as the square of the linear dimensions of the tadpole. However, the food the tadpole collects must nourish its entire body, and the volume of the body increases in proportion to the cube of the linear dimensions of the tadpole. The result of that relationship is a decreasing effectiveness of food collection as a tadpole grows; the body it must nourish increases in size faster than does its food-collecting apparatus. This mathematical stumbling block might explain why there are no paedomorphic anurans.

Caecilians

The reproductive modes of caecilians are as specialized as their body form and ecology. Fertilization is internal via a male intromittent organ that is protruded from the cloaca. Some species of caecilians lay eggs, and the female coils around the eggs, remaining with them until they hatch. Viviparity is widespread, however; about 75% of the species are viviparous and matrotrophic (receiving nutrition from the mother).

The fetuses of some species scrape lipid-rich cells from the walls of the oviducts with specialized embryonic teeth. The epithelium of the oviduct proliferates and forms thick beds surrounded by ramifications of connective tissue and capillaries. As the fetuses exhaust their yolk supply, these beds begin to secrete a thick, white, creamy substance that has been called uterine milk. When their yolk supply has been

(A)

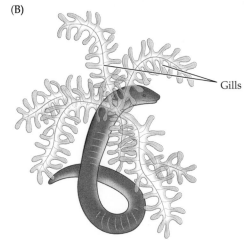

(B)

Gills

(C)

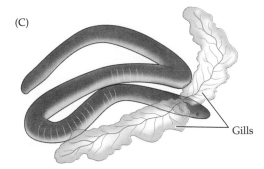

Gills

Figure 11.17 Reproduction by caecilians (A) The South American ringed caecilian (*Siphonops annulatus*) is an oviparous dermatophagous species. The young have spoon-shaped teeth in the lower jaw, each tooth bearing clawlike cusps. The young feed by seizing the lipid-rich skin on the mother's body and spinning about their body axis to tear off pieces. The mother remains calm during the feeding bout, which lasts only a few minutes and peels off the outer layer of the mother's skin. (B,C) Embryos of viviparous caecilians have gills that are pressed against the walls of the oviduct. All terrestrial species have a pair of three-branched gills (B), whereas embryos of species in the aquatic family Typhlonectidae have a pair of saclike gills (C). (A, courtesy of Carlos Jared; B, C, after Taylor 1968.)

in the stratum corneum (the outermost layer) of the skin of brooding females are thickened and contain vesicles filled with lipids. The young have a specialized fetal dentition that allows them to peel off single layers of these lipid-rich skin cells from their mother's body.

Gas exchange in developing caecilians appears to be achieved by close contact between the fetal gills and the walls of the oviducts. All the terrestrial species of caecilians have fetuses with a pair of triple-branched filamentous gills (**Figure 11.17B**). In preserved specimens, the fetuses frequently have one gill extending forward beyond the head and the other stretched along the body. The aquatic genus *Typhlonectes* has saclike gills (**Figure 11.17C**) that are positioned in the same way. Both the gills and the walls of the oviducts are highly vascularized, and it seems likely that exchange of gases, and possibly of small molecules such as metabolic substrates and waste products, takes place across the adjacent gill and oviduct. The gills are absorbed before birth, and cutaneous gas exchange may be important for fetuses late in development.

11.3   Amphibian Metamorphosis

The transition from larva to adult involves few conspicuous changes for salamanders and caecilians: the gills and lateral lines are lost, eyelids and skin glands develop, and the adult teeth appear. For anurans, however, the process involves a complete metamorphosis in which tadpole structures are broken down and their chemical constituents are rebuilt into the structures of an adult frog (**Figure 11.18**). These changes are stimulated by the actions of thyroxine, and the production and release of thyroxine are controlled by thyroid-stimulating hormone (TSH), a product of the pituitary gland. The action of thyroxine on larval tissues is both specific and local. In other words, it has a different effect in different tissues, and that effect is produced by the presence of thyroxine in the tissue; it does not depend on induction by neighboring tissues.

The final stage of metamorphosis begins with the appearance of the forelimbs and ends with the disappearance of the tail. This is the most rapid part of process, taking only a few days after a larval period that lasts for weeks or months. One reason for the rapidity of metamorphic climax may be the vulnerability of larvae to predators during this period. A larva with legs and a tail is neither a good tadpole nor a good frog; the legs inhibit swimming, and the tail interferes with jumping. As a result, predators are more successful at catching anurans during metamorphic climax than during prometamorphosis or following the completion of metamorphosis.

11.4   Exchange of Water and Gases

Both water and gases pass readily through amphibian skin, and all amphibians depend on cutaneous respiration for a significant part of their gas exchange. Under some circumstances, venous blood returning from the skin has a higher oxygen content than blood returning in the pulmonary veins from the lungs. Although the skin permits the passive movement of water and gases, it controls the movement of other compounds. Sodium is actively transported from the outer surface to the inner, and urea is retained by the skin. These characteristics are important in the regulation of osmotic concentration and in facilitating the uptake of water by terrestrial species.

Cutaneous respiration

The balance between cutaneous and pulmonary uptake of oxygen varies among species, and within a species it depends on body temperature and the animal's rate of activity. Amphibians show increasing reliance on the lungs for oxygen uptake as temperature and activity increase.

The patterns of blood flow within the hearts of adult amphibians reflect the use of two respiratory surfaces. The following description is based on the anuran heart (**Figure 11.19**). The atrium of the heart is divided anatomically into left and right chambers by a septum. Oxygenated blood from the lungs flows into the left side of the heart, and deoxygenated blood from the systemic circulation flows into the right side.

The spongy muscular interior of the undivided ventricle minimizes mixing of oxygenated and deoxygenated blood,

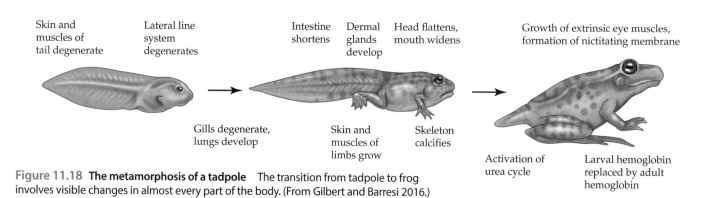

Skin and muscles of tail degenerate   Lateral line system degenerates   Intestine shortens   Dermal glands develop   Head flattens, mouth widens   Growth of extrinsic eye muscles, formation of nictitating membrane

Gills degenerate, lungs develop   Skin and muscles of limbs grow   Skeleton calcifies   Activation of urea cycle   Larval hemoglobin replaced by adult hemoglobin

Figure 11.18 The metamorphosis of a tadpole The transition from tadpole to frog involves visible changes in almost every part of the body. (From Gilbert and Barresi 2016.)

(A)

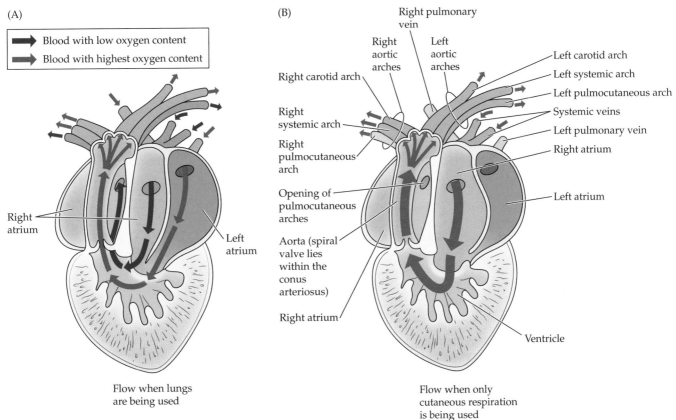

Blood with low oxygen content
Blood with highest oxygen content

Right atrium

Left atrium

Flow when lungs
are being used

(B)

Right pulmonary
vein

Right
aortic
arches

Left
aortic
arches

Right carotid arch

Right
systemic arch

Right
pulmocutaneous
arch

Opening of
pulmocutaneous
arches

Aorta (spiral
valve lies
within the
conus
arteriosus)

Right atrium

Left carotid arch

Left systemic arch

Left pulmocutaneous arch

Systemic veins

Left pulmonary vein

Right atrium

Left atrium

Ventricle

Flow when only
cutaneous respiration
is being used

Figure 11.19 Oxygenated and deoxygenated blood streams are kept separate in the anuran heart (A) The pattern of flow when lungs are being ventilated. (B) The flow when only cutaneous respiration is taking place.

and the position within the ventricle of a particular parcel of blood appears to determine its fate on leaving the contracting ventricle. The short conus arteriosus contains a spiral valve of tissue that differentially guides blood from the left and right sides of the ventricle to the aortic arches.

The anatomical relationships within the heart are such that oxygenated blood returning to the heart from the pulmonary veins enters the left atrium, which injects it on the left side of the common ventricle. Contraction of the ventricle tends to eject blood in laminar streams that spiral out of the pumping chamber, carrying the left-side blood into the ventral portion of the spirally divided conus. This half of the conus is the one from which the carotid and systemic aortic arches arise.

Thus, when the lungs are actively ventilated, oxygenated blood returning from them to the heart is selectively distributed to the tissues of the head and body. Deoxygenated venous blood entering the right atrium is directed into the dorsal half of the spiral valve in the conus. It goes to the pulmocutaneous arch, destined for oxygenation in the lungs.

However, when the skin is the primary site of gas exchange, as it is when a frog is underwater, the highest oxygen content is in the systemic veins that drain the skin. In this situation, the lungs may actually be net users of

oxygen, and because of vascular constriction, little blood passes through the pulmonary circuit.

Because the ventricle is undivided and the majority of the blood is arriving from the systemic circuit, the ventral section of the conus receives blood from an overflow of the right side of the ventricle. The scant left atrial supply to the ventricle also flows through the ventral conus. Thus, the most oxygenated blood coming from the heart flows to the tissues of the head and body during this shift in primary respiratory surface, a phenomenon possible only because of the undivided ventricle. Variability of the cardiovascular output in amphibians is an essential part of their ability to use alternate respiratory surfaces effectively.

Blood flow in larvae and adults

Larval amphibians rely on their gills and skin for gas exchange, whereas adults of species that complete full metamorphosis lose their gills and develop lungs. As the lungs develop, they are increasingly used for respiration. Late in development, tadpoles and partly metamorphosed froglets can be seen swimming to the surface to gulp air. As the gills lose their respiratory function, the carotid arches also change their roles (**Figure 11.20**).

Cutaneous permeability to water

The internal osmotic pressure of amphibians is approximately two-thirds that of most other vertebrates—about 200 milliosmoles (mOsm) per kilogram. The primary cause

(A)

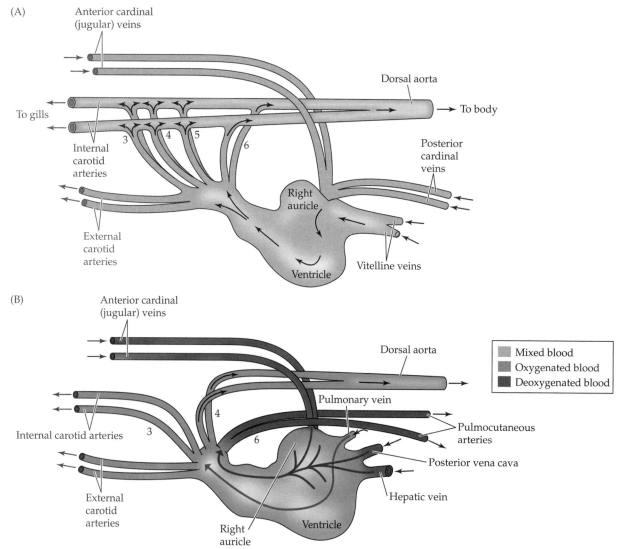

(B)

Figure 11.20 Changes in the circulatory system at metamorphosis   Blood flow through the aortic arches of (A) a larval amphibian (with gills) and (B) an adult (without gills). The head is to the left. Arches 1 and 2 are lost early in embryonic development. In tadpoles, blood is oxygenated in the gills and skin and the heart contains a mixture of oxygenated and deoxygenated blood. Arches 3 through 5 supply blood to the gills and thence to the internal carotid arteries that carry the blood to the head. Arch 6 carries blood to the dorsal aorta via a connection called the ductus arteriosus. After metamorphosis, blood is oxygenated in the lungs and skin. The pulmocutaneous veins return oxygenated blood which is distributed to the systemic circulation. Deoxygenated blood returning through the jugular veins, vena cava, and hepatic veins is directed to the lungs and skin via the pulmocutaneous arteries. At metamorphosis, arch 3 becomes the supply vessel for the internal carotid arteries. Initially, arches 4 and 5 supply blood to the dorsal aorta; however, arch 5 is usually lost in anurans, so arch 4 becomes the main route by which blood from the heart enters the aorta. Arch 6 primarily supplies blood to the lungs and skin via the pulmocutaneous arteries. The posterior cardinal veins return blood from the posterior body, the vitelline veins carry blood from the intestine, and the hepatic vein brings blood from the liver.

of the dilute body fluids of amphibians is low sodium content—approximately 100 millimoles (mM) per kilogram compared with 150 mM/kg in other vertebrates. Amphibians can tolerate a doubling of the normal sodium concentration, whereas an increase from 150 mM/kg to 170 mM/kg is the maximum humans can tolerate.

Evaporation from the skin of most amphibians occurs as rapidly as it does from a drop of water. Mucus glands secrete a watery solution of polysaccharides onto the surface of the skin, ensuring that evaporative water loss occurs from the skin surface, rather than from deeper layers.

Amphibians are most abundant and diverse in moist habitats, especially temperate and tropical forests. Nonetheless, a surprisingly large number of amphibian species live in dry regions. Anurans have been by far the most successful amphibian invaders of arid habitats. All but the harshest deserts have substantial anuran populations, and different families have converged on similar specializations. Avoiding the harsh conditions of the ground surface is the most common mechanism by which amphibians have managed to invade deserts and other arid habitats. Anurans and salamanders in deserts may spend 9–10 months of the year in moist retreat sites, sometimes more than a meter underground, emerging only during the rainy season and compressing feeding, growth, and reproduction into just a few weeks.

Many species of arboreal frogs have skin that is less permeable to water than the skin of terrestrial frogs, and a remarkable specialization is seen in a few treefrogs. The African rhacophorid *Chiromantis xerampelina* and the South American hylid *Phyllomedusa sauvagii* lose water through the skin at a rate only one-tenth that of most frogs. Several species of *Phyllomedusa* use their legs to spread lipid-containing secretions of dermal glands over the body surface in a complex sequence of wiping movements, and similar wiping behavior has been describe for an Indian rhacophorid, *Polypedates maculatus*, and an Australian hylid, *Litoria caerulea*.

The basis for the impermeability of *Chiromantis* is not yet understood but may involve a layer of specialized pigment cells (iridiphores) in the skin that contains crystals of guanine. These flat, shiny crystals give the frogs a chalky white appearance during the dry season, increasing reflectance and thereby reducing the heat load from solar radiation. The iridophores might form a barrier to water loss, but this function has not been demonstrated.

Chiromantis and *Phyllomedusa* are also unusual because they excrete nitrogenous wastes as precipitated salts of uric acid (as do lizards and birds) rather than as urea. This method of disposing of nitrogen provides still more water conservation.

Other species of amphibians in arid regions bury themselves in soil and slough off layers of skin to form a cocoon that is quite impermeable to water. Cocoons have evolved independently in several anuran families, including Hylidae, Ranidae, Myobatrachidae, and Leptodactylidae.

Behavioral control of evaporative water loss

For animals with skins as permeable as those of most amphibians, the main difference between rainforests and deserts may be how frequently they encounter a water shortage. The Puerto Rican coquí lives in wet tropical forests; nonetheless, it has elaborate behaviors that reduce

evaporative water loss during its periods of activity. Each male coquí has a calling site on a leaf in the understory vegetation from which it calls to attract a female. Male coquís emerge from their daytime retreats at dusk and move to their calling sites, remaining there until shortly before dawn, when they return to their daytime retreats.

The activities of the frogs vary from night to night, depending on rainfall. On nights after a rainstorm, when the forest is wet, the coquís begin to vocalize soon after dusk and continue until about midnight, when they fall silent for several hours. They resume calling briefly just before dawn. When they are calling, coquís extend their legs and raise themselves off the surface of the leaf (**Figure 11.21**). In this position, they lose water by evaporation from the entire body surface.

On dry nights, the frogs' behavior is quite different. Males move from their retreat sites to their calling stations, but they call only sporadically. Most of the time they rest in a water-conserving posture in which the body and chin are flattened against the leaf surface, the eyes are closed, and the limbs are pressed against the body. A frog in this posture exposes only half its body surface to the air, thereby reducing its rate of evaporative water loss. The effectiveness of the postural adjustments is illustrated by the water losses of frogs in the forest at El Verde, Puerto Rico, on dry nights. Frogs in one test group were placed individually in small wire-mesh cages on leaf surfaces. A second group consisted of unrestrained frogs sitting on leaves. The caged frogs spent most of the night climbing around the cages trying to get out. This activity, like vocalization, exposed the entire body surface to the air, and the caged frogs had an evaporative water loss that averaged 27.5% of their initial body mass. In contrast, the unrestrained frogs adopted water-conserving postures and lost an average of only 8% of their initial body mass by evaporation.

Experiments showed that the jumping ability of coquís was not affected by an evaporative loss of as much as 10% of the initial body mass, but a loss of 20% or more substantially decreased the distance frogs could jump. Thus, coquís

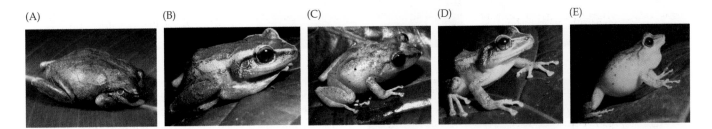

(A)   (B)   (C)   (D)   (E)

Figure 11.21 Male coquís change their posture to control evaporative water loss (A) In the water-conserving posture adopted on dry nights, the frog presses its chin and ventral surface against a leaf, tucks its toes under its body, and closes its eyes. In this posture, nearly half the frog's body surface is protected from evaporation. (B) In the chin up posture, the frog raises its head and opens its eyes. The ventral surface is still pressed against a leaf and the toes are beneath the body, but the frog can survey its surroundings and react to visual stimuli. In the low alert (C) and high alert (D) postures the frog partially extends its legs. Frogs in these postures respond to prey. (E) During vocalization, the frog extends its legs and nearly the entire body surface is exposed to evaporation. Pumping air between the lungs and vocal sac disrupts the boundary layer of air surrounding the frog, increasing evaporation. (Photos by Harvey Pough.)

can use behavior to limit their evaporative water losses on dry nights to levels that do not affect their ability to escape from predators or to capture prey. Without those behaviors, they would probably lose enough water by evaporation to affect their survival.

Uptake and storage of water

The mechanisms that amphibians use for obtaining water in terrestrial environments have received less attention than those used for retaining it. Amphibians do not drink water. Because of the permeability of their skins, species that live in aquatic habitats face a continuous osmotic influx of water that they must balance by producing urine. The impressive adaptations of terrestrial amphibians are ones that facilitate rehydration from limited sources of water. One such specialization is the **pelvic patch**. This area of highly vascularized skin in the pelvic region is responsible for a large portion of an anuran's cutaneous water absorption. Toads that are dehydrated and completely immersed in water rehydrate only slightly faster than those placed in water just deep enough to wet the pelvic area. In arid regions, water may be available only as a thin layer of moisture on a rock or as wet soil, and the pelvic patch allows an anuran to absorb this water.

Aquaporins (AQPs), tubular proteins that are inserted into the plasma membranes of cells in response to the release of the hormone vasotocin, are the channels for the movement of water through the pelvic patch. The aquaporin family of genes appears to have undergone an extensive radiation in anurans, producing genes that code for aquaporins with different sites of activity and different functions.

AQP1 and AQPa2 are localized in the skin and urinary bladder; AQP-h2 and AQP-h3 provide channels for water uptake through the pelvic patch; and AQP3 and AQP5 may control the movement of water, glycerol, and mucus to keep the skin moist.

The urinary bladder plays an important role in the water relations of terrestrial amphibians, especially anurans. Amphibian kidneys produce urine that is hyposmolal to the blood, so the urine in the bladder is dilute. Amphibians can reabsorb water from urine to replace water they lose by evaporation, and terrestrial amphibians have larger bladders than aquatic species. Storage capacities of 20–30% of the body mass of an animal are common for terrestrial anurans, and some species have still larger bladders: the Australian desert frogs *Notaden nichollsi* and *Neobatrachus wilsmorei* can store urine equivalent to about 50% of their body mass, and a bladder volume of 78.9% of body mass has been reported for the Australian moaning frog (*Heleioporus eyrei*).

11.5 Toxins, Venoms, and Other Defense Mechanisms

Amphibian skin has several important functions in addition to exchange of gases and water with the environment. The skin of amphibians often has colors and patterns that either conceal an animal from predators or advertise its distasteful properties. The skin of some amphibians is fluorescent or transparent (**Figure 11.22**).

Skin glands

The mucus that covers the skin of an amphibian has a variety of properties. It is mucus that makes some amphibians slippery and hard for a predator to hold, whereas other species have mucus that is extremely adhesive. Many salamanders have a concentration of mucus glands on the dorsal surface of the tail. When one of these salamanders is attacked by a predator, it bends its tail

(A)

(B)

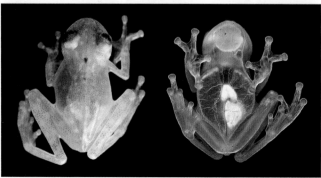

Figure 11.22 Amphibian skin can be fluorescent or transparent (A) The skin of the Brazilian treefrog *Hypsiboas punctatus* absorbs short light wavelengths and reemits the energy at longer wavelengths, creating a green fluorescence that increases the brightness of the frog. The wavelength of the emitted light matches the sensitivity of the retinal green rods that are a derived character of amphibians. (B) Glassfrogs (family Centrolenidae) are named for the transparent skin on the ventral surface of the body. Most species of glassfrogs have a window of transparent skin near the heart, but the entire ventral surface of *Hyalinobatrachium yaku* is transparent. (A, from Taboada et al. 2017, courtesy of Julian Faivovich and Carlos Taboada, Museo Argentino de Ciencias Naturales Bernardino Rivadavia; B, from Guayasamin et al. 2017, CC BY 4.0.)

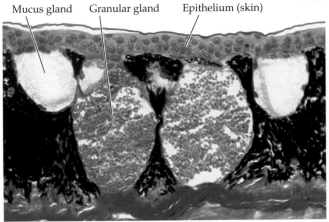

Mucus gland · Granular gland · Epithelium (skin)

Figure 11.23 Mucus and granular glands This cross section of skin is from a red-legged salamander *(Plethodon shermani)*. Mucus glands secrete mucopolysaccharides and proteoglycans that keep the skin moist, minimize abrasion, and may have antimicrobial properties. Some species of amphibians have sticky mucus, whereas other species have slippery mucus. Granular glands secrete biogenic amines, toxins, alkaloids, and a variety of proteins. Both mucus and granular glands contribute to the sticky secretions that red-legged salamanders use to deter predators. (From von Byern et al. 2015, courtesy of Janek von Byern.)

forward and slaps its attacker. The sticky mucus causes debris to adhere to the predator's snout or beak, and with luck the attacker soon concentrates on cleaning itself, losing interest in the salamander. When a California slender salamander *(Batrachoseps attenuatus)* is seized by a garter snake, the salamander curls its tail around the snake's head and neck. This behavior makes the salamander hard for the snake to swallow and also spreads sticky secretions on the snake's body. A small snake can find its body glued into a coil from which it is unable to escape.

Although secretions of the mucus glands of some amphibian species are irritating or toxic to predators, an amphibian's primary chemical defense system is located in its granular glands (**Figure 11.23**). These glands are concentrated on the dorsal surfaces of the animal, and defense postures of both anurans and salamanders present the glandular areas to potential predators.

A great diversity of pharmacologically active substances has been found in amphibian skins. Some of these substances are extremely toxic; others are less toxic but capable of producing unpleasant sensations when a predator bites an amphibian. Biogenic amines such as serotonin and histamine, peptides such as bradykinin, and hemolytic proteins have been found in frogs and salamanders belonging to many families. Many of these substances are named for the animals in which they were discovered—bufotoxin, epibatidine, leptodactyline, and physalaemin are examples.

Toxicity and diet

Cutaneous alkaloids are abundant and diverse among the Neotropical dart-poison frogs, Dendrobatidae. Most of these frogs are brightly colored and move about on the ground in daylight, making no attempt at concealment (**Figure 11.24A,B**). More than 200 alkaloids have been described from dendrobatids. Most of the alkaloids found in the skins of dart-poison frogs are similar to those found in ants, beetles, and millipedes that live in the leaf litter with the frogs, suggesting that frogs obtain alkaloids from their prey.

The *Mantella* frogs endemic to Madagascar (**Figure 11.24 C,D**) provide a striking example of evolutionary convergence with the New World dendrobatids. *Mantella* are small and brightly colored, with colors and patterns that are sometimes very similar to those of dendrobatids. Like dendrobatids, *Mantella* contain defensive alkaloids that they obtain from the ants and millipedes they eat. Ants that *Mantella* eat on Madagascar are not closely related to the ants that dendrobatids eat in Central and South America, and the convergent evolution of defensive alkaloids by ants on opposite sides of the world appears to have been

Figure 11.24 Aposematically colored frogs obtain toxins from the insects they eat Dendrobatid frogs in the Neotropics (A,B) and mantellid frogs on Madagascar (C,D) are protected by toxic alkaloids that they obtain from their prey. Frogs in both lineages are aposematically colored, and some of the patterns and colors are convergent. (A) Panamanian green poison frog *(Dendrobates auratus)*. (B) The golden poison frog *(Phyllobates terribilis)* from Colombia. (C) Baron's mantella *(Mantella baroni)* and (D) the golden mantella *(Mantella aurantiaca)* from Madagascar. (A, photo by Harvey Pough; B, courtesy of Charles W. Meyers, American Museum of Natural History; C, D, courtesy of Christopher Raxworthy, American Museum of Natural History.)

(A)

(B)

(C)

Granular glands

Rib tip

1 mm

(D)

Skin covers bony spikes

(E)

Bony spikes project from the maxillae

(F)

As the spike penetrates the skin, it punctures granular glands and carries venom into the wound

Bony spike

Figure 11.25 Some amphibians are venomous (A) The brightly colored glandular warts along the sides of the body may be an aposematic signal warning predators to avoid this Chinese crocodile newt (*Echinotriton andersoni*). (B) When a newt is attacked, the sharp ribs penetrate the body wall, passing through granular glands and carrying toxin into the wounds they create. (C) The warts are collections of granular glands, and a sharply pointed rib extends to just beneath the skin in the center of each wart. (D) The maxillae of Greening's frog (*Corythomantis greeningi*) are studded with bony spikes. (E) Normally the spikes are covered by skin, but when a frog is attacked it butts the predator with its head. (F) The bony spikes are surrounded by granular glands and inject venom into the predator. (A, courtesy of Axel Hernandez; B, courtesy of Edmund D. Brodie, Jr.; C after Brodie et al. 1984; D–F from Jared et al. 2015, photos by Carlos Jared, courtesy of E. D. Brodie, Jr.)

a prerequisite for the convergence between the frogs. Strangely, chemically protected lineages of frogs have a higher risk of extinction than unprotected species, and the probability of being classified as threated by the IUCN is 60% greater for species with chemical defenses than for those that rely only on mucus.

Some toxic frogs pass toxins on to their tadpoles. The trophic eggs that female strawberry poison frogs feed to their tadpoles contain alkaloids. By the time the tadpoles metamorphose, they have their toxic defenses in place.

Many amphibians, including dendrobatid and mantellid frogs, advertise their distasteful properties with conspicuous **aposematic** (warning) colors and behaviors. A predator that makes the mistake of seizing one is likely to spit it out because it is distasteful. The toxins in the skin may also induce vomiting that reinforces the unpleasant taste for the predator. Subsequently, the predator will remember its unpleasant experience and avoid the distinctly marked animal that produced it.

Venomous amphibians

A venomous animal injects a poisonous substance into a predator when it is attacked, whereas a toxic animal makes a predator sick after it has been eaten. A few species of amphibians have morphological specializations that qualify them as venomous animals:

- Blunt rib tips elevate brightly colored warts along the sides of the body of some Asian newts (*Tylototriton*). The warts contain granular glands and release toxins when a predator bites. Another genus of Asian salamandrid, *Echinotriton*, and a Spanish salamandrid, *Pleurodeles waltl*, have sharply pointed ribs that extend through the body wall (**Figure 11.25A–C**). You can imagine the shock for a predator that bites a salamander and finds its tongue and palate impaled by bony spikes! To make the predator's situation still worse, the ribs carry chemicals from the granular glands into the predator's wounds.

- Two South American hylid frogs have a similar method of defense; Greening's frog (*Corythomantis greeningi*) and Bruno's Casque-headed frog (*Aparasphenodon brunoi*) have bony spines on the skull that pierce the skin in areas with concentrations of granular glands (**Figure 11.25D–F**).

11.6 Why Are Amphibians Vanishing?

Amphibians are vanishing on a worldwide basis (**Figure 11.26**). More than 40% of amphibian species are classified as being at some level of risk by the IUCN Red List—a higher proportion than any other major group of vertebrates. The geographic distribution of declining populations is particularly distressing because the highest proportions of species with declining populations are located in Central America, sub-Saharan Africa, and Southeast Asia, which are the global hotspots for amphibian diversity.

Disease

In the past 50 years, disease-causing organisms have emerged as a major cause of the extinction of amphibians. This phenomenon is something new, because most amphibian species long ago established stable relationships with their local pathogens. But when animals carrying these pathogens were moved to regions where the pathogens were unknown, lethal epidemics spread, just as they did when Europeans carried previously unknown pathogens to the New World. Two well-documented epidemic diseases of amphibians are ranavirus disease and chytridiomycosis.

Ranaviruses   Ranaviruses infect frogs and salamanders and can lead to the extinction of local populations. Fishes are also vulnerable to ranaviruses, and the introduction of fishes to water bodies has accelerated the spread of ranaviruses among amphibians.

Ranaviruses are not new pathogens, and their distribution appears to have shaped the biology of some populations of tiger salamanders (*Ambystoma tigrinum*). In Arizona, tiger salamanders have two larval forms, the normal morph and a cannibalistic morph with an enlarged head and jaws. Some populations consist entirely of the normal morph, whereas in other populations some individuals develop into cannibalistic larvae that eat individuals of the normal morph. A survey of the occurrence of normal and cannibalistic forms revealed that cannibals are absent from populations that are infected by ranaviruses but occur regularly in populations without the viruses. Apparently in virus-free populations it is safe to eat other larvae, but when ranaviruses are present, a cannibal risks infection when it eats other larval salamanders.

The correspondence between the presence of ranaviruses and the life-history structure of salamander populations suggests that this host–pathogen association is ancient and that the salamander hosts and their ranavirus pathogens have evolved a stable relationship. This generalization probably applies to other amphibians that serve as hosts for ranaviruses, and it appears that ranaviruses cause fluctuations in amphibian populations but are not usually responsible for extinctions.

Chytridiomycosis   The chytrid fungus *Batrachochytrium dendrobatidis*—or Bd as it is usually called—is responsible for the disappearances of amphibians in the Americas, Europe, Australia, and New Zealand. Bd has motile reproductive zoospores that live in water and can penetrate the skin of an amphibian, causing a disease called chytridiomycosis. When a zoospore enters an amphibian, it matures into a spherical reproductive body, the zoosporangium, which has branching structures that extend through the skin. These structures interfere with respiration and control of water movement and at high densities kill adult frogs.

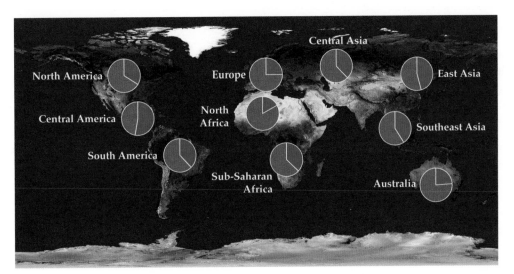

Figure 11.26 Global status of amphibian populations   On a worldwide basis, more than 40% of amphibian species have populations that are declining, and at least 168 species are known to have become extinct in recent years. The percentage of species from continental regions considered to be at risk by the IUCN (indicated in red on the pie charts) ranges from 24% (Australia) to 53% (Central America). (Data extracted from IUCN Red List 2016-3, iucnredlist.org; map courtesy of NASA Earth Observatory.)

Chytridiomycosis has played a large role in the worldwide decline of amphibians. The epidemic has been studied especially well in Central and South America. In the late 1980s, two iconic frog species, the golden toad (*Incilius periglenes*) and the harlequin frog (*Atelopus varius*), abruptly vanished from the Monteverde Cloud Forest Reserve in the mountains of northern Costa Rica. The disappearance of those two species was followed by others, and 20 of the 49 frog species that once lived in the reserve have vanished. The reason frogs were vanishing remained a mystery for several years, until Bd was detected and its pathogenic effect was recognized.

Initially, Bd was regarded as a novel pathogen that had only recently infected amphibians, but a more complicated picture has emerged. Surveys of amphibians in museum collections have revealed that Bd has been present in parts of Africa, North America, and Asia for more than a century. Genetic analyses have shown that Bd consists of at least four genetically distinct lineages that differ in geographic origin and virulence as well as a hypervirulent global pandemic lineage know as BdGPL. The worldwide spread of BdGPL is probably responsible for the massive die-offs and population extinctions that have been occurring for the last 50 years.

The role of chytrid fungi in amphibian declines took a horrifying turn in 2013 when a new fungus, *B. salamandrivorans* ("salamander eater") was identified in a population of fire salamanders (*Salamandra salamandra*) in the Netherlands. Within six months, 90% of the population had died, and in two years it was extinct. Bsal, as the new fungus is known, is substantially more threatening than Bd in several ways:

- Bsal has three times more genes that code for skin-destroying enzymes (metaloproteases) than Bd. These genes are expressed during infection, creating skin ulcers that speed fungal colonization.

- Salamanders do not develop an immune response to Bsal, suggesting that this fungus has an immune-suppressing property that is absent from Bd. Thus, it is unlikely that wild populations of amphibians will acquire immunity or that a vaccine can be developed.

- Unlike Bd, Bsal makes two kinds of zoospores, and one of them is hardier than the spores of Bd. These spores adhere to the feet of aquatic birds, which can transport them for long distances.

- A single spore of Bsal is sufficient to cause a lethal infection, whereas approximately 10,000 Bd zoosporangia are needed to cause chytridiomycosis.

In addition, Bsal spreads rapidly, even at very low population densities of salamanders, and it is lethal—no treatment or mitigation procedures are effective. Rigorous containment is the only method that offers hope of limiting the spread of Bsal. Genetic analyses indicate that this fungus arose in Asia and was transported to Europe via importation of salamanders for the pet trade.

Bsal would be devastating in North America, which has the world's greatest species diversity of salamanders. In January 2016, the United States prohibited importation of any species of salamander from a genus in which at least one species was a carrier of Bsal. This regulation affects 201 species of salamanders in 20 genera. Although it is a step in the right direction, it is a leaky net because many genera that have not been tested can probably carry Bsal. In May 2017, Canada adopted a stricter policy, prohibiting importation of any species of salamander. The Canadian ban covers both living and dead specimens, as well as eggs, sperm, embryos, and tissue cultures.

Synergisms

Extinctions rarely have a single cause; instead multiple factors simultaneously stress populations and species, and synergisms are likely to intensify the impact of some of the factors. For amphibians, habitat loss, agricultural chemicals, climate change, drought, and ultraviolet B radiation increase the vulnerability of amphibians to epidemic diseases. Acting together, these stresses have a greater impact than the sum of their individual effects.

Summary

The monophyly of lissamphibians is supported by a large number of shared derived characters.

A moist, permeable skin plays a major role in gas exchange, and mucus and granular glands in the skin are important in shaping the ecology and behavior of amphibians.

Pedicillate teeth is a character shared with a few temnospondyls, the lineage from which lissamphibians are derived.

A papilla amphibiorum in the inner ear and green rods in the retina are unique to lissamphibians.

Although Lissamphibia is monophyletic, its three major lineages are ecologically and morphologically distinct.

Salamanders (Urodela) have the most generalized body form—elongated with four limbs in most species. Some families of salamanders are paedomorphic, and paedomorphosis occurs as a life-history variant in some species that normally metamorphose. Aquatic salamanders use suction feeding to capture prey, whereas most terrestrial salamanders simply seize prey. Some plethodontid salamanders are able to project their tongues farther than their own body length to capture prey.

Frogs (Anura) have short bodies with nearly inflexible spines and lack tails as adults. All anurans have four well-developed legs. Species with long hindlegs leap and climb, whereas species with hindlimbs and forelimbs of similar length walk or hop. Aquatic species have extensive webbing on their toes.

Caecilians (Gymnophiona) are elongated, legless, burrowing or aquatic amphibians. They are predators on smaller animals, including insects, worms, and probably fishes and tadpoles.

Amphibians display a remarkable range of reproductive modes.

Aquatic eggs that hatch into free-living aquatic larvae and metamorphose into aquatic or terrestrial adults represent the ancestral reproductive mode of amphibians, but extant species have evolved many modifications of that pattern, including terrestrial eggs with or without aquatic larvae, direct development, and viviparity.

Parental care, in some species by the male and in other species by the female, occurs in many lineages. In some species with maternal care, the mother feeds her offspring.

Many species of salamanders have elaborate courtship behaviors that include colors, patterns, movements, and pheromones.

All but a few primitive lineages of salamanders have internal fertilization achieved via a spermatophore.

Aquatic eggs and larvae represent the ancestral reproductive mode of salamanders, but many species of plethodontids have terrestrial eggs with direct development, and a few species of salamandrids are viviparous. Egg yolk supports embryonic development of some species; embryos of other species consume some of their siblings.

Cannibalistic individuals appear in populations of some salamander species with free-living larvae.

Paedomorphosis appears as a life-history variant in places where the terrestrial habitat is hostile for adult salamanders, such as caves and arid regions.

Vocalizations are a prominent part of the mating behavior of most anurans.

Males give species-specific advertisement calls that may also contain information that females use for mate selection. Advertisement calls are energetically expensive and may attract predators as well as females.

Vocalizing males employ a variety of trade-offs to maximize attracting females while minimizing energetic cost and risk of predation. Some frog species have very brief

breeding seasons, whereas other species have prolonged breeding seasons.

Nearly all anurans that breed in water have external fertilization; internal fertilization is probably widespread among species that breed on land.

Many anuran species have aquatic larvae, allowing a species to exploit two different resource bases during a lifetime, but loss of the larval stage has occurred in many lineages.

Eggs are deposited in or over water, on land, or on the body of a parent, usually the mother. Parental care is widespread, especially among terrestrial species. Some species that deposit eggs on land have direct development, whereas others carry tadpoles to water. Females of some of the latter species return to feed their tadpoles on unfertilized eggs.

Tadpoles are morphologically and ecologically distinct from adult anurans.

All tadpoles filter food particles from a stream of water drawn through the mouth, and many tadpoles use keratinized beaks to rasp algae and bacteria from surfaces, thereby placing suspended particles into the water.

Some species of frogs have tadpoles that prey on other tadpoles. Cannibal morphs can develop in populations of some species of spadefoot toads; cannibal morph tadpoles prey on unrelated individuals more often than their own kin.

Most caecilians are viviparous and matrotrophic.

Males have an intromittent organ, and fertilization is internal. Viviparous species are matrotrophic; embryos of *Typhlonectes* scrape lipid-rich cells from the walls of their mother's oviducts. The young of three oviparous species are known to employ dermatophagy; they tear off and consume pieces of their mother's lipid-rich epidermis.

Thyroxin plays a key role in the metamorphosis of amphibian larvae.

Metamorphosis of salamanders and caecilians involves few conspicuous changes, but in anurans, tadpole structures must be broken down and their constituents recycled into adult structures. All anuran species undergo metamorphosis; there are no paedomorphic anurans.

Anurans are especially vulnerable to predation in the final stage of metamorphosis (metamorphic climax), when they are unable either to swim or to leap effectively. This stage is usually brief.

(Continued)

Summary *(continued)*

The skin of amphibians plays a major role gas exchange and water uptake and loss.

The spongy interior of the undivided ventricle separates oxygenated and deoxygenated blood streams in the heart while accommodating shifting reliance on cutaneous versus pulmonary gas exchange.

Patterns of blood flow in the circulatory systems of amphibians change during metamorphosis as the sites of gas exchange switch from skin and gills to skin and lungs.

The internal fluid concentrations of amphibians are about two-thirds those of most other vertebrates, primarily because of the body fluids' low sodium content. Amphibians do not drink; they absorb water through the skin. Aquaporins create channels for water movement through plasma membranes.

Water evaporates from the skin of most amphibians as rapidly as it does from a drop of water.

Evaporative water loss can be a problem even in humid environments, and amphibians use sheltered microhabitats and postural adjustments to control evaporation. The body water content and osmolality of terrestrial amphibians probably vary from hour to hour.

A surprising number of amphibian species—especially anurans—use a combination of physiological characters and behavioral adjustments to live in arid regions.

Some species of anurans spread waterproofing lipids across their skin, and others bury themselves in the soil. Some species form waterproof cocoons and rely on water stored in a large bladder to maintain normal tissue osmolality though the dry season; others allow their osmolality to increase and absorb water from the soil through their permeable skin.

Granular glands contain toxins that protect amphibians.

Some of these toxins are synthesized by the amphibian itself, whereas others are derived from toxins in prey. Most toxins deter predators by tasting unpleasant or producing distressing symptoms, but a few species of amphibians are lethally toxic, even to predators as large as humans. Some toxic frogs pass toxins on to their tadpoles.

More than 40% of amphibian species are at risk of extinction.

Virus and fungal diseases have decimated populations of amphibians and have driven some species to extinction. Ranaviruses infect fish as well as amphibians, and introduction of fish to new water bodies has accelerated the spread of the viruses.

A correspondence between the geographic distributions of ranaviruses and the occurrence of cannibal morphs in populations of tiger salamanders (*Ambystoma tigrinum*) suggests that this is an ancient host–pathogen association.

Two fungi, *Batrachochytrium dendrobatidis* (Bd) and *B. salamandrivorans* (Bsal), have emerged as major threats to amphibian populations. Bd has been linked to declines of amphibian populations on all continents, most dramatically in North America, Central America, and Australia.

Bsal, which was identified in Europe in 2013, is substantially more dangerous than Bd and has caused widespread extinction of European fire salamanders (*Salamandra salamandra*) in the Netherlands.

Multiple stresses are acting on amphibians, including habitat loss, climate change, agricultural chemicals, drought, increased ultraviolet-B radiation, and disease. When these factors act synergistically, they have an impact that is greater than the sum of their individual impacts.

Discussion Questions

1. Deserts appear to be the least likely habitats in which to find animals with skins that are very permeable to water, yet amphibians, especially anurans, inhabit deserts all over the world. How is this possible?

2. Four species of salamanders in the genus *Desmognathus* form a streamside salamander guild in the mountains of North Carolina. The four species are found at different distances from the stream, as listed in the table. Suggest one or more hypotheses to explain the relationship between the body sizes of these species and the distance each is found from water. What prediction can you make for each hypothesis, and how could you test that prediction?

Species	*D. quadramaculatus*	*D. monticola*	*D. ochrophaeus*	*D. aeneus*
Adult body length	100–175 mm	83–125 mm	70–100 mm	44–57 mm
Distance from stream	Less than 10 cm	Less than 1 m	1–4 m	4–7 m

3. Why do you think most of the male coquís in the population emerge from their daytime retreats on dry nights? Only female coquís that are seeking a mate are active on these nights; the remaining females remain in their shelters. Why don't males do that as well?

4. Aquatic eggs and larvae are ancestral characters of anurans, but terrestrial eggs that hatch into tadpoles or bypass the larval stage entirely have evolved independently in many lineages. Why didn't all anuran lineages evolve that way? That is, why might it be advantageous for a species of frog to have retained an aquatic egg and a larval stage?

5. There are about ten times as many species of anurans as there are species of salamanders. Can you think of a reason why anurans might be a more successful form of amphibian than salamanders are?

Additional Reading

Clark VC, Raxworthy CJ, Rakotomalala V, Sierwald P, Fisher BL. 2005. Convergent evolution of chemical defense in poison frogs and arthropod prey between Madagascar and the Neotropics. *Proceedings of the National Academy of Sciences USA* 102: 11617–11622.

Dodsworth S, Guignard MS, Hidalgo IJ, Pellicer J. 2016. Salamanders' slow slither into genomic gigantism. *Evolution* 70: 2915–2916.

Endlein T and 7 others. 2017. The use of clamping grips and friction pads by tree frogs for climbing curved surfaces. *Proceedings of the Royal Society B* 284: 20162867.

Farrer RA and 14 others. 2011. Multiple emergences of genetically diverse amphibian-infecting chytrids include a globalized hypervirulent recombinant lineage. *Proceedings of the National Academy of Science USA* 108: 18732–18736.

Hof C, Araujo MB, Jetz W, Rahbek C. 2011. Additive threats from pathogens, climate and land-use change for global amphibian diversity. *Nature* 480: 516–519.

Lips KR. 2011. Museum collections: Mining the past to manage the future. *Proceedings of the National Academy of Science USA* 108: 9323–9324.

Lips KR. 2016. Overview of chytrid emergence and impacts on amphibians. *Philosophical Transactions of the Royal Society B* 371: 20150465.

Roca IT and 9 others. 2016. Shifting song frequencies in response to anthropogenic noise: A meta-analysis on birds and anurans. *Behavioral Ecology* 27: 1269–1274.

Schloegel LM, Daszak P, Cunningham AA, Speare R, Hill B. 2010. Two amphibian diseases, chytridiomycosis and ranaviral disease, are now globally notifiable to the World Organization for Animal Health (OIE): An assessment. *Diseases of Aquatic Organisms* 92: 101–108.

Spitzen-van der Sluijs A and 24 others. 2016. Expanding distribution of lethal amphibian fungus *Batrachochytrium salamandrivorans* in Europe. *Emerging Infectious Diseases* 22: 1286–1288.

Stuart SN and 6 others. 2004. Status and trends of amphibian declines and extinctions worldwide. *Science* 306: 1783–1786.

Wells, K. D. 2007. *The Ecology and Behavior of Amphibians.* University of Chicago Press, Chicago.

Go to the Companion Website **oup.com/us/vertebratelife10e** for active-learning exercises, news links, references, and more.

Living on Land

The spread of plants and then invertebrates across the land in the middle Paleozoic provided a new habitat for vertebrates and imposed new selective forces. The demands of terrestrial life are quite different from those of aquatic life because water and air have different physical properties (see Table 4.1). Air is less viscous and less dense than water, so streamlining is a minor factor for terrestrial tetrapods, whereas a skeleton that supports the body against the pull of gravity is essential. Respiration, too, is different in air and water. Gills don't work in air because the gill filaments collapse on each other without the support of water, greatly reducing the surface area available for gas exchange. And because water is both dense and viscous, pumping water into and out of a close-ended gas-exchange structure is prohibitively expensive; aquatic animals have flow-through gas-exchange structures. Terrestrial animals need a gas-exchange organ that won't collapse, and given the low density and viscosity of air, terrestrial tetrapods can use a tidal flow of air into and out of a saclike lung. Heat capacity and heat conductivity are additional differences between water and air that are important to terrestrial animals. Terrestrial habitats can have large temperature differences over short distances, and even small terrestrial tetrapods can maintain body temperatures that are substantially different from air temperatures. As a result, terrestrial animals have far more opportunity than aquatic organisms for regulation of their body temperatures.

12.1 Support and Locomotion on Land

Perhaps the most important difference between living in water and on land is the effect of gravity on support and locomotion. Gravity has little significance for a fish living in water because the bodies of vertebrates are approximately the same density as water, and hence fish are essentially weightless in water. Gravity is a critically important factor on land, however, and the skeleton of a terrestrial tetrapod must be able to support the body.

Water and air also require different forms of locomotion. Because water is dense, a fish swims merely by passing a sine wave along its body—the sides of the body and the fins push backward against the water, and the fish moves forward. Pushing backward against air doesn't move a terrestrial animal forward unless it has wings. Most terrestrial tetrapods (including fliers, when they are on the ground) use their legs and feet to transmit a backward force to the substrate. Thus, both the skeletons and modes of locomotion of terrestrial tetrapods are different from those of fishes or specialized aquatic tetrapods, such as cetaceans.

The skeleton

The skeleton of terrestrial vertebrates is composed of cartilage and bone. It must be rigid enough to resist the force of gravity and the forces exerted as an animal starts, stops, and turns. The remodeling capacity of bone is of great importance for a terrestrial animal, as the internal structure of bone adjusts continually to the changing demands of an animal's life. In humans, for example, intense physical activity results in an increase in bone mass, whereas inactivity (prolonged bed rest or being in space) results in loss of bone mass. In addition, remodeling allows broken bones to mend, and the skeletons of terrestrial animals experience greater stress than those of aquatic animals and are more likely to break. (Chapter 2 provides a background on bone in vertebrates, including a contrast between the endoskeleton of endochondral bone and the "exoskeleton" of dermal bone.)

Amniotes have bone that is arranged in concentric layers around blood vessels, forming cylindrical units called **Haversian systems**, each of which encloses a capillary and a venule (**Figure 12.1A**). The structure of a bone is not uniform. If it were, animals would be very heavy. The external layers of a bone are formed of dense, compact or lamellar bone, but the internal layers are lighter, spongy (cancellous) bone. The joints at the ends of bones are covered by a smooth layer of articular cartilage that reduces friction as the joint moves. (Arthritis results from wear or damage to this cartilage.) The bone within the joint is composed of

(A)

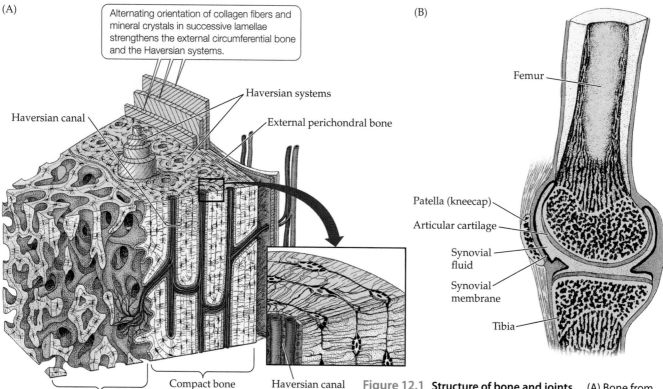

Alternating orientation of collagen fibers and mineral crystals in successive lamellae strengthens the external circumferential bone and the Haversian systems.

Haversian systems

Haversian canal

External perichondral bone

Compact bone

Haversian canal

Cancellous bone
(= spongy or trabecular bone)

(B)

Femur

Patella (kneecap)

Articular cartilage

Synovial fluid

Synovial membrane

Tibia

Figure 12.1 **Structure of bone and joints.** (A) Bone from a section of the shaft of a long bone of a mammal, showing Haversian systems. (B) Section through a human knee joint, showing the gross internal structure of a long bone and the structure of a joint capsule. Note that the patella (kneecap) is an example of a sesamoid bone (a bone formed within a tendon).

cancellous bone rather than dense bone, and the entire joint is enclosed in a joint capsule, containing synovial fluid for lubrication (**Figure 12.1B**).

The cranial skeleton

The skull of early tetrapods was much like that of early bony fishes, with extensive dermal skull bones. (These bones are retained in extant tetrapods—most of the bones of the human skull represent the legacy of this bony fish dermatocranium.) The dermal head bones of bony fishes include the opercular bones that protect the gills and play a vital role in gill ventilation. A series of dermal bones behind and above the operculum (the supracleithral bones) connect the head to the pectoral girdle. A fish cannot turn its head—it must pivot its entire body. Tetrapods have lost the bony connection between the head and pectoral girdle and, as a result, have a flexible neck region and can move the head separately from the rest of the body. This is also the condition in *Tiktaalik*, the fish most closely related to tetrapods (see Chapter 10).

The axial skeleton: Vertebrae and ribs

The vertebrae and ribs of fish stiffen the body so it will bend, rather than shorten, when muscles contract. In terrestrial tetrapods, the axial skeleton is modified for support on land. Processes called **zygapophyses** (singular *zygapophysis*) on the vertebrae of most tetrapods interlock and resist twisting (torsion) and bending (compression), allowing the spine to transfer the weight of the viscera to the limbs

(**Figure 12.2**). (Tetrapods that have permanently returned to the water, such as whales, have lost the zygapophyses, as did many of the extinct Mesozoic marine reptiles.)

The cervical (neck) vertebrae of tetrapods allow the head to turn from side to side and up and down relative to the trunk, and the muscles that support and move the head are attached to processes on the cervical vertebrae. The two most anterior cervical vertebrae, which allow the head to move on the neck, are the atlas and axis, and they are highly differentiated in mammals.

The trunk vertebrae are in the middle region of the body and bear the ribs. In mammals, the trunk vertebrae are differentiated into two regions: thoracic vertebrae (those that bear ribs) and lumbar vertebrae (those that have lost ribs).

The sacral vertebrae, which are derived from the trunk vertebrae, have a tight articulation with the pelvic girdle and allow the hindlimbs to transfer force to the axial skeleton. Extant amphibians have a single sacral vertebra (as did early tetrapods), lepidosaurs and crocodylians[1] usually have two sacral vertebrae, mammals usually have two to five, and some nonavian dinosaurs had a dozen or more.

---

[1]Birds, derived reptiles, have a synsacrum that is formed by fusion of the two sacral vertebrate of the ancestral condition with the lumbar vertebrae and often with a variable number of caudal vertebrae. In turn, the synsacrum fuses with the pelvic girdle forming a rigid structure that extends for most of the length of the trunk.

(A) *Eryops*, an early tetrapod

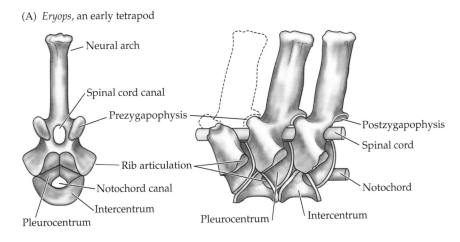

(B) Human, an extant tetrapod

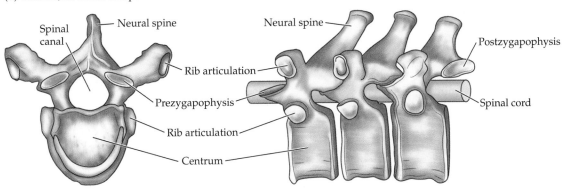

Figure 12.2 The vertebral column resists compression and the pull of gravity. Zygapophyses on the vertebrae of tetrapods interlock each vertebra with two adjacent vertebrae, thereby allowing the vertebral column to resist gravity. (A) The vertebral centrum of tetrapodomorph fishes and some early tetrapods was a two-part structure (pleurocentrum + intercentrum), as shown here. (B) Later tetrapods (including all extant ones) have a single centrum (corresponding to the pleurocentrum) fused to the neural arch, and the notochord is obliterated in the adult (although the embryonic notochord contributes to the intervertebral discs). (A after Rockwell et al. 1938; B after Cunningham 1903.)

The caudal vertebrae, found in the tail, are usually simpler in structure than the trunk vertebrae and bear only the weight of the tail.

The ribs of early tetrapods were fairly stout and more prominently developed than those of fish. They now played a role in body support a well as locomotion, and may have stiffened the trunk in animals that had not yet developed much postural support from the axial musculature (**Figure 12.3**, number 7). The trunk ribs are the most prominent ones in tetrapods in general, but many non-mammalian tetrapods have small ribs on their cervical vertebrae. Modern amphibians have almost entirely lost their ribs; in mammals, ribs are confined to the thoracic vertebrae.

Axial muscles

The axial muscles assumed two novel roles in tetrapods: postural support of the body and ventilation of the lungs. These functions are more complex than the side-to-side bending produced by the axial muscles of fish, and the axial muscles of tetrapods are highly differentiated in structure and function. Muscles are important for maintaining posture on land because the body is not supported by water; without muscular action, the skeleton would buckle and collapse. Likewise, the method of ventilating the lungs is different if the chest is surrounded by air rather than by water.

The axial muscles still participate in locomotion in basal tetrapods, producing the lateral bending of the backbone also seen during movement by salamanders, lepidosaurs, and crocodylians. In birds and mammals, however, limb movements have largely replaced the lateral flexion of the trunk by axial muscles. The trunks of birds are rigid, but dorsoventral flexion is an important component of mammalian locomotion. In secondarily aquatic mammals (e.g., whales), the axial muscles again assume a major role in locomotion in powering the tail, which moves up and down rather than side to side (as in fishes or aquatic reptiles).

The hypaxial muscles form two layers in bony fishes (the external and internal oblique muscles), but in tetrapods a third inner layer is added, the transversus abdominis (see Figure 12.3 [number 12]). This muscle is responsible for expiration of air from the lungs of modern amphibians (which, unlike amniotes, do not use their ribs to breathe), and it would have been essential for lung ventilation on land by early tetrapods. Air-breathing fishes use the pressure of the surrounding water to force air from the lungs, but land-dwelling tetrapods need muscular action.

The rectus abdominis is another novel hypaxial muscle in tetrapods. It runs along the ventral surface from the pectoral girdle to the pelvic girdle, and its role appears to be primarily postural. (This is the muscle responsible for the "six-pack" abdomen of human bodybuilders.) The costal muscles in the rib cage of amniotes are formed by all three layers of the hypaxial muscles and are responsible for inspiration as well as for expiration. The use of the ribs

(A) Generalized anatomy

Generalized lobe-finned fish

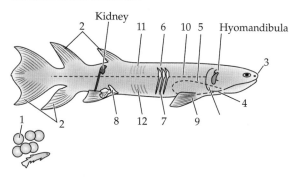

(B) Cross-section

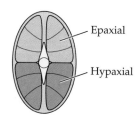

Epaxial

Hypaxial

Generalized primitive non-amniote

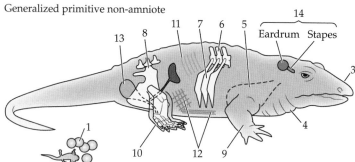

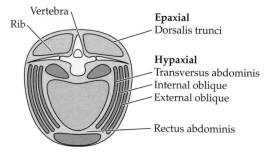

Vertebra

Rib

Epaxial
Dorsalis trunci

Hypaxial
Transversus abdominis
Internal oblique
External oblique

Rectus abdominis

Generalized early amniote

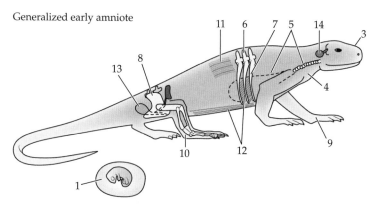

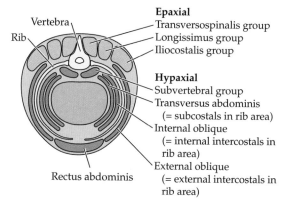

Epaxial
Transversospinalis group
Longissimus group
Iliocostalis group

Hypaxial
Subvertebral group
Transversus abdominis
    (= subcostals in rib area)
Internal oblique
    (= internal intercostals in
    rib area)
External oblique
    (= external intercostals in
    rib area)

Vertebra

Rib

Rectus abdominis

Figure 12.3 Morphological and physiological characteristics of sarcopterygian fishes, basal tetrapods, and amniotes.
(A) Numbers indicate the various systems that are referred to in the text. 1: Mode of reproduction—shelled egg in amniotes. 2: Presence of midline and tail fins—absent in post-Devonian tetrapods. 3: Length of snout—longer in tetrapods. 4: Length of neck—no neck in fish, longer neck in amniotes. 5: Form of lungs and trachea—complex lungs and trachea in amniotes. 6: Interlocking of vertebral column—seen in all tetrapods. 7: Form of ribs—stouter in tetrapods, used for ventilation in amiontes. 8: Attachment of pelvic girdle to vertebral column—one sacral attachment in basal tetrapods, two in amniotes. 9: Form of the limbs—digits in all tetrapods, limbs more gracile and more important in locomotion in amniotes. 10: Form of the ankle joint—distinct joint in middle of ankle in amniotes. 11: Differentiation of epaxial muscles—differentiated in amniotes. 12: Differentiation of hypaxial muscles—not differentiated in the basal fish condition (as shown here), the internal and external divisions are differentiated into multiple layers in tetrapods. 13: Presence of urinary bladder—in tetrapods. (The kidneys have been portrayed in a mammalian bean-shaped form, for familiarity and convenience. In fact, the kidneys of fishes and nonamniotic tetrapods are elongated structures lying along the dorsal body wall.) 14: Form of the acousticolateralis system and middle ear—lateral line in fish only, middle ear independently derived in non-amniotes and amniotes. (B) Cross-sectional view of body (tail in fish) showing axial musculature.

and their associated musculature as devices to ventilate the lungs was probably an amniote innovation.

In fish and modern amphibians, the epaxial muscles form an undifferentiated single mass. This was probably their condition in the earliest tetrapods (see Figure 12.3, number 11). In salamanders, both epaxial and hypaxial muscles contribute to the bending motions of the trunk while walking on land, much as they do in fish swimming in water. The epaxial muscles of extant amniotes are distinctly differentiated into three major components, and their role is now postural in addition to locomotory. In all amniotes, the transversopinalis muscles run between individual vertebrae and contribute to postural stability. In reptiles, the iliocostalis is the largest portion of the epaxials,

Figure 12.4 Evolution of the neck of tetrapods. The bones of the skull are connected to the pectoral girdle by the supracleithral bones in bony fishes, as shown here by the basal lobe-finned fish *Eusthenopteron* (A) and the derived lobe-finned fish *Tiktaalik* (B). Note that the posterior edge of the operculum of *Eusthenopteron* is not joined to the bones behind it: the operculum is mobile, and this is where water exits from the gills. The opercular and supracleithral bones have been lost in tetrapods, as shown here by the basal tetrapod *Acanthostega* (C) and the more derived tetrapod *Proterogyrinus* (D). (The anocleithrum of *Acanthostega* is one of the few examples of retention of one of the supercleithral bones by early tetrapods.) (From Schoch 2014.)

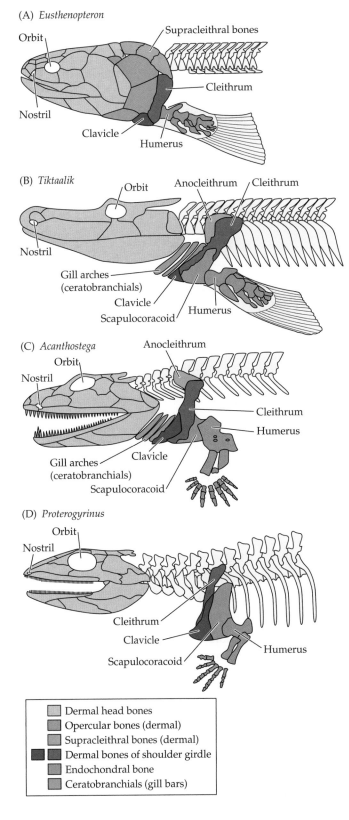

(A) *Eusthenopteron*

(B) *Tiktaalik*

(C) *Acanthostega*

(D) *Proterogyrinus*

☐ Dermal head bones
☐ Opercular bones (dermal)
☐ Supracleithral bones (dermal)
☐ Dermal bones of shoulder girdle
☐ Endochondral bone
☐ Ceratobranchials (gill bars)

running from the pelvis to the top of the rib cage, and it contributes to lateral flexion of the trunk. In mammals, the longissimus dorsi is the largest portion of the epaxials, running along the length of the spine, and it contributes to dorsoventral flexion of the trunk.

Thus, in terrestrial tetrapods the axial skeleton and its muscles assume very different roles from their original functions in aquatic vertebrates. The skeleton now participates in postural support and ventilation of the lungs, as well as in locomotion, and some of these functions are incompatible. For example, the side-to-side bending of the trunk that occurs when a lizard runs means that it has difficulty using its ribs for lung ventilation, creating a conflict between locomotion and respiration. More derived tetrapods such as mammals and birds have addressed this conflict by a change in posture from sprawling limbs to limbs that are held more directly underneath the body. These tetrapods are propelled by limb movements rather than by lateral bending.

The appendicular skeleton: limbs and limb girdles

In the ancestral gnathostome condition, illustrated by sharks, the pectoral girdle (supporting the front fins) is formed solely from the endoskeleton; it is a simple cartilaginous rod, called the coracoid bar, with a small ascending scapular process. In bony fishes, the endoskeletal portion of the pectoral girdle (the scapulocoracoid) joins with some of the supracleithral dermal bones (the clavicle and cleithrum), and these dermal bones are in turn connected to bones that form the posterior portion of the dermal skull roof (**Figure 12.4**). The pelvic girdle in both kinds of fishes is represented by the puboischiatic plate, which has no connection with the vertebral column but merely anchors the hindfins in the body wall. Neither arrangement works well on land.

The tetrapod limb is derived from the fin of fish. The basic structure of a fin consists of fanlike basal elements supporting one or more ranks of cylindrical radials, which usually articulate with raylike structures that support most of the surface of the fin web. The tetrapod limb is made up of the limb girdle and five segments that articulate end to end. All tetrapods have jointed limbs, wrist and ankle joints, and hands and feet with digits (see Figure 12.3, number 9). The feet of basal tetrapods were used mainly as holdfasts, providing frictional contact with the ground. Propulsive force was generated mainly by the axial musculature of

the body, as in salamanders today. In contrast, the feet of amniotes play a more complex role in locomotion. They are used as levers to propel the animal: the knee turns forward, the elbow turns backward, and the ankle forms a distinct hinge joint (mesotarsal joint; see Figure 12.3, number 10).

(A)

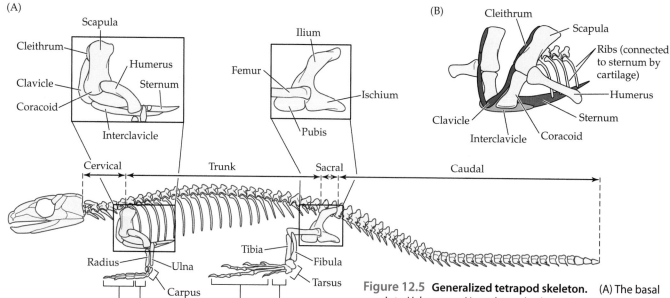

(B)

Figure 12.5 **Generalized tetrapod skeleton.** (A) The basal amniote *Hylonomus*. Note the arched vertebral column with articulations between adjacent vertebrae that transfer the force of gravity to the limbs; rib cage enclosing the thoracic and abdominal organs; and long tail. (B) The pectoral girdle rests in a muscular sling and does not articulate with the axial skeleton, unlike the pelvic girdle, which articulates tightly with the sacral vertebrae. (After Carroll and Baird 1972.)

Some non-amniotes (e.g., frogs) are like amniotes in this condition.

The basic form of the tetrapod skeleton is illustrated in **Figure 12.5A**. The pelvic girdle articulates with modified sacral vertebrae, and the hindlimbs are the primary propulsive mechanism. The pelvic girdle contains three paired bones on each side (a total of six bones): ilium (plural *ilia*), pubis (plural *pubes*), and ischium (plural *ischia*). The ilia on each side connect the pelvic limbs to the vertebral column, forming an attachment at the sacrum via one or more modified ribs (see Figure 12.3, number 8).

In fish, support of the forelimb is only a minor role of the pectoral girdle, which mainly serves to anchor the muscles that move the gill arches and lower jaw. In tetrapods the pectoral girdle is freed from its bony connection with the head, and support of the body is now its major role. In bony fishes, the pectoral girdle and forelimb are attached to the back of the head via a series of dermal bones (the supracleithral bones; see Figure 12.4). In tetrapods these bones are lost, and the pectoral girdle is freed from the dermal skull roof (**Figure 12.5B**). The endochondral bones are the scapula and the coracoid; the humerus (upper arm bone) articulates with the pectoral girdle where these two bones meet. However, some dermal bones (the cleithrum and clavicle) become incorporated into the pectoral girdle along the anterior border of the scapula, and a novel dermal bone, the interclavicle, forms a ventral midline structure, lying ventral to the sternum (see below). The cleithrum is seen only in some extinct tetrapods. The clavicle (collar bone of humans) connects the scapula to the interclavicle, or to the sternum if the interclavicle is absent. The interclavicle is seen today in lizards and crocodiles. It has been lost in birds and in most mammals but is still present in the monotremes.

Unlike the pelvic girdle, the pectoral girdle does not articulate directly with the vertebral column: this is why humans can shrug their shoulders but not their hips. (Only in some pterosaurs—extinct flying reptiles—is there an equivalent of a sacrum in the anterior vertebral column, a structure called the notarium.) In all other vertebrates, the connection between the pectoral girdle and the vertebral column consists of muscles and connective tissue (in addition to the clavicle) that hold the pectoral girdle to the sternum and the ribs. The sternum is a midventral structure, formed from cartilage or endochondral bone and usually segmented, that links the lower ends of right and left thoracic ribs in amniotes. It forms an important supportive role in vertebrates today, supplanting the interclavicle, but its original function may have involved ventilation via rib movements. The sternum is extensively ossified only in birds and mammals. Bones (sternal elements) are seen in the shoulder girdle of frogs and salamanders, but they may not be homologous with the amniote sternum.

The midline fins of fish—the dorsal and anal fins—help reduce roll, but they have no function on land and are not present in terrestrial animals (although a dorsal-fin equivalent, without any internal fin structure, is present in some secondarily aquatic tetrapods such as dolphins and the extinct marine reptiles known as ichthyosaurs). The pectoral and pelvic fins of fish become the limbs of tetrapods. The appendicular muscles of tetrapods are more complicated and differentiated than those of fish. The ancestral fish pattern of a major levator and a major depressor is retained, but derived fishes (humans, for example) have many additional muscles in the shoulder region alone—not to mention the muscles that move the other joints, including the fingers and toes.

Size and scaling

Body size is one of the most important things to know about an organism. Because all structures are subject to the laws of physics, the absolute size of an animal profoundly affects its anatomy and physiology. The study of scaling, or how shape changes with size, is known as **allometry** (Greek *allos*, "different"; *metron*, "measure"). If the features of an animal showed no relative changes with increasing body size (i.e., if a larger animal appeared just like a photo enlargement of a smaller one), then all of its component parts would be scaled with isometry (Greek *isos*, "similar"). However, animals are not built this way, and very few body components scale isometrically.

Underlying all scaling relationships is the issue of how the linear dimensions of an object relate to its surface area and volume: when linear dimensions double (a twofold change), the surface area increases as the square of the change in linear dimensions (a fourfold change) and the volume, and hence the weight, increases as the cube of the change in linear dimensions (an eightfold change). Thus, an animal that is twice as tall as another is eight times as heavy.

It is the cross-sectional area of a limb bone that actually supports an animal's weight on land. If an animal increased in size isometrically, its weight would rise as the cube of its linear dimensions, but the cross section of its bones would increase only as the square of their linear dimensions. As a result, the animal's limbs would be unable to support the increased weight. Thus, the limb bones must have greater cross-sectional area to keep up with increases in weight, and bone diameter scales with positive allometry; that is, bigger animals have proportionally thicker limb bones than smaller ones. The skeleton of a bigger animal can easily be distinguished from that of a smaller animal, even when they are shown at the same size, by the proportionally thicker bones of the large animal (**Figure 12.6**).

Although the cross-sectional area of the limb bones of terrestrial vertebrates does indeed scale with positive allometry, the area does not increase in proportion to the stress the bones experience. Consequently, the limbs of large animals are more fragile than the limbs of small animals, and large animals have a different posture than small animals. Small animals stand with flexed joints, whereas large animals stand more erect on straighter legs. (Compare the postures of the cat and the elephant in Figure 12.6.) The reason for this difference lies in the mechanical aspects of bone: A bone withstands compressive forces (forces exerted parallel to the long axis of the bone) much better than shearing forces (those exerted at an angle to the long axis). The larger the animal, the less flexed are its limb joints, and the less its bones are likely to be subjected to shearing forces. The pillarlike weight-bearing stance of very large animals, such as elephants and the huge sauropod dinosaurs, reduces the effect of shearing on their limb bones. But this posture also makes these animals less agile.

Locomotion

The basic form of quadrupedal locomotion was probably similar to that seen today in salamanders, combining lateral axial movements with diagonal pairs of legs moving together, but with the limbs used more as holdfasts than for propulsion. The right forelimb and left hindlimb move as one unit, and the left forelimb and right hindlimb move as another, in a type of gait known as the walking-trot.

Lizards retain a modified version of this mode of locomotion, although their limbs are more important for propulsion. Even though humans are bipedal, relying entirely on the hindlegs for locomotion, we retain this ancestral coupling of the limbs in walking, swinging the right arm forward when striding with the left leg, and vice versa. This type of coupled, diagonally paired limb movement is probably an ancestral feature for gnathostomes, because sharks also

(A)

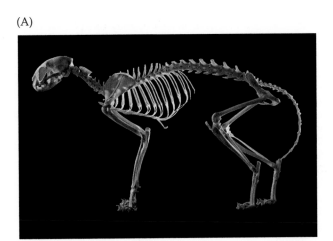

(B)

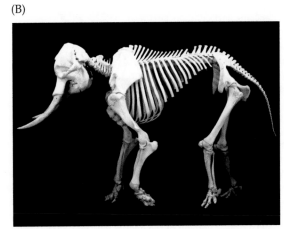

Figure 12.6 Which is bigger? Even though these animals are shown at the same size, the proportions of the bones, especially the limbs, make it instantly apparent that a cat (A) is smaller than an elephant (B). (A, Stock Up/ Shutterstock; B, Sklmsta/CC BY 1.0.)

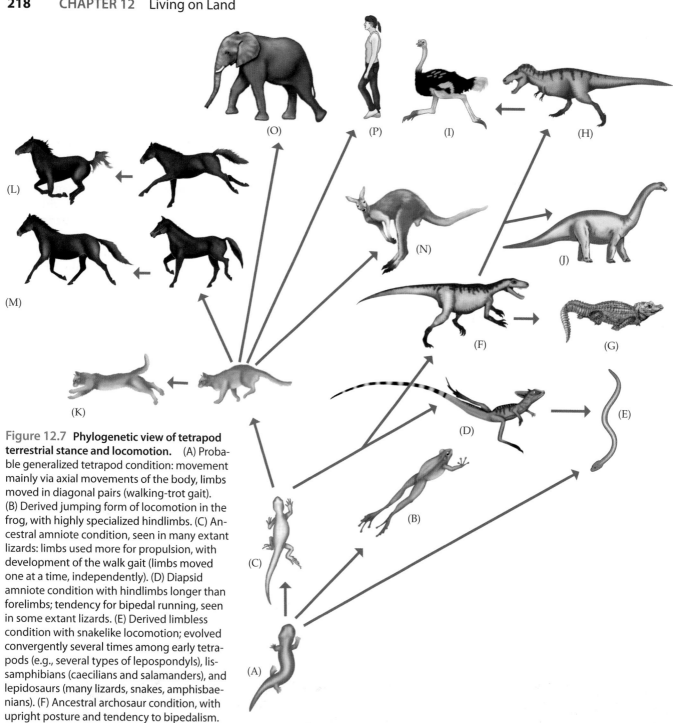

Figure 12.7 Phylogenetic view of tetrapod terrestrial stance and locomotion.   (A) Probable generalized tetrapod condition: movement mainly via axial movements of the body, limbs moved in diagonal pairs (walking-trot gait). (B) Derived jumping form of locomotion in the frog, with highly specialized hindlimbs. (C) Ancestral amniote condition, seen in many extant lizards: limbs used more for propulsion, with development of the walk gait (limbs moved one at a time, independently). (D) Diapsid amniote condition with hindlimbs longer than forelimbs; tendency for bipedal running, seen in some extant lizards. (E) Derived limbless condition with snakelike locomotion; evolved convergently several times among early tetrapods (e.g., several types of lepospondyls), lissamphibians (caecilians and salamanders), and lepidosaurs (many lizards, snakes, amphisbaenians). (F) Ancestral archosaur condition, with upright posture and tendency to bipedalism. (G) Secondary return to sprawling posture and quadrupedalism in crocodylians. (H) Obligate bipedalism in early dinosaurs and (I) birds. (J) Return to quadrupedalism several times within dinosaurs. (K) Ancestral therian mammal condition: upright posture and use of the bound as a fast gait with dorsoventral flexion of the vertebral column (all mammals use the walk as a slow gait). In the bound, the animal jumps off the two hindlimbs, flies through the air with limbs outstretched, and lands on the two forelimbs (or on one forelimb and then the other, as in this half-bounding cat). (L) Condition in larger mammals where the bound is turned into the gallop: the limbs move one at a time, and the period of suspension when all four feet are in the air occurs when the legs are bunched up, as shown. (M) The trot is used at intermediate speeds between the walk and gallop. The canter is essentially a slower version of the gallop. (N) The ricochet, a derived hopping gait of kangaroos and some rodents. (O) The amble, a speeded-up walk gait seen in the fast gait of elephants and in some horses. (P) The unique human condition of upright bipedal striding. Penguins can also walk with an upright trunk, but they waddle rather than stride.

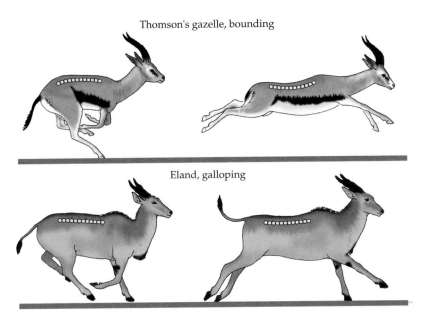

Thomson's gazelle, bounding

Eland, galloping

Figure 12.8 Bounding and galloping. Small species of antelopes, such as Thomson's gazelle (*Eudorcas thomsonii*; 15–35 kg), bound, whereas larger species, such as the eland (*Taurotragus oryx*); 350–900 kg), gallop. The position of the vertebral column is indicated. (After McMahon and Bonner 1983.)

move their fins in this fashion when bottom-walking over the substrate. **Figure 12.7** shows the modes of locomotion used by tetrapods in a phylogenetic perspective.

Amniotes employ the "walk" gait, in which each leg moves independently in succession, usually with three feet on the ground at any one time. The "trot" may be used by amniotes for faster movement. In this gait, seen predominantly in mammals, diagonal pairs of limbs move together (e.g., right front and left rear) as in the walking-trot, but it is a faster gait with a period of suspension when all four legs are off the ground. The fast gait that is characteristic of therian mammals (marsupials and placentals) is the "bound," which involves jumping off the hindlegs and landing on the forelegs, with flexion of the back contributing to the length of the stride. In larger mammals, the bound is modified into the "gallop"—as seen, for example, in horses—and there is less flexion of the backbone (**Figure 12.8**).

12.2 Eating on Land

The differences between water and air profoundly affect feeding by tetrapods. In water, most food items are nearly weightless, and aquatic animals can suck the food into their mouth and move it within the mouth by creating currents of water. Suction feeding is not an option on land, however, because air is much less dense than the food particles. (You can suck up the noodles in soup along with the liquid, but you cannot suck up the same noodles if you put them on a plate.)

The operculum and the bones that connect the pectoral girdle to the head in bony fish have been lost in all tetrapods. The skull of bony fishes has a short snout; movements of the jaws, hyoid apparatus (lower part of the hyoid arch), and operculum cause water to be sucked into the mouth for both gill ventilation and feeding.

Early tetrapods had wide, flat skulls and longer snouts than their fish ancestors, so that most of the tooth row was in front of the eye. Their flat heads and long snouts combined the functions of feeding and breathing, as do the heads of extant amphibians, which use movements of the hyoid apparatus to ventilate the lungs. This method of lung ventilation is a type of positive-pressure mechanism called buccopharyngeal pumping. The same expansion of the buccopharyngeal cavity is used for suction feeding in water.

The tongue of jawed fishes is small and bony, whereas the tongue of tetrapods is large and muscular. (The muscular tongues of lampreys and hagfishes are not homologous with the tongues of tetrapods, and they are innervated by different cranial nerves—V in cyclostomes and XII in tetrapods.) The tetrapod tongue works in concert with the hyoid apparatus and is probably a key innovation for feeding on land. Most tetrapods use the tongue to manipulate food in the mouth and transport it to the pharynx. Most terrestrial salamanders and lizards have sticky tongues that help capture prey and transport it into the mouth—a phenomenon called prehension. In addition, some tetrapods—such as frogs, salamanders, and the true chameleon lizards—can project their tongue to capture prey. (The mechanism of tongue projection is different in each group; tongue projection is an example of convergent evolution.)

Salivary glands are known only in terrestrial vertebrates, probably because saliva is not required to swallow food in water. The salivary glands of tetrapods lubricate food and have additional functions as well, such as secreting enzymes that begin the chemical digestion of carbohydrates while food is still in the mouth in mammals. Some insectivorous mammals, two species of lizards, and several lineages of snakes have elaborated salivary secretions into venoms that kill prey.

With the loss of gills in tetrapods, much of the associated branchiomeric musculature was also lost, but the gill levators are a prominent exception. In fish, these muscles are combined into a single unit, the cucullaris, and this muscle in tetrapods becomes the trapezius, which runs from the top of the neck and shoulders to the shoulder girdle. In mammals, the trapezius helps rotate and stabilize the scapulae in locomotion and humans use the trapezius to shrug the shoulders.

Understanding the original homologies of the trapezius muscle explains an important aspect of human spinal injuries. Because the trapezius is an old branchiomeric muscle, it is innervated directly from the brain by cranial nerves (cranial nerve XI, which is actually part of nerve X), not by the nerves exiting from the spinal cord in the neck. Thus, people who are paralyzed from the neck down by a spinal injury can still shrug their shoulders. Small muscles in the throat—for example, those powering the larynx and the vocal cords—are other remnants of the branchiomeric muscles associated with the gill arches. Ingenious biomedical engineering allows quadriplegic individuals to use this remaining muscle function to control prosthetic devices.

The major branchiomeric muscles that are retained in tetrapods are associated with the mandibular and hyoid arches and are involved solely in feeding. The adductor mandibulae remains the major jaw-closing muscle, and it becomes increasingly complex in more derived tetrapods. The hyoid musculature forms two novel important muscles in tetrapods. One is the depressor mandibulae, running from the back of the jaw to the skull and helping the hypobranchial muscles open the mouth. The other is the sphincter colli that surrounds the neck and aids in swallowing food. In mammals, the sphincter colli has become the musculature of facial expressions.

12.3 Breathing Air

In some respects, air is an easier medium for respiration than is water. The low density and viscosity of air make tidal ventilation of a saclike lung energetically feasible, and the higher oxygen content of air reduces the volume of fluid that must be pumped to meet an animal's metabolic requirements.

The lung is an ancestral feature of bony fishes, and lungs were not evolved for breathing on land. For many years it was assumed that lungs evolved in fishes living in stagnant, oxygen-depleted water where gulping oxygen-rich air would supplement oxygen uptake by the gills. However, although some lungfishes are found in stagnant, anoxic environments, other air-breathing fishes (e.g., the bowfin [*Amia calva*]) are active animals found in oxygen-rich habitats. An alternative explanation for the evolution of lungs is that air-breathing evolved in well-aerated waters in active fishes in which the additional oxygen is needed primarily to supply the heart muscle rather than the body tissues.

In contrast to non-amniote tetrapods, which use a positive-pressure buccal pump to inflate the lungs, amniotes use a negative-pressure aspiration pump. Expansion of the rib cage by the intercostal hypaxial muscles creates a negative pressure (i.e., below atmospheric pressure) in the abdominal cavity and sucks air into the lungs. Air is expelled by compression of the abdominal cavity, primarily through elastic return of the rib cage to a resting position and contraction of the elastic lungs, as well as by contraction of the transversus abdominis muscle.

The lungs of many extant amphibians are simple sacs with few internal divisions. They have only a short chamber leading directly into the lungs. In contrast, amniotes have lungs that are subdivided, sometimes in very complex ways,

to increase the surface area for gas exchange. Amniotes also have a long trachea (windpipe), strengthened by cartilaginous rings, that branches into a series of bronchi in each lung.

The form of lung subdivisions is somewhat different in mammals than in other amniotes, suggesting independent evolutionary origins of complex lungs. The combination of a trachea and negative-pressure aspiration has allowed many amniotes to develop longer necks than those seen in modern amphibians or in extinct non-amniote tetrapods. Amniotes also possess a larynx (derived from pharyngeal arch elements), at the junction of the pharynx and the trachea, that is used for sound production, most notably in mammals.

12.4 Pumping Blood Uphill

Blood is weightless in water, and the heart needs to overcome only fluid resistance to move blood around the body. Circulation is more difficult for a terrestrial animal because gravity causes blood to pool in low spots, such as the limbs, and it must be forced through the veins and back up to the heart by more blood pumped through the arteries. Thus, terrestrial tetrapods have blood pressures high enough to push blood upward through the veins against the pull of gravity, and they have valves in the limb veins that resist backflow.

The walls of capillaries are somewhat leaky, and high blood pressure forces some of the blood plasma (the liquid part of blood) out of the vessels and into the intercellular spaces of the body tissues. This fluid is recovered and returned to the circulatory system by the lymphatic system. The lymphatic system is a one-way system of blind-ended, veinlike vessels that parallel the veins and allow fluid in the tissues to drain into the venous system at the base of the neck. (A lymphatic system is also well developed in teleost fishes, but it is of critical importance on land, where the cardiovascular system is subject to the forces of gravity.) In tetrapods, valves in the lymph vessels prevent backflow, and contraction of muscles and tissues keeps lymph flowing toward the heart.

Lymph nodes—concentrations of lymphatic tissues—are found in mammals and some birds at intervals along the lymphatic channels. Lymphatic tissue is also involved in the immune system; white blood cells (macrophages) travel through lymph vessels, and the lymph nodes can intercept foreign or unwanted material, such as migrating cancer cells.

With the loss of the gills and the evolution of a distinct neck in tetrapods, the heart has moved posteriorly. In fish, the heart lies in the gill region in front of the shoulder girdle, whereas in tetrapods it lies behind the shoulder girdle in the thorax. The sinus venosus and conus arteriosus are reduced or absent in the hearts of tetrapods.

Tetrapods have a double circulation in which the pulmonary circuit supplies the lungs with deoxygenated blood and the systemic circuit supplies oxygenated blood to the body. The atrium of the heart is divided into left and right chambers (atria) in lungfishes and tetrapods, and the ventricle is divided either by a fixed barrier or by the formation of transiently separate chambers as the heart contracts. The right side of the heart receives deoxygenated blood returning from the body via the systemic veins, and the left side of the heart receives oxygenated blood returning from the lungs via the pulmonary veins. The double circulation of tetrapods can be pictured as a figure eight, with the heart at the intersection of the loops (**Figure 12.9**).

The aortic arches have undergone considerable change in association with the loss of the gills in tetrapods. Arches 2 and 5 are lost in most adult tetrapods (although arch 5 is retained in salamanders). Three major arches are retained: arch 3 (carotid arch) going to the head, arch 4 (systemic arch) going to the body, and arch 6 (pulmonary arch) going to the lungs (**Figure 12.10** and **Figure 12.11**).

Extant amphibians retain a fishlike condition, in which the aortic arches do not arise directly from the heart. This condition, with retention of a conus arteriosus and a ventral arterial trunk (the ventral aorta of fish and the truncus arteriosus of amphibians), was probably found among the earliest tetrapods. In amniotes, the pulmonary artery receives blood from the right ventricle, and the right systemic and carotid arches receive blood from the left ventricle, although details of the heart anatomy suggest that this condition may have evolved independently in mammals and in other amniotes.

A ventricular septum of some sort is present in all amniotes, but its form is different in the various lineages. A transient ventricular septum is formed during ventricular contraction in turtles and lepidosaurs, whereas a permanent ventricular septum is present in crocodylians, birds, and mammals. This phylogenetic pattern indicates that a permanent septum evolved independently in the sauropsid (reptiles, including birds) and synapsid (mammal) lineages. This interpretation is also supported by developmental evidence.

The heavy workload and divided ventricle of birds and mammals introduce another complication: how to supply oxygen to the heart muscle. Modern amphibians and nonavian reptiles have lower blood pressures than mammals and birds, their hearts don't work as hard, and their ventricles allow some mixing of oxygenated and deoxygenated blood. The hearts of these animals never evolved an extensive system of coronary arteries, presumably because enough oxygen diffuses into the cardiac muscle from the blood in the lumen of the ventricle. In contrast, the ventricular muscles of mammals and birds are thicker and must work harder than those of amphibians and nonavian reptiles to generate higher blood pressures. In addition, mammals and birds have a permanent ventricular septum, so the right ventricle contains only deoxygenated blood. Coronary arteries that supply oxygenated blood to the heart are seen in many gnathostomes, but the basal pattern appears to be for supply only to the conus arteriosus (or base of the aorta), which is composed of compact muscle and does not receive oxygen from within the heart. Among tetrapods, extensive coronary supply to muscles of the ventricles is seen only in those forms that have a permanent ventricular septum (which prevents oxygenated blood from circulating throughout the heart): crocodylians, birds, and mammals.

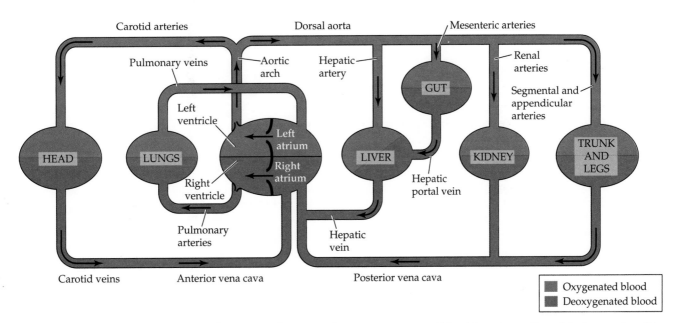

Figure 12.9 Double-circuit cardiovascular system of a tetrapod. Deoxygenated blood returns via the anterior vena cava and posterior vena cava to the right atrium and is pumped into the right ventricle. From there it goes via the pulmonary arteries to the lungs, where carbon dioxide is released and oxygen is bound. The oxygenated blood returns via the pulmonary veins to the left atrium and then enters the left ventricle. Contraction of the left ventricle sends oxygenated blood into the aortic arch. The carotid arteries carry oxygenated blood from the aortic arch to the head, and the dorsal aorta carries oxygenated blood to the rest of the body. The liver receives oxygenated blood via the hepatic artery and deoxygenated blood from the gut via the hepatic portal vein. Veins in the body return deoxygenated blood to the right atrium via the posterior vena cava, and the carotid veins return blood from the head to the anterior vena cava.

(B) Teleost or derived ray-finned fish

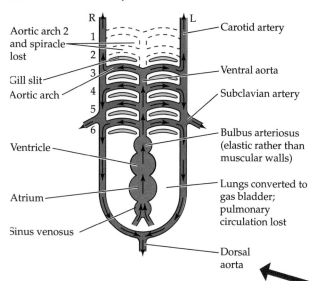

(C) Generalized non-coelanth lobe-finned fish

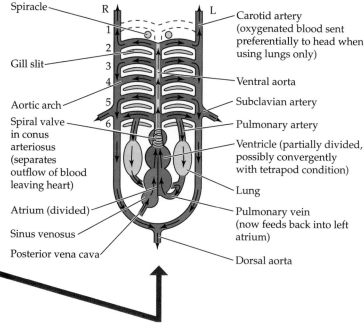

Figure 12.10 The heart and aortic arches in jawed fishes and lungfishes. (A) The ancestral bony fish condition is to have lungs and some sort of pulmonary circuit, as shown in an extant actinopterygian with lungs, *Polypterus*. This basal fish is shown as having five aortic arches (labeled 2–6; aortic arch 1 is present in the embryos of all vertebrates but not in any extant adult). (B) Teleosts have converted the lung into a gas bladder and have lost the pulmonary circuit and aortic arch 2. (C) The generalized sarcopterygian condition above the level of coelacanths. Some extant lungfishes have reduced the number of aortic arches from this condition. Here the pulmonary artery feeds back into the left atrium directly, and the ventricle is partially divided. R = right, L = left.

(A) Basal bony fish

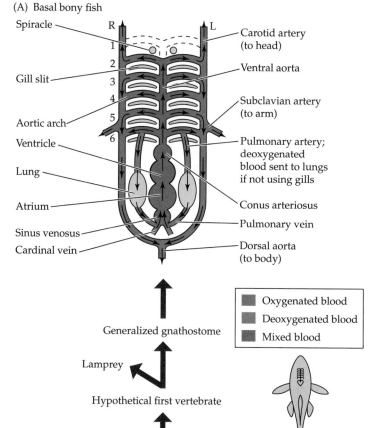

12.5 Sensory Systems in Air

Some of the sensory modes that are exquisitely sensitive in water are useless in air, whereas other modes work better in air than in water. Air is not dense enough to stimulate the mechanical receptors of the lateral line system, for example, and does not conduct electricity well enough to support electrosensation. Chemical systems work well on land, however, at least for molecules small enough to be suspended in air, and air offers advantages for both vision and hearing.

Vision

Vision is more acute in air than in water because light is transferred through air with less disturbance than through water. Air is rarely murky in the way that water frequently is, and thus vision is more useful as a distance sense in air than in water.

As explained in Chapter 4, terrestrial and aquatic animals focus images on the retina in different ways. The refractive index of the cornea is very similar to that of water, so the cornea of fish has little effect on focusing the eye. Fish focus light by moving the

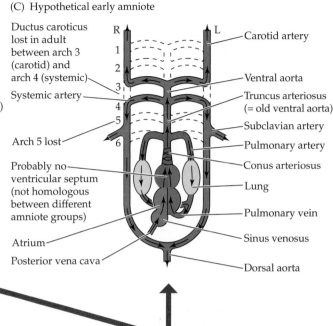

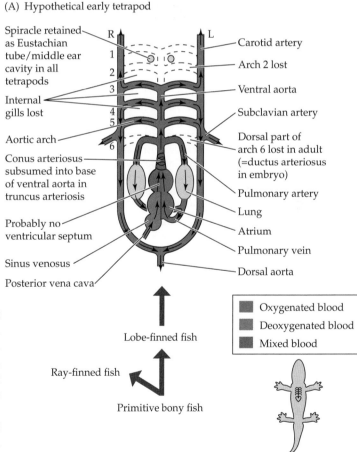

(B) Extant amphibian

- Ductus caroticus (lost in frogs)
- Arch 5 (lost in frogs)
- Frogs only: cutaneous artery (takes blood to skin)
- Some oxygenated blood comes into right side of heart from cutaneous vein
- Posterior vena cava
- Carotid artery
- Ventral aorta
- Truncus arteriosus (= old ventral aorta)
- Subclavian artery
- Pulmonary artery
- Conus arteriosus
- Lung
- Pulmonary vein
- Dorsal aorta

(C) Hypothetical early amniote

- Ductus caroticus lost in adult between arch 3 (carotid) and arch 4 (systemic)
- Systemic artery
- Arch 5 lost
- Probably no ventricular septum (not homologous between different amniote groups)
- Atrium
- Posterior vena cava
- Carotid artery
- Ventral aorta
- Truncus arteriosus (= old ventral aorta)
- Subclavian artery
- Pulmonary artery
- Conus arteriosus
- Lung
- Pulmonary vein
- Sinus venosus
- Dorsal aorta

(A) Hypothetical early tetrapod

- Spiracle retained as Eustachian tube/middle ear cavity in all tetrapods
- Internal gills lost
- Aortic arch
- Conus arteriosus subsumed into base of ventral aorta in truncus arteriosis
- Probably no ventricular septum
- Sinus venosus
- Posterior vena cava
- Carotid artery
- Arch 2 lost
- Ventral aorta
- Subclavian artery
- Dorsal part of arch 6 lost in adult (=ductus arteriosus in embryo)
- Pulmonary artery
- Lung
- Atrium
- Pulmonary vein
- Dorsal aorta

Lobe-finned fish

Ray-finned fish

Primitive bony fish

- Oxygenated blood
- Deoxygenated blood
- Mixed blood

Figure 12.11 The heart and aortic arches in tetrapods. (A) The inferred early tetrapod (above the *Acanthostega/Ichthyostega* level) is similar to lungfishes; however, the internal gills have been lost, and aortic arch 2 may also have been lost by this stage (as seen in all extant tetrapods). (B) In the frog, arch 5 and the connection of the dorsal aorta between arches 3 and 4 (the ductus caroticus) have been lost, but both of these features are retained in salamanders, so this condition cannot have been inherited from an early tetrapod ancestor. A derived condition in anurans is to have a cutaneous artery leading from the pulmonary artery taking blood to the skin to be oxygenated; the blood returns via the cutaneous vein that feeds into the subclavian vein. The ventricular septum is absent in modern amphibians; it is not clear if this is a primary or secondary condition, because a partial septum is present in lungfishes. (C) The proposed early amniote condition is similar to that of the frog, with the exception of the absence of a cutaneous circuit. The ventricular septum would have been simple, if it was present at all, and there would have been no division of the vessels leaving the heart because the more derived conditions in extant sauropsid and synapsid amniotes are not homologous (i.e., one cannot be derived from the other).

position of the lens within the eye. The refractive index of the cornea is greater than that of air, however, and in air the cornea *does* participate in forming a focused image. Terrestrial vertebrates have flatter lenses than do fish and marine mammals, and all tetrapods except snakes focus by changing the shape of the lens (**Figure 12.12**).

In air, the eye's surface must be protected and kept moist and free of particles. Novel features in terrestrial tetrapods include eyelids, glands that lubricate the eye and keep it moist (including tear-producing lacrimal glands), and a nasolacrimal duct that drains the tears from the eyes into the nose. Many species of mudskippers spend much of the day on land; these fishes retract their eyes into small water-filled chambers to moisten the corneal surface, as they lack the eyelids of tetrapods.

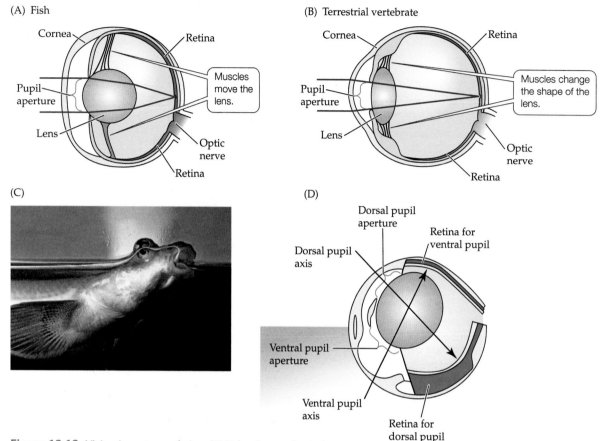

(A) Fish

Cornea
Retina
Pupil aperture
Muscles move the lens.
Lens
Optic nerve
Retina

(B) Terrestrial vertebrate

Cornea
Retina
Pupil aperture
Muscles change the shape of the lens.
Lens
Optic nerve
Retina

(C)

(D)

Dorsal pupil aperture
Retina for ventral pupil
Dorsal pupil axis
Ventral pupil aperture
Ventral pupil axis
Retina for dorsal pupil

Figure 12.12 Vision in water and air. (A) Fishes have spherical lenses. The entire lens is moved backward or forward by contraction or relaxation of muscles to adjust for objects at different distances. (B) Terrestrial vertebrates have oval lenses. The focus is adjusted for near or distant objects by muscular contractions that change the shape of the lens. (C) The four-eyed fish *Anableps anableps* swims at the surface of the water. (D) Each of its eyes has two pupils and two retinas. The dorsal pupil admits light from above the water surface, and the ventral pupil admits light from below. The lens is asymmetric, with a long dimension and a short dimension. The cornea of the dorsal pupil is in air and helps focus light entering through the dorsal pupil. The axis of the dorsal pupil passes through the short dimension of the lens. The cornea of the ventral pupil is in water and has no effect on focus. The axis of the ventral pupil passes through the long dimension of the lens. (C, Mark Conlin/Alamy Stock Photo; D, after Sivak 1976.)

Hearing

Sound perception is very different in air than in water. The density of animal tissue is nearly the same as the density of water, and sound waves pass freely from water into animal tissue. Because water is dense, movement of water molecules directly stimulates the hair cells of the lateral line system.

Air is not dense enough to move the cilia of hair cells, and tetrapods have a middle ear that amplifies sound vibrations and transmits them to the inner ear, where they are transformed to nerve impulses (**Figure 12.13**). The middle ear receives the relatively low energy of sound waves on its outer membrane, the tympanum (eardrum), and these vibrations are transmitted by the stapes (in sauropsids; see

Figure 12.13A) or by the ossicular chain (malleus, incus, and stapes in extant mammals; see Figure 12.13B) to the oval window, which links the middle and inner parts of the ear. The area of the tympanum is much larger than that of the oval window, and the difference in area (plus, for synapsids, the lever system of the ossicular chain) amplifies the sound waves. In-and-out movement of the oval window produces waves of compression in the fluids of the inner ear, and these waves stimulate the hair cells in the organ of Corti, which lies within a flask-shaped structure called the lagena in sauropsids and the cochlea in synapsids. The organ of Corti discriminates the frequency and intensity of vibrations it receives and transmits this information to the central nervous system. The semicircular canals provide information about orientation and acceleration.

The middle ear is not an airtight cavity—the auditory tube (called the Eustachian tube in mammals), derived from the spiracle of fish, connects the pharynx to the middle ear. Air flows into or out of the middle ear as air pressure changes. The Eustachian tubes sometimes become blocked. When that happens, changes in external air pressure can produce a painful sensation in addition to reduced auditory sensitivity. Anyone who has had a bad cold while traveling in an airplane knows about this.

The enclosed middle ear of tetrapods, with an eardrum (tympanum), has evolved convergently several times, although in each case the stapes (the old fish hyomandibula,

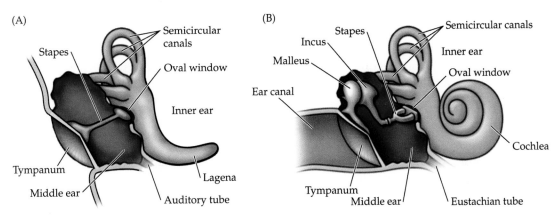

Figure 12.13 Anatomy of the middle and inner ear of tetrapods. (A) In sauropsids, vibrations received by the tympanum (eardrum) are transmitted through the middle ear by a single bone, the stapes. (B) In synapsids, vibrations are transmitted by the ossicular chain of three bones, the malleus, incus, and stapes. The tympanum is on the surface of the head in some sauropsids, and at the end of a short ear canal in others. All synapsids have relatively long ear canals. Nearly all extant synapsids have a pinna (external ear) that helps to channel sound to the ear canal.

often called the columella in non-mammalian tetrapods) transmits vibrations between the tympanum and inner ear (contained within the otic capsule). Modern amphibians have an organization of the inner ear that is different from that of amniotes, indicating an independent evolution of terrestrial hearing, and the anatomical evidence shows that an enclosed middle ear has evolved independently in mammals and other amniotes. Even in non-mammalian amniotes, differences in anatomy suggest that evolution of a fully enclosed middle ear occurred independently in turtles, lizards, and archosaurs, and that the definitive closed middle ear of modern mammals occurred independently in monotremes and therians.

Olfaction

Volatile odor molecules generally travel faster in air than in water, making olfaction an effective rapid-response distance sense for terrestrial vertebrates. Nonvolatile molecules deposited on objects in the environment are used to mark territories. The olfactory receptors responsible for the sense of smell are located in the olfactory epithelium in the nasal passages of tetrapods, and air passes over the olfactory epithelium with each breath. The receptors can be extraordinarily sensitive, and some chemicals can be detected at concentrations below 1 part in 1 million trillion (10^{15}) parts of air.

Among tetrapods, mammals probably have the greatest olfactory sensitivity; the area of the olfactory epithelium in mammals is increased by the presence of scrolls of thin bone called **ethmoturbinates** (Figure 12.14A). The turbinates of birds, which are analogous to those of mammals but probably not homologous, are formed of cartilage. Primates, including humans, have a relatively poor sense of smell because our snouts are too short to accommodate large turbinates and an extensive olfactory epithelium.

Tetrapods (except birds) have an additional chemosensory system located in a unique organ of olfaction in the anterior roof of the mouth—the **vomeronasal organ**. When snakes flick their tongues in and out of their mouth, they are capturing molecules in the air and

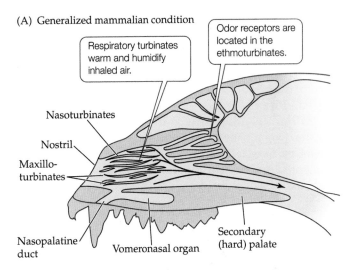

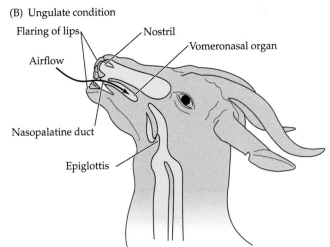

Figure 12.14 Olfactory system of a mammal. (A) Generalized mammal, showing the positions of the respiratory (naso- and maxilloturbinates) and olfactory turbinates (ethmoturbinates) and the vomeronasal organ. (B) The flehmen behavior of an ungulate: With the lips flared and the head lifted, a sharp inhalation draws air through the nasopalatine duct into the vomeronasal organ. (A, after Hillenius 1992.)

transferring them to this organ. Many male ungulates (hoofed mammals) sniff or taste the urine of a female, a behavior that permits them to determine the stage of her reproductive cycle. This sniffing is usually followed by "flehmen," a behavior in which the male curls his upper lip and often holds his head high, probably inhaling molecules of pheromones into the vomeronasal organ (**Figure 12.14B**). Primates, with their relatively flat faces, were thought to have lost their vomeronasal organs, but some recent work suggests the presence of a remnant of this structure in humans that is used for pheromone detection.

Proprioception

Aquatic vertebrates don't have long appendages, and their appendages have relatively little range of movement in relation to the body. Their heads are attached to their pectoral girdles, and their fins move either from side to side or forward and back. That is not true of terrestrial animals, which have necks and limbs that can move in three dimensions with respect to the body. It is important for a terrestrial vertebrate to know where all the parts of its body are, and **proprioception** provides that information. (It is the proprioceptors in your arm that enable you to touch your finger to your nose when your eyes are closed.) Proprioceptors include muscle spindles, which detect the amount of stretch in the muscle, and tendon organs, which convey information about the position of the joints. Muscle spindles are found only in the limbs of tetrapods, and they are important for determining posture and balance on land.

12.6 Conserving Water in a Dry Environment

Aquatic vertebrates are surrounded by water, but sources of water for terrestrial vertebrates are often scattered in time and space. Thus, gaining water and limiting loss of water are important considerations for terrestrial tetrapods. In air, water is evaporated from the body surface and respiratory system as water vapor and is lost through the kidneys and in feces as liquid water. The permeability of the skin of terrestrial vertebrates depends on its structure and varies from very high in most extant amphibians to very low in most amniotes.

The epidermal cells of vertebrates synthesize keratin (Greek *keras*, "horn"), an insoluble protein that ultimately fills those cells. The outer layer of the skin of vertebrates is composed of layers of keratinized epidermal cells, forming the stratum corneum (Latin *cornu*, "horn"). The stratum corneum is only a few cell layers deep in fish and amphibians, but it is many layers deep in amniotes. The primary function of these keratinized cells is to protect the skin against physical wear; keratin has some waterproofing effect, but lipids in the skin are the main agents that limit evaporative water loss.

The sauropsid and synapsid lineages developed different solutions to the problem of minimizing water loss through

the kidney: sauropsids excrete nitrogenous wastes primarily as semisolid mixtures of uric acid and ions, whereas synapsids excrete concentrated solutions containing those substances. (The urinary systems of the two groups will be compared in greater detail in Chapter 14.) What sauropsids and synapsids have in common, however, is a urinary bladder—a saclike structure that receives urine from the kidney and voids it to the outside. A bladder is a derived feature of tetrapods, although some bony fishes have a bladderlike extension of the kidney duct. A bladder was probably an ancestral character of tetrapods, and is a site for recovery of water in extant lissamphibians and some sauropsids.

Amniotes have a novel duct draining the kidney—the ureter—derived from the base of the archinephric duct. In most vertebrates the urinary, reproductive, and digestive systems reach the outside through a single common opening, the cloaca (**Figure 12.15**). Only in therian mammals (marsupials and placentals) is the cloaca replaced by separate openings for the urogenital and digestive systems.

The penis is a conduit for urine only in therian mammals. In all other amniotes, it is purely an intromittent organ, used to introduce sperm into the reproductive tract of the female to fertilize the egg before it is encased in a shell.

(A)

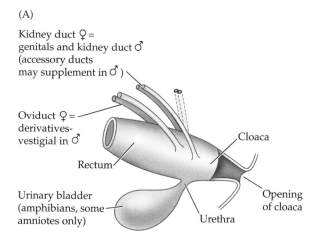

(B)

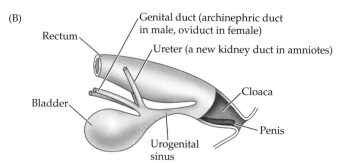

Figure 12.15 Anatomy of urogenital ducts of tetrapods. (A) Generalized condition in jawed vertebrates, including non-amniote tetrapods and some basal amniotes. (B) More derived amniote condition (except for therian mammals), as illustrated by a male monotreme. The urogenital sinus is the name given to the urethra past the point where the genital duct has joined the system. In this example, the entire structure between the bladder and the cloaca is the urogenital sinus, but this is variable in different species of amniotes.

An animal on land is in a physical environment that varies over small distances and can change rapidly. Temperature, in particular, varies dramatically in time and space in terrestrial environments, and changes in environmental temperature have a direct impact on terrestrial animals, especially small ones, because they gain and lose heat rapidly.

This difference in aquatic and terrestrial environments results from differences in the physical properties of water and air. The heat capacity of water is high, as is its ability to conduct heat. As a result, water temperatures are relatively stable and do not vary much at a given depth. A small aquatic animal has little capacity to maintain a body temperature different from the water temperature or to change its body temperature by moving from place to place.

As soon as an animal moves out onto the shore, however, it encounters a patchwork of warm and cool spots, with temperatures that may differ by several degrees within a few centimeters. Terrestrial animals can select favorable temperatures within this thermal mosaic, and the low heat conductivity of air means the animals can maintain body temperatures that are different from the air temperature.

In fact, **thermoregulation** (regulating body temperature) is essential for most tetrapods because they encounter temperatures that are hot enough to kill them or cold enough to incapacitate them. In general, tetrapods maintain body temperatures that are higher than the air temperature, and to do this they need a source of heat. The heat used to raise the body temperature to levels permitting normal activity can come from the chemical reactions of metabolism (**endothermy**) or from basking in the sun or being in contact with a warm object such as a rock (**ectothermy**). The convergent evolution of endothermy in synapsid and sauropsid lineages will be discussed in Chapter 14; here we focus on ectothermy, which is the ancestral form of thermoregulation by tetrapods.

Ectothermy

Ectothermy is the method of thermoregulation used by nearly all non-amniotes and by turtles, lepidosaurs, and crocodylians. Despite the evolutionary antiquity of ectothermy, it is a complex and effective way to control body temperature. Ectothermy is based on balancing the movement of heat between an organism and its environment. Many lizards can maintain stable body temperatures with considerable precision and have body temperatures very similar to those of birds and mammals.

A brief discussion of the pathways by which thermal energy moves between an organism and its environment is necessary to understand the thermoregulatory mechanisms employed by terrestrial ectotherms and how thermoregulation interacts with other activities. Ectotherms and endotherms gain or lose energy by several pathways: solar radiation, infrared (thermal) radiation, convection, conduction, evaporation, and metabolic heat production. Adjusting the flow of energy through these pathways allows an animal to warm up, cool down, or maintain a stable body temperature.

Figure 12.16 illustrates pathways of thermal energy exchange. Heat comes from both internal and external sources, and the flow in some of the pathways can be either into or out of an organism.

- For most ectotherms, the sun is the primary source of heat, and solar energy always results in heat gain. Solar radiation reaches an animal directly when it is in a sunny spot. In addition, solar energy is reflected from clouds and dust particles in the atmosphere and from other objects in the environment, and reflected solar radiation reaches the animal by these pathways. The wavelength distribution of solar radiation is the portion of the solar spectrum that penetrates Earth's atmosphere. About half this energy is in the visible portion of the spectrum (roughly 390–750 nanometers, nm), and most of the rest is in the infrared region of the spectrum (wavelengths longer than 750 nm).

- Infrared radiation is an important part of the thermal exchange. All objects, animate or inanimate, radiate energy at wavelengths determined by the absolute temperature of their surface. Objects in the temperature range of animals and Earth's surface (roughly –20°C to +50°C) radiate in the infrared portion of the spectrum. Animals continuously radiate energy to the environment and receive infrared radiation from the environment. Thus, infrared radiation can lead to either gain or loss of heat, depending on the relative temperature of the animal's body surface and the environmental surfaces, as well as on the radiation characteristics of the surfaces themselves. In Figure 12.16, the lizard is cooler than the sunlit rock in front of it and receives more energy from the rock than it loses to the rock. However, the lizard is warmer than the shaded rock behind it and has a net loss of energy in that exchange. The radiative temperature of a clear sky is about 20°C, so the lizard loses energy by radiation to the sky.

- Heat is exchanged between objects in the environment and the air via convection—the transfer of heat between an animal and a fluid. Convection can result in either gain or loss of heat. If the air temperature is lower than an animal's surface temperature, convection leads to heat loss—in other words, it's a cooling breeze. If the air is warmer than the animal, convection results in heat gain. In still air, convective currents formed by local heating produce convective heat exchange. When the air is moving—that is, when there is a breeze—forced convection replaces natural convection, and the rate of heat exchange is greatly increased. In Figure 12.16, the lizard is warmer than the still air and loses heat by convection.

- Heat exchange by conduction occurs where the body and the substrate are in contact—the transfer of energy between an animal and a solid material. Conductive heat exchange resembles convection in that its direction depends on the relative temperatures of

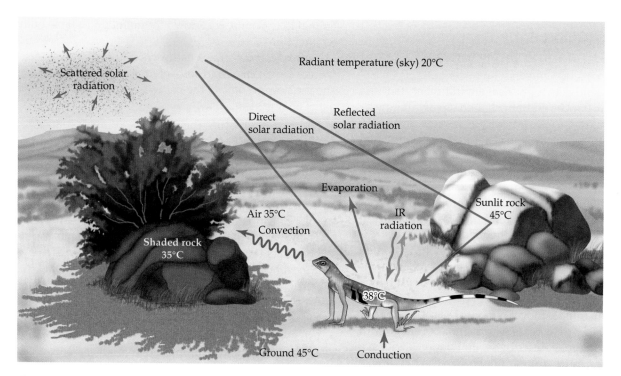

Figure 12.16 Pathways of energy exchange for a terrestrial animal. Energy is exchanged between a terrestrial organism and its environment via several pathways. These are illustrated in simplified form by a lizard resting on the floor of a desert arroyo. The direction of energy flow is indicated by the arrowheads. Small adjustments of posture or position can change the magnitude and even the direction of energy exchange in the various pathways, giving a lizard considerable control over its body temperature.

the animal and the environment. Conduction can be modified by changing the surface area of the animal in contact with the substrate and by changing the rate of blood flow in the parts of the animal's body that are in contact with the substrate. In Figure 12.16, the lizard gains heat by conduction from the warm ground.

- Evaporation of water occurs from the body surface and from the pulmonary system. Each gram of water evaporated represents a loss of about 2,450 joules (J) (the exact value changes slightly with the temperature). Evaporation of water transfers heat from the animal to the environment and thus represents a loss of heat. (The inverse situation—condensation of water vapor on an animal—would produce heat gain, but it rarely occurs under natural conditions.)

- Metabolic heat production is the final pathway by which an animal can gain heat. Among ectotherms, metabolic heat gain is usually trivial in relation to the heat gained directly or indirectly from solar energy.

Endothermy

Endotherms (birds and mammals for the purpose of this discussion) exchange energy with the environment by the same routes as ectotherms. Everyone has had the experience of getting hot in the sun (direct solar radiation) and starting to sweat (evaporation). When you were in that situation you probably moved into the shade, and that is the same behavioral thermoregulatory response that a lizard would use.

What is different about endotherms is the magnitude of their metabolic heat production. Endotherms have metabolic rates that are seven to ten times higher than those of ectotherms of the same body mass. During cellular metabolism, chemical bonds are broken, and some of the energy in those bonds is captured in the bonds of other molecules, such as adenosine triphosphate (ATP). Metabolism is an energetically inefficient process, however, and only a portion of the energy released when a bond is broken is captured—the rest is released as heat. This wasted energy from metabolism is the heat that endotherms use to maintain their body temperatures at stable levels.

Ectothermy, endothermy, and heterothermy

Neither ectothermy nor endothermy can be considered the better mode of thermoregulation, because each one has advantages and disadvantages:

- By producing heat internally, endotherms gain considerable independence from environmental temperatures. Endotherms can live successfully in cold climates and can be nocturnal in situations that would not be possible for ectotherms. These benefits come at the cost of high energy (food) requirements, however.

- Ectotherms save energy by relying on solar heating, and an ectotherm eats less than an endotherm with the same body mass. Because of that difference, ectotherms can live in places that do not provide enough energy to sustain an endotherm.

Endothermy and ectothermy are not mutually exclusive modes of thermoregulation, and many animals use them in combination. **Heterothermy** refers to variation in body temperature. Day-to-night variation in core body temperature is normal for birds and mammals; human rectal temperature has a daily cycle of 1.5–2°C. Some endotherms become torpid or hibernate. These species of birds and mammals allow their body temperatures to fall for periods ranging from hours to weeks before returning to the previous high set point—a phenomenon called **temporal heterothermy**.

Localized heat production or loss can lead to spatial variation in the temperature of tissues in different parts of the body, a phenomenon called **regional heterothermy**. Several large, active species of fishes are regional heterotherms, and during the winter the legs and feet of mammals and birds can be substantially colder than the core of the body.

In general, birds and mammals are primarily endothermal, but 5 species of birds and more than 20 species of mammals bask in the sun to warm themselves in the morning. For example, fat-tailed dunnarts (*Sminthopsis crassicaudata*) are small (10 g) Australian marsupials that become torpid at night during the fall and winter and bask to rewarm in the morning (**Figure 12.17A**). Using solar radiation to rewarm reduces their daily energy expenditure by about 25%.

The high surface area to volume ratios of newly weaned mammals may make basking a necessity rather than an energy-saving for some species. Newly weaned dunnarts must bask to maintain normal body temperatures during the first days or weeks after they leave their mother's pouch, and similar reliance upon basking is probably more common than we realize.

Snakes are normally ectothermal, but the females of several species of pythons coil around their eggs and produce heat by rhythmic contraction of their trunk muscles. The rate of contraction increases as air temperature falls, and a female Burmese python (*Python bivittatus*) is able to maintain her body temperature and the temperature of her

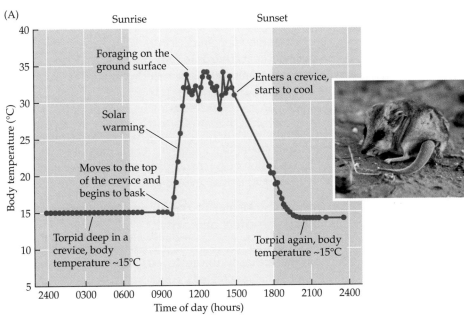

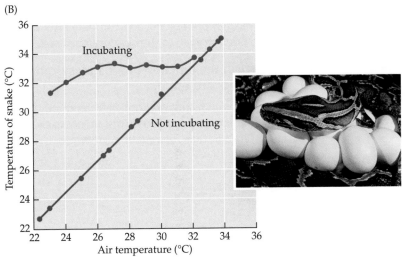

Figure 12.17 Ectothermal endotherms and endothermal ectotherms. Endotherms sometimes use external sources of heat to raise their body temperature, and ectotherms sometimes use metabolic heat production. (A) Fat-tailed dunnarts (*Sminthopsis crassicaudata*) spend fall and winter nights deep in crevices in the soil and allow their body temperature to fall to the temperature of the soil—about 15°C. In the morning (around 0900 h on the day shown here), a dunnart, its body temperature still near 15°C, climbs to the top of the crevice and basks in the sun; by 1100 h its body temperatures has reached between 30°C and 35°C. At this time, the dunnart begins to forage, and its body temperature rises and falls as it moves between sun and shade. Late in the day (at around 1500 h here), the dunnart moves back into a crevice and its body temperature cools to 15°C for the night. (B) When a female Burmese python (*Python bivittatus*) is incubating eggs, muscular contractions allow her to maintain the temperature of her body (and of her eggs) between 32°C and 34°C. When the python is not incubating, her body temperature does not differ from the ambient temperature. (A from Warnecke et al. 2008, photo © Chris Watson/Shutterstock; B from Van Mierop and Barnard 1978, photo © Paul Tessier/123RF.)

eggs at 31°C or above at air temperatures as low as 23°C (Figure 12.17B). Incubation speeds development of the embryos and avoids the developmental defects that occur at low temperatures, but it entails a substantial increase in the python's metabolic rate. At 23°C, a female python uses about 20 times as much energy when she is incubating as she does when she is not incubating.

Additional examples of endothermy among generally ectothermal lineages, ectothermy in endothermal lineages, and heterothermy will be described in later chapters. Generalizations about the body temperatures and thermoregulatory physiology of vertebrates must be made cautiously, and the actual mechanisms used to regulate body temperature must be determined.

Summary

Living on land differs from living in water in a host of ways because the physical properties of water and air are so different.

Aquatic vertebrates are nearly neutrally buoyant in water, so gravity is only a small factor in their lives. The vertebral column of a fish needs to resist only lengthwise compression as trunk muscles contract. In contrast, the skeleton of a terrestrial tetrapod has to support the animal's weight.

Zygapophyses on the vertebral column of terrestrial tetrapods transmit forces from one vertebra to the next, resisting the pull of gravity. In all but the earliest terrestrial tetrapods the vertebral centrum is a single, solid structure and the notochord is only present during embryonic development (although it contributes to the intervertebral discs in the adult).

The cranial skeleton of fishes is connected to the shoulder girdle via dermal bones, but these connecting bones are lost in tetrapods, which now have a mobile neck. The vertebral column of tetrapods is divided into cervical (neck), trunk, and caudal (tail) regions.

Ribs are found primarily on the trunk vertebrae, but small cervical ribs may be present in non-mammalian tetrapods. Mammals have divided their trunk vertebrae into thoracic (bearing ribs) and lumbar (without ribs) sections.

Tetrapods have specialized their axial musculature. In fishes, axial muscles are only used for locomotion, but in tetrapods they are also used for posture and lung ventilation. All tetrapods have differentiated hypaxial muscles, and amniotes have additionally differentiated their epaxial ones.

The limbs of a terrestrial tetrapod lift the body off the ground and push against the substrate as the animal moves. Tetrapod limbs differ from fish fins in having hands and feet with digits, and a pelvic girdle with ilia that connect to the vertebral column via one or more specialized vertebrae called sacral vertebrae. The pectoral girdle lacks a bony connection to the vertebral column (except in some extinct flying reptiles).

The weight of an organism increases as the cube of its linear dimensions, but the strength of an organism's bones increases only as the square of their cross-sectional area.

Because of this relationship, the limb bones of large terrestrial tetrapods are proportionally thicker than the limb bones of smaller tetrapods, and large tetrapods stand more erect than small ones.

Air has lower density, lower viscosity, and higher oxygen content than water.

Suction feeding is ineffective in air. Tetrapods use their mobile necks and heads to seize prey, and their muscular tongues to manipulate prey in the mouth.

Tidal respiration is energetically feasible for air-breathing vertebrates.

Non-amniote tetrapods use the buccal cavity to create positive pressure that forces air into the lungs. In amniotes, expansion of the rib cage creates a negative pressure in the abdominal cavity that draws air into the lungs.

The cardiovascular system of terrestrial tetrapods reflects the effect of gravity.

Venous blood must be forced upward from regions of the body that are below the heart. Tetrapods have high blood pressures that are created by thick-walled muscular hearts and valves in the vessels of the limbs prevent backflow.

In crocodylians, birds, and mammals, which have a permanent ventricular septum, the right side of the heart receives only oxygen-poor blood, and coronary arteries are needed to bring oxygen to the heart muscle itself.

The difference in the function of sensory systems in air and water is profound.

Light generally travels farther through air than through water, making vision a more effective distance sense for terrestrial vertebrates than for aquatic vertebrates.

Summary (continued)

In terrestrial vertebrates, the cornea participates in focusing light on the retina. In all terrestrial tetrapods except snakes, an image is focused by changing the shape of the lens rather than by moving the entire lens, as in fish.

Air is not dense enough to activate the cilia of hair cells. The hair cells of terrestrial vertebrates are found in the cochlea of the inner ear, and a lever system in the middle ear amplifies sound-pressure waves.

Olfaction is often an effective distance sense for terrestrial tetrapods, and nonvolatile molecules are used to mark objects in the environment.

Terrestrial tetrapods have long appendages, and proprioceptors provide information about the positions of the appendages relative to the body.

Evaporative water loss is a challenge in terrestrial environments.

In sauropsids and synapsids, lipids in the skin are the primary barriers to evaporative water loss.

The kidney and urine of sauropsids and synapsids represent different solutions to minimizing the amount of water needed to excrete metabolic waste products.

The bladder, a site for storing urine, is a derived character of tetrapods that has been lost in some lineages.

The urinary, reproductive, and digestive systems of most tetrapods discharge into a common opening, the cloaca. Only marsupial and placental mammals have separate openings for the urogenital and digestive systems.

Air temperature is more variable than water temperature.

Most terrestrial tetrapods encounter lethally hot and cold temperatures, and methods of regulating body temperature are crucial.

Two types of thermoregulation, ectothermy and endothermy, lie at the ends of a continuum and are not mutually exclusive. Many terrestrial tetrapods employ elements of both modes of thermoregulation.

The routes of energy gain and loss by terrestrial ectotherms and endotherms are the same: They gain or lose heat via solar radiation, infrared radiation, convection, conduction, evaporation, and metabolic heat production.

Ectothermy (using external sources of heat to raise body temperature) is the ancestral mode of thermoregulation for tetrapods. Endothermy, using heat from metabolism to warm the body, evolved independently in the sauropsid and synapsid lineages.

Ectothermy is energetically efficient, but it works only when external sources of heat are available. Endothermy provides greater independence but at a high energetic cost.

Both ectotherms and endotherms may be heterothermal—having body temperatures that vary from time to time or from one region of the body to another.

Discussion Questions

1. Why is bone such a useful structural material for tetrapods on land? Calcified cartilage is lighter than bone. Isn't that an advantage for a terrestrial vertebrate?

2. Why would we expect secondarily marine tetrapods (such as whales) to have lost the zygapophyses that interconnect vertebrae?

3. How would you determine whether an isolated vertebra of a mammal came from the thoracic region or the lumbar region?

4. We usually think of our ribs as being used for breathing. How is this different from their original function in fish, or even their original function in early tetrapods?

5. How can we be almost certain that the earliest tetrapods had a transversus abdominis component of their hypaxial musculature, as in extant tetrapods?

6. Why do you think it might be primarily mammals that evolved gaits (such as the bound) that involved dorsoventral (versus lateral) flexion of the backbone?

7. Why do large mammals stand with less-flexed joints than smaller mammals do?

Additional Reading

Ashley-Ross MA, Hsieh ST, Gibb AC, Blob RW. 2013. Vertebrate land invasions—past, present, and future: An introduction to the symposium. *Integrative and Comparative Biology* 53: 192–196.

Biewener AA. 2005. Biomechanical consequences of scaling. *Journal of Experimental Biology* 208: 1665–1676.

Clack JA. 2012. *Gaining Ground: The Origin and Evolution of Tetrapods,* 2nd Ed. Indiana University Press, Bloomington.

Farmer C. 1997. Did lungs and the intracardiac shunt evolve to oxygenate the heart in vertebrates? *Paleobiology* 23: 358–372.

Lillywhite HB. 2006. Water relations of tetrapod integument. *Journal of Experimental Biology* 209: 202–226.

McInroe B and 8 others. 2016. Tail use improves performance on soft substrates in models of early vertebrate land locomotors. *Science* 353: 154–158.

Michel KB, Heiss E, Aerts P, Van Wassenbergh S. 2015. A fish that uses its hydrodynamic tongue to feed on land. *Proceedings of the Royal Society B* 282: 20150057.

Neenan JM, Ruta N, Clack JA, Rayfield EJ. 2014. Feeding biomechanics in *Acanthostega* and across the fish–tetrapod transition. *Proceedings of the Royal Society B* 281: 20132689.

Reilly SM, McElroy EJ, Odum RA, Hornyak VA. 2006. Tuataras and salamanders show that walking and running mechanics are ancient features of tetrapod locomotion. *Proceedings of the Royal Society B* 273: 1563–1568.

Ruf T, Geiser F. 2015. Daily torpor and hibernation in birds and mammals. *Biological Reviews* 90: 891–926.

Sayer MDJ. 2005. Adaptations of amphibious fish for surviving life out of water. *Fish and Fisheries* 6: 186–211.

Wacker C, McAllen BM, Körtner G, Geiser F. 2017. The role of basking in the development of endothermy and torpor in a marsupial. *Journal of Comparative Physiology B*: DOI 10.1007/s00360-017-1060-2.

Wright PA, Turko AJ. 2016. Amphibious fishes: evolution and phenotypic plasticity. *Journal of Experimental Biology* 219: 2245–2259.

Go to the Companion Website **oup.com/us/vertebratelife10e** for active-learning exercises, news links, references, and more.

Geography and Ecology of the Mesozoic Era

The Mesozoic is famous for its dinosaurs, but all of the extant lineages of tetrapods had appeared well before the end of that era. Although their diversity increased, the rate of diversification was relatively slow, and the number of tetrapod species only doubled or tripled during the Mesozoic. The appearance and rapid radiation of angiosperms (flowering seed plants) during the Cretaceous were important floral changes.

Two lineages of tetrapods diversified during the Mesozoic:

- Synapsida (the lineage that includes extant mammals) had radiated during the Permian into a variety of carnivorous and herbivorous therapsids. In the Early Triassic, synapsids were the most abundant and diverse terrestrial tetrapods, but by the end of Mesozoic the lineage was represented only by basal mammals. Synapsids were not part of the marine fauna during the Mesozoic.

- Sauropsida (the lineage represented today by reptiles, including birds) included a diverse array of lineages: turtles, archosaurs (the lineage that includes extant crocodylians and birds as well as nonavian dinosaurs), and lepidosaurs (represented by the extant tuatara, lizards, and snakes). By the middle of the Triassic, sauropsids had replaced synapsids as the most abundant and diverse terrestrial tetrapods and had radiated into a variety of marine forms.

The details of how each of these lineages evolved to cope with life on land will be covered in Chapter 14.

Dinosaurs are often depicted as inhabiting a strange world, but in fact the Mesozoic ecosystems were not so different from those of today. The main difference was that the adaptive zones for large vertebrates were occupied by dinosaurs rather than by mammals. A mass extinction at the end of the Triassic established the dinosaur-dominated faunas of the later Mesozoic. Another mass extinction at the end of the Cretaceous is best known for the demise of nonavian dinosaurs, but many other types of organisms were also affected.

13.1 Continental Geography of the Mesozoic

By the early Mesozoic, Earth's entire land surface had coalesced into a single continent, Pangaea, that stretched from pole to pole (**Figure 13.1A**). Triassic faunas and floras showed some regional differentiation due to climate but had no oceanic barriers to dispersal. The fragmentation of Pangaea began in the Jurassic with the separation of Laurasia (the northern portion of Pangaea) and Gondwana (the southern portion) by a westward extension of the Tethys Sea. Laurasia rotated away from the other continents, ripping North America from its connection with South America and increasing the size of the newly formed Atlantic Ocean (**Figure 13.1B**).

Separation of the continents and rotation of the northern continents continued during the Jurassic and into the Cretaceous. Epicontinental seas spread during the Early Cretaceous and receded toward the end of the period. By the Late Cretaceous, continents were approaching their current positions. India was still close to Africa, but Madagascar had split from India, while Australia, New Zealand, and New Guinea, conjoined in the single landmass of Australasia, were still well south of their present-day positions (**Figure 13.1C**).

With the breakup of Pangaea in the Jurassic and Cretaceous, floras and faunas were geographically isolated and became distinct in different parts of the world. However, in the relatively warm world of the Mesozoic, there was little of the latitudinal zonation of floras and faunas that characterizes today's world.

13.2 Mesozoic Climates

Temperatures rose suddenly and profoundly at the end of the Paleozoic, as did levels of atmospheric carbon dioxide (see Figure 5.3). Temperatures remained high throughout the Mesozoic, apart from a dip in the mid-Jurassic that lasted into the Early Cretaceous. Carbon dioxide levels started

(A) Triassic

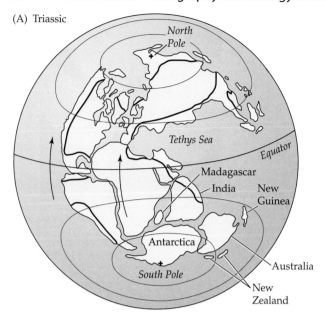

(B) Jurassic

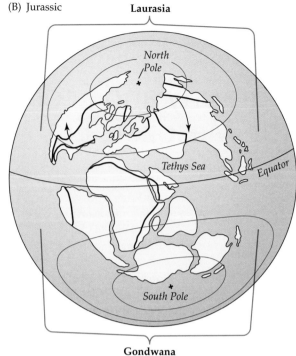

(C) Cretaceous

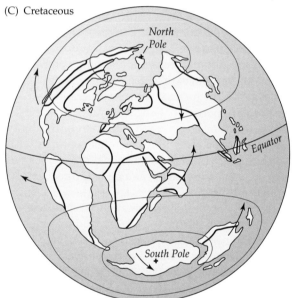

Figure 13.1 Continental blocks in the Mesozoic. Continental land masses are shown in beige, with epicontinental seas in light blue. Arrows indicate the direction of continental rotation. (A) Land surface in the early Triassic was a single mass, Pangaea, with no oceanic barriers to dispersal. (B) Westward expansion of the Tethys Sea began the separation of Laurasia and Gondwana. (C) The continents approached their current positions over the course of the Cretaceous, but with vast continental areas covered by epicontinental seas for much of the period.

extending into the polar regions. Arctic faunas contained a diversity of dinosaurs and large, aquatic sauropsid predators (champsosaurs) that would have required a temperate or subtropical climate. A reduction in the diversity of plants at higher latitudes at the end of the Cretaceous signaled an episode of worldwide cooling, with several periods of abrupt warming and cooling that destabilized marine and terrestrial ecosystems.

13.3 Mesozoic Aquatic Life

An increase in marine nutrients derived from the weathering of rocks on land led to a tremendous burst of animal life in the seas during the Mesozoic. A bloom of planktonic organisms that began in the Late Triassic and accelerated through the rest of the Mesozoic provided the base for the food chain that supported the radiation of large marine animals.

Mesozoic fish radiations occurred mainly among elasmobranch cartilaginous fishes and actinopterygian bony fishes. Chondrichthyans were the most abundant and diverse fishes in the late Paleozoic, but in the Mesozoic neopterygian osteichthyans assumed that position. Teleosts appeared and radiated in the Late Triassic, and neoselachian elasmobranchs radiated in the Jurassic. Fishes seem to have been relatively

to fall at the start of the Cretaceous, but were still high in comparison with present day levels.

Mesozoic climates were similar worldwide, without the type of latitudinal zonation seen today and with no polar ice caps at any time during the era. Fossils of large temnospondyls (aquatic non-amniote tetrapods) are found in high-latitude Triassic deposits, and coal deposits in both the Northern and Southern hemispheres point to moist climates. In contrast, low and middle latitudes were probably dry—either seasonally or year-round—until the Late Cretaceous and early Cenozoic, when coal deposits in middle latitudes indicate the presence of swamps and suggest that the climate had become wetter. These dry lower latitudes had a type of vegetation different from the equatorial vegetation of today, with the absence of wet tropical rainforests.

The Cretaceous plant record also suggests a relatively equable world for most of the period, with temperate forests unaffected by the end-Triassic and end-Cretaceous extinctions, although the end-Cretaceous extinction appears to have affected marine lineages more than freshwater ones.

All of the Mesozoic marine tetrapods were sauropsids, including ichthyosaurs, plesiosaurs, mosasaurs, placodonts, and three lineages of marine crocodylomorphs. None of these survived into the Cenozoic, but the lineages of marine sauropsids that include sea turtles and sea snakes also first appeared in the Mesozoic and are still around today. Two lineages of flying sauropsids evolved—pterosaurs in the Triassic and birds in the Jurassic. One lineage of Cretaceous birds became marine swimmers (although not as underwater-adapted as penguins), and some pterosaurs became adapted to fish from the sea, like seagulls or pelicans.

13.4 Mesozoic Terrestrial Ecosystems

The Mesozoic was marked by a series of large-scale changes in flora and fauna. Herbivory had been established among vertebrates and insects in the late Paleozoic, and tropical terrestrial ecosystems had achieved an essentially modern form by the end of the Permian, with a broad base of herbivores supporting several levels of carnivores (**Figure 13.2**).

The Triassic

The Triassic world was ice-free, with a broad equatorial arid zone. Vertebrates were confined to northern and southern temperate belts, resulting in different floral and faunal assemblages in the northern and southern regions that later separated as Laurasia and Gondwana. The end-Permian extinctions had left an impoverished terrestrial fauna and flora. The first 10 million years of the Triassic appear to have been a barren time, with an absence of forests on land and an absence of corals in the sea. Recovery to pre-extinction diversity required 10–15 million years.

The climate of the Early Triassic (252–247 Ma) was unstable. Plant fossils show two episodes of ecosystem

collapse within the first million years. Three or four episodes of potentially lethal global warming, accompanied by lower levels of O_2, high levels of rainfall and high levels of atmospheric CO_2, were interspersed with cooler and drier periods with lower levels of CO_2.

Early Triassic vertebrate faunas were dominated by the herbivorous dicynodont therapsid *Lystrosaurus* as well as by carnivorous therapsids and basal archosaurs. Although dinosaurs did not diversify until the end of the Triassic, they are first known from the beginning of the Middle Triassic, and footprints hint at an even earlier occurrence.

You saw in Chapter 5 that low levels of atmospheric oxygen in the Late Permian may have contributed to heightened levels of extinction. Oxygen levels continued to drop during the Triassic, reaching a low of 15% by the middle of the period, and did not approach present-day levels (around 21%) until the Late Cretaceous (see Figure 5.3). These low oxygen levels may help explain patterns of Triassic faunal turnover. The previously successful therapsids were now overshadowed by the rise of pseudosuchians (the lineage of archosaurs that includes the extant crocodiles). In the low oxygen levels of the Triassic, the flow-through respiratory

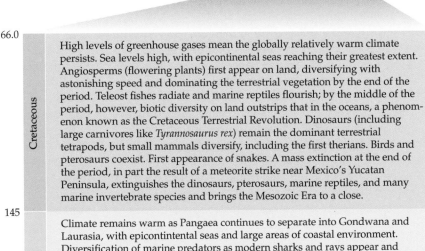

Figure 13.2 Key events during the Mesozoic. The Mesozoic was marked by large-scale changes in both flora and fauna, including the appearance and rapid radiation of angiosperms (flowering plants) and the diversification of herbivorous forms in many animal lineages.

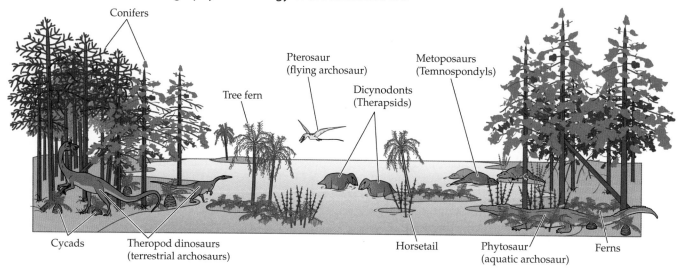

Figure 13.3 A scene from the Late Triassic. Based on fossils excavated from sites in what is now New Mexico, this reconstruction depicts the early part of the Late Triassic, a time that saw the first appearance of the dinosaurs (small, carnivorous theropods).

system of archosaurs may have been superior to the tidal-flow respiratory systems of therapsids.

Triassic vegetation included familiar modern groups of gymnosperms, such as conifers (pines and other cone-bearing trees), ginkgophytes (relatives of the extant ginkgo tree), cycads, and the now-extinct seed ferns. Other plants included ferns, tree ferns, and horsetails, all of which survive today but are less diverse than they were in the Mesozoic. Various kinds of insects first appeared in the Triassic, including cockroaches, aphids, moths, dipterans (flies), and stick insects. Beetles, first known from the Late Carboniferous, further radiated at this time.

Middle Triassic herbivorous vertebrates (including cynodont and dicynodont therapsids, pseudosuchians, and archosaur-related rhynchosaurs) ranged in body mass from 10 to nearly 1,000 kg (goat- to buffalo-size). All of these were generalized browsers that would have foraged within a meter of the ground (**Figure 13.3**). Arboreal vertebrates were rare; there were no arboreal herbivores, but there were several insectivorous sauropsids, including gliding forms.

A shift in plant communities occurred during the Late Triassic, especially in Gondwana, where conifers replaced seed ferns as the dominant plants. The fossil record shows faunal changes as well, with no evidence of simple competitive replacements (see Chapter 19).

Basal mammals (tiny insectivorous forms) made a first appearance in the Late Triassic, as did other extant groups such as sphenodontids (the sister group of snakes and lizards), turtles, and crocodyliforms (the group that now includes the semiaquatic extant crocodiles, but in the Mesozoic included many terrestrial forms, as well as fully aquatic marine forms), and now-extinct groups such as pterosaurs (flying archosaurs).

The Jurassic

The conifer- and fern-dominated vegetation of the Late Triassic continued into the Jurassic, although the diversity of seed ferns declined. Several new kinds of insects appeared in the Jurassic, including lepidopterans (butterflies and moths), and fleas that probably fed on feathered dinosaurs.

A major change occurred in terrestrial ecosystems during the Jurassic with the appearance of long-necked sauropod dinosaurs that could browse at treetop height. These large herbivores were prey for large carnivorous dinosaurs, which also appeared in the Jurassic. Ornithischian dinosaurs (such as stegosaurs and ornithopods) browsed in the 1- to 2-m height range. Although cause-and-effect links between changes in vegetation and dinosaur radiation during the Mesozoic have been proposed, none has stood up to scrutiny.

Jurassic mammals continued as small insectivores, with the addition of small omnivores (multituberculates). Lizards and the modern groups of amphibians first appeared in the Jurassic, and *Archaeopteryx* (considered to be the earliest bird; see Chapter 23) is known from the Late Jurassic.

Sauropod dinosaurs were gigantic, with body masses ranging from 10,000 kg (larger than an elephant) to 50,000 kg or more. The different reproductive strategies of dinosaurs (large number of young) and large mammals (single births or twins) would have led to a Mesozoic landscape dominated by large numbers of juvenile dinosaurs, which were probably the principal prey of carnivores.

The Cretaceous

The Cretaceous world had a greenhouse climate resulting from levels of CO_2 and other greenhouse gases that were 2–16 times higher than those of preindustrial levels of the 18th century. Pole-to-Equator temperature gradients were reduced, and polar regions were relatively warm. Sea levels were 170–250 m above present-day levels, and epicontinental seas extended across parts of the continents. At their height, these seas separated North America into three landmasses and Eurasia into two large regions and numerous smaller ones, split Africa into two masses, and made deep incursions into South America and Australia (see Figure 13.2C).

The vegetation of the Early Cretaceous was similar to that of the Late Jurassic. However, global vegetation had changed drastically by the Late Cretaceous. Angiosperms appeared in the fossil record in the middle of the Early Cretaceous,

Birds

Dinosaurs

Crocodylians

Turtles

Snakes

Lissamphibians

Mammals

Lepidosaurs

3 meters

Figure 13.4 Terrestrial tetrapods in the Late Cretaceous were a mix of the old and the new. The species present at the Lance fossil site in Wyoming and the Bug Creek Anthills site in Montana appear to have inhabited wooded, swampy habitats with large streams and some ponds. All of the extant tetrapod groups are present: mammals, crocodylians, birds, lepidosaurs (lizards and snakes), turtles, and lissamphibians. These had their origins in the Triassic or Jurassic. The number of individuals shown represents the number of genera in each group (e.g., 5 dinosaur genera), based on data in Estes and Berberian 1970. Dinosaurs were only a small portion of the fauna in terms of number of genera, but they made up much of the biomass, as indicated by the relative sizes of members of each group.

about 130 Ma, although molecular phylogenetic data suggest a Jurassic or earlier origin for the group. Angiosperms were the first plants to be pollinated by insects, although the radiation of the insects that feed on angiosperms and pollinate them today, including butterflies and bugs (hemipterans), preceded angiosperm diversification. Other types of insects appearing in the Cretaceous included termites (first known in the latest Jurassic), ants, and hive-forming bees.

The Late Cretaceous radiation of angiosperms, which formed up to 80% of the plants in many fossil assemblages by the end of the Cretaceous, was instrumental in the **Cretaceous Terrestrial Revolution** (**KTR**[1]), which marked the point at which diversity on land outstripped that in the sea.

Angiosperms were present mainly as small trees and low-growing shrubs and herbaceous (i.e., not woody) vegetation. Conifers were still the predominant large trees in the Cretaceous, and remain dominant today in high-latitude forests. Angiosperm-based tropical forest ecosystems may date back to the mid-Cretaceous. Fires may have helped the spread of angiosperms in the later part of the Cretaceous, because angiosperms grow faster than gymnosperms in burned areas. Ferns also underwent a major diversification in the Cretaceous, including lineages growing on the forest floor and other lineages in the tree canopy. Grasses (which are angiosperms) made an initial appearance in the latest Cretaceous but were not an important feature of the landscape until well into the Cenozoic.

Several new types of herbivorous dinosaurs appeared in the Late Cretaceous, including hadrosaurs (duck-billed dinosaurs) and ceratopsians (horned dinosaurs)—both low-level feeders with teeth or horny beaks capable of dealing with tough vegetation; preserved stomach contents reveal a diet of conifers. Other Cretaceous faunal changes included a radiation of birds, the appearance of snakes and modern types of crocodyliforms, and the appearance of mammals belonging to the three modern groups, monotremes, metatherians (stem marsupials), and eutherians (stem placentals). Cretaceous mammals remained small; most were shrew- to mouse-size, and none was larger than a raccoon (**Figure 13.4**). Although both birds and mammals diversified in the Cretaceous, the modern orders and families did not appear until the Cenozoic.

[1] Geologists use K to denote the Cretaceous, to avoid confusion with the Carboniferous.

The epicontinental seas that had penetrated deeply into the continents receded when sea levels fell near the end of the Cretaceous, changing the landscape as rivers eroded valleys in the exposed beds of the inland seas. Simultaneously, global temperatures became unstable, disrupting the hothouse conditions that had endured until this time and stressing ecosystems on a global basis.

13.5 Mesozoic Extinctions

The first round of Mesozoic extinctions occurred at the end of the Triassic and affected both terrestrial and marine vertebrates as well as marine invertebrates. Eighteen families of tetrapods became extinct at the end of the Triassic. The Triassic extinctions coincided with the initial breakup of Pangaea at the Triassic-Jurassic boundary. Two episodes of massive volcanism, which appear to have occurred in two pulses 20,000 years apart, released vast amounts of sulfur dioxide (SO_2, which is toxic to both plants and animals) and CO_2 (leading to rapid global warming and ocean acidification).

Some minor extinctions occurred at the end of the Jurassic, principally affecting larger tetrapods such as dinosaurs, crocodyliforms, pterosaurs, and marine reptiles. About 20% of genera became extinct, followed by a rapid radiation of new forms in the Cretaceous. The Jurassic extinctions appear to have been driven by a change in sea levels, with seas retreating from the continental shelf and causing massive changes in habitats.

The extinctions at the K–Pg (Cretaceous–Paleogene) boundary 66 Ma included the demise of the nonavian dinosaurs. This extinction event was larger than the end-Triassic one, but not nearly as large as the end-Permian extinction when almost all vertebrates larger than 10 kg became extinct. Nonetheless, the K–Pg extinction event was massive: 43% of tetrapod families disappeared.

Extinctions were not evenly distributed. Nonavian dinosaurs, plesiosaurs, and pterosaurs were wiped out by the K–Pg extinctions, and 75% of bird families became extinct. However, extinction rates for crocodyliforms and turtles were only twice baseline rates, and extinction rates of lizards, snakes, and amphibians were indistinguishable from baseline rates. Eutherian mammals (stem placentals) also suffered somewhat elevated extinctions, as did non-therian mammals (including multituberculates and monotremes), but metaterians (stem marsupials) were more profoundly affected (89%). There were extensive extinctions among marine invertebrates, but extinctions of chondrichthyans and osteichthyans were little higher than baseline rates. The angiosperm-dominated flora, which had been expanding throughout the Late Cretaceous, also continued into the Cenozoic relatively undisturbed.

The timescale of the K–Pg extinctions has long been debated: were the extinctions sudden or gradual? The answer appears to be "both." That is, some of the lineages that suffered massive extinctions at the end of the Cretaceous had already been declining for tens of millions of years. For example:

- Ichthyosaurs were not part of the end-Cretaceous extinctions; they had disappeared 30 Ma before the end of the Cretaceous, probably in two waves during the early part of the Late Cretaceous. An initial period of extinction that reduced the diversity and variation in ichthyosaur feeding methods was not followed by the evolution of new lineages to fill those adaptive zones. The final extinction corresponded with an increase in global temperatures that caused a massive disruption in marine food webs.

- Pterosaurs also experienced a dramatic decline in diversity before their final disappearance. By the Late Cretaceous, only two lineages remained, one of which appears to have been extremely rare.

- A growing body of evidence indicates that rates of speciation had slowed and were exceeded by extinction rates in all three dinosaur clades (ornithischians, sauropods, and theropods) tens of millions of years before the end-Cretaceous event. Other scientists dispute this conclusion, however, and the tempo and mode of dinosaur extinction remains a contentious issue.

What finally drove these declining lineages to extinction? The impact of an asteroid about 10 km in diameter has seized the imagination of the public as "the" cause of the extinction of dinosaurs. The site of the impact is marked by the Chicxulub crater, on the northern coast of Mexico's Yucatan Peninsula. This impact would have created a cloud of dust and aerosols, as evidenced by the layer of the rare earth element iridium found in the K-Pg boundary sediments around the world. Scientists hypothesize that this dust cloud would have reduced the intensity of sunlight reaching Earth's surface for several years, resulting in global cooling, decreased rainfall, and greatly reduced rates of primary production.

The drama and simplicity of the asteroid impact hypothesis have made it the favorite explanation in popular literature for the extinction of dinosaurs, and it has found its way into introductory biology texts, but many paleobiologists believe that postulating a single cause oversimplifies a complex event. The diverse pattern of extinction and survival of lineages at the K–Pg boundary points to multiple causes acting on lineages, some of which were already declining. This interpretation considers the Chicxulub impact to be just one among many factors that began several million years before the end of the Cretaceous.

- Sea levels had fallen, draining epicontinental seas and causing rivers to erode east-to-west-running valleys that could have impeded the north-to-south migratory movements of dinosaurs.

- The long-lived hothouse climate of the Cretaceous was interrupted near the K–Pg boundary by a series of cooling episodes, including temperatures that were 6–8°C lower than temperatures during the 100,000 years before the boundary. These temperature changes disrupted ecosystems on a global scale.

- Volcanic eruptions in the Deccan Traps volcanic province of western India that began about 66.3 Ma and continued across the K–Pg boundary released massive quantities of CO_2, Cl, SO_2, and toxic metals and caused a spike in global temperatures that further disrupted ecosystems.

Conservation biologists use the term **extinction vortex** to describe a situation in which positive feedback among multiple processes drives a population into an irreversible downward spiral to extinction. On a paleontological time-scale, dinosaurs, pterosaurs, and other Mesozoic lineages were in an extinction vortex. In the chaos of the Late Cretaceous, the Chicxulub impact and Deccan Traps volcanism were straws added to already stressed backs.

Summary

The Mesozoic was a time of great faunal and floral change.

All extant lineages of tetrapods had appeared by the end of the Mesozoic, including mammals, amphibians, squamates (snakes and lizards), and birds.

The angiosperms (flowering seed plants) that make up most of today's vegetation appeared and diversified in the Cretaceous.

Two lineages of tetrapods diversified during the Mesozoic.

- Synapsids, which had radiated during the Permian period of the Paleozoic, were still the most abundant and diverse terrestrial tetrapods during the Early Triassic.
- Sauropsids gradually replaced synapsids during the Triassic and dominated terrestrial ecosystems during the Jurassic and Cretaceous.

Three mass extinction events shaped the history of the Mesozoic.

The end-Permian extinction that brought the Paleozoic era to a close left a depauperate terrestrial fauna. Recovery initially was slow, with several types of herbivorous synapsids remaining as the dominant tetrapods of the early part of the Mesozoic.

The end-Triassic extinction resulted in establishment of the dinosaur-dominated faunas of the Jurassic and Cretaceous.

The end-Cretaceous (K–Pg) extinction wiped out nonavian dinosaurs and several other lineages, but left other lineages nearly unscathed.

By the start of the Mesozoic, the continents had coalesced into the single supercontinent Pangaea.

Separation into a northern continent (Laurasia) and southern continent (Gondwana) began in the Jurassic. By the Late Cretaceous, the continents were nearing their present-day positions.

Climates of the Mesozoic were relatively equable.

Polar regions were relatively warm and covered by forests. In general, the climate was warm and wet, although lower latitudes were dry for most of the era, and equatorial rainforests were lacking. However, the climate during the Early Triassic was unstable, with periods of lethally high temperatures.

Near the end of the Cretaceous, periods of abrupt warming and cooling disrupted ecosystems on a global scale.

Mesozoic marine life was rich.

Nutrient enrichment of the oceans resulted in a bloom of plankton, supporting radiations of marine invertebrates and the fishes that dominate the oceans today, elasmobranch cartilaginous fishes and teleost bony fishes.

Mesozoic lineages of marine tetrapods were all sauropsids—ichthyosaurs, plesiosaurs, mosasaurs, crocodyliforms, snakes, turtles, and birds.

Terrestrial ecosystems evolved to their present-day form during the Mesozoic.

Early Triassic faunas of therapsids and basal archosaurs included herbivores that browsed within 1 m of the ground.

By the end of the Triassic, many of the groups seen in today's world had appeared: mammals, sphenodontids, turtles, and crocodyliforms. Dinosaurs also first appeared at this time.

These lineages were joined in the Jurassic by lizards, modern types of amphibians, and birds. With the exception of crocodyliforms, members of these lineages were small in body size.

The largest tetrapods during the Jurassic were dinosaurs, including a large radiation of the enormous high-browsing, long-necked sauropods and carnivorous dinosaurs.

Angiosperms first appeared in the middle of the Early Cretaceous, and by the end of the period were the predominant vegetation.

(Continued)

Summary *(continued)*

Along with the new flora, new kinds of insects appeared, including termites, ants, and hive-forming bees. New types of dinosaurs also appeared, including the herbivorous hadrosaurs and ceratopsians and new types of predatory theropods, including tyrannosaurs.

The Mesozoic was punctuated by extinction events at the end of each period.

The end-Triassic extinction was moderately severe and was probably caused by volcanism related to the breakup of Pangaea at the Triassic-Jurassic boundary.

The end-Jurassic extinctions were relatively minor and were probably related to falling sea levels.

The end-Cretaceous extinctions came at a time that was characterized by changes in sea level and in terrestrial topography, climate instability and ecosystem disruption, profound volcanism, and the impact of an asteroid. Patterns of extinction varied greatly among lineages; those that were already in an extinction vortex became extinct, whereas others suffered few or no ill effects.

For Discussion

1. The Mesozoic is commonly thought of as the "Age of Dinosaurs." In what way might that concept be misleading when considering how similar—or dissimilar—Mesozoic ecosystems were to those of today?

2. What is a likely reason so many lineages of tetrapods returned to the sea during the Mesozoic?

3. What might be an environmental reason behind the rise to dominance of the archosaurs during the Triassic?

4. How did the floral changes in the Cretaceous (the radiation of angiosperms) affect the fauna?

5. Several studies indicate a gradual decline in the rate of appearance of new species of dinosaurs toward the end of the Cretaceous. Is this sufficient to account for their extinction?

Additional Reading

Archibald JD and 28 others. 2010. Cretaceous extinctions: Multiple causes. *Science* 328: 973.

Benson RBJ and 5 others. 2016. Near-stasis in the long-term diversification of Mesozoic tetrapods. *PLoS Biology* 14(1): e1002359.

Benton MJ. 2010. The origins of modern biodiversity on land. *Philosophical Transactions of the Royal Society B* 365: 3667–3679.

Courtillot V, Fluteau F. 2010. Cretaceous extinctions: The volcanic hypothesis. *Science* 328: 973–974,

Doyle JA. 2012. Molecular and fossil evidence on the origin of angiosperms. *Annual Review of Earth and Planetary Sciences* 40: 301–326.

Fischer V, Bardet N, Benson RBJ, Arkhangelsky MS. 2016. Extinction of fish-shaped marine reptiles associated with reduced evolutionary rates and global environmental volatility. *Nature Communications* 7: 10825

Mannion PD, Upchurch P, Benson RBJ, Goswami A. 2014. The latitudinal biodiversity gradient through deep time. *Trends in Ecology and Evolution* 29: 42–50.

Montani R. 2016. Paleobiology: Born and gone in global warming. *Current Biology* 26: R461–R480.

Petersen SV, Dutton A, Lohmann KC. 2016. End-Cretaceous extinction in Antarctica linked to both Deccan volcanism and meteorite impact via climate change. *Nature Communications* 7: 12079.

Sahney S, Benton MJ. 2008. Recovery from the most profound extinction of all time. *Proceedings of the Royal Society B* 275: 759–765.

Sakamoto M, Benton MJ, Venditti C. 2016. Dinosaurs in decline tens of millions of years before their final extinction. *Proceedings of the National Academy of Sciences USA* 113: 5036–5040.

Schulte P and 42 others. 2010. The Chicxulub asteroid impact and mass extinction at the Cretaceous–Paleogene boundary. *Science* 327: 1214–1218.

Whiteside JH, Grice K. 2016. Biomarker records associated with mass extinction events. *Annual Review of Earth and Planetary Sciences* 44: 581–612.

14

Synapsids and Sauropsids

Two Approaches to Terrestrial Life

The terrestrial environment provided opportunities for new ways of life that amniotes have exploited. The amniote egg may be a critical element of the success of synapsids and sauropsids because amniote eggs are larger than non-amniote eggs and produce larger hatchlings that grow into larger adults. Early in their evolutionary history, amniotes split into the two evolutionary lineages that dominate terrestrial habitats today, Sauropsida and Synapsida. Extant sauropsids include turtles, lepidosaurs (tuatara, lizards, and snakes), crocodylians, and birds, whereas mammals are the only extant synapsids. Both lineages underwent extensive radiations in the late Paleozoic and late Mesozoic that included animals that are now extinct and have no modern equivalents; the dinosaurs and pterosaurs were sauropsids, and the pelycosaurs and therapsids were synapsids.

By the mid-Carboniferous, the sauropsid and synapsid lineages had separated and amniotes had evolved few derived characters associated with terrestrial life. As a result, the sauropsid and synapsid lineages independently developed most of the derived characters necessary for terrestrial life, such as respiratory and excretory systems that conserve water and locomotor systems that are compatible with high rates of lung ventilation.

Both lineages eventually developed fast-moving predators that could pursue fleeing prey, as well as fleet-footed prey that could run away from predators, and both lineages included species capable of powered flight. Both lineages had members that became endothermal, evolving high metabolic rates and insulation to retain metabolic heat in the body, and both lineages evolved extensive parental care and complex social behavior.

Despite the parallel evolutionary trends in synapsids and sauropsids, differences in the way they carry out basic functions show that they evolved those derived characters independently:

- A terrestrial animal that runs for a sustained period must eliminate the conflict between respiratory movements and locomotion that is characteristic of early amniotes. Derived sauropsids (pterosaurs, dinosaurs, and birds) and derived synapsids (therapsids and mammals) both reduced the side-to-side bending of the rib cage by placing their legs more under the body (upright posture) and by relying more on movement of the legs than of the trunk. In both lineages, the ankles developed a hinged joint that could propel the body forward. Derived sauropsids became bipedal and retained expansion and contraction of the rib cage as the primary method of creating the pressure differences that move air into and out of the lungs. In contrast, derived synapsids remained quadrupedal and added a diaphragm to aid with lung ventilation.

- The high rates of oxygen consumption that are needed to sustain rapid locomotion require respiratory systems that can take up oxygen and release carbon dioxide rapidly. In many, perhaps most, extant sauropsids (birds, crocodylians, and some lizards), these functions are accomplished with a one-way flow of air through the lung (a through-flow lung), whereas synapsids retain the basic tetrapod condition of in-and-out airflow (a tidal-flow lung).

- A terrestrial animal requires an excretory system that eliminates nitrogenous wastes while conserving water. Sauropsids do this by having a waste product with low solubility (uric acid), kidneys that do not produce concentrated urine (some marine birds are an exception to that generalization), and glands that secrete salt. Synapsids excrete a highly soluble waste product (urea) through kidneys that can produce very concentrated urine, and they lack salt-secreting glands.

These differences in structural and functional characters of sauropsids and synapsids show that there is more than one way to succeed as a terrestrial amniote vertebrate.

14.1 The Conflict between Locomotion and Respiration

Running involves much more than just moving the legs rapidly. If an animal is to run very far, the muscles that

Figure 14.1   The demands of ventilation and locomotion can conflict. The effect of axial bending on the lung volume of a running lizard (top view) and a galloping dog (side view). The bending axis of the lizard's thorax is between the right and left lungs. As the lizard bends laterally, the lung on the concave side is compressed and air pressure in that lung increases (shown by +), while air pressure on the convex side decreases (shown by –). Air moves between the lungs (arrow), but little or no air moves into or out of the animal. In contrast, the bending axis of a galloping mammal's thorax is dorsal to the lungs. As the vertebral column bends, the volume of the thoracic cavity decreases; pressure in both lungs rises (shown by +), pushing air out of the lungs (arrow). When the vertebral column straightens, the volume of the thoracic cavity increases, pressure in the lungs falls (shown by –), and air is pulled into the lungs (arrow). (From Carrier 1987.)

locomotion and compressing the rib cage bilaterally to ventilate the lungs—and these activities cannot happen simultaneously. **Figure 14.1** illustrates the problem: the side-to-side bending of the lizard's rib cage compresses one lung as it expands the other, so air flows from one lung into the other, interfering with airflow in and out via the trachea.

Short sprints are feasible for animals that use lateral bending of the trunk for locomotion because the energy for a sprint is supplied initially by a reservoir of high-energy phosphate compounds, such as adenosine triphosphate (ATP) and creatine phosphate (CP), that are present in muscle cells. When these compounds are used up, the muscles switch to anaerobic metabolism, which draws on glycogen stored in the cells and does not require oxygen. The problem arises when rapid locomotion must be sustained beyond a minute or two, because at that point the supplies of high-energy phosphate compounds and glycogen in the muscles have been consumed. Because of this conflict between running and breathing, lizards that retain the ancestral modes of locomotion and ventilation are limited to short bursts of activity.

move the limbs require a steady supply of oxygen, and that is where the ancestral form of vertebrate locomotion encounters a problem. Early tetrapods moved with lateral undulations of the trunk, as do most extant salamanders and lizards. The axial muscles provide the power for this form of locomotion, bending the body from side to side. The limbs and feet are used in alternate pairs (i.e., left front and right rear, right front and left rear) to provide purchase on the substrate as the trunk muscles move the animal.

Tetrapodal locomotion based on the trunk muscles works only for short dashes. The problem with this ancestral locomotor mode is that the axial muscles are responsible for two essential functions—bending the trunk unilaterally for

Sustained locomotion requires a way to separate respiration from locomotion. Synapsids and sauropsids both developed modes of locomotion that allow them to hold the trunk rigid and use the limbs as a major source of propulsion, but the ways they did this were quite different.

Locomotion and lung ventilation of synapsids

Early non-mammalian synapsids (pelycosaurs) retained the short limbs, sprawling posture, and long tail that are ancestral characters of amniotes and are still seen in extant lizards and crocodylians. Later non-mammalian synapsids (therapsids) adopted a more upright posture with limbs held more underneath the trunk (although not as fully as in extant mammals) (**Figure 14.2**). Limbs in this position can move fore and aft with less bending of the trunk.

A second innovation in the synapsid lineage also contributed to resolving the conflict between locomotion and

(A) Early synapsid

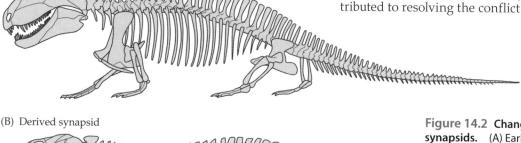

(B) Derived synapsid

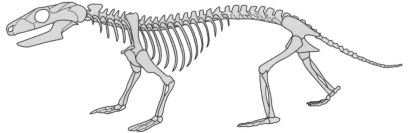

Figure 14.2   Changes in the anatomy of synapsids.   (A) Early synapsids such as *Mycterosaurus* (a pelycosaur) retained the ancestral conditions of ribs on all thoracic vertebrae, short legs, and a long tail. (B) Later synapsids such as *Massetognathus* (a cynodont therapsid) had lost ribs from the posterior vertebrae and had longer legs and a shorter tail. These changes probably coincided with the development of a diaphragm for respiration and fore-and-aft movement of the legs during locomotion. (After Kemp 2007.)

respiration. Ancestrally, contraction of the trunk muscles created the reduced pressure within the thorax that draws air into the lungs for inspiration, but this situation was modified with the development of a diaphragm in mammals, possibly as early as in some derived therapsids (cynodonts; see Figure 14.2B).

The diaphragm is a sheet of muscle that separates the body cavity into an anterior portion (the pulmonary cavity) and a posterior portion (the abdominal cavity). The diaphragm is convex anteriorly (i.e., it bulges toward the head) when it is relaxed and flattens when it contracts. This flattening increases the volume of the pulmonary cavity, creating a negative pressure that draws air into the lungs. Simultaneous contraction of the hypaxial muscles pulls the ribs forward and outward, expanding the rib cage—you can feel this change when you take a deep breath. Relaxation of the diaphragm permits it to resume its domed shape, and relaxation of the hypaxial muscles allows elastic recoil of the rib cage. These changes raise the pressure in the pulmonary cavity, causing air to be exhaled from the lungs.

Movements of the diaphragm do not conflict with locomotion, and in fact the bounding gait of therian mammals (marsupials and placentals) further resolved the conflicting demands of locomotion and respiration. The inertial backward and forward movements of the viscera (especially the liver) with each bounding stride work with the diaphragm to force air into and out of the lungs (see Figure 14.1). Thus, in derived mammals, respiration and locomotion work together in a synergistic fashion rather than conflicting.

Humans have little direct experience of this basic mammalian condition because our bipedal locomotion has separated locomotion and ventilation, but locomotion and respiration interact in many quadrupedal mammals, with a coupling of gait and breathing; that is, an animal inhales and exhales in synchrony with limb movements.

The evolution of the diaphragm in synapsids correlates with changes in the vertebral column. Synapsids differ from other tetrapods in having a vertebral column that is both more regionalized and more constrained in the number of vertebrae. The trunk vertebrae (rarely more than 20, but approaching 100 in some cetaceans) are differentiated into two regions, thoracic (with ribs) and lumbar (without ribs, although small lumbar ribs were retained in some early mammals). With very few exceptions (sloths and manatees), the number of cervical (neck) vertebrae is fixed at seven; mammals with longer necks simply have longer cervical vertebrae.

Mammals have transverse processes on their lumbar vertebrae where muscles that stabilize and flex the spine attach. These structures are not homologous to ribs but are projections of the vertebrae, and they may have evolved independently in different mammalian lineages.

These changes and stabilizations of vertebral form are first seen in cynodonts, and all of them may be linked to a common developmental cause, related to the evolution of the mammalian diaphragm. The diaphragm forms at the level of the thoraco-lumbar boundary, and the loss of lumbar ribs in cynodonts has long been thought to signal the evolutionary appearance of the diaphragm. Many vertebrates have a transverse septum (a sheet of connective tissue) in the peritoneal cavity that separates the lungs from the viscera and stabilizes the viscera. The mammalian diaphragm is basically a muscularization of this septum, as can be seen during mammalian development.

The muscles that form the diaphragm share a common developmental origin with the muscles of the forelimbs. Studies of the brachial plexus (nerves that exit from between the cervical vertebrae and combine to innervate the forelimb muscles) suggest that there has been a backward shift in both the exit of the plexus nerves and the position of the developing forelimb of mammals in comparison with other amniotes. A shift in Hox gene expression that changed the patterning of the vertebral column and the muscles forming in the anterior portion of the body may have allowed forelimb muscles on the medial side of the scapula to invade the diaphragm in mammals.

Thus, many key features of mammalian locomotion and respiration are linked by a common developmental change that probably happened at or near the base of the cynodont lineage, possibly in correlation with a higher metabolic rate allowing greater levels of activity.

Locomotion and lung ventilation of sauropsids

Early sauropsids were quadrupedal animals that moved with lateral undulations of the trunk, just as early synapsids did and as nearly all lizards do today. Derived sauropsids (birds and many dinosaurs) found a different solution to the problem of decoupling locomotion and respiration, however. They developed bipedal locomotion, using only the hindlimbs without movements of the trunk. Instead of developing a diaphragm, sauropsids appear to have incorporated pelvic movements and the ventral ribs (**gastralia**, bones in the ventral abdominal wall of some reptiles) into lung ventilation.

14.2 Limb-Powered Locomotion

The conflicting demands placed on the hypaxial trunk muscles by their dual roles in locomotion and respiration probably limited the ability of early amniotes to occupy many of the adaptive zones that are potentially available to a terrestrial vertebrate. If respiration nearly stops when an animal moves, as is the case for most extant lizards, both speed and distance are limited.

Separating locomotion and respiration allowed tetrapods to move farther and faster. That separation was arrived at in different ways in the synapsid and sauropsid lineages, as summarized in **Figure 14.3**. The synapsid solution—loss of the gastralia and the lumbar ribs and development of a muscular diaphragm—appeared fairly early in

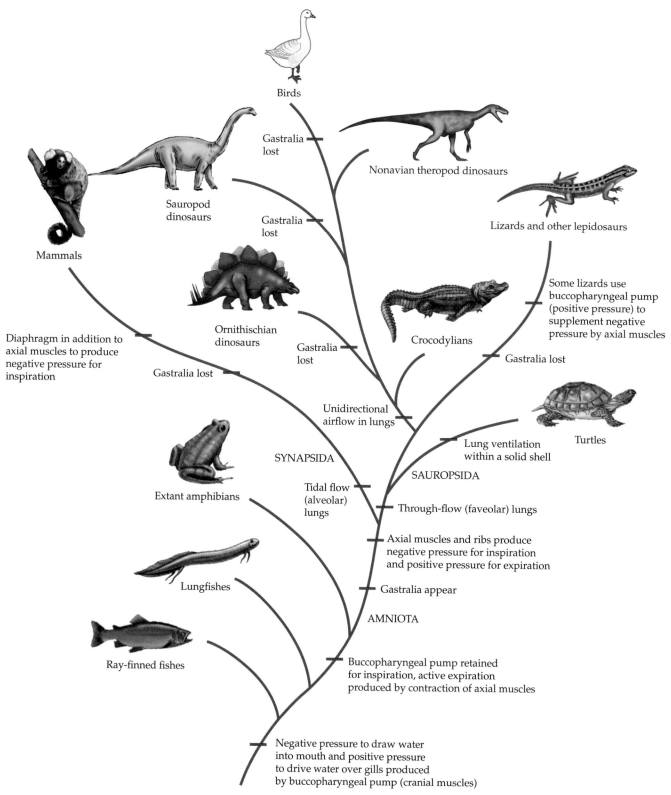

Birds

Gastralia lost

Nonavian theropod dinosaurs

Sauropod dinosaurs

Gastralia lost

Lizards and other lepidosaurs

Mammals

Some lizards use buccopharyngeal pump (positive pressure) to supplement negative pressure by axial muscles

Diaphragm in addition to axial muscles to produce negative pressure for inspiration

Ornithischian dinosaurs

Gastralia lost

Crocodylians

Gastralia lost

Gastralia lost

Unidirectional airflow in lungs

Lung ventilation within a solid shell

Turtles

SYNAPSIDA

SAUROPSIDA

Extant amphibians

Tidal flow (alveolar) lungs

Through-flow (faveolar) lungs

Lungfishes

Axial muscles and ribs produce negative pressure for inspiration and positive pressure for expiration

Gastralia appear

AMNIOTA

Ray-finned fishes

Buccopharyngeal pump retained for inspiration, active expiration produced by contraction of axial muscles

Negative pressure to draw water into mouth and positive pressure to drive water over gills produced by buccopharyngeal pump (cranial muscles)

Figure 14.3 Phylogenetic pattern of lung ventilation among tetrapods. The division of amniotes into sauropsids and synapsids was marked by a split into tidal-flow (alveolar) and through-flow (faveolar) lungs, as well as by gains and losses of gastralia. Fish use buccopharyngeal pumping to draw water into the mouth and expel it past the gills, and amphibians use buccopharyngeal pumping to inflate the lungs. Amniotes switched to negative-pressure inspiration, using the axial muscles and ribs to draw air into the lungs, and a diaphragm evolved in mammals. Synapsid lungs are alveolar and lung ventilation is tidal; sauropsid lungs are faveolar and lung ventilation appears to be flow-through in many (perhaps all) members of the lineage. Gastralia (ventral ribs) evolved in early amniotes and were retained in theropod dinosaurs and crocodylians, but were lost independently in mammals, ornithischian dinosaurs, sauropod dinosaurs, birds, and lepidosaurs.

the development of the lineage and is found in synapsids from the Early Triassic through extant mammals. In contrast, some archosaurs retained gastralia and used them to change the volume of the thorax, whereas lizards lost gastralia and emphasized rib movements and the increased flexibility of the trunk for lung ventilation.

The single solution adopted by synapsids and the multiple solutions of sauropsids probably reflect the diversity of body forms in the two lineages. Synapsids mostly remained quadrupedal and terrestrial throughout the Mesozoic, whereas sauropsids became enormously diverse, with bipedal, flying, limbless, and secondarily aquatic species in addition to quadrupeds.

The basal amniote ankle joint

As the limbs move beneath the body and swing in a parasagittal plane (i.e., parallel to the long axis of the body), controlling and ultimately limiting movement in the ankle joint becomes important. You saw in Chapter 10 that an ankle with a distinct plane of motion within the joint is a derived character of amniotes; basal tetrapods had no distinct plane of movement within the ankle, in contrast to basal amniotes. This change in the ankle joint was related to a change from axial to appendicular (limb-based) movement in locomotion.

In non-amniotes, the main propulsion of the body comes from the axial muscles; the feet function mainly as holdfasts to anchor the body to the ground, and as pivots from which to push. In amniotes, limb muscles provide more of the propulsion, with the feet acting more like levers than holdfasts and the axial muscles providing postural support. The lever action of the hindfoot, which pushes the body forward, is especially important, as reflected in the ankle joint (**Figure 14.4A**).

In the ankle (tarsus), a proximal (upper) row of elements articulates with the bones of the shin (tibia and fibula), and a distal (lower) row articulates with the upper bones of the foot (metatarsals). In basal amniotes, the original several proximal elements of the ankle became consolidated into two: the calcaneum, articulating with the fibula on the lateral (outer) side of the leg, and the astragalus, articulating with the tibia on the medial (inner) side. The original plane of articulation was through the middle of the tarsal joint, between the proximal and distal rows of bones. This is a **mesotarsal** ankle joint, sometimes called an intratarsal joint (**Figure 14.4B**).

The sauropsid ankle joint

The mesotarsal joint is retained in many extant reptiles, although various modifications may exist: for example, extant lizards (and also the tuatara) have fused the astragalus and calcaneum into a single element (**Figure 14.4C**).

Archosauromorphs have a complex history of changes in the ankle joint. The basal archosauromorph condition was a simple mesotarsal joint, but throughout archosauromorph evolution most of the tarsal elements have been lost or never ossified. A joint with a plane of motion between the astragalus and calcaneum (**Figure 14.4D**) developed at the base of Archosauria.[1] This is called a **crocodyloid joint** because a derived version is present in extant crocodiles. The true archosaurs are divided into two clades: Pseudosuchia, crocodiles and their relatives, including most of the Triassic archosaurs; see Chapter 19) and Ornithodira (pterosaurs, dinosaurs, and birds[2]).

The crocodyloid intratarsal articulation comprised a peg on the astragalus that fit into a socket on the calcaneum (the "crocodile-normal" condition), but some other archosaurs had the reverse condition (the peg was on the calcaneum—the "crocodile-reversed" condition). For many years these conditions were thought to have evolved independently, but it now seems that the morphology of the crocodyloid joint is plastic, and one type of joint can evolve from the other. The crocodyloid joint appears to allow the foot to turn inward, and it is thought to be associated with the ability of the hindlimbs to switch between a more sprawling stance and a more upright stance, as seen in extant crocodylians (see Chapter 18).

The crocodyloid joint of derived pseudosuchians is called a crurotarsal joint; in addition to the main joint between the astragalus and calcaneum, it includes a joint between a ball on the calcaneum and a socket on the fibula (**Figure 14.4E**). The calcaneum is now firmly integrated with the foot and has a tuber (heel) where the Achilles tendon inserts. The muscles that flex the foot (primarily the gastrocnemius) insert on the heel via the Achilles tendon and push the animal forward when the foot is on the ground; mammals have a similar structure that evolved independently.

In Ornithodira, the ankle joint became much more restricted, in association with the adoption of a more upright posture and an increasing tendency to become bipedal. In dinosaurs, the ankle joint became restricted to supporting this motion and lost any ability for rotation. The ornithodiran joint is a derived mesotarsal one, in which the proximal tarsals become closely appressed with the bones of the lower leg. The tibia is the main bone in these archosaurs (as it is in mammals), and the astragalus is large, with an ascending flange that interlocks with the base of the tibia. The calcaneum is reduced in size, fusing with the astragalus (**Figure 14.4F**).

In birds this condition is taken to an extreme. Adult birds appear to have no ankle bones at all because the astragalus and calcaneum have become fused with the lower limb, forming the tibiatarsus, while the distal tarsals have become fused with the foot, forming the tarsometatarsus.

The synapsid ankle joint

Pelycosaurs retained the basal amniote mesotarsal joint (see Figure 14.4B). A crocodyloid type of joint developed in

---

[1]The joint between the astragalus and calcaneum in archosauromorphs is often referred to as the "crurotarsal joint," but this terminology is incorrect. As the name suggests, a crurotarsal joint lies between the "crus" (lower leg) and the ankle (tarsus). Mammals have evolved a similar type of joint.
[2]An alternative pair of names for these lineages is based on their ankle joints: Crurotarsi (= Pseudosuchia) and Avemetatarsalia (= Ornithodira).

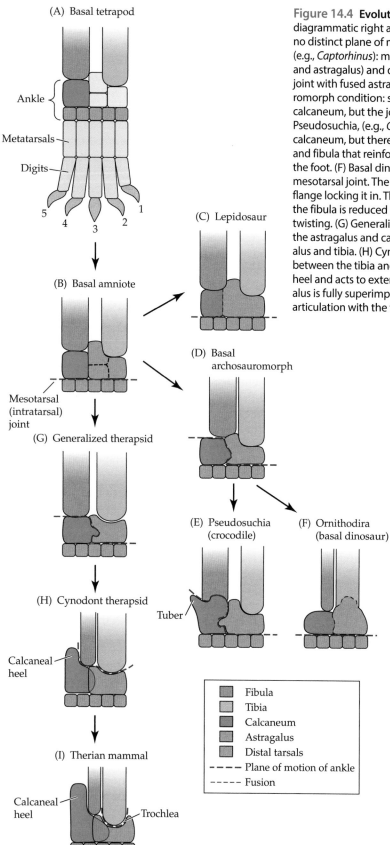

(A) Basal tetrapod

Ankle

Metatarsals

Digits

5 4 3 2 1

(B) Basal amniote

Mesotarsal
(intratarsal)
joint

(C) Lepidosaur

(D) Basal
archosauromorph

(G) Generalized therapsid

(E) Pseudosuchia
(crocodile)

Tuber

(F) Ornithodira
(basal dinosaur)

(H) Cynodont therapsid

Calcaneal
heel

(I) Therian mammal

Calcaneal
heel

Trochlea

Fibula
Tibia
Calcaneum
Astragalus
Distal tarsals
– – – Plane of motion of ankle
– – – – Fusion

Figure 14.4 Evolution of the ankle joint in amniotes. All figures show a diagrammatic right ankle viewed from the front. (A) Basal tetrapod condition: no distinct plane of movement within the ankle. (B) Basal amniote condition (e.g., *Captorhinus*): mesotarsal joint between the proximal tarsals (calcaneum and astragalus) and distal tarsals (C) Derived lepidosaur condition: mesotarsal joint with fused astragalus and calcaneum limiting twist. (D) Basal archosauromorph condition: some hinge-like motion between the astragalus and calcaneum, but the joint is mainly mesotarsal. (E) Derived archosaur condition, Pseudosuchia, (e.g., *Crocodylus*): the main joint is between the astragalus and calcaneum, but there is also a ball-and-socket joint between the calcaneum and fibula that reinforces the ankle. The calcaneum is strongly integrated with the foot. (F) Basal dinosaur condition (Ornithodira; e.g., *Lagosuchus*): derived mesotarsal joint. The astragalus is fused with the tibia, with an ascending flange locking it in. The calcaneum is small and fused with the astragalus, and the fibula is reduced in size. This joint functions only as a hinge, allowing no twisting. (G) Generalized therapsid condition (e.g., *Lycaenops*): joint between the astragalus and calcaneum, with rotation also possible between the astragalus and tibia. (H) Cynodont condition (e.g., *Cynognathus*: the main joint lies between the tibia and astragalus. The Achilles tendon inserts on the calcaneal heel and acts to extend the foot. (I) Therian condition (e.g., *Homo*): the astragalus is fully superimposed on the calcaneum and forms a gully (trochlea) for articulation with the tibia, allowing only a hingelike action at this joint.

therapsids, with a "crocodile-reversed" articulation between the astragalus and calcaneum, and a capacity for rotation between the tibia and astragalus (**Figure 14.4G**).

Cynodonts had a crurotarsal joint between the proximal tarsals and the lower limb (**Figure 14.4H**), with a distinct calcaneal heel for the attachment of the Achilles tendon (a condition approached by some other derived therapsids). The crurotarsal joint emphasizes fore-and-aft movement of the foot on the leg, which is advantageous for rapid terrestrial locomotion, but rotation was retained at the crocodyloid joint between the astragalus and calcaneum.

The peglike joint on the calcaneum was enlarged into a flange underlying the astragalus (the sustentacular process) so that the astralagus partially rested on top of the calcaneum. Early mammals had a somewhat similar condition, as do extant monotremes. This superposition of the astragalus on the calcaneum is developed to a much greater extent in therian mammals, where there is now a pronounced gully-like indentation, or trochlea, on the astragalus for the articulation of the tibia (**Figure 14.4I**; also see Chapter 24).

The original therapsid joint between the astragalus and calcaneum is retained in many mammals, including humans, as the "lower ankle joint." The original amniote mesotarsal joint may be retained as the "transverse tarsal joint," and there may also be joints between the distal tarsals and the metatarsals. These joints allow the foot to be turned inward and outward and are especially important in climbing mammals. However, these joints are fused in derived mammals specialized for running, such as horses, and motion is possible only between the tibia and astragalus.

14.3 Increasing Gas Exchange

Expiration powered by contraction of the transverse abdominal muscles is a derived character of tetrapods. Fish lack this muscle, and expiration of air by fish is a passive process that results from water pressure and elastic recoil of the gas bladder. Active expiration of air from simple lungs—basically internal sacs in which inhaled air could exchange oxygen and carbon dioxide with blood in capillaries of the lung wall—was probably sufficient for early tetrapods.

Rates of oxygen consumption would have increased as sustained high levels of muscular activity appeared, and a larger surface area would have been needed in the lungs for gas exchange. Complex lungs appeared in both synapsid and sauropsid lineages, but the additional surface area for gas exchange and the network of air-conducting tubes became organized in very different ways (**Figure 14.5**). In synapsids the conducting airways have a treelike dichotomous pattern of branching, where the walls of the last generations of the branches contain cuplike chambers, the alveoli, that are densely populated by blood capillaries. This type of lung is called an **alveolar lung**.

Sauropsids developed a multitude of different branching patterns in the lungs, and the size and distribution of the gas-exchange units vary enormously, depending on the

Figure 14.5 Synapsid and sauropsid lungs. (A) The ancestral tetrapod lung, exemplified by an amphibian, is a simple sac with a variable number of outgrowths (septa) around its periphery that increase its surface area. Airflow is tidal, and the lung contains a mixture of freshly inhaled air (high O_2, low CO_2) and air retained from previous breaths (lower O_2, higher CO_2). (B) The alveolar lungs of mammals have multiple branching airways (bronchi and bronchioles) that end in closed-ended alveoli, which are the sites of gas exchange. (C) An alveolus contains fresh air and retained air, and blood in capillaries exchanges O_2 and CO_2 with this mixed gas. (D) The faveolar lungs of sauropsids have air passages on their peripheries connected by thousands of parabronchi with faveoli in their walls. Gas exchange occurs in the faveoli, and airflow through the parabronchi and faveoli is unidirectional. (E) Fresh air flows in one direction through the parabronchi of a faveolar lung, and blood flows in the opposite direction, crossing the direction of airflow at an angle (i.e., cross-current exchange). (A, after Overton 1897.)

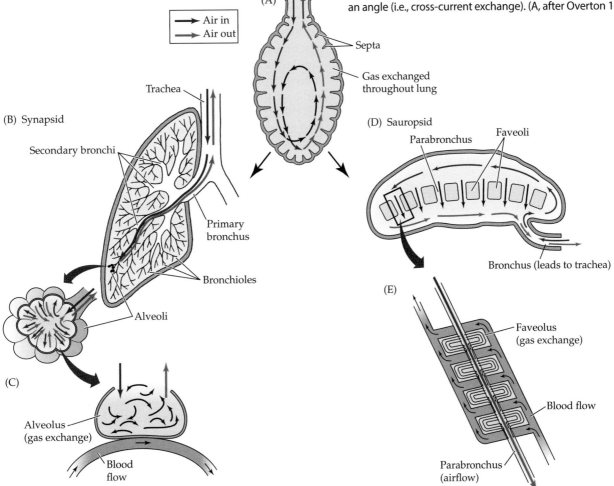

lineage. Gas exchange takes place in millions of cuplike chambers (faveoli) that line the walls of regions of the airways known as parabronchi. This type of lung is called a **faveolar lung**. The location and distribution of the faveoli vary, and in many lineages, such as snakes, the posterior region of the lung does not contain gas-exchange tissues but acts as a storage space and a bellows.

Synapsid lungs

The structure of the synapsid respiratory system is an elaboration of the saclike lungs of the earliest tetrapods. Air passes from the trachea through a series of progressively smaller passages, beginning with the primary bronchi and extending through many branch points; there are 23 levels of branching in the human respiratory system. Inhaled air ultimately reaches the respiratory bronchioles and alveolar sacs. Air flows into the alveoli and out again through the same tubes, a pattern called tidal ventilation.

The alveoli within the alveolar sacs are the sites of gas exchange. Alveoli are tiny (about 0.2 mm in diameter but varying with body size) and thin-walled. Blood in the capillaries of the alveolar walls is separated from air in the lumen of the alveolus by approximately 0.6–0.9 µm of tissue. This very short diffusion distance is critically important because during exercise a red blood cell passes through an alveolus in less than a second, and in that time it must release carbon dioxide and take up oxygen.

The alveoli expand and contract as the lungs are ventilated, and elastic recoil of alveoli in the mammalian lung helps expel air. The alveoli are so tiny that once they are emptied on expiration they could not be reinflated if it were not for the presence of a surfactant substance secreted by alveolar cells that reduces the surface tension of water. This substance is far more ancient than the alveolar lung and has been detected in the lungs of all vertebrates, as well as in the gas bladders of bony fishes. The total surface area of the alveoli is enormous—in humans it is 70 square meters, equal to the floor space of a large room.

Sauropsid lungs

The sauropsid respiratory system is more variable than that of synapsids. In some lineages it is very simple: the primary bronchi terminate at the lung and there are no air-conducting structures within the lung itself; the gas-exchange cups simply line the wall of the lungs. This morphology is characteristic of many kinds of lizards.

In other lineages of lizards, such as monitor lizards, and in some turtles, the primary bronchi continue into the lungs, where they give rise to secondary bronchi. These secondary bronchi can branch to give rise to tertiary bronchi. In birds and crocodylians, the secondary bronchi are connected to one another through this third level of tubes, called the **parabronchi**.

Thus, the lungs of birds and crocodylians have only three levels of branching rather than the 23 levels seen in humans and other synapsids. Furthermore, air flows in the same direction during inspiration and expiration in most of the secondary and tertiary bronchi. This unidirectional flow is facilitated by the alignment of the primary and secondary bronchi, and no valves are involved.

The respiratory system of birds

The respiratory system of birds differs from that of mammals in two ways. First, the gas-exchange structures are not cups but millions of interconnected small tubules known as **air capillaries** that radiate from the parabronchi. The air capillaries intertwine closely with vascular capillaries that carry blood. Airflow and blood flow pass in opposite directions, although they are not exactly parallel because the air and blood capillaries follow winding paths. This arrangement is called a **cross-current exchange system**. At less than 0.5 µm, the walls of avian parabronchi are even thinner than the walls of mammalian alveoli.

The presence of **air sacs** is the second characteristic of the respiratory system of birds. Two groups of air sacs, anterior and posterior, occupy much of the dorsal part of the body and extend into cavities (called pneumatic spaces) in many of the bones. The air sacs are poorly vascularized and do not participate in gas exchange, but they are large—about nine times the volume of the lung—and are bellows and reservoirs that store air during parts of the respiratory cycle to create a through-flow lung in which air flows in only one direction. Two respiratory cycles are required to move a unit of air through the lung (**Figure 14.6**).

The one-way passage of air in the bird lung may facilitate the extraction of oxygen from lung gases. The cross-current flow of air in the air capillaries and blood in the blood capillaries allows an efficient exchange of gases, like the countercurrent flow of blood and water in the gills of fish. In addition, the volumes of the secondary and tertiary bronchi and the air capillaries change very little during ventilation, so the blood vessels are not stretched at each respiratory cycle. As a result, the walls of the air capillaries and blood capillaries of birds can be thinner than the walls of mammalian alveoli, reducing the distance that carbon dioxide and oxygen must diffuse. Rapid diffusion of gas between blood and air is probably one of the mechanisms that allows birds to breathe at very high elevations.

Why are synapsid and sauropsid lungs so different?

The derived lung structures of synapsids and sauropsids probably evolved at different geological times, and differences in the atmospheric oxygen concentration may account for the differences in their lungs. Synapsids were the predominant large tetrapods in the late Paleozoic, when oxygen levels were high (higher than the current level of ~21%, and possibly as high as 30%). But when oxygen levels fell to about 15% in the Triassic, the archosaurian sauropsids radiated, with the rise of dinosaurs in the Jurassic completely eclipsing the synapsids.

Synapsids might have been at a respiratory disadvantage in the lower oxygen levels of the Mesozoic, and this situation may have contributed to mammals remaining small until the Cenozoic. This bottleneck of low oxygen levels may also explain why mammals evolved enucleate red blood cells—without a nucleus, more oxygen-carrying hemoglobin molecules could be packed into each cell.

Mammalian lungs combine a large internal surface area with relatively long diffusion distances between blood and air in the alveoli (that is, a relatively thick blood–gas barrier). The advantage of this thick barrier is that there is less chance of mechanical disruption to the capillaries during locomotion; this is important for the spongy, alveolar mammalian lung, which expands and contracts during ventilation. Under conditions of high atmospheric oxygen, there would be a large pressure gradient for diffusion, and there would be no need for a thinner blood–gas barrier.

Ancestral archosaurs probably had lungs like those of crocodiles, where the pumping of the heart promotes unidirectional airflow through the lung when the animal is not actively breathing. This unidirectional airflow may have facilitated partitioning of the respiratory system into air sacs that expand and contract during ventilation and compact, relatively immobile lungs used only for gas exchange. That separation of functions permits parabronchial lungs to have a thin barrier between air and blood that does not require high atmospheric oxygen levels to drive diffusion. Thus, the lung morphology of archosaurs may have given them an advantage in the low-oxygen conditions of the Mesozoic.

(A)

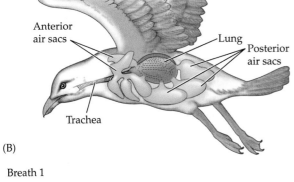

(B)

Breath 1

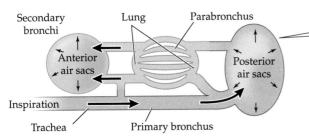

On the first inspiration, the volume of the thorax increases, drawing fresh air (blue) through the trachea and primary bronchi into the posterior air sacs, while the air that was in the parabronchial lung at the beginning of that inspiration (tan) is pulled into the anterior set of air sacs.

On the first expiration, the volume of the thorax decreases, forcing the air from the posterior sacs (blue) into the parabronchial lung and expelling the air from the anterior sacs (tan) to the exterior via the trachea.

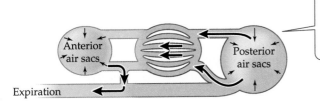

Breath 2

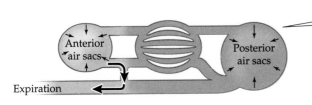

The second inspiration draws that unit of air (blue) into the anterior air sacs, and a new unit of fresh air (orange) into the posterior sacs.

The second expiration sends air (blue) out through the trachea and forces the air from the posterior sacs (orange) into the parabronchial lung.

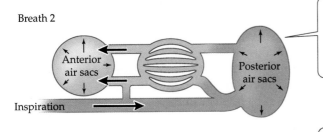

Figure 14.6 Pattern of airflow during inspiration and expiration by a bird. Note that air flows in only one direction through a parabronchial lung. Two respiratory cycles are required to move a unit of air through the lung. (After Sadava et al. 2016.)

14.4 Transporting Oxygen to the Muscles: The Heart

Changes in the mechanics of lung ventilation resolved the conflict between locomotion and breathing, and internal divisions of the lungs increased the surface area for gas exchange. These features were essential steps toward occupying adaptive zones that require sustained locomotion, but another element is necessary—oxygen must be transported rapidly from the lungs to the muscles and carbon dioxide from the muscles to the lungs to sustain high levels of cellular metabolism.

A powerful heart can produce enough pressure to move blood rapidly, but there is a complication: although high blood pressure is needed in the systemic circulation to drive blood from the heart to the limbs, high blood pressure would be bad for the lungs. Lungs are delicate structures because the tissue separating air and blood is so thin, and high blood pressure in the lungs would force plasma out of the thin-walled capillaries into the air spaces. When these spaces are partly filled with fluid instead of air—as happens in pneumonia, for example—gas exchange is reduced. Thus, amniotes must maintain different blood pressures in the systemic (body) and pulmonary (lungs) circuits while they are pumping blood at high speed. The solution that evolved in derived synapsids and sauropsids was separation of the ventricle into systemic and pulmonary sides with a permanent septum. Differences in the hearts of the two lineages indicate that this solution was reached independently in each lineage.

The ancestral amniote heart probably lacked a ventricular septum. The flow of blood through the ventricle was probably directed by a spongelike internal structure and perhaps by a spiral valve in a conus arteriosus, as in extant lungfishes and amphibians. Turtles and lepidosaurs do not have a permanent complete septum in the ventricle formed by tissue. Instead, during ventricular contraction the wall of the ventricle presses against a muscular ridge in the interior of the ventricle, keeping oxygenated and deoxygenated blood separated. This anatomy, which is undoubtedly derived and more complex than the ancestral amniote condition, plays an important functional role in the lives of turtles and lizards because it allows them to shunt blood between the systemic and pulmonary circuits in response to changing conditions.

Differences in resistance to flow in the pulmonary and systemic circuits are important in controlling the movement of blood through the hearts of turtles and lizards, and their blood pressures and rates of blood flow are low compared with those of birds and mammals. It may be that higher blood pressures and higher rates of blood flow made a permanent division necessary for derived synapsids (mammals; **Figure 14.7**) and derived sauropsids (crocodylians, birds, and probably nonavian dinosaurs **Figure 14.8**).

Inserting a complete anatomical separation between the left and right chambers of the ventricle requires that the same amount of blood flows through the pulmonary and systemic circuits. Additionally, high-oxygen-content blood can no longer be shunted from the left ventricle to the right ventricle (low oxygen content) to supply oxygen to the ventricular muscle. Instead, ventricular coronary vessels developed to oxygenate the heart muscle. All amniotes have some blood supply to the heart muscle, but an extensive ventricular coronary system evolved independently in mammals and derived sauropsids (archosaurs).

When the ventricle is permanently divided, the relative resistance to blood flow in the systemic and pulmonary circuits no longer determines where blood goes when it leaves the ventricle; instead, blood can flow only into the vessels that exit from each side of the ventricle: the right ventricle leads to the pulmonary circuit and the left ventricle to the systemic circuit. Synapsids and sauropsids have both reached this stage, but they must have done it independently because the relationship of the systemic arches to the left ventricle and the development of the septum are different in the two lineages.

While most sauropsids retain both aortic arches, in birds the left arch is lost entirely. In some nonavian sauropsids, notably turtles, crocodylians, and varanid lizards, the left systemic arch assumed a new function. In these animals, the celiac artery—the major artery supplying the stomach and anterior intestines—branches off from the base of the left systemic arch rather than from the dorsal aorta as in other vertebrates. This arrangement routes acidic, deoxygenated blood into the gut, perhaps aiding digestion.

Developmental studies show that mammalian and avian embryos both start off with two systemic arches and that one is subsequently lost (the right arch in mammals, the left in birds). However, there is a difference: mammals show no evidence of ever having had two separate arches leaving the heart; instead, they retain a remnant of the truncus arteriosus, and the two systemic arches branch from this structure, and a portion of the right systemic arch is retained as the brachiocephalic artery.

Why have birds and mammals each lost one of the systemic arches (or in the case of mammals, the bottom part of the right systemic arch)? The independent reduction to a single arch in each lineage suggests that one arch is somehow a better adaptive solution in these lineages than two arches, although two arches appear to be entirely functional for less derived sauropsids. Perhaps the advantage of a single arch is related to the high blood pressures and high rates of blood flow in the aortic arches of mammals and birds. One vessel with a large diameter creates less friction between flowing blood and the wall of the vessel than would two smaller vessels carrying the same volume of blood. In addition, turbulence may develop where the two arches meet, and that would reduce flow. Thus, a single arch may be the best conduit for blood leaving the heart at high pressure.

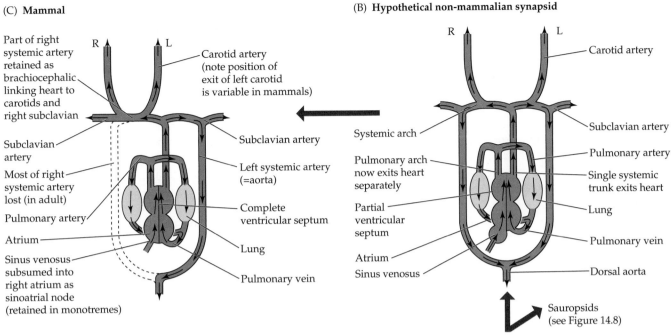

(C) Mammal

Part of right systemic artery retained as brachiocephalic linking heart to carotids and right subclavian

Carotid artery (note position of exit of left carotid is variable in mammals)

Subclavian artery

Most of right systemic artery lost (in adult)

Pulmonary artery

Atrium

Sinus venosus subsumed into right atrium as sinoatrial node (retained in monotremes)

Subclavian artery

Left systemic artery (=aorta)

Complete ventricular septum

Lung

Pulmonary vein

(B) Hypothetical non-mammalian synapsid

Carotid artery

Systemic arch

Pulmonary arch now exits heart separately

Partial ventricular septum

Atrium

Sinus venosus

Subclavian artery

Pulmonary artery

Single systemic trunk exits heart

Lung

Pulmonary vein

Dorsal aorta

Sauropsids (see Figure 14.8)

Figure 14.7 Diagrammatic view of the heart and aortic arches in synapsids. (A) Hypothetical early amniote condition. A conus arteriosus with a spiral valve and a truncus arteriosus are retained, and a ventricular septum is lacking. This condition, basically like that of extant amphibians, is proposed here to account for the differences in these structures between synapsids and sauropsids. (B) Hypothetical early synapsid condition. Mammal ancestors cannot have had the sauropsid pattern of dual systemic arches. Dual systemic arches are not seen in mammalian development, and retention of the left systemic arch as the aorta in mammals cannot be derived from the sauropsid double-arch condition, in which the right arch forms the aorta. Here a separation of the truncus arteriosus into separate pulmonary and (single) systemic trunks is proposed. Some degree of shunting between pulmonary and systemic circuits and within the heart may have been possible with an incomplete ventricular septum. The sinus venosus is shown as retained because a small sinus venosus is still present in monotremes. (C) Mammal (therian). The ventricular septum is complete, and the lower portion of the right systemic arch has been lost. (R = right, L = left.)

(A) Hypothetical early amniote

Ductus carotidus lost in adult between arch 3 (carotid) and arch 4 (systemic)

Systemic arch

Arch 5 lost

Probably no ventricular septum (not homologous between different amniote groups)

Atrium

Posterior vena cava

Carotid artery

Ventral aorta

Truncus arteriosus

Subclavian artery

Conus arteriosus

Pulmonary artery

Lung

Pulmonary vein

Sinus venosus

Dorsal aorta

Modern amphibian

Hypothetical early tetrapod

Lungfish

Teleost

Generalized bony fish

Generalized gnathostome

Lamprey

Hypothetical first vertebrate

Protovertebrate

Oxygenated blood

Deoxygenated blood

Mixed blood

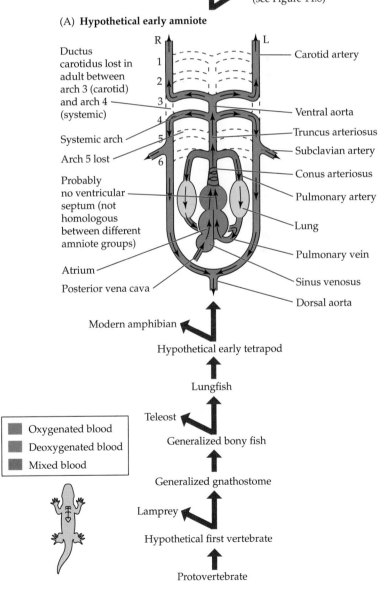

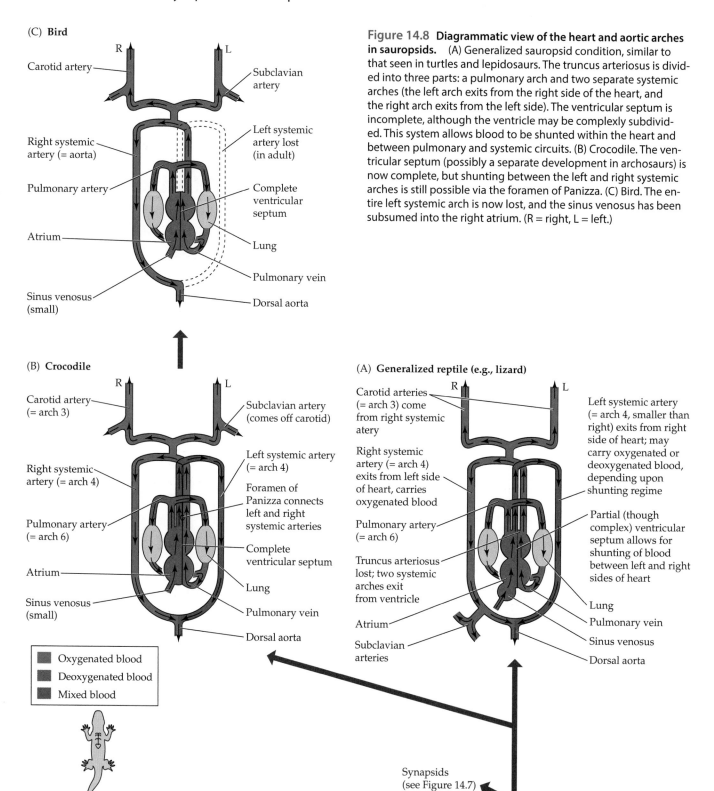

(C) Bird

Carotid artery

Subclavian artery

Right systemic artery (= aorta)

Left systemic artery lost (in adult)

Pulmonary artery

Complete ventricular septum

Atrium

Lung

Pulmonary vein

Sinus venosus (small)

Dorsal aorta

(B) Crocodile

Carotid artery (= arch 3)

Subclavian artery (comes off carotid)

Right systemic artery (= arch 4)

Left systemic artery (= arch 4)

Foramen of Panizza connects left and right systemic arteries

Pulmonary artery (= arch 6)

Complete ventricular septum

Atrium

Lung

Sinus venosus (small)

Pulmonary vein

Dorsal aorta

Oxygenated blood
Deoxygenated blood
Mixed blood

(A) Generalized reptile (e.g., lizard)

Carotid arteries (= arch 3) come from right systemic atery

Left systemic artery (= arch 4, smaller than right) exits from right side of heart; may carry oxygenated or deoxygenated blood, depending upon shunting regime

Right systemic artery (= arch 4) exits from left side of heart, carries oxygenated blood

Pulmonary artery (= arch 6)

Partial (though complex) ventricular septum allows for shunting of blood between left and right sides of heart

Truncus arteriosus lost; two systemic arches exit from ventricle

Atrium

Lung

Pulmonary vein

Subclavian arteries

Sinus venosus

Dorsal aorta

Synapsids (see Figure 14.7)

Hypothetical early amniote

Figure 14.8 Diagrammatic view of the heart and aortic arches in sauropsids. (A) Generalized sauropsid condition, similar to that seen in turtles and lepidosaurs. The truncus arteriosus is divided into three parts: a pulmonary arch and two separate systemic arches (the left arch exits from the right side of the heart, and the right arch exits from the left side). The ventricular septum is incomplete, although the ventricle may be complexly subdivided. This system allows blood to be shunted within the heart and between pulmonary and systemic circuits. (B) Crocodile. The ventricular septum (possibly a separate development in archosaurs) is now complete, but shunting between the left and right systemic arches is still possible via the foramen of Panizza. (C) Bird. The entire left systemic arch is now lost, and the sinus venosus has been subsumed into the right atrium. (R = right, L = left.)

14.5 The Evolution of Endothermy

Endotherms and ectotherms represent different ways to succeed as a tetrapod. Comparison of ectothermal tetrapods (amphibians and reptiles) and endotherms (birds and mammals) reveals two major physiological differences:[3]

- Endotherms continuously maintain body temperatures that are higher than ambient temperatures.
- The mass-specific metabolic rate (energy expended per gram of tissue) of an endotherm is about 10 times that of an ectotherm of the same body size when both are measured at the same temperature.

These two physiological differences lead to important behavioral and ecological differences:

- Endotherms can be active when ectotherms cannot—for example, on cool nights, during the winter, and in cold habitats.
- Endotherms can engage in activities that ectotherms cannot, such as sustained periods of foraging and, in derived forms, wing-powered flight or prolonged running to pursue prey or escape from a predator.

These enhanced abilities of endotherms come with an energetic cost. Not only are metabolic rates of endotherms an order of magnitude higher than those of ectotherms, but the daily energy requirement is even higher because endotherms maintain high metabolic rates at night, whereas ectotherms cool off at night and their metabolism falls. Because of that difference, an endotherm expends about 25 times as much energy daily as an ectotherm of the same body size; the food an endotherm consumes daily would support an ectotherm for a month. Endotherms use most of the energy from their food to stay warm, whereas ectotherms devote most of their ingested energy to growth and reproduction.

The evolution of endothermy was a major step in the history of tetrapods, and the question of how endothermy evolved has engaged biologists for decades. The benefits of endothermy as seen in birds and mammals are obvious, but birds and mammals are highly derived endotherms. Thus, comparing extant endotherms with extant ectotherms provides a great deal of information about how each mode of thermoregulation functions today, but offers little insight into how endothermy evolved. How could so energetically expensive an innovation succeed when it was competing with a preexisting, less expensive way to be a tetrapod? How did endothermy increase reproductive success enough to offset its energetic cost?

How did endothermy evolve?

Most hypotheses about the selective pressures that led to the evolution of endothermy focus either on the direct benefits of having a high and stable body temperature, or on the indirect evolution of endothermy as a by-product of selection for some other character. These hypotheses propose proximate benefits that could offset the energetic cost of endothermy. They are not mutually exclusive; more than one of them could apply at different stages during evolutionary change, and selective factors could have been different for mammals and birds.

Direct selection for a high and stable body temperature   The **thermogenic opportunity model** originated with the hypothesis that basal mammals were nocturnal. This proposal rests partly on characteristics of the visual pigments of mammals that suggest a "nocturnal bottleneck" in their evolution, and partly on the observation that extant mammals of the same body size as basal mammals are predominantly nocturnal (see Chapter 23). Even in the comparative warmth of the Mesozoic, nights were chilly and endothermy might have been essential for nocturnal activity.

The **warmer is better model** proposes that an organism benefits from having a high and stable body temperature because biochemical and physiological processes proceed faster or more forcefully as temperature increases. This observation, which can be quantified as the Q_{10} effect, means that everything an animal does from the cellular to the whole-organism level happens faster and more powerfully at warmer temperatures. Of course, these rates do not continue to increase indefinitely; eventually they level off and then begin to fall. At about 45°C many enzymes begin to lose catalytic effectiveness, and many birds and mammals regulate body temperature between 37.5°C and 42.5°C—that is, they are close to the maximum temperature for enzyme function but have a margin of safety.

Indirect selection for characters that depend on high and stable body temperature.   The **aerobic scope model** focuses on how changes in foraging mode could have affected physiology in the ancestors of birds and mammals. Both ectotherms and endotherms increase their rates of metabolism when they engage in high levels of activity. The aerobic scope model proposes that an increase in the maximum metabolic rate must be accompanied by an increase in the resting rate and, consequently, by higher rates of internal heat production when an animal is at rest as well as when it is active.

The changes (described earlier in the chapter) in postcranial skeletons of therapsids and nonavian dinosaurs suggest that these animals were developing increasing capacity for sustained locomotion. Active searching would require changes in the respiratory and circulatory systems that increased the rates at which oxygen and glucose were delivered to the muscles.

Increasing the metabolic rate would increase internal heat production because oxidative metabolism is not very efficient—less than half of the energy in chemical bonds of metabolic substrates is captured in high-energy molecules

---

[3] These are generalizations; exceptions will be described in subsequent chapters.

such as ATP. The rest of the energy is lost as heat, which is why you get hot when you exercise vigorously.

The **parental care model** is another hypothesis that attributes the evolution of endothermy to selection for another character. Endothermy could have several benefits for reproduction:

- A warmer temperature increases both the rate of embryonic development and the growth of neonates, resulting in shorter periods before neonates are no longer dependent on the parents.

- A stable temperature can increase the viability of embryos by preventing some developmental defects.

- Because some elements of developmental plasticity are temperature-sensitive, the mother can influence the phenotype of her offspring by controlling incubation temperature.

- Providing food for young probably required an increase in sustained activity.

Evaluating the models

The energetic cost of endothermy is enormous, and that cost must be offset by benefits that are at least as great. Incipient endotherms could have allocated the energy they used for thermogenesis to other activities, such as faster growth or having more or larger offspring. A model for the evolution of endothermy should present a convincing reason why investing energy in thermogenesis leads to a greater increase in fitness than would have resulted from other uses of the energy.

Although the aerobic scope model has been widely accepted, it has two major weaknesses:

1. It rests on the assumption that there is an obligate mechanistic link between resting and maximum metabolic rates, but most comparisons of extant species do not reveal the expected positive correlation between those rates. Failure to find that correlation undermines the hypothesis that selection acting on the maximum metabolic rate would change the resting rate.

2. A large proportion of the heat production by extant endotherms occurs in visceral organs, not in the skeletal muscles. Thus, it is not certain that an increase in skeletal muscle metabolism would have much effect on the resting metabolic rate or on thermogenesis when an animal was inactive.

The parental care model also has at least two weaknesses:

1. Thermogenesis is energetically expensive, and endotherms incur that expense all year. If the benefit of

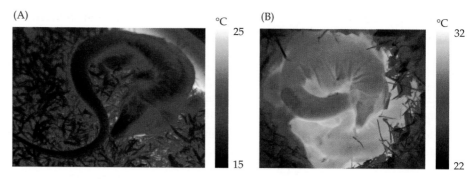

Figure 14.9 Endothermy of tegu lizards during the breeding season.   (A) A tegu in a burrow during the nonbreeding season is ectothermal. Its surface temperature (indicated by matching the color of its body in this infrared image to the scale to the right of the image) varies from 16°C to 18°C, which is the same as the substrate of the burrow. (B) During the breeding season, tegus can maintain body temperatures as much as 10°C higher than the burrow temperature. This image shows two lizards, one warmer than the other. (From Tattersall 2016.)

thermogenesis is realized during reproduction, why not be endothermal for only that part of the year? Tegu lizards have recently been shown to do exactly that. During reproduction, they raise their body temperatures as much as 10°C above ambient temperature (**Figure 14.9**).

2. Males play no role in embryogenesis in viviparous species of either synapsids or sauropsids, so why are males endothermal? Participation of males in incubating eggs and neonates appears to be an ancestral character of birds, but there is no evidence of an ancestral role of male mammals in caring for neonates.

As stated earlier, the four models we have discussed are not mutually exclusive. On the contrary, it is likely that an increase in thermogenesis resulting from one model would have changed the playing field, perhaps making additional heat from other models more beneficial and ultimately leading to a situation in which endothermy is beneficial for both sexes and at all times of the year. For example, selection for the ability to process food and assimilate energy to improve parental care could have favored higher rates of metabolism by the visceral organs and greater food consumption, which in turn would have established a selective advantage for greater ability to search for food via increases in the metabolic capacity of muscles, and expansion of foraging into the nocturnal predator adaptive zone.

When did endothermy evolve?

While the models that have been proposed to explain how endothermy evolved provide plausible explanations, the question of when endothermy evolved remains unresolved. The physiological characters that distinguish endotherms from ectotherms do not fossilize, and investigators are forced to depend on inferences drawn mostly from skeletal morphology.

Hard evidence: Bones   Two skeletal characters once appeared to provide solid evidence of endothermy—the microstructure of bone and the presence of respiratory turbinates in the nasal passages—but doubts have been expressed about the interpretation of both of these characters.

- *Lamellarzonal bone*, the primary bone type of most extant amphibians and nonavian reptiles, is poorly vascularized and forms relatively slowly. Consequently, lamellarzonal bone has been assumed to reveal low growth rates and ectothermy. Fibrolamellar bone, which is characteristic of extant mammals and birds, is well vascularized and is deposited during the rapid growth characteristic of juveniles. Thus, fibrolamellar bone has been assumed to reveal high growth rates and endothermy. However, recent studies have shown that both types of bone are present in fossil synapsids and sauropsids. Thus, bone histology is no longer believed to provide evidence of endothermy, although it probably does reflect higher growth rates.

- *Respiratory turbinates* reduce heat and water loss in extant mammals. Indications that respiratory turbinates may have been present in two lineages of non-mammalian synapsids in the Late Permian and Early Triassic were interpreted as evidence of an early appearance of endothermy in synapsids, whereas the absence of turbinates among nonavian dinosaurs was believed to show that these sauropsids were ectothermal. Reevaluation of the role of turbinates has cast doubt on their validity as markers of endothermy, however.

If bone histology and turbinate bones do not provide solid evidence of endothermy, can soft tissues help? Does the presence of hair or feathers indicate that an animal was an endotherm?

Soft evidence: Insulation   Producing heat is only half the battle for an incipient endotherm—retaining heat is equally important. Birds use feathers as insulation, and mammals use either hair or, in a few species, a thick layer of fat (blubber) under the skin. Feathers, hair, and blubber provide excellent insulation, but insulation is advantageous only after thermogenesis has evolved. Ectothermal tetrapods depend on external sources of heat to warm their bodies, and placing insulation between the external heat source and the core of the body prevents an ectotherm from warming.

Thus, the selective value of insulation creates a paradox in the evolution of endothermy: neither thermogenesis nor insulation is beneficial by itself; both characters must be present simultaneously. The solution to the paradox lies in hypothesizing a selective advantage that accounts for the evolution of insulation in an ectotherm before its

thermogenic capacity reaches the level of an endotherm. We have excellent fossil evidence of the presence of feathers in several lineages of nonavian theropod dinosaurs, as well as in the lineage that gave rise to birds. The nearly universal role of colored feathers in concealment from predators and in the social behavior of extant birds supports the hypothesis that feathers initially served these functions, and were later co-opted as insulation.

The earliest fossil evidence of hair in non-mammalian synapsids consists of probable hairs in coprolites (fossilized feces) of derived therapsids from Late Permian deposits in Russia and South Africa.

Both synapsids and sauropsids underwent a dramatic reduction in body size in the lineages leading to birds and mammals; the earliest mammals and birds were much smaller than their most recent non-mammalian and nonavian ancestors. Small animals have more surface area relative to their volume than large ones, and consequently heat loss from the body surface is a challenge for small endotherms. The predecessors of the first birds and mammals could have had body temperatures above ambient temperatures simply because large animals lose heat relatively slowly. The reduction in body size that occurred at the appearance of mammals and birds could have made it necessary to co-opt existing body coverings to serve a new role as insulation.

14.6 Getting Rid of Wastes: The Kidneys

Nitrogenous wastes are excreted primarily in urine, but urine is mostly water, and for a terrestrial animal water is too valuable to waste. The challenge is to excrete nitrogen while retaining water, and synapsids and sauropsids found different ways to do this (**Table 14.1**).

Metabolism of protein produces ammonia, NH_3. Ammonia is quite toxic, but it is very soluble in water and diffuses rapidly because it is a small molecule. Aquatic non-amniotes (bony fishes and aquatic amphibians) excrete a large proportion of their nitrogenous waste as ammonia, and ammonia is a nitrogenous waste product of terrestrial amniotes as well—human urine and sweat contain small amounts of ammonia.

Ammonia can be converted to urea, $CO(NH_2)_2$, which is less toxic and even more soluble than ammonia. Because it is

Table 14.1	Characteristics of the major nitrogenous waste products of vertebrates					
Compound	Chemical formula	Molecular weight	Solubility in water[a] (g/L)	Toxicity	Metabolic cost of synthesis	Water conservation efficiency[b]
Ammonia	NH_3	17	520	High	None	1
Urea	$CO(NH_2)_2$	60	1,080	Moderate	Low	2
Uric acid	$C_5H_4O_3N_4$	168	0.06	Low	High	4

[a]Solubility changes with temperature; these values refer to the normal body temperatures of vertebrates.
[b]The efficiency of water conservation is expressed as the number of nitrogen atoms per osmotically active particle; higher ratios mean more nitrogen is excreted per osmotic unit.

both soluble and relatively nontoxic, urea can be accumulated within the body and released in a concentrated solution in urine, thereby conserving water. Urea synthesis is an ancestral character of amniotes and probably of all gnathostomes. Synapsids retained the ancestral pattern of excreting urea, and the mammalian kidney has evolved to be extraordinarily effective in producing concentrated urine.

A complex metabolic pathway converts several nitrogen-containing compounds into uric acid, $C_5H_4O_3N_4$. Unlike ammonia and urea, uric acid is only slightly soluble, and it readily combines with sodium ions and potassium ions to precipitate as a salt of sodium or potassium urate. Synapsids normally synthesize and excrete small quantities of uric acid; excessive synthesis or failure to excrete uric acid can lead to accumulation of uric acid crystals in the joints, a painful condition called gout. Sauropsids synthesize and excrete uric acid and recover the water that is released when it precipitates.

Nitrogen excretion by synapsids: The mammalian kidney

The mammalian kidney is a highly derived organ composed of millions of nephrons, the basic unit of kidney structure that is recognizable in nearly all vertebrates (**Figure 14.10**). Each nephron consists of a glomerulus that filters the blood and a long tube in which the chemical composition of the filtrate is altered. A portion of this tube, the loop of Henle, is a derived character of mammals that is largely responsible for their ability to produce concentrated urine. The mammalian kidney can produce urine more concentrated than that of any non-amniote—and in most cases, more concentrated than that of sauropsids as well (**Table 14.2**).

Urine formation   Understanding how the mammalian kidney works is important for understanding how mammals can thrive in places that are seasonally or chronically short of water. Urine is concentrated by removing water from the ultrafiltrate that is produced in the glomerulus when water and small molecules are forced out of the capillaries. Because cells are unable to transport water directly, they manipulate the movement of water molecules by transporting ions to create osmotic gradients. In addition, the cells lining the nephron actively reabsorb substances important to the body's economy from the ultrafiltrate and secrete toxic substances into it.

The nephron's activity is a six-step process (**Figure 14.11**). Each step is localized in a region that has special cell characteristics and distinctive variations in the osmotic environment.

1. The first step is production of an ultrafiltrate at the glomerulus. The ultrafiltrate is isosmolal with blood plasma and resembles whole blood after the removal of (a) cellular elements, (b) substances with a molecular weight of 70,000 or higher (primarily proteins), and

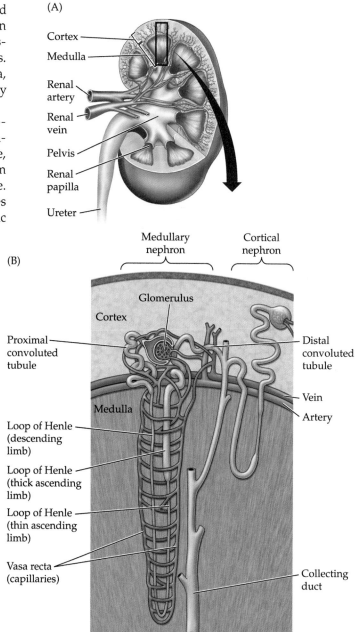

Figure 14.10 The mammalian kidney and nephron.   (A) Structural divisions of a human kidney and proximal end of the ureter. (B) A mammalian kidney contains millions of nephrons, the functional unit of the kidney. General nephron structure, including a glomerulus that filters the blood and tubules in which the blood filtrate is altered into urine, is recognizable in nearly all vertebrates. The Loop of Henle is a derived character of mammals that allows them to produced highly concentrated urine. This enlarged diagram is of a section extending from the outer cortical surface at the apex of the renal papilla. (From Sadava et al. 2016.)

(c) substances with molecular weights between 15,000 and 70,000, depending on the shapes of the molecules. Humans produce about 120 milliliters (mL) of ultrafiltrate per minute—that is 170 liters (L) of glomerular filtrate per day, twice the body mass of a large man! Humans excrete only about 1.5 L of urine because the

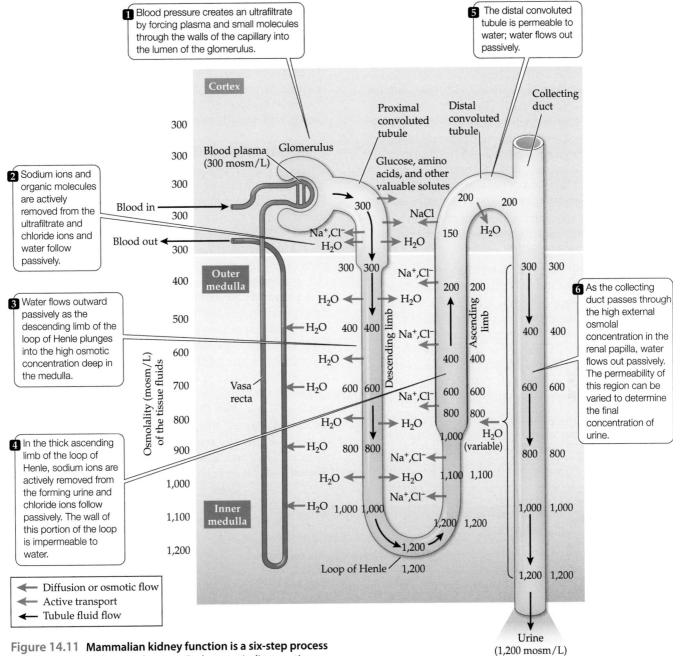

1 Blood pressure creates an ultrafiltrate by forcing plasma and small molecules through the walls of the capillary into the lumen of the glomerulus.

2 Sodium ions and organic molecules are actively removed from the ultrafiltrate and chloride ions and water follow passively.

3 Water flows outward passively as the descending limb of the loop of Henle plunges into the high osmotic concentration deep in the medulla.

4 In the thick ascending limb of the loop of Henle, sodium ions are actively removed from the forming urine and chloride ions follow passively. The wall of this portion of the loop is impermeable to water.

5 The distal convoluted tubule is permeable to water; water flows out passively.

6 As the collecting duct passes through the high external osmolal concentration in the renal papilla, water flows out passively. The permeability of this region can be varied to determine the final concentration of urine.

Cortex

Proximal convoluted tubule

Distal convoluted tubule

Collecting duct

Blood plasma (300 mosm/L)   Glomerulus

Glucose, amino acids, and other valuable solutes

Blood in

Blood out

Outer medulla

Inner medulla

Vasa recta

Descending limb

Ascending limb

Loop of Henle

Osmolality (mosm/L) of the tissue fluids

Urine (1,200 mosm/L)

← Diffusion or osmotic flow
← Active transport
← Tubule fluid flow

Figure 14.11 Mammalian kidney function is a six-step process that takes place in the nephrons. Red arrows indicate active transport, blue arrows indicate passive flow. Numbers represent the approximate osmolality of the fluids in the indicated regions. (From Sadava et al. 2016.)

kidney reabsorbs more than 99% of the ultrafiltrate produced by the glomerulus.

2. The second step in producing urine is the action of the proximal convoluted tubule (PCT) in decreasing the volume of the ultrafiltrate. The PCT cells have greatly enlarged surface areas that actively transport sodium ions, glucose, and amino acids from the lumen of the tubules to the exterior of the nephron. Chloride ions and water move passively through the PCT cells in response to the removal of sodium ions. By this process, about two-thirds of the salt is reabsorbed in the PCT,

and the volume of the ultrafiltrate is reduced by the same amount. Although the urine is still very nearly isosmolal with blood at this stage, the substances contributing to the osmolality of the urine after it has passed through the PCT are at different concentrations than in the blood.

3. The next alteration occurs in the descending limb of the loop of Henle. The thin, smooth-surfaced cells freely permit diffusion of water. Because the descending limb of the loop passes through tissues of increasing osmolality as it plunges deeper into the kidney, water is lost from the urine, which becomes more concentrated. In humans, the osmolality of the fluid in the descending limb may reach 1,200 millimoles per

kilogram (mmol/kg), and other mammals can achieve considerably higher concentrations.

4. The fourth step takes place in the thick ascending limb of the loop of Henle, which has cells with numerous large, densely packed mitochondria. The ATP produced by these organelles is used to remove sodium ions from the forming urine. Because these cells are impermeable to water, the volume of urine does not decrease as sodium ions are removed, and because the sodium ions were removed without loss of water, the urine is hyposmolal with the body fluids as it enters the next segment of the nephron. Although this sodium ion–pumping, water-impermeable, ascending limb does not concentrate or reduce the volume of the forming urine, it contributes to setting the stage for these important processes.

5. The last portion of the nephron changes in physiological character, but the cells closely resemble those of the ascending loop of Henle. This region, the terminal portion of the distal convoluted tubule (DCT), is permeable to water. The osmolality surrounding the DCT is the same as that of the body fluids, and water in the entering hyposmolal fluid flows outward and equilibrates osmotically. This process reduces the fluid volume to as little as 5% of the original ultrafiltrate.

6. The final step in producing a small volume of highly concentrated mammalian urine occurs in the collecting duct. Like the descending limb of the loop of Henle, the collecting duct courses through tissues of increasing osmolality, which withdraw water from the urine. The significant phenomenon associated with the collecting duct and with the terminal portion of the DCT is variable permeability to water. During excess fluid intake, the collecting duct demonstrates low water permeability: only half of the water entering it may be reabsorbed and the remainder excreted. In this way, copious dilute urine can be produced.

Antidiuretic hormone (**ADH**, also known as vasopressin) is released into the circulation in response to an increase in blood osmolality or when blood volume drops. ADH is a polypeptide that is produced in the hypothalamus and stored in the posterior pituitary. When ADH is carried by the circulation to the cells of the terminal portion of the DCT and the collecting duct, it causes them to insert proteins called aquaporins into the plasma membranes of the cells lining the collecting duct.

Aquaporins are tubular proteins with nonpolar (hydrophobic) amino acid residues on the outside and polar (hydrophilic) residues on the inside. When aquaporins are inserted into the plasma membranes of the cells, the hydrophobic residues on the outer surface anchor them in the phospholipid bilayer of the membrane and the hydrophilic residues on the inner surface form a channel through which water molecules can flow. When aquaporins are inserted into plasma membrane, the cells of the terminal portion of the DCT and the collecting duct become permeable to water, which flows outward following its osmotic gradient (**Figure 14.12**).

The water that is reabsorbed from the DCT and collecting duct enters the blood, reducing its osmolality and increasing its volume. As water is removed, the volume of the urine in the collecting duct decreases and its concentration increases. The final urine volume leaving the collecting duct may be less than 1% of the original ultrafiltrate volume. In some desert rodents, so little water remains that the urine crystallizes almost immediately upon urination. Alcohol inhibits the release of ADH, inducing a copious urine flow, and this can result in dehydrated misery the morning after a drinking binge.

By the time the filtrate enters the collecting duct, its volume has been reduced to about 5% of its initial volume, but up to this point the walls of the nephron have been impermeable to urea. As a result, the concentration of urea in the fluid that enters the collecting duct is very high. The walls of the terminal

Table 14.2	Maximum urine concentrations of some synapsids and sauropsids	
Species	Maximum observed urine concentration (mmol/kg)	Approximate urine:plasma concentration ratio
Synapsids		
Human (*Homo sapiens*)	1,430	4
Camel (*Camelus dromedarius*)	2,800	8
Marsupial mouse (*Dasycercus cristicauda*)	3,231	10
Cat (*Felis domesticus*)	3,250	10
Desert woodrat (*Neotoma lepida*)	4,250	12
Kangaroo rat (*Dipodomys merriami*)	6,382	18
Australian hopping mouse (*Notomys alexis*)	9,370	22
Sauropsids		
Desert tortoise (*Gopherus agassizii*)	337	0.7
American alligator (*Alligator mississippiensis*)	312	0.95
Desert iguana (*Dipsosaurus dorsalis*)	300	0.95
Gopher snake (*Pituophis catenifer*)	342	1
Savannah sparrow (*Passerculus sandwichensis*)	601	1.7
House sparrow (*Passer domesticus*)	826	2.4
House finch (*Carpodacus mexicanus*)	850	2.4

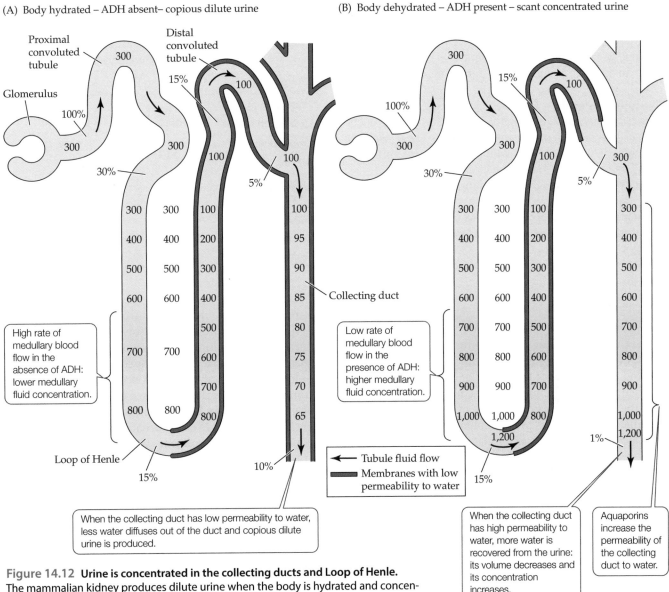

(A) Body hydrated – ADH absent– copious dilute urine

(B) Body dehydrated – ADH present – scant concentrated urine

High rate of medullary blood flow in the absence of ADH: lower medullary fluid concentration.

Low rate of medullary blood flow in the presence of ADH: higher medullary fluid concentration.

When the collecting duct has low permeability to water, less water diffuses out of the duct and copious dilute urine is produced.

When the collecting duct has high permeability to water, more water is recovered from the urine: its volume decreases and its concentration increases.

Aquaporins increase the permeability of the collecting duct to water.

← Tubule fluid flow
▬ Membranes with low permeability to water

Figure 14.12 Urine is concentrated in the collecting ducts and Loop of Henle. The mammalian kidney produces dilute urine when the body is hydrated and concentrated urine when the body is dehydrated. As in the previous figure, numbers show the approximate osmolality of the fluids in the region. Percentages are the volumes of the forming urine relative to the volume of the initial ultrafiltrate. (A) When blood osmolality drops below the normal concentration (~300 mmol/kg of body weight), levels of antidiuretic hormone (ADH, also known as vasopressin) drop and excess body water is excreted. (B) When osmolality rises above normal, ADH levels rise and water is conserved. (From Hill et al. 2016.)

portion of the collecting duct are very permeable to urea, and urea diffuses outward from the collecting duct into the extracellular fluid deep in the kidney. This urea contributes to the high osmolal concentration of the tissues at the lower end of the loop of Henle.

Structure of the nephron The structural characteristics of the cells that form the walls of the nephron are directly related to the processes that take place in that portion of the nephron. The cells of the PCT contain many mitochondria and have an enormous surface area produced by long, closely spaced microvilli. These structural features reflect the function of the PCT in actively moving sodium ions from the lumen of the tubule to the peritubular space and capillaries. Chloride follows the electric charge gradient created by the movement of sodium ions, and water moves in response to the concentration gradient produced by the movement of those two ions.

Farther along the nephron, the cells of the thin segment of the loop of Henle are waferlike and contain fewer mitochondria. The descending limb of the loop of Henle permits passive flow of water. Cells in the thick ascending limb are similar to those in the PCT and actively remove sodium ions from the ultrafiltrate.

Finally, cells of the collecting duct appear to be of two kinds. Most seem to be suited to the relatively impermeable state characteristic of periods of sufficient body water. Other cells are rich in mitochondria and have a larger surface area.

These are probably the cells that respond to the presence of ADH from the pituitary gland, triggered by insufficient body fluid. Under the influence of ADH, the collecting duct becomes permeable to water, which flows from the lumen of the duct into the concentrated peritubular fluids.

The loop of Henle   The key to producing concentrated urine is the passage of the loops of Henle and collecting ducts through tissues with increasing osmolality. These osmotic gradients are formed and maintained within the mammalian kidney as a result of its structure (see Figure 14.12).

The structural arrangements within the kidney medulla of the descending and ascending segments of the loop of Henle and its blood supply, the vasa recta, are especially important. These elements create a series of parallel tubes with flow passing in opposite directions in adjacent vessels (countercurrent flow). As a result, sodium ions pumped from the ascending limb of the loop of Henle diffuse into the medullary tissues to increase their osmolality, and this excess salt forms part of the steep osmotic gradient within the medulla.

The final concentration of a mammal's urine is determined by the concentration of sodium ions and urea accumulated in the fluids of the medulla. Physiological alterations in the concentration in the medulla result primarily from the effect of ADH on the rate of blood flushing the medulla. When ADH is present, blood flow into the medulla is retarded and salt accumulates to create a steep osmotic gradient. Another hormone, aldosterone, from the adrenal gland increases the rate of sodium-ion reabsorption into the medulla to promote an increase in medullary salt concentration.

In addition to having these physiological means of concentrating urine, a variety of mammals have morphological alterations of the medulla. Most mammals have two types of nephrons: those with a cortical glomerulus and abbreviated loops of Henle that do not penetrate far into the medulla, and those with juxtamedullary glomeruli, deep within the cortex, with loops that penetrate as far as the papilla of the renal pyramid. Obviously, the longer, deeper loops of Henle experience larger osmotic gradients along their lengths. The flow of blood to these two populations of nephrons seems to be independently controlled. Juxtamedullary glomeruli are more active in regulating water excretion; cortical glomeruli function in ion regulation.

Nitrogen excretion by sauropsids: Renal and extrarenal routes

All extant representatives of the sauropsid lineage, including turtles and birds, are uricotelic—that is, they can excrete nitrogenous wastes in the form of uric acid. They are not limited to uricotely, however, and many sauropsids can excrete nitrogenous wastes as ammonia or urea (**Table 14.3**).

The strategy for water conservation when uric acid is the primary nitrogenous waste is entirely different from

Table 14.3	Distribution of nitrogenous end products among sauropsids		
	Total urinary nitrogen (%)		
Group	**Ammonia**	**Urea**	**Salts of uric acid**
Tuatara	3–4	10–28	65–80
Lizards and snakes	Small	0–8	90–98
Crocodylians	25	0–5	70–75
Turtles			
Aquatic	4–44	45–95	1–24
Desert	3–8	15–50	20–50
Birds			
Omnivores	11–29	1–11	45–72
Carnivores	9–16	1	76–87

that required when urea is produced. Because urea is so soluble, concentrating urine in the kidney can conserve water, but concentrating uric acid would cause it to precipitate and block the nephrons. The kidneys of lepidosaurs are elongate and lack the long loops of Henle that allow mammals to reduce the volume of urine and increase its concentration (**Figure 14.13**). The osmolal concentration of urine from the kidneys of lepidosaurs is the same as blood plasma or even slightly lower. The kidneys of birds have two types of nephrons: short loop nephrons like those of lepidosaurs and long loop nephrons that extend down into the medullary cone. The long loop nephrons allow birds to produce urine that is two to three times more concentrated than the plasma. These ratios are lower than those of mammals, and even the highest urine to plasma ratio recorded for a bird is relatively low compared with mammalian ratios (see Table 14.2).

If sauropsids depended solely on the urine-concentrating ability of their kidneys, they would excrete all their body water in urine. This is where the low solubility of uric acid becomes advantageous: uric acid precipitates when it enters the cloaca or bladder. The dissolved uric acid combines with ions in the urine and precipitates as a light-colored mass that includes sodium, potassium, and ammonium salts of uric acid as well as other ions held by complex physical forces. (This mixture is familiar to anyone who has left a car beneath a tree where birds roost.) When uric acid and ions precipitate from solution, the urine becomes less concentrated and water is reabsorbed into the blood. In this respect, excretion of nitrogenous wastes as uric acid is even more economical of water than is excretion of urea because the water used to produce urine is reabsorbed and reused.

Water is not the only substance that is reabsorbed from the cloaca, however. Many sauropsids also reabsorb sodium and potassium ions and return them to the bloodstream. At first glance, that seems a remarkably inefficient thing to

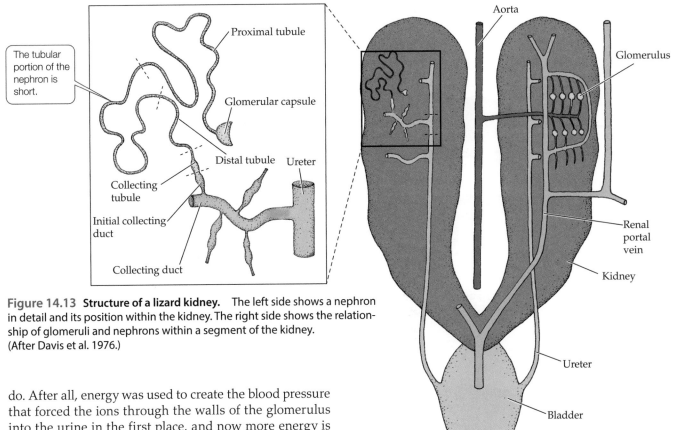

Figure 14.13 Structure of a lizard kidney.   The left side shows a nephron in detail and its position within the kidney. The right side shows the relationship of glomeruli and nephrons within a segment of the kidney. (After Davis et al. 1976.)

do. After all, energy was used to create the blood pressure that forced the ions through the walls of the glomerulus into the urine in the first place, and now more energy is used in the cloaca or bladder to operate the active transport system that returns the ions to the blood. The animal has used two energy-consuming processes, and it is back where it started—with an excess of sodium and potassium ions in the blood. Why do that?

The solution to this paradox lies in a water-conserving mechanism that is present in many sauropsids: salt-secreting glands that provide an extrarenal (i.e., in addition to the kidney) pathway that disposes of salt with less water than urine. Salt glands have evolved independently multiple times in sauropsids, and different glands have evolved a salt-secreting function in different lineages (**Figure 14.14**). Only the function of salt glands is novel, not the glands themselves. Homologous but non-salt-secreting glands are present in the sister taxa of all the lineages with salt glands. Apparently, salt glands have evolved by repeated co-option of existing unspecialized glands.

Salt glands are characteristic of many species that live in salty habitats (the sea or estuaries) or xeric habitats (deserts), as well as species that eat dry food (such as seeds) or animal flesh, but they are not limited to species in those categories. For example, salt glands are widespread among marine and shore birds (including gulls, sandpipers, plovers, pelicans, albatrosses, penguins), carnivorous birds (hawks, eagles, vultures, roadrunners), and upland game birds (pheasants, quail, grouse). More surprisingly, salt glands have been identified in freshwater birds (ducks, loons, grebes), three species of doves, and three species of hummingbirds.

Crocodiles, which have lingual salt glands, are more marine than alligators and gharials, which lack salt-secreting lingual glands. Among turtles, only sea turtles and one estuarine species have salt glands. The lacrimal glands of sea turtles are greatly enlarged (in some species each gland is larger than the turtle's brain) and secrete a salty fluid around the orbits of the eyes. Photographs of nesting sea turtles frequently show clear paths streaked by tears through the sand that adheres to a turtle's head. Those tears are salt-gland secretions. Terrestrial turtles, even those that live in deserts, do not have salt glands. Salt glands have evolved or been lost many times among lizards; they been described in eight families of lizards, but five of those families also include species without salt glands.

Despite their different origins and locations, the structural and functional properties of salt glands are quite similar. They secrete fluid containing primarily sodium or potassium cations and chloride or bicarbonate anions in high concentrations (**Table 14.4**). Sodium ions are the predominant cations in the salt-gland secretions of marine vertebrates, and potassium ions are present in the secretions of terrestrial lizards, especially herbivorous species such as the desert iguana (*Dipsosaurus dorsalis*). Chloride is the major anion, and herbivorous lizards may also excrete bicarbonate ions.

The total osmolal concentration of the salt-gland secretion may reach 2,000 mmol/kg, which is more than six

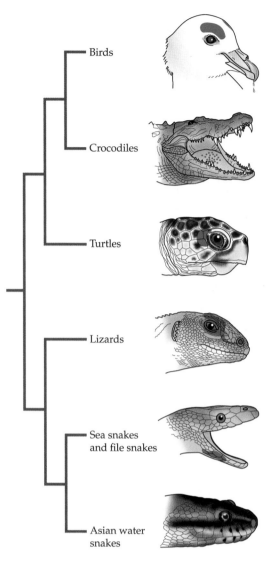

Figure 14.14   Salt-secreting glands of sauropsids.   The phylogenetic distribution of salt glands (shown in red in this illustration) reveals that at least five different glands have independently evolved salt-secreting properties. The nasal glands of birds excrete salt. These glands were ancestrally located on the snout, but in birds they have migrated backward and are located above the eyes. The secretions empty into the nasal passages and are expelled with a sneeze or head shake. The lingual glands of crocodiles secrete salts, but the lingual glands of alligators and gharials do not produce secretions more concentrated than blood plasma. In sea turtles and the diamondback terrapin (*Malaclemys terrapin*, an estuarine species), the lacrimal (tear) glands are the site of salt secretion. Lizards have specialized nasal glands that secrete salt. The salt-secreting ability of these glands evolved independently of the parallel specialization of the nasal glands of birds, and the scattered distribution of salt secretion among species within lizard families indicates multiple independent origins of the trait. Surprisingly, the salt-secreting glands of snakes are different from those of lizards, and two different glands have developed salt-secreting properties in snakes. In sea snakes and the marine elephant trunk snake (*Acrochordus javanicus*), the posterior sublingual gland has been modified to secrete salt, whereas in estuarine snakes in the Indo-Australian Homalopsidae it is the premaxillary gland that secretes salt.

times the concentration of urine that can be produced by the kidney. This efficiency of excretion is the explanation for the paradox of active uptake of salt from the urine. As ions are actively reabsorbed, water follows passively, so an animal recovers both water and ions from the urine. The ions can then be excreted via the salt gland at much higher concentrations, with a corresponding reduction in the amount of water needed to dispose of the salt. Thus, by investing energy to recover ions from urine, sauropsids with salt glands can conserve water by excreting ions through the more efficient extrarenal route.

Table 14.4	**Salt-gland secretions of sauropsids**		
	Ion concentration (mmol/kg)		
Species and conditions	**Na⁺**	**K⁺**	**Cl⁻**
Turtles			
Loggerhead sea turtle (*Caretta caretta*), seawater	732–878	18–31	810–992
Diamondback terrapin (*Malaclemys terrapin*), seawater	322–908	26–40	ND
Lizards			
Desert iguana (*Dipsosaurus dorsalis*), estimated field conditions	180	1,700	1,000
Fringe-toed lizard (*Uma scoparia*), estimated field conditions	639	734	465
Snakes			
Sea snake (*Pelamis platurus*), salt-loaded	620	28	635
Estuarine snake (*Cerberus rynchops*), salt-loaded	414	56	ND
Crocodylian			
Saltwater crocodile (*Crocodylus porosus*), natural diet	663	21	632
Birds			
Black-footed albatross (*Phoebastria nigripes*), salt-loaded	800–900	ND	ND
Herring gull (*Larus argentatus*), salt-loaded	718	24	ND

ND, No data.

In the phylogenetic tree (left): Birds, Crocodiles, Turtles, Lizards, Sea snakes and file snakes, Asian water snakes.

14.7 Sensing and Making Sense of the World

In many respects, synapsids and sauropsids perceive the world around them very differently; indeed, the lineages are quite different in some of their sensory capacities. Most synapsids are exquisitely sensitive to odors but have relatively poor vision. Scent plays an important role in the ecology and behavior of most mammals. (Primates in general—and humans in particular—are an exception to that generalization; primates and especially humans have lost many of the genes associated with olfactory reception.)

Sauropsids (especially lizards and birds) are strongly visually oriented, and colors, patterns, and movements are important components of their behavior. Birds have long been considered to have a poor sense of smell compared with mammals, but recent studies cast doubt on that assertion. Some birds use odor for navigation, predation, and in social interactions.

Vision

The vertebrate retina contains two types of cells that respond to light—rods and cones. **Rod cells** are sensitive to a wide range of wavelengths, and the electrical responses of many rod cells are transmitted to a single bipolar cell that sums their inputs. Because of these characteristics, rod cells are sensitive to low light levels but do not produce high visual acuity. Rods are the predominant cells in mammalian retinas, with cones restricted to a region called the fovea, but the retinas of other vertebrates are mainly cone-based.

The population of **cone cells** in the retina of a vertebrate includes subgroups of cells that contain pigments that are sensitive to different wavelengths of light. Blue light stimulates one set of cells, red light another, and so on. The variety of cone cells in vertebrate eyes is extensive and includes cells with pigments that are sensitive to wavelengths from deep red to ultraviolet light. Bipolar cells receive input from only a few cone cells, so cone cells require higher levels of illumination than rod cells and are capable of producing sharper images than rod cells. Most non-mammalian vertebrates have cones (and accompanying opsin genes) that have pigments sensitive to four different wavelengths of light, including very short wavelengths in the ultraviolet range.

The ability to distinguish four colors appears to be the basal condition for gnathostomes. At some point in the evolution of the synapsid lineage, the ability to perceive color was evidently reduced, possibly in connection with the adoption of a nocturnal lifestyle by early mammals. Most mammals have dichromatic (two pigments) vision—a long-wavelength pigment with maximum absorption in the green portion of the spectrum and a short-wavelength pigment with maximum sensitivity at blue wavelengths. Old World monkeys and apes have trichromatic vision, having duplicated a photopigment gene, adding red sensitivity.

The full ancestral condition of cones and color vision was retained in sauropsids. Birds have tetrachromatic (four pigments) vision, with red, green, and blue pigments and also an ultraviolet-sensitive pigment. Surprisingly, the ultraviolet-sensitive pigment in the avian retina is not the same as the ancestral short-wavelength pigment of vertebrates; instead it is a novel pigment that evolved when birds "reinvented" ultraviolet sensitivity.

Differences also exist in the way visual information is processed. Birds and other sauropsids retain the ancestral vertebrate condition of relying mainly on the optic lobes of the midbrain for visual processing. In mammals, the optic lobes are small and are used mainly for optical reflexes such as tracking motion, while a novel portion of the cerebrum (the visual cortex) interprets the information.

Chemosensation: Gustation and olfaction

Taste and smell are forms of chemosensation mediated by receptor cells that respond to the presence of chemicals with specific characteristics. Taste and smell seem to be distinct sensations from our perspective as terrestrial animals, and they are mediated by different cranial nerves. The functional distinction is blurred among fishes, however, which have taste buds that are widely dispersed across the body surface.

Taste buds are clusters of cells derived from the embryonic endoderm that open to the exterior through pores. Chemicals must be in contact with the sensory cells in a taste bud to elicit a response, and the sensory cells produce a relatively narrow range of responses. Amniotes have taste buds in the oral cavity (especially on the tongue) and in the pharynx. The taste buds of mammals are broadly distributed over the oral cavity, whereas in sauropsids they are only on the tongue and palate.

Humans perceive chemicals as combinations of salt, sweet, sour, bitter, and umami (savory) sensations, but other vertebrates have different repertoires of taste perception. Carnivorous mammals have lost the ability to taste sweet and umami compounds, and herbivorous mammals are less sensitive to bitter stimuli than are omnivores and carnivores. Much of the variation in sensitivity to different taste categories appears to be consistent with differences in diet, but not all of it. Some of the mismatches may result from multiple functions of taste genes. For example, the gene responsible for detecting sweet and umami tastes is also present in the testes of mice, and its deletion causes male sterility.

Olfactory cells are derived from neural crest tissue and are distributed in the epithelium of the olfactory turbinates that are adjacent to the nasal passages. Chemicals that stimulate olfactory cells are called odorants, and some mammals are exquisitely sensitive to olfactory stimuli. Even humans, who have poor olfactory sensitivity by mammalian standards, may be able to detect as many as 10,000 different odors. The chemical structures of odorant molecules allow them to bind only to specific receptor proteins in the membranes of olfactory cells. Genetically defined families of receptor proteins respond to different categories of odorants, but it is not clear how humans distinguish so many odors. Olfaction and gustation interact to produce

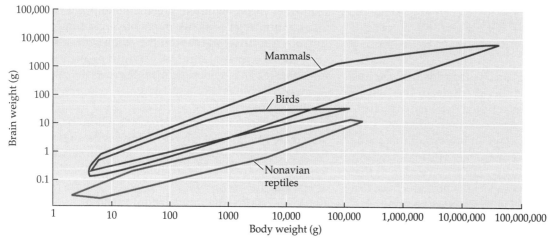

Figure 14.15 Relationship between body size and brain size.   The polygons enclose the values for 301 genera of mammals, 174 species of birds, and 62 species of nonavian reptiles. (After Jerison 1970.)

the sensation we call taste, and the process can be highly stylized, as in the sniffing, sipping, and swirling ritual of wine tasting.

Hearing

A hearing ear was not a feature of the earliest tetrapods. It evolved independently in amniotes and non-amniotes, and separately in different lineages of aminotes. The cochlea, the inner ear structure devoted to hearing, seems to be an ancestral feature for all amniotes. In contrast, the middle ear—which contains the bones that transmit and amplify vibrations from the tympanum (eardrum) to the oval window on the cochlea—appears to have evolved independently several times among amniotes. Mammals have a completely enclosed middle ear involving the stapes (the old fish hyomandibula) and two additional bones (malleus and incus), whereas all sauropsids have a single bone (the stapes) in the middle ear. Beyond those differences, details of the ear structure suggest that among sauropsids the final condition of the middle ear evolved independently in turtles, lepidosaurs, and archosaurs, and within synapsids a fully enclosed middle ear evolved independently in monotremes and therian mammals.

Brains

All amniotes have a relatively enlarged forebrain compared with amphibians, especially with respect to the forebrain (telencephalon). Furthermore, the brains of mammals and birds are larger in proportion to their body size than are the brains of reptiles (**Figure 14.15**). For a 1-kg animal, an average-size mammalian brain would weigh 9.9 g, a bird's brain 6.7 g, and a reptile's brain only 0.7 g. Thus, both birds and mammals are "brainy" compared with reptiles.

Both birds and mammals have larger brains in proportion to their body size than do (or did) basal sauropsids and synapsids, but the increase in size has been achieved differently. In both groups, a distinct cerebrum is formed

by the expansion of the dorsal portion of the telencephalon—the so-called dorsal pallium. In the original amniote condition, the dorsal pallium had two components, and these components had different fates in the sauropsid and synapsid lineages.

In sauropsids, one portion of the dorsal pallium became enlarged into what is called the dorsal ventricular ridge, while the other portion—the lemnopallium—remained small. In synapsids, the reverse occurred, with the cerebrum formed from the lemnopallium. In addition, the mammalian dorsal pallium formed a distinctive six-layered structure now known as the **neocortex**. This structure is highly convoluted in larger-brained mammals, resulting in the type of wrinkled surface that is so apparent in the human cerebrum.

We tend to think of birds as being less intelligent than mammals (the epithet "birdbrain" is a common one), but some birds (e.g., crows and parrots) are capable of highly complex behavior, learning, and probably cognition. Evaluating how "brainy" an animal is can be problematic. Brain size is not necessarily related to intelligence, cognitive ability, or even behavioral flexibility. The size of brains scales with negative allometry; that is, larger animals have proportionally smaller brains than do small animals. The reason for this negative scaling is not known for certain but may be related to the fact that bigger bodies don't need absolutely more brain tissue to operate. For example, if 200 nerve cells in the brain were needed to control the right hindleg of a mouse, there is no reason to suppose that more cells would be needed to control the right hindleg of an elephant. Of course, larger animals have brains that are absolutely larger than smaller ones; they are just not as much larger as one would predict for their size.

Although we think of the tendency to evolve a large brain as a natural outcome of mammalian evolution, the situation is not that simple. Pack-hunting carnivores have larger brains than solitary carnivores, for example, but many small-brained animals display complex behavior and substantial capacity for learning. Much of the enlargement of mammalian brains is related to increased sizes of gene families associated with specific functions, including chemosensation, cell-to-cell signaling, and immune response. Nonetheless, a large brain contains more neurons than a small brain and can make more different connections among neurons. A greater number of neural connections may confer greater behavioral flexibility, which might manifest itself in the ability to adjust to new habitats.

Indeed, some evidence supports this hypothesis: across the entire spectrum of tetrapods—amphibians, reptiles, and mammals—species that have successfully established themselves after being transported to a new area do have larger brains than species that were unsuccessful invaders.

Summary

Synapsids (mammals) and sauropsids (reptiles) are the major terrestrial vertebrate lineages.

The split between synapsids and sauropsids had occurred by the mid Carboniferous, and specializations for terrestrial life evolved independently in the two lineages.

Locomotion and lung ventilation are linked.

The ancestral mode of terrestrial locomotion relied on side-to-side bending of the trunk and created a conflict between locomotion and lung ventilation. Both synapsids and sauropsids resolved this conflict with changes in the limbs, trunk, and lungs.

Limb position of both synapsids and sauropsids became more vertical, with the legs moving more under the body. Changes in the ankle joint in both lineages shifted the feet from pivot points to levers that exert force on the substrate.

Progressive reduction in the ribs in the posterior region of the trunk of synapsids indicates that they developed a diaphragm that ventilated the lungs with a tidal airflow. In contrast, sauropsids continued to use expansion and contraction of the rib cage, but many (perhaps all) lineages developed a one-way airflow through at least some regions of the lung.

Synapsid and sauropsid lungs differ in anatomy and functional properties.

The alveolar structure of synapsid lungs evolved during the late Paleozoic, when atmospheric oxygen concentrations were substantially higher than they are now, whereas the faveolar structure of sauropsid lungs evolved in the early Mesozoic, when oxygen concentrations were lower than current levels.

The tidal airflow in a synapsid lung stretches the gas-exchange surfaces (the alveoli). The alveolar walls are thin, facilitating rapid gas exchange between air and blood.

The flow-through design of sauropsid lungs minimizes expansion and contraction of the gas-exchange surfaces, and may have been the key to sauropsid radiation during the low-oxygen conditions of the early Mesozoic.

A heart is a heart is a heart if you're a synapsid, but sauropsid hearts differ.

Derived synapsids and derived sauropsids have a ventricular septum that separates the pulmonary and systemic circuits of the circulatory system. This arrangement allows blood pressure to be lower in the pulmonary circuit than in the systemic circuit, but it requires flow to be the same in both circuits.

More basal sauropsids (turtles and lepidosaurs) lack a permanent ventricular septum. This cardiac anatomy has the advantage of allowing blood to be shifted between circuits.

The evolution of endothermy was a major event in vertebrate evolution, and came with advantages and disadvantages.

Endotherms continuously maintain body temperatures that are higher than ambient temperatures, allowing them to be active when the environment is cold.

The high metabolic rates of endotherms allow them to engage in activities that are not feasible for ectotherms, such as prolonged running and powered flight, but this benefit comes at a substantial energetic cost: an endotherm uses about 25 times as much energy daily as an ectotherm of the same body size.

Why and how endothermy evolved remain open questions.

Hypotheses about the evolution of endothermy focus either on the direct benefits of having a high and constant body temperature or on selection for another character that secondarily increased metabolic rate and thus metabolic thermogenesis.

Two kinds of fossil evidence that were once thought to distinguish endotherms from ectotherms are now regarded as questionable: bone histology and the presence or absence of respiratory turbinates.

Metabolic rate and insulation present a paradox in the evolution of endothermy, because neither a high metabolic rate nor insulation is beneficial unless the other component is already present. Resolving that paradox requires identifying an advantage for either an insulating body covering or a high metabolic rate that was independent of the presence of the other.

Some hypotheses propose that heat production came first. The elevated metabolic rates required for high levels of locomotor activity or for accelerating development of embryos and neonates might have created a

(Continued)

Summary *(continued)*

selective value for development of insulation to retain metabolic heat.

Other hypotheses propose that insulation came first. Many nonavian dinosaurs had feathery body coverings that probably played the same role in social behavior that the feathers of extant birds do. These feathery body coverings could have been co-opted secondarily to retain heat produced by elevated metabolic rates. Evidence that non-mammalian synapsids probably had hair suggests a similar resolution to the paradox of ectothermy and insulation in the synapsid lineage.

Synapsids and sauropsids evolved different methods of eliminating metabolic wastes while minimizing urinary water loss.

Nitrogenous wastes are excreted in urine, but because water is a limited resource for terrestrial tetrapods, the excretory system must minimize water loss.

All tetrapods excrete some ammonia, urea, and uric acid. The primary waste product of a species is related to its phylogeny and ecology.

- Ammonia (NH_3) is energetically cheap to produce and soluble in water, but it is toxic and cannot be allowed to accumulate in the body. It can be excreted as a dilute solution in urine and through the skin and gills.
- Urea ($CO[NH_2]_2$) is less toxic and more soluble than ammonia. It can be accumulated and excreted as a concentrated solution in urine.
- Uric acid ($C_5H_4O_3N_4$) is relatively nontoxic and only slightly soluble. It combines with ions of sodium and potassium and precipitates as sodium or potassium urate.

Synapsids excrete nitrogenous wastes primarily as urea, with some ammonia and uric acid. Nonavian sauropsids excrete all three forms of nitrogenous waste, and the proportion of each form varies with an animal's hydration status. Birds excrete primarily salts of uric acid.

The mammalian kidney is capable of producing concentrated urine containing primarily urea and ions.

Blood pressure forces water, ions, and small molecules through the walls of the glomerular capillaries, producing an ultrafiltrate that has the same osmolal concentration as blood plasma.

Sodium ions, glucose, and amino acids are removed from the ultrafiltrate in the proximal convoluted tubule (PCT), and water follows passively.

The descending limb of the loop of Henle is permeable to water, and the loop passes through tissue with a high osmolal concentration that draws water from the urine, leaving solutes behind.

Sodium ions are reabsorbed from the ascending limb of the loop of Henle, but the walls are impermeable to water. The sodium ions removed from the urine sustain the high osmolal concentration of the tissues surrounding the loop of Henle.

The walls of the distal convoluted tubule (DCT) are permeable to water, and water flows outward until the concentration of the urine has equilibrated with plasma concentration..

The collecting duct plunges back through the portion of the kidney with high osmolal concentrations, and the permeability of its walls can be altered to adjust the final osmolal concentration of urine.

Antidiuretic hormone (ADH) causes aquaporin molecules to be inserted into the plasma membranes of the collecting duct cells forming water channels and increasing the concentration of urine.

The sauropsid kidney operates on a different principle because uric acid, the primary form of nitrogenous waste of sauropsids, has very low solubility.

The low solubility of uric acid precludes the production of concentrated urine; concentrating urine in the kidney would cause urate salts to precipitate and clog the nephrons.

Nonavian reptiles have nephrons with only short loops of Henle and kidneys without regions of high osmolal concentration. These kidneys produce urine that has the same osmolal concentration as plasma.

Birds have two types of nephrons: short loop nephrons and long loop nephrons that extend into regions of higher osmolal concentration. The long loop nephrons allow birds to produce urine that is two to three times more concentrated than plasma.

After it leaves the kidney, the urine of sauropsids is concentrated in the bladder or cloaca. Ions are actively withdrawn from the urine, and water follows passively, raising the concentration of uric acid until some of it precipitates as urate salts. Precipitation reduces the concentration of urine, setting the scene for another round of active uptake of ions and precipitation of urate salts. Ultimately what remains is a slurry of urate salts with very little water.

Many sauropsids have an additional water-conserving mechanism: salt glands that excrete the ions that were removed from the bladder at concentrations five or more times the concentration of plasma.

Summary *(continued)*

Synapsids and sauropsids differ in their sensory modalities.

In general, synapsids rely more on olfaction than on vision, whereas sauropsids have excellent vision and variable olfactory sensitivity.

At some point in their evolution, synapsids lost the ancestral gnathostome ability to perceive four colors. Most mammals have dichromatic (two pigment) vision, with maximum sensitivity to green and blue pigments. Anthropoid primates are an exception; a gene duplication added a red-sensitive pigment, providing trichromatic vision.

Sauropsids retain the ancestral condition of four photosensitive pigments—red, green, blue, and a pigment sensitive to ultraviolet wavelengths.

Taste (gustation) and smell (olfaction) are forms of chemosensation, mediated by receptors that respond to chemicals with particular characteristics.

In synapsids, taste receptors are broadly distributed in the oral cavity, particularly on the tongue. In sauropsids, taste receptors are only on the tongue and palate.

Olfactory receptors located in the olfactory turbinates detect airborne molecules.

Sauropsids have been considered to be less sensitive to olfactory stimuli than mammals, but that view is changing as new studies reveal that some sauropsids use olfaction in navigation, predation, and social behavior.

A hearing ear evolved independently in different lineages of amniotes.

The cochlea of the inner ear is an ancestral character of amniotes. The middle ear, which contains the bones that transmit vibrations from the tympanum to the cochlea, evolved independently several times among amniotes.

The middle ear of mammals contains three bones (stapes, malleus, and incus), whereas the middle ear of sauropsids contains only the stapes.

The brains of mammals are larger in proportion to their body size than are the brains of sauropsids.

Both synapsids and sauropsids have an enlarged telencephalon (cerebrum) compared with amphibians, but these cerebra are elaborated from different portions of the dorsal telencephalon (the dorsal ventricular ridge in sauropsids and the lemnopallium in synapsids).

Brain size cannot be directly equated with intelligence, cognitive ability, or behavioral flexibility. Much of the enlargement of mammalian brains is linked to increased sizes of gene families associated with specific functions (chemosensation, cell-to-cell signaling, and immune response).

For Discussion

1. How does the ancestral locomotor system of tetrapods interfere with respiration?

2. Why is reabsorption of water from the bladder or cloaca essential for sauropsids?

3. By the time you feel thirsty, you are already somewhat dehydrated. What is happening to the ADH in your pituitary gland at that point? How does that response contribute to regulating the water content of your body?

4. Birds regularly fly at elevations higher than human mountain climbers can ascend without using auxiliary oxygen; radar tracking of migrating birds shows that they sometimes fly as high as 6,500 m. The alpine chough (*Pyrrhocorax graculus*) lives at elevations of about 8,200 m on Mount Everest, and migrating bar-headed geese (*Anser indicus*) pass directly over the summit of the Himalayas at elevations of 9,200 m. Explain the features of the sauropsid respiratory system that allow birds to breathe at such high elevations.

5. Understanding the evolution of endothermy requires resolving a paradox: increasing the resting metabolic rate can provide heat to raise the body temperature, but without insulation (hair, feathers, or blubber) that heat is rapidly lost to the environment. But a layer of insulation does not benefit a vertebrate unless the animal has a high metabolic rate. Discuss the evidence suggesting that Mesozoic synapsids and sauropsids had developed both high metabolic rates and insulation, and identify the puzzling gaps in the information currently available.

6. The nasal salt glands of insectivorous lizards respond to injections of chloride ions by increasing secretions of sodium or potassium chloride, whereas injections of sodium or potassium ions do not increase the activity of the nasal salt glands. What difference in the routes of excretion of sodium and potassium ions compared with chloride ions accounts for the different responses of the lizards?

Additional Reading

Brainerd EL. 2015. Major transformations in vertebrate breathing mechanisms. *In* KP Dial et al. (eds.), *Great Transformations in Vertebrate Evolution*, pp. 47–61. The University of Chicago Press, Chicago.

Cieri RL, Farmer CG. 2016. Unidirectional pulmonary airflow in vertebrates: A review of structure, function, and evolution. *Journal of Comparative Physiology B* 186: 541–552.

Clavijo-Baque S, Bozinovic F. 2012. Testing the fitness consequences of the thermoregulatory and parental care models for the origin of endothermy. *PLoS ONE* 7(5): e37069. doi:10.1371/journal.pone.0037069.

Dantzler WH. 2016. *Comparative Physiology of the Vertebrate Kidney*, 2nd Ed. Springer Science+Business Media, New York.

Farmer CG. 2015. Similarity of crocodylian and avian lungs indicates unidirectional flow is ancestral for archosaurs. *Integrative and Comparative Biology* 55: 962–971.

Hedrick MS, Hillman SS. 2016. What drove the evolution of endothermy? *Journal of Experimental Biology* 219: 300–301.

Hicks JW, Wang T. 2012. The functional significance of the reptilian heart: New insights into an old question. *In* D Sedmera and T Wang (eds.), *Ontogeny and Phylogeny of the Vertebrate Heart*, pp. 207–227. Springer, London.

Larsen EK and 6 others. 2014. Osmoregulation and excretion. *Comprehensive Physiology* 4: 405–573.

Owerkowicz T, Musinsky C, Middleton KM, Crompton AW. 2015. Respiratory turbinates and the evolution of endothermy in mammals and birds. *In* KP Dial et al. (eds.), *Great Transformations in Vertebrate Evolution*, pp. 143–165. The University of Chicago Press, Chicago.

Rakus K, Ronsmans M, Vanderplasschen A. 2017. Behavioral fever in ectothermic vertebrates. *Developmental & Comparative Immunology* 66: 84–91.

Tattersall GJ. 2016. Reptile thermogenesis and the origins of endothermy. *Zoology* 119: 403–405

Vandewege MW and 5 others. 2016. Contrasting patterns of evolutionary diversification in the olfactory repertoires of reptile and bird genomes. *Genome Biology and Evolution* 8: 470–480.

Go to the Companion Website **oup.com/us/vertebratelife10e** for active-learning exercises, news links, references, and more.

Ectothermy: A Low-Energy Approach to Life

Ectothermy is an ancestral character of vertebrates, and like many ancestral characters it is just as effective as its derived counterpart, endothermy. Furthermore, the mechanisms of ectothermy are as complex and specialized as those of endothermy. Here we consider the consequences of ectothermy in shaping broad aspects of the lifestyle of fishes, amphibians, and nonavian reptiles.

15.1 Vertebrates and Their Environments

Vertebrates manage to live in the most unlikely places. Amphibians live in deserts where rain falls only a few times a year, and several years may pass with no rainfall at all. Lizards live on mountains at elevations above 5,000 m, where the temperature falls below freezing nearly every night of the year and does not rise much above freezing during the day.

Of course, vertebrates do not seek out challenging places to live—birds, lizards, mammals, and even amphibians can be found on tropical beaches (sometimes running between the feet of surfers), and fishes cruise the shore. However, even these apparently benign environments present challenges. Examining the ways that vertebrates live in extreme environments has provided much information about how they function as organisms—that is, how morphology, physiology, ecology, and behavior interact and have been shaped during vertebrate evolution.

Elegant adaptations allow certain specialized vertebrates to colonize demanding habitats. More common and more impressive than these specializations, however, is the realization of how minor are the modifications of the ancestral vertebrate body plan that allow animals to endure air temperatures ranging from –70°C to +60°C or conditions ranging from complete immersion in water to complete independence of liquid water. No obvious differences distinguish animals from vastly different habitats; a lizard from sea level looks very much like one from the Andes Mountains (**Figure 15.1**). The adaptability of vertebrates lies in the combination of minor modifications of their ecology, behavior,

morphology, and physiology. A view that integrates these elements shows the startling beauty of organismal function of vertebrates.

15.2 Dealing with Dryness: Ectotherms in Deserts

A desert is defined as a region in which the potential loss of water (via evaporation and the transpiration of water by plants) exceeds the input of water via precipitation. Desert conditions are produced by various combinations of topography, air movements, and ocean currents and are found from the poles to the Equator. But whatever their cause, the feature all deserts have in common is a scarcity of liquid water.

Dryness is at the root of many features of deserts that make them difficult places for vertebrates to live. The dry air characteristic of most deserts seldom contains enough moisture to form clouds that would block solar radiation during the day or prevent radiative cooling at night. As a result, deserts in the temperate regions are hot by day and cold at night. Solar radiation is intense, and air does not conduct heat rapidly. A sunlit patch of ground can be lethally hot, whereas a shaded area just a few centimeters away can be substantially cooler. Underground retreats offer protection from both heat and cold. The annual temperature extremes at the surface of the ground in the Mohave Desert in the southwestern United States extend from a low that is below freezing to a maximum above 50°C; but just 1 m below the surface of the ground, the annual temperature range is only from 10°C to 25°C. Desert animals rely on the temperature differences between sunlight and shade and between the surface and underground burrows to escape both hot and cold.

Scarcity of water is reflected by sparse plant life. With few plants, deserts have low densities of insects for small vertebrates to eat and correspondingly sparse populations of small vertebrates that would be prey for larger vertebrates. Food shortages are chronic and are worsened by seasonal shortages and by unpredictable years of drought when the usual pattern of rainfall does not develop.

(A) *L. isabelae*: Rainfall <1 cm

(B) *L. chiliensis*: Rainfall >1.5 m

(C) *L. nigromaculatus*: Sea level

(D) *L. nigericeps*: Elevation >5,200 m

Figure 15.1 One of evolution's great continental radiations. Originating from a single ancestor, the lizard genus *Liolaemus* contains more than 260 species that occupy a wide variety of habitats in South America. These habitats cover the spectrum of extreme conditions. For example, *L. isabelae* (A) occurs in Andean deserts where precipitation occurs as snow and amounts to less than 1 cm of liquid water annually, whereas *L. chiliensis* (B) occurs in a region that receives 1.2 *meters* of rain annually. For another example, *L. nigromaculatus* (C) lives at sea level, while *L. nigericeps* (D) lives at elevations above 5,200 m. Despite the enormous differences in their habitats, these generalist species of *Liolaemus* are similar in body form. The only conspicuous differences are their colors and patterns, each of which is cryptic in the species' habitat. (Photos courtesy of Daniel Pincheira-Donoso.)

The low metabolic rates of ectotherms alleviate some of the difficulty caused by scarcity of food and water, but many desert ectotherms must temporarily extend the limits within which they regulate body temperatures or body fluid concentrations, become inactive for large portions of the year, or adopt a combination of these responses.

Desert tortoises

Tortoises (*Gopherus agassizii*) in Rock Valley, Nevada, construct shallow burrows that they use as daily retreat sites during the summer and deeper burrows for hibernation in winter. The tortoises in one study area emerged from hibernation in spring, and aboveground activity extended throughout the summer until they began hibernation again in November.

Figure 15.2 shows the annual cycle of time spent aboveground and in burrows by the tortoises in the study and the annual cycles of energy, water, and salt balance. Examination of the behavior and dietary habits of the tortoises throughout the year shows what was happening.

After they emerged from hibernation in the spring, the tortoises were active for about 3 hours (h) every fourth day; the rest of the time they spent in their burrows. From March through May the tortoises were eating annual plants that had sprouted after the winter rains. They obtained large amounts of water and potassium from this diet, and their water and salt balances were positive. Desert tortoises lack salt glands, and their kidneys cannot produce concentrated urine, so they have no way to excrete the salt without losing a substantial amount of water in the process. Instead they retain the salt, and the osmolality of the tortoises' body fluids increased by 20% during the spring. This increased concentration shows that they were osmotically stressed as a result of the high concentrations of potassium in their food. Furthermore, the energy content of the plants was not great enough to balance the metabolic energy expenditure of the tortoises, and they were in negative energy balance. During this period, the tortoises were using stored energy by metabolizing their body tissues.

As the temperature increased from late May through early July, the tortoises shortened their daily activity periods to about an hour every sixth day and spent the rest of the time in shallow burrows. The annual plants died, but the tortoises shifted to eating grass and achieved positive energy balances. They stored this extra energy as new body tissue. The dry grass contained little water, however, and

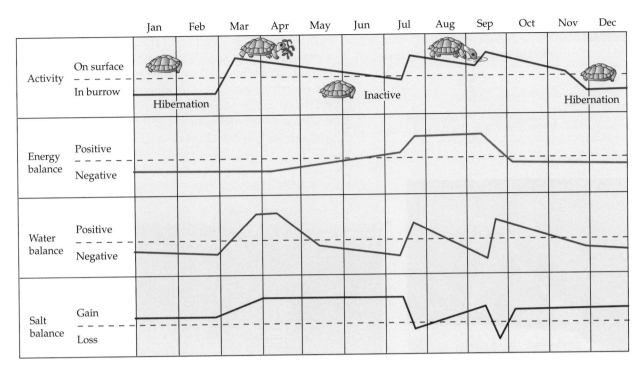

Figure 15.2 Annual cycle of desert tortoises. The energy balance, water balance, and salt balance of tortoises change through the year as the animals adjust feeding and activity. A positive balance means the animals show a net gain, whereas a negative balance represents a net loss. Positive energy and water balances indicate that conditions are good for the tortoises, but a positive salt balance means that ions are accumulating in the body fluids faster than they can be excreted. That situation indicates a relaxation of homeostasis and is probably stressful for the tortoises. This study showed that the animals were often in negative balance for water and energy, and that they accumulated salt during much of the year. (Based on data from Nagy and Medica 1986.)

the tortoises were in negative water balance. The osmolal concentrations of their body fluids remained at the high levels they had reached earlier in the year.

In mid-July, thunderstorms dropped rain on the study site, and most of the tortoises emerged from their burrows. They drank water from natural basins, and some of the tortoises constructed basins by scratching shallow depressions in the ground that could then trap rainwater. The tortoises drank large quantities of water (nearly 20% of their body weight) and voided the contents of their urinary bladders. The osmolal concentrations of their body fluids and urine decreased as the tortoises moved into positive water balance and excreted the excess salts they had accumulated when water was scarce. The behavior of the tortoises changed after the rain: they ate every 2–3 days and often spent their periods of inactivity aboveground instead of in their burrows.

August was dry, and the tortoises lost body water and accumulated salts as they fed on dry grass. They were in positive energy balance, however, and their weight increased. More thunderstorms in September allowed the tortoises to

drink again and to excrete the excess salts they had been accumulating. Seedlings sprouted after the rain, and in late September the tortoises started to eat them.

In October and November, the tortoises continued to feed on freshly sprouted green vegetation; however, low temperatures reduced the tortoises' activity, and they were in slightly negative energy balance. Salts accumulated and the osmolal concentrations of the body fluids increased slightly. In November, the tortoises entered hibernation. Hibernating tortoises had low metabolic rates and lost water and body tissue slowly. When they emerged from hibernation the following spring, they weighed only a little less than they had in the fall. Over the entire year, the tortoises increased their body tissues by more than 25% and balanced their water and salt budgets, but they did this by tolerating severe imbalances in their energy, water, and salt relations for periods that extended for months at a time.

The chuckwalla

The chuckwalla (*Sauromalus ater*) is an herbivorous iguanid lizard that lives in the rocky foothills of desert mountain ranges. The annual cycle of the chuckwalla, like that of the desert tortoise, is shaped by the interaction of food and temperature. The lizard faces many of the same problems that the tortoises encounter, but its responses are different. The lizard has nasal glands that allow it to excrete salt at high concentrations, and it does not drink rainwater but instead depends on water it obtains from the plants it eats.

Two categories of water are available to an animal from the food it eats: Preformed water and metabolic water. Preformed water corresponds to the water content of the

Table 15.1	Quantity of Water Produced by Metabolism of Different Substrates	
Compound	**Grams of Water/Gram of Compound**	
Carbohydrate	0.556	
Fat	1.071	
Protein	0.396 when urea is the end product; 0.499 when uric acid is the end product	

food—that is, it is in the form of molecules of water (H_2O) that are absorbed across the wall of the intestine. Metabolic water is a by-product of the chemical reactions of metabolism. Protons (H^+) are combined with oxygen (O_2) during aerobic metabolism, yielding a molecule of water for every two protons. The amount of metabolic water produced can be substantial; more than a gram of water is released by metabolism of a gram of fat (Table 15.1). For animals such as the chuckwalla that do not drink liquid

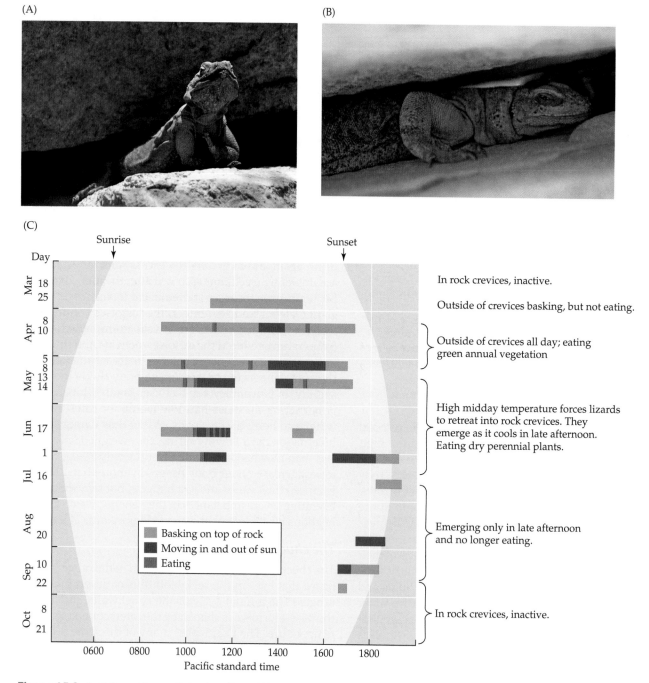

Figure 15.3 Activity patterns of chuckwallas. (A) In spring, chuckwallas (*Sauromalus ater*) feed on fresh vegetation that grows following winter rains. (B) The lizards retreat to rock crevices to escape heat and cold. To deter predators, they wedge themselves into crevices by inflating their lungs. (C) Daily behavior patterns of chuckwallas studied at Black Mountain in the Mohave Desert of California. (A, René C. Clark © Dancing Snake Nature Photography; B, courtesy of Julia Goebel; C from Nagy 1973, courtesy of the American Society of Ichthyologists and Herpetologists.)

water, preformed water and metabolic water are the only routes of water gain that can replace the water lost by evaporation and excretion.

Chuckwallas studied at Black Mountain in California's Mohave Desert spent the winter hibernating in rock crevices and emerged from hibernation in April. Individual lizards spent about 8 h a day on the surface in April and early May (Figure 15.3). By mid-May, air temperatures were rising above 40°C at midday, and the chuckwallas retreated into rock crevices for about 2 h during the hottest part of the day, emerging again in the afternoon when the desert was cooler. In April and May, annual plants that sprouted after the winter rains supplied both water and nourishment, and the chuckwallas gained weight rapidly. The average increase in body weight between April and mid-May was 18%. The water content of the chuckwallas increased faster than total body weight, indicating that the animals were storing excess water.

By early June, the annual plants had withered and the chuckwallas were feeding on perennial plants that contained less water and more ions than the annual plants. Both the body weight and the water contents of the lizards declined. The activity of the lizards decreased in June and July as midday temperatures increased. Lizards emerged in the morning or afternoon, but not during the middle of the day. By late June, the chuckwallas had reduced their feeding activity, and in July they stopped eating altogether. They spent most of the day in the rock crevices, emerging only in late afternoon to bask for an hour or so every second or third day. From late May through autumn, the chuckwallas lost water and body weight steadily. When they entered hibernation in October, they weighed an average of 37% less than they had in April when they emerged from hibernation.

The water budget of a 200-g chuckwalla is shown in Table 15.2. In early May, the annual plants it fed on contained more than 2.5 g of free water per gram of dry plant material, and the lizard showed a positive water balance, gaining 0.81 g of water per day. By late May, when the plants had withered, their free water content had dropped to just less than 1 g of water per gram of dry plant matter, and the chuckwalla was losing 0.84 g of water per day. The rate of water loss fell to 0.32 g per day when the lizard stopped eating.

Evaporation from the respiratory surfaces and from the skin accounted for about 61% of the total water loss of chuckwallas. As midday temperatures increased and lizards stopped eating, they spent most of the day in rock crevices. The body temperatures of inactive chuckwallas were lower than the temperatures of lizards that remained active. Because of their low body temperatures, the inactive chuckwallas had lower rates of metabolism. They breathed more slowly and lost less water from their respiratory passages. Also, the humidity was higher in the rock crevices than on the surface of the desert, and this reduction in the humidity gradient between the animal and the air further reduced evaporation. Most of the remaining water loss occurred in the feces (31%) and urine (8%). When a lizard stopped eating, it also stopped producing feces and reduced the amount of urine it excreted. The combination of these factors reduced the daily water loss of chuckwallas by almost 90%.

The food plants were always hyperosmolal with the body fluids of the lizards and had high concentrations of potassium. Despite this dietary salt load, osmotic concentrations of the chuckwallas' body fluids did not show the variation seen in tortoises, because the lizards' nasal salt glands were able to excrete ions at high concentrations. The concentration of potassium ions in the salt-gland secretions was nearly ten times the concentration in urine. The precipitation of potassium salts of uric acid was the second major route of potassium excretion by the lizards and was nearly as important in the overall salt balance as nasal secretion. The chuckwallas would not have been able to balance their salt budgets without these two extrarenal

Table 15.2	Seasonal Changes in the Water Balance of a 200-Gram Chuckwalla			
Water gain or loss (g/day)	Early May (eating green plants)	Late May (eating dry plants)	September (not eating)	Explanation
Total water intake	7.24	3.42	0.20	The lizards gain water even when they are eating dry plants. When they shift to dry plants, the contribution of metabolic water rises from less than 10% to 20% of total water gain.
Preformed water in food	6.56	2.74	0.00	Green plants provide more than twice as much preformed water as dry plants.
Metabolic water	0.68	0.68	0.20	Digesting green and dry plants yields the same amount of metabolic water.
Total water loss	6.41	4.26	0.52	Less time spent outside rock crevices in late May reduces water loss, and the further reduction in surface activity and fasting in September further reduce water loss.
Net water gain or loss	+0.81	−0.84	−0.32	Lizards are in positive water balance only in early May; the rest of the year, they are minimizing water loss until green plants become available again in the following spring.

routes of ion excretion, but with them they were able to maintain stable osmolal concentrations.

Both the chuckwallas and tortoises illustrate the interaction of behavior and physiology in responding to the characteristics of their desert habitats. The tortoises lack salt-secreting glands and store the salt they ingest, tolerating increased body fluid concentrations until a rainstorm allows them to drink water and excrete the excess salt. The chuckwallas are able to stabilize their body fluid concentrations by using their nasal glands to excrete excess salt, but they do not take advantage of rainfall to replenish their water stores. Instead they became inactive, reducing their rates of water loss by almost 90% and relying on energy stores and metabolic water production to see them through periods of drought.

Conditions for the chuckwallas were poor at the Black Mountain site during the study. Only 5 cm of rain had fallen during the preceding winter, and that low rainfall probably contributed to the early withering of the annual plants that forced the chuckwallas to cease activity early in the summer. Unpredictable rainfall is a characteristic of deserts, however, and the animals living there must be able to adjust to the consequences. Rainfall records from the weather station closest to Black Mountain showed that in 5 of the preceding 10 years, the annual total rainfall was about 5 cm. Thus, the year of the study was not unusually harsh; conditions are sometimes even worse—only 2 cm of rain fell during the winter after the study. However, conditions in the desert are sometimes good. Fifteen centimeters of rain fell in the winter of 1968, and vegetation remained green and lush throughout the following summer and fall. Chuckwallas and tortoises live for decades, and their responses to the boom-or-bust conditions of their harsh environments must be viewed in the context of their long lifespans. A temporary relaxation of the limits of homeostasis in bad years is an effective trade-off for survival that allows the animals to exploit the abundant resources of good years.

Desert amphibians

Permeable skins and high rates of water loss are characteristics that would seem to make amphibians unlikely inhabitants of deserts, but certain species are abundant in desert habitats. Most remarkably, these animals succeed in living in the desert because of their permeable skins, not despite them. Toads are the most common desert amphibians; tiger salamanders are found in the deserts of North America, and several species of plethodontid salamanders occupy seasonally dry habitats in California.

Western spadefoot toads (genus *Spea*) are the most thoroughly studied desert anurans (**Figure 15.4**). They inhabit the desert regions of North America—including the edges of the Algodones sand dunes in southern California, where the average annual precipitation is only 6 cm, and in some years no rain falls at all. An analysis of the mechanisms that allow an amphibian to exist in a habitat like that must include consideration of both water loss and gain. A desert anuran must control its water loss behaviorally by its choice of sheltered microhabitats free from solar radiation and wind movement. Different species of anurans use different microhabitats—a hollow in the bank of a desert wash, the burrow of a ground squirrel or kangaroo rat, or a burrow the anuran excavates for itself. All these places are cooler and wetter than exposed ground.

Desert anurans spend extended periods underground, emerging on the surface only when conditions are favorable. Spadefoot toads construct burrows about 60 cm deep, filling the shaft with dirt and leaving a small chamber at the bottom, which they occupy. In southern Arizona, the spadefoots construct these burrows in September, at the end of the summer rainy season, and remain in them until the rains resume the following July.

At the end of the rainy season when the spadefoots first bury themselves, the soil is relatively moist. The water tension created by the normal osmolal concentration of a spadefoot's body fluids establishes a gradient favoring movement of water from the soil into the toad. In this situation, a buried spadefoot can absorb water from the soil just as the roots of plants do. With a supply of water available, a spadefoot toad can afford to release urine to dispose of its nitrogenous wastes.

As time passes, the soil moisture content decreases, and the soil moisture potential (the driving force for movement of water) becomes more negative, until it equals the water potential of the spadefoot. At this point, there is no longer a gradient allowing movement of water into the toad. When its source of new water is cut off, a spadefoot stops excreting urine and instead retains urea in its body, increasing the osmotic pressure of its body fluids. Osmotic concentrations as high as 600 milliosmolal (mOsm) have been recorded in spadefoot toads emerging from burial at the end of the dry season. The low water potential produced by the high osmolal concentration of a spadefoot's body fluids may reduce the water gradient between the animal and the air in its underground chamber so that evaporative water loss is reduced. Sufficiently high internal concentrations would create potentials that would allow spadefoot toads to absorb water from even very dry soil.

Figure 15.4 A Mexican spadefoot toad (*Spea multiplicata*). These toads spend 9–10 months underground, digging their way into the soil backward, using their hindfeet. The spade that gives them their name is a keratinous structure on each hindfoot. (Photo © designpics/123RF.)

The ability to continue to draw water from soil enables a spadefoot toad to remain buried for 9–10 months without access to liquid water. In this situation, its permeable skin is not a handicap to the spadefoot—it is an essential feature of the toad's biology. If the spadefoot had an impermeable skin, or if it formed an impermeable cocoon as some other amphibians do, water would not be able to move from the soil into the animal. Instead, the spadefoot would have to depend on the water contained in its body when it was buried. Under those circumstances, spadefoot toads probably would not be able to invade the desert because their initial water content would not see them through a 9-month dry season.

15.3 Coping with Cold: Ectotherms in Subzero Conditions

Temperatures drop below freezing in the habitats of many vertebrates on a seasonal basis, and some animals at high elevations may experience freezing temperatures on a daily basis for a substantial part of the year. Birds and mammals respond to cold by increasing metabolic heat production and insulation, but ectotherms do not have those options. Instead, ectotherms show one of three responses—they retreat to shelters that do not cool to freezing, they experience freezing temperatures but avoid freezing by supercooling or by synthesizing antifreeze compounds, or they tolerate freezing and thawing by using mechanisms that prevent damage to cells and tissues.

Frigid fishes

The temperature at which water freezes is affected by its osmolal concentration: pure water freezes at 0°C, and increasing the osmolal concentration lowers the freezing point. Body fluid concentrations of marine fishes are 300–400 mOsm, whereas seawater has a concentration near 1,000 mOsm. The osmolal concentrations of the body fluids of marine fishes correspond to freezing points of –0.6°C to –0.8°C, and the freezing point of seawater is –1.86°C. The temperature of arctic and antarctic seas falls to or below –1.86°C in winter, yet the fishes swim in this water without freezing. How do they do that?

A classic study of freezing avoidance of fishes was conducted in Hebron Fjord in Labrador. In summer, the temperature of the surface water at Hebron Fjord is above freezing, but the water at the bottom of the fjord is –1.73°C (**Figure 15.5**). In winter, the surface temperature of the water also falls to –1.73°C, like the bottom temperature. Several species of fish live in the fjord; some are bottom-dwellers and others live near the surface. These two zones present

(A) Summer

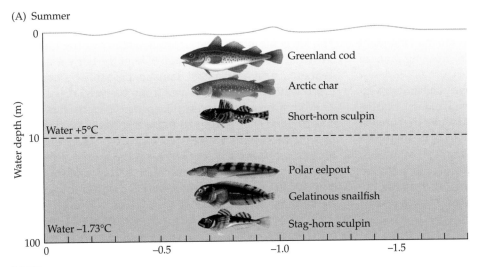

(B) Winter

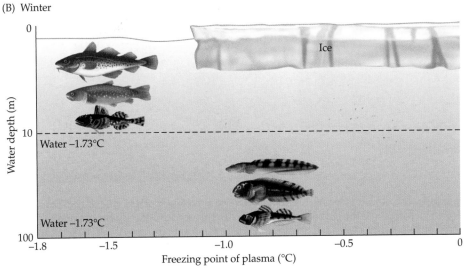

Figure 15.5 Water temperatures and freezing points of fishes in Hebron Fjord, Labrador. The water temperature and occurrence of ice are shown for summer (A) and winter (B). In summer, the temperature in the upper 10 m of water rises to about 5°C, but in winter it falls to –1.73°C. Deeper water remains at –1.73°C year-round. Fishes that live in deep water (polar eelpout, *Lycodes turneri*; gelatinous snailfish, *Liparis koefoedi*; and staghorn sculpin, *Gymnacanthus tricuspis*) have freezing points of –0.8°C in summer and winter, whereas the fishes that live in shallow water (Greenland cod, *Gadus ogac*; short-horn sculpin, *Myxocephalus scorpius*; and arctic char, *Salvelinus alpinus*) decrease their freezing points from about –0.8°C in summer to about –1.5°C in winter. (After Scholander et al. 1957.)

different challenges for the fishes. The temperature near the bottom of the fjord is always below freezing, but ice is not present because ice is lighter than water and remains at the surface. Surface-dwelling fishes live in water temperatures that rise well above freezing in the summer and drop below freezing in winter, and ice is present in the winter.

The body fluids of bottom-dwelling fishes in Hebron Fjord have freezing points of −0.8°C year-round. Because their body temperatures are −1.73°C, these fishes are supercooled; that is, the water in their bodies is in the liquid state even though it is below its freezing point. When water freezes, the water molecules become oriented in a crystal lattice. The process of crystallization is accelerated by nucleating agents that hold water molecules in the proper spatial orientation for freezing. In the absence of nucleating agents, pure water can remain liquid at −20°C. In the laboratory, the fishes from the bottom of Hebron Fjord can be supercooled to −1.73°C without freezing, but if they are touched with a piece of ice, which serves as a nucleating agent, they freeze immediately. At the bottom of the fjord there is no ice, and the bottom-dwelling fishes exist year-round in a supercooled state.

What about the fishes in the surface waters? They do encounter ice in winter when the water temperature is below the osmotically determined freezing point of their body fluids, and supercooling would not be effective in that situation. Instead, the surface-dwelling fishes synthesize antifreeze substances in winter that lower the freezing point of their body fluids to approximately the freezing point of the seawater in which they swim.

Antifreeze compounds are widely developed among vertebrates (and also among invertebrates and plants). Marine fishes have two categories of organic molecules that protect against freezing: glycoproteins with molecular weights of 2,600 to 33,000 and polypeptides and small proteins with molecular weights of 3,300 to 13,000. These compounds are extremely effective in preventing freezing. For example, the blood plasma of the Antarctic fish *Trematomus borchgrevinki*

contains a glycoprotein that is several hundred times more effective than sodium chloride in lowering the freezing point. The glycoprotein is adsorbed onto the surface of ice crystals and hinders their growth by preventing water molecules from assuming the proper orientation to join the ice-crystal lattice.

Frozen frogs

Terrestrial amphibians that hibernate in winter show at least two categories of responses to low temperatures. One group, which includes salamanders, toads, and aquatic frogs, buries deeply in the soil or hibernates in the mud at the bottom of ponds. These animals are not exposed to temperatures below the freezing point of their body fluids. As far as we know, they have no antifreeze substances and no capacity to tolerate freezing. However, other amphibians are exposed to temperatures below their freezing points. Unlike fishes, these amphibians freeze at low temperatures but are not killed by freezing (**Figure 15.6**). These species can remain frozen at −3°C for several weeks, and they tolerate repeated bouts of freezing and thawing without damage. However, temperatures lower than −10°C are lethal.

Freeze-tolerant animals draw water out of their cells before it can freeze, and restrict formation of ice crystals to the extracellular fluids. Low-molecular-weight cryoprotectant substances in the cells prevent intracellular ice formation. The ice content of frozen frogs is usually in the range of 34–48%. Freezing of more than 65% of the body water appears to cause irreversible damage, probably because too much water has been removed from the cells.

Wood frogs (*Rana sylvatica*), spring peepers (*Pseudacris crucifer*), and chorus frogs (*Pseudacris triseriata*) use glucose as a cryoprotectant, whereas gray treefrogs (*Hyla versicolor*) use glycerol. Glycogen in the liver is the source of the glucose and glycerol. The accumulation of these substances is stimulated by freezing and is initiated within minutes of the formation of ice crystals.

(A)

(B)

Figure 15.6 Freezing tolerance of wood frogs. (A) At normal temperature, a wood frog (*Rana sylvatica*) is light brown with darker brown markings, and the lens of the eye is clear. (B) The frog turns dark when it freezes, and the lens of the eye becomes white because ice crystals reflect light. The frog resumes its normal color when it thaws. To demonstrate that frogs do survive freezing, these photographs are shown in the reverse of the sequence in which they were taken—that is, (A) shows the frog after it had thawed out. (Courtesy of Janet M. Storey.)

Frozen frogs are, of course, motionless. Breathing stops, the heartbeat is exceedingly slow and irregular or may cease entirely, and blood does not circulate through frozen tissues. Nonetheless, the cells are not frozen; they have a low level of metabolic activity that is maintained by anaerobic metabolism. The glycogen content of frozen muscle and kidney cells decreases, and concentrations of lactic acid and alanine (two end products of anaerobic metabolism) increase.

15.4 Energetics of Ectotherms and Endotherms

Life as an animal is costly. In thermodynamic terms, an animal lives by breaking chemical bonds that were formed by a plant (if the animal is an herbivore) or by another animal (if it is a carnivore) and using the energy from those bonds to sustain its own activities. Vertebrates are particularly expensive animals because they generally are larger and more mobile than invertebrates. Big animals require more energy (i.e., food) than small ones, and active animals use more energy than sedentary ones.

In addition to body size and activity, an animal's method of temperature regulation (ectothermy, endothermy, or a combination of the two mechanisms) is a key factor in determining how much energy it uses and therefore how much food it requires. Because ectotherms rely on external sources of energy to raise their body temperatures to the level needed for activity, whereas endotherms use heat generated internally by their metabolism, ectotherms use substantially less energy than do endotherms. The metabolic rates (i.e., rates at which energy is used) of terrestrial ectotherms are only 10–14% of the metabolic rates of birds and mammals of the same body size. The lower energy requirements of ectotherms mean that they need less food than does an endotherm of the same body size.

Body size

Body size is another major difference between ectotherms and endotherms that relates directly to their mode of temperature regulation and affects their roles in terrestrial ecosystems. The smallest ectothermal tetrapods are much smaller than the smallest endotherms (**Figure 15.7**). The very high energetic cost of endothermy at small body sizes accounts for this difference. As body weight decreases, the mass-specific cost of living (energy expended per gram of body tissue) for an endotherm increases rapidly, becoming nearly infinite at very small body sizes (**Figure 15.8**). This is a finite world, and infinite energy requirements are just not feasible. Thus, energy apparently sets a lower limit to the body size possible for an endotherm.

The mass-specific energy requirements of ectotherms also increase at small body sizes, but because the energy requirements of ectotherms are about one-tenth those of

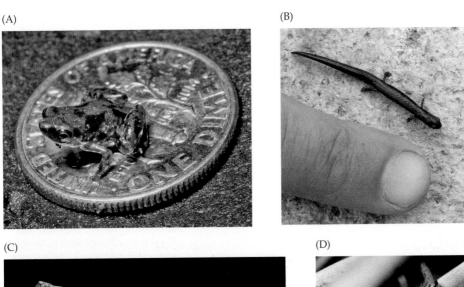

(A)

(B)

(C)

(D)

Figure 15.7 Miniaturization is a repeated phenomenon among amphibians and reptiles. Many species are no larger than insects and arachnids. (A) A frog from Papua New Guinea, *Paedophryne amanuensis*, that averages 7.7 mm in length and weighs < 1 g is believed to be the smallest species of vertebrate. (B) Adults of the arboreal salamander (*Thorius arboreus*) from Mexico are longer than the frog (17 mm), but also weigh less than 1 g. (C) The Ambohibooataba Forest chameleon (*Brookesia ramanantsoia*) from Madagascar is 25 mm long and weighs about 0.2 g. (D) Invertebrates, such as this wandering spider (Ctenidae) on Barro Colorado Island, Panama, are important predators of small amphibians and reptiles. (A, courtesy of Christopher C. Austin; B, courtesy of Sean Rovito; C, courtesy of Stephen Zozaya; D, courtesy of Robin Andrews.)

Figure 15.8 Weight-specific energy use by endotherms and ectotherms. Resting metabolic rate (in Joules/gram of body weight per hour) of ectothermal and endothermal tetrapods is shown as a function of body size. The metabolic rates of endotherms are seven to ten times higher than those of ectotherms of the same body size. Weight-specific energy requirements increase at small body sizes, and the rates for ectotherms are substantially lower than those for endotherms. As a result, ectotherms can be smaller than endotherms. The dashed portions of the lines for passerine birds and mammals show hypothetical extensions into body sizes below the minimum for adults of most species of birds and mammals. (From Pough 1980.)

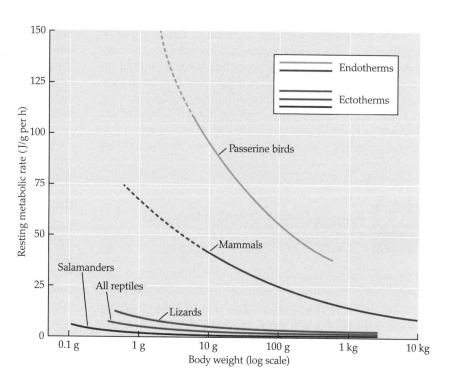

endotherms of the same body size, an ectotherm can be an order of magnitude smaller than an endotherm before its metabolic rate becomes impossibly large. Few adult birds and mammals have body weights less than 10 g. The smallest species of birds and mammals weigh about 2 g, but many ectotherms are only one-tenth that size (0.2 g) or less. Amphibians are especially small—20% of salamander species and 17% of frog species have adult body weights of less than 1 g, and 65% of salamanders and 50% of frogs are lighter than 5 g. Squamates are generally larger than amphibians, but 8% of lizard species and 2% of snake species weigh less than 1 g. At the opposite end of the spectrum, only 3–4% of species of amphibians and squamates weigh more than 100 g, whereas 31% of species of birds and 52% of mammals are heavier than that (**Figure 15.9**).

Body shape

Body shape is another aspect of vertebrate body form in which ectothermy allows more flexibility than does endothermy. An animal exchanges heat with the environment through its body surface, and the surface area of the body in relation to the volume of the body (the surface/volume ratio) is one factor that determines how rapidly heat is gained or lost. Small animals have higher surface/volume ratios than do large ones, and that is why endothermy becomes increasingly expensive at progressively smaller body sizes.

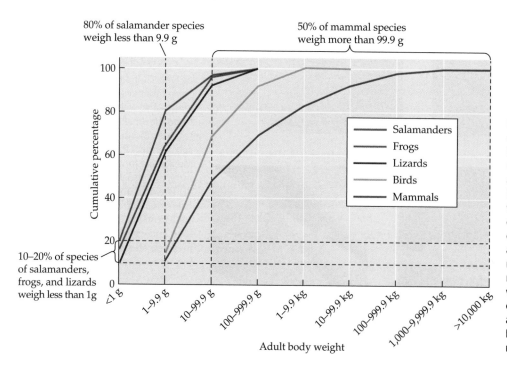

Figure 15.9 Adult body weights of ectotherms and endotherms. More than 80% of salamander species weigh less than 9.9 g (arrow), and more than half of frog and lizard species also weigh less than 9.9 g. Only 15% of birds and 12% of mammals are that small. The contrast in body sizes of ectotherms and endotherms is even more dramatic at the bottom of the weight range: no bird or mammal species weigh less than 1 g, but 10% to 20% of salamanders, frogs, and lizards are that small (bracket). (Data from Pough 1980, Smith et al. 2003, Dunning 2008.)

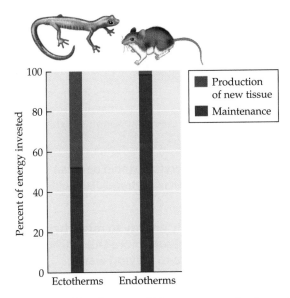

Figure 15.10 Efficiency of biomass conversion by endotherms and ectotherms. Ectotherms use an average of 52.8% of the energy in the food they eat for maintenance and the other 47.2% for production (i.e., growth of individuals and production of young). Endotherms use an average of 98.6% of the energy they consume for maintenance (primarily heat production) and devote only 1.4% to growth and reproduction. Bars show average values for 12 species of endotherms and 16 species of ectotherms. (Data from Pough 1980.)

Metabolic rates of small endotherms must be high enough to balance the heat lost across their large body surface areas.

Similarly, body shapes that increase the surface/volume ratio have an energy cost that makes them disadvantageous for endotherms. There are no highly elongated endotherms, whereas elongated body forms are widespread among fishes (true eels, moray eels, pipefishes, barracudas, and many more), amphibians (all caecilians and most salamanders, especially the limbless aquatic forms), and reptiles (many lizards and all snakes). Dorsoventral or lateral flattening is another shape that increases the surface/volume ratio. There are no flat endotherms, but flat fishes are common (dorsoventral flattening—skates, rays, flounders; lateral flattening—many coral reef and freshwater fishes). Some reptiles are also flat (dorsoventral flattening— aquatic turtles, especially soft-shelled turtles, horned lizards; lateral flattening—many arboreal lizards, especially chameleons). Small body sizes and specialized body forms allow ectotherms to fill ecological niches that are not available to endotherms.

Conversion efficiency

The amount of energy required by ectotherms and endotherms is not the only important difference between them; equally significant is what they do with that energy once they have it. Endotherms expend more than 90% of the energy they take in to produce heat to maintain their high body temperatures. Less than 10%—often as little as 1%—of the energy a bird or mammal assimilates is available for net conversion (that is, increasing the species' biomass by growth

of an individual or production of young). Ectotherms do not rely on metabolic heat. The solar energy they use to warm their bodies is free in the sense that it is not drawn from their food. Thus, most of the energy they ingest is converted into the biomass of their species. On average, ectotherms devote about 50% of the energy in their food to production compared with less than 2% for endotherms (**Figure 15.10**).

Because of the difference in how energy is used, a given amount of chemical energy invested in an ectotherm produces a much larger biomass return than it would in an endotherm. A study of salamanders in the Hubbard Brook Experimental Forest in New Hampshire showed that although their energy consumption was only 20% that of the birds or small mammals in the watershed, their conversion efficiency was so high that the annual increment of salamander biomass was equal to that of birds or small mammals. Similar comparisons can be made between lizards and rodents in deserts.

15.5 The Role of Ectotherms in Terrestrial Ecosystems

Small amphibians and squamates occupy key positions in the energy flow through an ecosystem. Because these animals are so small, they can capture tiny insects and arachnids that are too small to be eaten by birds and mammals. Because they are ectotherms, they are efficient at converting the energy in the food they eat into their own body tissues. As a result, the small ectothermal vertebrates in terrestrial ecosystems can be viewed as repackaging energy in a form that avian and mammalian predators can exploit. In other words, when a shrew or a bird searches for a meal in the Hubbard Brook Forest, the most abundant vertebrate prey it will find is salamanders. In this context, frogs, salamanders, lizards, and snakes occupy a position in terrestrial ecosystems that is important both quantitatively (in the sense that they constitute a substantial energy resource) and qualitatively (in that ectotherms exploit food resources that are not available to endotherms).

In a very real sense, small ectotherms can be thought of as living in a different world from that of endotherms. Interactions with the physical world may be more important in shaping the ecology and behavior of small ectotherms than are biological interactions such as competition. In some cases, these small vertebrates may have their primary predatory and competitive interactions with insects and arachnids rather than with other vertebrates. For example, orb web spiders (Areneidae) and *Anolis* lizards on some Caribbean islands are linked by both predation (spiders eat hatchling lizards, and adult lizards eat spiders) and competition (lizards and spiders eat many of the same kinds of insects). The competitive relationship between these distantly related animals was demonstrated by experiments. When lizards were removed from experimental plots, the abundance of insect prey increased, and the spiders consumed more prey and survived longer than those in control plots where lizards were present.

Ectothermy and endothermy thus represent fundamentally different approaches to the life of a terrestrial vertebrate. An appreciation of ectotherms and endotherms as animals requires understanding the functional consequences of the differences between them. Ectothermy is an ancestral character of vertebrates, but it is a very effective way of life in modern ecosystems.

Summary

The ancestral character of ectothermy is just as effective, and just as complex, as its derived counterpart, endothermy.

Although ectothermy is the ancestral mode of thermoregulation for vertebrates, it is the basis for a successful lifestyle that provides morphological, physiological, and ecological options that are not available to endotherms.

Deserts are characterized by chronically low resource bases and frequent periods of resource scarcity.

The low energy requirements of ectotherms and their ability to lower their needs still further by being inactive are advantageous characters in deserts.

Seasonal activity patterns of desert ectotherms are linked to the availability of food and water. Many species tolerate negative energy and water balances for months at a time. Some retreat to shelters and become inactive when even relaxation of homeostasis is inadequate, and others relax homeostasis and remain active when they are in negative energy and water balance.

Characters that at first glance appear to be deleterious can actually be keys to survival. For example, the high permeability of the skin of amphibians allows them to draw water from soil to supplement the body water content with which they entered hibernation.

Unlike endotherms, ectotherms cannot respond to cold by increasing metabolic heat production.

Nonetheless, some ectotherms succeed in seasonally cold environments. Many species hibernate in or retreat to shelters, such as burrows, that do not cool to freezing.

Other species, especially fishes, spend long periods—even their entire lives—in a supercooled state at temperatures below the freezing temperature of their body fluids. Other species have antifreeze compounds in their body tissues that inhibit freezing by preventing water molecules from assuming the lattice structure of ice.

Still other species tolerate repeated freezing and thawing of some of their body tissue by withdrawing water from cells, where formation of ice crystals would be lethal, and allowing extracellular (but not intracellular) fluids to freeze.

Endotherms use metabolic energy to raise their body temperature, and devote more than 90% of the energy they assimilate to staying warm.

Because an ectotherm relies on external sources of heat and has no metabolic cost of thermoregulation, it uses only one-seventh to one-tenth as much energy as an endotherm of the same body size.

Mass-specific metabolic rate increases as body size decreases. That relationship holds for both ectotherms and endotherms, but because the metabolic rate of an ectotherm is an order of magnitude lower than that of an endotherm, ectotherms can reach smaller body sizes than endotherms before the cost of living becomes prohibitive.

Few endotherms are smaller than 20 g, but most lizards, frogs, and salamanders are that small, and many weigh only 1 g or even less. Many ectotherms are so small that their important predators and competitors are insects and spiders.

Ectotherms have more scope for altering body shape than do endotherms, because energy loss through a large body surface area is not a concern for an ectotherm. Elongated and flattened body shapes are widespread among ectotherms but are not seen in endotherms.

Endotherms use most of the energy in the food they eat to stay warm, whereas ectotherms devote most of their energy to growth and reproduction.

The combination of small body sizes and efficient conversion of food into their own biomass creates a unique role for ectotherms in terrestrial ecosystems.

Ectotherms consume prey items that are too small for endothermal predators to capture, and efficiently convert them into packages of energy that are large enough for endotherms to eat.

Discussion Questions

1. The Lake Eyre dragon (*Ctenophorus maculosus*) is an agamid lizard that lives only on the barren, vegetation-free, salt crust of several dry lake basins in South Australia. The lizard burrows into moist sediment beneath the crust to escape the extreme heat of midday, and it feeds on insects and plant debris that blow onto the salt crust from the vegetation surrounding the lake basins. All of the lizard's water is derived from its food; there is no water for it to drink. Describe the ecological and physiological characters of the Lake Eyre dragon that allow it to survive in such an inhospitable environment.

2. Although most fishes are ectotherms, the efficiency of secondary production by fishes is substantially lower than the average of 50% measured for terrestrial ectotherms. What difference between aquatic and terrestrial environments might account for this difference?

3. The dwarf tegu (*Callopistes maculatus*) is an insectivorous teiid lizard that occurs in desert habitats in Chile. A study of its daily activity cycle in summer showed a decrease in foraging activity during midday, although the body temperatures of the lizards that were active at midday were not different from the body temperatures of lizards that were active in the morning or afternoon. That observation is not consistent with the hypothesis that midday temperatures are so high that the lizards must seek shelter to avoid overheating. Apparently dwarf tegus in this habitat are able to maintain their normal-activity body temperatures throughout the day. What alternative hypothesis can you suggest to account for the decrease in activity of dwarf tegus at midday? If your hypothesis is correct, what additional concern does it suggest about the potential effect of global climate change on lizards?

4. Spadefoot toads rely on the permeability of their skin to absorb water from the soil during the months they are underground. However, not all desert-dwelling anurans use this mechanism. Some anuran species form waterproof cocoons when they bury themselves. These cocoons reduce water loss, but also prevent the frogs from absorbing water from the soil. How can a cocooning species of frog survive for months underground?

Additional Reading

Burton TM, Likens GE. 1975. Energy flow and nutrient cycling in salamander populations in the Hubbard Brook Experimental Forest, New Hampshire. *Ecology* 56: 1068–1080.

Davie RH, Welsh HH Jr. 2005. On the ecological role of salamanders. *Annual Review of Ecology and Systematics* 35: 405–434.

Dunning JB Jr. 2008. *CRC Handbook of Avian Body Masses*, 2nd ed., CRC Press, Boca Raton, FL.

Evans CW and 5 others. 2011. How do Antarctic notothenioid fishes cope with internal ice? A novel function for antifreeze glycoproteins. *Antarctic Science* 23: 57–64.

Larson DJ, Barnes BM. 2016. Cryoprotectant production in freeze-tolerant wood frogs is augmented by multiple freeze-thaw cycles. *Physiological and Biochemical Zoology* 89: 340–346.

Parra-Olea G and 5 others. 2016. Biology of tiny animals: three new species of minute salamanders (Plethodontidae: *Thorius*) from Oaxaca, Mexico. *PeerJ* 4: e2694.

Pough FH. 1980. The advantages of ectothermy for tetrapods. *American Naturalist* 115: 92–112.

Rittmeyer EN, Allison A, Gründler MC, Thompson DK, Austin CC. 2012. Ecological guild evolution and the discovery of the world's smallest vertebrate. *PLoS ONE* 7(1): e29797. doi:10.1371/journal.pone.0029797.

Sharp KA. 2011. A peek at ice binding by antifreeze proteins. *Proceedings of the National Academy of Sciences USA* 108:781–782.

Smith FA and 8 others. 2003. Body mass of late Quaternary mammals. *Ecology* 84: 3403.

Storey KB, Storey JM. 2017. Molecular physiology of freeze tolerance in vertebrates. *Physiological Reviews* 97: 623–665.

Voituron Y, Barré H, Ramløv H, Douady CJ. 2009. Freeze tolerance evolution among anurans: Frequency and timing of appearance. *Cryobiology* 58: 241–247.

CHAPTER

16

Turtles

Harvey Pough

Turtles provide a contrast to the lepidosaurs, which we will describe in the next chapter, in the relative lack of diversity in their life histories. All turtles lay eggs, and probably only a few exhibit parental care of the nest or hatchlings. Turtles show morphological specializations associated with terrestrial, freshwater, and marine habitats, and marine turtles make long-distance migrations rivaling those of birds. Probably turtles and birds use many of the same navigation mechanisms to find their way. Most turtles are long-lived, with relatively poor capacities for rapid population growth. Many turtles, especially sea turtles and large tortoises, are endangered by human activities. Some efforts to conserve turtles have apparently been frustrated by a feature of the embryonic development of many species of turtles: an individual's sex is determined by the temperature to which it is exposed during embryonic development.

16.1 Everyone Recognizes a Turtle

Molecular data indicate that turtles are the sister lineage of archosaurs, and are among the most derived vertebrates. They are diapsids; the Middle Triassic stem turtle *Pappochelys rosinae* has a diapsid skull, but extensive emargination of the bone has obscured the diapsid condition of the skull in derived turtles. Furthermore, turtles are encased in a bony shell, with the limbs inside the ribs—if turtles had become extinct at the end of the Mesozoic, they would rival dinosaurs in their novelty. However, because they survived, turtles are regarded as commonplace and are often used inappropriately in comparative anatomy courses to represent basal amniotes.

Shell and skeleton

The shell is the most distinctive feature of a turtle. Beta-keratin, chemically similar to the beta-keratin in the scales of crocodylians and the feathers of birds, forms a layer of horny scutes on the exterior of the **carapace** (upper shell) and **plastron** (lower shell). These scutes do not coincide in

number or position with the underlying bones. The carapace has a row of five central scutes, bordered on each side by four lateral scutes. Ten to twelve marginal scutes on each side turn under the edge of the carapace. The plastron is covered by a series of six paired scutes (**Figure 16.1**).

The bony inner layer of the carapace is formed primarily by lateral expansion of the ribs (endochondral bone), with the incorporation of some dermal elements. The shell grows faster than the rest of the embryo, and both girdles are enclosed by the rib cage. The plastron is formed largely from dermal ossifications, but the entoplastron is derived from the interclavicle, and the paired epiplastra anterior to it are derived from the clavicles. Processes from the hyoplastron and hypoplastron fuse with the first and fifth pleurals, forming a rigid connection between the plastron and the carapace called the bridge. A stem turtle from the Middle Triassic, *Pappochelys rosinae*, provides evidence that the carapace and plastron are separate elements that evolved independently from each other.

Hinges are present in the shells of many turtles. The most familiar examples are the North American and Asian box turtles (*Terrapene* and *Cuora*, respectively), which have a hinge between the hyoplastral and hypoplastral bones. Mud turtles have two hinges in the plastron; the anterior hinge runs between the epiplastra and the entoplastron (which is triangular in kinosternid turtles rather than diamond shaped), and the posterior hinge is between the hypoplastron and the xiphiplastron. The African forest tortoises (*Kinixys*) have a hinge on the posterior part of the carapace. The margins of the epidermal shields and the dermal bones of the carapace are aligned, and the hinge runs between the second and third pleural scutes and the fourth and fifth costals. The presence of hinges is sexually dimorphic in some species of tortoises. The erratic phylogenetic occurrence of kinetic (hinged) shells and differences among related species indicate that shell kinesis has evolved many times in turtles.

Extant turtles have only ten vertebrae in the trunk and eight in the neck. The centra of the trunk vertebrae are elongated and lie beneath the dermal bones in the dorsal midline of the shell. The centra are constricted in their centers and fused to each other. The neural arches in the anterior two-thirds of the trunk lie between the centra as a result of anterior displacement, and the spinal nerves exit near the middle of the preceding centrum. The ribs are also shifted anteriorly; they articulate with the anterior part of the neurocentral boundary, and in the anterior part of the trunk, where the shift is most pronounced, the ribs extend onto the preceding vertebra.

The 346 extant species of turtles are distributed among 14 families (**Table 16.1**). The two lineages of extant turtles can be traced through fossils to the Mesozoic. **Cryptodires** (Greek *kryptos*, "hidden"; *deire*, "neck") retract the head into the shell by bending the neck in a vertical S shape; **pleurodires** (Greek *pleura*, "side") retract the head by bending the neck laterally (**Figure 16.2**). The cervical vertebrae of cryptodires have specialized articulating surfaces between vertebrae, called ginglymi, that permit formation of the S-shaped bend. In most families, the bend is formed by two successive ginglymoidal joints between the sixth and seventh and the seventh and eighth cervical vertebrae. The lateral bending of the necks of pleurodires is accomplished by simple ball-and-socket or cylindrical joints between adjacent cervical vertebrae.

Nearly three-quarters of the extant species of turtles are cryptodires, and cryptodires are the only turtles now found in most of the Northern Hemisphere. There are aquatic and terrestrial species of cryptodires in South America, Asia, and Africa. Australia has only a single species of cryptodire. Pleurodires now occur only in the Southern Hemisphere, although they were distributed worldwide in the late Mesozoic and early Cenozoic and included both freshwater and marine species. All extant pleurodires are at least semiaquatic, but the shells of some fossil pleurodires suggest that they may have been terrestrial.

Shell morphology reflects the ecology of turtle species (**Figure 16.3**). The most terrestrial forms, tortoises and box turtles, have a domed carapace. Box turtles are among several kinds of turtles that have flexible regions called hinges in the plastron, which allow the anterior and posterior lobes to be raised to close the openings of the shell. Tortoises use their heavily scaled forelimbs to block the front of the shell and retract their hindlegs so that only the soles of the feet are exposed, and the African hinge-backed tortoises (*Kinixys*) have a hinge across the carapace that allows them to close off the rear opening of the shell.

Aquatic turtles have low carapaces that offer little resistance to movement through water. Soft-shelled turtles (Trionychidae) are fast swimmers. Their feet are large with extensive webbing, and ossification of the shell is greatly reduced. The distal ends of the broadened ribs are embedded in flexible connective tissue, and the carapace and plastron

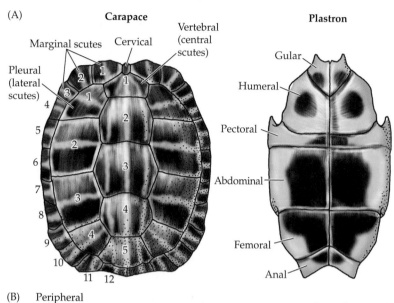

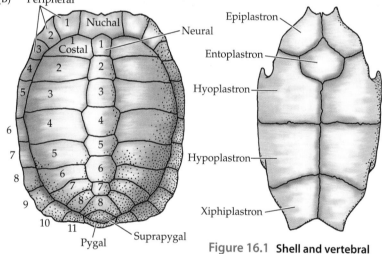

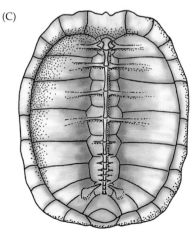

Figure 16.1 Shell and vertebral column of a turtle. (A) Epidermal scutes of the carapace and plastron. The carapace has a central (vertebral) row of 5 scutes with 4 lateral (pleural) scutes on each side and 10–12 marginal scutes. The plastron has 6 paired scutes. (B) Dermal bones of the carapace and plastron. (C) Vertebral column, seen from inside the carapace. Note that anteriorly, the ribs articulate with two vertebral centra. (After Zangerl 1969.)

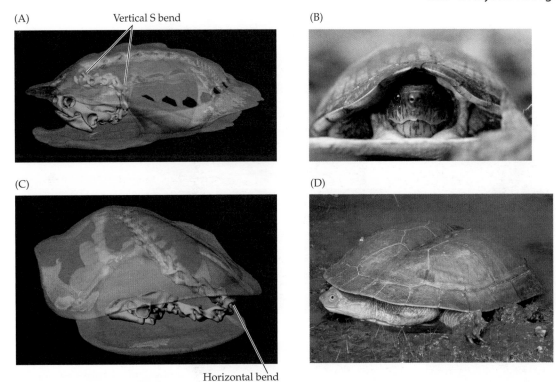

Figure 16.2 Neck retraction by cryptodires and pleurodires.
(A) Cryptodires retract the head into the shell by bending the neck in a vertical S shape. (B) A loggerhead musk turtle (*Stenotherus minor*) shows how this bend allows the head to be pulled directly back into the shell. In most families of cryptodires, the bend is formed by two successive ginglymoidal joints between the 6th and 7th and the 7th and 8th cervical vertebrae. (C) Pleurodires retract the head by bending the neck laterally. (D) Instead of being retracted into the shell, the head and neck are folded against the shell opening, leaving part of the neck exposed. (A, C from Werneburg et al. 2015; B, courtesy of Todd Pierson; D, courtesy of Stephen Zozaya.)

Table 16.1   Families of extant turtles

Cryptodira: 255 species

Dermochelyidae: A single species, the leatherback (*Dermochelys coriacea*), is the largest extant turtle (up to 240 cm in length and up to 900 kg). This is a marine turtle in which the shell is reduced to thousands of small bones embedded in a leathery skin. It occurs in oceans worldwide, and its range extends north and south into seas too cold for other sea turtles.

Cheloniidae: 6 species of large (70 cm) to very large (150 cm) sea turtles with bony shells covered with epidermal scutes and paddlelike forelimbs, found worldwide in tropical and temperate oceans.

Chelydridae: 5 species of large (*Chelydra*, 50 cm) to very large (*Macrochelys*, 80 cm and 100 kg) freshwater turtles in North and Central America.

Kinosternidae: 26 species of small (11 cm) to medium-size (40 cm) bottom-dwelling freshwater turtles in North, Central, and South America.

Dermatemydidae: A single medium-size species (50 cm), *Dermatemys mawii*, once inhabited large rivers from Mexico through Nicaragua; it is critically endangered and probably extinct in many parts of its former range.

Testudinidae: 60 species of small (15 cm) to very large (1 m) terrestrial turtles with a worldwide distribution in temperate and tropical regions. Although tortoises are clumsy swimmers, they float well and withstand long periods without food or water. These characteristics have allowed tortoises swept to sea by flooding rivers to populate oceanic islands, such as the Galápagos Islands.

Geoemydidae: 71 species of small (12 cm) to large (75 cm) freshwater, semiaquatic, and terrestrial turtles, primarily in Asia.

Emydidae: 51 species of small (12 cm) to large (60 cm) freshwater, semiaquatic, and terrestrial turtles, mostly in North America; 1 genus in Central and South America; and 1 genus in Europe, Asia, and North Africa.

Platysternidae: A single small (18 cm) species, the big-headed turtle (*Platysternon megacephalum*), inhabits mountain streams in Southeast Asia.

Carettochelyidae: A single large (70 cm) species, the pig-nosed turtle (*Carettochelys insculpta*) inhabits fresh water and lagoons in southern New Guinea and northern Australia. Its bony carapace and plastron are covered with soft skin instead of rigid scutes.

Trionychidae: 32 species of small (25 cm) to very large (130 cm) freshwater turtles in North America, Africa, and Asia. The carapace and plastron are covered with soft skin, and the underlying bone is reduced.

Pleurodira: 93 species

Podocnemididae: 8 species of aquatic turtles found in northern South America and Madagascar. *Podocnemis expansa*, which occurs in the Amazon and Orinoco rivers, is the largest extant pleurodire; females reach shell lengths of 90 cm. The extinct *Stupendemys* (from the late Neogene of Venezuela) was more than 2 m long.

Pelomedusidae: 27 species of small (12 cm) to medium-size (50 cm) aquatic turtles in Africa, Madagascar, and the Seychelles Islands.

Chelidae: 58 species of small (15 cm) to medium-size (50 cm) aquatic turtles in South America, Australia, and New Guinea.

(A)

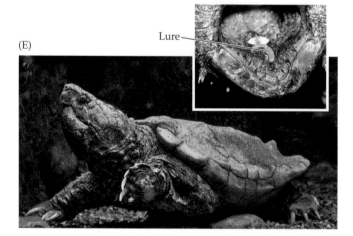

Hinge

(B)

(C)

(D)

(E)

Lure

(F)

Figure 16.3 Body forms of turtles. (A,B) Most terrestrial turtles have domed shells that the jaws of mammalian predators cannot easily crush. (A) An eastern box turtle (*Terrapene carolina*). A flexible hinge in the plastron (inset) allows box turtles to raise the anterior and posterior portions of the plastron when attacked. (B) Sonoran desert tortoise (*Gopherus morafkai*). Tortoises fold their heavily scaled forelimbs across the anterior opening of the shell and expose only the soles of their hindfeet in the shell's rear opening (inset). (C) Most aquatic turtles, such as the Australian saw-shelled turtle (*Myuchelys latisternum*), are streamlined and have large webbed feet. (D) Soft-shelled turtles such as this North American spiny softshell (*Apalone spinifera*) lack peripheral ossifications and epidermal scutes. Soft-shelled turtles are ambush predators; their flat, flexible shells allow them to bury into the substrate, leaving only the head exposed (inset). (E) The alligator snapping turtle (*Macrochelys temminckii*) of North America is among the largest extant freshwater turtles. Old adults can reach 80 cm in length and weigh more than 100 kg. Alligator snappers rest on the pond bottom with their mouths open and wiggle a wormlike lure (inset) to attract fish. (F) The forelimbs of sea turtles such as this leatherback (*Dermochelys coriacea*) are modified as flippers and the hindlimbs are used as rudders. (A, Harvey Pough, inset courtesy of Ken Dodd; B, René Clark © Dancing Serpent Nature Photography, inset Shellie R. Puffer, USGS; C, courtesy of Stephen Zozaya; D and inset, courtesy of Todd Pierson; E, © Maik Dobiey, inset Harvey Pough; F, Claudia Lombard, USFWS.)

are covered with skin. Soft-shelled turtles lie in ambush partly buried in debris on the bottom of the pond. Their long necks allow them to reach considerable distances to seize the invertebrates and small fishes on which they feed.

Marine turtles are cryptodires. Cheloniidae and Dermochelyidae show more extensive specialization for aquatic life than any freshwater turtle; for example, the forelimbs are modified as flippers in both families. The largest extant marine turtle, the leatherback sea turtle (*Dermochelys coriacea*), ranges far from land, and it has a wider geographic distribution than any other ectothermal amniote. Leatherbacks penetrate far into cool temperate seas and have been recorded in the Atlantic from Nova Scotia to Argentina and in the Pacific from Japan to Tasmania. They dive to depths of more than 1,000 m. One dive that drove the depth recorder off the scale is estimated to have reached 1,200 m, which exceeds the deepest dive recorded for a sperm whale (1,140 m). Nearly all marine turtles are carnivorous: leatherbacks feed largely on pelagic colonial invertebrates such as jellyfishes. Unfortunately, floating plastic bags look like jellyfishes to leatherbacks, and the bacteria that grow on the bags emit an odor that is a key feeding stimulus for many marine organisms. More than a third of dead leatherbacks autopsied had ingested plastic, and obstruction of the intestine was the probable cause of death for some of these turtles.

16.2 Turtle Structure and Function

Lung ventilation

As you have seen, basal amniotes probably used movements of the rib cage to draw air into the lungs and to force it out, and lizards still employ that mechanism. The fusion of the ribs of turtles with their rigid shells makes that method of breathing impossible. Only the openings at the anterior and posterior ends of the shell contain flexible tissues. The lungs of a turtle, which are large, are attached to the carapace dorsally and laterally. Ventrally, the lungs are attached to a sheet of non-muscular connective tissue that is itself attached to the viscera (**Figure 16.4**). The weight of the viscera keeps this sheet of tissue stretched downward.

Turtles draw air into the lungs by contracting muscles that increase the volume of the visceral cavity, allowing the viscera to settle downward. Because connective tissue attaches the viscera to the ventral surface of the lungs, downward movement of the viscera expands the lungs, drawing in air. Expiration is accomplished by contracting muscles that force the viscera upward, compressing the lungs and expelling air.

The basic problems of respiring within a rigid shell are the same for most turtles, but the mechanisms show some variation. For example, aquatic turtles use the hydrostatic pressure of water to help move air into and out of the lungs. In addition, many aquatic turtles are able to absorb oxygen from and release carbon dioxide to the water. The

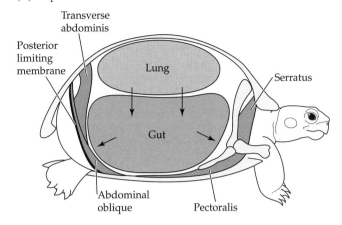

(A) Inspiration

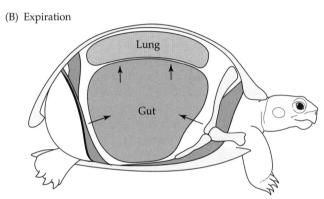

(B) Expiration

Figure 16.4 Schematic view of the lungs and respiratory movements of a tortoise. In turtles, the viscera are used as a piston to change the volume of the lungs. (A) On inspiration, the serratus and its associated muscles pull the pectoral girdle forward while the abdominal oblique pulls the posterior limiting membrane back. These movements allow the gut to drop downward, increasing the volume of the lungs. (B) On expiration, the pectoralis pulls the pectoral girdle backward while the transverse abdominis pulls the posterior limiting membrane forward. These movements force the gut upward, forcing air out of the lungs. The in-and-out movements of the forelimbs and of the soft tissue at the rear of the shell during breathing are conspicuous.

pharynx and cloaca appear to be the major sites of aquatic gas exchange. In 1860, in *Contributions to the Natural History of the United States of America*, Louis Agassiz pointed out that the pharynx of soft-shelled turtles (*Apalone*) contains fringelike processes and suggested that these structures are used for underwater respiration. Subsequent study has shown that soft-shelled turtles use movements of the hyoid apparatus to draw water into and out of the mouth and pharynx (oropharynx) when they are confined underwater, and that pharyngeal respiration accounts for most of the oxygen absorbed from the water. Musk turtles are also capable of aquatic gas exchange, and histological examination shows that their oropharyngeal region is lined by flat-topped papillae. These structures are highly vascularized and are probably the site of gas exchange.

The Australian turtle *Rheodytes leukops* uses cloacal respiration. Its cloacal orifice may be as large as 30 mm in diameter, and the turtle holds it open. Large bursae (sacs) open from the wall of the cloaca, and the bursae have a well-vascularized lining with numerous projections (villi). The turtle pumps water into and out of the bursae at rates of 15–60 times per minute. Captive turtles rarely surface to breathe, and experiments have shown that the rate of oxygen uptake through the cloacal bursae is very high.

The heart

The circulatory systems of tetrapods can be viewed as consisting of two circuits: the systemic circuit carries oxygenated blood from the heart to the head, trunk, and appendages, whereas the pulmonary circuit carries deoxygenated blood from the heart to the lungs. The blood pressure in the systemic circuit is higher than the pressure in the pulmonary circuit, and the two circuits operate in series. That is, blood flows from the heart through the lungs, back to the heart, and then to the body. The morphology of the hearts of derived synapsids (mammals) and sauropsids (crocodylians and birds) makes this sequential flow obligatory, but the hearts of turtles and lepidosaurs have the ability to shift blood between the pulmonary and systemic circuits.

The route of blood flow is flexible because the ventricular chambers in the hearts of turtles and lepidosaurs are in anatomical continuity, instead of being completely divided by a septum like the ventricles of birds and mammals. The flow of blood is controlled partly by the relative resistance to flow in the pulmonary and systemic circuits. The pattern of blood flow can best be explained by considering the morphology of the heart and how intracardiac pressure changes during a heartbeat. **Figure 16.5** shows a schematic view of the heart of a turtle. The left and right atria are completely separate, and as the ventricle contracts it forms three subcompartments. A muscular ridge in the core of the heart divides the ventricle into two spaces, the cavum pulmonale and the cavum venosum. The muscular ridge is not fused

to the wall of the ventricle, and thus the cavum pulmonale and the cavum venosum are only partly separated. A third subcompartment of the ventricle, the cavum arteriosum, is located dorsal to the cavum pulmonale and cavum venosum. The cavum arteriosum communicates with the cavum venosum through an intraventricular canal. The pulmonary artery opens from the cavum pulmonale, and the left and right aortic arches open from the cavum venosum.

The right atrium receives deoxygenated blood from the body via the sinus venosus and empties into the cavum venosum, and the left atrium receives oxygenated blood from the lungs and empties into the cavum arteriosum.

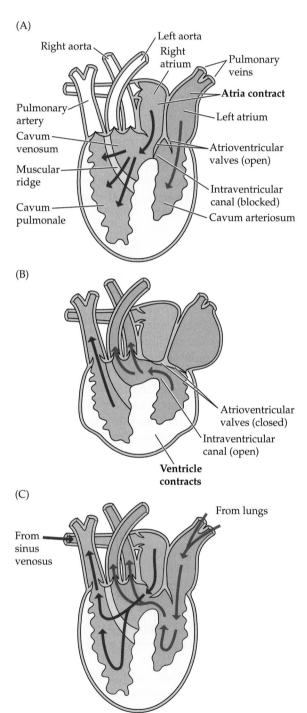

Figure 16.5 Blood flow in a turtle heart.   Although the ventricle is anatomically a single chamber, contraction transiently creates three ventricular chambers and produces considerable separation of oxygenated and deoxygenated blood. (A) As the atria contract, oxygenated blood (red) from the left atrium enters the cavum arteriosum while deoxygenated blood (blue) from the right atrium first enters the cavum venosum, then crosses the muscular ridge and enters the cavum pulmonale. The atrioventricular valves block the intraventricular canal and prevent mixing of the oxygenated and deoxygenated blood. (B) As the ventricle contracts, the deoxygenated blood in the cavum pulmonale is expelled through the pulmonary artery and travels to the lungs. The atrioventricular valves close and no longer obstruct the intraventricular canal, and oxygenated blood in the cavum arteriosum is forced into the cavum venosum and expelled through the left and right aortas. Contact between the wall of the ventricle and the muscular ridge prevents mixing of oxygenated and deoxygenated blood. (C) Summary of the pattern of blood flow through the heart. Deoxygenated blood passes from the right atrium through the cavum venosum, across the muscular ridge, into the cavum pulmonale, and out the pulmonary artery into the pulmonary circuit. Oxygenated blood passes from the left atrium into the cavum arteriosum, through the intraventricular canal to the cavum venosum, and out the left and right aortas into the systemic circuit. (From Pough et al. 2016.)

The atria are separated from the ventricle by flaplike atrio-ventricular valves that open as the atria contract and then close as the ventricle contracts, preventing blood from being forced back into the atria. The anatomical arrangement of the connections among the atria, their valves, and the three subcompartments of the ventricle is crucial because it is those connections that allow pressure differentials to direct the flow of blood and to prevent mixing of oxygenated and deoxygenated blood.

Patterns of circulation and respiration

The morphological complexity of the hearts of turtles (and of lizards) allows them to adjust blood flow through the pulmonary and systemic circuits to meet short-term changes in respiratory requirements by changing paths of blood flow within the heart (intracardiac shunting). Blood can be shifted from the pulmonary to the systemic circuit (a right-to-left shut) or in the opposite direction (a left-to-right shunt). The key to these adjustments is changing pressures in the systemic and pulmonary circuits.

Body size and temperature regulation

Turtles basking on a log in a pond are a familiar sight in many parts of the world; few pond turtles are large enough to maintain body temperatures higher than the temperature of the water surrounding them. Emerging from the water to bask is the only way most pond turtles can raise their body temperatures to speed digestion, growth, and the production of eggs.

Small terrestrial turtles, such as box turtles and small species of tortoises, thermoregulate by moving between sunlight and shade. Familiarity with a home range may assist this type of thermoregulation, and turtles can locate suitable microhabitats within an unfavorable area. A study conducted in Italy compared the thermoregulation of resident Hermann's tortoises (*Testudo hermanni*) living in their own home ranges with that of individuals that were brought to the study site and tested before they had learned their way around. The resident tortoises warmed faster and maintained more stable shell temperatures than did the strangers.

Turtles are unusual among extant reptiles in having a substantial number of species that reach large body sizes. The bulk of a large tortoise provides considerable thermal inertia, and large species such as the Galápagos tortoise (*Chelonoidis nigra*) and Aldabra giant tortoise (*Aldabrachelys gigantea*) of Aldabra Atoll heat and cool slowly. Aldabra giant tortoises, which weigh 60 kg or more, allow their body temperatures to rise to 32–33°C on sunny days and cool to 28–30°C overnight.

Large body size slows the rate of heating and cooling, but it can make temperature regulation more difficult. A small turtle can find shade beside a bush or even a clump of grass, but a giant tortoise needs a bigger object—a tree, for example. In open, sunny habitats, overheating can be a problem for giant tortoises. The difficulty is particularly acute for some tortoises on Grande Terre, an island in the Indian Ocean. During the rainy season each year, some of the turtles move from the center of the island to the coast. This movement has direct benefits because the migrant turtles gain access to a seasonal flush of plant growth on the coast. As a result of the extra food, migrant females are able to lay more eggs than females that remain inland.

There are risks to migrating, however, because shade is limited on the coast and the rainy season is the hottest time of the year. Tortoises on the coast must limit their activity to the vicinity of patches of shade, which may be no more than a single tree in the midst of a grassy plain. During the day, tortoises forage on the plain and move back toward the shade of the tree as their body temperatures rise. As the day grows hotter, tortoises try to get into the deepest shade, and the biggest individuals do this most successfully. As the big tortoises (which are mostly males) push their way into the shade, they force smaller individuals (most of which are females) out into the sunlight, and some of these tortoises die of overheating.

Most adult marine turtles are large enough to achieve a degree of endothermy. A body temperature of 37°C was recorded by telemetry from a green sea turtle swimming in water that was 20°C. The leatherback is the largest extant turtle; adults may weigh up to 900 kg. Leatherbacks range far from warm equatorial regions and in the summer can be found off the coasts of New England and Nova Scotia in water as cool as 8–15°C. Body temperatures of these turtles can be 12°C higher than the water temperature. A countercurrent exchange mechanism conserves the heat produced by muscular activity in the flippers, and a layer of fat insulates blood vessels in the neck.

16.3 Reproductive Biology of Turtles

All turtles are oviparous. Female turtles use their hindlimbs to excavate a nest in sand or soil and deposit a clutch that ranges from 4 or 5 eggs for small species to more than 100 eggs for the largest sea turtles. Most turtle species lay eggs that have soft, flexible shells, but some species have rigid shells. Embryonic development typically requires 40–60 days, and in general, soft-shelled eggs develop more rapidly than rigid-shelled eggs.

Some turtle species lay their eggs in late summer or fall, and the eggs have a period of arrested embryonic development (**diapause**) during the winter and resume development when temperatures rise in the spring. The northern snake-necked turtle (*Chelodina oblonga*) from Australia lays its eggs underwater during the wet season. The eggs remain in diapause until the floodwater recedes, begin development when they are exposed to air, and hatch at the start of the next wet season.

Moisture and egg development

The amount of moisture in the soil surrounding a turtle nest is an important variable during embryonic development of the eggs. Moist incubation conditions produce larger hatchlings than do dry conditions, apparently because water is needed for metabolism of the yolk. When water is limited, turtles hatch early and at smaller body sizes, and they retain a quantity of yolk that was not used during embryonic development. Hatchlings from nests under wetter conditions are larger and contain less remaining yolk. The large hatchlings that emerge from moist nests are able to run and swim faster than hatchlings from drier nests, and as a result, they may be more successful at escaping predators and catching food.

Temperature-dependent sex determination

We usually think of the sex of an individual as being genetically determined, and mammals and nearly all birds have **genetic sex determination (GSD)**. Beyond those two groups, however, sex determination is more varied. For some reptiles, including many turtles, sex is determined by the temperature an embryo experiences during development (**Figure 16.6**). **Temperature-dependent sex determination (TSD)** is widespread among turtles, is apparently universal among crocodylians, and is known for the tuatara, a few species of lizards, and one group of birds. The switch from one sex to the other can occur within a span of 3–4°C.

Three patterns of TSD have been described:

1. Type Ia produces males at low temperatures and females at high temperatures.
2. Type Ib produces females at low temperatures and males at high temperatures.
3. Type II produces females at both low and high temperatures and males at intermediate temperatures.

Some families of turtles have only one pattern, others have two patterns, and a few have species with both TSD and GSD (**Table 16.2**).

Allowing environmental factors to determine the sex of one's offspring sounds like a risky proposition, and several hypotheses have sought to find a benefit to TSD. An early suggestion proposed that a female turtle could select a nest site that would produce male or female young, depending on the ratio of the sexes in the adult population—presumably it would be beneficial to produce young of the rarer sex. A more recent hypothesis holds that TSD may be correlated with sexual size dimorphism of adults—that is, for each species high incubation temperatures produce the sex that benefits from being larger as an adult.

A third possibility is that selection acts on a different effect of incubation temperature and that sex is merely a by-product of that selection. Incubation temperature affects body size, growth rate, swimming and running speeds, and mode of escape behavior employed by hatchling turtles, and selection could act on any one of these traits or on a combination of them. For example, hatchling snapping turtles from eggs incubated at 28°C attempted to flee to escape predators, whereas hatchlings produced at 26°C or 30°C remained motionless and avoided detection. In a field enclosure, hatchlings from 26°C and 30°C had significantly higher survival rates after 1 year than did hatchlings from 28°C.

Temperatures of natural nests are not completely stable, of course. Daily temperature variation is superimposed on a seasonal cycle of

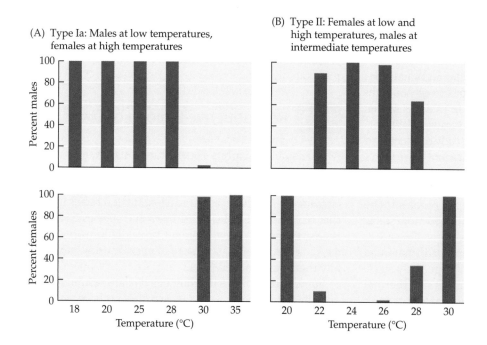

(A) Type Ia: Males at low temperatures, females at high temperatures

(B) Type II: Females at low and high temperatures, males at intermediate temperatures

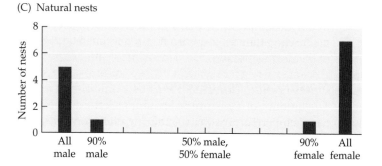

(C) Natural nests

Figure 16.6 **Temperature-dependent sex determination.**
(A) Type Ia: Eggs of the European pond turtle (*Emys orbicularis*) hatch into males when they are incubated at temperatures from 18°C to 28°C and into females at 30°C and above.
(B) Type II: Eggs of the North American snapping turtle (*Chelydra serpentina*) produce females at low and high temperatures and males at intermediate temperatures.
(C) Natural nests of the Ouachita map turtle (*Graptemys ouachitensis*) produce 90-100% males or 90–100% females. (A, data from Bull 1980; B, data from Yntema 1976; C, data from Bull and Vogt 1979.)

Table 16.2	Sex-determining mechanisms among turtles[a]
Group	**Mechanism(s)**
Dermochelyidae	Ia
Cheloniidae	Ia
Chelydridae	II
Kinosternidae	GSD (XY), Ia, II
Dermatemydidae	Ia
Testudinidae	Ia
Geomydidae	GSD (XY, ZW), Ia, II
Emydidae	GSD (heterogametic sex unknown), Ia, II
Platysternidae	unknown
Carettochelyidae	Ia
Trionychidae	GSD (ZW)
Podocnemididae	unknown
Pelomedusidae	Ia, II
Chelidae	GSD (XY)

[a]Genetic sex determination (GSD) can be male heterogametic (XY) or female heterogametic (ZW). Temperature-dependent sex determination (TSD) can be type Ia (males at low temperatures, females at high temperatures), type Ib (females at low temperatures, males at high temperatures), or type II (females at low and high temperatures, males at intermediate temperatures).

changing environmental temperatures. The middle third of embryonic development is the critical period for sex determination; the sex of the embryos depends on the temperatures they experience during those few weeks. When eggs are exposed to a daily temperature cycle, the high temperature in the cycle is most critical for sex determination.

In addition to external sources of temperature variation within a nest, a turtle embryo may have a limited capacity for thermoregulation while it is still in the egg. Embryos of a Chinese soft-shelled turtle (*Pelodiscus sinensis*) rotate within their eggshells so that their backs point toward the source of heat. Temperature measurements revealed a 1° C difference between the sides of the embryo facing toward and away from the source of heat.

Because of the narrowness of the thermal windows involved in sex determination and the variation that exists in environmental temperatures, both sexes are produced under field conditions but not necessarily in the same nests. A study of painted turtles found that two-thirds of the nests produced only males or only females while the remaining nests produced hatchlings of both sexes.

The wetness of a nest interacts with temperature in determining the sex of hatchlings. Dry substrates induced the development of some female painted turtles at low temperatures (26.5°C and 27°C) that would normally have produced only males. The wetness of the substrate did not affect the sex of turtles from eggs incubated at 30.5°C and 32°C: all the hatchlings from these eggs were females, as would be expected on the basis of TSD alone.

Parental care

Because fertilization is internal and egg deposition occurs days to months after insemination, male turtles are not present when the eggs are laid. Most species of turtles apparently provide no parental care; after a female covers a nest, she leaves. However, some female tortoises guard their nests and attack egg predators. Usually this behavior ends within a few days after nesting, but for two consecutive years a female desert tortoise was still guarding a nest that contained newly hatched young.

A South American river turtle, *Podocnemis expansa*, is the only species of turtle known to provide post-hatching care (**Figure 16.7**). Females nest communally on mid-river sandbars and then remain just off the nesting beach while the eggs develop. As the embryos approach hatching, they vocalize within the eggs to synchronize hatching. Continued vocalization by the hatchlings in the nest stimulates them dig their way out of the nest as a group. The hatchlings vocalize when they enter the river and the females respond by vocalizing in the water, attracting the hatchlings to them. Only after the hatchlings have entered the water do the females move away from the nest beach, migrating with the hatchlings to flooded forests.

Hatching and the behavior of baby turtles

Internal and external cues appear to be important in synchronizing the hatching of turtle embryos. Temperature variation within the nest causes some embryos to develop more rapidly than others, but vocalizations by these embryos stimulate slower embryos to increase their rates of development in a brief catch-up period shortly before hatching.

Emergence from the nest The first challenge a turtle faces after hatching is escaping from the nest, and in some instances interactions among all the hatchlings in a nest may be essential. Sea-turtle nests are quite deep; the eggs may be buried 50 cm beneath the sand, and the hatchling turtles must struggle upward through the sand to the surface. After several weeks of incubation, the eggs all hatch within a period of a few hours, and 100 or so baby turtles find themselves in a small chamber at the bottom of the nest hole. Spontaneous activity by a few individuals sets the whole group into motion, crawling over and under one another. The turtles at the top of the pile loosen sand from the roof of the chamber as they scramble about, and the sand filters down through the mass of baby turtles to the bottom of the chamber.

Periods of a few minutes of frantic activity are interspersed with periods of rest, possibly because the turtles' exertions reduce the concentration of oxygen in the nest and they must wait for more oxygen to diffuse into the nest from the surrounding sand. Gradually, the entire group of turtles moves upward through the sand as a unit until it reaches the surface. As the baby turtles approach the surface, high sand temperatures probably inhibit further activity, and the turtles wait a few centimeters below the surface until night falls, when a decrease in temperature triggers emergence.

(A)

(B)

(C)

(D)

Figure 16.7 Parental care by giant Amazon river turtles.
(A) Females arrive at the nesting beaches in groups that once were numbered in thousands, but now are only hundreds. The females have a distinct vocalization they use as they leave the water.
(B) Adult females and hatchlings are equipped with radio transmitters allowing them to be tracked. (C) Hatchlings vocalize as they enter the water, and females respond by vocalizing. (D) Hatchlings and adult females depart from the nesting beaches in groups containing hundreds of females and thousands of hatchlings. Maintaining contact with vocalizations, the turtles travel 100 km or more into the flooded forests where they feed. (Photos by Richard C. Vogt.)

All the babies emerge from a nest in a very brief time, and all the babies in different nests that are ready to emerge on a given night leave their nests at almost the same time, probably because their behavior is cued by temperature. The result is the sudden appearance of hundreds or even thousands of baby turtles on the beach, each one crawling toward the ocean as fast as it can.

Simultaneous emergence is an important feature of the hatching of sea turtles because the babies suffer high mortality crossing the few meters of beach and surf. Terrestrial predators—crabs, foxes, raccoons, and other predators—gather at the turtles' breeding beaches at hatching time and await their appearance. Some of the predators come from distant places to prey on the baby turtles. In the surf, sharks and bony fishes patrol the beach. Few, if any, baby turtles would get past that gauntlet were it not for the simultaneous emergence that brings them all out at once and temporarily swamps the predators.

The early years Turtles are self-sufficient at hatching, and with the single known exception of the South American river turtle *Podocnemis expansa*, adults provide no parental care to hatchlings. Baby turtles are secretive and are rarely encountered in the field. Probably small turtles spend most of their time in concealment because they are vulnerable to predators. A hard shell is a turtle's defense, and the adults of most species have shells that predators cannot crush. Baby turtles are in a very different position, however: they are bite-size and their shells are not rigid enough to resist crushing—baby turtles are Oreo cookies from a predator's perspective.

We know even less about the biology of baby sea turtles than we do about the hatchlings of terrestrial and freshwater species. Where the turtles go in the period following hatching has been a long-standing puzzle in the life cycle of sea turtles. For example, green sea turtles that hatch in the late summer in the Caribbean disappear from sight as soon as they are at sea, and they are not seen again until they weigh 4–5 kg, some 3 years later. They apparently spend the intervening period floating in ocean currents, and transatlantic dispersal is probably common. Material drifting on the surface of the sea accumulates in areas where currents converge, forming drift lines of flotsam that include sargassum (brown algae) and the vertebrate and invertebrate fauna associated with it. These drift lines are probably important resources for juvenile sea turtles.

16.4 Social Behavior, Communication, and Courtship

Although turtles are popularly viewed as solitary and not very interactive, this impression is probably erroneous. Turtles have a variety of social interactions during which they use tactile, visual, and olfactory signals, and social behavior among turtles is probably more common than has been realized.

Many pond turtles have distinctive stripes of color on their heads, necks, and forelimbs and on their hindlimbs and tail. These patterns are used by herpetologists to distinguish the species, and they may be species-isolating mechanisms for the turtles as well. During the mating season, male pond turtles swim in pursuit of other turtles, and the color and pattern on the posterior limbs may enable males to identify females of their own species. At a later stage of courtship, when the male turtle swims backward in front of the female and vibrates his claws against the sides of her head, both sexes can see the patterns on their partner's head, neck, and forelimbs. This titillation is characteristic of young male pond turtles, whereas larger males use coercive behaviors, such as biting and holding a female under water. Males of other species of turtles, especially tortoises, ram a female with their shell during courtship.

Among terrestrial turtles, the behavior of tortoises is best known. Many tortoises vocalize while they are mating; the sounds they produce have been described as grunts, moans, and bellows. The frequencies of the calls that have been measured range from 500 to 2,500 Hz. Some tortoises have glands that become enlarged during the breeding season and appear to produce pheromones. The secretion of the subdentary gland on the underside of the jaw of tortoises in the North American genus *Gopherus* appears to identify both the species and the sex of an individual. During courtship, males and females of the Florida gopher tortoise (*G. polyphemus*) rub their subdentary gland across one or both forelimbs and then extend the limbs toward the other individual, which may sniff at them. Males also sniff the cloacal region of other tortoises, and male tortoises of some species trail females for days during the breeding season. Fecal pellets may be territorial markers; fresh fecal pellets from a dominant male tortoise have been reported to cause dispersal of conspecifics.

Movements of the head appear to act as social signals for tortoises, and elevating the head is a signal of dominance in some species. Herds of tortoises have social hierarchies that are determined largely by aggressive encounters. Ramming, biting, and hooking are employed in these encounters, and the larger individual is usually the winner—although experience may play some role. These social hierarchies are expressed in the priority of different individuals in access to food or forage areas, mates, and resting sites. Dominance relationships also appear to be involved in determining the sequence in which individual tortoises move from one place to another. The social structure of a tortoise herd can be a nuisance for zookeepers trying to move the animals from an outdoor pen into an enclosure for the night because the tortoises resist moving out of their proper rank sequence.

16.5 Navigation and Migrations

Pond turtles and terrestrial turtles usually lay their eggs in nests that they construct within their home ranges. The mechanisms of orientation that they use to find nesting areas are probably the same ones they use to find their way among foraging and resting areas. Familiarity with local landmarks is an effective method of navigation for these turtles, and they may also use the sun for orientation. Sea turtles have a more difficult time, partly because the open ocean lacks conspicuous landmarks but also because feeding and nesting areas are often separated by hundreds or thousands of kilometers. Most sea turtles are carnivorous. The leatherback feeds on jellyfishes, the loggerhead sea turtle (*Caretta caretta*) and Kemp's Ridley sea turtle (*Lepidochelys kempii*) eat crabs and other benthic invertebrates, and the hawksbill sea turtle uses its beak to scrape encrusting organisms (sponges, tunicates, bryozoans, mollusks, and algae) from reefs. Juvenile green sea turtles are carnivorous, but the adults feed on vegetation, particularly turtle grass, which grows in shallow water on protected shorelines in the tropics. The areas that provide food for sea turtles often lack the characteristics needed for successful nesting, and many sea turtles move long distances between their feeding grounds and their breeding areas.

The ability of sea turtles to navigate over thousands of kilometers of ocean and find their way to nesting beaches that may be no more than tiny coves on a small island is astonishing. The migrations of sea turtles, especially the green sea turtles at Tortuguero on the Caribbean coast of Costa Rica, have been studied for more than 50 years, and tag returns from turtle catchers and fishing boats have allowed the major patterns of population movements to be established (**Figure 16.8**).

Four major nesting sites of the green sea turtle have been identified in the Caribbean and South Atlantic: one at Tortuguero on the coast of Costa Rica, one on Aves Island in the eastern Caribbean, one on the coast of Suriname, and one on Ascension Island between South America and Africa. Male and female green sea turtles congregate at these nesting grounds during the nesting season. The male turtles remain offshore, where they court and mate with females, and the female turtles come ashore to lay eggs on the beaches. A typical female green sea turtle at Tortuguero produces three clutches of eggs about 12 days apart. Turtles often return to the same part of a nesting beach repeatedly during a nesting season and again in later seasons. About a third of the female turtles in the Tortuguero population nest in alternate years, and the remaining two-thirds follow a 3-year breeding cycle.

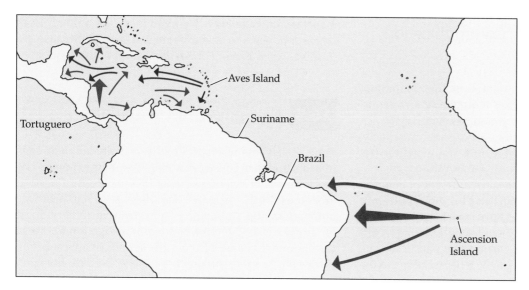

Figure 16.8 Migratory movements of green sea turtles (*Chelonia mydas*). The turtles nest on open beaches, but feed on turtle grass that grows in protected bays. The nesting sites of green sea turtles in the Caribbean and South Atlantic are located at considerable distances from their feeding areas. The Tortuguero and Aves Island populations (red and blue arrows, respectively) feed at sites throughout the Caribbean; the Ascension Island population (green arrows) feeds on the coast of Brazil.

The nesting grounds of green sea turtles in the Caribbean lack the beds of turtle grass on which the turtles feed, and the turtles disperse between breeding periods before returning to their natal beaches to nest. The precision with which they home is astonishing; female green sea turtles at Tortuguero return to the same kilometer of beach to deposit each of the three clutches of eggs they lay in a breeding season and to the same portion of the beach in successive breeding seasons.

Probably the most striking example of the ability of sea turtles to home to their nesting beaches is provided by the green sea turtle colony that has its feeding grounds on the coast of Brazil and nests on Ascension Island, a small volcanic peak that emerges from the ocean. The island is 2,200 km east of Brazil and less than 20 km in diameter—a tiny target in the vastness of the South Atlantic.

Navigation by adult turtles

How do adult turtles migrating to or from breeding sites find their way across thousands of kilometers of ocean?

Other migratory animals use a variety of cues for navigation, and sea turtles probably do so as well. Chemosensory information may be one important component of their navigation. For example, the South Atlantic equatorial current flows westward, washing past Ascension Island and continuing toward Brazil, and the odor plume of the island may help guide female turtles back to the island to nest. That is, a female turtle leaving the coast of Brazil may swim upstream in the South Atlantic equatorial current (i.e., up the odor gradient) to locate Ascension Island.

It is impractical to locate female turtles off the coast of Brazil as they are about to begin their journey to the island, but it is easy to find turtles that have completed nesting at Ascension Island and are ready to start back to Brazil. Five female green sea turtles were tracked on their return trip using the Argos satellite system (**Figure 16.9**). The turtles traveled 1,777 to 2,342 km and reached Brazil in 33–74 days. For the first 500 km, they followed a west-southwest heading that carried them slightly south of a direct route toward the bulge of Brazil. At this stage, they were following the

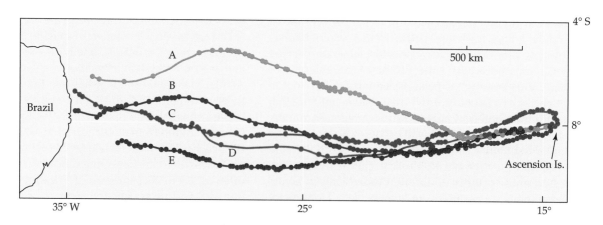

Figure 16.9 Turtle tracks. The paths of five green sea turtles (*Chelonias mydas*) were tracked by satellite as the turtles returned from Ascension Island to Brazil. Each dot is a location detected by the satellite. (From Luschi et al. 1998.)

route of the South Atlantic equatorial current. Perhaps the turtles were simply being carried off course by the current, but they may have been using the same guidance system that they rely on for their outward journey—that is, staying within the plume of the island's scent. Even though the current carried them slightly south of a direct route to Brazil, they may have saved energy and moved faster by initially staying in the current.

If they had remained in the current for the entire trip, they would have been carried too far south, so the turtles made a midcourse correction. The new course headed west-northwest on a nearly direct route toward the bulge of Brazil. The shift in direction may have been triggered by the waning strength of the scent plume. The turtles spent more than 90% of the journey underwater, suggesting that they might have been sampling the plume in three dimensions.

Navigation by hatchling and juvenile sea turtles

Several studies of navigation by hatchling loggerhead sea turtles have shown that they use at least three cues for orientation: light, wave direction, and magnetism. These stimuli played sequential roles in the turtles' behavior. When they emerged from their nests, the hatchlings crawled toward the brightest light they saw. The sky at night is lighter over the ocean than over land, so this behavior brought them to the water's edge. (Lights at shopping centers, streetlights, and even porch lights on beachfront houses can confuse loggerheads and other species of sea turtles and lead them inland, where they are crushed on roads or die of overheating the next day.)

In the ocean, the loggerhead hatchlings swam into the waves. This response moved them away from shore and ultimately into the Gulf Stream and then into the North Atlantic Gyre, the current that sweeps around the Atlantic Ocean in a clockwise direction, northward along the coast of the United States and then eastward across the Atlantic. Off the coast of Portugal, the gyre divides into two branches. One turns north toward England, and the other swings south past the bulge of Africa and eventually back westward across the Atlantic. It is essential for the baby turtles to turn right at Portugal; if they fail to make that turn, they are swept past England into the chilly North Atlantic, where they perish. If they do turn southward off the coast of Portugal, they are eventually carried back to the coast of tropical America—a round-trip that takes 5–7 years.

Magnetic orientation appears to keep the turtles in the North Atlantic Gyre and to tell them when to turn right to catch the current that will carry them to the South Atlantic. We usually think of Earth's magnetic field as providing two-dimensional information—north–south and east–west—but it's more complicated than that. The field loops out of Earth's north and south magnetic poles. At the Equator, the field is essentially parallel to Earth's surface (in other words, it forms an angle of 0 degrees), and at the poles it intersects the surface at an angle of 90 degrees. Thus, the three-dimensional orientation of Earth's magnetic field provides both directional information (which way is magnetic

north?) and information about latitude (what is the angle at which the magnetic field intersects Earth's surface?).

When loggerheads in a wading pool in a laboratory were exposed to an artificial magnetic field at the 57° angle of intersection with Earth's surface that is characteristic of Florida, they swam toward artificial east—even when the magnetic field was changed 180°, so that the direction they thought was east was actually west. That is, they were able to use a compass sense to determine direction. But that wasn't all they could do: subsequent studies showed that as the three-dimensional magnetic field was changed to match points around the North Atlantic Gyre, the hatchlings turned in the appropriate direction at each location to remain in the gyre and ultimately to return to the coast of Florida (**Figure 16.10**). Thus, it appears that young loggerheads can use magnetic sensitivity to recognize both direction and latitude.

16.5 The Fateful Life-History Characteristics of Turtles

Turtles are long-lived animals, and their life history depends on adult survival. Maturity comes late, but many species live as long as humans. Even small species such as the painted turtle do not mature until they are 7 or 8 years old, and females can live for more than 60 years. Many turtles have indeterminate growth—that is, growth continues after adult size is reached—and clutch size increases with body size. Reproductive senescence is delayed; female painted turtles more than 50 years old are still reproducing, although their eggs have lower hatching success than the eggs of younger turtles.

Long lifetimes are associated with low replacement rates of individuals in a population. In the absence of humans, juvenile mortality is high but adults are relatively safe from most natural predators. Humans, however, prey on adult turtles, short-circuiting a crucial element of the life-history strategy of turtles. As long-lived animals with low reproductive rates and high juvenile mortality, turtles have exactly the wrong characteristics to withstand heavy predation. When adults do not survive, populations decline and low replacement rates make recovery problematic.

The plight of large tortoises and sea turtles is particularly severe, not only because these species are among the largest and slowest growing of all turtles, but also because other aspects of their biology expose them to additional risks. For example, adult sea turtles are by-catch (nontarget species) in gill net, trawl, and longline fisheries. In a single year, more than 200,000 loggerhead sea turtles and 50,000 leatherbacks were captured by longline gear, and a large proportion of these turtles died.

Pollution of nesting beaches can reduce or eliminate recruitment, sometimes for years following a pollution event. The impact of the Deepwater Horizon spill in the Gulf of Mexico in 2010 included residual oil on beaches and oil balls

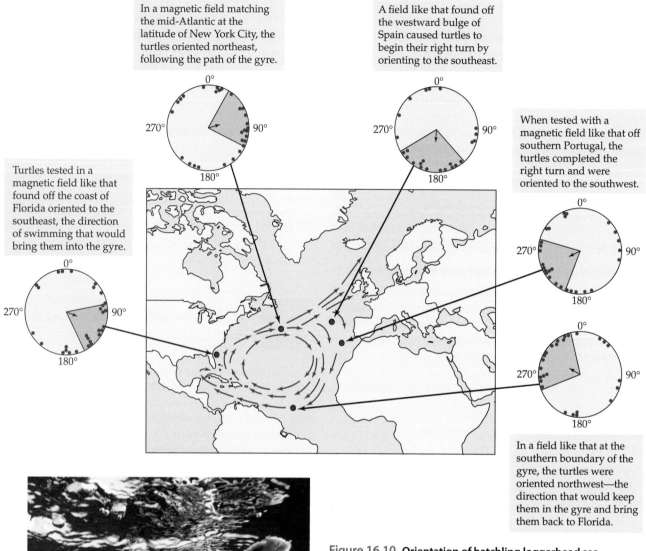

In a magnetic field matching the mid-Atlantic at the latitude of New York City, the turtles oriented northeast, following the path of the gyre.

A field like that found off the westward bulge of Spain caused turtles to begin their right turn by orienting to the southeast.

When tested with a magnetic field like that off southern Portugal, the turtles completed the right turn and were oriented to the southwest.

Turtles tested in a magnetic field like that found off the coast of Florida oriented to the southeast, the direction of swimming that would bring them into the gyre.

In a field like that at the southern boundary of the gyre, the turtles were oriented northwest—the direction that would keep them in the gyre and bring them back to Florida.

Figure 16.10 Orientation of hatchling loggerhead sea turtles (*Caretta caretta*) in magnetic fields along their migratory route.   The experiment tested the turtles in magnetic fields duplicating locations along their migratory routes. The North Atlantic Gyre (the predominant ocean current) is shown by arrows. In the orientation circles, each dot represents a single hatchling, the arrow in the center of each circle indicates the mean angle of the group, and the pie-shaped segment shows the 95% confidence intervals of the mean angle. (From Lohmann et al. 2001, additional data from Fuxjager et al. 2011; photo, WaterFrame/Alamy Stock Photo.)

in the water, and the effects extended southward to Brazil and across the Atlantic to Africa.

The largest extant land turtles are found on the Galápagos and Aldabra islands. The relative isolation of these small and (for humans) inhospitable landmasses was long an important factor in the survival of tortoises. In the past two centuries, however, human colonization of the islands has brought with it domestic animals such as goats and donkeys, which compete with tortoises for the limited quantities of vegetation found in these arid habitats, as well as dogs, cats, and rats that prey on tortoise eggs and hatchlings.

Entire turtle faunas are threatened in some areas. Nearly all species of turtles in Southeast Asia are now at risk because of economic and political changes in the region. Turtles have traditionally been used in China for food and for their supposed medicinal benefits. A 35-month survey recorded nearly 1 million turtles passing through markets in just three cities in China. In three years, China imported nearly 32 million turtles from the United States, and 700,000 of these were wild-caught adults.

Very little is known about the natural history of Chinese turtles. In fact, some species, such as McCord's box turtle (*Cuora mccordi*), are known scientifically only from specimens purchased in markets, and no wild population has ever been described. These species may never be known in the wild; specimens have not been seen in markets for several years, and the species may be extinct.

While China is the deepest black hole for turtles, it is by no means the only one. Madagascar is home to endangered species of many kinds of animals, including tortoises. Although these species are protected by international treaty, they are smuggled out of the country by the score for sale as pets in Japan, Europe, and North America, where they fetch high prices. The United States imports and exports more live reptiles for the pet trade than any other country. In four years, more than 8 million reptiles were imported and more than 36 million exported, and the most endangered species command the highest prices. Commercial exploitation is a threat to many species of vertebrates, but the combination of high juvenile mortality, late maturity, and long reproductive lifespans that makes adult survival the key to population survival renders turtles especially vulnerable.

Summary

If it has four legs and a shell, it's a turtle.

The morphology of turtles is unique; no other extant tetrapod has a bony external shell.

The highly specialized morphology of turtles—the trunk and limb girdles enclosed within a shell and the absence of skull fenestrae—has long obscured the phylogenetic affinities of this remarkable lineage.

Molecular analyses place turtles as the sister group of archosaurs.

The method of retracting the head into the shell distinguishes two groups of turtles: cryptodires bend the neck vertically, whereas pleurodires bend the neck laterally.

The upper shell (carapace) is formed primarily by expansion of the ribs, which are endochondral bone. The lower shell (plastron) is composed largely of dermal bone with some endochondral elements.

Beta-keratin, chemically similar to the beta-keratin in the scales of crocodylians and the feathers of birds, forms the horny scutes on the exterior of the carapace and plastron.

Because the ribs are incorporated into the carapace, turtles cannot expand and contract the rib cage to ventilate the lungs.

Instead, turtles ventilate by moving the viscera to compress and expand the lungs.

Some aquatic turtles employ vascularized tissue in the pharynx or cloaca for gas exchange while they are submerged.

The hearts of turtles are anatomically three chambered but functionally five-chambered.

The right atrium receives oxygenated blood returning from the body, and the left atrium receives oxygenated blood returning from the lungs, as in other tetrapods.

The turtle ventricle is anatomically undivided, but during ventricular contraction three transient subcompartments form. Valves respond to differences in pressure in the pulmonary and systemic circuits to direct oxygenated blood to the head and body via the aorta and deoxygenated blood to the lungs via the pulmonary artery.

The absence of an anatomical ventricular septum allows turtles to create intracardiac blood shunts that shift blood between the pulmonary and systemic circuits.

Small terrestrial turtles bask in the sun or move into the shade to control body temperatures. Large species of turtles, especially the enormous leatherback, retain heat produced by muscular activity, using countercurrent exchange of heat between veins and arteries in the flippers, and an insulating layer of fat in the neck.

All turtles are oviparous, depositing flexible-shelled or rigid-shelled eggs in nests that females excavate with their hindlimbs.

Nest moisture and temperature play important roles during embryonic development.

Eggs must take up water from the soil if the embryo is to metabolize all of its yolk. Hatchlings from dry nests are smaller than those from moist nests, and cannot run or swim as rapidly.

A few families of turtles include species that exhibit genetic sex determination (GSD), but most species exhibit temperature-dependent sex determination (TSD)—meaning sex is determined by the temperatures that embryos experience during the middle third of development.

Males can develop from eggs incubated at low temperatures (type Ia TSD), high temperatures (type Ib TSD), or intermediate temperatures (type II TSD). The switch in sexes typically occurs over a very narrow temperature range.

Spatial variation in nest temperature and interaction between temperature and moisture in sex determination ensure that hatchlings of both sexes are produced.

(Continued)

Summary *(continued)*

Few species of turtles provide parental care, and even in those species care is limited to brief periods of nest attendance. Only one species, a South American river turtle, is known to provide post-hatching care.

As eggs near hatching, vocalizations by the embryos synchronize their emergence from the shell.

Sea turtles emerge nearly simultaneously from many nests on a beach. The emergence of large numbers of baby turtles rushing en masse toward the water helps reduce their mortality by temporarily swamping predators.

Hatchling and juvenile turtles are difficult to find, and little is known about the early years of most turtle species.

Turtles employ a variety of sensory modalities—including tactile, visual, and olfactory signals—during social interactions.

Social behavior is probably more extensive and phylogenetically widespread among turtles than is yet appreciated.

The species-specific patterns of stripes and colors on the heads, necks, and tails of many species of pond turtles are probably used by the turtles to identify conspecifics.

Some turtles vocalize during courtship and mating, emitting sounds that have been described as grunts, moans, and bellows.

Herds of tortoises have social hierarchies that are determined largely by aggressive encounters.

Turtles can travel considerable distances between their feeding and breeding sites; some sea turtles travel thousands of kilometers.

Sea turtles return to breed and lay eggs at the beaches where they hatched, often retuning to the same part of a beach repeatedly during a single season and again in subsequent breeding seasons.

Sea turtles use multiple senses to navigate, including olfaction and magnetic orientation. They can sense both compass directions and their location on Earth's surface, and make appropriate changes in course to reach their goal.

The life-history characteristics of turtles are a recipe for disaster in a human-dominated world.

Most turtles require years to mature, and juvenile mortality is high. As result, replacement rates are low, and longevity of adults is a critical factor in the viability of populations.

Adult turtles are relatively safe from natural predators, but not from humans. Anthropogenic mortality of adult turtles strikes at most the vulnerable element of turtle life history—survival of adults.

Discussion Questions

1. Considering the life-history characteristics of most turtles, which of these approaches to conservation of a turtle species is likely to be more effective, and why: protecting eggs and nests so that hatchlings can reach the water, or protecting adult females?

2. Why is it so difficult to determine the phylogenetic relationships of turtles to other sauropsids?

3. Does temperature-dependent sex determination offer any benefits to turtles, or is it better regarded as a possibly ancestral system of sex determination that has persisted in most lineages of turtles because it works well enough?

4. What simple experiment could you conduct to test the hypothesis that hatchling sea turtles use orientation to Earth's magnetic field to swim in the correct direction when they reach the ocean?

5. The absence of a complete ventricular septum allows turtles to reduce the blood flow to the lungs by shunting blood from the right (pulmonary) circuit to the left (systemic) circuit. When and why would a right-to-left blood shunt be advantageous?

Additional Reading

Benhamou S and 5 others. 2011. The role of geomagnetic cues in green turtle open sea navigation. *PLoS ONE* 6(10): e26672.

Casey JP, James MC, Willard AS. 2014. Behavioral and metabolic contributions to thermoregulation in freely swimming leatherback turtles at high latitudes. *Journal of Experimental Biology* 217: 2331–2337.

Chiari Y, Cahais, Galtier, Delsuc F. 2012. Phylogenomic analyses support the position of turtles as the sister group of birds and crocodiles (Archosauria). *BMC Biology* 10: 65.

Cordero CA, Telemeco RS, Gangloff EJ. 2018. Reptile embryos are not capable of behavioral thermoregulation in the egg. *Evolution & Development* 20: 40–47.

Davenport J, Jones TT, Work TM, Balaz GH. 2016. Topsy-turvy: Turning the countercurrent heat exchange of leatherback turtles upside down. *Biology Letters* 11: 20150592.

Du W-G, Zhao B, Chen Y, Shine R. 2011. Behavioral thermoregulation by turtle embryos. *Proceedings of the National Academy of Sciences USA* 108: 9513–9516.

Ferrara CR, Vogt RC, Sousa-Lima RS. 2013. Turtle vocalizations as the first evidence of posthatching parental care in chelonians. *Journal of Comparative Psychology* 127: 24–32.

Field DJ and 5 others. 2014. Toward consilience in reptile phylogeny: miRNAs support an archosaur, not lepidosaur, affinity for turtles. *Evolution and Development* 16: 189–196.

Jensen MP and 7 others. 2018. Environmental warming and feminization of one of the largest sea turtle populations in the world. *Current Biology* 28: 154–159.

Joyce WG. 2016. The origin of turtles: A paleontological perspective. *Journal of Experimental Zoology* 324: 181–193.

Kashon EAF, Carlson BE. 2018. Consistently bolder turtles maintain higher body temperatures in the field but may experience greater predation risk. *Behavioral Ecology and Sociobiology* 72: 9.

Lyson TR and 7 others. 2014. Origin of the unique ventilatory apparatus of turtles. *Nature Communications* 5: 5211.

McGlashan JK, Spencer RJ, Old JM. 2012. Embryonic communication in the nest: Metabolic responses of reptilian embryos to developmental rates of siblings. *Proceedings of the Royal Society B* doi:10.1098/rspb.2011.2074

Mitchell-Campbell MA, Campbell HA, Cramp RL, Booth DT, Franklin CE. 2011. Staying cool, keeping strong: Incubation temperature affects performance of a freshwater turtle. *Journal of Zoology* 285: 266–273.

Mrosovsky N, Ryan GD, James MC. 2009. Leatherback turtles: the menace of plastic. *Marine Pollution Bulletin* 58: 287–289.

Olmo E. (ed.). 2010. *Sex Determination and Differentiation in Reptiles*. Karger, Basel, Switzerland.

Van Dijk PP, Stuart BL, Rhodin AGJ. 2000. *Asian Turtle Trade. Chelonian Research Monographs* No. 2. Chelonian Research Foundation.

Go to the Companion Website **oup.com/us/vertebratelife10e** for active-learning exercises, news links, references, and more.

CHAPTER

17

Lepidosaurs

Tuatara, Lizards, and Snakes

© Dancing Snake Nature Photography

Lepidosaurs (Greek *lepisma*, "scale"; *sauros*, "lizard") form the largest and most diverse group of nonavian reptiles, containing a single rhynchocephalian (the tuatara, *Sphenodon punctatus*) and more than 10,000 species of squamates—nearly 6,500 species of lizards and more than 3,600 species of snakes (**Table 17.1**). Lepidosaurs are predominantly terrestrial tetrapods with some secondarily aquatic species. They are found on all continents except Antarctica.

The skin of lepidosaurs is covered by scales, and the outer layer of the epidermis is shed at intervals (**Figure 17.1**). Tuatara and most lizards have four limbs, but reduction or complete loss of limbs is widespread among some groups of lizards, and all snakes are limbless. Lepidosaurs have a transverse cloacal slit rather than the longitudinal slit that characterizes other tetrapods.

Lepidosaurs are the sister lineage of archosaurs (crocodylians and birds). Within Lepidosauria, Rhynchocephalia (the tuatara) is the extant sister group of Squamata (lizards and snakes). Snakes are nested within lizards; lizards and snakes can be distinguished in colloquial terms but not phylogenetically. Thus, "lizards" in the colloquial sense is a paraphyletic group because it does not include all the descendants of a common ancestor. Nonetheless, lizards and snakes are distinct in morphology and in many aspects of their ecology and behavior, and the colloquial separation is useful in discussing them.

17.1 Rhynchocephalians and the Biology of Tuatara

Rhynchocephalians were a diverse group of diapsids with a worldwide distribution in the Mesozoic. Triassic rhynchocephalians were small, with body lengths of only 15–35 cm, and were probably insectivores. During the Jurassic and Cretaceous, rhynchocephalians included herbivorous and marine forms, some of which reached lengths of 1.5 m.

The tuatara is the only extant rhynchocephalian (**Figure 17.2**). (*Tuatara* is a Maori word meaning "spines on the back"; no *s* is added to form the plural.) The tuatara has several derived characters that disqualify it from being considered a "living fossil." Tuatara formerly inhabited the North and South Islands of New Zealand, but the advent of the Maori and the Pacific rats that arrived with them exterminated tuatara on the main islands about 800 years ago. Now the species is restricted to 32 natural populations on small islands off the coast of New Zealand, 9 translocated populations on islands, and 5 translocated populations in sanctuaries on the North and South Islands.

Adult tuatara are about 60 cm long. They are nocturnal, and in the cool, foggy nights that characterize their island habitats, they cannot raise their body temperatures by basking in sunlight as lizards do. Body temperatures ranging from 6°C to 16°C have been reported for active tuatara, and these are low compared with the temperatures of most lizards. During the day, tuatara do bask, raising their body temperatures to 28°C or higher.

Tuatara feed largely on invertebrates but occasionally eat a frog, lizard, or seabird. The jaws and teeth of tuatara produce a shearing effect during chewing: The upper jaw contains two rows of teeth on each side, one on the maxilla and the other on the palatine. The teeth of the lower jaw fit between the two rows of upper teeth, and the lower jaw closes with an initial vertical movement, followed by an anterior sliding movement. As the lower jaw slides, the food item is bent or sheared between the triangular cusps on the teeth of the upper and lower jaws.

Tuatara live in burrows that they may share with nesting seabirds. The burrows are spaced at intervals of 2–3 m in dense colonies, and both male and female tuatara are territorial. They use vocalizations, behavioral displays, and color change in their social interactions.

The ecology of tuatara rests to a large extent on exploitation of resources provided by seabird colonies. Tuatara occasionally feed on adult birds, which are most vulnerable to predation at night. More important, the quantities of guano produced by the birds, the scraps of food they bring to their nestlings, and the bodies of dead nestlings attract huge numbers of arthropods that, in turn, tuatara

Table 17.1	Major Lineages of Extant Lepidosaurs[a]

Rhynchocephalia
One extant species, the tuatara (*Sphenodon punctatus*), in New Zealand

Squamata: Lizards
A total of 42 lizard families are recognized, containing from 1 species to more than 1,600 species

Dactyloidae: 416 species of arboreal lizards, mostly in the genus *Anolis*, found in the Americas and West Indies and previously included in Iguanidae. Repeated species radiations have made *Anolis* models for studies of evolution and ecomorphology. From 10 to 50 cm.[b]

Iguanidae: 42 species of herbivorous lizards—iguanas—found primarily in the New World but with representatives on the Galápagos Islands and Fiji. Most species are terrestrial or arboreal, but the Galápagos marine iguana (*Amblyrhynchus cristatus*) dives as deep as 12 m to scrape algae from submerged rocks. From ~25 cm to >1 m.

Phrynosomatidae: 153 species of terrestrial, rock-dwelling, and arboreal lizards, mostly in North America and previously included in Iguanidae. Includes the wide-ranging genus *Sceloporus* (spiny swifts) and *Phrynosoma* (horned lizards). Most phrynosomatids are oviparous, some are viviparous, and *Sceloporus aeneus* includes both oviparous and viviparous populations. From 10 to 30 cm.

Agamidae: 480 species found in the Middle East and parts of Asia, Africa, the Indo-Australian Archipelago, and Australia. Most species are terrestrial or arboreal, and a few are semiaquatic. From ~10 cm to 1 m.

Chamaeleonidae: 203 species found in Africa and Madagascar and extending into southern Spain and along the west coast of the Mediterranean. Primarily arboreal, but includes a few grassland and terrestrial species. From ~3 cm to 0.5 m.

Varanidae: 79 species found in Africa, Asia, and the East Indies; about half the species are limited to Australia. *Varanus* has the greatest size range of any extant genus of vertebrates, extending from 20 cm (*V. brevicauda*) to 3 m (*V. komodoensis*).

Anguidae: 77 species found in North and South America, Europe, the Middle East, and southern China. Most species are terrestrial, but legless lizards (*Anniella*) are burrowers, and all *Abronia* are arboreal. From 20 cm to >1 m.

Helodermatidae: 2 heavy-bodied, venomous species found in the southwestern U.S. and Mexico. From 50 cm to ~1 m.

Amphisbaenia: 6 families, 197 species of elongated, legless, burrowing lizards found in the West Indies, South America, sub-Saharan Africa, and around the Mediterranean. All are limbless except for the three species of *Bipes* (Bipedidae), which have stout forelimbs. Some evidence suggests that the ancestors of *Bipes* were limbless and that forelimbs re-evolved; if that interpretation is correct, it represents a remarkable evolutionary reversal. From 10 to 80 cm.

Lacertidae: 323 species of small to medium-size terrestrial lizards found in Europe, Africa, and Asia. Lacertids are oviparous, except for the genera Eremias and Zootoca, which have both oviparous and viviparous representatives. From <10 cm to ~80 cm.

Teiidae: 155 species of terrestrial lizards ranging from just south of the Canadian border in North America to central Argentina, and including the West Indies. Several genera include parthenogenetic species. From 20 cm to ~1 m.

Scincidae: With 1,612 species, the most species-rich family of lepidosaurs. Skinks occur on all continents except Antarctica. Body forms vary from stout to elongated, and many lineages have reduced limbs. Most species are terrestrial, but arboreal, semifossorial, and semiaquatic species are represented. Oviparity is the dominant reproductive mode, but some skinks are viviparous and two species of Australian skinks contain both oviparous and viviparous populations. From ~10 to 40 cm.

Gekkonidae: 1,103 species, found on every continent except Antarctica. Some have modified scales (setae) on the bottom of the toes that allow them to climb vertical surfaces and even to hang by a single toe. Many species are nocturnal. From ~3 to 30 cm.

Pygopodidae: 46 species of elongated lizards that move through grass and leaf litter with sinusoidal locomotion. The forelimbs are absent, and the hindlimbs are reduced to flaps. Most species are insectivorous, but two species of *Lialis* feed on skinks and have highly kinetic skulls and hinged teeth. From ~6 to 30 cm.

Squamata: Serpentes (snakes)
A total of 24 families of snakes are recognized, containing from 1 species to more than 1,800 species

Elapidae: 359 venomous species with hollow fangs near the front of relatively immobile maxillae. Elapids occur on all continents except Antarctica and are most diverse in Australia. Sea snakes (Hydrophiinae) are elapids. Most species are small (25 cm–1 m), but the king cobra (*Ophiophagus hannah*) of Asia reaches 5 m.

Colubridae: 1,866 species, found on all continents except Antarctica. Many colubrids have glands that secrete venom that kills prey, but they lack hollow teeth specialized for injecting venom. From 20 cm to ~3 m.

Viperidae: 341 venomous species in which the maxillae rotate about their attachment to the prefrontals, allowing the fangs to rest horizontally when the mouth is closed and swing erect when the mouth is opened. True vipers (Viperinae, 100 species) are found in Eurasia and Africa, pit vipers (Crotalinae, 239 species) in the New World and Asia, and fea vipers (Azemiopinae, 2 species) in China and Southeast Asia. Viperids are absent from Australia and Antarctica. From ~75 cm to 2 m.

Pythonidae: 40 species of terrestrial and arboreal snakes found in Africa, Asia, and Australia. All pythons are oviparous, and females of some species coil around their eggs and use heat generated by muscular contractions to warm the eggs during incubation. From ~1 to almost 10 m.

Boidae: 61 species of terrestrial, arboreal, and semiaquatic snakes from western North America through subtropical South America and the West Indies. Most boids are viviparous, but 3 species are oviparous, and oviparity might have re-evolved from a viviparous ancestral state. From ~50 cm to almost 9 m.

Typhlopidae: 406 species of small to medium-size fossorial snakes with reduced eyes, found on all continents except Antarctica. Typhlopids have teeth only on the maxillae, which are oriented horizontally and rake small insect prey into the mouth. The flowerpot snake (*Indotyphlops braminus*) is the only snake known to be parthenogenetic. It has been accidentally transported in soil and now occurs on every continent except Antarctica. From 20 to 75 cm.

Leptotyphlopidae: 139 species of extremely small, slim (Greek *leptos*, "thin, small") fossorial snakes, known as threadsnakes, found in the Americas, Africa, and the Middle East. The Barbados threadsnake (*Leptotyphlops carlae*), with an adult length of 10 cm, may be the smallest extant species of snakes. Leptotyphlopids have teeth only on the dentary bone and feed on soft-bodied prey, including termites and ant pupae. From 10 to 30 cm.

[a] Only major families of squamates are listed, with the most basal species at the bottom of each list and the most derived species at the top. Number of species in each family are from the Reptile Database (www.reptile-database.org) as of March 2017.
[b] Sizes are given as total body length (i.e., from nose to tail tip).

(A)

(B)

Figure 17.1 **Lepidosaurs shed their skins.** A distinguishing characteristic of lepidosaurs is the production of successive generations of epidermal cells that are shed at intervals over the course of their lives. (A) Tuatara and lizards shed the skin in pieces, as seen for this lesser earless lizard (*Holbrookia maculata*). (B) In contrast, snakes such as the kingsnake (*Lampropeltis getula*) shed the entire skin intact. (Photos by René Clark © Dancing Snake Nature Photography.)

eat. These arthropods are largely nocturnal and must be hunted when they are active. Thus, the nocturnal activity of tuatara and the low body temperatures resulting from being active at night are probably derived characters that stem from the association of tuatara with colonies of nesting seabirds.

Figure 17.2 **Henry the tuatara (*Sphenodon punctatus*).** Henry has lived at the Southland Museum in Invercargill, New Zealand, since 1970 and was the model for the New Zealand 5-cent coin. He is estimated to be at least 80 years old, and possibly as old as 100. For many years Henry was too aggressive to breed, but removal of a tumor from his cloaca in 2009 reignited his ardor, and he finally added some diversity to the museum's genetic breeding stock. The dominant breeding male, Albert, had sired so many offspring that he was retired from the breeding program to prevent his genes from becoming overrepresented in the limited gene pool of captive tuatara. By 2017, Henry had sired more than a dozen offspring each from two female tuatara. (Photos by Harvey Pough.)

17.2 Radiation of Squamates

Determinate growth may be the most significant derived character of squamates, especially lizards, because it allows many species to remain small enough to eat insects. Crocodylians and turtles continue to grow throughout their lives, but about 80% of extant species of lizards weigh less than 20 g as adults—smaller than a white mouse. Generalized lizards of this size can readily capture insects, whereas large insect-eating vertebrates, such as mammalian anteaters, require morphological or ecological specializations to capture tiny prey.

Lizards

Lizards range in size from diminutive geckos and chameleons less than 3 cm long to the Komodo monitor (*Varanus komodoensis*), which is 3 m long at maturity and can weigh as much as 80 kg. Giant varanids were widespread in Eurasia and Australasia during the Neogene. A reconstruction of the skeleton of a fossil monitor, *Varanus priscus* from the Pleistocene of Australia, is 5.5 m long, and in life the lizard may have weighed 600 kg. *Varanus priscus* may well have preyed on humans; fossil dates indicate that the two species coexisted for at least 10,000 years.

Insectivores Lizards display an enormous variety of body forms (**Figure 17.3**). Spiny swifts (*Sceloporus*) and the Australian lizards known as dragons (*Ctenophorus*) are examples of small, generalized insectivores. Other small lizards have specialized diets: the North American horned lizards (*Phrynosoma*) and the Australian thorny devil (*Moloch horridus*) feed on ants. Many geckos are nocturnal, and some species are so closely associated with humans that they have been accidentally transported around the world.

Herbivores Most large lizards are herbivores. Many iguanas are arboreal inhabitants of the Neotropics. Large terrestrial iguanas live only on islands in the West Indies

(A)

(B)

(C)

(D)

(E)

(F)

(G)

(H)

Figure 17.3 Body forms of lizards. (A) Most small lizards, like this netted dragon (*Ctenophorus nuchalis)* from Australia, are generalized insectivores and eat a wide variety of insects and spiders. In contrast, North American horned lizards, such as the Texas horned lizard (*Phrynosoma cornutum*; B) and the Australian thorny devil (*Moloch horridus*; C) eat only ants. Both have dorsoventrally flattened bodies, short legs and spines that deter predators. (D) Most lizards are diurnal, but geckos are nocturnal. This orange-spotted gecko (an undescribed viviparous species of *Mokopirirakau*) is from the mountains of southern New Zealand. (E) Most herbivorous lizards, such as the Galápagos land iguana (*Conolophus subcristatus*), have short legs and bulky bodies that accommodate the long digestive system needed to digest plants. (F) In contrast, highly predatory lizards are slim, with long legs. The Australian racehorse goanna (*Varanus gouldii*) can sprint faster than 20 km/h to escape from predators. (G) Chameleons, the most specialized arboreal lizards, have zygodactylous feet, prehensile tails, and eyes in turrets that can swivel independently. The giant chameleon of Madagascar (*Furcifer oustaleti*), which reaches a length of nearly 70 cm, is probably the largest extant species. (H) Limb reduction or loss has evolved many times among lizards. This Australian skink (*Lerista planiventralis*) has tiny forelegs with only two digits and the hindlegs have only three toes. They use their wedge-shaped snout to burrow through loose soil. (A,F,G,H, courtesy of Stephen Zozaya; B, courtesy of Todd Pierson; C, photo by Harvey Pough; D, courtesy of Alison Cree; E, courtesy of Robert Rothman.)

(*Cyclura*) and the Galápagos Islands (*Conolophus*), probably because the absence of predators on the islands has allowed these iguanas to spend a large part of their time on the ground. Smaller terrestrial herbivores like the black iguanas (*Ctenosaura*) live on the mainland of Mexico and Central America, and still smaller relatives such as chuckwallas (*Sauromalus*) and desert iguana (*Dipsosaurus dorsalis*) range as far north as the southwestern United States.

Carnivores Monitors (Varanidae) and tegus (Teiidae) are exceptions to the generalization that lizards too large to subsist on insects are herbivores. Both feed on a variety of vertebrates and invertebrates, including birds and mammals. Varanids circumvent the conflict between locomotion and lung ventilation that constrains the activity of other lizards by using a positive-pressure gular pump to assist the axial muscles, and they are able to sustain high levels of activity. The Komodo monitor can kill adult water buffalo, but deer and feral goats are its usual prey. Large monitors were widely distributed on the islands between Australia and Indonesia during the Pleistocene and may have preyed on pygmy elephants that also lived on the islands at that time.

The effectiveness of monitors as predators is reflected in reciprocal geographic distributions of monitors and small mammalian carnivores in the Indo-Australian Archipelago, Australia, and New Guinea. Wallace's Line is a zoogeographic boundary that extends between the islands of Borneo on the west and Sulawesi on the east. It traces the location of a deep oceanic trench, and areas to the west of the line have never had a land connection with areas to the east. Because of this ancient separation, Wallace's Line marks the easternmost boundary of the occurrence of many species from the Asian mainland and the westernmost boundary of species from Australia and New Guinea. In particular, small carnivorous placental mammals are found west of the line and small carnivorous marsupials to the east. Large species of monitors occur on both sides of the line, but small species are found on only the eastern side. West of the line, the adaptive zone for small carnivores is filled by mammals (such as small cats, civets, mongooses, and weasels), but east of the line—especially in Australia and New Guinea—most of the small carnivores are monitors.

Arboreal lizards   Many lizard species are arboreal, with the Old World chameleons (Chamaeleonidae) being the most specialized of these. The toes on their **zygodactylous** (Greek *zygos*, "a joining"; *daktylos*, "finger or toe") feet are arranged in two opposable groups that grasp branches firmly, and additional security is provided by a prehensile tail. The tongue and hyoid apparatus of chameleons are specialized, and the tongue can be projected forward more than a body's length to capture insects that adhere to its sticky tip. This feeding mechanism requires good eyesight, especially the ability to gauge distances accurately so that the correct trajectory can be employed. Chameleons' eyes are elevated in small cones and can move independently. When a chameleon is at rest, its eyes swivel back and forth, viewing its surroundings. When it spots an insect, the lizard fixes both eyes on it and cautiously stalks to within shooting range.

Legless lizards   Limb reduction has evolved more than 60 times among lizards, and every continent (except Antarctica) has one or more families with legless, or nearly legless, species. Two ecomorphs characterize legless lizards: surface-dwelling species have long tails, and burrowers have short tails. Legless surface-dwellers usually live in dense grass or shrubbery, where a slim, elongated body can maneuver more easily than a short one with functional legs and where a long tail can contribute to sinusoidal locomotion. In contrast, a tail creates friction in a burrow but makes little contribution to locomotion.

Amphisbaenians are **fossorial** (Latin *fossor*, "digger") lizards with specializations that differ from those of other squamates. The earliest amphisbaenian known is a fossil from the Late Cretaceous. Most amphisbaenians are legless, but the three species in the Mexican genus *Bipes* have well-developed forelimbs that they use to assist entry into the soil but not for burrowing underground (see Figure 17.4D).

The skin of amphisbaenians is distinctive. The **annuli** (rings; singular *annulus*) that pass around the circumference of the body are readily apparent from external examination, and dissection shows that the integument is nearly free of connections to the trunk. Thus, it forms a tube within which the amphisbaenian's body can slide forward or backward. The separation of trunk and skin is employed during locomotion through tunnels. Integumentary muscles run longitudinally from annulus to annulus. The skin over this area of muscular contraction is then telescoped and buckles outward, anchoring that part of the amphisbaenian against the walls of its tunnel. Next, contraction of muscles that pass anteriorly from the vertebrae and ribs to the skin slide the trunk forward within the tube of integument. Amphisbaenians can move backward along their tunnels with the same mechanism by contracting muscles that pass posteriorly from the ribs to the skin. The name "amphisbaenian" is from the Greek (*amphis*, "double"; *baino*, "to walk") and refers to the ability of amphisbaenians to move forward and backward with equal facility.

The dental structure of amphisbaenians is also distinctive: they possess a single median tooth in the upper jaw, a feature unique to this group of vertebrates (see Figure 17.9). The median tooth is part of a specialized dental battery that makes amphisbaenians formidable predators, capable of subduing a wide variety of invertebrates and small vertebrates. The upper tooth fits into the space between two teeth in the lower jaw and forms a set of nippers that can bite out a piece of tissue from a prey item too large for the mouth to engulf as a whole.

The skulls of amphisbaenians are used for tunneling, and they are rigidly constructed. The burrowing habits of amphisbaenians make them difficult to study, but three basic body forms and functional categories can be recognized (**Figure 17.4**):

1. *Blunt-snouted* forms burrow by ramming their head into the soil to compact it. Sometimes an oscillatory rotation of the head, which has heavily keratinized scales, is used to shave material from the face of the tunnel.

2. *Shovel-snouted* forms ram their snout into the end of the tunnel and then lift the head to compact soil into the roof.

3. *Keel-snouted* forms ram the snout into the end of the tunnel and then use the snout or the side of the neck to compress the material into the walls of the tunnel.

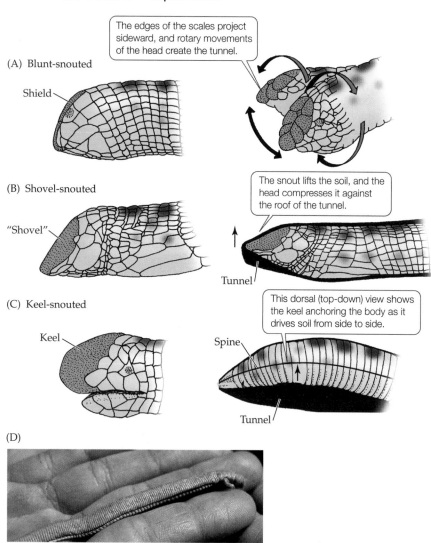

(A) Blunt-snouted

Shield

The edges of the scales project sideward, and rotary movements of the head create the tunnel.

(B) Shovel-snouted

"Shovel"

The snout lifts the soil, and the head compresses it against the roof of the tunnel.

Tunnel

(C) Keel-snouted

Keel

Spine

This dorsal (top-down) view shows the keel anchoring the body as it drives soil from side to side.

Tunnel

(D)

Figure 17.4 Body forms of amphisbaenians. Blunt-snouted amphisbaenians are found near the surface, where the soil is loose. More specialized shovel- and keel-snouted amphisbaenians have modifications of the head that allow them to burrow deeper into more compacted soil. (A) *Agamodon anguliceps*, a blunt-snouted species from Africa, has heavy scales forming a shield on the front of the head (left). The edges of the scales project sideward, and *Agamodon* uses rotary movements of the head to shave soil from the working face of its tunnel (right). (B) Shovel-snouted forms (e.g., *Rhineura floridana*) have flat, wedge-shaped heads (left). The snout is driven into the soil at the front of the tunnel, and the head is lifted to compress loosened soil against the roof of the tunnel (right). (C) The heads of keel-snouted amphisbaenians (e.g., *Amphisbaena kingii*) have a vertically oriented sharp-edged scale on the snout (left). The scale is driven into the soil at the front of the tunnel and anchors the anterior part of the body as it pushes loose soil from side to side in its tunnel (right; note that this view is from above the animal; all other views are lateral). (D) The five-toed Mexican worm lizard (*Bipes biporus*) has short front legs. The three species of *Bipes* are the only amphisbaenians with limbs, and they use them only to enter the soil. Once underground, they burrow by forcing their blunt snouts through the soil. (A–C adapted from Gans 1974; D, photo by Harvey Pough.)

Snakes

Snakes range in size from diminutive burrowing species, which feed on termites and grow to only 10 cm, to large constrictors, which approach 10 m in length. The phylogenetic affinities of snakes are hotly debated. Clearly they are nested within the squamates, and they appear to have evolved in a terrestrial environment. The leptotyphlopids and typhlopids are probably similar to the ancestral condition for snakes. These burrowing snakes have shiny scales and reduced eyes. Traces of the pelvic girdle remain in most species, but the braincase is snakelike. Their anatomical and ecological characters appear to be consistent with a long-standing hypothesis that snakes evolved from a subterranean lineage of lizards with greatly reduced eyes. The hypothesis that the eyes of extant surface-dwelling snakes were redeveloped after nearly disappearing during a fossorial stage in their evolution explains some puzzling differences in the eyes of snakes and lizards.

Unlike surface-dwelling legless lizards, surface-dwelling snakes have long trunks and relatively short tails. Elongation of the trunk probably results from a change in expression of the regulatory factor Oct4 during development. A second regulatory factor, ZRS, which controls expression of the protein Sonic hedgehog (Shh), has been mutated in snakes. These mutations reduce the activity of ZRS and the expression of Shh. Replacing mouse ZRS with snake ZRS causes mice to be born with stumps instead of legs.

The morphology of snakes is highly specialized, and correlated suites of characters identify parallel evolution of ecomorphs in different families (**Figure 17.5**). Generalized terrestrial snakes are elongated and have long tails; vipers that prey on mammals that are large in relation to the size of the snake are stout; arboreal snakes are long and thin; burrowing snakes have short tails and smooth, shiny scales; and specialized aquatic snakes are laterally flattened and have nostrils that are closed by valves.

Snakes use four methods of locomotion, depending on the body form of the species and the substrate over which the snake is moving (**Figure 17.6**):

1. In *lateral undulation,* the body is thrown into a series of curves. All snakes employ this mode of locomotion.

(A)

(B)

(C)

(D)

Eye

(E)

Nasal valves

Figure 17.5 Body forms of snakes. (A) The body length of a generalized terrestrial snake, such as the North American fox snake (*Pantherophis vulpinus*) is 10 to 15 times its maximum circumference, and these snakes feed on mice and other small, slim mammals. (B) Some vipers, especially African vipers in the genus *Bitis* such as this puff adder (*B. arietans*), are exceedingly stout, with body lengths only 5 times their circumference. These snakes are capable of swallowing bulky prey, up to and including small antelope. Vipers have long fangs that inject proteolytic venom deep into the prey, killing it and also initiating digestion from the inside out while the snake's digestive enzymes work from the outside in. (C) Arboreal specialists like this Central American blunt-headed tree snake (*Imantodes cenchoa*) are exceeding slim with body lengths 20 to 25 times their diameter. These snakes hunt in vegetation at night (note the large eyes). Their long, thin bodies allow them to spread their weight over many twigs and they can extend their heads and necks like construction cranes to seize small lizards that sleep on leaves at the tips of branches where other predators cannot reach them (inset). (D) Fossorial snakes have slim bodies, smooth scales, and short tails. The most specialized fossorial snakes, such as this western threadsnake (*Rena humilis*), have extremely reduced eyes. The head and tail are similar in appearance; a threadsnale attacked by a predator curls into a ball with its head inside its coil and waves its tail to direct the predator's attack to this non-vital site. Some threadsnakes live in colonies of army ants and travel in their raiding columns. Live threadsnakes have also been found in the nests of owls, apparently escaping after having been brought as food for nestlings and living on the insects attracted to the owls' nests by fecal matter and uneaten scraps of prey. (E) Specialized sea snakes, such as the olive sea snake (*Aipysurus laevis*), never leave the water. They lack ventral scales and have laterally compressed bodies, flattened tails, and valvular nostrils located on the top of the snout. (A,C, courtesy of Todd Pierson; B, photo by Harvey Pough; D, René Clark © Dancing Snake Nature Photography; E, WaterFrame/Alamy Stock Photo; inset by Harvey Pough.)

2. *Concertina locomotion* is used in narrow passages such as rodent burrows that do not provide space for the broad curves of lateral undulation.

3. *Rectilinear locomotion* is used primarily by heavy-bodied snakes such as large vipers, boas, and pythons. Rectilinear locomotion is slow, but it is effective even when there are no surface irregularities strong enough to resist the sideward force exerted by lateral undulation.

4. *Sidewinding* is used primarily by snakes in deserts, where windblown sand substrate slips away during lateral undulation. Because the snake's body is extended nearly perpendicular to its line of travel, sidewinding is an effective means of locomotion only for small snakes that live in habitats with few plants or other obstacles.

(A)

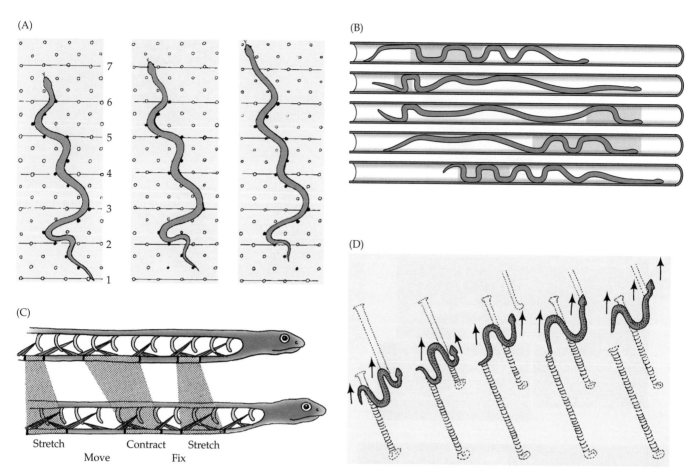

(B)

(C)

Stretch   Contract   Stretch
        Move   Fix

(D)

Figure 17.6 The four modes of snake locomotion. (A) Lateral undulation is the most generalized form of locomotion. The body may curve irregularly, as shown in this illustration of a snake crawling across a board with fixed pegs. Each curve presses backward; the pegs against which the snake is exerting force are shown in solid color. The lines numbered 1 to 7 are at 3-inch intervals, and the positions of the snake are shown at 1-second intervals. (B) In concertina locomotion, a snake anchors the posterior part of its body by pressing several loops against the walls of a burrow and extends the front part of its body. When the snake is fully extended, it forms new loops anteriorly and anchors itself with these while it draws the rear end of its body forward. (C) In rectilinear locomotion, alternate sections of the ventral skin are lifted off the ground and pulled forward by muscles that originate on the ribs and insert on the ventral scales. The intervening sections of the body rest on the ground and support the snake's body. Waves of contraction pass from anterior to posterior. Because the snake moves slowly and in a straight line, it is inconspicuous, and rectilinear locomotion is used by some snakes when stalking prey. (D) A sidewinding snake raises its body in loops, resting its weight on two or three points that are the only body parts in contact with the ground. The loops are swung forward through the air and placed on the ground, with the points of contact moving smoothly along the body. Force is exerted downward; the lateral component of the force is so small that the snake does not slip sideways. This downward force is shown here by imprints of the ventral scales in the tracks. (B from Astley and Jayne 2009.)

17.3 Foraging Modes

The methods that snakes and lizards use to find, capture, subdue, and swallow prey are diverse, and they are important in determining the interactions among species in a community. Among the many astonishing specializations that have evolved are blunt-headed snakes with long lower jaws that can reach into a shell to winkle out a snail; nearly toothless snakes that swallow bird eggs intact and then slice them open with sharp ventral processes (hypapophyses) on the neck vertebrae; and chameleons that project their tongues to capture insects or small vertebrates on the sticky tips.

The activity patterns of lizards span a range from extremely sedentary species that spend hours in one place to species that are in nearly constant motion. Field observations of the curly tail lizard *Leiocephalus schreibersi* and the ameiva *Pholidoscelis chrysolaema* in the Dominican Republic revealed two extremes of behavior. *Leiocephalus* rested on elevated perches from sunrise to sunset and were motionless for more than 99% of the day. Their only movements were short, rapid dashes to capture insects or chase away other lizards. These periods of activity never lasted longer than 2 seconds, and the frequency of movements averaged 9.6 per hour. In contrast, *Pholidoscelis* were active for only 4–5 hours in the middle of the day, but they were moving more than 70% of that time, and their velocity averaged one body length every 2–5 seconds.

For convenience, the extremes of this spectrum of activity patterns are frequently called sit-and-wait predators and actively foraging predators (Table 17.2), and the intermediate condition has been called a cruising forager. Other field studies have shown that this spectrum of locomotor behaviors is widespread in lizard faunas. In North America, for example, tree lizards (*Urosaurus*) are sit-and-wait predators and whiptail lizards (*Aspidoscelis*) are actively foraging predators (Figure 17.7). Snakes display a range of foraging modes similar to those of lizards, from ambush predators to active foragers. A spectrum of foraging modes is not unique to squamates; it probably applies to nearly all kinds of mobile animals, including fishes, mammals, birds, frogs, insects, and zooplankton.

Correlates of foraging mode

The ecological, morphological, and behavioral characters that are correlated with the foraging modes of different species of squamates appear to define many aspects of the biology of these animals. For example, sit-and-wait predators and actively foraging predators consume different kinds of prey and fall victim to different kinds of predators. They have different social systems, probably emphasize different sensory modes, and differ in some aspects of their reproduction and life history.

A weakness of the generalizations about lizard foraging behavior summarized in Table 17.2 should be emphasized: sit-and-wait species (at least the ones that have been studied most) are primarily phrynosomatids and agamids, whereas actively foraging species are mostly skinks and teiids. That phylogenetic split raises the question of whether the differences observed between sit-and-wait and actively foraging lizards are really the consequences of differences in foraging behavior or if they are ancestral characters of the lineages of lizards. If the latter is true, the association with different foraging modes may be misleading. In either case, Table 17.2 integrates a large quantity of information about the biology of lizards.

Prey detection Squamates with different foraging modes use different methods to detect prey. Sit-and-wait lizards normally remain in one spot from which they can survey a broad area. These motionless lizards detect the movement of prey visually and capture it with a quick dash from their

Table 17.2	Ecological and Behavioral Characters Associated with Foraging Modes of Lizards [a]	
	Foraging Mode	
Character	**Sit-and-wait**	**Actively foraging**
Foraging behavior		
No. movements/hour	Few	Many
Distance moved/hour	Low	High
Sensory modes	Vision	Vision and olfaction
Exploratory behavior	Low	High
Types of prey	Mobile, large	Sedentary, often small
Predation		
Risk of predation	Low	Higher
Types of predators	Actively foraging	Sit-and-wait and actively foraging
Body form		
Trunk	Stocky	Elongated
Tail	Often short	Often long
Physiological characters		
Endurance	Limited	High
Sprint speed	High	Intermediate to low
Aerobic metabolic capacity	Low	High
Anaerobic metabolic capacity	High	Low
Heart size	Small	Large
Hematocrit	Low	High
Energetics		
Daily energy expenditure	Low	Higher
Daily energy intake	Low	Higher
Social behavior		
Size of home range	Small	Large
Social system	Territorial	Not territorial
Reproduction		
Volume of eggs or embryos relative to the volume of the mother	High	Low

[a] Foraging modes of lizards form a continuum from sit-and-wait predators to actively foraging predators. In most cases, however, data are available only for species at the extremes of the continuum.

observation site. Sit-and-wait lizards may be most successful in detecting and capturing relatively large insects such as beetles and grasshoppers. Vipers are sit-and-wait predators. Rattlesnakes wait in ambush beside trails that rodents use, and many vipers that prey on birds, lizards, or frogs twitch their tail to lure prey within reach, as do Australian death adders (*Acanthophis*), elapids that have converged on the stout body form of vipers. The African puff adder (*Bitis arietans*) uses its tongue as well as its tail to lure prey, and the tail of the aptly named spider-tailed viper (*Pseudocerastes urarachnoides*) terminates in bristles that have a striking resemblance to the legs of a camel spider.

(A)

Figure 17.7 **Sit-and-wait versus actively foraging squamates.**   Both extremes of the spectrum of foraging modes can be found among lepidosaurs. (A) This ornate tree lizard (*Urosaurus ornatus*) is a sit-and-wait predator that perches in an elevated site and uses vision to detect moving prey. Its blotched pattern makes it inconspicuous when it is motionless and its stout body accommodates large prey in relation to its own size. (B) The giant spotted whiptail lizard (*Aspidoscelis stictogramma*) forages actively, moving continuously and searching for hidden prey in crevices and under leaves. It has the slim, elongated body and long tail that are characteristic of actively foraging species of lizards. Longitudinal stripes are characteristic of actively foraging lizards, probably because they make movement hard to detect. Similar combinations of colors, patterns, body forms, and behaviors are found in snakes. (C) Rattlesnakes, such as this western diamondback (*Crotalus atrox*), wait in ambush beside trails that rodents use regularly, sometimes remaining in the same ambush position for days at a time. Blotched patterns make them difficult to see in daylight, and scent (transferred via the tongue) and heat-sensitive pit organs allow rattlesnakes to detect and strike prey in the dark. (D) The Sonoran whipsnake (*Coluber bilineatus*) is an actively foraging predator that moves continuously, exploring places where prey might be hiding. Its slim, elongated body moves rapidly and its striped pattern confuses predators. (A–C, René Clark, ©Dancing Snake Nature Photography; D, Harvey Pough.)

(B)

(C)

Tongue

Nostrils

Pit organs

(D)

that are sit-and-wait predators. Thus, the different foraging behaviors of lizards lead to differences in their diets, even when the two kinds of lizards occur in the same habitat. Actively foraging snakes patrol a home range, examining crevices where prey might be hiding.

Different foraging modes result in differences in the exposure of squamates to their own predators. A lizard that spends the bulk of its time resting motionless is relatively inconspicuous, whereas a lizard that spends most of its time moving is more easily seen. Sit-and-wait lizards are probably most likely to be discovered and captured by predators that are active searchers, whereas actively foraging lizards are likely to be caught by sit-and-wait predators. Because of this difference, foraging modes may alternate at successive levels in the food chain (**Figure 17.8**). Insects that move about may be captured by lizards that are sit-and-wait predators, and those lizards may be eaten by actively foraging predators. Insects that are sedentary are more likely to be discovered by an actively foraging lizard, and that lizard may be picked off by a sit-and-wait predator.

Color and pattern   The body forms and color patterns of sit-and-wait predators may reflect selective pressures different from those that act on actively foraging species. Sit-and-wait lizards are often stout-bodied, short-tailed, and cryptically colored. Many of these species have dorsal patterns formed by blotches of different colors that probably obscure the outlines of the body as the lizard rests motionless on a rock or tree trunk. Actively foraging species of lizards are usually slim and elongated with long tails, and they often have patterns of stripes that may produce optical illusions as they move. However, one predator-avoidance mechanism, the ability to break off the tail when it is seized by a predator—caudal **autotomy** (see Figure 17.15A)—does not differ among lizards with different foraging modes.

Metabolism and foraging mode   What physiological characters are necessary to support different foraging modes? The energy requirements of a dash that lasts for only a second or two are quite different from those of

Actively foraging lizards spend most of their time on the ground, moving steadily and poking their snouts under fallen leaves and into crevices in the ground. These lizards apparently rely largely on chemical cues to detect insects, and they probably seek out local concentrations of patchily distributed prey such as termites. Actively foraging species of lizards appear to eat more small insects than do lizards

(A)

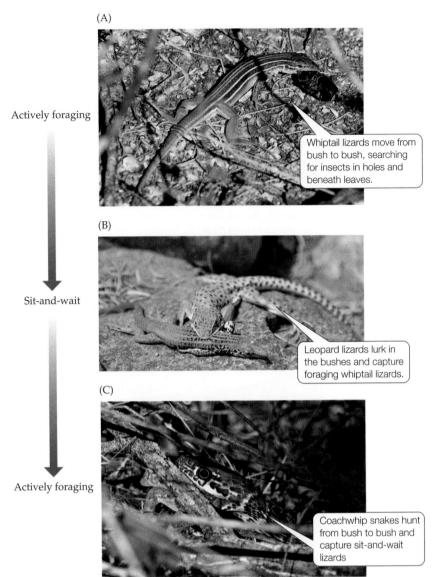

Actively foraging

> Whiptail lizards move from bush to bush, searching for insects in holes and beneath leaves.

(B)

Sit-and-wait

> Leopard lizards lurk in the bushes and capture foraging whiptail lizards.

(C)

Actively foraging

> Coachwhip snakes hunt from bush to bush and capture sit-and-wait lizards

Figure 17.8 Foraging modes alternate at successive trophic levels. In the desert Southwest, most insects live in the leaf litter in the shade of bushes, where they find food and hiding places. (A) Whiptail lizards (*Aspidocelis*) actively forage the insects' niche, investigating crevices and searching beneath leaves. (B) Longnose leopard lizards (*Gambelia wislenzenii*) hide in bushes and ambush actively foraging lizards. (C) Sonoran coachwhips (*Coluber flagellum*) actively forage for sit-and-wait lizards. (A, C, René Clark, © Dancing Snake Nature Photography; B, Leon Werdinger/ Alamy Stock Photo.)

locomotion that is sustained nearly continuously for several hours. Sit-and-wait and actively foraging species of squamates differ in their relative emphasis on the two metabolic pathways that provide adenosine triphosphate (ATP) for activity and in how long that activity can be sustained:

- *Sit-and-wait* lizards move in brief spurts, and they rely largely on anaerobic metabolism to sustain their movements. Anaerobic metabolism uses glycogen stored in the muscles as a metabolic substrate and produces lactic acid as its end product. It is a way to synthesize ATP quickly (because the glycogen is already in the muscles and no oxygen is required), but it is not good for sustained activity because the glycogen is soon exhausted and lactic acid inhibits cellular metabolism. Lizards that rely on anaerobic metabolism can make brief sprints but become exhausted when they are forced to run continuously.

- *Actively foraging* lizards rely on aerobic metabolism, using glucose that is carried to the muscles by the

circulatory system as a metabolic substrate. Aerobic metabolism produces carbon dioxide and water as end products, and can continue for long periods because the circulatory system brings more oxygen and glucose and carries carbon dioxide away. As a result, actively foraging species can sustain activity for long periods without exhaustion.

The differences in exercise physiology are associated with differences in the oxygen transport systems of the lizards: Actively foraging species have larger hearts and more red blood cells in their blood than do sit-and-wait species. As a result, each beat of the heart pumps more blood, and that blood carries more oxygen to the tissues in an actively foraging species of lizard than it does in a sit-and-wait species.

Sustained locomotion is probably not important to a sit-and-wait lizard that makes short dashes to capture prey or to escape from predators, but sprint speed might be vitally important in both these activities. Speed may be relatively unimportant to an actively foraging lizard that moves slowly, methodically looks for prey under leaves and in cavities, and can hide under a bush to confuse a predator. As you might predict from these considerations, sit-and-wait lizards generally have high sprint speeds and low endurance, whereas actively foraging species usually have lower sprint speeds and greater endurance.

The continuous locomotion of actively foraging species of lizards is energetically expensive. Measurements of energy expenditure of lizards in the Kalahari showed that the daily energy expenditure of an actively foraging species averaged 1.5 times that of a sit-and-wait species. However, the energy that the actively foraging species invested in foraging was more than repaid by its greater success in finding prey. The daily food intake of the actively foraging species was twice that of the sit-and-wait predator. As a result, the actively foraging species had more energy available to use for growth and reproduction than did the sit-and-wait species, despite the additional cost of its mode of foraging.

17.4 Skull Kinesis

Many of the feeding specializations of squamates are related to changes in the structure of the skull and jaws (**Figure 17.9**). The ancestral lepidosaur skull was diapsid, with a lower temporal arch formed by a connection between the jugal and quadrate and an upper arch formed by the postorbital and squamosal. The tuatara is fully diapsid, albeit secondarily so, having re-evolved the connection forming the lower arch. Lizards have lost the lower arch, and snakes have lost the upper arch as well, increasing the gape and lateral spread of the jaw articulation during feeding. In addition, lizards and snakes have developed flexible joints between some of the bones of the skull (skull kinesis). Highly kinetic skulls with wide gapes accommodate large prey items, but the loss of rigidity at the joints diminishes the crushing force of the jaws.

The skulls of amphisbaenians have a completely different sort of specialization. These burrowing lizards use their heads to construct tunnels in the soil (see Figure 17.4). Their akinetic skulls are heavy and many of the bones are fused. The jaws of amphisbaenians have small gapes, but their unusual dentition allows them to bite pieces from large prey.

17.5 Feeding Specializations of Snakes

In popular literature, snakes are sometimes described as "unhinging" their jaws during feeding. That's careless writing and rather silly—unhinged jaws would merely flap back and forth. What those authors are trying to say is that snakes have extremely kinetic skulls that allow extensive movement of the jaws.

The skulls of derived snakes are much more flexible than the skulls of lizards. A snake skull contains multiple elements that have flexible joints between them, allowing a staggering degree of complexity in the movements of the skull (**Figure 17.10**). To complicate things, the elements are paired, and each side of the head acts independently. In addition, the tips of the mandibles are connected only by elastic skin and muscle, so the sides of the lower jaw are able to separate widely—to as much as 20 times the resting distance in some species.

Whereas the mandibles of lizards are joined at the front of the mouth by a rigid bony connection, the mandibles

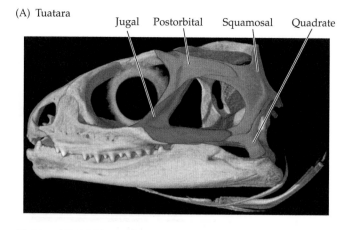

(A) Tuatara

Jugal   Postorbital   Squamosal   Quadrate

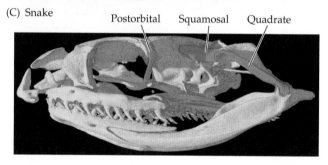

(B) Lizard (iguana)

Jugal   Postorbital   Squamosal   Quadrate

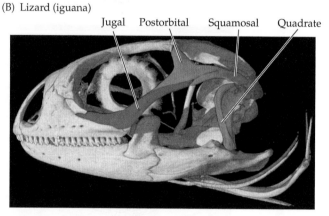

(C) Snake

Postorbital   Squamosal   Quadrate

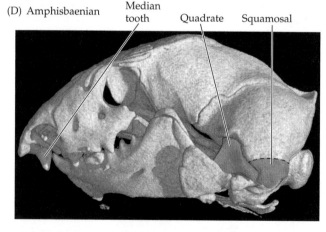

(D) Amphisbaenian   Median tooth   Quadrate   Squamosal

Figure 17.9 Modifications of the diapsid skull among lepidosaurs. (A) The tuatara (*Sphenodon punctatus*) has reverted to the ancestral condition for lepidosaurs, with two temporal arches. The upper arch is formed by the postorbital and squamosal, and the lower arch by the jugal and quadrate. This fully diapsid condition is not characteristic of all sphenodontids; some of the Mesozoic forms did not have a complete lower temporal arch. (B) Crown group lizards have developed some kinesis by eliminating the lower arch; there is a gap between the jugal and quadrate, as shown in the desert iguana (*Dipsosaurus dorsalis*). (C) Skull kinesis is further developed in snakes by loss of the upper temporal arch, as in the black racer (*Coluber constrictor*) shown here. (D) Amphisbaenians such as Zarudnyi's worm lizard (*Diplometopon zarudnyi*) use their heads for burrowing through soil. They have derived akinetic skulls in which many of the bones have fused. (3D HRXCT reconstructions courtesy of DigiMorph.org.)

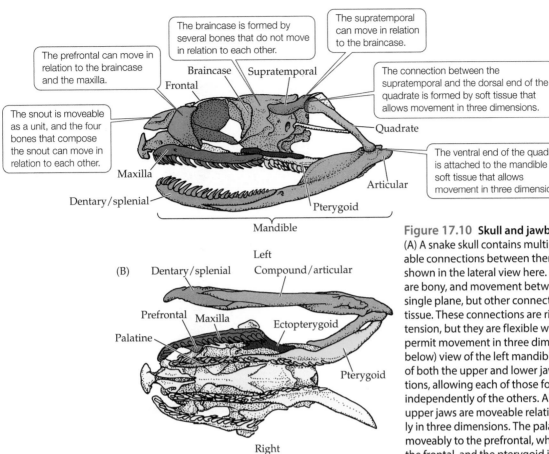

The braincase is formed by several bones that do not move in relation to each other.

The supratemporal can move in relation to the braincase.

The prefrontal can move in relation to the braincase and the maxilla.

The connection between the supratemporal and the dorsal end of the quadrate is formed by soft tissue that allows movement in three dimensions.

The snout is moveable as a unit, and the four bones that compose the snout can move in relation to each other.

The ventral end of the quadrate is attached to the mandible by soft tissue that allows movement in three dimensions.

Braincase Supratemporal
Frontal
Quadrate
Maxilla
Articular
Dentary/splenial
Pterygoid
Mandible

(B) Left
Dentary/splenial Compound/articular
Prefrontal Maxilla
Palatine Ectopterygoid
Pterygoid
Right

Figure 17.10 Skull and jawbones of a snake.
(A) A snake skull contains multiple elements with moveable connections between them, some of which are shown in the lateral view here. Most of the elements are bony, and movement between them occurs in a single plane, but other connections are formed by soft tissue. These connections are rigid when they are under tension, but they are flexible when they are relaxed and permit movement in three dimensions. (B) Ventral (from below) view of the left mandible The left and right sides of both the upper and lower jaws have flexible connections, allowing each of those four elements to move independently of the others. All of the bones of the upper jaws are moveable relative to each other, typically in three dimensions. The palatine and maxilla attach moveably to the prefrontal, which is itself moveable on the frontal, and the pterygoid is movably attached to the quadrate and the compound/articular. In the lower jaws, the left and right mandibles can be protracted and retracted independently of each other and independently of the upper jaws. (From Gans 1961.)

of snakes are attached by only muscles and skin, so they can spread laterally and move forward or backward independently. Loosely connected mandibles and flexible skin in the chin and throat allow a snake's jaw tips to spread, so that the widest part of the prey passes ventral to the articulation of the jaw with the skull.

Snakes usually swallow prey headfirst, perhaps because that approach presses the prey's limbs against its body, out of the snake's way (**Figure 17.11**). Swallowing movements take place slowly enough to be easily observed. The mandibular and pterygoid teeth of one side of the head are

anchored in the prey, and the head is rotated to advance the opposite jaw as the mandible is protracted and grips the prey ventrally. As this process is repeated, the snake draws the prey item into its mouth. Once the prey has reached the esophagus, it is forced toward the stomach by contraction of the snake's neck muscles. Usually the neck is bent sharply to push the prey along.

(A)
(B)

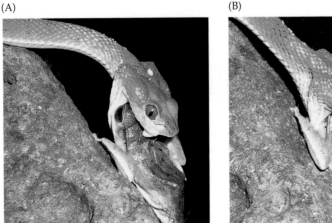

Figure 17.11 Movements of a snake during swallowing. A ruby-eyed green tree viper (*Trimeresurus rubeus*) from Southeast Asia illustrates the combination of head movements and protraction and retraction of the jaws that allow snakes to swallow prey. (A) The snake holds the prey with its left maxilla and mandible and spreads the right maxilla and mandible. (B) Bending its neck to the left, the snake advances its jaws over the prey. The backward-curving teeth slide across the surface of the prey like the runners of a sled; closing the jaws embeds the teeth in the prey. (The fangs, visible in A, are not used during swallowing.) The swallowing process continues with alternating left and right movements until the entire prey has passed through the snake's jaws. (Photos by Wayne Van Devender.)

Many snakes seize prey and swallow it as it struggles. The risk of damage to the snake during this process is a real one, and various features of snake anatomy seem to give some protection from struggling prey. The frontal and parietal bones of a snake's skull extend downward, entirely enclosing the brain and shielding it from the protesting kicks of prey being swallowed. Possibly the kinds of prey that can be attacked by snakes without a specialized feeding mechanism are limited by the snake's ability to swallow the prey without being injured in the process.

Constriction and venom are predatory specializations that permit a snake to tackle large prey with little risk of injury to itself. Constriction is characteristic of boas and pythons as well as some colubrid snakes. Despite travelers' tales of animals crushed to jelly by a python's coils, the process of constriction involves very little pressure. A constrictor seizes prey with its jaws and throws one or more coils of its body around the prey. The loops of the snake's body press against adjacent loops, and friction prevents the prey from forcing the loops open. Each time the prey exhales, the snake takes up the slack by tightening the loops slightly until the increased internal pressure eventually stops the heart.

Venom and fangs

Early colubrids may have used venom to immobilize prey. Duvernoy's gland, found in the upper jaw of many extant colubrid snakes, is homologous with the venom glands of viperids and elapids and produces a toxic secretion that immobilizes prey. The presence of Duvernoy's gland appears to be an ancestral character for colubrid snakes, and some extant colubrids have venom that is dangerously toxic; humans have died as a result of being bitten by the African boomslang (*Dispholidus typus*).

In this context, then, the front-fanged venomous snakes (Elapidae and Viperidae) are not a new development, but represent specializations of an ancestral venom-delivery system. A variety of snakes have enlarged teeth (fangs) on the maxillae. Three categories of venomous snakes are recognized (**Figure 17.12**):

1. **Opisthoglyphous** (Greek *opisthen*, "behind"; *glyphe*, "a carving") snakes have one or more enlarged teeth near the rear of the maxillae, with smaller teeth in front. In some forms the fangs are solid; in others, there is a groove on the surface of the fang that may help conduct venom into the wound. The fangs do not engage the prey until it is well into the mouth. Birds and lizards are the primary prey of opisthoglyphs, and the snakes often hold the prey in their mouth until the venom takes effect and the prey has stopped struggling.

2. **Proteroglyphous** (Greek *pro*, a prefix meaning "earlier or before") snakes include cobras, mambas, coral snakes, and sea snakes (Elapidae). The hollow fangs of proteroglyphous snakes are located at the front of the maxillae, and there are often several small, solid teeth

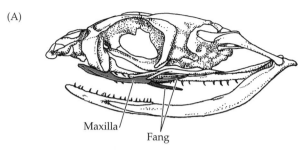

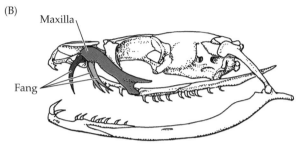

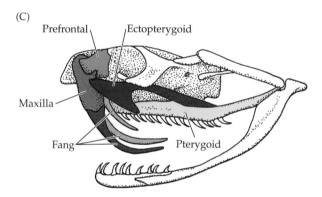

Figure 17.12 Dentition of snakes. (A) Opisthoglyphous snakes (e.g., the African boomslang, *Dispholidus typus*) have grooved fangs near the rear of the maxillae that conduct venom. (B) Proteroglyphous snakes (e.g., the African mamba, *Dendroaspis jamesoni*) have permanently erect hollow fangs at the front of the maxillae. (C) Solenoglyphous snakes (e.g., the African puff adder, *Bitis arietans*) have hollow fangs on rotating maxillae. When the mouth is closed, the fangs of solenoglyphs lie against the roof of the mouth. When the mouth is opened to strike, the fangs are erected by anterior movement of the pterygoid that is transmitted through the ectopterygoid to the maxilla, causing it to rotate about its articulation with the prefrontal.

behind the fangs. The fangs are permanently erect and relatively short.

3. **Solenoglyphous** (Greek *solen*, "pipe") snakes include the pit vipers of the New World and the true vipers of the Old World. The hollow fangs are the only teeth on the maxillae, which rotate so that the fangs are folded against the roof of the mouth when the jaws are closed. This folding mechanism permits solenoglyphous snakes to have long fangs that inject venom deep into the tissues of prey. The venom first kills the prey and then speeds its digestion after it has been swallowed.

Snake venom is a complex mixture of proteins and polypeptides, including a family of small polypeptides called "three-finger toxins" because they consist of three loops extending from a central core. Three-finger toxins interfere with communication between cells, and consequently have a variety of neurotoxic, cardiotoxic, and anticoagulant effects. The actions of three-finger toxins from venom are more specific than those of conventional pharmacological agents. In biomedical applications, three-finger toxins from venom are more effective and have fewer undesirable side-effects than conventional drugs, such as opioid painkillers. These properties are being exploited to develop new adrenergic agents based on snake venoms.

The toxicities of some venoms are matched to the primary prey of a species of snake. For example, juvenile Australian brown snakes (several species in the genus *Pseudonaja*) feed on lizards and their venom is primarily neurotoxic, whereas the venom of adults, which feed on mammals, is primarily cytotoxic. Each property is maximally effective against the appropriate prey. Northern Pacific rattlesnakes (*Crotalus o. oreganus*) are engaged in an "arms race" with the California ground squirrels that are their principal prey; as the squirrels evolve resistance, the snakes evolve increasingly lethal venom.

Venomous snakes synthesize and store multiple lethal doses of venom in a form that is ready for immediate use but is not toxic to the snake. Acidity in the main venom gland of rattlesnakes inactivates the venom during storage, and secretions from the accessory gland activate the venom when a snake strikes.

Snakes that can inject a disabling dose of venom into their prey have evolved a very safe prey-catching method. A constrictor is in contact with its prey while it is dying and runs some risk of injury from the prey's struggles, whereas a solenoglyphous snake needs only to inject venom and allow the prey to run off to die; the snake can then follow the scent trail of the prey to find its corpse. This is the prey-capture pattern of most vipers, and experiments have shown that a viper can distinguish the scent trail of a mouse it has bitten from the scent trails left by uninjured mice.

Several features of the body form of vipers allow them to eat larger prey in relation to their own body size than can most nonvenomous snakes or elapids. Many vipers, including rattlesnakes (*Crotalus*), the jumping pit viper (*Atropoides nummifer*), the African puff adder, and the Gaboon viper (*Bitis gabonica*), are very stout snakes. The triangular head shape usually associated with vipers is a result of the outward extension of the rear of the skull, especially the quadrate bones. The wide-spreading quadrates allow bulky objects to pass through the mouth, and even a large meal makes little bulge in the stout body and thus does not interfere with locomotion. Vipers have specialized as sedentary predators that wait in ambush and can prey even on quite large animals. Elapids are primarily slim-bodied snakes that actively forage for prey, but one genus of elapids, the Australian death adders (*Acanthophis*), has converged on the stout body form and ambush mode of foraging that is typical of vipers.

Hearts and stomachs

Kinetic skulls and safe ways to kill prey enable snakes to consume meals that are large in relation to their own body size, but having swallowed the prey they must digest it. Binge-feeding snakes and lizards—species that eat large meals at infrequent intervals—conserve energy by allowing the intestine to shrink between meals and to hypertrophy rapidly following a meal. The small intestine of a Burmese python (*Python bivittatus*) doubles its size while a meal is being digested and shrinks again when digestion is completed. Digestion requires a large increase in blood flow to the intestine, and the size of the ventricle of the heart increases by 40% within 48 hours after a python swallows a meal. The weights of the liver, pancreas, and kidney also increase, as do the activities of several digestive enzymes. The bacteria in the gut also show a dramatic post-feeding increase in overall species diversity and a shift in the dominant species of bacteria.

17.6 Predator Avoidance and Defense

Squamates are preyed on by birds, mammals, and larger species of squamates, and they have evolved a variety of ways to avoid being attacked or to evade a predator after they are attacked. Many of these mechanisms are size-specific; small animals can hide more easily than large ones, and large individuals can fight back more effectively than small ones. Temperature also plays a role; many species use one type of defense when they are cold and sluggish and a different one when they are warm and agile. The defensive behavior of the tegu lizard *Salvator merianae* combines the effects of both body temperature and ontogenetic increases in size; juvenile *S. merianae* run from predators at all body temperatures, whereas adults (which are large enough to defend themselves) attempt to bite a predator at low body temperatures, but flee when they are warm and can run rapidly.

Crypsis

Avoiding detection is a sure way to avoid predation, and crypsis is the first line of defense for many squamates. Many arboreal lizards and snakes are green or brown, colors that blend with leaves and twigs, and chameleons can change their patterns to increase crypsis on some backgrounds. Most sit-and-wait species of lizards have blotched patterns that obscure their body outlines when they are motionless. The striped patterns of most actively foraging lizards and snakes conceal their movements. Especially when only a part of the body can be seen, as is the case when an animal is moving through vegetation, stripes give the impression that the animal is motionless. Body form can enhance crypsis; some arboreal lizards have crests on their heads or other ornaments that render them hard to identify as lizards when seen in profile, and

Figure 17.13 Antipredator mechanisms: Crypsis. The North American short-tailed horned lizard (*Phrynosoma modestum*) remains motionless when it detects a predator and escapes detection because it looks like a small stone. (Photo by Wade C. Sherbrooke.)

some small species of horned lizards resemble pebbles (**Figure 17.13**).

"Eavesdropping"

Although most squamates cannot vocalize, many have good hearing, and some lizards "eavesdrop" on birdcalls. Mexican spiny-tailed iguanas (*Ctenosaura similis*) pause in

their foraging or flee from calls of Harris's hawks, with female iguanas responding more strongly than males (perhaps because they are smaller and more vulnerable to attack). Western girdled lizards (*Zonosaurus laticaudatus*) of Madagascar eavesdrop on the alarm calls of paradise flycatchers, pausing for five minutes or more after an alarm call before resuming activity.

Deterrence

Once a snake or lizard has been detected by a predator, deterring attack is a secondary line of defense. Some lizards have alternating black and white markings on the underside of the tail that are not visible when the tail is in its normal position, but become conspicuous when the tail is curled over the lizard's back and waved from side to side (**Figure 17.14A**). This behavior, which appears to signal that the lizard is aware of the predator and ready to flee, may cause a predator to move on in search of a less alert meal.

Spines (which are enlarged, sharp-pointed scales) can be effective deterrents if a predator attacks. Many lizards, such as North American horned lizards and the Australian thorny devil (see Figure 17.3B,C), have spines on the head. Other lizards, including the Neotropical spiny-tailed iguanas (*Ctenosaura*, **Figure 17.14B**) and the African girdled lizards (*Cordylus*) and mastigures (*Uromastyx*) have spiny tails. Small species use their tails to prevent predators from following them into retreat sites, and a slap from the tail of one of the larger species can cause a predator to back off.

Figure 17.14 Antipredator mechanisms: Deterrence. (A) The dorsal pattern of the North American gridiron-tailed lizard (*Callisaurus draconoides*) makes it hard to detect against a background of desert soil. When a lizard knows that it has been seen by a predator, it curls its tail over its back and waves it from side to side, displaying a conspicuous pattern of black and white bars. This behavior apparently tells the predator that it has been detected, because predators rarely pursue lizards that are waving their tails. (B) Some small lizards, such as the Yucatán spiny-tailed iguana (*Ctenosaura defensor*), use their tails to block access to their retreat sites. Some larger species use the tail to slap predators. (C) The spines on the tail of the southwestern spiny-tailed gecko from Australia (*Strophurus krisalys*) are hollow. Glands beneath the spines can spray a bad-smelling adhesive substance as far as 50 cm to entangle a predator. (A, René Clark © Dancing Snake Nature Photography; B, courtesy of Catherine Malone and Joseph Burgess; C, courtesy of Stephen Zozaya.)

Two lineages of geckos deter predators by squirting a liquid from hollow spines on the tail (e.g., *Strophurus krisalys*; **Figure 17.14C**). The ejected fluid can travel as far as 50 cm, and forms long, sticky filaments that are very difficult to remove. Soil particles and bits of plant debris adhere to the filaments, distracting a predator while the gecko makes its escape.

Autotomy

Many lizards and a few snakes can autotomize the tail—that is, voluntarily break the tail (caudal autotomy) and run off, leaving part of the tail in a predator's mouth. Tail muscle has a high anaerobic metabolic capacity, and the broken tail writhes and twists for many minutes, distracting the predator and allowing the lizard to escape. The break occurs in the middle of a vertebra, not between vertebrae, and autotomized tails are regenerated with a rod of cartilage replacing the lost vertebrae (**Figure 17.15A**). The cartilage does not break, so subsequent autotomies must be closer

(A)

(B)

Figure 17.15 Antipredator mechanisms: Autotomy. (A) Many lizards can autotomize their tail and subsequently regenerate it. The regenerated portion of the tail of this Madrean alligator lizard (*Elgaria kingii*) has a distinctly different pattern from the original tail, the stump of which can still be seen. (B) Some geckos autotomize portions of their skin. This fish-scaled gecko (*Geckolepis megalepis*) has autotomized nearly all of the skin from its back. Although the injury looks horrifying, the skin and scales will be regenerated within weeks (inset). (A, Harvey Pough; B, courtesy of Frank Glaw.)

to the body. At least some lizards are conservative in the portion of the tail they autotomize; when they are cold and unable to run fast, lizards leave a larger part of the tail for the predator than they do when they are warm.

In addition to caudal autotomy, some geckos use dermal autotomy as a mode of escape, sacrificing large areas of skin to a predator (**Figure 17.15B**). The skin splits between the integument (epidermis, connective tissue, and subcutaneous fat) and the underlying dermis. Capillaries in the dermis constrict to prevent blood loss, and the skin regenerates in a few weeks without a scar.

Venomous and poisonous snakes

Although venom evolved as a predatory mechanism, some specializations enhance its value in defense. Many venomous snakes have distinctive markings or displays, such as the rattles of rattlesnakes (**Figure 17.16A**) and the hoods of cobras. These deterrents may prevent a predator from attacking a snake. Venom is a second line of defense, and many venoms contain substances that produce intense pain without causing tissue damage. Probably these substances make the predator flee before it has injured the snake.

About one-third of the 31 species of cobras have a modification of the fang that makes venom a defensive weapon. The venom channel in the fangs of these "spitting cobras" makes a right-angle bend near the tip, so that venom is directed forward and released as a spray from the front surface of the fang (**Figure 17.16B**). Spraying venom is a defensive behavior; a spitting cobra bites like any other cobra when it strikes prey, and releases more venom when it bites than when it spits. Experimental manipulations using photographs showed that a face with eyes elicited spitting most reliably; a face without eyes was not as effective. The snake aims at the eyes, and venom causes immediate pain and at least temporary blindness. Spitting has evolved three times independently in cobras, once in Asia and twice in Africa.

Mimicry of a dangerous species, such as a venomous snake, can protect a nonvenomous snake from attack, and mimicry of coral snakes by rear-fanged and nonvenomous snakes is a widespread phenomenon in Central and South America (**Figure 17.16C, D**). Field studies have shown that predatory birds avoid artificial snakes with the patterns of coral snakes, even when the mimetic resemblance is imprecise.

Some snakes are poisonous rather than venomous—that is, they contain toxic substances derived from the snakes' prey that sicken a predator. Some Asian keelback snakes (*Rhabdophis*) store toxins from toads in a series of paired glands on the neck. When a *Rhabdophis* is attacked by a predator, it arches its neck and butts the portion containing the glands against the predator's face. The skin over the glands ruptures, spraying the contents of the glands into the predator's mouth and eyes. Red-sided garter snakes (*Thamnophis sirtalis parietalis*) from populations that prey on rough-skinned newts retain enough of the newts' protective tetrodotoxin to incapacitate birds and mammals that prey on the snakes.

(A)

(B)

(C)

(D)

Figure 17.16 Antipredator mechanisms: Venoms, toxins, and mimicry. Although venom's first evolutionary functions involved predation, it can also function as a defense. (A) The buzz produced by the rattle of the venomous prairie rattlesnake (*Crotalus viridus*) warns away potential predators. (B) Many cobras, such as this Mozambique cobra (*Naja mossambica*), can spray venom as far as a meter, and aim for a predator's face. (C) Many elapids, especially the New World coral snakes such as this western coral snake (*Micruroides euryxanthus*) are aposematically colored, warning predators away. (D) Some non-venomous snakes, such as this Arizona mountain kingsnake (*Lampropeltis pyromelana*), are mimics of coral snakes, thus taking advantage of the aposematic defense. (A, courtesy of Todd Pierson; B, courtesy of Wolfgang Wüster; C, René Clark © Dancing Snake Nature Photography; D, courtesy of Andreas Kettenburg.)

17.7 Social Behavior

Squamates employ a variety of visual, auditory, chemical, and tactile signals in the behaviors they use to maintain territories and to choose mates. The various sensory modalities employed by animals have biased the amount of information we have about the behaviors of different species. Because humans are primarily visually oriented, we perceive the visual displays of other animals quite readily. Because of our own sensory bias, the extensive repertoires of visual displays of *Anolis* lizards figure heavily in the literature of behavioral ecology, but much less is known about the chemical and tactile signals that are probably important for other lizards and for snakes. Many geckos, for example, are nocturnal and use vocalizations during territorial defense and courtship.

Territoriality

Diurnal lizards employ visual assertion displays during social interactions, whereas nocturnal geckos use vocalizations. The genus *Anolis* includes more than 400 species of small to medium-size diurnal lizards that occur primarily in tropical America. Male *Anolis* have dewlaps, areas of skin beneath the chin that can be distended by the hyoid apparatus during visual displays. The brightly colored scales and skin of the gular fans of many *Anolis* species are conspicuous signaling devices, and they are used in conjunction with movements of the head and body.

Figure 17.17 shows the color and size of the gular fans of three *Anolis* species that occur together in Costa Rica. Since no two species of *Anolis* in a given habitat have the

same combination of colors on their gular fans, it is possible to identify a species solely by seeing the colors it displays.

In addition to its unique gular fan, each species also has a unique behavioral display that consists of raising the body by straightening the forelegs (called a push-up), bobbing the head, and extending and contracting the gular fan. The combination of these three sorts of movements creates a display action pattern. No two species have the same display action pattern, so it is possible to identify any of the co-occurring *Anolis* species by seeing its display action pattern. This redundancy ensures that a lizard species can be identified in an environment in which vegetation moving in the wind and shifting patterns of light and dark penetrating the leaf canopy may obscure parts of the signal.

The value of the visual signal transmitted by a dewlap would be low in the dim interior of the forests where many species of *Anolis* live, were it not for an unusual property of these dewlaps: when they are illuminated from behind, the dewlaps of forest-dwelling *Anolis* transmit diffused light and appear as bright, glowing spots in the forest (**Figure 17.18**).

Some populations of the side-blotched lizard (*Uta stansburiana*, a common species of western North America) have a complex association between gular color and territorial and reproductive behavior. Male lizards in these populations have one of three colors in the gular region—blue, orange, or yellow, corresponding to different levels of testosterone (**Figure 17.19**). Males with blue throats are territorial, maintaining small territories that overlap with the home ranges of one or two females and mating with those females. Males with orange throats are more aggressive

(A) Simple (*A. carpenteri*)

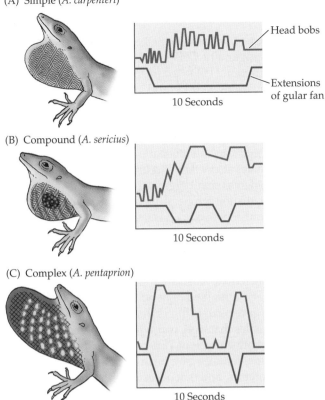

Head bobs

Extensions of gular fan

10 Seconds

(B) Compound (*A. sericius*)

10 Seconds

(C) Complex (*A. pentaprion*)

10 Seconds

Figure 17.17 Species-specific displays of *Anolis* lizards. Nine species of Costa Rican *Anolis* lizards occupying the same habitat signal their species identity redundantly. The species can be separated into three groups based on the size and color of their gular fans. Simple fans are unicolored (A), compound fans are bicolored (B), and complex fans are bicolored and very large (C). In addition, the aggression display of each species has a unique action pattern of push-ups, head bobs, and extensions of the gular fan. The horizontal axis is time (the duration of these displays is about 10 seconds), and the vertical axis is vertical height. Blue lines graph the number and height of head bobs; the red lines graph extensions of the gular fan. (Adapted from Echelle et al. 1971.)

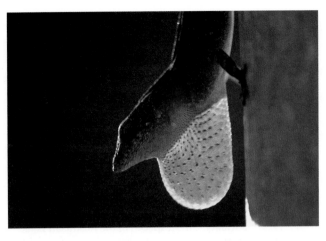

Figure 17.18 The dewlaps of some *Anolis* glow. When they are illuminated from behind, the dewlaps of some species of *Anolis*, such as *A. sagrei* seen here, transmit diffuse light. The glowing dewlaps are easy to distinguish from the background and increase the ability of a lizard to signal its species to other lizards, even from a distance and at low light levels. (Photograph by Manuel Leal.)

than blue-throated males; they displace blue-throated males from their territories and mate with the females. Yellow-throated males mimic the colors and behaviors of females and try to sneak into the territories of blue-throated and orange-throated males to steal a mating before they are chased away by the territorial male.

The most effective mating tactic depends on the number of lizards using each tactic. Yellow sneakers beat orange males, which beat blue males, which beat yellow sneakers—like a game of rock-paper-scissors. The fitness of any one of the color forms relative to the others depends on the proportions of the three forms in the population; the rarest form always has an advantage, and the proportion of the three colors in a population shifts from generation to generation.

Males with orange throats displace blue-throated males and establish large territories containing several females.

Males with blue throats guard small territories with one or two females and chase off yellow-throated males.

Orange

Blue

Yellow

Males with yellow throats are non-territorial, and try to sneak into territories of blue- and orange-throated males.

Figure 17.19 The mating system of male side-blotched lizards (*Uta stansburiana*) resembles a game of rock-paper-scissors. Yellow-throated males have the lowest testosterone levels and mimic the behavior of non-receptive females to sneak into the territories of blue- and orange-throated males. Blue-throated males, with intermediate testosterone levels, have small territories that contain only one or two females. They are usually successful at chasing yellow-throated sneaker males away. Orange-throated males have the highest testosterone levels and are aggressive; they take over the territories of several blue-throated males to create large territories with multiple females, but because their territories are large, they are vulnerable to yellow-throated sneaker males. (Background photo by Harvey Pough; photos of *U. stansburiana* by Barry Sinervo.)

Figure 17.20 Sociality. Most species of the Australian skink genus *Egernia*, such as *E. epsisolus*, spend their lives in stable family groups that consist of parents and several generations of offspring. (Courtesy of Stephen Zozaya).

Sociality

In general, lizards and snakes are not considered to be social animals; that is, they generally do not appear to live in groups. The *Egernia* lineage of Australian skinks is a well-studied exception to that generalization, however (Figure 17.20). Social behavior in this group ranges from solitary species with only transient contacts between individuals to species that form stable, long-term family groups consisting of a monogamous pair of adults and their offspring. Group members share a common shelter site, bask close to one another, and defecate on a common scat pile.

Sociality based on long-term associations among parents or siblings may be more common than currently realized. For example, a cross-fostering experiment with desert night lizards (*Xantusia vigilis*) found that juvenile lizards released in the field with their biological siblings were twice as likely to form aggregations as juveniles that were released with unrelated individuals, and the aggregations formed by juveniles released with their siblings were more than three times larger than the aggregations formed by unrelated juveniles.

17.8 Reproductive Modes

Squamates show a range of reproductive modes from **oviparity** (development occurs outside the female's body and is supported entirely by the yolk—i.e., lecithotrophy) to **viviparity** (eggs are retained in the oviducts, and development is supported by transfer of nutrients from the mother to the fetuses—matrotrophy). Intermediate conditions include retention of the eggs for a time after they have been fertilized and the production of **precocial** young that were nourished primarily by material in the yolk.

Sex determination

Squamates exhibit a spectrum of sex-determining mechanisms. Genotypic sex determination (GSD) characterizes many species; some are male heterogametic (XY), others are female heterogametic (ZW), and temperature-dependent sex determination (TSD) occurs in the tuatara and in a few species of agamids, skinks, and geckos.

Characterizing species as having GSD or TSD oversimplifies the situation, however. For example, the sex of veiled chameleons (*Chamaeleo calyptratus*) is determined by the interaction of egg size and incubation temperature: large eggs produce females at low temperatures and males at high temperatures, whereas small eggs have the opposite relationship between incubation temperature and hatchling sex.

TSD among lepidosaurs has a feature that is not present in turtles: some viviparous species of lizards have TSD, and a pregnant female lizard can determine the sex of her embryos by adjusting her thermoregulatory behavior. Free-ranging female Australian snow skinks, (*Niveoscincus ocellatus*) that maintained high body temperatures for many hours a day gave birth to litters that were about 65% female, whereas females that were warm for shorter periods of time had litters that were about 75% male.

The warmer females gave birth earlier in the season than the cooler females, and that observation suggests an advantage for producing females at high incubation temperatures: birth date influences the adult body size of snow skinks—individuals that are born early in the year are larger when they reproduce than individuals that are born later. Large body size is probably more advantageous for female snow skinks than for males because large females produce more young than do small females. A small body size may not be a disadvantage for a male snow skink because males of this species do not compete with other males to hold territories, and even a small male can fertilize the eggs of a large female.

Oviparity and viviparity

Oviparity is assumed to be the ancestral reproductive mode for squamates, and viviparity has evolved more than 100 times; about 20% of squamates are viviparous. Some viviparous skinks have specialized chorioallantoic placentae; more than 99% of the weight of the fetus of a Brazilian skink (*Mabuya heathi*) results from transport of nutrients across the placenta. A few species of lizards are reproductively bimodal—that is, they include both viviparous and oviparous populations. The Arabian sand boa (*Eryx jayakari*) probably represents a reversal from viviparity to oviparity, and several other reversals have been tentatively identified.

Viviparity is a high-investment reproductive strategy. Females of most viviparous squamates produce relatively small numbers of large young, although there are exceptions to that generalization. Viviparity is not evenly distributed among lineages of squamates: nearly half the origins of viviparity in the group have occurred in skinks, whereas it is unknown in teiid lizards and occurs in only two genera of lacertids. Viviparity has advantages and disadvantages as a reproductive mode. The most commonly cited benefit is the opportunity it provides for a female snake or lizard to use her own thermoregulatory behavior to control the temperature of the embryos during development. This hypothesis is appealing in an ecological context because a relatively short period of retention of the eggs by the female might substantially reduce the total amount of time required for development, especially in a cold climate.

Viviparity potentially lowers reproductive output because a female that is retaining one clutch of eggs cannot produce another. Lizards in warm habitats may produce more than one clutch of eggs in a season, but that is not possible for a viviparous species because development takes too long. In a cold climate, lizards are not able to produce more than one clutch of eggs in a breeding season anyway, and viviparity would not reduce the annual reproductive output of a female lizard. Phylogenetic analyses of the origins of viviparity suggest that it has evolved most often in cold climates.

Viviparity has other costs. The agility of a female lizard is substantially reduced when her embryos are large. Experiments have shown that pregnant female lizards cannot run as fast as nonpregnant females and that snakes find it easier to capture pregnant lizards than nonpregnant ones. Females of some lizard species become secretive when they are pregnant, perhaps in response to their vulnerability to predation. They reduce their activity and spend more time in hiding places. This behavioral adjustment may contribute to the reduction in body temperature seen in pregnant females of some lizard species, and it probably reduces their rate of prey capture as well.

Parthenogenesis

Parthenogenesis—reproduction by females without fertilization by males—has been identified among squamates in six families of lizards and one species of snake. The phenomenon of all-female species is particularly widespread in teiids and lacertids and occurs in several species of geckos. However, parthenogenesis is probably more widespread among squamates than this list indicates because parthenogenetic species are not conspicuously different from bisexual species. Parthenogenetic species are usually detected when a study undertaken for an entirely different purpose reveals that a species contains no males.

Many parthenogenetic species appear to have had their origin as interspecific hybrids. These hybrids are diploid (2*n*), with one set of chromosomes from each parental species. Some parthenogenetic species are triploids (3*n*). These forms are usually the result of a backcross of a diploid parthenogenetic individual to a male of one of its bisexual parental species or, less commonly, the result of hybridization of a diploid parthenogenetic species with a male of a bisexual species different from its parental species. The first known tetraploid species (4*n*) of a lizard, *Aspidoscelis neavesi*, resulted from fertilization of a female of the triploid parthenogenetic species *A. exsanguis* (3*n*) by sperm (1*n*) of a male of the diploid species *A. inornata*.

It is common to find the two bisexual parental species and a parthenogenetic species living in overlapping habitats. Parthenogenetic species of *Aspidoscelis* often occur in habitats such as the floodplains of rivers that are subject to frequent disruption. Disturbance of the habitat may bring together closely related bisexual species, fostering the hybridization that is the first step in establishing a parthenogenetic species. Once a parthenogenetic species has become established, its reproductive potential is twice that of a bisexual species because every individual of a parthenogenetic species is capable of producing young. Thus, when a flood or other disaster wipes out most of the lizards in an area, a parthenogenetic species can repopulate a habitat faster than a bisexual species.

Parental care

Parental care has been recorded for more than 100 species of squamates. A few species of snakes and a larger number of lizards remain with the eggs, protecting them from predators. Some female skinks rotate the eggs to prevent the embryos from adhering to the shell, and some remove dead eggs from the clutch. Some species of pythons brood their eggs: the female coils tightly around the eggs, and in some species, muscular contractions of the female's body produce sufficient heat to raise the temperature of the eggs to about 30°C, which is substantially warmer than air temperature (see Figure 12.17B). One unconfirmed report exists of baby pythons returning at night to their empty eggshells, where their mother coiled around them and kept them warm. Little interaction between adult and juvenile lizards has been documented. In captivity, female prehensile-tailed skinks (*Corucia zebrata*) have been reported to nudge their young toward the food dish, as if teaching them to eat. Maternal care of young may be the ancestral condition for pit vipers; newborn rattlesnakes (*Crotalus*) remain with their mother for 2 weeks or more after birth, not dispersing until they have shed their skins and begun to feed (**Figure 17.21**).

Figure 17.21 Parental care by a snake.
A female timber rattlesnake (*Crotalus horridus*) with four young. The young snakes are 3 or 4 days old and have not yet shed their skins. The mother remains at the natal site until the young have shed for the first time. Juveniles continue to associate with their mothers and siblings or with close relatives after they disperse. Juveniles that were separated from their mothers immediately after birth had lower first year survival than those that were allowed to remain with their mothers for a few days. (Photo by Matthew G. Simon, courtesy of William Brown.)

17.9   Thermal Ecology

The behavioral mechanisms involved in ectothermy are quite straightforward and are employed by insects, birds, and mammals (including humans) as well as by ectothermal vertebrates. Lizards, especially desert species, are particularly good at behavioral thermoregulation. Movement back and forth between sunlight and shade is the most obvious thermoregulatory mechanism they use. Early in the morning or on a cool day, lizards bask in the sun, whereas in the middle of a hot day, they retreat to shade and make only brief excursions into the sun. Sheltered or exposed sites may be sought out. In the morning, when a lizard is attempting to raise its body temperature, it is likely to be in a spot protected from the wind. Later in the day, when it is getting too hot, the lizard may climb into a bush or onto a rock outcrop where it is exposed to the breeze and its convective heat loss is increased.

Body size influences the thermal ecology of lizards. Small species (or hatchlings of large species) have a high surface/volume ratio; as a result, small ectotherms warm and cool faster than large ones. In addition, because small ectotherms lose heat rapidly by convection, they reach lower equilibrium temperatures than large individuals. A small lizard can sit in full sun without overheating when a larger lizard would have to retreat to shade.

Color change can further increase a lizard's control of radiative exchange. Objects look dark because they are absorbing energy in the visible part of the solar spectrum, and the radiant energy they absorb warms them. The lightness or darkness of a lizard affects the amount of solar radiation it absorbs, and lizards can darken or lighten by moving dark pigment in their skin. Melanophores are cells that contain the pigment melanin. They are shaped rather like mushrooms, with a broad upper portion connected by a stalk to a lower section. When melanin granules are dispersed into the upper part of the cell, close to the skin surface, the skin appears dark; when the granules are drawn away from the surface into the lower section of the cell, the skin appears light. Lizards heat 10–75% faster when they are dark than they do when they are light.

A new dimension was added to studies of ectothermy in the 1960s with the discovery that ectotherms can use physiological mechanisms to adjust their rate of temperature change. The original observations showed that several different species of large lizards were able to heat faster than they cooled when exposed to a 20°C difference between body and ambient temperatures. The basis of this control of heating and cooling rates lies in changes in peripheral circulation. Heating the skin of a lizard causes a localized expansion of dermal blood vessels (vasodilation) in the warm area. Dilation of the blood vessels, in turn, increases the blood flow through them, and the blood is warmed in the skin and carries the heat into the core of the body. Thus, in the morning, when a cold lizard orients its body perpendicular to the sun's rays and the sunlight warms its back, local vasodilation in that region speeds heat transfer to the rest of the body.

The thermoregulatory mechanisms employed by ectotherms allow many species of lizards and snakes to keep their body temperature within a range of a few degrees during the active part of their day. Many species of lizards have body temperatures between 33°C and 38°C while they are active (the **activity temperature range**), and snakes often have body temperatures between 28°C and 34°C. This is the region of temperature in which an ectotherm carries out its full repertoire of activities—feeding, courtship, territorial defense, and so on.

Organismal performance and temperature

Minimizing variation in body temperature greatly simplifies the coordination of biochemical and physiological processes. An organism's body tissues are the sites of a tremendous variety of biochemical reactions, proceeding simultaneously and depending on one another to provide the proper quantity of the proper substrates for successive

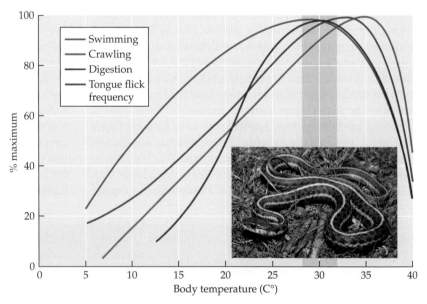

Figure 17.22 Effect of temperature on performance of the western garter snake (*Thamnophis elegans*). Crawling and swimming speeds, tongue-flick frequency, and speed of digestion all reach their maximum within the species' activity temperature range. The vertical axis shows the percentage of maximum performance achieved at each temperature. The shaded bar indicates the temperature range 28–32°C. (From Stevenson et al. 1985.) (Photo courtesy of Todd Pierson.)

steps in biochemical pathways. Each reaction has a different sensitivity to temperature, and coordination is greatly facilitated when temperature variation is limited.

In most cases, the body temperatures that ectotherms maintain during activity are the temperatures that maximize organismal performance. The western garter snake (*Thamnophis elegans*) provides examples of the effects of body temperature on physiological and behavioral functions. This species prowls the shorelines of ponds, moving in and out of the water and hunting fishes and amphibians. Chemosensation, accomplished by flicking the tongue, is an important mode of prey detection. The snakes spend the night in shelters, where their body temperatures fall to ambient levels (4–18°C), and emerge in the morning to bask. During activity on sunny days, the snakes maintain body temperatures between 28°C and 32°C.

Temperature affects the speed of crawling and swimming, frequency of tongue flicks, and rate of digestion, and all of the functions are close to their maximum between 28°C and 32°C (**Figure 17.22**). Anywhere within that range of body temperatures, snakes are able to crawl, swim, tongue-flick, and digest prey at rates that are at least 80% of maximum.

The relationship between the body temperatures of active garter snakes and the temperature sensitivity of various behavioral and physiological functions is probably representative of ectotherms in general.

17.10 Lepidosaurs and Climate Change

Lepidosaurs are ectotherms, and that characteristic lies at the heart of the effects of global climate change for this group. Both ectothermic and endothermic mechanisms allow vertebrates to control their body temperatures, but there is an important difference between the two types of mechanisms: by using metabolic heat to raise their body temperature, endotherms achieve substantial independence from the temperature of their immediate surroundings. If a mountain lion needs to crouch in the shadow of a rock while it waits to ambush a deer, it can do that. Life is not so simple for a lepidosaur because it must select an ambush site that not only allows it to see and attack prey but also allows it to thermoregulate. As the global climate changes, lepidosaurs in some habitats are having difficulty integrating thermoregulation with their other activities.

A study of populations of 48 species of spiny swifts (*Sceloporus*) in Mexico revealed the vulnerability of these lizards to climate change. Spiny swifts bask in the sunlight to raise their body temperatures to activity levels and then move to shaded crevices to avoid overheating. As global temperatures rise, the lizards spend more time in crevices, thereby reducing the time they can spend feeding, defending territories, and searching for mates. The limitation on energy intake may be exacerbated if higher nighttime temperatures increase the lizards' metabolic rates when they are sleeping, causing the lizards to consume energy that could otherwise be devoted to reproduction.

Increasing temperatures may also affect embryonic development directly. Many high-elevation species of *Sceloporus* are viviparous, and female *Sceloporus* maintain lower body temperatures when they are pregnant than they do at other times of the year. Higher temperatures could force pregnant females to choose between maintaining body temperatures that are higher than the optimum for their embryos or consuming less food and potentially depriving the embryos of nutrients. Oviparous species of *Sceloporus* may have difficulty finding nest sites with suitable temperature or water conditions for egg development. Temperature-dependent sex determination is yet another potential complication; TSD is not known to occur in *Sceloporus*, but it is hard to detect without detailed study and we cannot say that it is absent from the genus.

Data from Mexican weather stations show that the maximum daily air temperature has increased most during the period from January to May, which corresponds to the breeding season for many *Sceloporus* species in Mexico. The increase in temperature has been greatest in northern and central Mexico and at high elevations. This information was used to make predictions about the risk of extinction of lizard populations (**Figure 17.23**).

In 2009, researchers tested their predictions by returning to 200 sites that had populations of *Sceloporus* in 1975. They found that 12% of the populations had become extinct, and that the change in maximum temperature at the 200 sites was positively correlated with extinctions of local populations. For example, two populations of *S. serrifer* were extinct and two others were still present. A lizard at the sites where populations were extinct would have had to spend most of the daylight hours in a retreat site to escape high temperatures; at the sites where populations were still present, only 4 hours a day were too hot.

The risk experienced by mountain-dwelling species is counterintuitive because one would think that these species could move upward to escape higher temperatures. In practice, that response does not work, for two related reasons. First, mountains end, and once a lizard population has been forced to the top of a mountain, it has nowhere else to go. Second, mountains grow narrower as one moves upward, so there is less area available—a problem that is exacerbated by interspecific competition if other species from lower on the mountain are also moving upward.

Extending the model to 2050 and 2080 expands the area of Mexico in which extinctions are expected to occur and raises the probability of extinction for many populations above 50%. Applying the model on a global level and extending it to include 34 families of lizards, researchers predict that extinctions of local populations in Mexico will reach 39% by 2080.

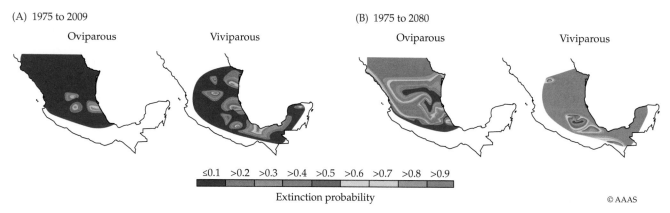

Figure 17.23 Climate change is already responsible for extinctions of *Sceloporus* lizards in Mexico. (A) A survey conducted in 2009 of 200 sites that had been occupied by *Sceloporus* species in 1975 found that 12% of those populations had become extinct. Extinctions were concentrated in the areas most affected by rising temperature. Populations at high elevations showed the highest levels of extinction (red and yellow areas), and viviparous species were more strongly affected than oviparous species. (B) The temperature increase predicted by global climate models is projected to lead to widespread extinctions of *Sceloporus* by 2080. Once again, viviparous species will feel the greatest impact. (From Sinervo et al. 2010.)

Summary

Lepidosaurs are the largest and most diverse group of nonavian reptiles.

Extant lepidosaurs include two lineages: a single species of rhynchocephalian (the tuatara) and more than 10,000 species of squamates (lizards and snakes). The phylogenetic relationship of snakes is nested within lizards, making lizards in the colloquial sense paraphyletic. Nonetheless, the groups popularly called snakes and lizards are very different in their morphology, ecology, and behavior.

Rhynchocephalians were diverse in the Mesozoic, but only the tuatara remains today.

The tuatara is a New Zealand endemic that once occurred on the North and South Islands. Today, natural populations are restricted to 32 small islands. Although sometimes called a "living fossil," the tuatara has several derived characters, including its nocturnal activity, low body temperature, and skull morphology.

Determinate growth is characteristic of squamates, especially lizards.

More than 80% of lizard species weigh less than 20 g as adults. Lizards of this size can subsist on insects without requiring morphological or ecological specializations to capture them, and most lizards are insectivores.

Most of the species of lizards that are too large to eat insects are herbivores. Their bulky bodies accommodate the large intestines needed to digest plants.

Monitors and tegus are exceptions to the generalization that large lizards are herbivores. Some species of monitors and tegus are large enough to prey on other vertebrates.

Many species of lizards are arboreal. Chameleons are the most specialized arboreal lizards; they have a prehensile tail, zygodactyl feet, eyes that swivel independently, and a projectile tongue used to capture prey.

Leglessness has evolved more than 60 times among lizards. Surface-dwelling legless lizards have long tails, whereas fossorial species are short-tailed.

Amphisbaenians are fossorial and the most specialized legless lizards. Their skin is nearly free of connections to the trunk, and they are able to burrow forward and backward with ease. A single median tooth in the upper jaw occludes with two teeth in the lower jaw, allowing amphisbaenians to nip pieces from prey.

Changes in the expression of regulatory factors during development elongate the trunk and suppress development of the limbs of snakes.

Snakes differ from legless lizards in having long trunks and relatively short tails.

Despite their specialized body form, ecomorphs of snakes are associated with different ways of life. Burrowing snakes are small with short tails, and some have vestigial eyes; arboreal snakes are elongated and large-eyed, and some have prehensile tails; sea snakes have laterally flattened tails and valvular nostrils; and vipers that feed on bulky prey have stout bodies and long fangs.

Lateral undulation is the generalized form of locomotion employed by all snakes. Concertina locomotion is used in narrow passages such as burrows; a snake presses the sides of its body against the walls, alternately anchoring and extending the body. Heavy-bodied snakes (vipers,

Summary (*continued*)

boas, and pythons) use rectilinear locomotion, successively raising portions of the body from the substrate and moving the skin forward. Snakes that move over loose sand use sidewinding, which exerts force downward rather than laterally.

Foraging modes of squamates extend over a spectrum from sit-and-wait predators to actively foraging predators.

Many aspects of the morphology and physiology of squamates are correlated with foraging mode.

Sit-and-wait species rely primarily on vision to detect prey on the surface; actively foraging species employ chemoreception to detect hidden prey. Some sit-and-wait snakes use their tail, and sometimes tongue, as lures to attract prey within reach.

Most sit-and-wait species of lizards have stout bodies, whereas actively foraging species are slim. Viviparity is more common among sit-and-wait species than actively foraging species. Many sit-and-wait species are territorial; most actively foraging species are not.

Sit-and-wait species, which remain motionless for long periods, often have cryptic colors and patterns that conceal them from predators. Many actively foraging species have striped patterns that obscure movement.

Sit-and-wait species make short, rapid sprints to capture prey or escape predators and have high anaerobic metabolic capacities. Actively foraging species spend much of their time moving and have high aerobic capacities and greater endurance than sit-and-wait species.

Modifications of the skulls of most lepidosaurs allow kinesis of the skull and jaws during feeding.

Tuatara have reverted to an akinetic, fully diapsid skull, with two temporal arches. Lizards have lost the lower arch, imparting a degree of kinesis to most species. Snakes have lost both the lower and upper arch; their highly kinetic skulls have eight elements that can move independently.

Snakes swallow prey intact, in some cases while it is still struggling, and usually head first.

Snakes draw prey into the mouth with alternating movements of the upper and lower jaws on the left and right sides of the head. The recurved teeth slide forward across the prey as the jaws are advanced and are embedded in the prey as the jaws are retracted.

Constriction and venom reduce the risk of injury to the snake by killing prey before it is swallowed. Duvernoy's gland, which is present in the upper jaw of many colubrid snakes, is homologous with the venom gland of elapids and vipers.

Opisthoglyphous snakes have one or more enlarged fangs near the rear of the maxillae that conduct venom into prey. Proteroglyphous snakes (elapids) have short, permanently erect fangs at the front of the maxillae with closed channels that conduct venom. Solenoglyphous snakes (vipers and pit vipers) have long fangs on the maxillae, which rotate so that the fangs lie parallel to the roof of the mouth when it is closed and are erected when the snake strikes.

Many snakes are binge eaters that consume large prey items at infrequent intervals. Between meals, the intestine and some internal organs shrink.

Squamates have a variety of mechanism for avoiding and deterring predators.

Many squamates have colors and patterns that make them hard to detect. Some lizards eavesdrop on birds, becoming motionless in response to avian alarm calls.

Some squamates deter predators by signaling that they are aware of the predator's presence, by threatening a predator, by mimicking the colors and patterns of venomous species, or by spraying a predator with blood or with sticky secretions.

Many lizards and a few snakes can break off a portion of the tail if it is seized by a predator. The break occurs within a vertebra, and the missing vertebra is regenerated with a cartilaginous rod. Some geckos have skin that rips easily, allowing the lizard to escape; the skin subsequently regenerates.

Venom is a predatory mechanism that has been modified to serve a protective function in some snakes. Many venoms contain compounds that cause intense pain rather than damaging tissues.

Squamates employ visual, auditory, chemical, and tactile signals to maintain territories and choose mates.

Diurnal species of lizards employ visual displays; males challenge intruding males, often by displaying a colored dewlap combined with doing push-ups and head bobs. Nocturnal geckos use vocalizations rather than visual displays.

Mating systems of a few lizard species are based on differences in testosterone levels of males (as reflected in their throat color) and corresponding mating strategies. The least common strategy has an advantage, and the frequency of each strategy changes from one generation to the next.

In general, squamates are believed to display little sociality, but that impression may reflect a lack of information.

(*Continued*)

Summary (continued)

Some skinks form monogamous relationships, and others live in family groups.

Squamates exhibit genotypic and temperature-dependent sex determination.

Genotypic sex determination includes male and female heterogamety. Temperature-dependent sex determination is known only in the tuatara and in a few species of agamids, skinks, and geckos.

For some species, sex is determined by the interaction of incubation temperature and egg size. Some viviparous species of lizards have temperature-dependent sex determination.

Oviparity is the ancestral reproductive mode for squamates, but viviparity has evolved more than 100 times, primarily in cold climates.

Viviparity allows a female to control the incubation temperature of her embryos, but pregnancy reduces a female's ability to flee from predators and limits reproduction to a single event in a season.

Parthenogenesis is widespread among lizards but apparently rare among snakes.

Parthenogenesis is known in six families of lizards and one species of snake, but parthenogenesis is hard to detect and probably more widespread than realized.

Interspecific hybridization of two bisexual species appears to be the starting point for parthenogenesis, producing a parthenogenetic diploid species. A backcross to one of the parental species produces triploid parthenogenetic species; one tetraploid parthenogenetic whiptail lizard species is known.

More than 100 species of squamates are known to exhibit parental care.

Females of some lizard species remain with their clutches, protecting them from predators, rotating the eggs, and removing dead eggs.

Some python species brood their eggs, using heat generated by muscular contractions to raise the temperature of the clutch. Newborn rattlesnakes remain with their mothers for 2 weeks or more.

Moving between sun and shade, changing color, and changing patterns of blood flow allow squamates to control their body temperatures.

Speed of locomotion, prey detection and capture, rate of digestion, and many other physiological functions normally reach their maximum within a species' activity temperature range.

Body size is an important component of thermoregulation; small individuals heat and cool faster than large ones and reach lower equilibrium temperatures.

Because lepidosaurs are ectotherms, climate change threatens their ability to carry out their daily activities.

Increasing temperatures limit the time that squamates can be outside their retreat sites, thereby reducing the time they can devote to feeding, defending territories, and seeking mates.

Females of viviparous species are especially at risk as they attempt to carry out essential activities while simultaneously maintaining their embryos at the temperature required for normal development.

Discussion Questions

1. Durophagy (eating hard-shelled prey, such as beetles or mollusks) appears as a dietary specialization among lizards, but it is rare among snakes. Why? What specializations of the teeth would you expect to find in durophagous lizards and in snakes?

2. *Anolis* lizards on the island of Jamaica do not show a reduction in signaling behavior during periods of visual "noise" created by the movement of windblown vegetation as *Anolis* on Puerto Rico do. The dewlaps of Jamaican *Anolis* are more brightly colored than those of the Puerto Rican lizards, and Jamaican *Anolis* expand and contract their dewlaps more rapidly than the Puerto Rican species do. What difference between the signaling systems of the Jamaican and Puerto Rican *Anolis* might allow the signals of the Jamaican species to be perceived even in visually noisy conditions?

3. Both the North American horned lizards (*Phyrnosoma*, Phrynosomatidae) and the Australian thorny devil (*Moloch horridus*, Agamidae) eat ants almost exclusively. Suggest an explanation for the convergence of ant-eating specialists in these two families on a short-legged, stocky body form and a spine-covered body.

4. The text mentioned several specializations that allow vipers to eat large prey. The large, heavy-bodied vipers in the genus *Bitis*, such as the African puff adder (*Bitis arietans*), provide particularly clear examples of these specializations. List as many of these specializations as you can, and explain how they work together.

5. Lizards that display sociality are all viviparous; in contrast, oviparous lizards do not appear to form long-term associations with other individuals of their species. What differences between these two groups of lizards might account for the difference in behavior?

Additional Reading

Bateman PW, Fleming PA. 2009. To cut a long tail short: A review of lizard caudal autotomy studies carried out over the last 20 years. *Journal of Zoology* 277: 1–14.

Chen Q, Liu Y, Brauth SE, Fang G, Tang Y. 2017. The thermal background determines how infrared and visual systems interact in pit vipers. *Journal of Experimental Biology* 220: 3103–3109.

Cree A. 2014. *Tuatara: Biology and Conservation of a Venerable Survivor.* Canterbury University Press, Christchurch, NZ.

Gardner MG, Pearson SK, Johnston GR, Schwarz MP. 2015. Group living in squamate reptiles: a review of evidence for stable aggregations. *Biological Reviews* 91: 925–936.

Greene HW, May PG, Hardly DL Sr., Sciturro JM, Farrell TM. 2002. Parental behavior by vipers. *In* Schuett GW, Hoggren M, Douglas ME, Greene HW (eds.). *Biology of the Vipers*, pp. 179–206. Eagle Mountain Publishing, Eagle Mountain, UT.

Halperin T, Carmel L. Hawlens D. 2017. Movement correlates of lizards' dorsal pigmentation patterns. *Functional Ecology* 31: 370–376.

Huey RB, Losos JB, Moritz C. 2010. Are lizards toast? *Science* 328: 832–833.

Jones MEH, Cree A. 2012. Tuatara. *Current Biology* 22: R986–R987.

Kaltcheva MM, Lewandoski M. 2017. Evolution: Enhanced footing for snake limb development. *Current Biology* 26: R1237–R1240.

Knight K. 2017. Thermal imaging and vision are equally important for hunting pit vipers. *Journal of Experimental Biology* 220: 3001.

Leal F, Cohn MJ. 2017. Loss and re-emergence of legs in snakes by modular evolution of *Sonic hedgehog* and *HoxD* enhancers. *Current Biology* 26: 2966–2973.

Lillywhite HB. 2014. *How Snakes Work.* Oxford University Press, New York.

Pla D and 6 others. 2017. What killed Karl Patterson Schmidt? Combined venom gland transcriptomic, venomic and antivenomic analysis of the South African green tree snake (the boomslang), *Dispholidus typus. Biochimica et Biophysica Acta* 1861: 814–823.

Sinervo B, Miles DB, Frankino WA, Klukowski M, DeNardo DF. 2000. Testosterone, endurance, and Darwinian fitness: Natural and sexual selection on the physiological basis of alternative male behaviors in side-blotched lizards. *Hormones and Behavior* 38: 222–233.

Crocodylians

Crocodylians include the largest extant species of reptiles, and the only ones that regularly prey on humans. The largest of these species, the saltwater crocodile (*Crocodylus porosus*), grows to 6 m or longer and weighs more than 1,000 kg. Not all crocodylians are large, however. Both Africa and South America have forest-dwelling species that are less than 2 m long.

A review of the biology of extant crocodylians dispels the mistaken impression of crocodylian behavior that comes from observing well-fed captive alligators and crocodiles resting inert for hours. In fact, crocodylians have extensive and complex predatory and social behaviors. They are highly social animals, communicating with vocalizations in the audible part of the sound spectrum as well as with subsonic frequencies below the range of human hearing. Crocodylians have been reported to use lures to attract prey and to engage in play—both activities traditionally considered to occur only in birds and mammals.

All extant crocodylians are semiaquatic predators. The Mesozoic precursors of crocodylians, however, included terrestrial lineages and herbivores, as well as enormous species that may have preyed on dinosaurs. Because crocodylians and birds form the extant phylogenetic bracket for pterosaurs and dinosaurs, they provide a basis for inferences about the behavior of those extinct Mesozoic reptiles. For example, crocodylians and birds both exhibit extensive parental care, and the inference that the bracketed groups also exhibited parental care is obvious.

18.1 Diversity of Extant Crocodylians

Only 24 extant species of crocodylians have been described, although the actual number may be greater, as molecular studies suggest the presence of several cryptic species of crocodiles. Recognition of cryptic species of crocodylians is an essential component of conservation programs designed to maintain the genetic diversity of Crocodylia.

Most of the extant species are found in the tropics or subtropics, but three species—the American alligator (*Alligator mississippiensis*), Chinese alligator (*A. sinensis*), and American crocodile (*Crocodylus acutus*)—have ranges that extend into the northern temperate zone, and the range of the Yacare caiman (*Caiman yacare*) extends into the southern temperate zone in Paraguay and Argentina. Alligators are more cold-tolerant than crocodiles. Chinese alligators hibernate in burrows during the winter; American alligators sometimes remain active even in below-freezing conditions, and have been reported to maintain a breathing hole through thin ice covering the water surface.

Molecular and morphological characters separate extant crocodylians (Crocodylia) into three lineages: Alligatoridae, Crocodylidae, and Gavialidae (**Figure 18.1**). Alligatoridae, a family of freshwater forms, includes the two alligator species and six species of caimans. Except for the Chinese alligator, alligatorids are found only in the New World. However, an extensive fossil record reveals their presence in Europe, and to a lesser extent in Asia, through the Oligocene; their restriction to the Americas dates from the middle Miocene. The American alligator occurs from the Gulf Coast states to North Carolina and west to Oklahoma, and several species of caimans range from Mexico to South America and occur on the island of Trinidad and Tobago.

Among Crocodylidae, some crocodiles (such as the saltwater crocodile) inhabit mangrove swamps and the estuaries of large rivers. The saltwater crocodile occurs widely in the Indo-Pacific region and penetrates the Indo-Australian Archipelago to northern Australia. In the New World, the American crocodile is quite at home in the sea and occurs in coastal regions from the southern tip of Florida through the Caribbean to northern South America.

Gavialidae contains only two species. The gharial (*Gavialis gangeticus*) once lived in large rivers from northern India to Burma but now is restricted to a small portion of the Ganges River drainage. The false gharial (*Tomistoma schlegelii*) inhabits the Malay Peninsula, Sumatra, Borneo, and Java. Although morphological phylogenies place the false gharial among crocodiles, molecular phylogenies ally

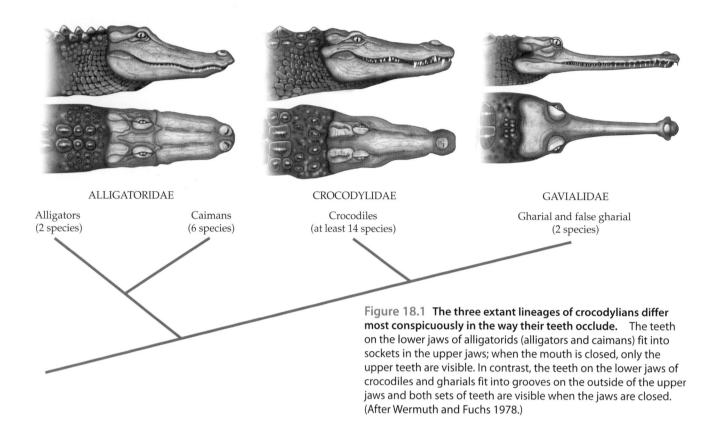

ALLIGATORIDAE

Alligators
(2 species)

Caimans
(6 species)

CROCODYLIDAE

Crocodiles
(at least 14 species)

GAVIALIDAE

Gharial and false gharial
(2 species)

Figure 18.1 The three extant lineages of crocodylians differ most conspicuously in the way their teeth occlude. The teeth on the lower jaws of alligatorids (alligators and caimans) fit into sockets in the upper jaws; when the mouth is closed, only the upper teeth are visible. In contrast, the teeth on the lower jaws of crocodiles and gharials fit into grooves on the outside of the upper jaws and both sets of teeth are visible when the jaws are closed. (After Wermuth and Fuchs 1978.)

(A)

(B)

(C)

(D)

Figure 18.2 Crocodylians have multiple modes of locomotion. (A) Lateral undulations of the tail propel swimming crocodylians. (B) On land, crocodylians use a belly crawl. Sinusoidal movements of the trunk and tail are coordinated with leg movements. (C) For longer movements on land, crocodylians use the high walk, with the legs directly beneath the animal. (D) Several species of crocodiles can gallop, moving the forelegs and hindlegs synchronously and flexing the vertebral column vertically. (A, USFWS/CC BY 2.0; B, Marianne Serra/CC BY 2.0; C, cuatrok77/CC BY-SA 2.0; D, from Grigg and Kirshner 2015, photo by Gordon Grigg courtesy of David Kirshner.)

it with the gharial. The gharial has the narrowest snout of any crocodylian; the mandibular symphysis (the point where the mandibles meet at the anterior end of the lower jaw) extends back for half the length of the lower jaw, and the snout of the false gharial is nearly as slender.

In the water, crocodylians swim with lateral sweeps of the tail, steering with their hindfeet (**Figure 18.2A**). A swimming crocodylian can be extremely inconspicuous while it is stalking prey, and it can move with frightening speed to attack a swimming animal. When not in the water, crocodylians have three modes of terrestrial locomotion:

1. *Crawl.* In the belly crawl (**Figure 18.2.B**), the legs are extended to the sides of the body and the ventral surface slides across the ground. The belly run is a faster version of the belly crawl. These gaits may look awkward, but they allow crocodylians to attain speeds of 12–14 km/h for short distances.

2. *Walk.* For longer terrestrial movements, crocodylians adopt the *high walk*, with the legs vertically under the body and moving in an anterior-posterior plane rather than extended to the side (**Figure 18.2C**). The high walk is slow, but it allows crocodylians to travel overland for many kilometers.

3. *Gallop.* Crocodiles (but not alligators and caimans) can gallop, holding the limbs vertically beneath the body and moving the fore- and hindlegs as pairs while vertical flexion of the vertebral column increases the stride length (**Figure 18.2D**). Galloping crocodiles can reach speeds of 17 km/h or more. Extant crocodiles gallop to water to escape from predators, moving only a few tens of meters, but predatory terrestrial crocodyliforms of the Mesozoic and Paleocene may have waited in ambush and galloped during a sprint to attack prey.

18.2 The Crocodylomorph Lineage

Basal crocodylomorphs had narrow skulls and short snouts, unlike the broad skulls and elongated snouts of extant crocodylians. During the Triassic, basal crocodylomorphs were the most diverse lineage of middle- and upper-level terrestrial predators. *Terrestrisuchus* (Latin *terra*, "earth"; Greek *souchos*, "crocodile"), a 0.5-m-long predator from the Late Triassic, was lightly built and had hindlegs that were longer than its forelegs, suggesting it could have run bipedally for short distances. *Protosuchus* (1 m) and *Sphenosuchus* (1.4 m), both from the Early Jurassic, were quadrupedal terrestrial predators.

The evolution of crocodylians was not a steady progression; rather, it proceeded in a series of diversification events that were limited to particular lineages and environments (**Figure 18.3**). The Jurassic saw the appearance of more derived forms in several lineages that included terrestrial and semiaquatic species. The notosuchian and neosuchian lineages separated in the Triassic and diversified during the Jurassic and Cretaceous—notosuchians primarily in terrestrial habitats and neosuchians mostly in semiaquatic habitats. That ecological division was not absolute, however—some notosuchians were semiaquatic, and some neosuchians were terrestrial.

Notosuchia

Notosuchians were diverse in the Southern Hemisphere during the Late Cretaceous, filling adaptive zones that were occupied by small dinosaurs in the Northern Hemisphere.

- *Araripesuchus wegeneri* (**Figure 18.4A**) was about 1 m long and had long legs. It appears to have been an omnivore.

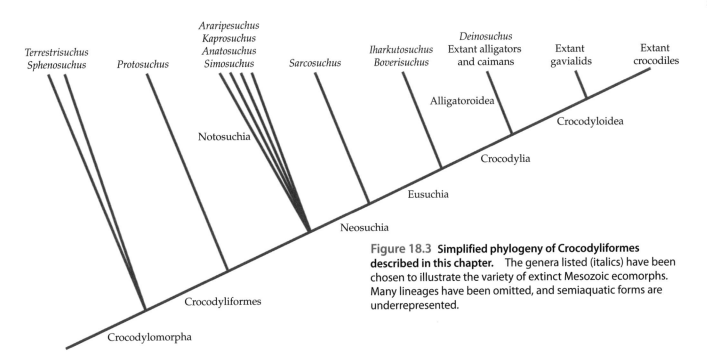

Figure 18.3 Simplified phylogeny of Crocodyliformes described in this chapter. The genera listed (italics) have been chosen to illustrate the variety of extinct Mesozoic ecomorphs. Many lineages have been omitted, and semiaquatic forms are underrepresented.

Retroarticular process

Figure 18.4 Terrestrial notosuchians of the Mesozoic.
(A) *Araripesuchus wegeneri* was an agile omnivore about 1 m long.
(B) *Kaprosuchus saharicus*, about 6 m long, was a formidable predator. The orientation of the eye sockets suggests that it had some binocular vision, and the long retroarticular process of the lower jaw indicates that the jaws had a wide gape that allowed *Kaprosuchus* to drive its teeth deeply into prey. (C) *Anatosuchus minor*, about 1 m long, had a broad snout rimmed with small teeth. Although it had long legs and an upright stance, it may have been semiaquatic, probing in shallow water to capture fishes and frogs. (D) The herbivorous *Simosuchus clarki*, only 60 cm long, was heavily armored and probably slow-moving. (A,C, Todd Marshall/CC BY 3.0; B, Carol Abraczinskas CC BY 3.0; D, from Krause et al. 2010, courtesy of David Krause.)

- *Kaprosuchus saharicus*, 6 m long, was a formidable predator. It had fanglike teeth in the upper and lower jaws, and the teeth were so long that they projected above and below the snout when the mouth was closed (**Figure 18.4B**).

- *Anatosuchus minor*, about 1 m long, may have been semiaquatic (**Figure 18.4C**); it gets its name from its broad, elongated snout (Latin *anas*, "duck").

- *Simosuchus clarki* (Greek *simos*, "flat-nosed"), only 60 cm long, was a terrestrial species with a blunt snout and a heavy covering of bony plates (**Figure 18.4D**). Its teeth had multiple cusps and closely resembled the teeth of extant herbivorous lizards.

During the Late Cretaceous in South America, lineages with flattened, serrated teeth arose. Some were up to 3.5–4 m long and were probably apex terrestrial predators, filling the ecological role played by smaller and medium-size carnivorous dinosaurs (theropods) in the Northern Hemisphere. One such group, Sebecidae, survived the Cretaceous–Paleogene extinctions and persisted until the Miocene in South America.

Neosuchia

Neosuchia was ancestrally a Laurasian group, although it included some Gondwanan lineages. This is the group that gave rise to the extant semiaquatic ambush predators, but during the Mesozoic it included both aquatic and terrestrial species. A group called hylaeochampsids had short snouts and heterodont dentition, with large crushing teeth in the rear of the jaw. These characters suggest an herbivorous diet, and the teeth of *Iharkutosuchus makadii,* an 80-cm-long

terrestrial species from the Late Cretaceous found near the town of Iharkút in western Hungary, had multiple cusps like the teeth of extant herbivorous lizards. The pattern of wear on the teeth indicates that the lower jaw moved from side to side to grind food.

Enormous body size—**gigantism**—is a recurrent theme in the evolution of neosuchians. *Sarcosuchus imperator* (Greek *sarkos*, "flesh"; Latin *imperator*, "ruler"), an Early Cretaceous crocodyliform from Africa, had an adult length of 11–12 m and weighed an estimated 8,000 kg. *Sarcosuchus* appears to have been a generalized predator, probably waiting in ambush at the water's edge, and it was large enough to have preyed on small dinosaurs.

Deinosuchus (Greek *deinos*, "terrible"), a Late Cretaceous crocodylian from North America related to extant alligators, was up to 10 m long and weighed 5,000 kg (**Figure 18.5**). *Deinosuchus* was semiaquatic, with a body form like that of extant alligators. It was undoubtedly an apex predator and probably preyed on small species of dinosaurs; bite marks matching the teeth of *Deinosuchus* have been found on bones of hadrosaurs (duck-billed dinosaurs).

Boverisuchus magnifrons (Greek *bous*, "ox"; *frons*, "brow"), about 3 m long, was probably an apex terrestrial carnivore during the Paleogene. *Boverisuchus* had laterally compressed teeth with serrations on the cutting edges, a deep snout, a tail that was rounded in cross section, and hooflike toes. With related forms, collectively known as ziphodonts (Greek *xiphos*, "sword"; *odous*, "tooth"), *Boverisuchus* may have occupied the large terrestrial predator adaptive zone left vacant by the extinction of nonavian theropod dinosaurs at the end of the Cretaceous.

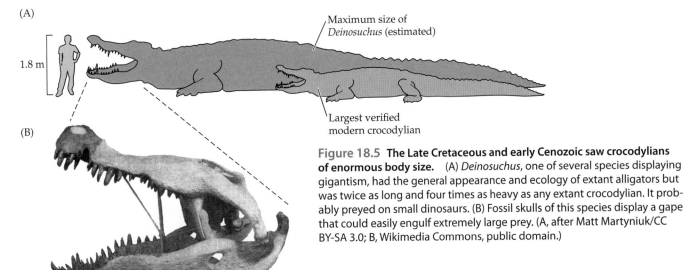

Figure 18.5 The Late Cretaceous and early Cenozoic saw crocodylians of enormous body size. (A) *Deinosuchus*, one of several species displaying gigantism, had the general appearance and ecology of extant alligators but was twice as long and four times as heavy as any extant crocodylian. It probably preyed on small dinosaurs. (B) Fossil skulls of this species display a gape that could easily engulf extremely large prey. (A, after Matt Martyniuk/CC BY-SA 3.0; B, Wikimedia Commons, public domain.)

During the Pliocene and Pleistocene, *Crocodylus thorbjarnarsoni* (7.5 m) inhabited the Lake Turkana basin, located in modern-day Kenya, and *C. anthropophagus* (5–6 m) (Greek *anthropos*, "man"; *phago*, "to eat") lived in the region that is now Olduvai Gorge in Tanzania. Both of these crocodiles were contemporaries and predators of early humans; fossils of early human bones with marks from crocodile teeth have been found at Olduvai Gorge.

18.3 Predatory Behavior and Diet of Extant Crocodylians

The head shapes of extant crocodylians vary from short and broad to long and slender (**Figure 18.6**). Conventional wisdom equates broad snouts with generalist diets and slender snouts with piscivory, but field data do not support this generalization. Among the slender-snouted forms—the gharial and false gharial, African slender-snouted crocodile (*Mecistops cataphractus*), Australian freshwater crocodile (*Crocodylus johnstoni*), and Orinoco crocodile (*C. intermedius*)—only the gharial is a specialized piscivore. Even the false gharial is a generalized predator, and there is a record of a large false gharial killing and eating a human.

All extant crocodylians are semiaquatic. Although they have well-developed limbs and some species make extensive overland movements (see Figure 18.2), they appear to be primarily aquatic predators. The heads of alligatorids are covered with small bulges, called **integumentary sensory organs**, that are exquisitely sensitive pressure receptors. In crocodylids and gavialids, these organs extend over the entire body. Patterns of the nerves associated with integumentary sensory organs are present in fossils of semiaquatic Mesozoic crocodyliforms, but not in terrestrial species.

Sensing water movement is probably part of the predatory repertoire of crocodylians; in complete darkness,

American alligators can lunge toward the point of impact of a single drop of water falling on the water surface. Some crocodylians float motionless on the water surface with their mouths open and their forelegs extended and swipe their head sideways to capture fishes detected by the integumentary sensory organs. In addition to detecting prey, integumentary sensory organs probably play a role in the social interactions of crocodylians.

Some of the predatory stealth of extant crocodylians results from the presence of a secondary palate, which allows crocodylians to breath with just their nostrils and eyes breaking the surface of the water (**Figure 18.7**). The choanae (internal nostrils) lie behind the gular valve, a flap of soft tissue that prevents water from entering the trachea. Gentle movements of the tail can propel a crocodylian through the water with barely detectable ripples, and a powerful tail stroke can propel it from the water before surprised prey can react.

After seizing a bird or mammal, a crocodylian drags it underwater to drown it. When the prey is dead, the crocodylian bites off large pieces and swallows them whole. Alternatively, crocodylians can use the inertia of a large prey item to pull off pieces: the crocodylian bites the prey and then rotates rapidly around its own long axis (a behavior called the "death roll"), tearing loose the portion it is holding. Sometimes crocodylians wedge a dead animal into a tangle of submerged branches or roots to hold it as the crocodylian pulls chunks of flesh loose. Some crocodylians leave large prey items to decompose for a few days until they can be dismembered easily.

Hunting behavior extends beyond simply waiting in ambush to attack prey. Crocodylians have been reported to hunt in groups and to use lures to attract birds. Groups of American alligators and caimans have been observed feeding side by side at the mouths of streams where the current sweeps fish toward them (**Figure 18.8A**). This behavior falls short of true cooperation, but the presence of so many individuals probably increases the capture rate for each of the participating caimans, because a fish dodging away from one set of jaws is likely to swim into the mouth of an adjacent caiman.

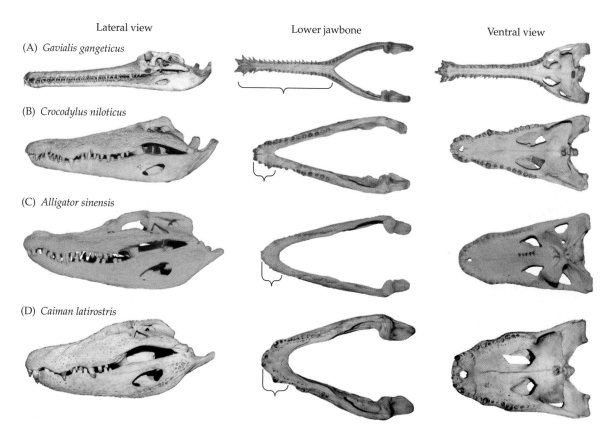

Lateral view   Lower jawbone   Ventral view

(A) *Gavialis gangeticus*

(B) *Crocodylus niloticus*

(C) *Alligator sinensis*

(D) *Caiman latirostris*

Figure 18.6 The snouts of extant crocodylians range from slender to broad.   The mandibular symphysis (brackets) extends for 50% of the length of the slender lower jaw of the gharial (*Gavialis gangeticus*; top) and less than 10% of the lower jaw of the broad-snouted caiman (*Caiman latirostris*; bottom). Species with intermediate snout widths include the Nile crocodile (*Crocodylus niloticus*) and Chinese alligator (*Alligator sinensis*). (From Walmsley et al. 2013, CC BY 4.0; courtesy of Christopher Walmsley.)

(A)

External nostril
Nasal passage
Internal nostrils (choanae)
Esophagus
Trachea
Secondary palate
Gular valve

(B)

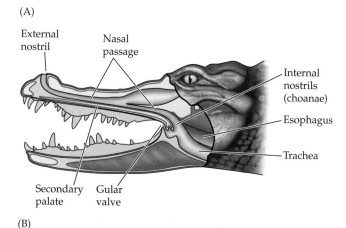

Some crocodylians have been reported to augment their ambush hunting strategy with lures. During the breeding season of wading birds, mugger crocodiles (*Crocodylus palustris*) and American alligators lie motionless for hours in shallow water with twigs and sticks balanced on their snouts (**Figure 18.8B**). This behavior is most common near rookeries during the period when birds are gathering sticks to build their nests. A bird that tries to take a stick is at risk of being seized (**Figure 18.8C**).

Although extant crocodylians are basically semiaquatic, hunting on land appears to be a regular behavior for large species. American alligators and several species of crocodiles have been observed waiting in ambush beside trails at night, and humans, dogs, and other mammals have been attacked (**Figure 18.8D**). Prior to the arrival of humans, crocodylians may have been the apex predators on some islands.

Figure 18.7 A secondary palate allows crocodylians to breathe with only their nostrils and eyes above the water.   (A) The secondary palate separates the nasal passage from the oral cavity. The internal nostrils (choanae) open into the pharynx behind the gular valve, which forms a seal that prevents water from entering the trachea. (B) Floating with only its nostrils and eyes above the water, a crocodylian is a stealthy predator. (A, after Gregg and Kirshner 2015; B, Vladimir Dinets.)

(A)

(B)

(C)

(D)

Figure 18.8 Predatory behaviors of crocodylians. (A) Yacare caimans (*Caiman yacare*) at the mouth of a stream in Brazil are facing into the current, capturing fish that must dodge between the closely spaced open mouths. (B) A mugger crocodile (*Crocodylus palustris*) with sticks on its head is waiting to ambush nest-building marsh birds. (C) Sticks can be effective lures for birds; here an American alligator (*Alligator mississippiensis*) is consuming a snowy egret that was attracted by a stick on the alligator's head. (D) An American alligator waiting in ambush at the edge of a road at night. (A, Carlos Yamashita, courtesy of F. Wayne King; B,D Vladimir Dinets; C from Dinets et al. 2015, photo by Don Specht.)

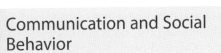

18.4 Communication and Social Behavior

Adult crocodylians, like birds, use sound extensively in their social behavior. Male crocodylians emit a variety of vocalizations during courtship and territorial displays and also slap their heads and tails against the water (**Figure 18.9**). Vocal displays are especially important for crocodylians such as American alligators that live in dense swamps, because males' territories are often out of sight of other males and females. The bellow of a male alligator includes frequencies between 20 and 250 Hz and resounds through the swamp, announcing his presence to other alligators up to 160 m away. Female alligators also roar, but only males produce a subsonic vibrations (~10 Hz, a frequency below the range of human hearing) that causes drops of water to dance on the water surface and travels for long distances underwater.

The displays may have different functions: the sharp onset of the sound produced by a head slap facilitates location of its source, but the sound does not travel far. Thus, head slaps may identify the position of a displaying male to nearby individuals. Bellows can be heard at distances up to 160 m and may stimulate vocalizations from other individuals, creating a crocodylian chorus. Subsonic vibrations travel a kilometer or more underwater, and probably convey information about the size of the crocodylian producing the sound.

As many as 30 American alligators have been reported to assemble in courtship gatherings during the breeding season, swimming within a small area; one such gathering included about 30 alligators in an area only 20 m in diameter, splashing, hissing, head slapping, and forming pairs. These gatherings may establish dominance hierarchies and appear to be preludes to mating. Courting crocodylians rub their chins over the head and body of their partner, engage in a variety of visual signals such as raising the snout, emit musky odors, and sometimes produce subsonic vibrations. Crocodylians are polygamous, and a female may mate with more than one male during a breeding season.

Do crocodylians play? "Play" can be defined as an activity that consumes time and energy with no immediate gain in fitness for the individual or individuals participating in it. Play is generally considered to be characteristic of mammals, but three activities have been interpreted as play by crocodylians:

(A)

(B)

Figure 18.9 **Male American alligators communicate with audible vocalizations and subsonic vibrations.** (A) During the audible portion of its display, a male alligator raises its head and emits a bellow containing sound frequencies between 20 and 250 Hz. These frequencies travel through the air. (B) After bellowing, the male sinks until it is partly submerged and produces subsonic vibrations that travel through the water. (A, Tristan Loper/CC BY-SA 2.0; B, Vladimir Dinets.)

- *Solitary locomotor play* consists of vigorous activity that is typically performed alone. Familiar mammalian examples include running, twisting, leaping, and sliding. Young American alligators have been observed repeatedly sliding down slopes into water, and a 2.5-m-long saltwater crocodile alarmed beachgoers by repeatedly body surfing in the waves at a beach near Port Douglas, Australia.

- *Object play* incorporates an inanimate object and can be a solitary or group activity. Dolphins play with floating objects, repeatedly tossing them into the air and retrieving them, and canids—from domestic dogs to coyotes and wolves—play with sticks. Adult crocodylians appear to be especially attracted to pink flowers floating on the water surface, repeatedly picking them up, pushing them, and carrying them in their teeth while ignoring other floating objects. Captive crocodylians play with floating balls, repeatedly seizing and releasing them, and will continue those behaviors for years with no reward other than the activity itself.

- *Social play* includes the chasing and wrestling bouts that can be observed among any litter of canids or felids, domestic or wild. Juvenile crocodylians have been observed swimming in tight circles as if chasing each other, wrestling, and repeatedly riding on another individual's back.

18.5 Reproduction and Parental Care

All crocodylians are oviparous, depositing hard-shelled eggs in a nest that is guarded by the female. Alligators and caimans are mound nesters; the female uses her hindlegs to build a pile of soil and vegetation and deposits her eggs in a depression in the center of the mound. About half of the extant species of crocodiles and the false gharial are also mound nesters. The remaining crocodiles and the gharial deposit eggs in holes that the female excavates with her hindlegs. Fossilized crocodylian nests have been found in Mesozoic and Cenozoic deposits, some of them adjacent to the nests of nonavian dinosaurs.

Temperature-dependent sex determination

All species of crocodylians that have been studied display type II temperature-dependent sex determination, with females produced at low and high temperatures and males at intermediate temperatures. Considering the elaborate nests that crocodylians prepare and the large numbers of eggs a female produces, natural nests probably have enough temperature variation within the nest to produce hatchlings of both sexes. For example, average temperatures ranged from 33°C to 35°C in the top center of nests of American alligators in marshes, and males hatched from eggs in this area. At the bottom and sides of the same nests, average temperatures ranged from 30°C to 32°C, and eggs from those regions hatched into females.

Human activities can distort the ratio of male to female hatchlings. Levees have been constructed in much of the habitat of the American alligator, and alligators have adopted these artificial sites for nest construction. Nests on levees (34°C) are hotter than nests in marshes (30°C), and nearly 100% of the hatchlings from nests on levees are males. Global climate change has been shown to affect sex determination of a lizard and of sea turtles, and it is likely to affect crocodylians as well.

Parental care

Crocodylians provide parental care that is as extensive as the care provided by birds. Adult crocodylians have few predators except for humans, but eggs and hatchlings are vulnerable to predation by fishes, frogs, lizards, turtles, other species of crocodylians, birds, and mammals. Most often the female is the caregiver, but males of some species participate. In every species that has been studied, the female remains near the nest throughout incubation and defends the clutch against egg-eating predators.

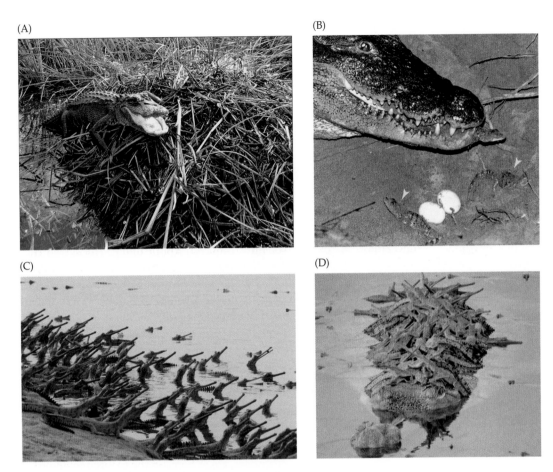

Figure 18.10 Crocodylians have extensive parental care. (A) An adult female American alligator guarding her nest. (B) An adult male mugger crocodile is carrying one hatchling from the nest to the water, while two more hatchlings (arrowheads) await transport. (C) Hatchling gharials (*Gavialis gangeticus*) assemble in crèches containing hundreds of individuals. (D) Adult males guard the crèches, assisted by young males. (A, NPS/Lori Oberhofer, NPS; B,C,D, courtesy of Jeffrey Lang.)

Baby crocodylians begin to vocalize before they have fully emerged from their eggs, and these vocalizations are loud enough to be heard some distance away. These calls synchronize the hatching of all the eggs in a clutch and stimulate one or both parents to excavate the nest (**Figure 18.10**). The female opens the nest and transports the hatchlings in every species of crocodylian for which information is available, and males of some species participate in this behavior. In some species, the parents gently break the eggshells with their teeth to help the young escape. The sight of a Nile crocodile (*Crocodylus niloticus*), with jaws that can crush the leg of a zebra, delicately cracking the shell of an egg little larger than a hen's egg and releasing the hatchling unharmed is truly remarkable.

Young crocodylians stay together in a group called a crèche for a considerable period—2 years for the American alligator and 3 years for the spectacled caiman (*Caiman crocodilus*) of Central and South America. A crèche contains young of many parents, and several adults serve as guards. The crèches of gharials can contain up to 1,000 hatchlings, and young males participate in guarding the crèches. Contact calls help the juveniles in a crèche maintain cohesion and alert them to the presence of food.

Frightened crocodylian hatchlings emit a distress call that stimulates adult male and female crocodylians to come to their defense. In addition to summoning the parents, these vocalizations may attract unrelated adults. When staff members at a crocodile farm in Papua New Guinea rescued a hatchling New Guinea crocodile (*Crocodylus novaeguineae*) that had strayed from the pond, the hatchling's distress call brought 20 adult crocodiles surging toward it—and toward the staff members! The dominant male head slapped the water repeatedly and then charged into the chain-link fence where the staff members were standing, while the females swam about, giving deep guttural calls and head slapping the water.

18.6 The Skin Trade

The major threats to crocodylian populations are probably habitat loss and alteration, and their effects are compounded by the biological characters of crocodylians. For example, temperature-dependent sex determination has turned unshaded levees into an ecological trap for American alligators. Alien species can be destructive; toxic cane toads have caused declines in populations of Australian freshwater crocodiles, and in Florida, introduced tegu and monitor lizards are nest predators of American alligators and American crocodiles.

In addition to these stresses, overexploitation of crocodylians for their skin and meat is a problem on a global level. World trade in crocodylian skins averaged 1.3 million skins per year between 2001 and 2012, and 400,000–1,000,000 kg of meat were shipped *each year* between 2000 and 2008. A dozen species of crocodylians were represented among the skins (crocodylian skins can be identified to species by structural characters, including the pattern of integumentary sensory organs), and most of them were species included in the protection supposedly afforded by the Convention on Trade in Endangered Species. Crocodylians are late-maturing and long-lived—both life-history characters that cannot withstand high adult mortality.

Farming of crocodiles and alligators is a mixed blessing at best. As with farmed fishes and sea turtles, large numbers of captive animals that are frequently poorly fed and housed are sources of diseases that can spread to wild populations. Crocodylians are hosts for mosquito-borne viruses, and West Nile virus is suspected as the cause of death of hundreds of captive alligators in Florida. Large species of crocodylians need a lot of food, and that can create ancillary problems; farmed crocodiles in Cambodia are fed snakes that are collected from the wild, consuming between 2.7 and 12.2 million snakes per year.

The recovery of the American alligator, which in the 1950s was close to extinction because of overharvesting and habitat loss, demonstrates the potential resilience of crocodylians. After well-designed regulations were established and enforced, American alligator populations rebounded, and in 1987 the U.S. Fish and Wildlife Service declared the species to be fully recovered. Unfortunately, most countries with threatened species of crocodylians lack the resources or will to provide the protection needed. Success stories such as that of the American alligator require well-informed policies and robust enforcement of regulations.

Summary

Crocodylians are the largest extant reptiles.

The largest species, the saltwater crocodile, grows to 6 m or more and weighs more than 1,000 kg. The smallest species are less than 2 m long.

Crocodylians and birds form the extant phylogenetic bracket for dinosaurs and pterosaurs.

Characters shared by crocodylians and birds—vocalizations, social behavior, and parental care—are probably ancestral characters that were present in some of the extinct lineages of Mesozoic nonavian reptiles.

Only 24 species of extant crocodylians are recognized, although there are probably several cryptic species that await description.

All extant crocodylians are semiaquatic predators. Mesozoic forms were more diverse.

Nearly all extant crocodylians live in the tropics or subtropics. The two species of alligators have the northernmost geographic ranges, and both encounter freezing temperatures in the winter.

Molecular and morphological phylogenies group crocodylians into three lineages.

Alligators and caimans (Alligatoridae) are freshwater species and, except for the Chinese alligator, are confined to the New World. The teeth on the lower jaws of alligatorids fit into grooves in the upper jaws and are not visible when the mouth is closed.

Crocodiles (Crocodylidae) are tolerant of estuarine habitats, and several species, including the American and saltwater crocodiles, are at home in the sea. The teeth on the lower jaws of crocodylids are visible when the mouth is closed.

Gavialids (Gavialidae) have extremely slender snouts, with the mandibular symphysis extending back half the length of the lower jaw.

Crocodylians swim with lateral undulations of the tail and steer with their hindfeet. On land, crocodylians have three gaits.

The belly crawl and its speedier version, the belly run, are used for short distances; the limbs extend horizontally, and the ventral surface slides across the ground.

The high walk, with the legs held vertically under the body, is used for longer terrestrial movements. Crocodiles can gallop, moving the fore- and hindlegs as pairs and flexing the vertebral column vertically to increase the stride length.

Several radiations of crocodyliforms during the Mesozoic gave rise to a variety of ecomorphs.

Basal crocodylomorphs were terrestrial, and radiated widely in terrestrial and semiaquatic habitats during the Mesozoic.

Summary (*continued*)

Notosuchians and neosuchians diverged in the Triassic and diversified during the Jurassic and Cretaceous.

In the Southern Hemisphere, notosuchians radiated into terrestrial predatory and herbivorous ecological niches and as semiaquatic predators. By the Late Cretaceous, sebecid notosuchians were probably the apex terrestrial predators in South America, filling the adaptive zone occupied in the Northern Hemisphere by carnivorous dinosaurs. Large terrestrial sebecids survived into the Miocene.

Neosuchia, an initially Laurasian group that spread to Gondwana, included terrestrial and semiaquatic predators. Gigantism evolved repeatedly among neosuchians, and some Cretaceous and early Cenozoic species exceeded 10 m in length and weighed as much as 8,000 kg. In the late Pleistocene, semiaquatic neosuchians preyed on early humans.

All extant crocodylians are primarily aquatic predators, although several species sometimes hunt on land, waiting in ambush beside trails.

The shape of the snout does not reveal the dietary habits of extant crocodylians. Among the slender-snouted forms, only the gharial is a piscivore; all of the other species are generalist predators.

The secondary palate allows a crocodylian to breathe with just its nostrils and eyes above the water. The gular valve prevents water from entering the trachea.

Crocodylians drag terrestrial prey underwater to drown it, then bite off pieces of the prey. Sometimes they seize a limb and spin rapidly to pull it off. Large prey items may be wedged underwater and left until they can be dismembered easily.

Collaborative feeding by groups of crocodylians falls short of true cooperation but probably increases the capture rate among participants.

Twigs and sticks on the snouts of crocodylians may act as lures, attracting nest-building birds that are seized when they reach for a stick.

Adult crocodylians use sound extensively in social interactions.

Bellows emitted by males travel through the air, while vibrations from head slaps and subsonic vibrations are transmitted through the water. Bellows can be heard from distances up to 160 m, while subsonic vibrations may travel farther than 1 km.

During the mating season, large numbers of individuals of some species of crocodylians assemble in small areas, vocalizing and head slapping. These aggregations may establish dominance hierarchies and male–female pairings.

Juvenile and adult crocodylians appear to engage in solitary and mutual play behaviors.

All crocodylians are oviparous, laying hard-shelled eggs in nests constructed by the female.

Alligators and caimans, about half of all crocodile species, and the false gharial build nest mounds of soil and vegetation and deposit the eggs in a hole the female digs in the top of the mound. In the remaining species of crocodylians, the female digs a nest in a sandy substrate.

Crocodylians have type II temperature-dependent sex determination, with females produced at low and high temperatures and males at intermediate temperatures.

Parental care by crocodylians includes guarding the nest, assisting hatchlings to emerge and carrying them to the water, and guarding crèches of hatchlings.

Females are the primary caregivers at nests, but males of some species help open the nest, transport young to water, and guard crèches.

Contact calls by juveniles maintain cohesion within a crèche, and distress calls bring adults to the rescue.

Habitat loss and alteration and the introduction of alien species are threats to crocodylians, but overharvesting strikes at the most vulnerable element of crocodylian life history.

An average of 1.3 million skins and up to 1 million kg of crocodylian meat are shipped annually in international trade. Many of the skins are from species that are supposedly protected by the Convention on Trade in Endangered Species.

Crocodylians are late maturing and long-lived and thus are poorly positioned to withstand severe adult mortality.

Crocodile farming introduces new problems, including disease and the harvesting of other threatened vertebrates to feed captive crocodylians.

Discussion Questions

1. Some crocodylians construct nest mounds of soil and vegetation, whereas other species dig holes in the ground. The occurrence of mound and hole nests is shown in the phylogeny below. What is the ancestral nest type for crocodylians? What hypotheses can you propose to explain the distribution of those two nest types? What additional information would you gather to test those hypotheses?

2. Crèche formation is an uncommon form of parental care, although examples are known among birds and fishes.

What conditions might predispose hatchlings of a crocodylian species to form a crèche? What benefits might hatchlings derive from being in a crèche? A guarding adult is related to only a small number of the young in a crèche, so is guarding a crèche an example of altruistic behavior? What costs does an adult incur by guarding a crèche? What benefits might an adult derive from guarding a crèche? What other questions can you pose about crèche formation by crocodylians?

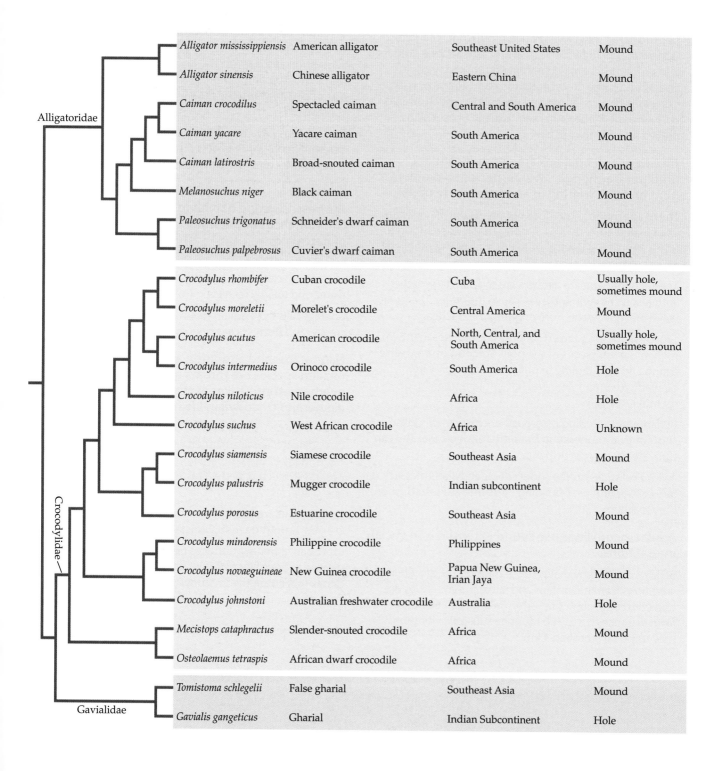

Alligator mississippiensis	American alligator	Southeast United States	Mound
Alligator sinensis	Chinese alligator	Eastern China	Mound
Caiman crocodilus	Spectacled caiman	Central and South America	Mound
Caiman yacare	Yacare caiman	South America	Mound
Caiman latirostris	Broad-snouted caiman	South America	Mound
Melanosuchus niger	Black caiman	South America	Mound
Paleosuchus trigonatus	Schneider's dwarf caiman	South America	Mound
Paleosuchus palpebrosus	Cuvier's dwarf caiman	South America	Mound
Crocodylus rhombifer	Cuban crocodile	Cuba	Usually hole, sometimes mound
Crocodylus moreletii	Morelet's crocodile	Central America	Mound
Crocodylus acutus	American crocodile	North, Central, and South America	Usually hole, sometimes mound
Crocodylus intermedius	Orinoco crocodile	South America	Hole
Crocodylus niloticus	Nile crocodile	Africa	Hole
Crocodylus suchus	West African crocodile	Africa	Unknown
Crocodylus siamensis	Siamese crocodile	Southeast Asia	Mound
Crocodylus palustris	Mugger crocodile	Indian subcontinent	Hole
Crocodylus porosus	Estuarine crocodile	Southeast Asia	Mound
Crocodylus mindorensis	Philippine crocodile	Philippines	Mound
Crocodylus novaeguineae	New Guinea crocodile	Papua New Guinea, Irian Jaya	Mound
Crocodylus johnstoni	Australian freshwater crocodile	Australia	Hole
Mecistops cataphractus	Slender-snouted crocodile	Africa	Mound
Osteolaemus tetraspis	African dwarf crocodile	Africa	Mound
Tomistoma schlegelii	False gharial	Southeast Asia	Mound
Gavialis gangeticus	Gharial	Indian Subcontinent	Hole

Alligatoridae

Crocodylidae

Gavialidae

3. Phylogenetic hypotheses for crocodylians based on molecular data differ from those based on morphological and paleontological data in the placement of the gharial (*Gavialis gangeticus*) and the false gharial (*Tomistoma schlegelii*). Molecular data place the two species as sister lineages within Gavialidae, whereas morphological and paleontological data place the gharial as the sister lineage of Alligatoridae + Crocodylidae and place the false gharial in Crocodylidae. How does the difference in these phylogenies (shown below) affect our interpretation of the derived versus ancestral condition of crocodylian characters? For example, integumentary sensory organs (ISOs) are present on the heads and bodies of crocodylids and gavialids, but only on the heads of alligatorids. What is the ancestral condition?

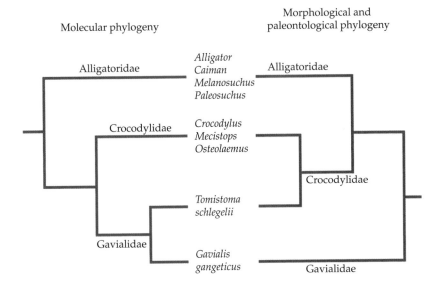

Molecular phylogeny

Alligatoridae

Crocodylidae

Gavialidae

Alligator
Caiman
Melanosuchus
Paleosuchus

Crocodylus
Mecistops
Osteolaemus

Tomistoma schlegelii

Gavialis gangeticus

Morphological and paleontological phylogeny

Alligatoridae

Crocodylidae

Gavialidae

Additional Reading

Brazaitis P, Watanabe ME. 2011. Crocodylian behaviour: A window to dinosaur behaviour? *Historical Biology* 23: 73–90.

Brochu CA. 2001. Crocodylian snouts in space and time: Phylogenetic approaches toward adaptive radiation. *American Zoologist* 41: 564–585.

Erickson GM, Brochu CA. 1999. How the "Terror crocodile" grew so big. *Nature* 398: 205–206.

Farmer CG. 2015. Similarity of crocodylian and avian lungs indicates unidirectional flow is ancestral for archosaurs. *Integrative and Comparative Biology* 55: 1–10.

Finger JW Jr., Gogal RM Jr. 2013. Endocrine-disrupting chemical exposure and the american alligator: a review of the potential role of environmental estrogens on the immune system of a top trophic carnivore. *Archives of Environmental Contamination and Toxicology* 65: 704–714.

Grigg G, Kirshner D. 2015. *Biology and Evolution of Crocodylians.* Cornell University Press, Ithaca, NY.

Lang SDJ, Farine DR. 2017. A multidimensional framework for studying social predation strategies. *Nature Ecology & Evolution* 1: 1230–1239.

Reber SA and 5 others. 2017. Formants provide honest acoustic cues to body size in American alligators. *Scientific Reports* 7: 1816.

Schwimmer D. 2002. *King of the Crocodylians: The Paleobiology of Deinosuchus.* Indiana University Press, Bloomington, IN.

Shirley MH, Villanova WL, Vliet KA, Austin JD. 2015. Genetic barcoding facilitates captive and wild management of three cryptic African crocodile species complexes. *Animal Conservation* 18:322-330.

Soares D. 2002. An ancient sensory organ in crocodylians. *Nature* 417:241-242.

Vergne AL, Pritz MB, Mathevon N. 2009. Acoustic communication in crocodylians: from behaviour to brain. *Biological Reviews* 84:391-411.

Walmsley CW and 8 others. 2013. Why the long face? The mechanics of mandibular symphysis proportions in crocodiles. *PLoS ONE* 8(1): e53873.

The Mesozoic extended for 186 million years, from the close of the Paleozoic 252 Ma to the beginning of the Cenozoic 66 Ma. Throughout this vast span of time, a worldwide fauna evolved, diversified, drifted with the moving continents, and radiated into most of the adaptive zones occupied by extant terrestrial vertebrates—and into some zones that no longer exist, such as those occupied by the enormous herbivorous and carnivorous tetrapods we call dinosaurs.

The Mesozoic saw the first modern vertebrate fauna, with all extant evolutionary lineages represented and with a full array of tetrapod herbivores and carnivores in the sea. Most of these animals were diapsids, the lineage that includes extant nonavian reptiles and birds. Mammals (synapsids), although present and diversifying, were a small part of the Mesozoic terrestrial fauna and, as far as we know, were absent from marine habitats.

19.1 Characteristics of Diapsids

The taxonomic categories Diapsida and Reptilia are confusing. The crown group Reptilia includes the common ancestor of all extant taxa of diapsids—that is, turtles, crocodylians, birds, tuatara, and squamates (snakes and lizards). This chapter, however, focuses on extinct lineages of diapsids that are not among the groups encompassed by Reptilia, and thus we refer to these animals as diapsids rather than as reptiles.

The name *diapsid* means "two arches" and refers to the presence of an upper and a lower fenestra in the temporal region of the skull (**Figure 19.1**). More distinctive than the openings themselves are the bones forming the arches that border the openings. The upper temporal arch is composed of a three-pronged postorbital bone and a three-pronged squamosal. The jugal and quadratojugal bones form the lower arch. The connection between the jugal and quadratojugal that forms the lower arch has been lost repeatedly in the radiation of diapsids, and the upper arch

(postorbital-squamosal) has also been lost in some forms. Extant lizards and snakes clearly show the importance of those modifications of the skull in permitting increased skull kinesis during feeding, and the same significance may attach to the loss of the arches in some extinct forms.

Diapsids comprise two groups: Archosauromorpha (Greek *archon*, "ruler"; *sauros*, "lizard"; *morphe*, "form") and Lepidosauromorpha (Greek *lepisma*, "scale") (**Figure 19.2**). Lepidosauromorphs include Lepidosauria (rhynchocephalians, squamates, and their extinct relatives) as well as specialized marine tetrapods that are now extinct—the ichthyosaurs and sauropterygians.

Archosauromorphs include Testudines (turtles) and Archosauria (crocodylians, pterosaurs, nonavian dinosaurs, and birds). Two lineages of archosaurs are recognized, based on the articulation of the ankle joint. In Pseudosuchia, the ankle is a **crurotarsal (crocodyloid) joint** (see Chapter 14), meaning it can twist sideways as well as flex forward and backward; in Ornithodira, the ankle is a more rigid **mesotarsal joint**, with a straight-line hinge that can bend only forward and back (**Figure 19.3**).

Ecological relationships provide a perspective on Mesozoic diapsids (**Table 19.1**). Similar ecomorphs evolved in different lineages, and in some cases members of these lineages lived side by side and preyed on, or competed with one another.

19.2 Diversity of Mesozoic Diapsids

The end-Permian mass extinction described in Chapter 5 wiped out 70% of the terrestrial species of vertebrates (including synapsids, which had diversified in the Permian) and 95% of all marine species. Diapsids entered this nearly empty world and diversified into an enormous variety of animals, many of which were entirely different from anything that had preceded them.

- *Marine diapsids* (ichthyosaurs, placodonts, nothosaurs, pachypleurosaurs, plesiosauroids, and mosasaurs) first

Figure 19.1 *Tyrannosaurus rex* **illustrates the characters of the diapsid skull.** Two temporal fenestrae are a distinguishing character of diapsids. In addition, archosaurs (crocodylians, pterosaurs, dinosaurs, and birds) have an antorbital fenestra on each side of the head, anterior to the eye, that houses an air sinus, and derived forms have a still more anterior maxillary fenestra. (EncycloPetey/CC BY-SA 3.0.)

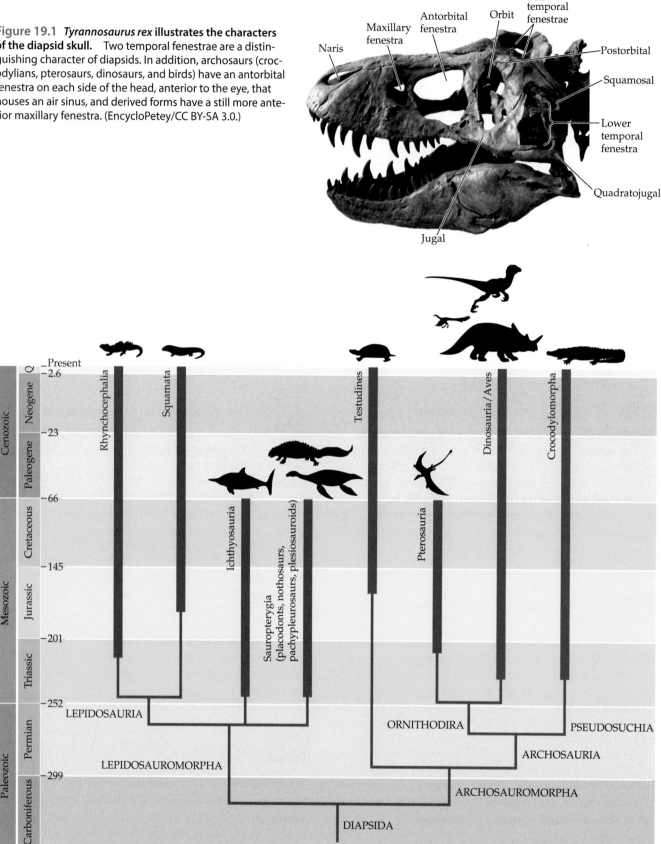

Figure 19.2 **Phylogenetic relationships of Diapsida.** Probable relationships among the major groups of diapsids; only the best-corroborated relationships are shown. Within archosauromorphs, turtles (Testudines) are the sister lineage of archosaurs. Within archosaurs, Pseudosuchia is the crocodile lineage (crocodylomorphs) and Ornithodira is the nonavian dinosaur and bird lineage. Three major lineages of lepidosauromorphs are distinguished. Ichthyosaurs and sauropterygians are extinct marine lineages that flourished during the Mesozoic. The lepidosaur lineage includes rhynchocephalians (diverse in the Mesozoic but with only one extant species, the tuatara) and squamates (lizards and snakes, with more than 10,000 extant species).

(A) Crurotarsal (B) Mesotarsal

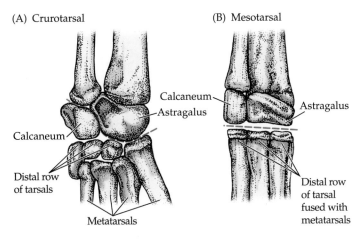

Figure 19.3 The way ankle joints bend defines the major lineages of archosaurs. (A) The axis of bending in the ankle joint of crocodylomorphs runs between the astragalus and calcaneum bones. This is called a crurotarsal or crocodyloid joint, and it allows the foot to twist sideward as well as to flex forward and back. Taxa belonging to Pseudosuchia have this type of joint. (B) In dinosaurs and birds, the ankle joint can flex only backward and forward. This is called a restricted mesotarsal joint; derived Ornithodira (nonavian dinosaurs and birds) have this type of joint. (By permission of Jeffrey Martz.)

appeared in the Triassic and diversified into the adaptive zones now occupied by sharks, seals, sea lions, walruses, and toothed whales, as well as some groups that have no extant parallels.

- On land, *crocodylomorphs* and several lineages of basal archosaurs were the earliest diapsid lineages to diversify as predators. Crocodylomorphs continued to diversify through the Mesozoic and into the Cenozoic. In the Southern Hemisphere, crocodylomorphs appear to have remained more abundant and diverse than dinosaurs, at least into the Early Cretaceous.

- *Flying diapsids* evolved twice—pterosaurs in the Late Triassic, followed by birds in the Jurassic. The two groups were contemporaneous from the Jurassic through the Cretaceous, a span of about 100 million years.

- The earliest *dinosaurs* appeared in the Middle Triassic, but dinosaurs were not very diverse until the Jurassic. The diversification of dinosaurs appears to have progressed in three stages. Their initial appearance 235–228 Ma was followed by a major radiation in the Middle to Late Jurassic (174–164 Ma) and then several pulses of diversification during the Cretaceous.

Table 19.1	Ecological Diversity of Mesozoic Diapsids	
Group	**Habitat**	**Description**
LEPIDOSAUROMORPHA		
Ichthyosaurs	Marine	Predators with body forms resembling those of sharks and cetaceans
Placodonts	Marine	Coastal herbivores
Pachypleurosaurs	Marine	Small predators
Nothosaurs	Marine	Medium-size predators
Plesiosaurs	Marine	Two types of large predators: long-necked plesiosaus and short-necked pliosaurs
Mosasaurs	Marine	Predators related to extant varanid lizards
Rhynchocephalians	Terrestrial	Small to medium-size herbivores
ARCHOSAUROMORPHA		
Turtles	Terrestrial, semi-aquatic, and marine	Carnivores and herbivores (see Chapter 16)
Crocodylomorphs: Metriorhynchids	Marine	Coastal predators related to crocodyliforms
Crocodylomorphs: Crocodyliforms	Semi-aquatic to terrestrial	Small herbivores and small to enormous carnivores (see Chapter 18)
Basal archosaurs (Rhynchosaurs)	Terrestrial	Herbivores
Ornithischian dinosaurs	Terrestrial	Herbivores
Saurischian dinosaurs: Sauropods	Terrestrial	Herbivores
Saurischian dinosaurs: Theropods	Terrestrial	Carnivores and omnivores
Pterosaurs	Aerial; secondarily terrestrial and aquatic	Filter feeders, predators, scavengers, mollusk-eaters, and omnivores
Birds	Aerial; secondarily terrestrial and aquatic	Small to very large carnivores, omnivores, and herbivores (see Chapters 21 and 22)

19.3 Lepidosauromorphs: Marine Diapsids

Extant lepidosaurs (lizards, snakes, and the tuatara) are terrestrial, but during the Mesozoic lepidosauromorphs dominated the seas.

Terrestrial lepidosauromorphs

Rhynchocephalia (Greek *rhynchos*, "beak"; *kephale*, "head") was the only lineage of lepidosauromorphs that radiated extensively on land.[1] Triassic rhynchocephalians were small (15–35 cm), light-bodied insectivores with small conical teeth. These early rhynchocephalians were lizardlike in body form and probably in ecology as well. Rhynchocephalians from the Jurassic and Early Cretaceous were up to 1.5 m long, and variation in their teeth indicates that they had radiated into new feeding niches: there were bladelike teeth for cutting and slashing, long pointed teeth for capturing fish, broad teeth for crushing armored prey, grinding teeth for shredding plants, and a remarkable species with jaws lined by massive, continuously growing tooth plates.

Although the tuatara is a member of an ancient lineage, the genus *Sphenodon* is no more than 100,000 years old. Except for reestablishment of a complete lower temporal arch, the tuatara is morphologically similar to some Mesozoic rhynchocephalians, but at the molecular level the tuatara has the highest rate of evolution known among vertebrates.

Lizards and snakes radiated during the Mesozoic, and some snakes grew large enough to prey on dinosaur eggs and hatchlings, but these lineages have reached their greatest diversity in the Cenozoic.

Marine lepidosauromorphs

Lepidosauromorphs were well represented in the complex marine ecosystems of the Early Triassic and included coastal herbivores as well as both coastal and pelagic predators.

Ichthyosaurs   Ichthyosaurs probably are the most familiar of the Mesozoic aquatic diapsids. Basal ichthyosaurs retained the lizardlike body form of their terrestrial ancestors, but the derived ichthyosaurs that lived during the Jurassic were more streamlined (**Figure 19.4A,B**). Ichthyosaurs had a hypocercal tail with the vertebral column bending sharply downward into the ventral lobe of the caudal fin. The upper lobe was formed of stiff tissue, and the lobes were nearly symmetrical. Ichthyosaurs also had a dorsal fin that was supported only by stiff tissue, not by a bony skeleton, and the paddles were extended by skin and connective tissue. We know about these soft tissues because ichthyosaur fossils in fine-grained sediments near Holzmaden in southern Germany contain an outline of the entire body preserved as a carbonaceous film.

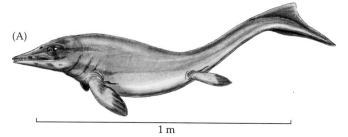

(A)

1 m

(B)

1 m

Figure 19.4 Evolutionary changes in ichthyosaurs.   (A) Basal ichthyosaurs, such as the Early Triassic form *Grippia longirostris*, probably swam with undulations of the body. (B) Later ichthyosaurs, represented by *Temnodontosaurus eurycephalus* from the Late Triassic, had a sharklike body form and swam with lateral undulations of the caudal fin. (Dmitry Bogdanov/CC BY 3.0.)

Ichthyosaurs had both forelimbs and hindlimbs, unlike extant cetaceans (whales and porpoises), which retain only the forelimbs. The limbs of ichthyosaurs were modified into paddles by both **hyperdactyly** (addition of extra fingers and toes) and **hyperphalangy** (addition of extra bones lengthening the fingers and toes).

The streamlining of later forms may have been associated with the development of carangiform locomotion (see Chapter 8) and rapid pursuit of prey. The Jurassic was the high point of ichthyosaur diversity. Ichthyosaurs were less abundant in the Early Cretaceous, and only a single family remained by the Late Cretaceous. Ichthyosaurs disappeared 30 million years before the end-Cretaceous extinction, perhaps victims of the global warming that characterized the Late Cretaceous.

Fossil ichthyosaurs with embryos in the body cavity indicate that these animals were viviparous. Some fossils may be those of females that died in the process of giving birth, as they preserve the young emerging tail-first, as in extant cetaceans and sirenians (manatees and sea cows). The terrestrial ancestors of ichthyosaurs appear to have been viviparous, but tail-first birth is a derived character associated with marine life.

Most ichthyosaurs had large heads with long, pointed jaws that were armed with sharp teeth. Stomach contents preserved in some ichthyosaur specimens include cephalopods and fishes, and the remains of a hatchling sea turtle and a bird have been found in fossil ichthyosaurs from the Cretaceous. The aptly named *Thalattoarchon saurophagis* (Greek *thallasa*, "sea"; *archon*, " ruler"; *phago*, "eat") was nearly 9 m long with a skull estimated to have been more than 1.2 m

---

[1]Two lineages of diapsids with beaklike snouts have confusingly similar names. Rhynchocephalia ("beak heads," with one extant species, the tuatara) are lepidosauromorphs, whereas the extinct Rhynchosauria ("beak lizards") were archosauromorphs.

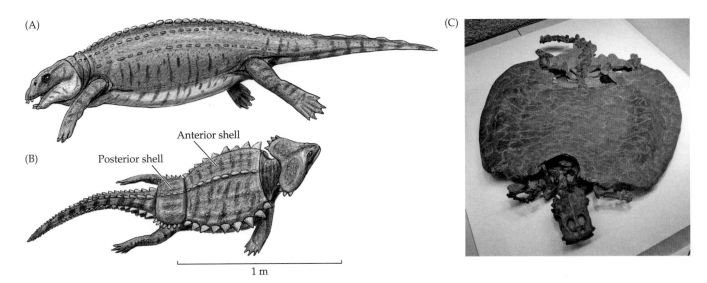

Figure 19.5 **Placodonts were coastal species.** (A) *Placodus* was relatively unspecialized and was probably herbivorous. *Cyamodus* (B) and (C) *Henodus* probably ate mollusks. Both of these species had dorsal shells composed of bony elements. *Cyamodus* had a two-part shell; the anterior portion extended from the neck to the hips and the posterior portion covered the hips and the base of the tail. (A,B, Dmitry Bogdanov/CC BY 4.0; C, Ghedoghedo/CC BY-SA 3.0.)

long and with bladelike teeth with serrations on both edges, and was an apex predator during the Middle Triassic.

Species of the Triassic genus *Shastasaurus* had short snouts and were toothless; they lacked the characters of the hyobranchial apparatus that characterizes suction feeders. Instead, they may have been ram-feeders, engulfing prey by swimming over it.

Derived ichthyosaurs had very large eyeballs that were supported by a ring of sclerotic bones. A *Temnodontosaurus* 9 m long had an eyeball with a diameter of 253 mm—roughly equivalent to an eye diameter of 50 mm for a human. These ichthyosaurs are believed to have hunted at great depths—500 m or more—and to have detected light emitted by the photophores of their prey. Traces of melanin in the skin of these deep-diving ichthyosaurs suggest that they were a uniform dark color, without the countershading (dark dorsal surface, light ventral surface) that is characteristic of most aquatic vertebrates.

Placodonts The Triassic placodonts were the least specialized of the marine lepidosauromorphs, retaining many characteristics of their terrestrial ancestors. They lived in shallow water habitats in the Tethys Sea (the body of water that spread from east to west as Laurasia and Gondwana separated during the Mesozoic). Basal forms, such as *Placodus*, were stocky with short legs and land feet modified as paddles. The tail was not laterally flattened and was probably not used in swimming. Broadened gastralia (bones in the ventral abdominal wall; see Chapter 14) covered the ventral surface of placodonts, and the dorsal surface of some species was covered by polygonal bony plates (**Figure 19.5**).

Placodonts had forward-projecting anterior teeth, large flat maxillary teeth, and enormous teeth on the palatine bones. The anterior teeth of fossils of *Placodus* are heavily worn, but the posterior teeth show little wear, and some scientists have proposed that *Placodus* was an herbivore—the Mesozoic equivalent of modern sea cows, which use their anterior teeth to dislodge seagrass from the seafloor, and which have greatly reduced posterior teeth.

Nothosaurs and pachypleurosaurs Nothosaurs (0.2–1 m) and pachypleurosaurs (1–4 m) were elongated marine predators with laterally flattened tails and slender jaws armed with short, pointed teeth. These characters suggest that nothosaurs and pachypleurosaurs were pursuit predators, akin to modern-day long-snouted dolphins, that chased and captured fish.

Plesiosauroids Plesiosauroids appeared in the Late Triassic and persisted to the very end of the Cretaceous. Basal forms had slightly elongated necks, with heads were proportional to their body size. Two ecological specializations—pliosaurs and plesiosaurs—were represented among derived plesiosauroids. Pliosaurs had long skulls (more than 3 m in some forms) and short necks (about 13 cervical vertebrae), whereas plesiosaurs had small skulls and exceedingly long necks with 32–76 vertebrae (**Figure 19.6**). The necks of plesiosaurs could bend downward, but not readily upward or to the side.

Both types of plesiosauroids had heavy, rigid trunks and appear to have rowed through the water with limbs that acted like oars and may also have served as hydrofoils, increasing the efficiency of swimming. Hyperphalangy increased the size of the paddles, and some plesiosaurs had as many as 17 phalanges per digit. In both types of plesiosauroids, the nostrils were located high on the head just in front of the eyes.

Pliosaurs developed an increasingly streamlined body form during their evolution as the neck became shorter and the paddles larger, whereas plesiosaurs became less

(A)

1 m

(B)

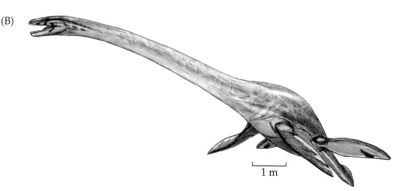

1 m

Figure 19.6 Pliosaurs and plesiosaurs were pelagic predators. Two radically different forms of plesiosauroids had evolved by the Jurassic. (A) Pliosaurs such as *Polyptychodon* were short-necked and fast-swimming. (B) Long-necked plesiosaurs such as *Elasmosaurus* were probably slow swimmers. (A, Dmitry Bogdanov/CC BY 3.0; B, Dmitry Bogdanov/CC BY-SA 4.0.)

streamlined as their necks lengthened and the paddles became smaller in proportion to body size. Pliosaurs were probably speedy swimmers that might have captured swimming cephalopod mollusks and fish by pursuing them the way seals and sea lions hunt their prey; the large species may have preyed on other marine diapsids. The downward-bending neck suggests that plesiosaurs may have fed on bottom-dwelling prey.

Plesiosauroids were viviparous. A fossil of the Late Cretaceous plesiosaur *Polycotylus latippinus* contains a single embryo, not yet at full term but already 32% of the body length of its mother. That is remarkably large, and the embryo's probable size at birth, based on the extent of ossification, was 35–50% of its mother's length. Other marine diapsids were viviparous, but those species produced litters consisting of several juveniles, each only 15–30% of the mother's body length. Production of a single, large offspring appears to have been unique to plesiosaurs among Mesozoic diapsids.

Mosasaurs Mosasaurs were a Late Cretaceous radiation of varanid lizards into the shallow epicontinental seas that spread across North America and Europe at that time. Early mosasaurs, such as *Aigialosaurus dalmaticus* from Texas, had

body proportions like those of modern varanid lizards, and their limbs ended in digits (**Figure 19.7A**). They probably swam with lateral undulations of the body and tail. The limbs of more derived mosasaurs had elongated digits connected by webbing that converted them into paddles, but their limbs were small and probably played little role in locomotion, unlike the much larger paddles of plesiosauroids. The most derived mosasaurs, including *Tylosaurus poriger*, had a hypocercal tail (**Figure 19.7B**). These forms probably used a modified carangiform swimming mode and would have been faster than the earlier forms.

The earliest mosasaurs were about 1 m long, but they increased in size during their evolutionary history, and the largest species reached lengths of 17 m or more. Examination of their teeth suggests that mosasaurs had radiated into a variety of adaptive zones. Their skulls were highly kinetic, an ancestral feature of varanid lizards, and the teeth of most species were sharp and conical—effective for seizing and holding prey. Mosasaurs probably ate whatever they were able to catch, and we know something of their diets from fossils that were preserved with the animals' last meals still intact. A single specimen contains bones from a fish, a smaller mosasaur, a flightless seabird, and possibly a shark. A turtle and a plesiosaur have been found in the stomachs of other mosasaurs. Fossil shells of ammonites (pelagic cephalopod mollusks) with marks that appear to match the shape and arrangement of mosasaur teeth have

(A)

(B)

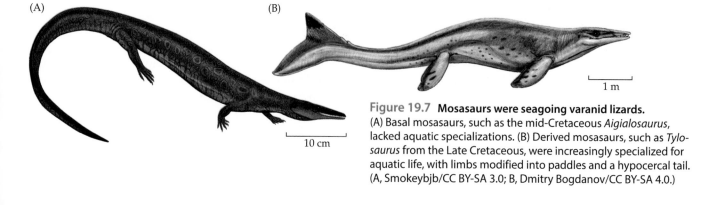

10 cm

1 m

Figure 19.7 Mosasaurs were seagoing varanid lizards. (A) Basal mosasaurs, such as the mid-Cretaceous *Aigialosaurus*, lacked aquatic specializations. (B) Derived mosasaurs, such as *Tylosaurus* from the Late Cretaceous, were increasingly specialized for aquatic life, with limbs modified into paddles and a hypocercal tail. (A, Smokeybjb/CC BY-SA 3.0; B, Dmitry Bogdanov/CC BY-SA 4.0.)

been described, and a fossil of *Globidens*, a specialized mosasaur with massive blunt teeth, has the remains of clam shells in its abdomen.

Mosasaurs were viviparous, giving birth to litters of four or five young, each about 15% of the body length of the mother. A fossil of *Carsosaurus marchesetti*, a mid-Cretaceous mosasaur, contains at least four advanced embryos. The orientation of the embryos suggests that they were born tail-first, like ichthyosaurs and extant cetaceans and sirenians, reducing the risk that the babies would drown before they had fully emerged.

19.4 Metriorhynchid Crocodylomorphs

Metriorhynchid crocodylomorphs constituted the only successful marine radiation by archosaurs during the Mesozoic. Metriorhynchids appeared in the Early Jurassic and persisted through the Early Cretaceous. Even the basal members of the group displayed a suite of adaptations to marine life, and the derived forms that lived in the Cretaceous were highly specialized (**Figure 19.8**). Their heads were streamlined, and air spaces in the bones made their skulls light, probably allowing them to float at the surface with only their heads protruding. The limbs of metriorhynchids were paddlelike, and the posteriormost vertebrae turned sharply downward to create a hypocercal tail with an upper lobe supported by only stiff tissues, as in derived ichthyosaurs and mosasaurs.

Most metriorhynchids were probably fish eaters, but species in the Cretaceous genera *Dakosaurus* and *Geosaurus* were probably apex predators that could attack and kill prey larger than themselves. *Dakosaurus* had bone-crunching jaws that could exert enormous force, whereas the teeth in the upper and lower jaws of *Geosaurus* formed a pair of blades that could slice through tissue. In addition, species

(A)

(B)

|_____|
        1 m

Figure 19.8 Metriorhynchids were marine crocodylomorphs. (A) *Geosaurus*, from the Cretaceous, had a short snout, suggesting that it fed on other large marine reptiles rather than on fish. (B) The slender snout of *Metriorhynchus superciliosus*, a Late Jurassic species, suggests that it fed on fish. Both species were about 3 m long. (A, Dmitry Bogdanov/CC BY 2.5; B, Dmitry Bogdanov/CC BY 3.0.)

in both genera probably employed the crocodylian death roll—seizing prey and then rotating rapidly around their own long axis, twisting off portions of the prey.

19.5 Pterosaurs: The First Flying Vertebrates

Archosaurs gave rise to two independent radiations of fliers: pterosaurs and birds. Pterosaurs first appeared in the Late Triassic, some 80 million years before the earliest birds. Pterosaurs persisted until the end of the Cretaceous and diversified into the major adaptive zones occupied by extant birds (**Figure 19.9**).

"Rhamphorhynchoids" were basal pterosaurs (probably a paraphyletic grouping) that retained a long tail stiffened by bony projections extending anteriorly, overlapping half a dozen vertebrae and preventing the tail from bending horizontally or dorsoventrally. A leaf-shaped expansion at the end of the tail may have acted as a rudder. Derived (pterodactyloid) pterosaurs, which were larger than rhamphorhynchoids and lacked tails and teeth, appeared in the Middle Jurassic and persisted until the end of the Cretaceous.

The structure of pterosaurs

The mechanical demands of flight are reflected in the structure of flying vertebrates, and it is not surprising that pterosaurs and birds show a high degree of convergent evolution. In derived pterosaurs, the teeth were reduced or entirely absent; the tail was lost; the sternum developed a keel that was the origin of flight muscles; the thoracic vertebrae became fused into a rigid structure; the bones were thin-walled; **postcranial pneumatization** (the development of open spaces in bone) was extensive; and the eyes, the parts of the brain associated with vision, and the cerebellum (which is concerned with balance) were large while the olfactory areas were small. Pterosaurs probably had a unidirectional flow of air through the lungs.

Limbs and locomotion   The limbs of pterosaurs were large in relation to the trunk and abdomen. The wings were formed by skin stiffened by internal fibers and were entirely different from the feathered wings of birds. The primary wing of pterosaurs, the **brachiopatagium** (arm wing), was supported anteriorly by the upper and lower arm bones and an extremely elongated fourth finger. A small splintlike bone, the pteroid, was attached to the front edge of the fourth finger and may have supported a membrane that ran forward to the neck. The brachiopatagium extended posteriorly to the hindlegs.

A posterior wing, the **uropatagium** (tail wing), provided additional lift at the rear of the body. The uropatagium of rhamphorhynchoid pterosaurs extended between the hindlegs, limiting their capacity for independent movement. In pterodactyloids, the uropatagium was reduced, extending from the pelvis to the ankle of each leg and allowing the hindlimbs to move independently.

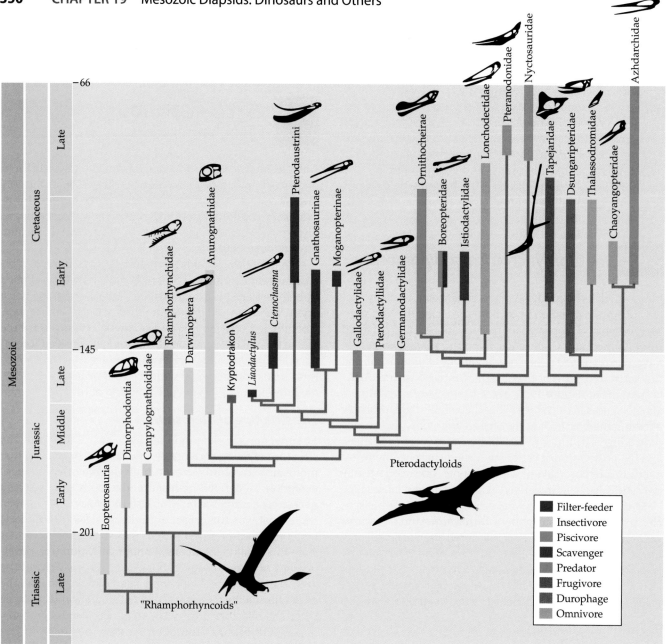

Figure 19.9 Pterosaurs filled the adaptive zones for flying vertebrates that birds occupy today. Rhamphorhynchoids included insectivores and piscivores. Morphological and ecological diversity expanded in the Middle Jurassic with the appearance of pterodactyloids, which included filter feeders, predators, scavengers, frugivores, durophagous mollusk eaters, and omnivores. (Modified from Zhou et al. 2017.)

Pterosaurs walked quadrupedally (Figure 19.10). The uropatagium of ramphorhynchoids probably required them to move the hindlegs together in a hopping gait, but pterodactyloids were able to stride. Trackways of walking pterodactyloids have been found in Jurassic and Cretaceous sediments on most continents. No tracks of ramphorhynchoids have been discovered, and basal pterosaurs may have climbed on trees and cliffs instead of walking across flat surfaces.

Aerodynamic tests and modeling indicate that all pterosaurs could fly, despite the fact that the largest species had

wingspans of 10 m and weighed 200 kg or more. The long, narrow wings of large species of pterosaurs were similar to those of birds such as albatrosses that glide at high speeds for great distances. The smallest pterosaur, the pterodactyloid *Nemicolopterus crypticus*, was a forest-dweller with a wingspan of only 25 cm. It probably had short, broad wings like those of birds that fly through woodlands.

How a pterosaur took flight has long been a subject of debate. Through most of the 20th century, reconstructions of pterosaurs showed them posed on cliffs, reflecting a belief that launching from a height was the only way a pterosaur could get into the air. More recently, a bipedal running takeoff was proposed, but both mechanical and aerodynamic arguments have discredited that hypothesis. The most likely method is a quadrupedal leap, which would have allowed the wing and hindlimb muscles to act simultaneously to produce high launch speeds and

Figure 19.10 Pterosaurs walked quadrupedally. A reconstruction of a walking *Pterodactylus* shows the erect stance and limb movement of a pterodactyloid pterosaur. Only the first three fingers of the forefeet touched the ground; the long fourth finger that supported the brachiopatagium was apparently folded against the body. The wrist was rotated outward, so impressions of the fingers of the forefoot are directed to the side and backward. The toes and soles of the hindfeet were in contact with the ground. (Copyright © Mark Witton.)

enough height to prevent the first wingbeats from striking the ground.

Jaws and teeth   Variation in body size and in the jaws and teeth of pterosaurs appears to have corresponded to differences in diet, although direct evidence of the diets of pterosaurs—such as fossilized food in the stomach—is scarce. Small species with short jaws containing numerous small, sharp teeth are believed to have been insectivorous, whereas long jaws with pointed teeth may have been characteristic of piscivores. Several lineages of pterodactyloids

in the Jurassic and Cretaceous that combined long jaws with extremely long, closely set teeth are believed to have been filter feeders. Dsungaripterids had long, narrow jaws with no anterior teeth; they may have used a pincerlike motion to pluck mollusks from rocks at low tide, then crushed them with broad, flat teeth at the rear of the jaws. Tapejarids had deep skulls with toothless beaks and may have been frugivorous, while istiodactylids had broad snouts with interlocking lancet-shaped teeth and were probably scavengers.

Azhdarchids (Uzbek *azhdarkho*, "a mythical dragon") may have been the most remarkable pterosaurs. Most species in this Cretaceous lineage of derived pterydacyloids were very large; *Arambourgiania philadelphiae* could have stood eye-to-eye with a giraffe (**Figure 19.11**).

Azhdarchids were terrestrial stalkers that foraged quadrupedally, using their long necks and beaks to seize prey. Described relatively recently, azhdarchids are turning out to be more diverse than was originally realized. The smallest species yet discovered was about the size of a house cat and had a wingspan of about 1.5 m, whereas the heads of the largest species were 4 m above the ground when they stood erect, and their wingspans were 10–11 m. Azhdarchids have no extant analogues, but in behavior and ecology they probably resembled ground hornbills and marabou storks, except that they were much larger than the birds.

Two general types of giant azhdarchids have been distinguished: gracile and robust. *Arambourgiania philadelphiae* is representative of gracile azhdarchids, which had long necks and relatively light skulls. *Hatzegopteryx thambema*, a robust form, had a shorter neck than *A. philadelphiae* but a larger, heavier bill. *A. philadelphiae* may have preyed on relatively small animals—up to the size of a human—whereas *H. thambema* was probably an apex predator, capable of killing dinosaurs. Indeed, it may be significant that fossils of large predatory dinosaurs have not been found in the deposits that contain azhdarchid pterosaurs.

Body covering and crests   Fossils of *Sordes pilosus* and *Jeholopterus ningchengensis* from fine-grained sediments show that their skin was covered by fine, hairlike fuzz called **pycnofibers** that probably provided insulation. A fuzzy body covering was probably widespread among pterosaurs; in addition to providing insulation, pycnofibers were probably colorful, forming patterns used for species and sex recognition.

Figure 19.11 The largest azhdarchid pterosaurs were apex predators. *Arambourgiania philadelphiae* (center) could have confronted an adult giraffe eye-to-eye. *Hatzegopteryx thambema* (right) was not quite as tall as *A. philadelphiae*, but it was more robust. Both of these pterosaurs were large enough to have preyed on animals the size of humans, and *H. thambema* may have preyed on dinosaurs. (Copyright © Mark Witton.)

Many pterosaurs had crests on their heads. These crests appeared early in the evolution of pterosaurs and included an enormous variety of sizes and shapes. Crests were formed by bones alone, by bones with soft tissue, or by soft tissue alone. The crests grew allometrically, becoming larger as an individual matured, and in many species they were sexually dimorphic—large in males and small or absent in females. Both of these characteristics suggest that the crests were used during intraspecific interactions, such as courtship and territorial disputes. Tests with models of pterosaurs in wind tunnels indicate that even large crests had little aerodynamic effect.

Reproduction, eggs, and parental care

Pterosaurs were oviparous, like all other archosaurs, but their eggs had flexible shells like those of many squamates, instead of the rigid shells that characterize the eggs of crocodylians, nonavian dinosaurs, and birds. The eggs were small in relation to the size of the adults. Some species of pterosaurs hatched at an advanced stage of development and, like some species of brush turkeys (megapode birds; see Chapter 22), were probably able to run and fly soon after emerging from the nest.

Did the evolution of birds doom the pterosaurs?

Pterosaurs had diversified widely by the Late Jurassic when the first birds appeared, and they filled a large number of adaptive zones that appear to be occupied now by birds. Thus, it is tempting to propose that competition with birds drove pterosaurs to extinction, but available evidence does not clearly support that hypothesis. A comparison of the body forms of pterosaurs and birds indicates that they occupied statistically different morphospaces—that is, they had different combinations of anatomical features, and as a result they did not overlap in the ways they lived. Thus, this analysis does not support the hypothesis of competition between pterosaurs and birds.

However, the appearance of birds coincided with an increase in the wingspan of pterosaurs from an average of about 1.5 m to about 7 m, while the wingspans of birds decreased to between 10 cm and 1 m. Character displacement of that sort might result from competition, but the timing of the increase in body size of pterosaurs also corresponds to the appearance of pterodactyloids and the gradual disappearance of "rhamphorhynchoids."

If competitive replacement did occur, it was remarkably slow. Birds appeared in the Late Jurassic, and pterosaurs persisted until the end of the Cretaceous; thus, they lived side by side for nearly 100 million years.

19.6 Triassic Faunal Turnover

Although dinosaurs are the iconic organisms of the Mesozoic, they appeared only in the Middle Triassic and did not begin to diversify until the Late Triassic and Early Jurassic.

Figure 19.12 Three diapsid faunas radiated during the Triassic. ▶ From the Middle Permian to the Middle Triassic, carnivorous and herbivorous synapsids were the most abundant tetrapods in terrestrial habitats. In the Early Triassic, synapsids achieved a worldwide distribution of rabbit- to cow-size herbivores (dicynodonts, traversodontids) that were preyed on by fox- to lion-size carnivores (cynodonts). Diapsids replaced synapsids as the most diverse and abundant terrestrial amniotes in the Early and Middle Triassic. Rhynchosaurs were the first diapsids to radiate. Initially a minor component of terrestrial faunas, rhynchosaurs replaced synapsids as the most abundant herbivores in the Middle to Late Triassic. A variety of peudosuchians composed the major assemblage of carnivores, and the heavily armored aetosaurs were the first herbivorous archosaurs. Ornithodirans displaced pseudosuchians during the Middle and Late Triassic. The first dinosauromorphs appeared early in the Middle Triassic and diversified during the Triassic turnover, when climates shifted from damp to arid. By the Late Triassic, dinosauromorphs had replaced synapsids and rhynchosaurs as the most abundant herbivores. Carnivorous dinosaurs (theropods) were diverse and abundant in the Late Triassic. The end-Triassic mass extinction wiped out all pseudosuchians except for crocodyliforms, and the Jurassic saw a major expansion of herbivorous and carnivorous dinosaurs. (Modified from Benton et al. 2014.)

Two earlier radiations preceded the appearance of dinosaurs (Figure 19.12):

- From the Middle Permian to the Middle Triassic, synapsids were the most abundant tetrapods in terrestrial habitats. Cynodonts were carnivores that ranged from the size of foxes to wolves. Several lineages, including dicynodonts and traversodontids, were herbivores, ranging in size from a few kilograms to as much as 600 kg. Most of these lineages were extinct by the Late Triassic.

- Rhynchosaurs, basal archosauromorphs, were small herbivores in the Early Triassic, but by the Middle Triassic some species were up to 2 m long. They fed on tough vegetation with a shearing dentition and jaw action.

During the Middle Triassic, basal pseudosuchians diversified into predators (ornithosuchids, rauisuchians, and phytosaurs) and herbivores (aetosaurs). Herbivorous lineages of dinosaurs began to diversify in the Late Triassic. Initially the herbivorous sauropodomorph dinosaurs expanded, replacing herbivorous synapsids, aetosaurs, and rhynchosaurs. Carnivorous lineages of dinosaurs did not diversify until after phytosaurs, ornithosuchids, and rauisuchians disappeared during the end-Triassic mass extinction.

Several lineages of basal ornithodires radiated in the Triassic. These animals can be described as "almost dinosaurs" because they had some, but not all, of the derived characters of dinosaurs. Our understanding of relationships among these lineages and of these lineages to dinosaurs (Dinosauria) is in flux as new fossils are discovered and older specimens are reanalyzed. The basal lineages were diverse in body form and ecology: *Marasuchus lilloensis* was a bipedal carnivore, only 30–40 cm long; *Sacisaurus agudoensis*, *Silesaurus opolensis*, and *Asilisaurus kongwe* were lightly built quadrupedal omnivores or herbivores, 1.5–2.5 m long.

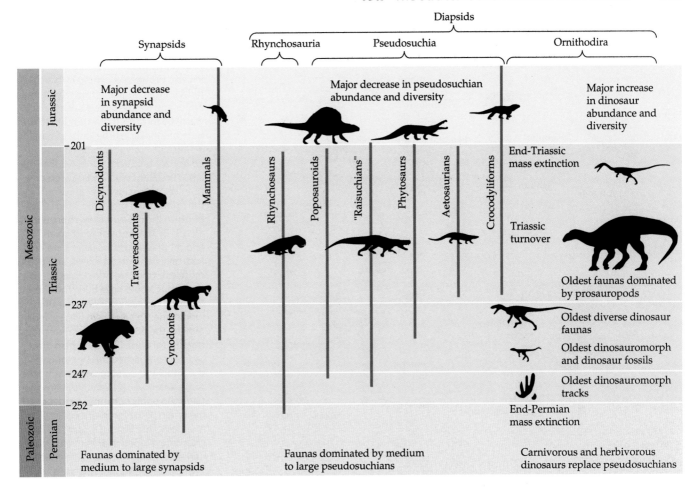

19.7 The Structure and Function of Dinosaurs

When most people hear the word "dinosaur," the image that comes to mind is a large animal—either a fearsome predator like *Tyrannosaurus rex* or an enormous, long-necked herbivore like *Apatosaurus*. However, the diversity of sizes and body forms of nonavian dinosaurs extends far beyond those two examples. The huge species were indeed spectacular, but there were also dinosaurs that were about the size of a chicken. In addition, traditional reconstructions of dinosaurs as a uniform gray or tan are probably also incorrect. Species that relied on camouflage to avoid detection by predators probably had colors and patterns that obscured their outlines, and it is likely that many feathered dinosaurs were as flamboyant and colorful as birds are today.

Mobility and social interaction were important components of the biology of many dinosaurs. Their mobility owed much to the structure of their hindlimbs, and their social interactions involved colors, movements, and sounds.

The skeletons of dinosaurs combined the strength required to support animals that weigh thousands or tens of thousands of kilograms with features that minimized the weight of the skeleton. Postcranial pneumatization was widespread in dinosaurs and pterosaurs; the open spaces in bone are believed to be the traces of air sacs like those in extant birds.

Hips and legs

Dinosaurs originated from small (about 1 m), agile terrestrial archosaurs. Some of these animals were bipedal and already had morphological changes in the hips that allowed the hindlimbs to be held beneath the body in an erect stance. In addition, the articulation of the ankle joint had been simplified so that it formed a straight-line hinge (mesotarsal joint; see Figure 19.3) rather than the complex articulation of several bones that characterized basal archosaurs. The novel ankle structure allowed the hindfeet to thrust backward forcefully without twisting. These two changes set the stage for the increase in body size that was so prominent in dinosaurs, but they required additional changes in the pelvis, and these changes distinguish the ornithischian and saurischian lineages of dinosaurs.

Early tetrapods had a sprawling posture, with the humerus and femur projecting more or less horizontally and a sharp bend at the elbow and knee. This posture is sufficient for small and medium-size tetrapods, but it does not work for large animals. Bones are far more resistant to force exerted parallel to their long axis (compressive force) than to force exerted at an angle to the long axis (shearing force). Thus, large extant mammals, such as rhinoceroses and elephants, support their weight on vertical limbs. Many dinosaurs were far larger than elephants, and the limbs of dinosaurs were also oriented vertically.

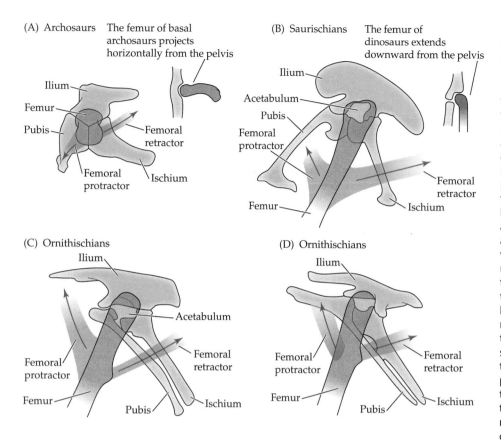

(A) Archosaurs   The femur of basal archosaurs projects horizontally from the pelvis

Ilium
Femur
Pubis
Femoral retractor
Femoral protractor
Ischium

(B) Saurischians   The femur of dinosaurs extends downward from the pelvis

Ilium
Acetabulum
Pubis
Femoral protractor
Femur
Femoral retractor
Ischium

(C) Ornithischians

Ilium
Acetabulum
Femoral protractor
Femoral retractor
Femur
Pubis
Ischium

(D) Ornithischians

Ilium
Femoral protractor
Femoral retractor
Femur
Pubis
Ischium

Figure 19.13 **The pelvic girdles of dinosaurs combined upright stance with long strides.** The femur of the leg was protracted (swung forward) by muscles that originated anterior to the femur and was retracted (swung backward) by muscles that extended to the base of the tail. (A) The limbs of basal archosaurs such as *Euparkeria* extended horizontally from the pelvis, like the limbs of extant crocodylians. Femoral protractor muscles originating on the pubis could produce a long stride because the leg swung in a horizontal arc. (B–D) The limbs of saurischian and ornithischian dinosaurs were held vertically beneath the pelvis, and the muscles extending from the pubis to the femur would have been too short to give a vertically oriented femur a long stride. This problem was solved by moving the origin of the femoral protractors forward. (B) Saurischian dinosaurs accomplished this by elongating the pubis and ischium and rotating the pubis anteriorly. (C) Ornithischians rotated the pubis posteriorly and moved the origin of the femoral protractor muscles to the ilium. (D) Some ornithischians developed a forward extension of the pubis (the prepubis) on which the femoral protractors inserted.

Changing the angle of the limbs solved the problem of weight-bearing but introduced a new problem. Among early tetrapods, muscles originating on the pubis and inserting on the femur protracted the leg (moved it forward), and muscles originating on the tail retracted the femur (moved it backward). The ancestral tetrapod pelvis, little changed from *Ichthyostega* through basal archosauromorphs, was platelike (**Figure 19.13A**). The ilium articulated with one or two sacral vertebrae, and the pubis and ischium did not extend far anterior or posterior to the hip socket (acetabulum).

Because these animals had a sprawling posture, the femur projected horizontally and the pubofemoral muscles extended outward from the pelvis to insert on the femur. Thus, they were long enough to swing the femur through a large arc relative to the ground. But moving the legs under the body made those muscles shorter and less effective in moving the femur, because a muscle's maximum contraction is about 30% of its resting length. The shorter muscles would have swung the vertical femur through a smaller arc and reduced the stride to a shuffle.

Dinosaurs solved that problem by moving the origins of the muscles to make them longer and combine upright stance with long strides. The saurischian (lizard-hipped) and ornithischian (bird-hipped) dinosaurs did that in different ways. In saurischians (**Figure 19.13B**), the pubis and ischium both became elongated and the pubis was rotated anteriorly, so that the pubofemoral muscles ran back from the pubis to the femur and were long enough to protract it. Ornithischians (Figure 19.13C) rotated the pubis posteriorly to lie against the ischium and pubofemoral muscles

originated on the ilium. Derived ornithischians (**Figure 19.13D**) developed an anterior projection of the pubis (the prepubis) that extended beyond the anterior part of the ilium, providing an anterior origin for protractor muscles.

Although these changes were anatomically different, they were functionally the same—both produced hip articulations that allowed the legs to be held vertically beneath the pelvis to support heavy bodies with stride lengths that enabled dinosaurs to move rapidly and over long distances.

Dinosaur lineages

An estimated 1,500 to 2,500 species of dinosaurs roamed Earth during the Mesozoic, and a new species is named about every two weeks. Since the 1980s, Dinosauria has been considered to be a monophyletic lineage containing two monophyletic sister lineages, the bird-hipped dinosaurs (Ornithischia, Greek *ornis*, "bird"; *ischion*, "hip joint") and the lizard-hipped dinosaurs (Saurischia) (**Figure 19.14**).

- All ornithischian dinosaurs were herbivores. They included hadrosaurs (duck-billed dinosaurs), stegosaurs, and ceratopsians (horned dinosaurs), among others. Ornithischians reached their greatest diversity in the Late Cretaceous.

- Saurischian dinosaurs include two lineages: Sauropoda (Greek *pous* "foot") were long-necked, long-tailed

(A)

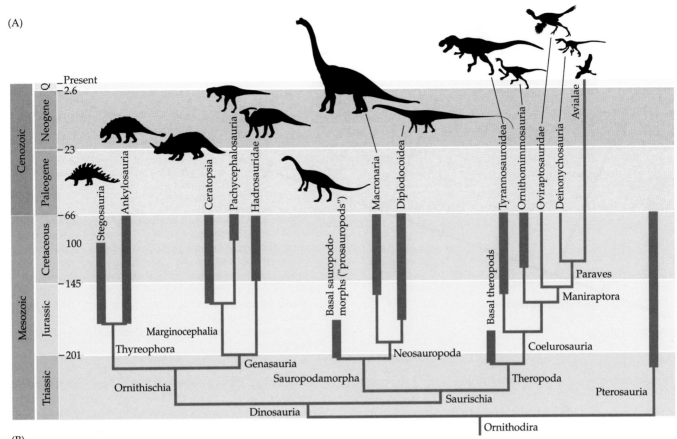

(B)

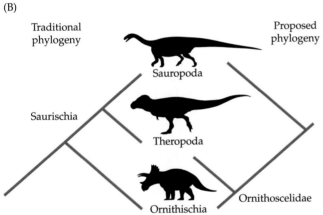

Traditional phylogeny

Proposed phylogeny

Saurischia

Sauropoda

Theropoda

Ornithischia

Ornithoscelidae

Figure 19.14 Phylogenetic relationships of Dinosauria. (A) This tree depicts the probable relationships among the major groups of dinosaurs, including birds. The lines show interrelationships only; they do not indicate times of divergence or the unrecorded presence of taxa in the fossil record. Only the best-corroborated relationships are shown. (B) In 2017, a major revision of dinosaur phylogeny was proposed that unites Ornithischia and Theropoda as terminal sister groups of a new lineage, Ornithoscelidae, that is the sister grap of the remaining Saurischia (see Baron et al. reference in Additional Readings). Extensive discussions of this hypothesis can confidently be anticipated. (A, data from Fastovsky and Weishampel 2016.)

herbivores that reached their maximum diversity in the Late Jurassic; Theropoda (Greek *ther,* "wild beast") were carnivores and reached their maximum diversity in the Late Cretaceous.

19.8 Ornithischian Dinosaurs

Body forms of ornithischians were diverse, and the elaborate crests, frills, and horns that decorated the heads of many species suggest that ornithischians had complex social behaviors. Many species probably lived in family groups or formed herds, and some species nested in colonies.

All ornithischians were herbivores, feeding on low-growing vegetation (mostly within 2 m of the ground). All had horny beaks, and all processed food orally—that is, they chewed. Chewing (reducing food to a pulp in the mouth before swallowing) is the norm for mammals, but outside of mammals, chewing is rare; most vertebrates either swallow prey whole or tear off and swallow chunks of their food.

Three major lineages of ornithischian dinosaurs are distinguished: the armored dinosaurs (Thyreophora), the horned dinosaurs (Marginocephalia), and the duck-billed dinosaurs (Ornithopoda).

Thyreophora

Thyreophorans (Greek *thyreos,* "shield"; *phoros,* "a bearer") were the armored dinosaurs (**Figure 19.15**). Their name refers to the parallel rows of **osteoderms** (bones embedded in the skin) that extended from the neck to the tail, armoring the back. The two major lineages, Stegosauria and Ankylosauria, were quadrupedal, although some basal forms were bipedal.

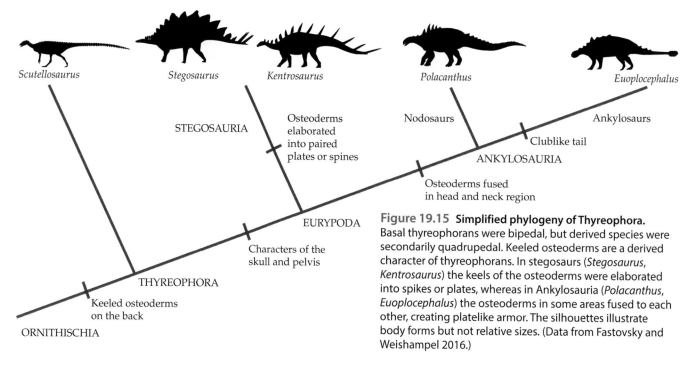

Figure 19.15 **Simplified phylogeny of Thyreophora.** Basal thyreophorans were bipedal, but derived species were secondarily quadrupedal. Keeled osteoderms are a derived character of thyreophorans. In stegosaurs (*Stegosaurus*, *Kentrosaurus*) the keels of the osteoderms were elaborated into spikes or plates, whereas in Ankylosauria (*Polacanthus*, *Euoplocephalus*) the osteoderms in some areas fused to each other, creating platelike armor. The silhouettes illustrate body forms but not relative sizes. (Data from Fastovsky and Weishampel 2016.)

Stegosauria   Stegosaurs (Greek *stegos*, "roof") were most abundant in the Late Jurassic, although some species persisted to the end of the Cretaceous. Stegosaurs were medium-size to large—up to 9 m—quadrupedal herbivores with small heads, short and sturdy forelegs, and long, columnar hindlegs. The short forelegs kept the head close to the ground, and stegosaurs may have browsed on ferns, cycads, and other low-growing plants. The skull was small for such a large animal and had a horny beak at the front of the jaws. The teeth were shaped like those of extant herbivorous lizards, and show none of the specializations seen in some other ornithischians, which appear to have been able to grind or cut plant material into small pieces that could be digested efficiently. Stegosaurs may have eaten large quantities of food without much chewing and used stone **gastroliths** (Greek *gastros*, "stomach"; *lithos*, "stone") in a muscular gizzard to pulverize plant material.

The distinctive feature of stegosaurs was a double row of spines or leaf-shaped plates that extended along the vertebral column. In addition, all stegosaurs (except for *Stegosaurus*) had a long spine that projected upward and backward from the shoulder. These spines and plates were hypertrophied keels of the osteoderms that are characteristic of thyreophorans.

The function of the plates of *Stegosaurus* has been a matter of contention for decades. Initially they were assumed to have provided protection from predators, and some reconstructions have shown the plates lying flat against the sides of the body as shields. A defensive function is not very convincing, however. Whether the plates were erect or flat, they left large areas on the sides of the body and the belly unprotected.

A role for the plates as heat exchangers initially seemed plausible. Grooves on the surface of the plates were interpreted as channels for blood vessels that could carry a large flow of blood to be warmed or cooled according to the needs of the animal. However, a reanalysis of the microstructure of the plates concluded that the channels merely carried nutrients to the overlying sheath that covered the plates, and would have had little, if any, role in thermoregulation.

The combinations of plates and spines were species-specific and dimorphic within a species. These characters suggest that the plates and spines identified species and sex during intraspecific behavioral interactions. The plates and spines increased the apparent size of a stegosaur when viewed from the side, and may have deterred both rivals and potential predators.

The tails of stegosaurs had large spines near their tips. These spines were almost certainly defensive structures, and a biomechanical analysis of the whipping motion of the tail of *Kentrosaurus* supports that interpretation. Using the tail muscle of an alligator as the basis for calculation, this study concluded that the tail tip of an adult *Kentrosaurus* could strike a predator at a velocity of 20–40 m/sec. At those speeds, the tip of a spine would have exerted enough force to penetrate skin and muscle to a depth of 30 cm or to pierce bony armor.

Ankylosauria   Ankylosaurs (Greek *ankylos*, "stiffening") were a group of heavily armored dinosaurs found in Jurassic and Cretaceous deposits in North America and Eurasia; the timing of their appearance may have been linked to the increase in size of carnivorous dinosaurs at that time. Ankylosaurs were quadrupedal ornithischians that ranged from 2 to 6 m long. They had short legs and broad bodies, with osteoderms that were fused on the neck, back, hips, and tail to produce large shieldlike pieces. Bony plates also covered the skull and jaws, and in *Euoplocephalus* even the eyelids had bony armor.

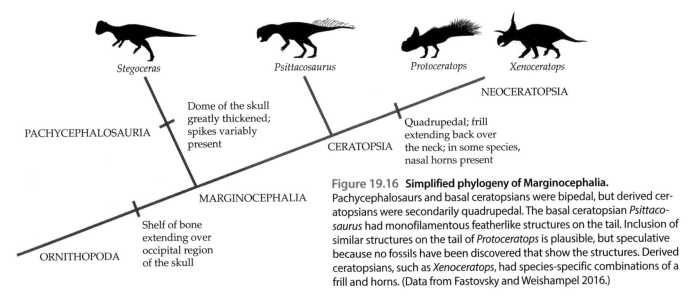

Dome of the skull
greatly thickened;
spikes variably
present

NEOCERATOPSIA

Quadrupedal; frill
extending back over
the neck; in some species,
nasal horns present

CERATOPSIA

MARGINOCEPHALIA

Shelf of bone
extending over
occipital region
of the skull

ORNITHOPODA

Figure 19.16 Simplified phylogeny of Marginocephalia.
Pachycephalosaurs and basal ceratopsians were bipedal, but derived ceratopsians were secondarily quadrupedal. The basal ceratopsian *Psittacosaurus* had monofilamentous featherlike structures on the tail. Inclusion of similar structures on the tail of *Protoceratops* is plausible, but speculative because no fossils have been discovered that show the structures. Derived ceratopsians, such as *Xenoceratops*, had species-specific combinations of a frill and horns. (Data from Fastovsky and Weishampel 2016.)

Ankylosaurs must have been difficult animals to attack. Some species of nodosaurids had spines projecting from the back and sides of the body, while derived species of ankylosaurids had a lump of bone at the end of the tail that could be swung like a club. Even species of ankylosaurs without clubs or spines had broad, flat bodies with armored backs and sides, and the act of merely lying flat on the ground might have been an effective defensive tactic.

Marginocephalia

Marginocephalians (Latin *margo*, "border") take their name from the bony shelf at the rear of the skull. Two lineages of marginocephalians are distinguished: ceratopsians and pachycephalosaurs (**Figure 19.16**).

Ceratopsia Ceratopsians (Greek *keras*, " horn"; *opsis*, "appearance"), the horned dinosaurs, were the most diverse marginocephalians. Ceratopsians appeared in the Late Jurassic or Early Cretaceous. The distinctive features of ceratopsians were the frill over the neck, which was formed by an enlargement of the parietal and squamosal bones; a parrotlike beak; and teeth that were arranged in batteries in each jaw, where they formed a series of knifelike edges rather than a solid surface. The feeding method of ceratopsians probably consisted of shearing vegetation into short lengths.

Although ceratopsians were quadrupedal, they were derived from a bipedal ancestor. Early ceratopsians,

such as *Protoceratops*, were small (cat-size to sheep-size), had a simple frill, and lacked nasal horns (**Figure 19.17A**). About 90 Ma, a lineage of larger animals (rhinoceros-size to elephant-size) appeared sporting a more elaborate frill, long brow horns, and nasal horns. Two groups of these

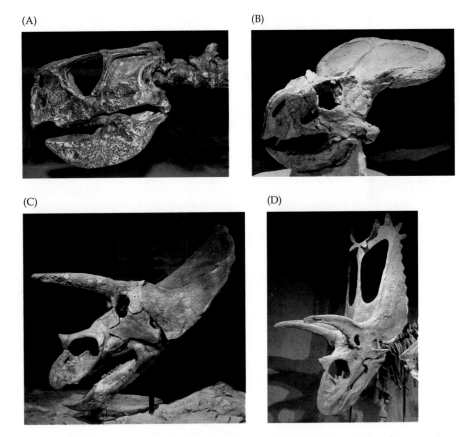

Figure 19.17 Elaboration of the frill and horns of ceratopsians. (A) Basal ceratopsians such as *Psittacosaurus* were largely bipedal. (B) The skulls of early ceratopsians such as *Protoceratops* had small frills and lacked horns. Later ceratopsians had both horns and more elaborate frills. (C) The frill of *Triceratops*, a short-frilled form, extended over the shoulders. (D) Long-frilled ceratopsians, such as *Titanoceratops*, had frills extended half the length of the trunk. (A, Matt Martyniuk(Dimoguy2)/CC BY-SA 4.0; B, Luis Miguel Bugallo Sanchez/CC BY-SA 3.0; C, Ed T/CC BY-SA 2.0; D, Kurt McKee/CC BY-SA 2.0.)

(A)   (B)   (C)

Figure 19.18 Pachycephalosaurs had greatly thickened skulls. (A) The bony dome on the skull of an adult male *Phrenocephale prenes* was as much as 25 cm thick. (B) The skull of *Pachycephalosaurus wyomingensis* combined a thick dome with short spikes. (C) The newly described *Dracorex hogwartsia* ("Dragon king of Hogwarts"), which had a skull adorned by many spikes, is controversial; some authorities believe that *Dracorex* is a juvenile *Pachycephalosaurus*. (A, Ghedoghedo/CC BY-SA 3.0; B, Balista/CC BY-SA 3.0; C, Ealdgyth/CC BY 3.0.)

derived ceratopsians can be distinguished. In short-frilled ceratopsians, such as *Triceratops*, the frill extended back over the neck, whereas in the long-frilled forms, such as *Titanoceratops*, the frill extended half the length of the trunk (**Figure 19.17C,D**).

Pachycephalosauria

Pachycephalosaurs were a group of small to medium-size (1–5 m) bipedal dinosaurs that lived during the Cretaceous. A distinctive feature of this lineage was a pronounced thickening of the skull roof (Greek *pachys*, "thick") in some species, and an array of bony projections on the skull of other species (**Figure 19.18**).

Ornithopoda

Ornithopods lived from the Middle Jurassic to the end of the Cretaceous (**Figure 19.19**). Most species were medium-size, although the largest species rivaled sauropods in

size. Small species were largely bipedal, and the largest species probably moved mostly on four legs.

Hadrosauridae   Hadrosaurs (Greek *hadros*, "bulky") were the last group of ornithopods to evolve, appearing in the mid-Cretaceous, and they were also the most speciose lineage. These were large animals, some reaching lengths of more than 15 m and weights greater than 13,000 kg. The anterior portion of the jaws was toothless and sheathed in a horny beak, and a remarkable battery of teeth occupied the rear of the jaws. On each side of the upper and lower jaws were four tooth rows, each containing about 40 teeth packed closely side by side to form a massive tooth plate. Several sets of replacement teeth lay beneath those in use, so a hadrosaur had several thousand teeth in its mouth, of which several hundred were in use simultaneously—perhaps the most advanced vertebrate chewing apparatus ever. Fossilized stomach contents of hadrosaurs consist of pine needles and twigs, seeds, and fruits of terrestrial plants.

Social behavior of ornithischian dinosaurs

Although it is impossible to know what a dinosaur was doing in the last moments before it died and embarked on the long trail to becoming a fossil, multiple discoveries of fossil beds that contain many individuals of the same species of ornithischian dinosaur support the hypothesis that they formed

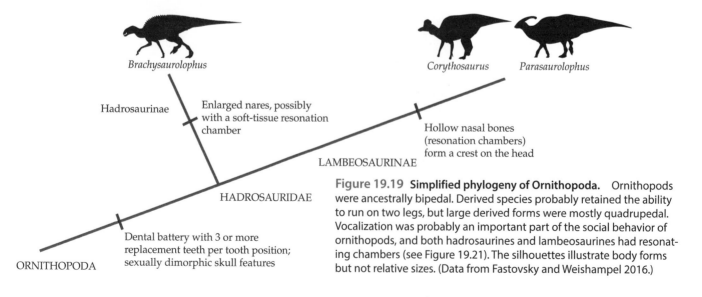

Brachysaurolophus   *Corythosaurus*   *Parasaurolophus*

Hadrosaurinae

Enlarged nares, possibly with a soft-tissue resonation chamber

Hollow nasal bones (resonation chambers) form a crest on the head

LAMBEOSAURINAE

HADROSAURIDAE

Dental battery with 3 or more replacement teeth per tooth position; sexually dimorphic skull features

ORNITHOPODA

Figure 19.19 Simplified phylogeny of Ornithopoda.   Ornithopods were ancestrally bipedal. Derived species probably retained the ability to run on two legs, but large derived forms were mostly quadrupedal. Vocalization was probably an important part of the social behavior of ornithopods, and both hadrosaurines and lambeosaurines had resonating chambers (see Figure 19.21). The silhouettes illustrate body forms but not relative sizes. (Data from Fastovsky and Weishampel 2016.)

herds. In many cases, the material surrounding the fossils represents a sudden fatal event such as a lahar (volcanic mudflow) or deep deposits of volcanic ash, reinforcing the impression that the fossils represent a group of individuals that lived and died together.

Ceratopsians   The frills and horns of ceratopsians (see Figure 19.17) were probably elements of social behavior, especially in male-male competition. Observations of modern horned mammals suggest that differences in the size of the frill and the length and arrangement of the horns were associated with different behaviors:

- Species of antelopes with small horns engage in side-on displays with other males. Competing males swing their heads sideways against the flank of their opponent in a comparison of strength. *Protoceratops* and other early ceratopsians may have used displays like this.

- Deer and elk have large antlers, and males engage with each other head-on, interlocking their antlers and twisting their necks as each individual attempts to knock the other off its feet. The sturdy horns of *Triceratops* and other short-frilled ceratopsians would have been suitable for trials of strength of this sort. Some fossils of *Triceratops* have wounds on the face and frill that match the size and spacing of the large brow horns.

- Moose have enormous antlers that consist mostly of flat surfaces. Rival males face each other head-on, twisting and shaking their heads to emphasize the breadth of their antlers. This comparison of antler size is often sufficient to determine dominance, and when moose do engage in trials of strength they are back-and-forth shoving contests. Long-frilled ceratopsians, such as *Chasmosaurus*, may have engaged in these sorts of displays.

The nasal and brow horns of ceratopsians would have been formidable weapons and were probably used for defense against carnivorous dinosaurs. A herd of ceratopsians might have used the same defense as some modern horned mammals, forming a defensive ring when a predator approached, with the adults on the outside, facing the predator, and the juveniles sheltered in the center of the ring.

Pachycephalosaurs   Mature pachycephalosaur males are believed to have used their domed heads in male-male combat for territories or access to mates, with different configurations of the dome and its projections corresponding to different methods of combat (**Figure 19.20**).

Hadrosaurs   The nasal regions of hadrosaurines were greatly enlarged, and may have been surrounded by soft

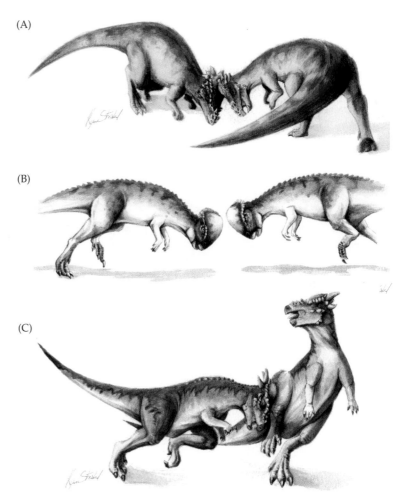

Figure 19.20 Pachycephalosaurs probably used their heads in male-male combat.   Analysis of injuries seen in fossilized skulls (see Figure 19.18) provides information about the combat methods of pachycephalosaurs. (A) The broad heads and marginal spines of adult male *Pachycephalosaurus wyomingensis* may have been used for head-to-head shoving contests, like those of some extant mammals, such as moose. (B) The thick, domed skulls of *Prenocephale prenes* could have protected their brains during head-on impacts of the sort seen in bighorn sheep. (C) The spine-covered heads of *Dracorex hogwartsia* might have been used to butt the flank of an opponent, as seen in extant goats. (From Peterson et al. 2013.)

tissue that could be inflated by exhaled air, rather like the nasal regions of male elephant seals. (**Figure 19.21A–C**). The heads of lambeosaurine hadrosaurs were crowned by hollow crests. These crests were sexually dimorphic—much larger in males than in females (**Figure 19.21D,E**). The crest of a male *Parasaurolophus* was a curved structure that extended back over the shoulders. Air in the hollow crest flowed from the external nares backward through the crest and then forward to the internal nares, which were located in the palate just anterior to the eyes (**Figure 19.21F**).

The inflatable proboscis of hadrosaurines and the hollow crests of lambeosaurines were probably used for visual displays and for vocalizations. Acoustic analysis of the ears of lambeosaurines indicates that they would have been able

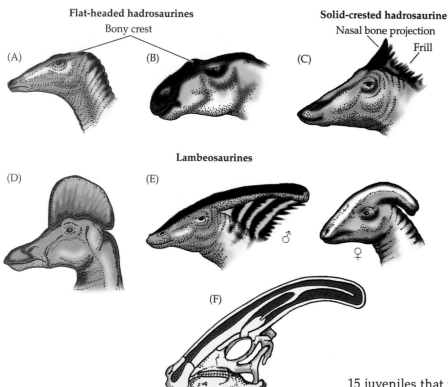

Flat-headed hadrosaurines
Bony crest

(A)

(B)

(C)

Solid-crested hadrosaurine
Nasal bone projection
Frill

Lambeosaurines

(D)

(E)

♂

♀

(F)

Figure 19.21 **Cranial ornamentation of hadrosaurs.** The hadrosaurines *Anatosaurus* (A) and *Kritosaurus* (B) lacked bony crests on their heads but may have had inflatable resonating structures formed by soft tissues. Solid-crested hadrosaurines such as *Saurolophus* (C) had a projection formed by a backward extension of the nasal bones that might have supported a frill of skin. In the lambeosaurines, such as *Corythosaurus* (D) and *Parasaurolophus* (E), the extensions of the nasal bones were hollow. These animals were sexually dimorphic, with the crests of males being larger than those of females. (F) The nasal passages of lambeosaurines (shown here for *Parasaurolophus*) ran from the external nares through the crests and down to openings in the roof of the mouth. (A–F modified from various sources, including Charig 1979.)

to hear the sound frequencies produced by the resonating columns of air in the nasal passages.

Nesting and parental care by ornithischians

Inferences about parental care by dinosaurs rest on a firmer basis than inferences about their social behavior. Not only do we have observations of parental care by the extant bracketing groups, crocodylians and birds, but thousands of nests of nonavian dinosaurs have been discovered.

Ornithischian dinosaurs laid eggs in an excavation that might have been filled with rotting vegetation to provide both heat and moisture for the eggs. Similar methods of egg incubation are used by crocodylians and by the brush turkeys of Australia.

Some ornithischians probably provided an extended period of parental care. For example, a nest of the hadrosaur *Maiasaura* (Greek *maia*, "good mother") uncovered in the Late Cretaceous Two Medicine Formation in Montana contained 15 juveniles that were about 1 m long (approximately twice the size of hatchlings found in the same area), indicating that the group remained together after they hatched. The teeth of the baby hadrosaurs showed that they had been feeding; some teeth were worn down to one-fourth of their original length. It seems likely that a parent remained with the young.

Other fossils suggest that *Maiasaura* and another hadrosaur, *Hypacrosaurus*, grew to one-fourth of adult size before they left the nesting grounds, and that another ornithopod found at the same site grew to half its adult size. A nest of the ceratopsian *Protoceratops andrewsi* from a Late Cretaceous fossil deposit in Mongolia contained

15 juveniles that were larger than a hatchling from the same area, again suggesting that hatchlings remained at the nest and were cared for by a parent. Discovery of a single adult *Psittacosaurus* with 34 babies suggests that juveniles from several clutches may have assembled in a crèche, a behavior that also occurs in several species of crocodylians.

19.9 Herbivorous Saurischians

Saurischians radiated into two very different lineages: the herbivorous sauropods and the carnivorous theropods. The sauropods included the gigantic long-necked, long-tailed dinosaurs that are the centerpieces of paleontology halls at many museums, but those are the derived forms. Although their ancestors were bipedal, derived sauropods were quadrupedal. Basal sauropodomorph dinosaurs (prosauropods), were most diverse in the Late Triassic and Early Jurassic. Prosauropods were not as large as their descendants; *Plateosaurus*, with a length of 6 m, was among the largest prosauropods.

The structure of sauropods

Prosauropods had long necks and small heads; *Plateosaurus* had 10 cervical vertebrae, 15 trunk vertebrae, 3 sacral vertebrae (the vertebrae that connect the hips to the vertebral column), and about 46 caudal vertebrae. The long necks of prosauropods suggest that they were able to browse on plant material at heights up to several meters above the ground. The ability to reach tall plants might have been a significant advantage during the shift from the low-growing *Dicroidium* flora to the taller bennettitales (extinct

(A)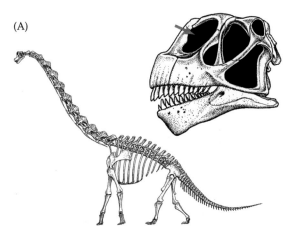

(B)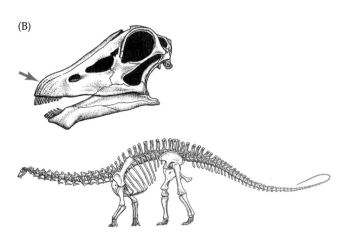

Figure 19.22 Macronarian and diplodocoid sauropods differed in body proportions and skull shapes. (A) Macronarians had long forelegs, and their backs sloped downward from the shoulders. With extremely long necks, they may have browsed from treetops; some reconstructions even show them standing on their hindlegs to extend their reach. The skulls of macronarians were boxy, and spatulate teeth extended the length of the jaws. The external nares of macronarians (arrow) were high on the face and were as large as, or larger than, the eyes. (B) Diplodocoids had short frond-bearing plants allied to cycads) and conifers that occurred in the Late Triassic.

forelegs, so their backs sloped down from the hips to the shoulders. Diplodocoid necks were long, although not as long as those of macronarians, and their heads were closer to the ground. They may have fed on low-growing vegetation, swinging the head in an arc to reach a large area. Diplodocoids had long, flat snouts; peglike teeth in the front of the jaws; and long tails with a whiplike tip. Current interpretations place the external nares near the front of the snout (arrow). (From Fastovsky and Weishampel 2005, illustrations by John Siddick.)

The derived sauropods of the Jurassic and Cretaceous—neosauropods—were enormous. Gigantism evolved independently in different lineages, which suggests that the ability to grow to a large body size is an ancestral character of the sauropodomorph lineage. Neosauropods were the largest terrestrial vertebrates that have ever existed. The largest of them may have exceeded 30 m in length and weighed 50,000 kg. (For comparison, a large African elephant is about 5 m long and weighs 4,000–5,000 kg.)

Long necks increased the area in which a sauropod could feed. The length of the necks of neosauropods resulted both from lengthening the cervical vertebrae and from increasing their number, which rose from 10 in prosauropods to a minimum of 12 and a maximum of 19 in neosauropods. The back became shorter in neosauropods as the number of dorsal vertebrae decreased from 15 in prosauropods to as few as 9 in neosauropods. The connection between the vertebral column and the hips was increased to 5 sacral vertebrae in neosauropods, and their tails contained 80 or more caudal vertebrae.

Neosauropods had remarkably small heads in proportion to the size of their bodies, perhaps because a large head at the end of a long neck would exert too much leverage. One major group of neosauropods, macronarians ("big nostrils"), had short, deep snouts with enormous nasal openings near the top of the skull, and spatulate teeth, which extended the full length of the jaws. *Camarasaurus* had a maximum of three replacement teeth in position for each active tooth, and teeth were replaced every 62 days. The forelimbs of macronarians were longer than their hindlimbs, so their backs sloped downward.

The other major lineage of neosauropods, diplodocoids, had long, flat snouts. Current interpretations place the nasal openings near the front of the snout. Diplodocoids had peglike teeth that were limited to the front of the jaws. Each *Diplodocus* tooth had 5 replacement teeth in position, and its active teeth were replaced every 35 days. The forelimbs of diplodocoids were shorter than the hindlimbs and they had long tails with a whiplike tip (**Figure 19.22**).

Simple teeth were sufficient for neosauropods because they were bulk feeders, consuming enormous quantities of food and digesting it slowly. They used their teeth to strip twigs and leaves from trees, but they did little oral processing of that material. Instead, their large gastrointestinal tracts allowed slow rates of passage, giving symbiotic microorganisms time to ferment the food and releasing volatile fatty acids that were absorbed across the wall of the intestine.

Sauropods were enormously heavy, and the skeletons of large sauropods clearly reveal selective forces favoring a combination of strength with light weight. The axial skeleton, vertebrae, and ribs were strongly pneumatized (i.e., contained many air spaces known as pleurocoels). The pneumatization moved progressively backward in neosauropods, eventually including the hips and tail vertebrae. The arches of the vertebrae acted like flying buttresses on a large building, while the neural spines held a massive and possibly elastic ligament that helped support the head, neck, and tail (**Figure 19.23**). In cross section, the trunk was deep, like that of an elephant.

Fossil trackways of sauropods show that the legs were held under the body with the left and right feet only a single foot width apart. The limbs were held straight in an elephantlike pose and moved fore and aft parallel to the midline of the body. This morphology produces the straight-legged locomotion familiar in elephants, and sauropods probably walked with an elephantlike gait, holding their

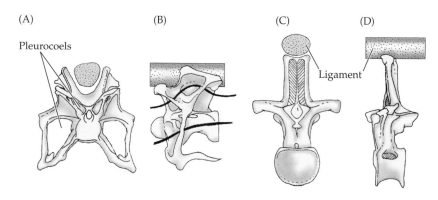

Figure 19.23 **Structural features of sauropods.** The skeletons of large sauropods combined lightness with strength. (A,B) Cervical vertebrae seen from the rear and side show pleurocoels that are thought to have contained air sacs. The black line in B indicates the positions of the pleurocoels. (C,D) The neural arches of the dorsal region seen from the front and side show a U-shaped depression that accommodated a large dorsal ligament (blue) that supported the head, neck, and tail. (A,B redrawn based on Osborn and Mook 1921; C,D redrawn based on Hatcher 1901.)

neck erect and their tails in the air. Sauropod trackways preserved in many parts of the world reveal the immense size of these animals (**Figure 19.24**).

Social behavior of sauropods

Sauropods lacked frills and other sexually dimorphic display structures of the sort seen among ornithischian dinosaurs, but that does not mean that social behavior was entirely absent. After all, extant crocodylians lack sexually dimorphic ornaments, yet they have an extensive repertoire of social behaviors. Still, the evidence for sociality by sauropods is so sparse that we can guess that sauropods had less extensive social interactions than ornithischians did.

An analysis of oxygen isotope ratios in the bones of *Camarasaurus* suggests that these sauropods made seasonal migrations of several hundred kilometers between upland and lowland environments. We cannot tell whether the dinosaurs traveled as individuals moving along the same route at the same time of year or as a cohesive group containing both adults and juveniles. Evidence of possible herd behavior by sauropods may be revealed by a series of tracks found in Early Cretaceous sediments at Davenport Ranch in Texas. These tracks show the passage of 23 sauropods in a group that appears to have moved in a structured fashion, with young animals in the center, surrounded by adults.

Another line of evidence comes from sites that preserve multiple individuals of a restricted age range of a single species of dinosaur. These associations suggest that juveniles lived in groups apart from adults, a phenomenon that has been observed in some extant species of grazing mammals. Several age-restricted sites of this sort have been discovered—for example, the Mother's Day Quarry, a Late Jurassic site in Montana, contains numerous fossils of juveniles of a *Diplodocus*-like sauropod, and the Late Cretaceous Big Bend *Alamosaurus* site contains only juveniles of the sauropod *Alamosaurus sanjuanensis*.

Nesting and parental care by sauropods

Concentrations of nests and eggs ascribed to sauropods in an Early Jurassic site in South Africa and in Cretaceous deposits in southern France and Patagonia suggest that these animals had well-defined nesting grounds to which they returned year after year. Eggs thought to be those of the large sauropod *Hypselosaurus priscus* have been found in association with fossilized vegetation similar to that used by alligators to construct their nests. The orientation of the nests suggests that each female dinosaur probably deposited about 50 eggs. The eggs had an average volume of 1.9 L (about 40 times the volume of a chicken egg). Fifty of these eggs together would have weighed about 100 kg, or 1% of

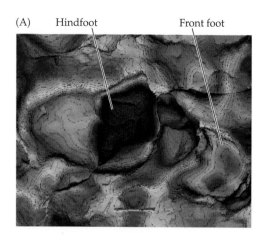

Figure 19.24 **A sauropod footprint from Early Cretaceous sandstone in Western Australia.** Trackways in rocks along the coast of Western Australia, north of the town of Broome, contain footprints of 16 different species of dinosaurs, including theropods, ornithopods, thyreophorans, and sauropods. The footprints shown here are designated as "Broome sauropod morphotype A." (A) Digital analysis uses colors to reveal the depth of the imprint of a hindfoot and a front foot. (B) Goolarabooloo Maja Richard Hunter beside the print of a hindfoot, revealing its size. The scale bar he is holding is 40 cm. (From Salisbury et al. 2016, courtesy of Steven W. Salisbury.)

the estimated body weight of the mother. Crocodylians and large turtles have egg outputs that vary from 1% to 10% of the adult body mass, so this estimate for *Hypselosaurus* seems reasonable.

Some nesting sites of sauropods have revealed astonishing quantities of eggs. During the Late Cretaceous, the Auca Mahuevo fossil site in Patagonia was a flood plain drained by shallow stream channels. Thousands of individuals of an unidentified species of large sauropod dinosaur journeyed here to construct nests, creating five layers of egg beds that extended for several kilometers and reached densities of 11 eggs per m². The eggs, which were roughly spherical, had diameters between 12 and 15 cm, with 20–40 eggs in each clutch. The eggs in a clutch were stacked on top of one another in two or three layers. There is no evidence of parental care. Indeed, considering the size of an adult sauropod and the dense spacing of the clutches, any loitering by an adult would have been more likely to crush eggs than protect them.

We do not yet have direct evidence of parental care by sauropods, but a nest of the prosauropod *Massospondylus carinatus* in South Africa may hint at parental care. The embryos in this nest appeared to be close to full term, but the ventral portions of the pelvic girdles were poorly developed, the heads were enormous in relation to the bodies, and teeth were virtually absent. This combination of characters would have made it difficult for the hatchlings to move about or feed themselves, and supports the inference that adults of this species of sauropod might have cared for their young.

19.10 Carnivorous Saurischians

In contrast to the herbivorous sauropods, the theropod lineage of saurischians was primarily carnivorous. Basal theropods were small, lightly built predators with long arms and legs. The hands contained three fingers and were held in the same position as human hands—palms inward. These theropods used their hands in a clapping motion to seize prey and hold it while they dismembered it with their jaws.

Derived theropod dinosaurs (coelurosaurs) included three general types of animals (**Figure 19.25**): predators that attacked large prey using their jaws as weapons (tyrannosauroids), fast-moving predators that seized small prey with their forelimbs (ornithomimisaurians), and fast-moving predators (maniraptorans). All theropods were fleet-footed; computer models suggest that ornithomimosaurs and maniraptors could reach speeds of 50–60 km/h, and even tyrannosaurs may have been able to sprint at 40 km/h.

Tyrannosauroides

Early tyrannosauroids were small and lightly built with long arms and legs. Although derived tyrannosaurs were very large, the distinctive features of the lineage appeared in the early small species. The evolutionary history of tyrannosaurs proceeded in three stages starting in the Middle Jurassic. In the first stage, the skull was strengthened, the anterior upper teeth were serrated and the jaw muscles were powerful. The second stage, in the Early Cretaceous, produced a miniature tyrannosaur: *Raptorex kriegsteini* was only 3 m long (compared with 15 m for *Tyrannosaurus rex*), but it had nearly all the distinctive features of tyrannosaurs, including a large skull, tiny forelimbs, and long hindlimbs. In the third stage, which extended throughout the rest of the Cretaceous, tyrannosaurs increased in size.

The tiny forelegs of tyrannosauroids are puzzling. The arms were too short to reach the mouth, and the third finger had been lost, yet the arm bones were robust and the fingers were tipped with large claws. These characters suggest that the arms and hands were not vestigial; they had a function, but we don't know what it was.

The teeth of large tyrannosauroids were as long as 15 cm, dagger shaped with serrated edges, and driven by powerful jaw muscles. Marks from the teeth of predatory dinosaurs are sometimes found on fossilized dinosaur bones, and these records of prehistoric predation provide a way to estimate the force of a dinosaur's bite. The pelvis of a horned dinosaur (*Triceratops*) found in Montana bears dozens of bite marks from a *Tyrannosaurus rex*, some as deep as 11.5 mm. Fossilized feces (**coprolites**) deposited by a *Tyrannosaurus rex* in Saskatchewan, Canada, contained fragments of crushed bone from a juvenile ornithischian, indicating that tyrannosaurs could crush bone—a phenomenon unique to derived tyrannosaurs among diapsids. (Some mammals can crush bones, but no other diapsids are known to do so.) The distribution of teeth in the jaws of large tyrannosauroids allowed them to apply enormous bite forces (8,526–34,552 newtons) to localized areas to crack bones that were then pulverized by repeated bites.

Featherlike structures (filamentous protofeathers) appeared long before birds (the evolution of feathers is discussed in Chapter 21). New discoveries are pushing the origin of protofeathers back, and feathers may even be an ancestral character of Dinosauria. Not all dinosaurs had feathers, however; they are known from only a few ornithischians, and they are not universally present in saurischians. No featherlike structures have been identified in sauropods, but protofeathers and feathers are widespread among theropods, from coelurosaurs (the lineage that gave rise to birds; see Chapter 21) onward.

The occurrence of feathers among tyrannosauroids is perplexing. Filaments described as protofeathers are present in *Dilong* and *Yutyrannus*, two basal species of tyrannosauroids from the Early Cretaceous, but the skins of *Tyrannosaurus rex* and several other derived tyrannosaurids from the Late Cretaceous were covered by scales.

Ornithomimisauria

The ornithomimisaurians were lightly built, cursorial (specialized for running) coelurosaurs of the Cretaceous. Despite their name, which means "bird mimic," the

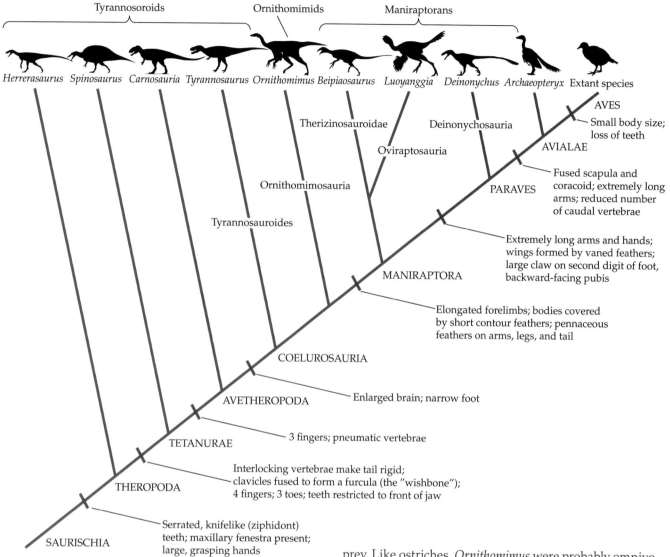

Figure 19.25 Simplified phylogeny of theropods. Bipedality, an ancestral character of theropods, was retained throughout the lineage. Feathers (protofeathers) may also have been ancestral, and some (but probably not all) species in the lineages shown here had feathers, at least on the arms, legs, or tail. Birdlike characters (reduction or complete loss of teeth and development of a beak) appeared independently in ornithomimids, oviraptosaurs, *Archaeopteryx*, and Aves (birds). The silhouettes illustrate body forms but not relative sizes. (Data from Fastovsky and Weishampel 2016.)

ornithomimisaurians are not closely related to birds, but they had evolved into birdlike forms. *Ornithomimus* was ostrichlike in size, shape, and probably ecology as well. The legs were long with three weight-bearing toes, and the upper limb (femur) was more than twice as long as the lower limb (tibia), a ratio that is characteristic of cursorial tetrapods.

Ornithomimus had a small skull on a long neck, and its toothless jaws were covered with a horny bill. The forelimbs were long, and three digits were retained on the hands. The inner digit was opposable and the wrist was flexible, making the hand an effective organ for capturing small

prey. Like ostriches, *Ornithomimus* were probably omnivorous and fed on fruits, insects, small vertebrates, and eggs. Quite possibly they lived in groups, as do ostriches, and their long legs suggest that they inhabited open regions rather than forests.

Maniraptora

Maniraptorans were active predators with long, slender legs and long arms with grasping fingers (their name means "predators that grasp with hands"). They were feathered—the body had a covering of downy feathers, and long pennaceous feathers were present on the arms, legs, and tail of many species. (Pennaceous feathers, present on the wings and tail of extant birds, have a central shaft with a vane on each side; see Figure 21.4.)

Oviraptosaurids were a group of mostly small maniraptorans (1–2.5 m) in the Late Cretaceous. They were toothless, had horny beaks, and were probably herbivorous or omnivorous. Oviraptosaurids had boxlike skulls, and most species had prominent crests on the head that appear to have been sexually dimorphic (presumably larger in males than in females). These crests were probably brightly colored and identified the species and sex of an individual.

(A)

(B)

Figure 19.26 Changing views of deinonychosaurs.
Deinonychus was about 3.5 m long, and its head would have been at chest height to an adult human. (A) *Deinonychus* was described in 1969, before the discovery of feathered dinosaurs, so its skin was assumed to be scaled. The impressive talon on the second toe led to the suggestion that *Deinonychus* hunted in packs and attacked large dinosaurs, disemboweling them with their talons. (B) We are now confident that *Deinonychus* was feathered, and probably hunted individually, preying on animals smaller than itself (such as the small ornithopod *Zephyrosaurus* in this illustration). It probably pinned prey down with its claws, as modern falcons and hawks do, using its jaws to kill the prey and flapping its feathered arms for balance as the prey struggled. (A, Universal Images Group North America LLC/DeAgostini/Alamy Stock Photo; B, Emily Willoughby/ CC BY-SA 3.0.)

Derived maniraptorans (Paraves) had a huge claw on the second digit of the hind foot. *Deinonychosaur* means "terrible claw lizard," and these maniraptorans deserved that name. Ranging in size from less than 1 m to at least 6 m, they were consummate predators, with large eyes and large brains. The claw on the second toe was sicklelike and was held off the ground during locomotion; only the third and fourth toes touched the ground. The arms and hands were long, with three long, claw-tipped fingers, one of which was a semiopposable thumb.

Although deinonychosaurs retained grasping hands, they used their feet to subdue prey. Focusing on the sickle-shaped claws, early interpretations proposed that *Deinonychus* used its claw to slash prey, perhaps hunting in packs and attacking sauropods and ornithopods many times its size. Tests with fossilized claws have shown that they are not effective at slicing, however, so the claws probably were not used to slash. Extant falcons and hawks have enlarged talons on the second toe that they use to pin down prey as they tear it apart with their beaks. It seems likely that deinonychosaurs were solitary hunters, attacking prey smaller than themselves. Like falcons, they may have used their body weight to pin prey to the ground while tearing at it with their jaws. Their feathered arms may have helped them balance on struggling prey (**Figure 19.26**).

Social behavior of theropods

Large terrestrial mammalian carnivores, such as tigers and large bears, are generally solitary. Adults of these species hunt individually, and groups form only when prey are concentrated in a small area—for example, Kodiak bears gather during the upstream migrations of salmon. Twelve large tyrannosaurs, *Albertosaurus sarcophagus*, found amid flood debris in the Late Cretaceous Horseshoe Canyon Formation in Alberta, Canada, might have been scavenging carcasses of animals killed in an earlier flood.

Some medium-size mammalian predators are more social; wolves, coyotes, and African hunting dogs live and hunt in family groups, for example, and smaller species of theropods, especially ornithomimids, might have been social. More than 20 juvenile *Sinornithomimus dongi* ranging from 1 to 7 years old were trapped in mud in a drying pond in western Mongolia, and 160 juvenile and adult individuals of *Avimimus* appear to have drowned while crossing a river in southern Mongolia. *Avimimus* was probably herbivorous, making it unlikely that these animals were drawn together by a temporary abundance of food.

Nesting and parental care by theropods

Recognition of parental care by theropods lagged behind discoveries of nests of ornithischian dinosaurs because of a mistaken identification in 1923. The fossil of a theropod dinosaur that apparently died while attending a nest of eggs was discovered in the Gobi Desert, but its significance was not recognized until 70 years later. The eggs, which were about 12 cm long and 6 cm in diameter, were thought to have been deposited by the small ceratopsian *Protoceratops andrewsi* because adults of that species were by far the most abundant dinosaurs at the site. The theropod was assumed to have been robbing the nest and was given the name *Oviraptor philoceratops*, which means "egg seizer, lover of ceratops."

In 1993, paleontologists from the American Museum of Natural History, the Mongolian Academy of Sciences, and the Mongolian Museum of Natural History discovered a fossilized embryo in an egg identical to the supposed

Protoceratops eggs. To their surprise, the embryo was an *Oviraptor* nearly ready to hatch. With the benefit of hindsight, it is apparent that the adult dinosaur found in 1923 had been resting on its own nest, apparently trying to shelter its eggs from the sandstorm that buried the adult and the

(A)

(B)

(C)

Figure 19.27 Parental care.   (A) Reconstruction of an adult *Oviraptor philoceratops* brooding a nest of eggs. (B) Male ostriches (*Struthio camelus*), palaeognath birds from Africa, brood their eggs in the same way. Uniparental care by the male is characteristic of palaeognaths and is believed to be the condition for coelurosaurs. (C) Parental care by palaeognaths continues after the chicks have hatched. With only minor changes, this image of an ostrich with its young could be a family group of *Oviraptor*. (A, © Julius T. Csotonyi/Science Source; B, Christophe Eyquem/CC BY 3.0; C, Bernard Dupont/CC BY-SA 2.0.)

nest. Additional fossils of adult *Troodon* and *Deinonychus*, which like *Oviraptor* were maniraptorans, have subsequently been found sitting on eggs with their legs folded, arms extended, and bellies in contact with the eggs—the same posture that extant ground-nesting birds use when incubating eggs (**Figure 19.27**).

Male parental care appears to be ancestral for coelurosaurs, and is retained in the most primitive extant birds, paleognaths such as emus and ostriches. Thus, it is likely that male maniraptorans were the caregivers. Young maniraptorans may have remained with their male parent for extended periods, as do young emus and ostriches.

19.11 Gigantothermy and the Body Temperature of Dinosaurs

The popularization of "hot-blooded" dinosaurs in the 1970s initiated a controversy about body temperatures of dinosaurs and their mechanisms of thermoregulation that continued for several decades. The discovery that many dinosaurs had feathers has added a new dimension to the discussion, because insulation is one of the functions of feathers.

With a half-century of hindsight, it has become clear that the relationship between high metabolic rates and high body temperatures that is characteristic of extant vertebrates is not applicable to dinosaurs. Body temperatures and thermoregulatory mechanisms are closely related to body size, and most dinosaurs were much larger than any extant tetrapods. As a result, the link between high body temperatures and high metabolic rates seen among extant vertebrates does not apply to dinosaurs.

Gigantothermy is a form of thermoregulation characteristic of large animals that have low metabolic rates but nonetheless maintain body temperatures higher than their surroundings as a result of having low surface/volume ratios.

Gigantotherms lose heat to the environment slowly. A biophysical model that assumes that dinosaurs had metabolic rates like those of extant crocodylians predicts that medium-size to large dinosaurs would have had body temperatures above 30°C with a day–night variation of less than 2°C even in far northern and southern latitudes. If hatching occurred in the spring, juveniles would have grown large enough to be gigantotherms before winter.

Thus, gigantothermy would have allowed large dinosaurs (both ornithischians and saurischians) to maintain stable core body temperatures with the low metabolic rates that are characteristic of extant turtles and crocodylians. Feathers would not have been necessary for thermoregulation; indeed, avoiding overheating would have been the major problem facing large dinosaurs. Skin imprints show that the bodies of large dinosaurs (tyrannosaurs, sauropods, and ornithopods) were scaly, not feathered. Repeated proposals that archosaurs had high metabolic rates ignore the physics of heat exchange and are incorrect.

Smaller dinosaurs (adult weights <100 kg) living at latitudes more than 45° north or south would not have had stable body temperatures without metabolic heat production, but heat production without insulation is ineffective. Thus, it is among these dinosaurs that we would expect to find feathers providing insulation. A dramatic reduction in body size occurred during the evolution of birds, from about 160 kg for basal maniraptorans to 100 g for the first birds. Fossils show that members of the Avialae (see Figure 19.25) were feathered, and they are assumed to have had high metabolic rates that maintained high and stable body temperatures, as will be discussed in Chapter 20.

Summary

Diapsids dominated the Mesozoic.

All extant evolutionary lineages of vertebrates were present in the Mesozoic. Diapsida was the most abundant and diverse Mesozoic lineage.

Diapsids are characterized by a skull with two temporal fenestrae. The postorbital and squamosal bones form the base of the upper fenestra, and the jugal and quadratojugal bones form the lower arch.

Two lineages of diapsids are recognized. Extant lepidosauromorphs include the tuatara, lizards, and snakes, and archosauromorphs include the extant crocodylians and birds. Turtles (Testudines) are the sister group of Archosauria, which includes two major lineages: Pseudosuchia (now represented by crocodylians) and Ornithodira (birds, plus extinct pterosaurs and nonavian dinosaurs).

In the sea, lepidosauromorphs (ichthyosaurs, plesiosaurs, and others) diversified into adaptive zones occupied now by sharks, cetaceans, and other marine mammals. There were also Mesozoic diapsids (such as the long-necked plesiosaurs) that have no modern equivalents.

On land, archosaurians (crocodylomorphs, pterosaurs, and dinosaurs) filled the adaptive zones now occupied by mammals and birds. There were also forms (such as the enormous sauropod, therapod, and ornithopod dinosaurs) that have no present-day equivalents.

In addition to these now-extinct diapsids, the lineages that include extant lizards, snakes, turtles, and the tuatara began to diversify in the Mesozoic. Only mammals were poorly represented during the Mesozoic, although they were present and diversifying in terrestrial habitats.

Lepidosauromorphs were a major component of the Mesozoic marine fauna.

Lepidosaurs were not a prominent component of terrestrial faunas in the Mesozoic. Rhynchocephalians were the most diverse terrestrial lepidosauromorphs during this time. Initially small terrestrial insectivores, they radiated into terrestrial and aquatic carnivores and terrestrial herbivores.

Ichthyosaurs evolved from lizardlike early forms to superbly streamlined aquatic predators with a body form that converged on that of fast-swimming cetaceans and sharks. Ichthyosaurs were viviparous, and the young were born tail-first, like those of extant cetaceans and sirenians.

Placodonts retained many characters of their terrestrial ancestors and were probably slow-swimming coastal herbivores, perhaps rather like extant sea cows.

Nothosaurs and pachypleurosaurs were elongated marine predators that probably pursued and captured fish.

Plesiosauroids radiated into two lineages: Pliosaurs had large heads and short necks and were probably pursuit predators. Plesiosaurs had small heads and long necks that could bend downward but not upward or laterally; plesiosaurs may have preyed on bottom-dwelling fishes and invertebrates.

Mosasaurs were a lineage of varanid lizards that took to the sea, and early forms looked much like their terrestrial ancestors. Derived forms were streamlined and preyed on other marine diapsids.

Crocodylomorphs also radiated in Mesozoic seas.

Metriorhynchids were far more specialized for life in water than is any extant crocodylian; the limbs were modified as paddles, and the tail was hypocercal.

Pterosaurs were the first flying vertebrates.

Pterosaurs appeared in the Late Triassic, 80 million years before the earliest birds, and persisted until the end of the Cretaceous.

Pterosaurs radiated into all of the adaptive zones occupied by extant birds, and some that are not occupied by extant birds. Azhdarchids, which were probably terrestrial predators, may have been the most remarkable pterosaurs. Some azhdarchids could stand eye-to-eye with a giraffe, and the largest of them could have preyed on dinosaurs.

The primary wings of pterosaurs were formed by skin that was supported by the arm bones and an elongated fourth finger. Some pterosaurs had a posterior wing that extended between the pelvis and the hind legs. The skin of at least some pterosaurs was covered by fine, hairlike fuzz called pycnofibers.

All pterosaurs could fly, probably using a four-legged leap to get into the air. Tracks show that pterosaurs walked quadrupedally.

(Continued)

Summary *(continued)*

Pterosaurs were oviparous, and hatchlings were probably able to run and fly soon after leaving the nest.

Birds and pterosaurs coexisted for about 100 Ma. The evolution of birds coincided with an increase in body size of pterosaurs, suggesting that birds may have outcompeted pterosaurs at small body sizes.

Two wholesale replacements of terrestrial vertebrate faunas occurred in the Triassic.

Herbivorous and carnivorous synapsids were the most abundant terrestrial vertebrates at the start of the Triassic.

Basal pseudosuchians diversified into predators and herbivores in the Middle Triassic. Basal archosauromorph and ornithodiran herbivores had largely replaced herbivorous synapsids by the end of the period.

Beginning in the Late Triassic, pseudosuchians and basal archosauromorphs were replaced by dinosaurs. Herbivorous dinosaurs radiated in the Late Triassic, and carnivorous dinosaurs radiated in the Jurassic, after the end-Triassic extinction wiped out most pseudosuchian predators.

An increase in body size in dinosaurs required changes in the pelvis and associated musculature.

Large tetrapods hold their limbs vertically beneath the body because bone is more resistant to compressive force than to shearing force.

As the limbs of dinosaurs moved to a vertical position, the origin of muscles that protract the femur shifted to maintain stride length.

The anatomical changes in the pelvis characterize the two major lineages of dinosaurs.

- Saurischian dinosaurs had a long pubis that provided an anterior site for the origin of femoral protractors.
- Ornithischian dinosaurs rotated the pubis backward to lie parallel to the ischium and shifted the origin of femoral protractors to an anterior extension of the ilium or to an anterior projection of the pubis.

Ornithischian dinosaurs radiated into a wide variety of body forms.

The lineage included bipedal and quadruped species; all were herbivorous, had horny beaks, and chewed their food.

Thyreophorans (armor-bearing dinosaurs, the stegosaurs and ankylosaurs) had rows of osteoderms on the back.

Osteoderms were elaborated into leaf-shaped plates or spines in stegosaurs. Some ankylosaurs had spines, but they probably relied mostly on thick osteoderms for protection. Thyreophorans fed on low-growing vegetation.

Marginocephalians include the horned dinosaurs (ceratopsians) and the thick-headed pachycephalosaurs.

Although derived from a bipedal ancestor, derived ceratopsians were quadrupedal. Derived forms developed a frill that extended backward from the skull and horns that, like the horns and antlers of extant mammals, were probably used for defense and in social interactions with other individuals. Pachycephalosaurs remained bipedal; their distinctive feature was a thick skull roof that was probably advantageous during male-male combat.

Ornithopods appeared in the Middle Jurrasic. Small species were largely bipedal, but the largest species were probably mostly quadrupedal.

Hadrosaurs (duck-billed dinosaurs) were the most speciose group of ornithopods. The largest species reached lengths of more than 15 m and weighed upward of 13,000 kg.

Hadrosaurs may have run bipedally to escape predators. They had dental batteries formed by thousands of teeth packed together to form tooth plates.

Many species of ornithischians probably formed herds. Social interactions probably included male-male competition for access to mates and territories. The plates and spines of stegosaurs, the frills and horns of ceratopsians, and the crests of lambeosaurines were probably visual and auditory signals in these interactions.

Hundreds of fossilized ornithischian nests have been discovered containing eggs, embryos, hatchlings, or older juveniles.

It appears that some ornithischians attended their nests, and hatchlings of some species probably remained with a parent for extended periods.

Saurichians radiated into two very different lineages.

Sauropods were quadrupedal herbivores, although they were derived from a bipedal lineage. Theropods were bipedal carnivores.

Sauropods had long necks, small heads, and long tails.

The earliest sauropods (prosauropods) were considerably smaller than derived sauropods (neosauropods). The two lineages of neosauropods, macronarians and diplodocoids, differed in the length of their necks and the structure of their skulls, teeth, and limbs. Neosauropods were the largest terrestrial animals that have ever existed.

Macronarians had boxy skulls with spatulate teeth running the full length of the jaws. Their forelegs were longer than their hindlegs. Macronarians probably browsed on high-growing vegetation, using their long necks for vertical reach.

Diplodocoids had long, flat snouts with peglike teeth at the front of the jaws. Their forelegs were shorter than their hindlegs, and they may have browsed on low-growing vegetation, using their long necks to cover a wide arc.

Some sauropods may have formed herds.

Fossilized tracks of sauropods indicate that they may have moved in herds with juveniles in the center, surrounded by adults. Other fossil sites preserve the remains of several subadult sauropods, all the same size, suggesting that juveniles of some species formed age-restricted herds, as do some extant mammals.

At least some sauropods had nesting grounds that were used by generation after generation.

Some species of sauropods may have constructed nests, but evidence for parental care of nests or hatchlings is weak.

Theropods included three general types of animals: tyrannosauroids, ornithomimids, and maniraptorans.

All theropods were carnivorous or omnivorous, and all were bipedal and fleet-footed. Many, but not all, were feathered, and the presence of feathers varied among species within some lineages.

Derived tyrannosaurs were enormous, heavily built predators.

Early tyrannosaurs were small and lightly built with long arms that they used to seize prey.

The forelimbs of derived tyrannosaurs were greatly reduced, and only two fingers remained. The arms were too short to lift food to the mouth, but they were robust and the fingers were clawed, suggesting that the arms and hands were not vestigial. Their function remains a mystery.

Derived tyrannosaurs had their weapons concentrated in their heads. Strong skulls, powerful jaw muscles, and serrated teeth allowed them to rip flesh from prey and crush bones.

Basal species of tyrannosaurs had filaments that are described as protofeathers, but fossilized skin impressions show that several derived tyrannosaurs were covered by scales.

Ornithomimids were ostrichlike in appearance and perhaps in their ecology.

Long hindlimbs with the femur longer than the tibia indicate that ornithomimids were cursorial. The forelimbs

were long and had three grasping fingers that were used to seize prey.

Ornithomimids had long necks, small skulls, and toothless jaws. They were probably omnivorous and may have lived in groups, as ostriches do.

Maniraptorans were fast-moving predators with an enlarged talon on the second toe.

Maniraptorans retained the ancestral characters of long legs and long arms with grasping fingers. Feathers were extensively developed and included downy feathers covering the body and pennaceous feathers on the arms, legs, and tail of some species.

Oviraptosaurids were toothless and had horny jaws; they were probably herbivorous or omnivorous.

Deinonychosaurs were pursuit predators. Current hypotheses propose that they were solitary predators that attacked animals smaller than themselves and used their talons to immobilize their prey while they tore at it with their teeth.

There is less evidence of social behavior by theropods than there is by saurischians.

Analogy with large mammalian carnivores suggests that large theropods were probably solitary. Some fossil deposits contain multiple individuals of the same species, raising the possibility that these represent mass-mortality events that overwhelmed a group of individuals.

At least some small species of theropods brooded their eggs in nests.

Fossils of maniraptorans show that adults brooded eggs as birds do, sitting on the clutch with their legs folded and their arms extended to cover the eggs. It is not known if parental care was present in tyrannosaurs.

The concepts of endothermy and ectothermy that describe thermoregulation of extant vertebrates cannot be transferred to dinosaurs.

Dinosaurs larger than about 100 kg would have been gigantotherms, meaning their low surface/volume ratios would have reduced the rate at which they lost heat to the environment. Computer models show that metabolic rates equivalent to those of extant crocodylians would have kept most medium-size to large dinosaurs warm; with higher metabolic rates, they would have overheated.

Only small dinosaurs would have required high metabolic rates and insulation. The transition from low to high metabolism in the avian lineage occurred during the extreme reduction in body size that characterized the evolution of birds.

Discussion Questions

1. Use the extant phylogenetic bracket method to test the hypothesis that male coelurosaur dinosaurs were the primary providers of care for eggs and young.

2. The necks of giraffes are the closest extant equivalent to the necks of sauropod dinosaurs. Two hypotheses have been proposed to explain the functional significance of the necks of giraffes, and those hypotheses can be applied to the necks of sauropods:

    a. Competing browsers hypothesis: The long neck makes more food resources available to a giraffe or sauropod by allowing it to reach up to browse on vegetation that is not available to short-necked browsers.

    b. Necks-for-sex hypothesis: Male giraffes and sauropods use their necks and heads to batter opposing males during intraspecific combat. Thus, long necks increase the fitness of males by increasing their opportunities to mate.

    What predictions can you make about the relative lengths of the necks of male and female sauropods to test these hypotheses?

3. Extensive dinosaur faunas inhabited Arctic regions well beyond the polar circle. We know that northern regions were warmer during the Mesozoic than they are now; clearly they were warm enough for dinosaurs. But Arctic winters would have had long periods of dim light (twilight) in the Mesozoic, as they do now. How might dinosaurs have responded to the seasonal changes in day length in the Arctic?

4. Compare the use of inferences about the ecology, behavior, and physiology of extinct animals based on extant phylogenetic brackets with those based on analogies with extant forms—for example, the inferences in this chapter about the social behaviors of marginocephalian dinosaurs that were based on analogies with modern herbivorous mammals with horns or antlers. How might you test your conclusions in each case?

5. When a giraffe holds its neck in a vertical position, its brain is about 2 m above its heart. Contraction of the ventricle must create enough pressure to raise the blood to the level of the head and perfuse the brain (i.e., circulate blood through the brain). The arterial pressure at the aorta of a giraffe is approximately 200 mm Hg, twice that of a human.

    The brain of a sauropod dinosaur such as *Barosaurus* would have been about 9 m above the level of its heart, assuming it held its neck vertically. Calculate the aortic pressure required to perfuse the brain using the following information:

    a. The density of blood is 1.055 g/cm^3 and the density of mercury 13.554 g/cm^3.

    b. An aortic pressure of 10 mm Hg is needed to overcome resistance to blood flow in the carotid arteries leading from the heart to the brain.

    c. A blood pressure of 50 mm Hg at the head is required to perfuse the brain.

    What inferences about the biology of sauropods can you draw from your calculation?

6. The concentration of oxygen in the atmosphere is currently ~21%, but it has been both higher and lower. At the start of the Permian it reached a maximum of 30–32%, then declined throughout the Permian and Triassic, reaching its minimum of 13–15% at the end of the Triassic. Dinosaurs and mammals both appeared in the Triassic, but dinosaurs flourished and radiated into a variety of large-bodied lineages, whereas mammals remained small and did not radiate into multiple lineages. Recall the characteristics of the respiratory systems of sauropsids and synapsids described in Chapter 14 to explain how the low concentration of atmospheric oxygen in the Permian may have favored the development of dinosaurs over mammals.

Additional Reading

Baron MG, Norman DB, Barrett PM. 2017. A new hypothesis of dinosaur relationships and early dinosaur evolution. *Nature* 543: 501–506. *See also* Langer MC and 8 others. 2017. Untangling the dinosaur family tree. *Nature* 551: E1–E3; and Baron MG, Norman DB, Barrett PM. Baron et al. reply. *Nature* 551: E4–E5.

Benton MJ, Twitchett RJ. 2003. How to kill (almost) all life: The end-Permian extinction event. *Trends in Ecology & Evolution* 18: 358–365.

Fastovsky DE, Weishampel DB. 2016. *Dinosaurs: A Concise Natural History*, 3rd ed. Cambridge University Press, Cambridge.

Fowler DW, Freedman EA, Scannella JB, Kambic RE. 2011. The predatory ecology of *Deinonychus* and the origin of flapping in birds. *PLoS ONE* 6(12): e28964.

Gignac PM, Erickson GM. 2017. The biomechanics behind extreme osteophagy in *Tyrannosaurus rex*. *Scientific Reports* 7: 2012.

Lee MSY, Cau A, Naish, Dyke GJ. 2014. Sustained miniaturization and the anatomical innovation in the dinosaurian ancestors of birds. *Science* 345: 562–566.

McGowan AJ, Dyke GJ. 2007. A morphospace-based test for competitive exclusion among flying vertebrates: Did birds, bats, and pterosaurs get in each other's space? *Journal of Evolutionary Biology* 20: 1230–1236.

Naish D, Barrett P. 2016. *Dinosaurs: How They Lived and Evolved*. Smithsonian Books, Washington, DC.

Sander PM and 15 others 2011. Biology of the sauropod dinosaurs: The evolution of gigantism. *Biological Reviews* 86: 117–155.

Seebacher F. 2003. Dinosaur body temperatures: The occurrence of endothermy and ectothermy. *Paleobiology* 29: 105–122.

Watson T. 2017. Beasts from the deep. *Nature* 543: 603–607.

Witton MP. 2013. *Pterosaurs: Evolution, Natural History, Anatomy*. Princeton University Press, Princeton, NJ.

Go to the Companion Website **oup.com/us/vertebratelife10e** for active-learning exercises, news links, references, and more.

Endothermy: A High-Energy Approach to Life

David Ellis, USFWS

Endothermy lies at one end of a continuum of thermoregulatory modes. Ectothermy and endothermy are not mutually exclusive, and many vertebrates use a combination of ectothermal and endothermal mechanisms of thermoregulation. Mammals (synapsids) and birds (sauropsids) evolved endothermy independently, but the benefits and costs are the same for both.

On the *benefit* side, endothermy allows birds and mammals to maintain high body temperatures when solar radiation is not available or is insufficient to warm them—at night, for example, or in winter. The thermoregulatory capacities of birds and mammals allow them to live in climates that are too cold for ectotherms.

On the *cost* side, endothermy is energetically expensive. The metabolic rates of birds and mammals are about an order of magnitude greater than those of amphibians and nonavian reptiles of the same body size. The energy to sustain those high metabolic rates comes from food, and gram-for-gram endotherms need more food than do ectotherms.

20.1 Balancing Heat Production with Heat Loss

Endotherms regulate their body temperatures by matching their rate of heat production to their rate of heat loss.

- Oxidative phosphorylation releases more than half of the energy in the chemical bonds of substrate molecules as heat. Thus, resting metabolism produces a basal level of heat, and increasing or decreasing the metabolic rate changes the production of metabolic heat.

- Insulation (the air trapped between hair or feathers) retards loss of heat to the environment, and endotherms can modify their rate of heat loss by adjusting their insulation.

- Evaporation of water by sweating or panting is an effective cooling mechanism, but as a long-term strategy it works only when water is readily available.

These factors must be considered simultaneously to understand how endotherms maintain their body temperatures at stable levels in the face of environmental temperatures that range from −70°C to +60°C (the approximate range of air temperatures on Earth; ground surface temperatures can go much higher). The ability of endotherms to live in polar regions, on mountains, and in deserts results primarily from interactions among of generalized anatomical, physiological, and behavioral traits rather than from novel adaptations. Normal body temperature (**normothermia**) varies somewhat for different mammals and birds, but most mammals and birds conform to the generalized diagram in **Figure 20.1**.

Endotherms usually live under conditions in which air temperatures are lower than the regulated body temperatures of the animals themselves, and in this situation heat is lost to the environment. Adjusting insulation—by fluffing feathers to increase the thickness of the layer of trapped air, for example—allows endotherms to remain at their resting metabolic rate for a time as environmental temperature falls, but there is a limit to how much insulation alone can accomplish. The **lower critical temperature (LCT)** is the point at which an animal must increase its metabolic heat production to balance its heat loss. Although some birds and mammals have impressively low LCTs, most species reach their LCT at moderate temperatures. Body size is the most important determinant of a species LCT. The modal body weight of mammals is 21 g, and the predicted LCT for a mammal of this size is 29°C. Nowhere on Earth does the minimum temperature remain above 29°C all year (**Figure 20.2**). Birds are only a little better off; the calculated LCT of a bird weighing 18 g (the modal weight of birds) is 21°C, and only in equatorial regions do minimum temperatures remain above that level all year. Thus, most birds and all mammals must at times increase heat production to maintain stable body temperatures. How do they do that?

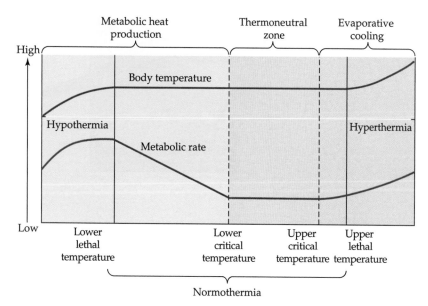

Figure 20.1 **Endotherms can maintain stable body temperatures over a range of environmental temperatures.** The thermoneutral zone is the range of environmental temperatures within which an endotherm can control its body temperature by adjusting its rate of heat loss without increasing heat production. The boundaries of the thermoneutral zone are the lower and upper critical temperatures. At temperatures below the lower critical temperature, an animal increases heat production to compensate for heat loss. The lower lethal temperature is the coldest environmental temperature at which heat production can keep up with heat loss; at lower temperatures, heat loss exceeds production, body temperature falls, metabolism and heat production decrease as a result of the Q_{10} effect, and hypothermia (a body temperature below normothermia) leads to death. Above the upper critical temperature, the animal employs evaporative cooling. The upper lethal temperature is the highest environmental temperature at which evaporative cooling can balance heat gain; at higher environmental temperatures, body temperature rises, metabolic heat production increases as a result of the Q_{10} effect, and hyperthermia (a body temperature above normothermia) leads to death.

Whole-body metabolism

The metabolic heat production of an organism is the sum of the heat production of its different tissues and organs, and the mass-specific basal metabolic rates (kilojoules per kilogram [kJ/kg] per day) of these tissues and organs vary widely. When an animal is at rest, the heart and kidneys are the most metabolically active tissues, consuming twice as much oxygen (and releasing twice as much heat) as the liver and brain. The metabolism of the remaining organs, of adipose tissue, and of skeletal muscle is still lower.

Thus, the basal metabolic heat production of an endotherm is derived primarily from the heart and kidneys. Muscular activity is not a significant source of heat when an animal is at rest, but intense activity can increase the heat production of skeletal muscle 10–15 times above its basal rate. In cold environments, this muscular heat can compensate for heat loss, but only when an animal is active.

Shivering and non-shivering thermogenesis

Mammals and birds use both shivering and non-shivering thermogenesis to increase heat production. Shivering consists of uncoordinated contraction of postural muscle fibers and transforms chemical energy (ATP) to elastic energy (a contracted muscle fiber) and then to heat energy (when the fiber relaxes). The heat is released in the muscle tissue and distributed by the circulatory system. Both birds and mammals shiver, and differences in the details of muscle fiber recruitment during shivering indicate that shivering evolved independently in the two groups.

Non-shivering thermogenesis is an increase in oxidative phosphorylation achieved by short-circuiting a cellular process. One form of non-shivering thermogenesis occurs in the mitochondria, where the electron transport chain pumps protons (H^+) from the mitochondrial matrix into the intermembrane space, creating a proton gradient. Most of the protons return to the matrix via ATP synthase, converting ADP to ATP, but some protons leak back into the matrix without passing through ATP synthase. This process is called a proton leak, and its effect is to increase oxidative phosphorylation without synthesizing ATP.

Proton leak has two components, basal and inducible proton leak. Basal proton leak, which accounts for 20–30% of the resting metabolic rate of laboratory rats, results from the permeability of the inner mitochondrial membrane and is constant. Inducible proton leak is produced by uncoupling proteins (UCPs), which increase the permeability of the inner mitochondrial membrane, allowing additional protons to bypass ATP synthase and thereby increasing heat production. Mammalian UPCs occur in brown fat, a specialized thermogenic tissue, and in a subset of cells in subcutaneous white fat, as well as in skeletal muscle. An avian-specific UPC is present in birds. In both mammals and birds, proton leakage induced by the UPCs is signalled by the nervous system in response to cold.

The sarcoplasmic reticulum is a second site of non-shivering thermogenesis. Contraction of skeletal muscle fibers is initiated by neural activity that releases Ca^{2+} ions from the sarcoplasmic reticulum. These ions are subsequently pumped back into the sarcoplasmic reticulum. A protein called sarcolipin increases the permeability of the sarcoplasmic reticulum, allowing Ca^{2+} ions to leak out, thereby increasing oxidative phosphorylation and releasing more heat.

Insulation

Hair and feathers reduce the rate of heat loss by trapping air, and a mammal or bird can increase its insulation by raising its hair or feathers to increase the thickness of the layer of trapped air. (We humans have goose bumps on our arms and

Mammals

Birds

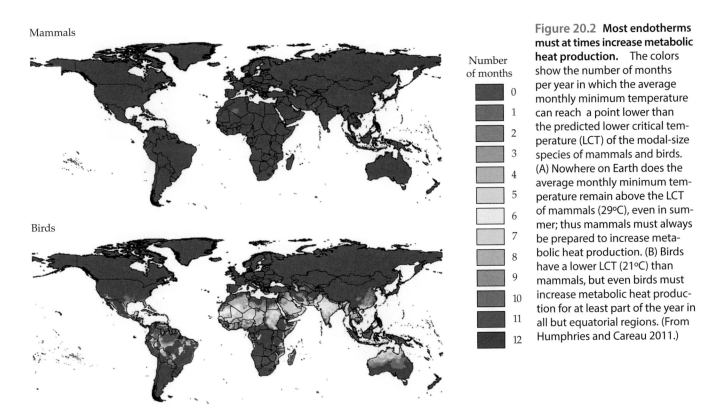

Figure 20.2 Most endotherms must at times increase metabolic heat production. The colors show the number of months per year in which the average monthly minimum temperature can reach a point lower than the predicted lower critical temperature (LCT) of the modal-size species of mammals and birds. (A) Nowhere on Earth does the average monthly minimum temperature remain above the LCT of mammals (29°C), even in summer; thus mammals must always be prepared to increase metabolic heat production. (B) Birds have a lower LCT (21°C) than mammals, but even birds must increase metabolic heat production for at least part of the year in all but equatorial regions. (From Humphries and Careau 2011.)

Number of months

0
1
2
3
4
5
6
7
8
9
10
11
12

legs when we are cold because our few remaining hairs rise to a vertical position in an attempt to increase our insulation.)

Heat loss can be increased by allowing hair or feathers to lie flat, minimizing the thickness of insulation, and by exposing thinly insulated parts of the body (called thermal windows), as a dog does when it lies on its back. Heat loss also can be enhanced by redirecting peripheral blood circulation to the skin, and especially to appendages with large surface areas, such as the ears of jackrabbits and elephants. In cold environments, retaining blood in the body core conserves heat.

Evaporative cooling

Evaporative cooling is another method of increasing heat loss. The skin of all vertebrates is permeable to water vapor, and this baseline evaporation, called insensible water loss, has a cooling effect. Humans and some other mammals increase evaporative heat loss by sweating, a process in which water is released from sweat glands and evaporates from the surface of the body. Many animals pant, breathing rapidly and shallowly so that evaporation of water from the respiratory system provides a cooling effect. Birds use a rapid fluttering movement of the gular (throat) region to evaporate water for thermoregulation.

Sweating has the advantage of requiring little energy, but it cools the surface of the skin, which is also the area that is being heated by the environment; thus, sweating may have a limited effect on the temperature of the body core. Panting and gular fluttering cool internal tissues, but both require muscular activity, which is an added source of metabolic heat.

20.2 Endotherms in the Cold

Endotherms expend most of the energy they consume just keeping themselves warm—even in the moderate conditions of tropical and subtropical climates. Nonetheless, endotherms have proven themselves very adaptable in extending their thermoregulatory responses to allow them to inhabit even polar and high-mountain regions.

The lower critical temperatures of arctic birds and mammals dramatically illustrate how effective insulation can be (**Figure 20.3A**). The effect of the fur of arctic mammals and of the layer of down beneath the feathers of arctic birds can be understood by comparing the LCTs of arctic and tropical animals. Tropical birds and mammals have LCTs between 20°C and 30°C. As air temperatures fall below their LCTs, these animals must increase their metabolic rates to maintain normal body temperatures. For example, at an environmental temperature of 25°C, a tropical raccoon has already increased its metabolic rate approximately 50% above its basal level.

In contrast, arctic birds and mammals maintain basal metabolic rates at environmental temperatures well below freezing, and they show only slow increases in metabolism (i.e., flatter slopes) below the LCT. The Arctic fox (*Vulpes lagopus*), for example, has an LCT of −40°C, and at −70°C (approximately the lowest air temperature ever recorded) a fox has elevated its metabolic rate only 50% above its basal level. Under those conditions the Arctic fox is maintaining a body temperature approximately 110°C higher than air temperature. Arctic birds are equally impressive. The

(A)

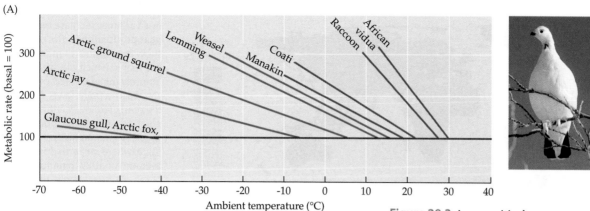

(B)

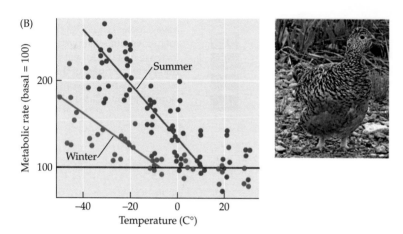

Figure 20.3 Lower critical temperatures for birds and mammals.   (A) Arctic species (graphed in blue) have lower LCTs than tropical species (graphed in red). The basal metabolic rate for each species is considered to be 100 units to facilitate comparisons among species. The lower end of the thermoneutral zone lies at the intersection of the basal metabolic rate (black horizontal line) and the increased metabolic rate needed to maintain a stable body temperature. The glaucous gull and Arctic fox don't have to increase heat production until the ambient temperature falls below –40°C. The steepness of the slope shows how rapidly metabolic heat production increases as ambient temperature falls. The slopes for tropical animals are steeper than those for arctic animals because the bodies of tropical animals have less insulation. (B) The white winter plumage of the willow ptarmigan (*Lagopus lagopus*) has a higher insulative value than its summer plumage; in winter, even the toes are feathered. This species' LCT is +10°C in summer and –5°C in winter. During summer, the metabolic rate has doubled at -20°C, but in winter it does not double until the ambient temperature has fallen below -40°C. (A, after Scholander et al. 1950; B, after West 1972; summer plumage photo by Alan Schmierer; winter plumage photo by Katie Thoresen/NPS.)

glaucous gull (*Larus hyperboreus*), like the Arctic fox, has an LCT near –40°C and can withstand –70°C with only a modest increase in metabolism.

Most mammals and birds replace their body covering in autumn and spring, a process called shedding for mammals and molting for birds. Winter plumage and fur are denser than the summer coverings and provide better insulation (**Figure 20.3B,C**).

Avoiding cold and sharing heat

Small mammals can buffer the effects of winter conditions by reducing the time they are outside shelter or by sharing a nest with other individuals. Snow provides insulation; the ground surface beneath a layer of snow can be more than 20°C warmer than the air temperature. Small mammals can forage in this protected subnivean (beneath the snow) zone. When they are inactive they can retreat to nests that provide insulation, and by huddling in groups small mammals can reduce their own heat loss and capture some of the heat that other individuals lose.

Modifying activity patterns and seeking favorable microclimates   Small mammals such as shrews and voles have such high surface/volume ratios that their **thermoneutral zones** (range of environmental temperatures within which an endotherm can control its body temperature by adjusting its rate of heat loss without increasing heat

production) are very narrow. Nonetheless, these species are active through the winter in northern regions. They confine their winter activity to sheltered microhabitats, such as spaces within the layer of leaves on the forest floor where the temperature is –4°C and tunnels that have a comparatively balmy temperature of 1°C. They also reduce their exposure to cold by spending less time outside their nests; northern short-tailed shrews (*Blarina brevicauda*) forage for less than 3 hours (h) per day in winter compared with 8 h per day in summer.

Nests and social thermoregulation   Many small mammals construct nests underground or in hollows in trees and form groups that occupy them together during the winter. White-footed deer mice (*Peromyscus leucopus*), for example, construct elaborate nests during the winter, lining them with dried grasses and herbaceous vegetation. Like voles, deer mice spend more time in their nests in winter than in summer. Several individuals huddle together in a nest, and the group has a smaller surface/volume ratio than an individual mouse has, thus reducing heat loss. The mice in

Figure 20.4 Small mammals use behavioral adjustments to avoid cold. (A) Voles (*Microtus*) are too small to grow long hair and have a lower critical temperature of 28.9°C. (B) Voles minimize their exposure to cold in the winter by foraging in tunnels beneath the snow; these tunnels are revealed as the snow melts. (A, Tomi Tapio K/CC BY 2.0; B, Ellen Snyder.)

the center of the pile are warmer than those on the outside, of course, and there is a continuous shifting of positions as mice from the outside of the pile grow cold and push their way into the center.

Social thermoregulation has a substantial effect on the temperatures that taiga voles (*Microtus xanthognathus*) experience in winter. Groups of 5–10 individuals nesting together can maintain the nest temperature between 4°C and 10°C when the soil around the nest is between –3°C and –6°C and air temperatures on the surface are between –10°C and –20°C. Furthermore, individual voles forage on different schedules, so the nest is never empty. As a result, a vole returns to a warm nest after foraging.

20.3 Facultative Hypothermia

Winters pose a challenge to endotherms because low environmental temperatures increase the need for heat production—and for the food to sustain higher rates of heat production—during the season when food becomes scarce. Birds and mammals that weigh less than 100 g are particularly vulnerable to low temperatures for three reasons:

1. They have large surface/volume ratios and consequently high rates of heat loss.

2. They have large food requirements because their mass-specific metabolic rates (i.e., energy consumed per gram of body weight) are higher than those of larger species.

3. Their layers of insulation are thinner than those of larger animals. (An animal the size of a reindeer can have hair 3 cm long, but if a mouse had hair that long, its legs would not reach the ground!)

The rate at which an animal loses heat to the environment can be reduced by decreasing the heat gradient between the animal and the environment. The sheltered microclimates described in the preceding section adjust the environmental side of that relationship, and intentionally lowering body temperature—**facultative hypothermia**—is another way to reduce the temperature gradient between an animal and its environment.

Daily cycles of body temperature are probably universal among endotherms, including humans, and body temperature also varies with physiological status. For example, African lions (*Panthera leo*) lower their body temperatures when they are pregnant and control body temperature more precisely. As a result, the frequency of hyperthermia—body temperatures above 39.5°C for this species, which can damage developing fetuses—is reduced during pregnancy.

Seasonal reductions in body temperature form a continuum from species that have slightly lower body temperatures in winter than in summer while continuing their normal activities (**seasonal hypothermia**), through those that allow their body temperature to drop a few degrees and become inactive for hours (rest-phase hypothermia), to those that have profound reductions of body temperature that extend for weeks or months (**hibernation**).

Seasonal hypothermia

Some animals lower their body temperature as much as 5°C in winter while maintaining normal activity. In North America, eastern gray squirrels (*Sciurus carolinensis*), Abert's squirrels (*S. aberti*), and red squirrels (*Tamiasciurus hudsonicus*) have body temperatures in winter that are 1–6°C lower than in summer. In late summer and autumn, red squirrels accumulate larders of conifer cones in tunnels and construct well-insulated nests. In winter, they limit the time they spend outside their nests. These behaviors, combined with seasonal hypothermia, keep the daily winter energy expenditures of red squirrels at summer levels.

Rest-phase hypothermia

Rest-phase hypothermia is a decrease in body temperature that coincides with an animal's normal period of inactivity—at night for most birds and during the day for most small

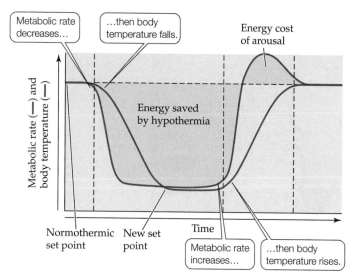

Figure 20.5 Changes in body temperature and metabolic rate during rest-phase hypothermia. As an animal enters rest-phase hypothermia, a decrease in metabolic rate (green trace) precedes a fall in body temperature to a new set point (red trace). On rewarming, an increase in metabolism precedes the return to normothermia. The metabolic rate during arousal briefly overshoots the resting rate, but the energy cost of arousal is less than the energy saved by the period of hypothermia.

mammals. Metabolic rate and thus heat production decrease, and body temperature falls 5–10°C lower than normal (**Figure 20.5**). The normal thermoregulatory mechanisms—thermogenesis and insulation—balance heat gain and heat loss at this new level, saving energy because the temperature gradient between the animal and its environment is smaller. At the end of the period of hypothermia, an increase in metabolic rate and heat production initiates warming, and the metabolic rate rises briefly above the resting rate. This period of elevated metabolism is the energy cost of arousal, and it is small compared with the energy that was saved during the time the animal was hypothermic. Calculations indicate that even entering hypothermia and immediately arousing saves energy for a small bird or mammal.

Chickadees on winter nights   Rest-phase hypothermia is widespread among small birds and mammals and has been reported in more than two dozen species of passerine birds. Black-capped chickadees (*Poecile atricapillus*) are an example (**Figure 20.6**). These small (10–12 g) birds are year-round residents in northern latitudes, where they regularly experience temperatures during winter that remain below freezing for days or weeks.

In winter, black-capped chickadees around Ithaca, New York, maintain body temperatures of 40–42°C during the day and cool to 29–30°C at night. This reduction in body temperature permits a 30% reduction in energy consumption. The chickadees rely primarily on fat stores they accumulate as they feed during the day to supply the energy needed to carry them through the following night. Thus, the energy available to them and the energy they use at night can be estimated by measuring the fat content of

birds as they go to roost in the evening and as they begin activity in the morning. The chickadees have an average of 0.80 g of fat when they cease activity in the evening, and by morning the fat store has decreased to 0.24 g. The fat metabolized during the night (0.56 g) corresponds to the metabolic rate expected for a bird with a body temperature of 30°C.

Nighttime hypothermia is essential for a chickadee because it would require 0.92 g of fat to maintain a body temperature of 40°C through the night—more fat than it has when it goes to roost in the evening. If chickadees did not become hypothermic, they would starve before morning. Even with hypothermia, they use 70% of their fat reserve overnight. They do not have an energy supply to carry them far past sunrise, and chickadees are among the first birds to begin foraging in the morning. They also forage in weather so foul that other birds, which are not in such a precarious energy balance, remain on their roosts. The chickadees must reestablish their fat stores each day if they are to survive the next night.

Migratory stopovers   Another function of rest-phase hypothermia may be to speed the rebuilding of energy stores during migratory stopovers. The Old World garden warbler (*Sylvia borin*) and icterine warbler (*Hippolais icterina*) normally have daytime body temperatures near 39.5°C, but they allow their temperatures to fall to 33.5°C at night while they remain at a stopover site, feeding and replenishing their fat stores. Migrating Eurasian blackcaps (*Sylvia atricapilla*) allow their body temperatures to fall from a daytime average of 42.5°C to 35.5°C at night, and that reduction in body temperature corresponds to an energy savings of more than 30%.

Hibernation

The deep hypothermia that occurs during hibernation is a comatose condition, much more profound than the deepest

Figure 20.6 The black-capped chickadee. The black-capped chickadee (*Poecile atricapillus*) is one of the smallest birds that lives year-round in Canada and the northern United States, and is in precarious energy balance during the winter. (David Ellis, US Fish & Wildlife Service.)

sleep. Hibernation is nearly confined to mammals; at most only a handful of bird hibernate and hibernation has been demonstrated only for the common poorwill (*Phalaenoptilus nuttallii*).

During hibernation, voluntary motor responses are reduced to sluggish postural changes, but some sensory perception of powerful auditory and tactile stimuli and environmental temperature changes is retained. Perhaps most dramatically, a hibernating animal can arouse spontaneously from this state using heat production by brown fat. Some endotherms can rewarm under their own power from the lowest levels of hibernation; others must warm passively with an increase in environmental temperature until some threshold is reached at which arousal starts.

Body size and hibernation The largest mammals that hibernate are marmots, which weigh about 5 kg, and deep hypothermia and body size are closely related. Deep hypothermia is not as advantageous for a large animal as for a small one. In the first place, the energy cost per gram of maintaining a high body temperature is lower for a large animal than for a small one, and as a consequence, a small animal has more to gain from hypothermia. Second, a large animal cools more slowly than a small animal, so its metabolic rate does not decrease as rapidly. Furthermore, large animals have more body tissue to rewarm on arousal, and their costs of arousal are correspondingly greater than those of small animals. Bears in winter dormancy, for example, lower their body temperatures only about 5°C from normal levels, and their metabolic rate decreases about 50%. That small reduction in body temperature, combined with the large fat stores bears accumulate before retreating to their winter dens, is sufficient to carry them through the winter, but their hypothermia is not deep enough to be considered hibernation.

Physiological changes during hibernation Profound changes occur in a variety of physiological functions of animals in deep hypothermia. Body temperature drops to within 1°C or less of the surrounding temperature, and in some cases (bats, for example) extended survival is possible at body temperatures just above the freezing point of the tissues. Arctic ground squirrels (*Urocitellus parryii*) allow the temperature of parts of their bodies to fall as low as −2.9°C. Active metabolic suppression (temperature-independent changes in gene expression and post-translational modifications of enzymes in oxidative metabolic pathways) starts before body temperature falls and reduces oxidative metabolism to as little as one-twentieth the rate at normal body temperatures.

Although body temperatures fall dramatically during hibernation, temperature regulation does not entirely cease. Instead the hypothalamic thermostat is reset. If the body temperature of a hibernating animal falls below the new set point, thermogenic processes bring the hibernating animal back to the regulated level.

Respiration is slow during hibernation, and an animal's overall breathing rate can be less than one breath per minute. Heart rate is drastically reduced, and blood flow to peripheral tissues is virtually shut down, as is blood flow posterior to the diaphragm. Most of the blood is retained in the body's core.

Arousal from hibernation Hibernation is an effective method of conserving energy during long winters, but hibernating animals rewarm at intervals. Heat production by brown fat between the shoulder blades is distributed by the circulatory system. The head, heart, and anterior part of the body warm first, followed several hours later by the posterior regions.

Periodic arousals are normal, and these arousals consume a large portion of the total energy used by hibernating mammals. An example of the magnitude of the energy cost of arousal is provided by Richardson's ground squirrels (*Urocitellus richardsonii*) in Alberta, Canada (**Figure 20.7**).

The activity season for ground squirrels in Alberta is short: about 100 days for adults. Emerging from hibernation in March, males establish territories and court females when they emerge about 2 weeks later. Males enter hibernation in June, and the females follow about 2 weeks later. Juveniles remain active, increasing their body weight as much as possible before entering hibernation in September. When the squirrels are active, they have body temperatures of 37–38°C; when they are hypothermic, their temperatures fall to as low as 3–4°C. Hibernation consists of a cycle of different phases, with periods of hypothermia alternating with periods of arousal and normothermy throughout the winter.

The periods of arousal account for most of the energy used by Richardson's ground squirrel during hibernation (**Table 20.1**). The energy costs associated with arousal include the cost of warming from the hibernation temperature to normothermy (37°C), the cost of sustaining normothermy for several hours in a cold burrow, and the metabolism as the body temperature slowly decreases during reentry into hypothermia. For the entire hibernation season, the combined metabolic expenditures for these three phases of the hibernation cycle account for an average of 85% of the total energy used by the squirrel.

Table 20.1	Energy use during the hibernation cycle of a Richardson's ground squirrel			
	Percent of total energy per month			
Month	**Entry**	**Hypothermia**	**Arousal**	**Normothermia**
July	17.8	8.5	17.2	56.5
September	15.7	19.2	15.2	49.9
November	13.0	20.8	23.1	43.1
January	11.2	24.8	24.1	40.0
March	6.3	3.3	14.1	76.4

Source: From Wang 1978.

(A)

Figure 20.7 Hibernation in Richardson's ground squirrel. (A) Richardson's ground squirrel (*Urocitellus richardsonii*). In Alberta, Canada, this species spends most of the year in hibernation. (B) Body temperature of a Richardson's ground squirrel from September through mid-March. This squirrel entered hibernation in September, when the temperature in the burrow was about 13°C. As winter progressed and the burrow's temperature fell, the intervals between arousals lengthened and the squirrel's body temperature decreased. By late December, the burrow temperature had dropped to 0°C and the periods between arousals were 14–19 days. In late February, the periods of hypothermia became shorter, and in early March the squirrel emerged from hibernation. (C) Body temperature of a Richardson's ground squirrel during a single hypothermia cycle. Entry into hypothermia began shortly after noon on February 16; 24 hours later, the body temperature had stabilized at 3°C. This period of hypothermia lasted until late afternoon on March 7, when the squirrel started to arouse. In 3 hours, the squirrel warmed from 3°C to 37°C. It maintained that body temperature for 14 hours and then entered hypothermia again (not shown). (A, Doug Collicutt; B,C, from Wang and Hudson 1978.)

(B)

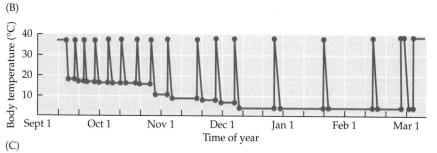

(C)

Communal hibernation Many species of ground squirrels and marmots hibernate in groups. For example, alpine marmots (*Marmota marmota*) hibernate in groups consisting of a male, female, and their young from the past several years—some groups contain as many as 20 individuals. The animals huddle together, producing a heap of animals that has a smaller surface/volume ratio than a single individual. The body temperatures of the marmots in the heaps remain 3–4°C above freezing, although the air temperature in the burrow is below freezing. Communal hibernation increases survival, especially of the youngest individuals (i.e., the ones born the previous spring).

Bouts of arousal by communally hibernating marmots are synchronized, which saves energy because heat is shared by all of the individuals. The older individuals begin arousal first, and heat transferred from them assists the younger individuals in warming up. The effectiveness of this synchrony in arousal is demonstrated by comparing weight loss during hibernation: marmots in highly synchronized groups lose 20–25% of their initial body weight, whereas those in poorly synchronized groups lose 40–45%—almost twice as much.

Surprisingly, we have no clear understanding of why a hibernating ground squirrel undergoes these arousals that increase its total winter energy expenditure nearly fivefold. Ground squirrels do not store food in their burrows, so they are not using the normothermal periods to eat. They do urinate during normothermy, so eliminating accumulated nitrogenous wastes may be the reason for arousal, and spending some time at a high body temperature may be necessary to carry out other physiological or biochemical activities, such as resynthesizing glycogen, redistributing ions, or synthesizing serotonin. Arousal may also allow a hibernating animal to determine when environmental conditions are suitable for emergence. Whatever their function, the arousals must be important because the squirrel pays a high price for them during a period of extreme energy conservation.

20.4 Endotherms in the Heat

Hot environments place more severe physiological demands on endotherms than do cold environments. Endotherms encounter two problems in regulating their body temperature in hot deserts. The first results from a reversal of the normal relationship of an animal to its environment. In most environments, an endotherm is warmer than the air temperature; heat flows from the animal to its environment, and thermoregulatory mechanisms achieve a stable body temperature by balancing heat production and heat loss. Very cold environments merely increase the gradient between an animal's body temperature and the environment. The example of the Arctic fox with a lower critical temperature of −40°C illustrates the success that endotherms have had in providing sufficient insulation to cope with enormous gradients between high core body temperatures and low environmental temperatures.

In a hot desert, the temperature gradient is not increased—it is reversed. Desert air temperatures can climb to 40–50°C during summer, and the ground temperature may exceed 60–70°C. Instead of losing heat to the environment, an animal is continuously absorbing heat, and that heat plus metabolic heat must somehow be dissipated to maintain the animal's body temperature in its normal range. Maintaining a body temperature 10°C below the ambient temperature can be a greater challenge for an endotherm than maintaining it 100°C above the ambient temperature.

Temperature stress and scarcity of water

Temperature stress and scarcity of water interact to challenge endotherms in deserts. Evaporating a kilogram of water dissipates approximately 2,400 kJ. (The exact value varies slightly with temperature.) Thus, evaporative cooling is an effective cooling mechanism as long as an animal has an unlimited supply of water. In a hot desert, however, where thermal stress is greatest, water is a scarce commodity and its use must be carefully rationed. Calculations show, for example, that if a kangaroo rat (*Dipodomys*) were to venture out into the desert sun, it would have to evaporate 13% of its body water per hour to maintain a normal body temperature. Most mammals die when they have lost 10–20% of their body water, so evaporative cooling is of limited use in deserts except as a short-term response to a critical situation.

Disposing of waste products in the desert is another problem. Birds and mammals must eat a lot to maintain their high metabolic rates, and they produce correspondingly large volumes of waste. Mammalian feces contain about 50% water by weight, a value that varies little with body size or habitat. The water content of the mixture of feces and urine excreted by birds is often higher—50% to 70%—but this value includes nitrogenous waste from the kidneys and cannot be compared directly with the waste products of mammals. Birds probably have less difficulty conserving urinary and fecal water than mammals do because the basal sauropsid excretory system reabsorbs water in the bladder or cloaca and excretes ions as sodium and potassium urates.

Form and function are intimately related in mammalian kidneys. A thick medulla forms a long renal papilla (= renal pyramid) with great concentrating power, and the maximum urine osmolalities of mammals are proportional to the relative medullary thickness of their kidneys (the ratio of the depth of the medulla to the thickness of the cortex). Some desert rodents with exceptionally long renal papillae have relative medullary thicknesses of 15–20, and the maximum urine concentrations of these animals exceed 7,000 millimoles (mmol)/kg. (The relative medullary thickness of human kidneys is an unimpressive 3.0, and the maximum urine concentration is only 1,430 mmol/kg.) **Figure 20.8** shows the strong correlation

between relative medullary thickness and maximum urine concentration, but substantial variation is apparent, indicating that other anatomical or physiological factors are involved. For mammals as a group, relative medullary thickness accounts for 59% of interspecific variation in maximum urine concentration.

Strategies for desert survival

Deserts are challenging environments, especially for endotherms, but they contain a mosaic of microenvironments that animals can use to find the conditions they need. Broadly speaking, endotherms have two general methods for coping with desert conditions:

- *Avoidance.* Some endotherms manage to avoid desert conditions by behavioral means. They live in deserts but are rarely exposed to the full rigors of desert life.

Coypu
RMT = 3.5
Urine osmolarity = 741

Pocket mouse
RMT = 17.6
Urine osmolarity = 8,600

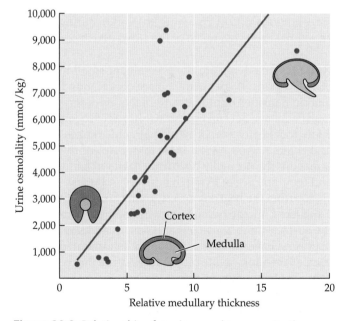

Figure 20.8 Relationship of maximum urine concentration to relative medullary thickness for 29 species of rodents. The sketches illustrate kidneys with different relative medullary thicknesses (RMTs). The medulla of species from moist habitats, such as coypu (*Myocastor coypus*), is small (low RMT), whereas species from arid habitats (Pacific pocket mouse, *Perognathus longimembris*) have high much larger medullas and larger RMTs. (Photos: Coypu, Richard Griffin; Pocket mouse, Joanna Gilkeson, US Fish & Wildlife Service.)

• *Relaxation of homeostasis.* Some endotherms have relaxed the limits of homeostasis. They survive in deserts by tolerating hyperthermia, hypothermia, or dehydration (low body-water content).

These two categories are not mutually exclusive; many desert endotherms combine elements of both responses.

Avoidance

Cloudless skies allow the desert surface to become very hot during the day and to grow cold at night as that heat is radiated into space. Shelters—natural rock crevices or burrows constructed by animals—make it possible to escape temperature extremes. Indeed, ecosystem engineering is a critical ecological service provided by burrowing animals, whose burrows serve as refuges for a wide variety of invertebrates as well as other vertebrates. In California's Mojave Desert, horned larks (*Eremophilus alpestris*), common poorwills (*Phalaenoptilus nuttallii*), and burrowing owls (*Athene cunicularia*) seek shelter during the heat of the day in the burrows of desert tortoises, and in Saudi Arabia four species of larks shelter in the burrows of large *Uromastyx* lizards. Moving higher off the ground is another way to escape heat. Climbing a meter into a bush can allow an animal to reach substantially cooler air, and hawks and vultures that soar above the desert during the day do not face thermal stress.

Similarly, deserts are not dry everywhere. Relative humidity is higher in burrows than it is on the desert floor, and oases are within the reach of many desert birds. Thus, behavior provides a first line of responses that mitigate desert conditions, and by changing its location during the day, an animal can avoid much of the stress of a desert.

Rodents   Rodents are the preeminent small mammals of arid regions. It is a commonplace observation that population densities of rodents are higher in deserts than in moist habitats. Several ancestral features of rodent biology allow rodents to extend their geographic ranges into hot, arid regions. Among the most important of these characters are the normally nocturnal habits of many rodents and their practice of living in burrows. A burrow provides escape from the heat of a desert, giving an animal access to a sheltered microenvironment while soil temperatures on the surface climb above lethal levels. In addition, rodents in general have kidneys that produce concentrated urine—even a laboratory white rat produces urine that is twice as concentrated as human urine.

Kangaroo rats are among the most specialized desert rodents in North America (**Figure 20.9**). Merriam's kangaroo rat (*Dipodomys merriami*) occurs in desert habitats from central Mexico to northern Nevada. A population of this species lives in extreme conditions in the Sonoran Desert of southwestern Arizona. In summer, daytime temperatures on the ground approach 70°C, and even a few minutes of exposure would be deadly. Kangaroo rats spend the day in burrows 1–1.5 m underground, where air temperatures do not exceed 35°C even during the hottest parts of the year.

Figure 20.9 Merriam's kangaroo rat (*Dipodomys merriami*). Kangaroo rats are nocturnal, and because deserts cool off rapidly at night, low environmental temperatures are a greater problem for them than heat. (Bcexp/CC BY-SA 4.0.)

In the evening, when the kangaroo rats emerge to forage, external air temperatures have fallen to about 35°C.

Birds   Because they can fly, birds are much more mobile than mammals. A kangaroo rat or ground squirrel is confined to a home range less than 100 m in diameter, but it is quite possible for a desert bird with the same body size as those rodents to fly many kilometers to reach water. For example, mourning doves (*Zenaida macroura*) in the deserts of North America congregate at dawn at water holes, with some individuals flying 60 km or more to reach them.

The normally high and variable body temperatures of birds give them another advantage that is not shared by mammals. With a body temperature of 40°C, a bird faces the problem of a reversed temperature gradient between its body and the environment for a shorter time each day than does a mammal. Furthermore, birds' body temperatures are normally variable, and birds tolerate moderate hyperthermia without apparent distress. These are all ancestral characters that are present in virtually all birds. Neither the body temperatures nor the upper lethal temperatures of desert birds are higher than those of related species from nondesert regions.

The mobility provided by flight does not extend to fledgling birds, and the most conspicuous adaptations of birds to desert conditions are those that ensure a supply of water for the young. Altricial fledglings (those that need to be fed by their parents after hatching) receive the water they need from their food. One pattern of adaptation in desert birds ensures that reproduction will occur at a time when succulent food is available for fledglings. In the arid central region of Australia, bird reproduction is precisely keyed to rainfall. The sight of rain is apparently sufficient to stimulate courtship, and mating and nest building commence within a few hours of the start of rain. This rapid response ensures that the baby birds will hatch in the flush of new vegetation and insect abundance stimulated by the rain.

A different approach, very like that of mammals, has evolved among columbiform birds (pigeons and doves), which are widespread in arid regions. Fledglings are fed

(A)

(B)

Figure 20.10 Short-term cycles of activity and body temperature of an antelope ground squirrel. (A) White-tailed antelope ground squirrels (*Ammospermophilus leucurus*) are diurnal, even during the summer when ambient temperatures exceed the species' upper lethal temperature. (B) Early in the day, a squirrel can dump heat by pressing its lightly furred belly against a shaded rock that is still cool from the night. Later in the day it must seek its burrow to find a cool substrate. (A, Marshal Hedin/CC BY 2.0; B, Renee/CC BY 2.0.)

on "pigeon's milk," a liquid substance produced by the crop under the stimulus of prolactin, which is a hormone secreted by the anterior pituitary gland. The chemical composition of pigeon's milk is very similar to that of mammalian milk; it is primarily water plus protein and fat, and it simultaneously satisfies both the nutritional requirements and the water needs of fledglings. This approach places the water stress on the adult, which must find enough water to produce the pigeon's milk as well as meet its own water requirements.

Seed-eating desert birds with precocial young, such as the sandgrouse (more than a dozen species of *Pterocles*) inhabiting the deserts of Africa and the Near East, face particular challenges in providing water for their young. Sandgrouse chicks begin to find seeds for themselves within hours of hatching. However, they are unable to fly to water holes as their parents do, and seeds do not provide the water they need. Instead, adult male sandgrouse transport water to their broods. The breast feathers of males have a unique structure in which the proximal portions of the barbules are coiled into helices. When the feather is wetted, the barbules uncoil and trap water. The feathers of male sandgrouse hold 15–20 times their weight with water, and the feathers of females hold 11–13 times their weight.

Male sandgrouse in the Kalahari Desert of southern Africa fly to water holes just after dawn and soak their breast feathers, absorbing 25–40 mL of water. Some of this water evaporates on the flight back to their nests, but calculations indicate that a male sandgrouse can fly 30 km and arrive with 10–28 mL of water still adhering to its feathers. As the male sandgrouse lands, the chicks rush to him and, seizing the wet breast feathers in their beaks, strip the water from them with downward jerks of their heads. In a few minutes, the young birds have satisfied their thirst, and the male rubs himself dry on the sand.

Relaxation of homeostasis by hyperthermia

Temporarily allowing the body temperature to rise above a species' normal set point allows an animal to conserve water that would otherwise have been used for evaporative cooling. In a desert, conserving water can be more critical than strictly regulating body temperature. Body size determines the way hyperthermia can be employed, and small mammals differ from large mammals.

Small mammals   Not all desert rodents are nocturnal. Ground squirrels forage during the day, running across the desert surface. The almost frenetic activity of desert ground squirrels on intensely hot days is a result of the thermoregulatory problems that small animals experience under these conditions. Studies of the white-tailed antelope ground squirrel (*Ammospermophilus leucurus*) at Deep Canyon, California (near Palm Springs) illustrate the short-term relaxation of body temperature homeostasis (**Figure 20.10**).

The heat on summer days at Deep Canyon is intense, and the squirrels are exposed to high heat loads for most of the day. They have a bimodal pattern of activity that peaks in the midmorning and again in the early evening. Relatively few squirrels are active in the middle of the day. The body temperatures of antelope ground squirrels are labile, and the body temperatures of individual squirrels can vary as much as 7.5°C (from 36.1°C to 43.6°C) during a day. The squirrels use this variability of body temperature to store heat during their periods of activity.

Because the squirrels are small and heat rapidly, high temperatures limit their bouts of activity to no more than 9–13 minutes. They sprint furiously from one patch of shade to the next, pausing only to seize food or look for predators. The squirrels minimize exposure to the highest temperatures by running across open areas, and they seek shade or their burrows to cool off; here their small size is an advantage because they cool rapidly. In the middle of a hot summer day, a squirrel can maintain a body temperature lower than 43°C (the maximum temperature it can tolerate) only by retreating every few minutes to a burrow deeper than 60 cm, where the soil temperature is 30–32°C.

Large mammals Large animals, including humans, have advantages and disadvantages in desert life that are directly related to body size. A large animal has nowhere to hide from desert conditions. It is too big to burrow underground, and few deserts have vegetation large enough to provide useful shade to an animal much larger than a jackrabbit. However, large body size offers some options not available to smaller animals. Large animals are mobile and can travel long distances to find food or water, whereas small animals may be limited to home ranges only a few meters or tens of meters in diameter. Large animals have low surface/volume ratios and can be well insulated. Consequently, they absorb heat from the environment slowly. A large body weight gives an animal a large thermal inertia; that is, it can absorb a large amount of heat energy before its body temperature rises to dangerous levels.

The two extant species of camels, the Asian (two-humped) bactrian camel (*Camelus ferus*) and the Arabian (one-humped) dromedary camel (*C. dromedarius*) are classic large desert animals, with adult body weights of 400–450 kg for females and up to 500 kg for males.

Behavioral mechanisms and the distribution of hair on the body aid camels in reducing their heat load. In summer, camels have hair 5–6 cm long on the neck and back and up to 11 cm long over the hump. On the ventral surface and legs, the hair is only 1.5–2 cm long. Early in the morning, camels lie down on surfaces that have cooled overnight by radiation of heat to the night sky. They tuck their legs beneath the body, placing the ventral surface, with its short covering of hair, in contact with the cool ground (**Figure 20.11A**). In this position, a camel exposes only its well-protected neck and back to the sun and places its lightly furred legs and ventral surface in contact with cool sand. Camels assemble in small groups and lie pressed closely together throughout the day. Spending a day in the desert sun squashed between two sweaty camels may not be your idea of fun, but in this posture a camel reduces its heat gain because it keeps its sides in contact with other camels (both at about 40°C) instead of allowing solar radiation to raise its fur surface temperature to 70°C or higher.

The camel's adjustments to a the shortage of water in deserts are revealed by comparing the daily cycle of body temperature in a camel that receives water daily with the cycle of one that has been deprived of water (**Figure 20.11B**). The body temperature of a watered camel is 36°C in the early morning and rises to 39°C in midafternoon. When a camel is deprived of water, the daily temperature variation doubles. Body temperature falls to 34.5°C at night and climbs to 40.5°C during the day.

The significance of this increased daily fluctuation in body temperature can be assessed in terms of the water that the camel would expend in evaporative cooling to prevent the 6°C rise. With a specific heat of 3.5 kJ/kg per degree Celsius, a 6°C increase in body temperature for a 500-kg camel represents storage of 10,500 kJ of heat. Evaporation of a liter of water dissipates approximately 2,400 kJ. Thus, a camel would have to evaporate about 4.5 L of water daily to

(A)

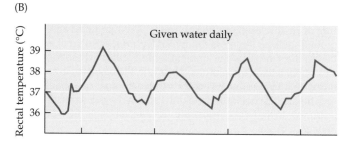

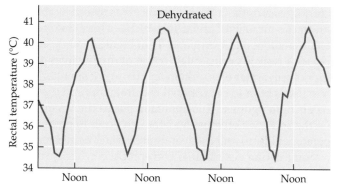

(B)

Figure 20.11 Daily cycles of body temperature of camels. (A) In the heat of the day, dromedary camels (*Camelus dromedarius*) face into the sun so that solar radiation falls on their well-insulated necks, shoulders, and backs. They are pressed together, so solar energy does not strike their sides, and their lightly insulated bellies and legs are in contact with sand that is still cool from the previous night. (B) When a dromedary camel has daily access to water (top), its body temperature cools to about 36°C at night and rises to 38–39°C during the day. When the camel is dehydrated (bottom), its body temperature falls below 35°C at night and rises above 40°C during the day. (A, Marcel Peek; B, after Schmidt-Nielsen et al. 1956.)

maintain a stable body temperature at the nighttime level, and it can conserve that water by tolerating hyperthermia during the day.

In addition to saving the water that was not used for evaporative cooling, the camel receives an indirect benefit from hyperthermia via a reduction of energy flow from the air to the camel's body. As long as the camel's body temperature is lower than the air temperature, a gradient exists that causes the camel to absorb heat from the air. At a body temperature of 40.5°C, the camel's temperature is

Table 20.2	Daily water loss in a 250-kg camel			
	Water loss (L/day) by different routes			
Condition	Feces	Urine	Evaporation	Total
Drinking daily (average of 8 days)	1.0	0.9	10.4	12.3
Not drinking (average of 17 days)	0.8	1.4	3.7	5.9

equal to that of the air for much of the day, and no net heat exchange takes place. Thus, the camel saves an additional quantity of water by eliminating the temperature gradient between its body and the air. The combined effect of these measures on water loss is illustrated by data from a young dromedary camel (**Table 20.2**). When deprived of water, the camel reduced its evaporative water loss by 64% and reduced its total daily water loss by more than 50%

The strategy a camel uses is basically the same as that employed by an antelope ground squirrel: saving water by allowing the body temperature to rise until the heat can be dissipated passively. The difference between the two animals is a consequence of their difference in body size. A camel weighs 500 kg and can store heat for an entire day and cool off at night, whereas an antelope ground squirrel weighs about 100 g and heats and cools many times during the course of a day.

Keeping a cool head   The brain is more sensitive to hyperthermia than other parts of the body are, and brain temperatures higher than 43°C rapidly cause damage in most mammals. Camels and many other large desert mammals, however, can maintain rectal temperatures well above 43°C for hours. They keep their brain temperatures lower than their body temperatures by using a countercurrent heat exchange to cool blood before it reaches the brain. The blood supply to the brain passes via the external carotid arteries, and at the base of the brain the arteries break into a rete mirabile that lies within a structure called the cavernous sinus (**Figure 20.12**). The blood in the sinus is returning from the walls of the nasal passages where it has been cooled by the evaporation of water. This chilled venous blood cools the warmer arterial blood before it reaches the brain.

A mechanism that produces selective cooling of the brain is widespread among vertebrates, but the anatomical details vary. Many ungulates (antelope, sheep, goats) have a carotid retia mirabile within a cavernous sinus, but horses cool the blood in the internal carotid arteries by passing it through the guttural pouches—outgrowths from the auditory tubes that envelope the internal carotid arteries and

are filled with air that is cooler than the blood. Birds use the ophthalmic retia to cool blood flowing to the brain. Blood in the ophthalmic retia, which lie behind the eyes, is cooled by evaporation of water from the surface of the eyes.

Hypothermia in the desert

Many desert rodents have the ability to reduce their body temperature in response to a shortage of food. When the food ration of the Arizona pocket mouse (*Perognathus amplus*) is reduced slightly below its daily requirements, the mouse allows its body temperature to fall for a part of the day. The duration of hypothermia is proportional to the severity of food deprivation. As its food ration is reduced, a pocket mouse spends more time each day in hypothermia and conserves more energy.

Adjusting the time spent in hypothermia to match the availability of food may be a general phenomenon among seed-eating desert rodents. These animals appear to assess the rate at which they accumulate food supplies during foraging rather than their actual energy balance. Species that accumulate caches of food become hypothermic even with large quantities of stored food on hand if they are unable to add to their stores by continuing to forage. When seeds are deeply buried in the sand, and thus hard to find, pocket mice spend more time in hypothermia than they do when the same quantity of seed is close to the surface. This behavior is probably a response to the chronic food shortage that desert rodents face because of the low primary productivity of desert communities and the effects of unpredictable variations from normal rainfall patterns, which may almost completely eliminate seed production by desert plants in dry years.

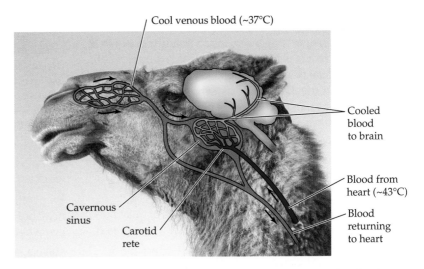

Figure 20.12 A countercurrent heat-exchange mechanism cools the blood going to a camel's brain.   Venous blood is cooled by evaporation in the nasal passages. When this blood enters the cavernous sinus, it has a temperature of about 37°C. Blood flowing to the brain via the internal carotid arteries has a temperature of about 43°C when it enters the vessels of a rete mirabile in the cavernous sinus. Here it is surrounded by the cool venous blood and has cooled to approximately 40°C by the time it reaches the brain. (Photo by Victoria Reay.)

Summary

Endothermy allows birds and mammals substantial independence from low environmental temperatures.

Endotherms can be active at night and the winter and can live in polar and high-mountain environments.

The benefits of endothermy come at an energetic cost. The metabolic rates of endotherms are about ten higher than those of ectotherms of the same body size, and on a gram-per-gram basis endotherms must find and consume more food than ectotherms of the same body weight.

The ability of endotherms to live in polar regions and deserts stems more from combinations of general anatomical, physiological, and behaviorial characters than it does from novel adaptations.

Endotherms regulate their body temperatures by matching their rate of heat production to their rate of heat loss.

The heart and kidneys are the sources of most metabolic heat production when an endotherm is at rest.

Metabolism of skeletal muscle makes little contribution to metabolic heat production at rest but can increases 10- to 15-fold during intense physical activity. In this situation, skeletal muscle activity can contribute to thermoregulation.

Birds and mammals use shivering and non-shivering thermogenesis to increase metabolic heat production.

Shivering consists of uncoordinated contraction of muscle fibers that releases heat and evolved independently in birds and mammals.

Non-shivering thermogenesis consists of mechanisms that increase the rate of oxidative phosphorylation by short-circuiting a cellular process. Uncoupling proteins, which increase proton leakage in the mitochondria, are found in brown fat. Sarcolipin increases the permeability of the sarcoplasmic reticulum, allowing Ca^{2+} ions to leak out.

Hair and feathers insulate an endotherm by trapping air. Their insulative value can be increased by raising the hair or feathers to deepen the layer of trapped air and decreased by allowing the hair or feathers to lie flat.

Evaporation of water—by sweating, panting, or gular fluttering—cools the surface from which water is evaporated. In the case of sweating, that surface is the skin, which is also the surface being heated by the environment. Panting and gular fluttering cool internal tissues but require muscular activity, which releases additional heat.

Some endotherms can maintain stable body temperatures in very cold environments.

Tropical endotherms must increase heat production at temperatures between 20°C and 30°C, whereas the insulation of polar birds and mammals allows them to maintain basal metabolic rates at temperatures well below 0°C.

The depth of insulation depends partly on the size of an animal. Very small birds and mammals have relatively thin layers of insulation and depend on behavioral mechanisms to avoid cold stress.

Many small mammals, such as squirrels, reduce their activity in winter; they spend most of their time in insulated nests and rarely venture into the cold. Shrews, mice, and voles forage beneath the snow, where temperatures are warmer. Some species share nests, pooling their heat production and huddling together to reduce the surface/volume ratio of the group.

Hypothermia reduces energy requirements.

Squirrels and some other small mammals exhibit seasonal hypothermia, lowering their body temperatures in winter and living for months at a time with body temperatures a few degrees lower than they are in summer.

Many birds and mammals reduce energy consumption by engaging in rest-phase hypothermia, allowing their body temperatures to fall several degrees when they are inactive, and warming to their usual temperatures when they resume activity.

Some mammals weighing no more than about 5 kg and one bird hibernate, allowing body temperatures to fall dramatically and remaining hypothermic for months at a time. These periods of hypothermia are interrupted by brief periods of warming, and those periods of normothermy account for most of the energy used by a hibernating animal.

Hot environments place more severe physiological demands on endotherms than cold environments do.

When the environmental temperature is higher than an animal's body temperature, heat flows from the environment into the animal.

Evaporative cooling is a short-term response to excess heat, but it is not a long-term strategy, especially in deserts where water shortage is combined with heat.

Many animals that live in deserts avoid heat stress. Most small rodents are nocturnal, spending the day underground and emerging at night when it is cool. Some birds shelter during the day in burrows constructed by other animals. Hawks, eagles, and vultures soar in cool air.

Summary (*continued*)

Birds are more mobile than mammals, and adults can fly to distant water sources. Fledgling birds cannot follow their parents, and getting water to nestlings is a problem. Many desert birds time their reproduction to occur at a time when succulent food is available for their nestlings. Pigeons and doves feed their young "pigeon's milk"—a watery mixture of lipids and proteins produced in the crop and similar in composition to mammalian milk. Male sandgrouse use their modified breast feathers to transport water tens of kilometers to their young.

Storing heat by tolerating periods of hyperthermia allows some desert animals to continue activity in extreme heat.

Small diurnal rodents, such as ground squirrels, allow their body temperatures to rise while they forage on the ground surface. They then retreat to burrows to cool off before resuming surface activity.

When they are unable to drink, camels allow their body temperatures to rise during the day and fall at night, avoiding the need for evaporative cooling by tolerating periods of hyperthermia.

The brain is more sensitive to high temperature than other parts of the body are, and many mammals and birds use countercurrent heat exchange to cool blood before it reaches the brain.

Although hypothermia in a hot desert is counterintuitive, mammals become hypothermic to conserve energy and water.

Food and water shortages stimulate periods of hypothermia by seed-eating desert rodents. The duration of hypothermia is proportional to the magnitude of the shortage.

Discussion Questions

1. Why does an endotherm's metabolic rate increase above the upper lethal temperature? And why does the metabolic rate fall below the lower lethal temperature?

2. What arctic animals use countercurrent circulation? Where in those animals is it most conspicuous?

3. Compare the advantages and disadvantages of large body size for mammals that live in cold environments. Go beyond the information presented in the chapter.

4. Most mammals have a layer of cutaneous fat beneath the skin, but dromedary camels concentrate fat storage in their hump. What is the significance of that difference for the thermoregulation of camels?

5. How can you estimate the concentration of the urine that a species of rodent produces if the only information you are given is a photograph of an intact kidney from that species? Explain.

6. Are birds more like humans or kangaroo rats in the challenges they face in deserts and the way they cope with these challenges?

Additional Reading

Blix AS. 2016. Adaptations to polar life in mammals and birds. *Journal of Experimental Biology* 219: 1093–1105.

Busiello RA, Savarese S, Lombardi A. 2015. Mitochondrial uncoupling proteins and energy metabolism. *Frontiers in Physiology* 6: article 36.

Gaudry MJ and 9 others. 2017. Inactivation of thermogenic UCP1 as a historical contingency in multiple placental mammal clades. *Science Advances* 3 (7): e1602878; doi: 10.1126.

Hetem PS and 7 others. 2012. Selective brain cooling in Arabian oryx (*Oryx leucoryx*): A physiological mechanism for coping with aridity? *Journal of Experimental Biology* 215: 3917–3924.

McGaugh S, Schwartz TS. 2017. Here and there, but not everywhere: Repeated loss of uncoupling protein 1 in amniotes. *Biology Letters* 13: 20160749.

Mitchell D and 6 others. 2002. Adaptive heterothermy and selective brain cooling in arid-zone mammals. *Comparative Biochemistry and Physiology Part B* 131: 571–585.

Newman SA. 2011. Thermogenesis, muscle hyperplasia, and the origin of birds. *Bioessays* 33: 653–656.

Nowack J, Geiser F. 2016. Friends with benefits: The role of huddling in mixed groups of torpid and normothermic animals. *Journal of Experimental Bioogy* 219: 590–596.

Rezende EL, Bacigalupe LD. 2015. Thermoregulation in endotherms: Physiological principles and ecological consequences. *Journal of Comparative Physiology B* 185: 709–727.

Ruf T, Geiser F, 2015. Daily torpor and hibernation in birds and mammals. *Biological Reviews* 90: 891–926.

Staples JF. 2014. Metabolic suppression in mammalian hibernation: The role of mitochondria. *Journal of Experimental Biology* 217: 2032–2036.

Walter I, Seebacher F. 2009. Endothermy in birds: Underlying molecular mechanisms. *Journal of Experimental Biology* 212: 2328–2336.

21

The Origin and Radiation of Birds

Reconstruction by Carl Buell.

Birds are derived theropod dinosaurs, and few evolutionary transitions are as clearly recorded in the fossil record as the appearance of birds. More than 150 years ago, Thomas Henry Huxley was an ardent advocate of that relationship, writing that birds are nothing more than "glorified reptiles." Huxley, in fact, was so impressed by the many similarities of birds and reptiles that he placed them together in the class Sauropsida. Traditional systematics, with its emphasis on strict hierarchical categories, obscured that evolutionary relationship by placing reptiles and birds at the same taxonomic level, as the classes Reptilia and Aves. Now, because phylogenetic systematics emphasizes monophyletic evolutionary lineages rather than hierarchies, birds are once again seen as the most derived theropod dinosaurs.

Over the past 25 years, spectacular new discoveries of fossils of nonavian dinosaurs and early birds have turned our understanding of the evolution of birds on its head. Although the origin of birds from the theropod dinosaur lineage was proposed by Huxley in 1867, the material available as recently as 25 years ago suggested that birds evolved feathers, wings, and the ability to fly within the last 10 million years. Now we know that avian characters extend back for at least 50 million years, and perhaps more. The stepwise evolution of avian characters provides a classic illustration of macroevolution.

21.1 Avian Characters in Nonavian Theropods

Although the large—in some cases enormous—sauropod dinosaurs attract attention in museum exhibits, it was a lineage of theropods, the coelurosaurs, that gave rise in the Late Jurassic to birds (**Figure 21.1**). Fossils from the fine-grained limestone of the Jehol formation in northeastern China have been especially important in tracing the evolution of birds in the Late Jurassic and Early Cretaceous.

Skeletal characters

Extant birds have four distinctive skeletal features: (1) a trunk that is held in a horizontal position, with a short tail; (2) highly modified forelimbs; (3) hindlimbs with extensive fusion of bones and digitigrade foot posture (i.e., walking on the toes, with the heels off the ground); and (4) a paedomorphic skull. Discoveries of derived coelurosaurs have revealed that all of these characters appeared long before birds. The developmental mechanisms responsible for some of these changes in skeletal features have been identified:

- *Furcula.* The **furcula** (wishbone), formed by fusion of the clavicles, can be traced to basal theropods and appears in nearly all theropod lineages. For birds, the furcula is a component of the flight mechanism, but the widespread occurrence of a furcula among theropods shows that the origin of this structure had nothing to do with flight.

- *Postcranial pneumatization.* Early in the evolution of theropods, air spaces developed in the cervical and anterior dorsal vertebrae (i.e., they became pneumatized). Expansion of pneumatization to the posterior dorsal vertebrae, ribs, and long bones appeared independently in as many as a dozen lineages of theropods as well as in sauropods and pterosaurs. Thus, skeletal pneumatization evolved independently of flight.

- *Wrists.* The wrist joint of extant birds allows the carpometacarpus (the fused bones that form the outer portion of the wing skeleton) to bend inward to rest against the ulna and radius when the wing is folded. (The human wrist bends inward only to a 90-degree angle; a bird's wrist folds all the way down.) This extreme mobility results from modifications of the radiale (one of the wrist bones) and appeared in coelurosaurs long before the evolution of wings or flight.

- *Ribs.* The ribs of birds have bony projections, the **uncinate processes**, that extend upward and backward from each rib. These structures provide a mechanical advantage for respiratory muscles.

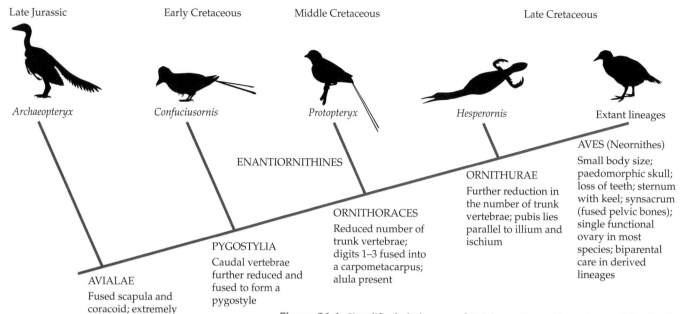

Figure 21.1 Simplified phylogeny of Avialae. Current knowledge of the fossil record reveals the evolution of the extant bird lineages over the latter half of the Mesozoic. (This general overview does not indicate actual times of divergence; see Figure 21.12.) Progressive elongation of the forelimbs (arms and hands) was accompanied by loss of digits 4 and 5, and ultimately by the fusion of digits 1, 2, and 3 into the carpometacarpus (familiar as the outer part of "buffalo wings"). The fibula became progressively reduced, and the tibia ("drumstick") fused to some of the tarsal (ankle) bones, forming the tibiotarsus. The remaining tarsal bones fused with the metatarsals, forming the tarsometatarsus. The last several caudal vertebrae fused into the pygostyle, and the sacrum fused with more anterior caudal vertebrae to form the synsacrum. The skulls of extant birds are paedomorphic, with a large cranium. Enantiothornine species were the dominant birds of the Middle Cretacous; however, modern birds (Aves) are believed to have arisen from the Ornithurae.

Oviraptors and dromaeosaurs also had uncinate processes, and these similarities in the structure of the rib cage strengthen the inference that nonavian theropods had a flow-through pattern of respiration like that of extant crocodylians, birds, and some lizards.

Feathers

A fossil imprint of a single feather found in the Solnhofen Limestone in southern Germany in 1860 was followed a year later by the discovery of a complete fossil of *Archaeopteryx lithographica* with impressions of wing and tail feathers (**Figure 21.2**). All of the 11 currently known specimens of *A. lithographica* were found in the Solnhofen Limestone. This fine-grained sedimentary rock retains detailed impressions of the feathers, which would not be visible in a coarser mineral matrix. These discoveries reinforced evidence from extant species—taxonomists believed that only birds have feathers, and therefore *Archaeopteryx* must be a bird. This view prevailed for

Figure 21.2 *Archaeopteryx lithographica.* The fossil pictured was the second specimen to be found. It was collected in 1874 or 1875 and is in the Museum für Naturkunde, Berlin. Feathers on the wings and tail are clearly visible in the specimen as it exists now (A), but a photo made in 1880 (B) shows feathers on the legs that were removed during preparation. (C) An isolated feather collected in 1860 or 1861 shows that *Archaeopteryx* had asymmetrically vaned feathers, suggesting that it could fly. (A, © Petr Tkachev/Shutterstock.com; B, from Vogt, 1880; C, from Carney et al. 2012.)

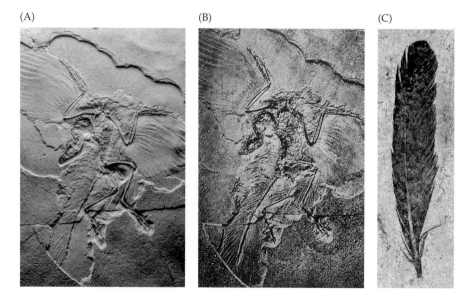

(A)

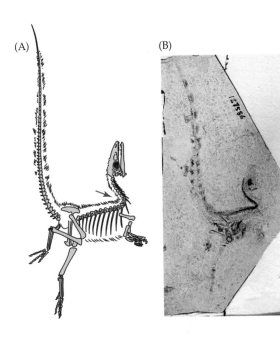

(B)

(C)

Figure 21.3 The earliest feathers were probably keratinaceous bristles. The basal coelurosaur theropod *Sinosauropteryx primus* from the Jehol formation in China. As seen in the sketch (A) and the fossil it is based on (B), traces of filamentous protofeathers can be seen on the head, spine, and both sides of the tail. The filaments increase in length from 13 mm on the head to 40 mm on the tail. (C) A detail of the neck region on the fossil (at the arrow in A) shows individual filaments. (From Chen et al. 1998.)

more than 150 years, but in the past 25 years a multitude of impressions of feathers and featherlike structures have shown that feathers appeared long before birds.

Featherlike structures called protofeathers have been found with theropod fossils. The simplest protofeathers, sometimes known as "dinofuzz," were single hollow filaments 1–5 cm long (**Figure 21.3**). These filaments bear little resemblance to the derived feathers of extant birds, and some paleontologists doubt that they are homologous with feathers. The question hangs on the embryonic origin of the structures. If they are epidermal, they are indeed protofeathers; but if they are dermal, they are merely part of the skin and their filamentous appearance results from decay of the skin during fossilization. Evidence for an epidermal origin has been provided by a fossil of a coelurosaur, *Shuvuuia deserti*, in which the filaments retained enough organic material for chemical tests. The results showed that the filaments were composed of a type of beta-keratin that is unique to feathers.

Filamentous projections were not restricted to coelurosaurs; filaments 15–20 cm long have been described on the head, neck, limbs, and tail of a large basal tyrannosaurid, *Yutyrannus huali,* and similar-looking filaments have been described in fossils of three ornithischian dinosaurs. One of the ornithischians, *Kulindadromeus*, had both scales and filaments.

More complex feathers from nonavian theropods consisted of downlike tufts of filaments and feathers with symmetrical vanes. Their locations varied among species—along the back, on the forelimbs or hindlimbs or both, and on the tail. Protofeathers and feathers were widely distributed among coelurosaurs (**Figure 21.4**). These feathers were probably used for species and sex identification, in courtship displays, and during male-male displays. For example, juvenile *Ornithomimus* had only filamentous feathers on their winglike structures, but adults had both filaments and symmetrical **pennaceous feathers**—feathers having a central shaft with a vane on each side. This ontogenetic change

suggests that the feathers of adults had social or reproductive functions, such as display, courtship, or brooding.

The excellent preservation of several feathered dinosaur fossils has led some paleontologists to speculate about the colors of feathers. The shapes and distribution of melanosomes (melanin-containing structures) in the feathers were identified by scanning electron microscopy, and their presumed colors were determined by matching the shapes and density of the melanosomes from a specimen of *Anchiornis huxleyi* to the melanosomes that produce black, gray, and reddish-brown colors in the feathers of extant birds. This analysis proposes that *Anchiornis* was gray, with a crest of reddish-brown feathers on top of the head and speckles of that color on the face. The long feathers on the wings and hindlimbs were white with black bands (**Figure 21.5**).

Some paleontologists view assignment of colors to fossils with skepticism. Bacterial decay may have led to postmortem changes in the soft tissues, and it has even been suggested that some of the structures interpreted as pigment cells are in fact bacteria. However, immunological tests have produced positive responses to the presence of both keratin and melanosomes in a fossil of *Eoconfuciusornis*, an Early Cretaceous bird.

Caution about assigning colors to fossils is warranted, but there is no reason to doubt that the feathers were pigmented and that their colors and patterns were important. In addition to providing color, patterns, and textures, early feathers could have provided insulation. Extant birds keep their feet and eggs warm at night by crouching to cover them with their body feathers, and tuck their thinly feathered heads and uninsulated beaks beneath their wings. Feathered coelurosaurs could have done these things as well. Furthermore, even protofeathers could have retained the heat produced by a theropod's resting metabolism and muscular activity, starting theropods on the path to endothermy, as described in Chapter 14.

The levels of structural complexity we have been considering are not simply a stepwise evolutionary progression

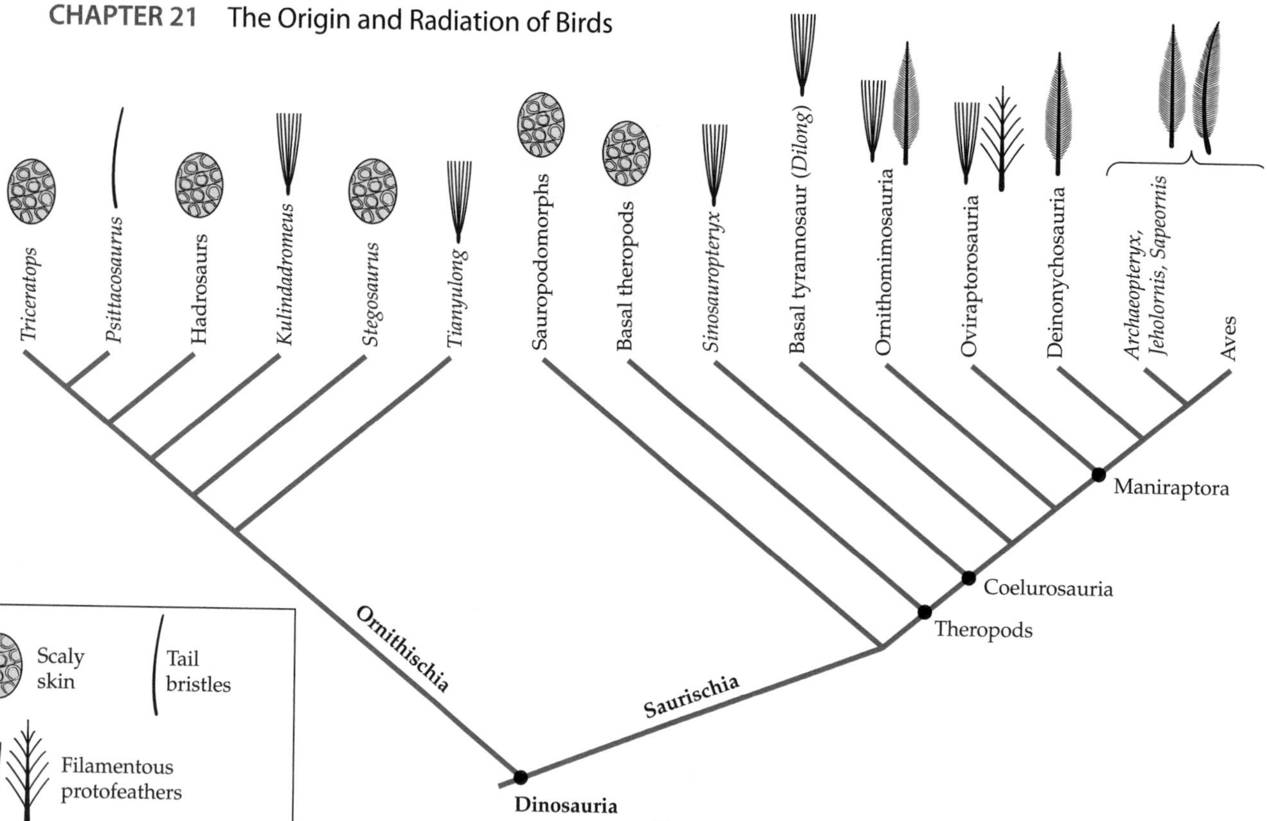

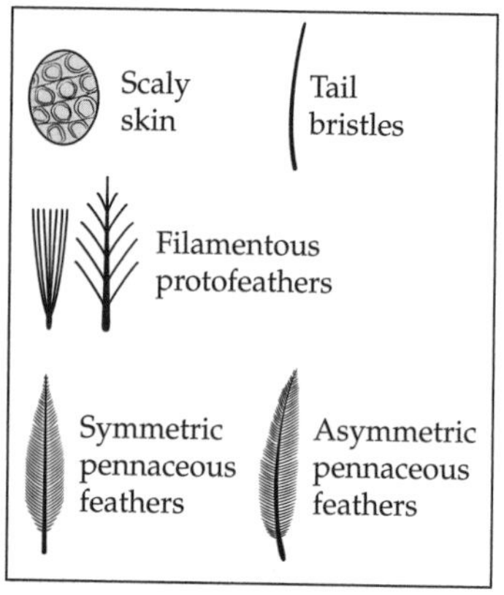

Scaly skin | Tail bristles

Filamentous protofeathers

Symmetric pennaceous feathers | Asymmetric pennaceous feathers

Figure 21.4 Distribution of feather types among dinosaurs. Adult, but not juvenile, *Psittacosaurus* had stiff bristles on the tail. Filamentous protofeathers have been described from the ornithischians *Kulindadromeus* and *Tianyulong*; from *Sinosauropteryx*, a basal coelurosaur; and from *Dilong*, a basal tyrannosaur. Oviraptorosaurs are currently known to have had various forms of filamentous protofeathers, and perhaps symmetrical pennaceous feathers as well. Deinonychosaurs had symmetrical pennaceous feathers. Asymmetrical pennaceous feathers appeared at least with *Archaeopteryx*, and perhaps earlier. (Tree modified from Witmer 2009; feather drawings from Xu et al. 2010.)

in which simple feathers come first and are progressively replaced by more complex feathers. Different types of feathers have different functions, and extant birds have feathers that consist of single strands and tufts of down as well as symmetrical and asymmetrical pennaceous feathers, each with a different function. Diversification of feathers resulted from extensive duplication and modification of beta-keratin genes. Thus, the widespread distribution of feathers among nonavian dinosaurs emphasizes that the evolution of feathers and the evolution of avian flight were entirely separate phenomena.

Reproduction and parental care

Crocodylians and extant birds share important reproductive characters. Both have assembly-line oviducts that successively surround a zygote with yolk, albumen, and finally a hard shell. But there are differences as well. Crocodylians, which are assumed to represent the ancestral archosaur condition, have two functional ovaries and oviducts and produce many small eggs that are deposited in a single laying event. Most extant birds have a single functional ovary and oviduct and produce fewer and larger eggs that are deposited one at a time over a period of days. The eggs

Figure 21.5 *Anchiornis huxleyi*. This Late Jurassic maniraptor was about the size of a small chicken. It had pennaceous feathers on its wings and legs and a feathery crest on its head. Comparison of the melanosomes in the fossilized feathers of an exceptionally well-preserved *Anchiornis* specimen with those of extant birds suggests the coloring shown in this recreation, including the black and white spotted bands and rufous crest and facial spots. (Illustration by Carl Buell.)

(A)

(B)

Figure 21.6 Nest and eggs of a Late Cretaceous coelurosaur, *Heyuannia huangi*. (A) Within the nest, the eggs were deposited in pairs, with the blunt ends pointed toward the center of the clutch. (B) A detail of two eggs reveals the texture and color of the eggshells. (From Wiemann et al. 2017, CC BY 4.0.)

of crocodylians are symmetrical, whereas the eggs of extant birds have a narrow end and a blunt end.

Fossils of nonavian coelurosaurs and their nests have shed light on the sequence in which the derived reproductive characters of birds appeared. Coelurosaurs deposited eggs in pairs within a roughly circular depression surrounded by a rim constructed by the adult (**Figure 21.6A**). The pairwise production of eggs, which was initially inferred by their positions in the nest, has been confirmed by the discovery of a female coelurosaur with two shelled eggs in the pelvic region. The eggs of coelurosaurs were asymmetrical like those of extant birds and were deposited with the blunt ends facing toward the center of the clutch. Analysis of the eggshells of a Late Cretaceous coelurosaur from China revealed the pigments protoporphyrin

and biliverdin, which would have given the eggs a blue-green color (**Figure 21.6B**); these pigments are also present in the eggs of extant birds. The final step to the extant avian condition—loss of a functional right ovary and oviduct and the consequent reduction of egg production to one at a time—is a derived character of Neornithes.

The extensive parental care that characterizes many extant birds also originated among nonavian dinosaurs. Uniparental care by the male appears to have been characteristic of coelurosaurs, and thus ancestral for birds, and the basal lineages of extant birds (Palaeognathae: ostriches, the emu [*Dromaius novaehollandiae*], and their relatives) also have predominantly male parental care. The biparental care that characterizes 90% of extant bird species is a derived character of Neognathae.

Body size

A substantial reduction in body size characterized the evolution of birds (**Figure 21.7**). Starting from basal coelurosaurs, which averaged about 25 kg, the body sizes of

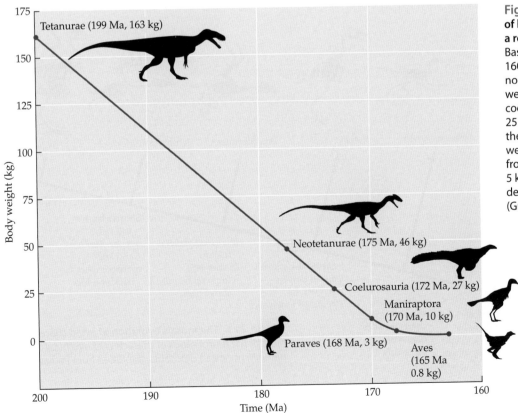

Figure 21.7 The evolution of birds was characterized by a reduction in body size. Basal theropods weighed 900 to 1600 kg, and the oldest tyrannosaurids and ornithomimids weighed about 100 kg. Basal coelurosaurs weighed about 25 kg. The remaining points on the graph are for avialians; body weights in this lineage declined from a starting point of about 5 kg to 100 g or less in the most derived Early Cretaceous forms. (Graph from Lee et al. 2014.)

Tetanurae (199 Ma, 163 kg)

Neotetanurae (175 Ma, 46 kg)

Coelurosauria (172 Ma, 27 kg)

Maniraptora (170 Ma, 10 kg)

Paraves (168 Ma, 3 kg)

Aves (165 Ma 0.8 kg)

Body weight (kg)

Time (Ma)

proavians (derived nonavian theropods) shrank to about 6.5 kg (the size of a large turkey) for *Caudipteryx*, to about 800 g (a small chicken) for *Archaeopteryx*, to 300 g (pigeon-size) for *Epidexipteryx*, and to 30 g (sparrow-size) for *Scansoriopteryx*.

Reduction in body size was not a general feature of coelurosaur biology. On the contrary, lineages of coelurosaurs basal to Avialae grew larger. From a starting point of 4 kg or less, oviraptorosaurs increased to about 50 kg, dromaeosaurs to 350 kg, and tyrannosaurs to 6,000–8,000 kg.

21.2   The Mosaic Evolution of Birds

Major evolutionary changes, such as the transition from nonavian dinosaurs to birds, do not occur all at once. Instead, one character changes, then another, and then another, so the transitional forms present mosaics of ancestral and derived characters. The mosaic nature of evolution is superbly illustrated by the transition from nonavian dinosaurs to birds (**Figure 21.8**).

Dromaeosaurs (Greek *dromas*, "running"), a derived group of coelurosaurs, had many birdlike characters, including a wrist structure that permitted them to flex the wrists sideways while rotating them. This motion probably allowed dromaeosaurs to use their hands to seize prey, and it is recognized in the name Maniraptora (Latin *manus*, "hand"; *raptor*, "robber"). Birds use the same wrist motion

to produce a flow of air over the **primaries** (the feathers on the outermost portion of the wing, corresponding to the hand) to create lift during flapping flight.

How—and why—birds got off the ground

The proavians of the Jurassic shared basic features of their body form and ecology. They were terrestrial predators, and their long legs suggest that they ran to pursue prey and to escape from predators. They competed with each other and preyed on each other; it must have been a tough place to make a living, and the earliest flying theropods did not have to be particularly adept; even a brief glide could have foiled an attack by a terrestrial predator, or surprised unsuspecting prey. Thus, enhanced agility is a plausible hypothesis for the advantage gained by winged proavians. That is far from the only hypothesis, however, and other scenarios also offer plausible ideas about advantages that could be gained from feathered wings. These hypotheses fall into two categories, known as "from the trees down" and "from the ground up," depending on the starting point for the aerial locomotion.

From the trees down   The ability to glide has evolved many times among vertebrates and is a plausible starting point for the evolution of powered flight. The small proavians that lie near the transition to flight were probably capable of climbing trees, and may have been arboreal. A proavian ambusher might have used its wings to steer as it swooped down on prey. Alternatively, arboreal

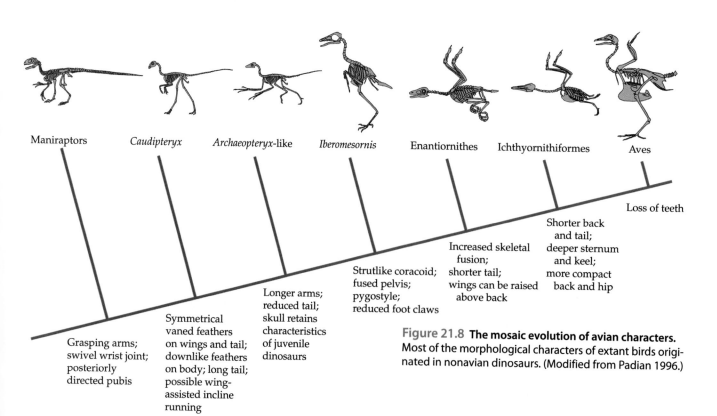

Figure 21.8 **The mosaic evolution of avian characters.** Most of the morphological characters of extant birds originated in nonavian dinosaurs. (Modified from Padian 1996.)

Figure 21.9 *Microraptor gui.* Both the fore- and hindlimbs of *M. gui* were feathered, and a fan of feathers sprouted from the end of the tail. This Early Cretaceous proavian is often restored as a glider, but because its feathers lacked barbules they would not have formed an airfoil and it probably could not glide. The feathers may have been used for social interactions, such as this threat display in which a *M. gui* is spreading its wing, leg, and tail feathers to appear as large as possible. (Reconstruction by Emily Willoughby/Stocktrek Images/Getty Images.)

proavians that climbed into trees to forage for prey could have glided from one tree to another to avoid descending to the ground, as do extant lizards in the genus *Draco*. *Microraptor gui*, a small Early Cretaceous proavian that had feathers on all four limbs and the tail, illustrates this hypothesis (**Figure 21.9**). However, the feathers of this four-winged proavian lacked barbules, the small hooks and catches that allow a feather to form a solid airfoil. Without barbules, *M. gui* would not have been able to glide, and it may have been a ground-dweller, using its feathers for social displays.

From the ground up The strongly **cursorial** (specialized for running) morphology of proavian ancestors suggests that the origin of flight should be sought in terrestrial locomotion. For example, an early hypothesis proposed that

a proavian cursorial predator might have combined an upward leap with wing flapping to swat a flying insect to the ground. No extant birds do this, however, and the feathers on the wings of proavians appear to have been as tightly packed as those on extant birds, making the proavian wing more like a Ping-Pong paddle than a tennis racquet. Thus, air resistance to a swing might have rotated the predator away from its prey.

Wing-assisted incline running (WAIR) may provide a more plausible scenario for evolving flight from a cursorial starting point. Juveniles and adults of some species of ground-feeding birds (e.g., grouse, partridges, chickens) use their wings to ascend steep slopes. For example, on level ground and on slopes of 45 degrees or less, chukars (*Alectoris chukar*) walk or run without flapping their wings, but on steeper slopes they run and flap their wings (**Figure 21.10**). Using WAIR, these partridges are able to ascend vertical slopes, such as tree trunks (see Figure 21.10D). Thus, WAIR might have been a starting point for powered flight.

The WAIR hypothesis has a weakness, however. Analysis of the scapular orientation in theropods and basal birds suggests that the shoulder joint of proavians and basal birds such as *Archaeopteryx* prevented the humerus from being lifted higher than the back, thus ruling out a downward wingbeat. Only in Ornithoraces (see Figure 21.1) does the shoulder joint permit the wing to be raised high enough for flapping flight.

The appearance of powered avian flight

Symmetrical pennaceous wing feathers are adequate for balancing, gliding, and even leaping into the air in pursuit of insects, but powered flight requires asymmetrical primaries to provide thrust. In extant birds, the primaries are asymmetrical. Symmetrical feathers were widespread among coelurosaurs, but so far asymmetrical feathers have been described only from *Archaeopteryx* and more derived forms.

Thus, there are two conflicting lines of evidence about the time at which powered flight evolved. Avialae (*Archaeopteryx* and more derived forms; see Figure 21.1) had asymmetrical wing feathers, which should indicate that they engaged in powered flight. But the shoulder joint

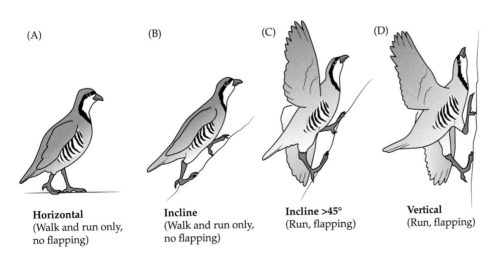

(A) **Horizontal** (Walk and run only, no flapping)

(B) **Incline** (Walk and run only, no flapping)

(C) **Incline >45°** (Run, flapping)

(D) **Vertical** (Run, flapping)

Figure 21.10 Wing-assisted incline running. Chukars (*Alectoris chukar*) walk or run on flat ground (A) and up slopes as steep as 45 degrees (B) without flapping their wings. On steeper slopes, they use their wings in addition to their legs (C), and can even ascend vertical objects such as tree trunks (D). (From Dial 2003.)

appears to have prevented the wings from being lifted above the back in all forms earlier than Ornithoraces. These contradictory interpretations have not yet been reconciled.

21.3 Early Birds

Feathered coelurosaurs such as *Caudipteryx* were not very birdlike in appearance. They had a long body, with the center of gravity near the hindlegs, a shallow trunk (because they lacked the sternal keel of extant birds), a long bony tail, forelegs that retained claws, and a full set of teeth. If you were out walking and saw a *Caudipteryx*, your immediate reaction would *not* be "Oh, there's a bird." By the Early Cretaceous, however, several anatomical changes had taken place, and if you saw a *Sinornis* or a *Confuciusornis* you would have no doubt that you were looking at a bird (**Figure 21.11**).

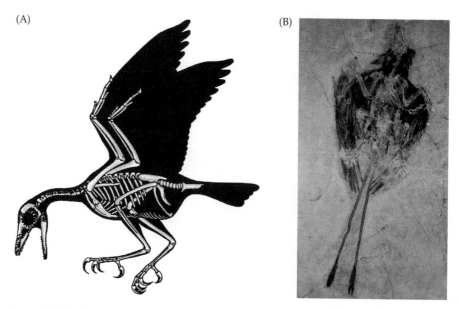

Figure 21.11 Early Cretaceous birds.   (A) *Sinornis* was a sparrow-size bird that has been found in Early Cretaceous lakebed deposits in China. (B) *Confuciusornis* was a crow-size bird, also from the Early Cretaceous of China. This fossil is thought to be that of a male because of the forked tail; other fossils of *Confuciusornis* lack a forked tail, suggesting that it was a sexually dimorphic character. (A from Sereno and Chenggang 1992; B from Naturhistorisches Museum Wien/CC BY 2.0.)

The most obvious changes were in general body form. The center of gravity had shifted forward toward the wings as in extant birds, the bony tail was greatly shortened, and fused vertebrae at the end of the tail formed a **pygostyle** as they do in extant birds. Less conspicuous changes were the strutlike coracoids (bones that help the shoulder girdle resist the pressures exerted on the chest by the wing muscles), a reduction in the size of the claws on the feet (making them better suited to perching in trees), and a larger sternum (giving more area for the origin of flight muscles).

During the Late Jurassic and throughout the Cretaceous, several lineages of proavians lived side by side with the earliest birds as well as with a diverse assemblage of pterosaurs. Amber from the Cretaceous of France and Canada contains feathers from both proavians and birds. The proavians were carnivores, and we have occasional direct evidence of predation; for example, a fossil of *Microraptor gui* from China contains the remains of a enantiornithine bird. Probably both proavians and the earliest birds were generalized predators, eating anything smaller than themselves that they could catch.

Although we cannot step back in time to directly observe feeding behavior, morphological evidence such as beak shape and the presence of a crop (an enlarged portion of the esophagus specialized for temporary food storage), and occasionally even fossilized stomach contents, indicates that later birds developed many dietary specializations seen in extant birds. By the Cretaceous, some birds (*Longipteryx, Jianchangornis, Ichthyornis, Confuciusornis,* and *Hesperornis*) were eating fish. Insectivores were represented by *Shenqiornis* and

Sulcavis, which ate beetles and other hard-bodied insects, and by *Pengornis*, which ate caterpillars and other soft-bodied invertebrates. *Longirostravis* probed for buried invertebrates such as worms and larvae, and *Archaeorhynchus* ate plant leaves and buds. *Sapeornis, Jeholornis,* and *Hongshanornis* appear to have been seed eaters, and the role of birds in seed dispersal may have begun at this time. Except for *Bohaiornis*, which may have been the first avian raptor, hawklike birds appear to have been rare or absent during the Cretaceous, possibly because that feeding niche was filled by pterosaurs in the air and by proavians on the ground.

21.4 The Mesozoic Radiations of Birds

The Mesozoic saw two independent radiations of birds (**Figure 21.12**). The birds in the earlier radiation are known as Enantiornithes (Greek *enantios*, "opposite"; *ornis* "bird") because the articulation between the scapula and coracoid is the reverse of the arrangement in extant birds. Enantiornithines were the dominant birds of the Cretaceous, and they ranged from small to large; *Sinornis* was the size of a sparrow, and *Enantiornis* was the size of a turkey vulture. Most enantiornithines were small to medium-size and probably lived in trees, but some had long legs and appear to have been wading birds, and others had powerful claws like those of extant hawks.

The second radiation of birds, Ornithurae (Greek *oura*, "a tail"), appeared later in the Cretaceous and radiated into a wide variety of ecological types. Early ornithurines were small, finchlike arboreal species, but by the Late Cretaceous

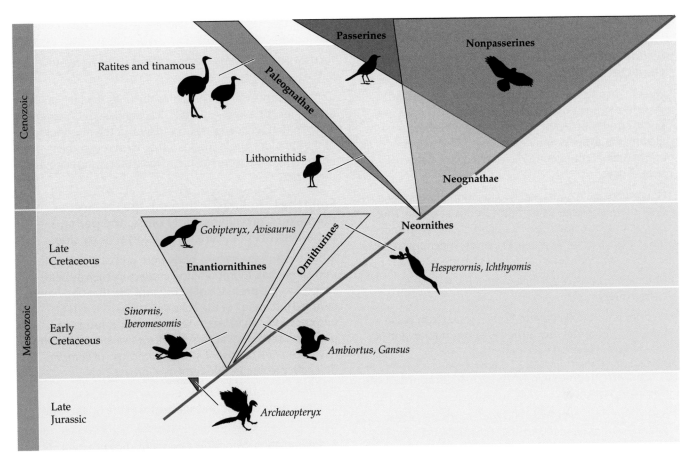

Figure 21.12 Mesozoic and extant birds. Enantiornithes was the first lineage to radiate and was the dominant group of birds during the Cretaceous. A second radiation, basal ornithurines, completed the Mesozoic avian fauna. Neither of these lineages survived the end of the Mesozoic, and extant birds (Neornithes) are modern ornithurines. The shaded areas indicate the changes in diversity of the lineages; the areas representing passerines and nonpasserines overlap. (From Feduccia 2003.)

the lineage had expanded to include waders, perchers, and secondarily flightless foot-propelled swimmers and divers. Neornithes probably originated among the ornithurines during the Late Cretaceous, and the last common ancestor of Neornithes may have been a seed eater.

Estimates based on molecular evidence have consistently placed the second radiation of birds within the Cretaceous, and *Vegavis iaai*, a member of the lineage that includes extant ducks and geese, provides fossil evidence that diversification of modern birds began before the Cenozoic. Clearly, there was some radiation of modern forms during the Mesozoic, because in addition to *V. iiai*, fossils of gallinaceous birds (fowl), and the flightless palaeognaths have been found in Cretaceous deposits. So far the fossil record is too sparse to say whether the big bang of neornithine radiation occurred before or after the end of the Cretaceous.

Summary

Birds provide one of the clearest examples of macroevolution.

The derived skeletal characters of birds evolved independently of flight among coelurosaurs, a lineage of bipedal theropod dinosaurs.

A furcula is present in nearly all theropod lineages, as is pneumatization of the postcranial skeleton. A wrist joint that allows the hands to be folded against the ulna and radius appeared in basal coelurosaurs, while uncinate processes on the ribs appeared with oviraptors.

Feathers are older than birds.

Filamentous protofeathers appeared with tyrannosaurs, and symmetrical pennaceous feathers with ornithomimosaurs. Asymmetrical pennaceous feathers, which are associated with powered flight, appeared with *Archaeopteryx*.

(Continued)

Summary (continued)

Asymmetrical eggs and parental care are ancestral characters of birds.

Extant crocodylians have two functional ovaries and deposit a large number of small, symmetrical eggs in a single egg-laying event, whereas extant birds have a single functional ovary and deposit one large, asymmetrical egg at a time, completing the clutch over a period of several days.

Nonavian coelurosaurs had two functional ovaries and deposited two large, asymmetrical eggs at a time. Thus, completion of a clutch over a period of days is an ancestral character of birds, whereas a single functional ovary is a derived character.

Uniparental care by the male was characteristic of coelurosaurs.

Predominantly male parental care is retained in palaeognaths. Biparental care is the derived condition and is characteristic of most neognaths.

A dramatic decrease in body size accompanied the evolution of birds.

Basal theropods weighed 1000 kg or more, the oldest tyrannosaurids were 100 kg, basal coelurosaurs weighed about 25 kg, *Archaeopteryx* 0.5 kg kg, and early birds 100 g or less.

From the trees down, or from the ground up?

One group of hypotheses about the origin of powered flight proposes that gliding set the stage. Another group focuses on the advantages that wings could have provided for a terrestrial runner.

Contradictory evidence obscures the time at which powered flight appeared. Asymmetrical pennaceous feathers suggest that *Archaeopteryx* could fly, but the morphology of the shoulder girdle suggests that forms older than enantiornithines were unable to raise their wings above the back.

From the Late Jurassic through the Cretaceous, pterosaurs, proavians, early birds, and perhaps the ancestors of extant birds lived side by side.

Proavians and the earliest birds were probably generalized carnivores. Morphological characters such as beak shape and the presence of a crop indicate that birds soon radiated into many of the adaptive zones that extant birds occupy: fish eaters, insect eaters, and substrate probers. Only flying raptors (hawk equivalents) appear to have been rare or absent, possibly because of competition with pterosaurs and terrestrial proavians.

The initial radiations of birds, enantornithines and ornithurines, were diverse and successful in the Cretaceous but did not extend into the Paleogene.

A later radiation, Neornithes, contains the ancestors of extant birds. Molecular dating indicates that many of these lineages originated in the Cretaceous, and a few fossils of basal members of extant lineages have been found. It is not currently known how much of the radiation of extant lineages occurred before or after the end of the Cretaceous.

Discussion Questions

1. Unidirectional airflow through the avian pulmonary system contrasts starkly with the tidal flow in the pulmonary system of mammals. Combine information from previous chapters to assemble evidence indicating that unidirectional lung ventilation is an ancestral character of the avian lineage.

2. Figure 21.4 shows the importance of fossils for interpreting the evolution of characters such as feathers. It also illustrates the paradox that having a great deal of information can complicate interpretation rather than clarify it. What parts of Figure 21.4 do you feel complicate the interpretation of the evolution of feathers, and how do you interpret them? Suppose the alternative phylogeny of Dinosauria shown in Figure 19.14B is correct? How would that change your interpretation?

3. Extant megapode birds such as the Australian brush turkey (*Alectura lathami*) are Neornithes that bury their eggs in soil as crocodylians do, rather than brooding them in a nest like other birds. How could you determine whether this mode of incubation is a retained ancestral trait or a derived condition that happens to have reverted to the crocodylian mode?

Additional Reading

Benton MJ. 2014. How birds became birds. *Science* 345: 508–509.

Botelho JF and 5 others. 2016. Molecular development of fibular reduction in birds and its evolution from dinosaurs. *Evolution* 70: 543–554.

Clarke JA, Tambussi CP, Noriega JI, Erickson GM, Ketcham RA. 2005. Definitive fossil evidence for the extant avian radiation in the Cretaceous. *Nature* 430: 305–308.

Godfroit P and 7 others. 2014. A Jurassic ornithischian dinosaur from Siberia with both feathers and scales. *Science* 345: 451–455.

Heers AM, Dial KP. 2015. From extant to extinct: Locomotor ontogeny and the evolution of avian flight. *Trends in Ecology and Evolution* 27: 296–305.

Lee MSY. Can A, Naish D, Dyke GJ. 2014. Sustained miniaturization and anatomical innovation in the dinosaurian ancestors of birds. *Science* 345: 562–566.

Lefèvre U and 6 others. 2017. A new Jurassic theropod from China documents a transitional step in the macrostructure of feathers. *The Science of Nature* 104: 74.

Lindgren J. 2016. Fossil pigments. *Current Biology* 26: R445–R460.

Mayr G, Peters DS, Plodowski G, Vogel O. 2002. Bristle-like integumentary structures at the tail of the horned dinosaur *Psittacosaurus*. *Naturwissenschaften* 89: 361–365.

Pan Y and 9 others. 2016. Molecular evidence of keratin and melanosomes in feathers of the Early Cretaceous bird *Eoconfuciusornis*. *Proceedings of the National Academy of Sciences, USA* 113: E7900–E7907.

Varracchio DJ and 5 others. 2008. Avian paternal care had dinosaur origin. *Science* 322: 1826–1828.

Xing L and 12 others. 2016. Mummified precocial bird wings in mid-Cretaceous Burmese amber. *Nature Communications* 7: 12089.

Go to the Companion Website **oup.com/us/vertebratelife10e** for active-learning exercises, news links, references, and more.

CHAPTER

22

Extant Birds

René C. Clark © Dancing Snake Nature Photography

Flight is a central characteristic of birds. At the structural level, the mechanical requirements of flight shape many aspects of the anatomy of birds. In terms of ecology and behavior, flight provides options for birds that terrestrial animals lack. The ability of birds to make long-distance movements is displayed most dramatically in their migrations. Even small species such as hummingbirds travel thousands of kilometers between their summer and winter ranges.

Not all birds fly through the air, however. Wing-propelled aquatic birds such as penguins "fly" through the water, whereas others (ducks and cormorants are familiar examples) use their feet to swim. Some birds, especially island species, are secondarily flightless, and many species spend most of their time on the ground and fly only short distances to escape predators.

A second characteristic of birds is diurnality—most species are active only during the day. In addition, most birds have excellent vision, and color and movement play important roles in their lives. Because humans also are diurnal and sensitive to color and movement, birds have been popular subjects for behavioral and ecological field studies. In terms of sound perception, noise pollution affects birds living in cities, interfering with their ability to detect predators and forcing them to alter the pitch of their songs. In short, important areas of biology draw heavily on studies of birds for information that can be generalized to other vertebrates.

Birds are a complex group of organisms—so complex that even the number of extant species is in question. The 2016 edition of the International Ornithological Congress's World Bird List (version 6.4) includes 10,660 species, but other authorities have argued that application of the phylogenetic species concept to birds would double that number. Phylogenetic relationships within Neoaves are controversial. **Figure 22.1** presents one hypothesis about the phylogenetic relationships of birds, and **Table 22.1** lists major lineages of extant birds.

22.1 The Structure of Birds: Specialization for Flight

Birds are variable in many respects. Their beaks and feet are specialized for different modes of feeding and locomotion, the morphology of their intestinal tract is related to dietary habits and can change from one season to another as their diets change, and their wing shapes reflect their specific flight characteristics. Despite that variation, the morphology of birds is more uniform than that of mammals. Much of this uniformity is a result of the specialization for flight.

As an example, consider body size. Flight imposes a maximum body size on birds. The muscle power required for takeoff increases by a factor of 2.25 for each doubling of body weight. That is, if species B weighs twice as much as species A, species B will require 2.25 times as much power to fly at its minimum speed. If the proportion of the total body weight allocated to flight muscles is constant, the muscles of a large bird must work harder than those of a small bird.

In fact, the situation is even more complicated, because the power output is a function of both muscular force and wingbeat frequency, and large birds have lower wingbeat frequencies than small birds. As a result, if species B weighs twice as much as species A, it will develop only 1.59 times as much power from its flight muscles, although it needs 2.25 times as much power to fly. Therefore, large birds require longer takeoff runs than do small birds, and a bird could ultimately reach a body weight at which any further increase in size would move it into a realm in which its leg and flight muscles were not able to provide enough power to take off.

Taking off is particularly difficult for large birds, which have to run and flap their wings to reach the speed needed for liftoff. The largest flying bird known is *Pelagornis chilensis*, a seabird that lived 5–10 Ma in what is now northern Chile. This gigantic predator had an estimated wingspan of 6.4 m, half again the wingspan of the wandering albatross (*Diomedea exulans*), which has the largest wingspan of any extant bird.

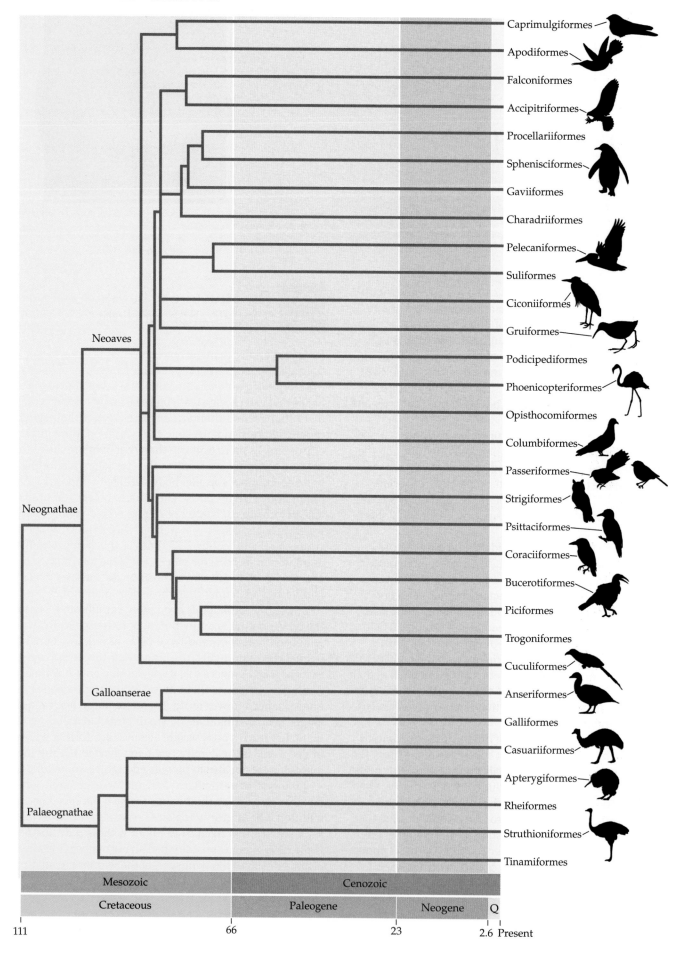

Caprimulgiformes
Apodiformes
Falconiformes
Accipitriformes
Procellariiformes
Sphenisciformes
Gaviiformes
Charadriiformes
Pelecaniformes
Suliformes
Ciconiiformes
Gruiformes
Podicipediformes
Phoenicopteriformes
Opisthocomiformes
Columbiformes
Passeriformes
Strigiformes
Psittaciformes
Coraciiformes
Bucerotiformes
Piciformes
Trogoniformes
Cuculiformes
Anseriformes
Galliformes
Casuariiformes
Apterygiformes
Rheiformes
Struthioniformes
Tinamiformes

Neoaves

Neognathae

Galloanserae

Palaeognathae

Mesozoic   Cenozoic
Cretaceous   Paleogene   Neogene   Q
111   66   23   2.6 Present

Table 22.1	Major lineages of extant birds[a]

NEOGNATHAE

Neoaves

Caprimulgiformes: A possibly paraphyletic grouping of nearly 100 species of nocturnal insectivores (nighthawks, nightjars, frogmouths, poor-wills) and the oilbird, a South American nocturnal fruit-eater.

Apodiformes: Hummingbirds and swifts. More than 400 species of arboreal birds with specialized flight. Swifts feed on insects they capture in flight, and hummingbirds are specialized nectar feeders that also capture flying insects.

Falconiformes: Falcons, caracaras, and their relatives. About 60 species of scavengers and birds of prey.

Accipitriformes: Hawks, eagles, sea eagles, kites, buzzards, New and Old World vultures, and relatives. More than 200 species of scavengers and birds of prey.

Procellariiformes: Albatrosses, petrels, and their relatives. More than 100 species of seabirds. Many are pelagic, flying across thousands of kilometers of open ocean before returning to land.

Sphenisciformes: Penguins. 17 species of wing-propelled divers that feed on fish. Largely confined to the cold Southern Hemisphere, including Antarctica, although one penguin species lives in the Galápagos Islands.

Gaviiformes: Loons. 5 species of diving birds.

Charadriiformes: Stilts, plovers, oystercatchers, auks, and their relatives. Fewer than 100 species. Most are long-billed shorebirds that probe for invertebrates buried in mud or sand. The auks are diving birds.

Pelecaniformes: A possibly paraphyletic grouping made up of the pelicans (8 species of water birds with large throat pouches), the shoebill stork (a large African wading bird), and the hammerkop (a wading bird found in Africa, Madagascar, and Arabia).

Suliformes: Frigatebirds, boobies, cormorants, gannets, anhingas, and relatives. About 60 species of water birds with fully webbed feet.

Ciconiiformes: Storks, herons, ibises, spoonbills, and relatives. About 100 species of mostly predatory birds, long-legged waders found near water.

Gruiformes: Cranes, rails, and coots. More than 300 species of mostly herbivorous birds that are associated with freshwater habitats.

Podicipediformes: Grebes. About 20 species of freshwater diving birds.

Phoenicopteriformes: Flamingos. 5 species of highly specialized aquatic filter feeders found in tropical regions.

Opisthocomiformes: A single species (*Opisthocomus hoazin*, the hoatzin) of South American forest bird. Fledglings have two claws on each wing that they use to clamber through vegetation; the claws are lost as the birds mature. The hoatzin is the only bird with a fermentative digestive system.

Columbiformes: Pigeons. More than 300 species of mostly rock dwellers, primarily seed eaters.

Passeriformes: Perching birds. Nearly 6000 species of birds with feet specialized for holding onto perches.

> *Suboscines:* Ovenbirds, antbirds, tyrant flycatchers, and their relatives. Six families with about 1000 species. Suboscines have limited control of the syrinx muscles and do not produce complex vocalizations.
>
> *Oscines:* Songbirds. About 70 families containing about 5000 species. Oscines have great control of their syrinx muscles and produce complex birdsongs.

Strigiformes: Owls. About 200 species of nocturnal predatory birds that hunt on the wing.

Psittaciformes: Parrots. Nearly 400 species of arboreal birds, mostly fruit- and seed-eaters.

Coraciiformes: Kingfishers, kookaburras, and relatives. More than 100 species of mostly arboreal, carnivorous, or insectivorous birds.

Bucerotiformes: A possibly paraphyletic grouping of hornbills (about 55 species of omnivorous birds with very large bills from Africa, Asia, and Melanesia), the hoopoe (an insectivore found in Europe and Africa), and the wood hoopoes and scimitarbills (8 species of insectivores from Africa).

Piciformes: Woodpeckers and relatives. More than 200 species of mostly arboreal insectivorous birds.

Trogniformes: Trogons and quetzals. About 40 species of colorful omnivores found in tropical forests worldwide.

Cuculiformes: Cuckoos. About 150 species of arboreal and ground-dwelling birds. There are both herbivorous and predatory forms.

Galloanserae

Anseriformes: Ducks, geese, and relatives. More than 150 species of semiaquatic birds.

Galliformes: Fowl, quail, and megapodes. More than 200 species of ground-dwelling birds.

PALAEOGNATHAE

All of the palaeognaths, also known as the ratites, are flightless.

Casuariiformes: 3 species of cassowary in Papua New Guinea and neighboring islands, and in northeastern Australia; and the single species of emu, endemic to Australia.

Apterygiformes: 5 species of kiwis in New Zealand

Rheiformes: 2 or 3 species of rheas living in the pampas and open woodland of South America.

Struthioniformes: 2 species of ostriches in Africa.

Tinamiformes: About 50 species of tinamous in South America.

[a] Note that although these are widely accepted divisions, many other names are also in use.

Flightless birds are spared the mechanical constraints associated with producing power for flight, but they still do not approach the body sizes of large mammals. The heaviest extant birds are the two species of flightless ostriches, which weigh about 150 kg, and the largest bird known, one of the extinct elephant birds, weighed an estimated 450 kg. In contrast, the largest terrestrial mammal, the African elephant, may weigh as much as 5,000 kg.

◀ **Figure 22.1 Phylogenetic relationships of extant birds (Neornithes).** Descriptions of the various groups are given in Table 22.1. Little agreement exists about the relationships within Neoaves, and attempts to resolve the nodes in this tree are controversial. This tree was created from data in Mayr 2014 and Prum et al. 2015 (see Further Reading); other recent studies have proposed different relationships.

The structural uniformity of birds is seen even more clearly if their body shapes are compared with those of other sauropsids. There are no quadrupedal birds, for example, nor any with horns or bony armor. Even those species of birds that have become secondarily flightless retain the basic body form of birds.

Feathers

Feathers develop from follicles in the skin, generally arranged in tracts, or **pterylae**, which are separated by patches of unfeathered skin, the **apteria**. Some birds—such as ratites, penguins, and mousebirds—lack pterylae, and the feathers are uniformly distributed over the skin.

For all their structural complexity, the chemical composition of feathers is remarkably simple and uniform. More than 90% of a feather consists of a specific type of

(A)

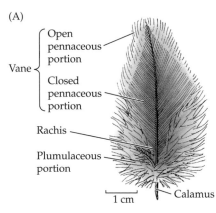

Vane {
  Open pennaceous portion
  Closed pennaceous portion
}

Plumulaceous portion

Rachis

Calamus

1 cm

Figure 22.2 Structural features of contour feathers. (A) A body contour feather. (B) A wing feather (quill), showing its main structural features. (After Lucas and Stettenheim 1972.)

(B)

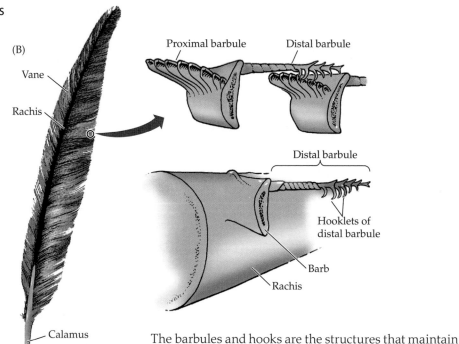

Vane

Rachis

Calamus

Proximal barbule   Distal barbule

Distal barbule

Hooklets of distal barbule

Barb

Rachis

beta-keratin, a protein related to the keratin that forms the scales of lepidosaurs. About 1% of a feather consists of lipids, about 8% is water, and the remaining fraction consists of small amounts of other proteins and pigments, such as melanin. The colors of feathers are produced by structural characters and pigments.

Each feather is anchored to a follicle in the skin by a short, tubular base, the **calamus**, which remains firmly implanted within the follicle until molt occurs (**Figure 22.2**). A long, tapered **rachis** extends from the calamus. The rachis of a flight feather changes from round to square in cross section as it is subjected to stress; the square shape maintains the feather's stiffness and prevents it from bending. The rachises of other feathers do not have this property; they remain round, and fail under stress. Closely spaced side branches called **barbs** project from the rachis of pennaceous (vaned) feathers, and proximal and distal **barbules** branch from opposite sides of the barbs. The ends of the distal barbules bear hooks that insert into grooves in the proximal barbules of the adjacent barb. The hooks and grooves hold adjacent barbs together, forming a flexible **vane**. The hooks and barbs can be pulled apart and then reattached by lightly stroking the feather, as a bird does when it preens.

A body contour feather (see Figure 22.2A) has several regions that reflect differences in structure. Near the base of the rachis, the barbs and barbules are flexible and the barbules lack hooks. This portion of the feather has a soft, loose, fluffy texture called plumulaceous or downy. It gives the plumage of a bird its properties of thermal insulation.

Farther from the base, the barbs form the tight surface called the vane, which has a pennaceous (sheetlike) texture. This is the part of the feather that is exposed on the exterior surface of the plumage, where it serves as an airfoil (lifting surface), protects the downy undercoat, sheds water, and reflects or absorbs solar radiation.

The barbules and hooks are the structures that maintain the pennaceous character of the feather vanes. They are arranged in such a way that any physical disruption to the vane is easily corrected by preening, as the bird realigns the barbules by drawing its slightly separated bill over them. Ornithologists distinguish five feather types:

1. **Contour feathers** include the outermost feathers on a bird's body, wings, and tail. The **remiges** (wing feathers; singular *remex*) and **rectrices** (tail feathers; singular *rectrix*) are large, stiff, mostly pennaceous contour feathers that are modified for flight. For example, the remiges are asymmetrical, and the distal portions of the outer remiges (the **primaries**) are abruptly tapered so that when the wings are spread, the tips of the primaries are separated by conspicuous gaps, called slots (**Figure 22.3**). The slots reduce drag on the wing and, in association with the marked asymmetry of the vanes, allow the feather tips to twist as the wings are flapped.

2. **Semiplumes** are intermediate in structure between contour feathers and down feathers. They combine a large rachis with entirely plumulaceous vanes. Semiplumes are hidden beneath the contour feathers, where they provide insulation and help streamline the contours of a bird's body.

3. **Down feathers** of various types are entirely plumulaceous; the rachis is short or entirely absent. Down feathers provide insulation, and powder down feathers disintegrate into fine particles of keratin that help waterproof the contour feathers.

4. **Bristles** are specialized feathers with a stiff rachis and without barbs or with barbs only on the proximal portions. Bristles occur most commonly around the base of the bill, around the eyes, and on the head and toes. Bristles screen out foreign particles from the eyes and nose, act as tactile sense organs, and aid in capturing flying insects.

5. **Filoplumes** are fine, hairlike feathers with a few short barbs or barbules at the tips and numerous nerve

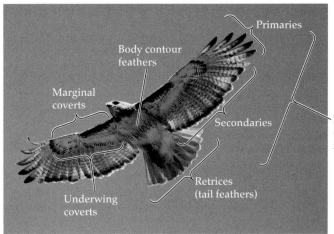

Figure 22.3 **Groups of feathers.** A soaring red-tailed hawk (*Buteo jamaicensis*) displays the major groups of feathers. The primaries and secondaries make up the wing feathers (remiges); the tail feathers are the retrices. The coverts (layers of smaller feathers that overlie the bases of the remiges and smooth airflow) include underwing coverts on the ventral surface of the wing, overwing coverts on the dorsal surface, and marginal coverts on the leading edge. Contour feathers fill out the body form and smooth the transitions from head to neck and neck to shoulders. Narrowing of the distal portions of the primaries creates slots that reduce turbulence at the wing tips. (Courtesy of the Missouri Department of Conservation.)

endings in their follicle walls. These nerves connect to pressure and vibration receptors around the follicles and transmit information about the position and movement of the contour feathers. This sensory system probably plays a role in keeping the contour feathers in place and adjusting them properly for flight, insulation, bathing, and display.

Streamlining and weight reduction

Birds are the only vertebrates that move fast enough in air for wind resistance and streamlining to be important factors in their lives. Many passerines are probably able to fly 50 km/h or even faster when they must, although their normal cruising speeds are lower. Ducks and geese can fly at 80–90 km/h, and peregrine falcons (*Falco peregrinus*) reach speeds as high as 200 km/h when they dive on prey. Fast-flying birds have many of the same structural characters as those seen in fast-flying aircraft. Contour feathers make smooth junctions between the wings and the body and often between the head and the body as well, eliminating sources of turbulence that would increase wind resistance. The feet are tucked close to the body during flight, further improving streamlining.

At the opposite extreme, some birds are slow fliers. Many of the long-legged, long-necked wading birds, such as spoonbills and flamingos, fall into this category. Their long legs trail behind them as they fly, and their necks are extended. They are far from streamlined, although they may be strong fliers.

Characteristics of some of the organs of birds reduce body weight. For example, birds lack urinary bladders, and most species have only one ovary (the left) as adults. All of the kiwis and a few other species of birds retain two functional ovaries. Males of most species of birds lack a phallus. The gonads of both male and female birds are usually small; they hypertrophy during the breeding season and regress when breeding has finished.

Skeleton

Structural modifications can be seen in several aspects of the skeleton of birds. The avian skeleton is not lighter in relation to the total body weight of a bird than is the skeleton of a mammal of similar size, but the distribution of weight is different.

Many bones are pneumatic (air-filled; **Figure 22.4**), and the skull is especially light. However, the leg bones of birds are proportionally heavier than those of mammals. Thus, the total weight of a bird's skeleton is similar to that of a mammal, but more of a bird's weight is concentrated in its hindlimbs.

Except for the specializations associated with flight, the skeleton of a bird is very much like that of a small coelurosaur (**Figure 22.5**). The center of gravity is beneath the wings, and the sternum is greatly enlarged compared with that of proavian coelurosaurs. The sternum of flying birds bears a keel from which the pectoralis major and supracoracoideus muscles originate. Strong fliers have a well-developed keel and large flight muscles. The scapula extends posteriorly above the ribs and is supported by the coracoid, which is fused ventrally to the sternum. Additional bracing is provided by the clavicles, which in most birds are fused at their ventral ends to form the furcula (wishbone).

The hindfoot of birds is greatly elongated, and the ankle joint is within the tarsals (a mesotarsal joint) as it was in Mesozoic theropods. The fifth toe is lost, and the metatarsals of the remaining toes are fused with the distal tarsals to form a bone called the **tarsometatarsus**. In other words, birds walk with their toes (tarsals) flat on the ground and with the tarsometatarsus projecting upward at an angle. The actual knee joint is concealed within the contour feathers on the body, and what looks like the lower part of a bird's leg is actually the tarsometatarsus. That is why birds appear to have knees that bend backward; the true knee lies between the femur and lower leg (i.e., between the thigh and the "drumstick"). The outer (distal) end of the tibia fuses with the uppermost (proximal) tarsal bones, creating

Figure 22.4 **Interior of an avian long bone.** The hollow core and reinforcing struts in the avian humerus combine strength with weight reduction. (The Natural History Museum/Alamy Stock Photo.)

(A)

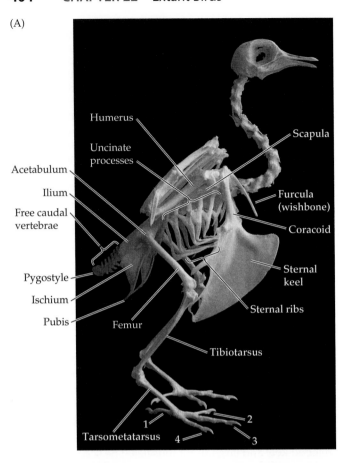

Humerus

Scapula

Uncinate processes

Acetabulum

Ilium

Furcula (wishbone)

Free caudal vertebrae

Coracoid

Pygostyle

Ischium

Sternal keel

Pubis

Femur

Sternal ribs

Tibiotarsus

1

2

Tarsometatarsus 4 3

(B)

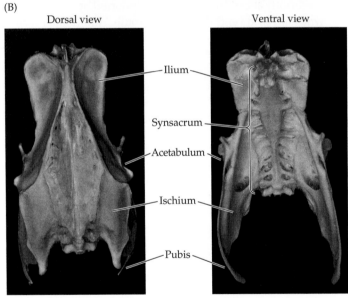

Dorsal view

Ventral view

Ilium

Synsacrum

Acetabulum

Ischium

Pubis

Figure 22.5 Skeleton of a bird. The skeletons of birds reveal their coelurosaurian ancestry, with modifications that are related to the demands of flight. (A) The skeleton of a rock pigeon (*Columba livia*). The long tail of coeluroesaurs has shortened to about five free caudal vertebrae and a pygostyle formed by the fusion of the remaining vertebrae. (B) The pelvic girdle of birds (that of a chicken, *Gallus gallus*, is shown here in dorsal and ventral views) is elongated, and the ischium and ilium have broadened into thin sheets that are firmly united with a synsacrum formed by the fusion of multiple vertebrae, as seen in the ventral view (right). The remaining thoracic vertebrae are connected by strong ligaments that are often ossified, and thus relatively immobile. These immobile thoracic vertebrae and the synsacrum combine with the elongated, rooflike pelvis to produce a nearly rigid vertebral column. Flexion is possible only in the neck, at the joint between the thoracic vertebrae and the synsacrum, and at the base of the tail. (A, © photowind/Shutterstock.com; B, M. H. Siddall.)

(*Apus apus*) fly nearly continuously for the 10-month duration of their nonbreeding season. These species have small legs; the leg muscles account for as little as 2% of the body weight. Predatory birds such as eagles, hawks, and owls use their legs to capture prey. In these species, the flight muscles make up about 25% of the body weight and the leg muscles 10%. Swimming birds such as ducks and grebes have an even division between leg and flight muscles; the combined weight of these muscles may be 30–60% of the total body weight. Birds such as rails, which are primarily terrestrial and rely on their legs to escape from predators, have leg muscles that are larger than flight muscles.

Muscle-fiber types and metabolic pathways also distinguish running birds from fliers. The familiar distinction between the light meat and dark meat of a chicken reflects those differences. Fowl, especially domestic breeds, rarely fly, but they are capable of walking and running for long periods. The dark color of the leg muscles reveals the presence of myoglobin in the tissues and indicates a high capacity for aerobic metabolism in the leg muscles of these birds. The white muscles of the breast lack myoglobin and have little capacity for aerobic metabolism. The flights of fowl (including wild species such as pheasants, grouse, and quail) are of brief duration and are used primarily to evade predators. These birds use an explosive takeoff, fueled by anaerobic metabolic pathways, followed by a long glide back to the ground. Birds that are capable of strong, sustained flight have dark breast muscles with high aerobic metabolic capacities.

a composite bone called the **tibiotarsus** that forms most of the lower leg. The fibula is reduced to a splint of bone.

Muscles

The relative size of the leg and flight muscles of birds is related to their primary mode of locomotion. The massive pectoralis major and supracoracoideus muscles, which produce the downstroke of the wings, originate on the keel of the sternum and make up 25% of the total body weight of strong fliers such as hummingbirds and swifts. Common swifts

22.2 Wings and Flight

Flapping flight is remarkable for its automatic, unlearned performance. A young bird learning to fly uses a form of locomotion so complex that it nearly defies analysis in physical and aerodynamic terms. Even nestlings that develop in confined spaces, such as terrestrial burrows and tree cavities in which it is impossible for them to spread their wings to practice flapping before they leave the nest, still manage to fly soon after fledging.

Many hole- and burrow-nesting birds are capable of covering considerable distances on their first flights. South Georgia diving-petrels (*Pelecanoides georgicus*) may fly as far as 10 km the first time out of their burrows. Young birds that are reared in open nests, especially large birds such as albatrosses, storks, vultures, and eagles, frequently flap their wings vigorously for several days before flying. This flapping may help develop muscles, but it is unlikely that these birds are learning to fly. However, a bird's flying abilities do improve with practice for a period of time after leaving the nest.

There are so many variables involved in flapping flight that it is difficult to understand exactly how it works. A beating wing is both flexible and porous, and it yields to air pressure. The shape, camber, angle relative to the body, and position of the individual feathers all change as a wing moves through its cycle. This is a formidable list of variables, far more complex than those involved in the fixed wing of an airplane, and it is no wonder that the aerodynamics of flapping flight are not yet fully understood. However, the general properties of a flapping wing can be described.

We can begin by considering the flapping cycle of a small bird in flight. Unlike the fixed wings of an airplane, the wings of a bird function both as airfoils and as propellers for forward motion. A bird cannot continue to fly straight and level unless it can develop a thrust to balance the drag operating against forward momentum. The downward stroke of the wings produces this thrust. The primary feathers at the outer end of the wing are the site where thrust is generated, whereas the feathers on the inner wings—the **secondaries**—generate lift. It is easiest to consider the forces operating on the inner and outer wing separately (**Figure 22.6**).

As the wings move downward and forward on the downstroke, the trailing edges of the primaries bend upward under air pressure, and each feather acts as an individual propeller biting into the air and generating thrust (**Figure 22.7**). During this downbeat, the thrust is greater than the total drag, and the bird accelerates. In small birds,

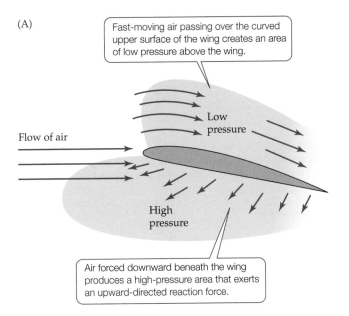

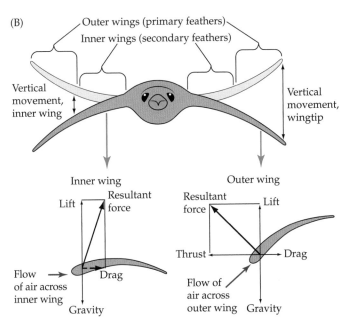

Figure 22.6 Aerodynamic aspects of bird flight. (A) A simplified representation of a bird's wing as an airfoil shows some of the aerodynamic forces involved in flapping flight. The combination of these forces acting together produces lift. (B) The arc of movement at the inner end of the wing is much smaller than it is at the outer end of the wing. Thus, the inner wing (the area of the secondaries) acts as if the bird were gliding and generates an upward-directed force (below left) that provides most of the lift need to counteract gravity. The inner wing produces little or no thrust (forward-directed force), but the outer wing (the area of the primaries; below right) is tilted during the downstroke, angling the resulting force forward to produce thrust. (A after Gill 2007.)

Figure 22.7 The primaries act as propellers. A bushtit (*Psaltriparus minimus*) in flight shows the primary feathers twisting and opening during the downstroke, causing these feathers to act as propellers and produce forward thrust. The primaries twist only on the downstroke of small birds, and during both the downstroke and the upstroke of large birds. (Photo by Bridget Spencer.)

(A)

(B)

(C)

Figure 22.8 **A flying bird creates vortices that form a wake behind it.** (A) Vortices (shown in red) formed at the wingtips create drag, but vortices that form at the tail reduce drag. (B) Large birds often fly in V-shaped formations. Following in the wake of another bird allows the following bird to take advantage of the updraft from the wingtip vortices of the bird ahead of it. (C) When birds fly in formation they either match the leader wingbeat for wingbeat, or (as in this photograph of pelicans) flap in regular succession from the leader backward. Pelicans realize an energy saving of 11%–14% by flying in formation. (A, © Frank Schulenburg/CC BY-SA 4.0; B, Mohamed Nuzrath/CC0 1.0 Public Domain; C, Tony Fischer/CC BY 2.0.)

the return stroke, which is upward and backward, provides little or no thrust. It is mainly a passive recovery stroke, and the bird slows during this part of the wingbeat cycle.

Vortices—eddies of swirling air—play a major role in avian flight. Vortices formed at the leading edge of the wings contribute lift to a flying bird, but they also increase drag. The tail does not create significant lift, but it reduces drag by keeping airflow close to the body surface. The wing vortices, and a pair of vortices shed from the tail, are left in a bird's wake, where they can provide lift to birds following another bird in the formation (**Figure 22.8**). The leading bird works harder than following bird, and leaders and followers switch positions within a formation.

A bird can change the aerodynamics of its wings from one wingbeat to the next and even during a single wingbeat by sweeping its wings back or extending them, a process called morphing. Instantaneously variable wing morphing modulates speed, turning rate, and glide angle.

Landing is a delicate maneuver because a bird must reduce its velocity during the approach, which increases the risk of stalling prematurely, then it must deliberately stall as it reaches the perch site. As a bird comes in for a landing, it rotates its body and wings, spreads its tail and extends its feet to create drag, and raises feathers called underwing coverts that lie on the lower surface of the leading edge of the wings. These feathers act like the Krueger flaps that rotate out from the lower surface of the leading edge of the wings of airplanes during landing; they maintain lift at low speed by increasing the camber of the wing. Birds also extend the **alula**, the feather-covered second digit that normally rests against the fused third and fourth digits, creating a vortex that maintains lift by holding airflow close to the surface of the wing. (**Figure 22.9**).

Wing muscles

Two large breast muscles, the pectoralis major and the supracoracoideus, are the major flight muscles (**Figure 22.10**).

(A)

Figure 22.9 **The alula creates vortices that stabilize airflow at low speeds.** The alula is the feather-covered remnant of the second digit of the hand. It normally rests against the fused third and fourth digits. (A) The alula is extended as a bird slows its flight in

(B)

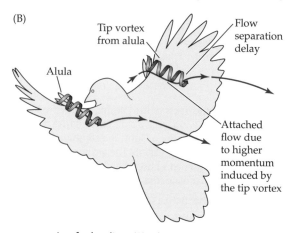

preparation for landing. (B) When the alula is extended, it creates a vortex (blue) that maintains lift at low speed by delaying the separation of airflow from the surface of the wing. (A, David Hypes/U.S. National Park Service; B, from Lee et al. 2015, CC BY 4.0.)

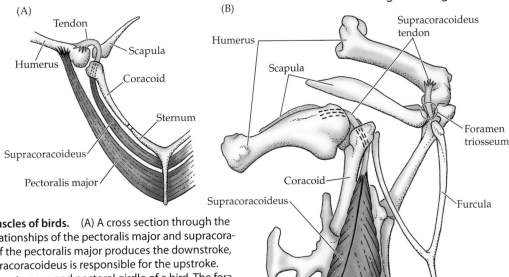

Figure 22.10 Major flight muscles of birds. (A) A cross section through the sternum of a bird shows the relationships of the pectoralis major and supracoracoideus muscles. Contraction of the pectoralis major produces the downstroke, whereas contraction of the supracoracoideus is responsible for the upstroke. (B) Frontal and lateral view of the sternum and pectoral girdle of a bird. The foramen is formed by the articulation of the furcula, coracoid, and scapula. The supracoracoideus tendon passes through the foramen triosseum onto the dorsal head of the humerus and inserts on the scapula. (A, modified from Storer 1943; B, after Feduccia 1980.)

Contraction of the pectoralis major produces the forceful downstroke. A powered upstroke results mainly from contraction of the supracoracoideus, which underlies the pectoralis major and attaches directly to the sternal keel. The supracoracoideus inserts on the dorsal head of the humerus by passing through the foramen triosseum, formed where the coracoid, furcula, and scapula join. In most species of birds, the supracoracoideus is a relatively small, pale muscle with low myoglobin content and is easily fatigued. In species that rely on a powered upstroke—for fast, steep takeoffs, for hovering, or for fast aerial pursuit—the supracoracoideus is larger. The ratio of weights of the pectoralis major and the supracoracoideus is a good indication of a bird's reliance on a powered upstroke; such ratios vary from 3:1 to 20:1. The total weight of the flight muscles shows the extent to which a bird depends on powered flight. Strong fliers such as pigeons and falcons have breast muscles that make up more than 25% of total body weight, whereas the flight muscles of some owls are only 10% of total body weight.

Wing shape and flight characteristics

Ornithologists recognize four structural and functional types of wings (**Figure 22.11**). Wings are often differentiated by two major characteristics: aspect ratio and wing loading. **Aspect ratio** refers to the length of the wing divided by its area; long, narrow wings have high aspect ratios. **Wing loading** refers to the weight of a bird divided by

its wing area; the larger the wing relative to the weight of the bird, the lower the wing-loading value. High-aspect-ratio wings provide the high lift-to-drag ratios needed by oceanic birds such as albatrosses that rely on dynamic soaring (a type of soaring in which energy is gained by repeatedly crossing between air masses with different wind speeds). Aerial foragers such as swallow and falcons have wings with high aspect ratios and employ wing morphing to adjust speed and

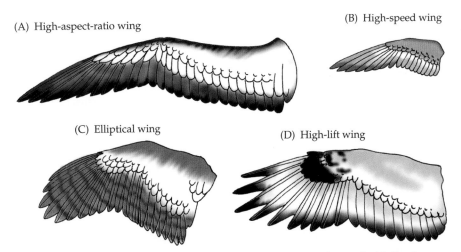

Figure 22.11 Flight characteristics can be deduced from the shape of a bird's wings. (A) High-aspect-ratio wings are long and narrow and lack slots in the outer primaries. These wings are characteristic of birds such as albatrosses and shearwaters that employ dynamic soaring over oceans with strong winds. The pointed tips minimize wingtip drag and give high-aspect-ratio wings their high lift-to-drag ratios. (B) High-speed wings are characteristic of birds that feed in flight, such as swallows. These wings have high aspect ratios, low camber, no slotting of the primaries, and a swept-back shape that can rapidly morph. (C) Short, elliptical wings are characteristic of birds such as grouse and quail that live in complicated, three-dimensional habitats where they must maneuver through and around vegetation. These wings have low aspect ratios and slotting of the primaries. (D) The high-lift wings of birds that soar in rising air currents, such as hawks and vultures, have low wing-loading values, intermediate aspect ratios, and pronounced slotting of the primaries. (After Peterson 1978.)

sinking rate and to make sharp turns. These wings have a flat profile (little camber) and often lack slots (spaces between the outer primaries). In flight, these wings show the swept-back attitude of jet fighter plane wings. All fast-flying birds have converged on this form.

In contrast, many forest and woodland birds have elliptical wings with low aspect ratios and a high degree of slotting in the outer primaries. While these birds may not be able to soar, their wings allow rapid takeoff and excellent maneuverability in forested environments that have many obstacles. These birds generally have high wing-loading values with relatively small wings, and thus have rapid flapping flight.

Birds that are static soarers, such as vultures, eagles, and storks, have high-lift wings that are intermediate between the elliptical wing and the high-aspect-ratio wing. These wings generally have a deep camber and marked slotting in the primaries. Static soarers also have broad wings with low wing-loading values. They remain airborne mainly by gliding on air masses that are rising at a rate faster than a bird's sinking speed, and thus they rarely need to flap their wings. The highly slotted wings of static soarers enhance maneuverability because the birds can respond to changes in wind currents using individual feathers instead of the entire wing.

In regions where topographic features and meteorological factors provide currents of rising air, static soaring is an energetically cheap mode of flight. By soaring rather than flapping, a large bird can decrease the energy required for flight by a factor of 20 or more, whereas the saving is only one-tenth that much for a small bird. Some condors and vultures cover hundreds of kilometers each day soaring in search of food.

22.3 Feet

A foot with four toes is the ancestral condition for birds, and most extant species retain four toes, although the arrangement of the toes has been modified in some species (**Figure 22.12**). The coelurosaur ancestors of birds were terrestrial runners. As the body size of proavians decreased, they became more arboreal, and the ability to perch in trees was an early feature of bird evolution. By the Cretaceous, both the enantiornithine and ornithurine lineages included perching birds.

Hopping, walking, and running

When they are on the ground, birds hop (make a succession of jumps in which both legs move together), walk (move the legs alternately with at least one foot in contact with the ground at all times), or run (move the legs alternately with both feet off the ground at some times). Modifications of these basic gaits include climbing, wading in shallow water or on insubstantial surfaces such as lily pads, and supporting heavy bodies.

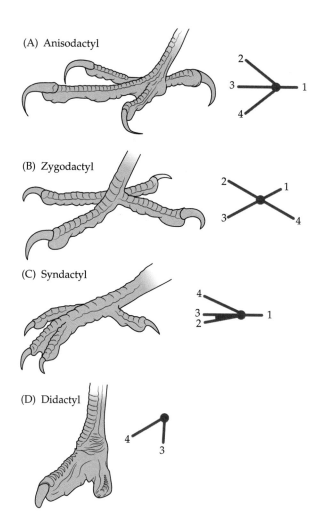

Figure 22.12 The arrangement of toes reflects locomotor specialization.   (A) In anisodactyly, toe 1 points backward and toes 2, 3, and 4 point forward. The terrestrial ancestors of birds were anisodactyl, and this arrangement is retained by most terrestrial, arboreal, and aquatic species of extant birds. (B) In zygodactyly, toes 1 and 4 point backward and toes 2 and 3 point forward. Zygodactyly has evolved independently in different lineages. Some climbing birds, such as woodpeckers and parrots, are zygodactyl, as are roadrunners (terrestrial runners); owls and the osprey (aerial raptors that seize prey with their feet); and some swifts (rapid fliers). (C) Syndactyly, in which 2 or 3 toes are fused for much of their length, is characteristic of several passerine families. (D) Didactyly is distinguished by having only toes 3 and 4 and is characteristic of the cursorial ostriches.

Hopping is a special form of locomotion found mostly in perching, arboreal birds. It is most highly developed in passerines, and only a few nonpasserine birds regularly hop. Many passerines cannot walk, and hopping is their only mode of terrestrial locomotion. In Corvidae, for example, ravens, crows, and rooks are walkers, whereas jays and magpies are hoppers.

Running is a modification of walking in which both feet are off the ground at the same time for a portion of each step cycle. In general, vertebrates that are fast runners have long, thin legs and small feet. The significance of long legs is obvious: they allow an animal to cover more distance with each stride. The reduction in leg and foot weight involves the physics of momentum. When a foot is in contact with the ground, it is motionless; after it pushes off from the ground, the foot and the lower part of the leg must be accelerated to a speed faster than the trunk of the running animal. Reducing the weight of the foot and leg reduces the energy needed for this change in momentum and allows the animal to run faster.

Reducing the number and length of the toes makes the foot smaller and lighter. No bird has reduced the length and number of toes in contact with the ground to the extent that the hoofed mammals have, but the large, fast-running ostriches have only two toes on each foot, and emus and rheas have only the three forward-directed toes in contact with the ground.

Climbing

Birds climb on tree trunks or other vertical surfaces by using their feet, tail, beak, and rarely, their forelimbs. Several distantly related groups of birds have independently acquired specializations for climbing and foraging on vertical tree trunks. Species such as woodpeckers and woodcreepers, which use their tails as supports, begin foraging near the base of a tree trunk and work their way vertically upward, headfirst, clinging to the bark with strong feet on short legs. The tail is used as a prop to brace the body against the powerful pecking exertions of the head and neck, and the pygostyle and free caudal vertebrae in these species are much enlarged and support strong, stiff tail feathers. A similar modification of the tail is found in certain swifts that perch on cave walls and inside chimneys.

Nuthatches and similarly modified birds climb on trunks and rock walls in both head-upward and head-downward directions while foraging. In these species, which do not use their tails for support, the claw on the backward-directed toe 1 is larger than the claws on the forward-directed toes and is strongly curved.

Swimming

Although no birds have become fully aquatic, nearly 400 species are specialized for swimming. Nearly half of these aquatic species also dive and swim underwater.

Swimming on the surface Modifications of the hindlimbs are the most obvious avian specializations for foot-propelled swimming on the water's surface. Other changes include a wide body that increases stability in water, dense plumage that provides buoyancy and insulation, a large preen gland which produces oil to waterproof the plumage, and structural modifications of the body feathers that retard penetration of water to the skin. The legs of foot-propelled swimmers are at the rear of the body, where the weight of leg muscles interferes least with streamlining and where the best control of steering can be achieved.

The feet that propel aquatic birds are either webbed or lobed (**Figure 22.13**). Webbing between toes 2, 3, and 4 (palmate webbing) has been independently acquired at least four times in the course of avian evolution. Totipalmate

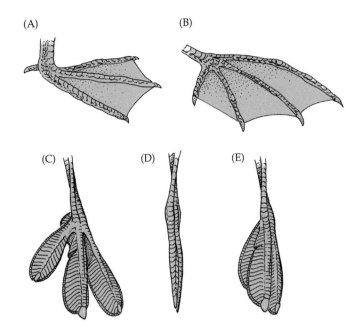

Figure 22.13 Webbed and lobed feet of some aquatic birds. (A) A duck's foot shows palmate webbing. (B) Totipalmate webbing, seen in the foot of the cormorant. (C–E) The lobed foot of a swimming grebe. During the backward power stroke (C), the lobes flare outward, maximizing the surface area of the foot. During the forward recovery stroke (D, front view; E, side view), toes 3 and 4 rotate 90 degrees and toe 2 is moved behind the tibiotarsus, minimizing the area presented to the water. (After Peterson 1978.)

webbing—webbing of all four toes—is found in pelicans, cormorants, and boobies. The hydrodynamic forces acting on the foot of a swimming bird are complex. At the beginning of the power stroke, when the foot is in its forward-most position, the web is nearly parallel to the water surface and moving downward. At this stage, the effect of foot movement is to lift the bird and propel it forward. Later in the stroke, when the web passes through the point of being perpendicular to the water surface, hydrodynamic drag is produced on the forward-facing (dorsal) side of the web by the vortex that develops as water flows around the web, and drag is the major force propelling the bird forward.

Lobes on the toes have evolved convergently in several lineages of aquatic birds. The lobes are flexible flaps that flare open to present a maximum surface on the backward stroke and fold back against the toes during the recovery stroked when the foot is moving forward through water. Grebes are unique in that the lobes on the outer sides of their toes are rigid and do not fold back as the foot moves forward. Instead of folding the lobes, grebes rotate the third and fourth toes 90 degrees at the beginning of the recovery stroke and move the second toe behind the tarsometatarsus. In this position, the sides of the toes slice through the water like knife blades, minimizing resistance.

Diving and swimming underwater The transition from a surface-swimming bird to a subsurface swimmer has occurred in two fundamentally different ways: (1) by further specialization of a hindlimb already adapted for swimming, or (2) by modification of the wing for use as a flipper. Highly specialized foot-propelled divers have evolved independently among grebes, cormorants, loons, and the extinct hesperornithids. All these families except the loons include some flightless forms. Wing-propelled divers include diving-petrels and shearwaters (Procellariiformes), penguins (Sphenisciformes), and auks (Charadriiformes). Only among waterfowl are there both foot-propelled and wing-propelled diving species, and none of these is as highly modified for diving as are specialists such as loons and auks. The American dipper (*Cinclus mexicanus*) is a passerine that dives and swims in mountain streams with great facility using its small wings, but it lacks any other morphological specializations.

22.4 Feeding and Digestion

With the specialization of the forelimbs as wings, which largely precludes any substantial role in capturing prey, birds have concentrated their predatory mechanisms in their beaks and feet. Modifications of the beak, tongue, and intestines are often associated with dietary specializations.

Beaks, skulls, and tongues

The presence of a horny beak in place of teeth is not unique to birds; turtles have beaks, and so did rhynchosaurs, many dinosaurs, pterosaurs, and dicynodonts. Birds' beaks are versatile structures. In addition to feeding, birds use their beaks to groom their feathers, manipulate objects (as when constructing a nest), in social interactions with other individuals of their species, and even for thermoregulation. The range of morphological specializations of beaks defies complete description, but some categories can be recognized.

Figure 22.14 A familiar generalist.   The American robin (*Turdus migratorius*) is a member of the thrush family (Turdidae), a group of passerine birds with worldwide distribution. Thrushes are generalists, feeding on insects, annelid worms, and other invertebrates as well as fruits and seeds. (Photo © René C. Clark, Dancing Snake Nature Photography.)

- Many birds are generalists, with beaks that can seize and process both animal and plant food (**Figure 22.14**).
- Probers and gleaners—birds that extract food from small or covered spaces—typically have pointed beaks that they use like forceps to seize specific items (**Figure 22.15**).
- Many carnivorous birds, including gulls, ravens, crows, and roadrunners, use their heavy, pointed

(A)

(B)

Figure 22.15 Probers and gleaners are characterized by their beaks.   (A) The white-breasted nuthatch (*Sitta carolinensis*). Nuthaches probe for insects or seeds hidden in the bark of tree trunks and branches. (B) The yellow warbler (*Setophaga petechia*). Warblers are gleaners, searching leaf surfaces for insects. (C) The southern bald ibis (*Geronticus calvus*) is a ground-dwelling bird that probes for insects and small vertebrates in dry grasslands. Other species of ibis probe for invertebrates in soft mud. (D) A toucan, the chestnut-eared aracari (*Pteroglossus castanotus*). Toucans pluck fruit from the trees of their Neotropical forest habitats, using their long beaks to reach among leaves and twigs. (A, Bill Thomas/USFWS; B, courtesy of U.S. National Park Service; C, © René C. Clark, Dancing Snake Nature Photography; D, Bernard Dupont/CC BY-SA 2.0.)

(C)

(D)

(A)

(B)
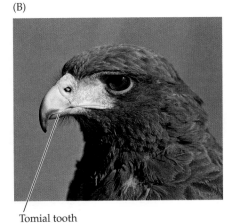

Tomial tooth

Figure 22.16 Carnivores. Carnivorous birds kill prey using heavy beaks and/or pointed talons. (A) Chihuahuan raven (*Corvus cryptoleucus*). The omnivorous ravens and crows (Corvidae) are predators of other birds and also eat recently killed carrion. (B) Harris's hawk (*Parabuteo unicinctus*). The tomial tooth is a projection in the center of the upper beak that severs the spinal cord when the hawk bites the neck of its prey. (Photos © René C. Clark, Dancing Snake Nature Photography.)

beaks as bludgeons to kill prey (**Figure 22.16**). By contrast, most hawks, owls, and eagles kill prey with their talons and use their beaks to tear off pieces small enough to swallow. Falcons stun prey with the impact of their dive, then bite the neck of the prey to disarticulate the cervical vertebrae.

- Many fish-eating birds have long beaks, often with a hooked tip (**Figure 22.17**).
- Filter feeding birds use a variety of mechanisms to capture prey and filter prey from the water (**Figure 22.18**).
- Seeds are usually protected by hard coverings (husks) that must be removed before the nutritious contents can be eaten. Specialized seed-eating birds use several methods to husk seeds before swallowing them. One group, including birds such as cardinals and grosbeaks, holds the seed in a strong beak and slices into it

with fore-and-aft movements of the lower jaw (**Figure 22.19A**). Another group, including blackbirds, holds the seed against ridges on the palate and cracks the husk by exerting an upward pressure with the robust lower jaw. After the husk has been opened, both these groups use their tongue to remove the contents. Other birds, such as the conifer-feeding crossbills (**Figure 22.19B**), have different specializations for eating seeds. Woodpeckers, nuthatches, and chickadees may wedge a nut or acorn into a hole in the bark of a tree and then hammer at it with their sharp bill until it cracks.

The skulls of most birds consist of four bony units that can move in relation to each other. This skull kinesis is important in some aspects of feeding. The upper jaw flexes upward as the mouth is opened, and the lower jaw expands laterally at its articulation with the skull (**Figure 22.20A**). The flexion of the upper and lower jaws increases the bird's

(A)

(B)

(C)

Figure 22.17 Fish-eaters. (A) The merganser (*Mergus merganser*) is a diving duck, a group that dives to capture prey under water rather than foraging in shallow water. Diving ducks have straight, narrow beaks with serrated margins and hooked tips. (B) Tufted puffin, *Fratercula cirrhata*). Puffins are diving birds of the world's Arctic and subarctic oceans. The short, deep beaks of puffins can hold previously captured fish while the bird makes additional captures. (C) American white pelican (*Pelecanus erythrorhynchos*). Pelicans feed mainly in near-surface ocean waters, dipping the lower beak into the water to scoop prey into an extensible gular pouch. They are gregarious and often hunt cooperatively in flocks. (A, Stan Bousson/USFWS/CC BY 2.0; B, Kuhnmi/CC BY 2.0; C, USFWS/CC BY 2.0.)

(A)

Lamellae

(B)

(C)

Figure 22.18 **Filter feeders use their beaks as sieves.**
(A) The mallard (*Anas platyrhynchos*) is a dabbling duck, a group that feeds in shallow water, upending (but not diving) to seize clods of aquatic vegetation and insects. Dabbling ducks have broad, flat beaks with rounded tips and soft edges. Lamellae along the edges of the upper and lower beak act as sieves, allowing mud and water to be expelled while insects, seeds, and other plant material is retained. (B) American flamingo (*Phoenicopterus ruber*). Flamingos immerse the head in water and feed with the head upside down. The thick tongue lies in a groove in the lower beak; movement of the tongue pumps water into the front of the mouth and out the sides of the beak, where complex rows of horny plates capture food. (C) Roseate spoonbills (*Platalea ajaja*) have broad-tipped beaks that they immerse and move from side to side. The convex upper surface and flat lower surface of the beak create vortex currents that lift prey from the bottom into the water column where the spoonbill seizes it. (A, Brad Sutton/NPS, inset Nic Redhead/CC BY SA 2.0; B, P. E. Hart, inset © René C. Clark, Dancing Snake Nature Photography; C and inset, James Diedrick/CC BY 2.0.)

gape in both the vertical and horizontal planes and probably assists in swallowing large items. Many birds use their beaks to search for hidden food. Long-billed shorebirds probe in mud and sand to locate worms and crustaceans. These birds display a form of skull kinesis called distal rhynchokinesis (Greek *rhynchos*, "beak") in which the flexible zone in the upper jaw has moved toward the tip of the beak, allowing the tip of the upper jaw to be lifted without opening the mouth (**Figure 22.20B**). This mechanism enables long-billed waders to grasp prey under the mud.

Tongues are an important part of the food-gathering apparatus of many birds. Woodpeckers drill holes into dead trees and then use their long tongues to investigate passageways made by wood-boring insects. The tongue of the European green woodpecker (*Picus viridis*), which extracts ants from their tunnels in the ground, extends four times the length of the bird's beak. The hyoid bones that support the tongue are elongated and housed in a sheath of muscles that passes around the outside of the skull and rests in the nasal cavity (**Figure 22.21**). When the muscles of the sheath contract, the hyoid bones are pushed around the back of the skull and the tongue is projected from the bird's mouth. The tip of a woodpecker tongue has barbs that impale insects and allow them to be pulled from their tunnels. Nectar-eating birds, such as hummingbirds and sunbirds, also have long

(A)

(B)

Figure 22.19 **Seed eaters.** (A) Blue grosbeak (*Passerina caerulea*). Grosbeaks slice seeds open using fore-and-aft movements of the lower jaw, then use the tongue to remove the contents. (B) Red crossbill (*Loxia curvirostra*). Crossbills extract the seeds of conifers from between the scales of the cones, using the diverging tips of their bills to pry the scales apart and a prehensile tongue to capture the seed inside (A, © René C. Clark, Dancing Snake Nature Photography; B, courtesy of Bridget Spencer.)

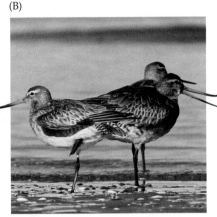

Figure 22.20 Avian skull and jaw kinesis. (A) A yawning herring gull (*Larus argentatus*) shows the kinetic movements of the skull and jaws that occur during feeding. Orange arrows show the positions of outward flexion, and green arrows show inward flexion. (B) Long-billed shorebirds such the bar-tailed godwit (*Limosa lapponica*) that probe for worms and crustaceans in soft substrates employ distal rhynchokinesis to raise the tip of the upper bill to seize prey without opening their mouth (A, Tim Lenz/CC BY-2.0; B, photo by Anne Collins.)

tongues and a hyoid apparatus that wraps around the back of the skull. The tip of the tongue of nectar-eating birds is divided into a spray of hair-thin projections, and capillary force causes nectar to adhere to the tongue.

The digestive system

The digestive systems of birds show some differences from those of other vertebrates. The absence of teeth prevents birds from doing much processing of food in the mouth, and the gastric apparatus takes over some of that role.

Esophagus and crop   Birds often gather more food than they can process in a short time and hold the excess in the esophagus. Many birds have a **crop**, an enlarged portion of the esophagus that is specialized for temporary food

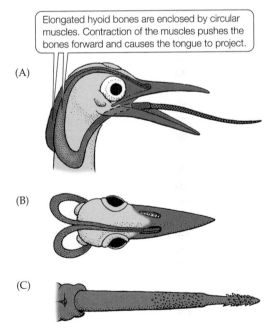

Elongated hyoid bones are enclosed by circular muscles. Contraction of the muscles pushes the bones forward and causes the tongue to project.

(A)

(B)

(C)

Figure 22.21 Tongue projection mechanism of a woodpecker. (A,B) The tongue itself is about the length of the beak, but it can be extended well beyond the tip of the beak by muscles that squeeze the posterior ends of the elongated hyoid apparatus, forcing the tongue forward. (C) A detail of the tongue shows the barbs on the tip that impale prey (insects and other small invertebrates). (A,B, after Leiber 1907; C after Eckstorm 1901.)

storage, with longitudinal folds that allow it to expand. An additional function of the crop is transportation of food for nestlings. The foraging adult gathers stores food in the crop, then returns to the nest where it regurgitates the material from its crop and feeds it to the young.

Many birds are omnivorous, and some change the proportions of plant and animal food in their diets seasonally. Plants are harder to digest than animal tissue, and the length of the intestine increases when plants are a major component of the diet (**Figure 22.22A**). The South American hoatzin (*Opisthocomus hoazin*) is the only bird known to employ fermentative digestion (**Figure 22.22B**).

In doves and pigeons, the hormone prolactin—the same hormone that stimulates lactation by mammals—stimulates the crop of both sexes to produce a nutritive fluid that is fed to the young. This crop milk (or "pigeon's milk") is produced by fat-laden cells that detach from the squamous epithelium of the crop and are suspended in an aqueous fluid. Crop milk is rich in lipids and proteins but contains no sugar. Its chemical composition is similar to that of mammalian milk, although it differs in containing intact cells.

Stomach   The form of a bird's stomach is related to its dietary habits. Carnivorous and piscivorous birds need expansible storage areas to accommodate large volumes of soft food, whereas birds that eat insects or seeds require a muscular organ that can contribute to the mechanical breakdown of food. Typically, the gastric apparatus of birds consists of two relatively distinct chambers: an anterior glandular stomach (**proventriculus**), and a posterior muscular stomach (**gizzard**). The proventriculus contains glands that secrete acid and digestive enzymes. The proventriculus is especially large in species that swallow large items such as intact fruit.

The gizzard has several functions, including food storage while the chemical digestion that was begun in the proventriculus continues, but its most important function is the mechanical processing of food. The thick, muscular walls of the gizzard squeeze the contents, and small stones that are held in the gizzards of many birds help grind the food. In this sense, the gizzard performs the same function as a mammal's teeth. The pressure that can be exerted on food in the gizzard is intense—a turkey's gizzard can grind two dozen

(A) Budgerigar

(B) Hoatzin

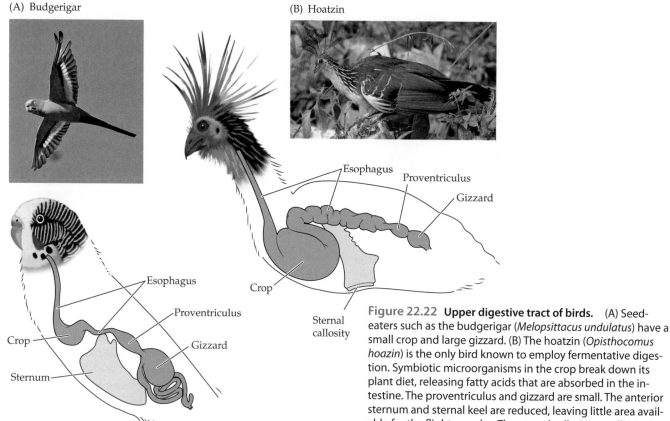

Figure 22.22 Upper digestive tract of birds. (A) Seed-eaters such as the budgerigar (*Melopsittacus undulatus*) have a small crop and large gizzard. (B) The hoatzin (*Opisthocomus hoazin*) is the only bird known to employ fermentative digestion. Symbiotic microorganisms in the crop break down its plant diet, releasing fatty acids that are absorbed in the intestine. The proventriculus and gizzard are small. The anterior sternum and sternal keel are reduced, leaving little area available for the flight muscles. The sternal callosity, an elliptical patch of horny skin, lies above the sternal keel. When its crop is full, a perching hoatzin supports itself by resting the callosity against a branch. (A, Jim Bendon/CC BY-SA 2.0; B, Bridget Spencer.)

walnuts in as little as 4 hours and can crack hickory nuts that require as much as 150 kilograms of pressure to break.

Intestine, ceca, and cloaca   The small intestine is the principal site of chemical digestion, as enzymes from the pancreas and intestine break down the food into small molecules that can be absorbed across the intestinal wall. The mucosa of the small intestine is modified into a series of folds, lamellae, and villi that increase its surface area. The large intestine is relatively short, usually less than 10% of the length of the small intestine. Passage of food through the intestines of birds is quite rapid: transit times for carnivorous and frugivorous (fruit-eating) species are in the range of a few minutes to a few hours. Passage of food is slower in herbivores and may require a full day. Birds generally have a pair of ceca at the junction of the small and large intestines. The ceca are small in carnivorous, insectivorous, and seed-eating species, but they are large in herbivorous and omnivorous species such as cranes, fowl, ducks, geese, and ostriches. Symbiotic microorganisms in the ceca apparently ferment plant material.

Birds respond to seasonal changes in diet with changes in the morphology of the gut. Many species of birds feed on insects and other animal prey during the summer and switch to plant food (such as berries) during the winter. Plant material takes longer to digest than animal food, so it must pass through the gut more slowly. To accommodate differences in passage time, the intestine changes length in response to changes in diet. European starlings (*Sturnus vulgaris*) feed mainly on insects from March through June and add progressively more plant material to their diet from late summer through winter. The length of the intestine shows a corresponding cycle, increasing in length by about 20% during fall and winter and decreasing by the same amount during spring and early summer. In addition to the anatomical changes in the intestine, the digestive enzymes change to match the reduction in protein and fat and the increase in simple sugars in fruit compared with animal food.

The cloaca temporarily stores waste products while water is being reabsorbed. Birds precipitate their nitrogenous waste in the form of potassium and sodium salts of uric acid, which requires very little water and thus is highly effective at water conservation. The white urate salts mixed with dark fecal material are familiar to anyone who has washed an automobile. Some species of birds have extra-renal salt-secreting glands that accomplish further conservation of water by removing ions directly from the blood.

22.5 Sensory Systems

A flying bird moves rapidly through three-dimensional space and requires a continuous flow of sensory information about its position and the presence of obstacles in its path. Vision is the sense best suited to provide this sort of information on a rapidly renewed basis, and birds have a well-developed visual system that remains active even

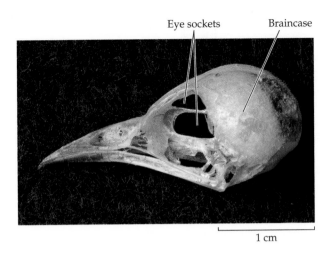

Eye sockets   Braincase

1 cm

Figure 22.23 The effects of a bird's large eyeballs. The eyeballs of many birds, such as this American yellow warbler (*Setophaga petechia*), meet in the middle of the face. The large eye sockets can occupy more than half the skull in some species, and in most birds displace the braincase caudally (toward the rear) and ventrally (downward). (Photo by M. H. Siddall.)

when a bird is asleep. The importance of vision is reflected in the brain: the optic lobes are large, and the midbrain is an important area for processing visual and auditory information. The cerebellum, which coordinates body movements, is large. The cerebrum is less developed in birds than it is in mammals and is dominated by the corpus striatum.

Vision

The eyes of birds are so large that the brain is displaced dorsally and caudally, and in many species the eyeballs meet in the midline of the skull (**Figure 22.23**). In its basic structure, the eye of a bird is like that of any other vertebrate, but the shape varies from a flattened sphere to something approaching a tube. An analysis of the optical characteristics of birds' eyes suggests that these differences are primarily the result of fitting large eyes into small skulls. The eyes of a starling are small enough to be contained within the skull, whereas the eyes of an owl bulge out of the skull. An owl would require an enormous, unwieldy head to accommodate a flat eye like that of a starling. The tubular shape of the owl's eye allows it to fit into a reasonably sized skull.

Although birds of prey are popularly considered to be "eagle-eyed" (i.e., to have very high visual acuity), their ability to resolve images is not much better than that of humans. The visual acuity of the wedge-tailed eagle (*Aquila audax*) is about 2.5 times that of humans, and visual acuity of falcons is similar to that in humans. The visual acuity of blue jays (*Cyanocitta cristata*) and pigeons is approximately the same as, or inferior to, that of humans. Birds do excel, however, in their ability to see fast—that is, to resolve visual stimuli that change rapidly. Flicker-fusion frequency is the rate of change that an eye can resolve; for humans it is about 60 hertz (Hz), but several species of passerines are able to resolve alternating light–dark cycles at up to 145 Hz. Ultrarapid vision may be important to birds as they

fly through complex three-dimensional environments, and for swallows and other aerial foragers that capture flying insects.

The pecten is a conspicuous structure in the avian eye. It is formed by blood capillaries surrounded by pigmented tissue and covered by a membrane; it lacks muscles and nerves. The pecten arises from the retina at the point where the nerve fibers from the ganglion cells of the retina join to form the optic nerve. In some species of birds the pecten is small, but in other species it extends so far into the vitreous humor that it almost touches the lens. The function of the pecten remains uncertain after 200 years of debate. Proposed functions include reduction of glare, a mirror to reflect objects above the bird, production of a stroboscopic effect, and a visual reference point, but none of these hypotheses is very plausible. The large blood supply flowing to the pecten suggests that it may provide nutrition for the retinal cells and help remove metabolic waste products that accumulate in the vitreous humor.

The world must look different to birds than it does to us because most birds have four photosensitive pigments in their retinal cells, whereas humans and other primates have only three. In other words, birds have tetrachromatic vision, while that of primates is trichromatic. Three of the avian retinal pigments absorb maximally at wavelengths equivalent those humans perceive—the red, green, and blue regions of the spectrum. The fourth pigment present in birds ancestrally responds to wavelengths of about 400 nanometers, or deep blue. In some bird species, a mutation in the gene encoding the deep blue pigment shifted its maximum sensitivity into the ultraviolet, thereby extending the visual sensitivity of these birds beyond the range of the human eye. Ultraviolet sensitivity has evolved independently several times among passerines, although it is rare among seabirds.

Sensitivity to wavelengths in the ultraviolet portion of the spectrum is important in many aspects of bird behavior. Ultraviolet reflectance distinguishes males from females in some species of birds in which the sexes are indistinguishable to human eyes, indicates the physiological status of prospective mates, and demonstrates the health of nestlings. The colors and patterns of eggs may help parent birds distinguish their own eggs from the eggs of brood parasites, such as cuckoos, and ultraviolet reflectance may enhance the effectiveness of the warning colors of toxic prey, such as dart-poison frogs.

Oil droplets are found in the cone cells of the avian retina, as they are in other sauropsids. The droplets range in color from red through orange and yellow to green. The oil droplets are filters, absorbing some wavelengths of light and transmitting others. The function of the oil droplets is unclear; it is certainly complex because the various colors of droplets are combined with the four visual pigments. Birds such as gulls, terns, gannets, and kingfishers that must see through the surface of water have a preponderance of red droplets. Aerial hawkers of insects have predominantly yellow droplets.

Hearing

In birds, as in other sauropsids, the columella (stapes) and its cartilaginous extension, the extracolumella, transmit vibrations of the tympanum to the oval window of the inner ear. The cochlea of birds appears to be specialized for fine distinctions of the frequency and temporal pattern of sound. The cochlea of a bird is about one-tenth the length of the cochlea of a mammal, but it has about ten times as many hair cells per unit of length. The space above a bird's basilar membrane (the scala vestibuli) is nearly filled by a folded, glandular tissue. This structure may dampen sound waves, allowing a bird's ear to respond very rapidly to changes in sounds.

The sensitivity of the avian auditory system is approximately the same as that of humans, despite the small size of birds' ears. Most birds have tympanic membranes that are large in relation to the head size. A large tympanic membrane enhances auditory sensitivity, and owls (which have especially sensitive hearing) have the largest tympani relative to their head size among birds. Sound pressures are amplified during transmission from the tympanum to the oval window of the cochlea because the area of the oval window is smaller than the area of the tympanum. The ratio for birds ranges from 11:1 to 40:1. High ratios suggest sensitive hearing, and the highest values are found in owls; songbirds have intermediate ratios (20:1 to 30:1). (The ratio is 36:1 for cats and 21:1 for humans.)

Owls are the most acoustically sensitive birds, with sensitivity as great as a cat's. In addition to their large tympanic membranes, owls have large cochleae and well-developed auditory centers in the brain. In an experimental test of their capacities for acoustic orientation, barn owls (*Tyto alba*) were able to seize mice in total darkness. If the mice towed a piece of paper across the floor behind them, the owls struck the rustling paper instead of the mouse, showing that sound was the cue they were using.

Anatomical features that are key to the acoustic abilities of owls include a facial ruff and an asymmetrical skull. The ruff is formed by stiff feathers (**Figure 22.24A**) and acts as a parabolic sound reflector, focusing sounds with frequencies higher than 5 kilohertz (kHz) and amplifying them by 10 decibels. The skulls of many owl species are markedly asymmetrical, and these are the species that have the greatest auditory sensitivity (**Figure 22.24B**). The asymmetry ends at the external auditory meatus; the middle and inner ears of owls are bilaterally symmetrical. The asymmetry of the external ear openings assists with localization of prey in the horizontal and vertical axes. Minute differences in the time and intensity at which sounds are received by the two ears indicate the direction of the source. The brains of owls integrate time and intensity information with extraordinary sensitivity to produce a map of their environment that integrates auditory and visual information.

Olfaction

Although birds have long been considered to rely primarily on vision and hearing, evidence that scent plays important roles in their lives is accumulating. In the 1970s, studies of homing pigeons (a domesticated strain of the rock pigeon, *Columba livia*) demonstrated that olfaction is a critical component of successful homing. Surprisingly, when the right nostrils of homing pigeons were plugged, they had more difficulty finding their way back to their home lofts than they did when their left nostrils were plugged. Tube-nosed seabirds, such as shearwaters, employ olfaction during migration and also to locate areas where food is abundant, as well as for homing. Unfortunately, detection of prey by odor now leads seabirds, and probably other marine organisms as well, to consume plastic flotsam because microorganisms growing on marine plastic debris release dimethyl sulfide, which is a key feeding stimulus for marine organisms.

(A)

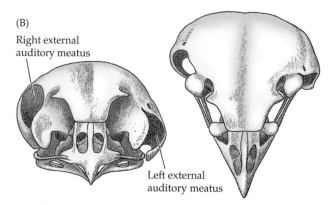

(B)

Right external auditory meatus

Left external auditory meatus

Figure 22.24 An owl's ruff and asymmetrical skull increase directional sensitivity of hearing. (A) A boreal owl (*Aegolius funereus*). The ruff of feathers that surrounds the face of many owls increases the directional sensitivity of their hearing. (B) Frontal and dorsal views of the skull of a boreal owl show the pronounced asymmetry in the position of the left and right ear openings (the external auditory meatus). (A, courtesy of U.S. National Park Service; B, from Norberg 1978.)

(A) 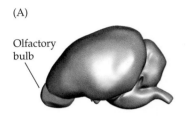 (B) 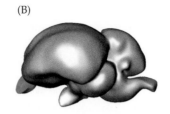 (C) 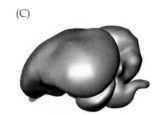 (D)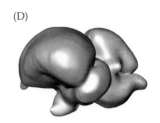

Olfactory
bulb

Figure 22.25 The size of birds' olfactory bulbs varies greatly. Three-dimensional models show the size of olfactory bulbs relative to the size of the brain. (A,B) Basal species such as the North Island brown kiwi (*Apteryx mantelii*; A) and aquatic species such as the purple swamphen (*Porphyrio porphyrio*; B) have large olfactory bulbs. (C,D) More derived species, such as the Australian magpie (*Cracticus tibicen*; C), a passerine, and terrestrial species such as the wild turkey (*Meleagris gallopavo*; D) have small lobes. (Modified from Corfield et al. 2015.)

Vultures locate carcasses by scent, and at least some insectivorous birds respond to the volatile chemicals released by plants that have been attacked by herbivorous insects. Kiwis have nostrils at the tip of their long bill and find earthworms underground by smelling them. Great tits (*Parus major*) are attracted to trees infested by caterpillars and ignore visually similar trees without caterpillars. This is a learned behavior; naive birds do not discriminate between caterpillar-infested and uninfested trees.

Odors also influence the social behaviors of birds, playing a role in species recognition and perhaps also in sex recognition, in the choice of partners, and in care of eggs and nestlings. The odors responsible for these behaviors probably originate in the uropygial gland secretions—waxy fluids that birds spread on their feathers when they preen.

The olfactory bulbs of birds vary greatly in size, ranging from more than 25% of the total brain volume in vultures and tube-nosed seabirds to less than 5% in some passerines (**Figure 22.25**). Basal species and species that live and forage in aquatic habitats have larger olfactory bulbs than more derived and terrestrial species. Small olfactory bulbs do not necessarily correspond to lack of sensitivity, however; even passerines have olfactory capacities that are similar to those of other birds.

Olfactory receptor (OR) genes code for the transmembrane proteins that detect odorant molecules. OR genes can be grouped in families, which originate with a single gene and diversify by gene duplication and mutations. A family can contain hundreds of variants, and OR genes constitute one of the largest multigene families in vertebrates. The OR genes within a subfamily generally respond to similar chemicals, suggesting that a subfamily may be responsive to a particular class of odors. In birds, expansions of certain OR gene subfamilies are associated with specific ecological characteristics. For example, the OR subfamilies OR51 and 52, which respond to hydrophilic (water-soluble) compounds, are common in aquatic birds, whereas OR14, which responds to hydrophobic compounds, is common in terrestrial birds. OR5, 8, and 9 are well represented in raptors, and OR6 and 10 in species with large vocal repertoires.

22.6 Social Behavior

Vision and hearing are important to birds, as they are to humans, and as a result of this similarity birds play an important role in behavioral studies. Furthermore, most birds are active during the day and are thus relatively easy to observe. A tremendous amount of information has been accumulated about the behavior of birds under natural conditions. This background has contributed to the design of experimental studies in the field and in the laboratory.

Plumage colors and patterns

The colors and patterns of a bird's plumage identify its species, and frequently its sex and age as well (**Figure 22.26**). Three types of pigments are widespread among birds and produce their distinctive plumages:

1. *Melanins* produce dark colors. Eumelanins produce black, gray, and dark brown, while phaeomelanins are responsible for reddish brown and tan shades.

2. *Carotenoids* are fat-soluble pigments that are responsible for most red, orange, and yellow colors. Birds obtain carotenoids from their diet, and in some cases the intensity of color can be used to gauge the fitness of a prospective mate.

3. *Porphyrins* are metal-containing compounds that are chemically similar to the pigments in hemoglobin and liver bile. Ultraviolet light causes porphyrins to emit a red fluorescence. Porphyrins are destroyed by prolonged exposure to sunlight, so they are most conspicuous in fresh plumage.

The structural colors of plumage—purples, blues, and greens—result from tiny air-filled structures in the cells on the surface of feather barbs. When daylight is scattered by these structures, some wavelengths show constructive interference (i.e., an intensification of that wavelength), whereas other wavelengths show destructive interference (reduction in intensity). The spacing of the air-filled structures determines which colors are intensified. Structural colors can be combined with pigments; green parakeets, for example, combine a structural blue with a yellow

(A)

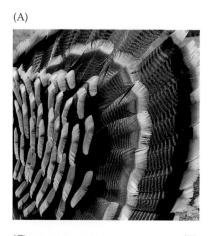

(B)

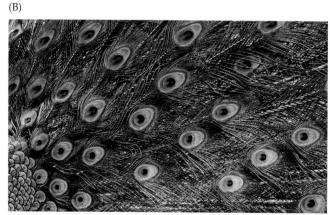

(C)

(D)

Figure 22.26 Pigments and structures combine to produce the colors and patterns of feathers. (A) The black, brown, tan, and cream regions in the tail feathers of a male wild turkey (*Meleagris gallopavo*) are produced by eumelanins and phaeomelanins. (B) Melanins are also responsible for the dramatic colors of the tail feathers of a male peacock (*Pavo cristatus*), but this is a structural color caused by interference. Cylinders of melanin in the barbules create interference at wavelengths that correspond to the spacing between cylinders—150 nm in the green regions, 140 nm in the blue areas. (C, D) The magenta gorget of a male Anna's hummingbird (*Calypte anna*) is also produced by interference. The gorget is bright and highly visible from some angles of view (C), and disappears when the bird turns slightly. (A, courtesy of U.S. National Park Service; B, C, D © René C. Clark, Dancing Snake Nature Photography.)

(A)

(B)

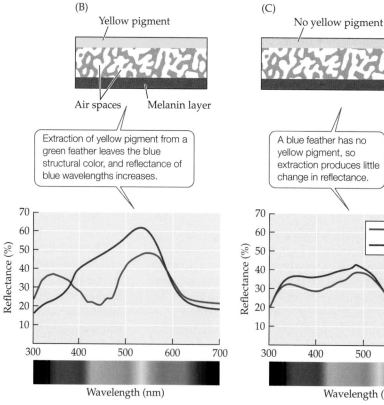

(C)

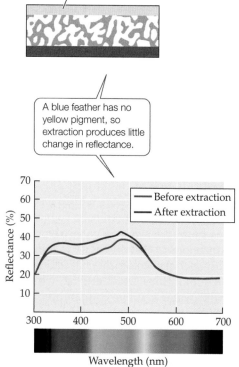

carotenoid, while blue parakeets are homozygous for an allele that blocks formation of the carotenoid (**Figure 22.27**). That allele is a simple recessive, so two blue parakeets can produce only blue offspring.

In addition to producing most yellow, orange, and red colors, carotenoids have beneficial physiological functions, and diverting carotenoids to ornamentation may signal that an individual is robust, and hence a desirable mate—an example of a phenomenon known as the handicap principle.

Vocalization, sonation, and visual displays

Birds use sounds, colors, and movements to recognize other individuals as well as their species, sex, and age. Two categories of sounds play a role in bird behavior: vocalization and sonation. Birds produce vocalizations by passing air through the **syrinx**, a unique avian structure that lies at the junction of the trachea and the two bronchi. A syrinx may be a derived character of crown group birds; a fossil syrinx has been described from a Late Cretaceous bird, but the structure has not been found in nonavian dinosaurs. **Sonations** are nonvocal sounds that are usually created either by feathers knocking against each other or by air moving between or across feathers.

Birdsong and SCRs The term "birdsong" has a specific meaning that is distinct from a bird call. Whereas bird calls are typically simple and brief, songs are often long and very complex vocalizations. Songs are generally associated with reproduction and territoriality, and in some species are only produced by mature males during the breeding season. Song is controlled by a series of song control regions (SCRs) in the brain. During the period of song learning, which occurs early in life, neural connections form between a part of the SCR that is associated with song learning and a region of the brain that controls the vocal muscles. Thus, song learning and song production are closely linked in birds.

Avian SCRs are under hormonal control, and in many bird species, the SCRs of males are larger than those of females, with more and larger neurons and longer dendritic processes. The vocal behavior of female birds varies greatly across taxonomic groups. Females of some species produce only simple calls, whereas in other species the females engage with males in complex song duets. Among the latter species, the SCRs of females are similar in size to those of the males. In species in which females do not vocalize, the SCRs of females apparently play a role in species-specific mate recognition: when the SCRs of female canaries were inactivated, the birds no longer distinguished the vocalizations of male canaries from those of sparrows.

A birdsong consists of a series of notes with intervals of silence between them. Changes in frequency (frequency modulation) are conspicuous components of the songs of many birds, and the avian ear may be very good at detecting rapid changes in frequency. Birds often have more than one song type, and some species may have repertoires of several hundred songs.

Birdsongs identify the species of bird that is singing, and they often show regional dialects (**Figure 22.28**). These dialects are transmitted from generation to generation as young birds learn the songs of their parents and neighbors. In the indigo bunting (*Passerina cyanea*), one of the best-studied species, song dialects that were characteristic of small areas persisted for as long as 15 years, which is substantially longer than the life of an individual bird. Birdsongs also show individual variation that allows birds to recognize the songs of residents of adjacent territories and to distinguish the songs of these neighbors from those of intruders. Male hooded warblers (*Setophaga citrina*) remember the songs of neighboring males and recognize them as individuals when they return to their breeding sites in North America after spending the winter in Central America.

The songs of male birds identify their species, sex, and occupancy of a territory. When a recording of a male's song is played back through a speaker that has been placed in the territory of another male, the territorial male responds with vocalizations and aggressive displays, and may even attack the speaker. These behaviors repel intruders, and even hearing the song of a territorial male keeps intruders at a distance; broadcasting recorded songs in a territory from which the territorial male has been removed delays occupation of the vacant territory by a new male.

Nonvocal displays In contrast to the precise definition of "birdsong," the words "drumming," "clapping," and "whistling" have been used to describe the sounds that birds make by striking objects with their beaks or wings and the sounds made as air passes over or through their feathers. These sounds play central roles in the social behavior of many bird species.

Woodpeckers and some other bird species make a drumming sound by pounding with their bills on a resonant object. Drumming could be a territorial signal, informing other woodpeckers that a stand of trees already has a resident, and it could advertise the presence of a male to unmated female woodpeckers. Storks and herons clap their

◀ **Figure 22.27 Many green birds are blue underneath.** Green feathers are often the combination of a blue structural color overlain by a yellow pigment, as in budgerigars (*Melopsittacus undulatus*). Wild budgerigars are yellow and green, and the white and blue varieties of domesticated budgies result from loss of the yellow pigment. (A) The bird in this photo is probably a mosaic, resulting from a mutation at its 2-cell stage of development that inactivated synthesis of yellow pigment in the cell that developed into the bird's left side. With the yellow pigment gene inactivated, the feathers reveal the blue structural color produced by refraction of light in the air spaces. (B) When the yellow pigment is experimentally extracted from a green feather, reflection in the blue region of the spectrum increases. (C) There is no yellow pigment in a blue feather, so subjecting it to the extraction process scarcely changes its color. (A, 2il org/CC BY 2.0; B,C, modified from D'Alba et al. 2012.)

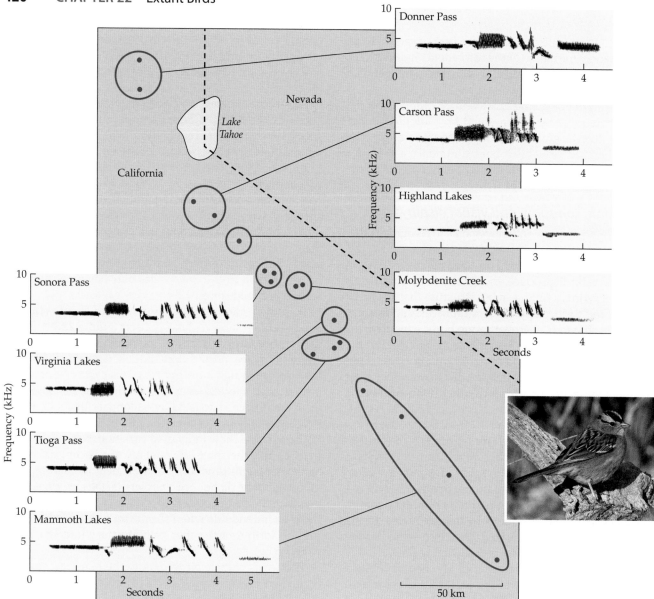

Figure 22.28 Regional song dialects of white-crowned sparrows (*Zonotrichia leucophrys*). Sound spectrograms show geographic variation in the songs of male white-crowned sparrows from different populations in the Sierra Nevada of California. The density of the spectrogram is proportional to the sound energy at that point. The difference between dialects increases with the distance between populations. (After MacDougall-Shackleton and MacDougall-Shackleton 2001; photo © René C. Clark, Dancing Snake Nature Photography.)

mandibles together, and wing-clapping has been described in many species of birds.

Flutter of primaries 6 and 7 of African broadbills produces a startlingly loud sound with frequencies between 1 and 2 kHz. Male Anna's hummingbirds (*Calypte anna*) emit a loud *chirp* by spreading the tail as they dive toward a female; air flowing across the surface of the vane of the outermost tail feathers causes them to flutter, producing a sound with a fundamental frequency of about 4 kHz. The feathers of broadbills and hummingbirds have no visible modifications for sound production, but some species do

have modified feathers. The rachi of secondaries 6 and 7 of male club-winged manakins (*Machaeropterus deliciosus*) are thickened toward their bases and bent, and the rachi of secondaries 1–5 are also thickened. During a male's display, these seven feathers vibrate as a unit, producing a vibration with a fundamental frequency of about 1.5 kHz and a harmonic with a frequency of about 3 kHz.

Visual displays Visual displays are frequently associated with songs; for example, a particular body posture that displays colored feathers may accompany singing. Male birds are often more brightly colored than females and have feathers that have become modified as the result of sexual selection. In this process, females mate preferentially with males that have certain physical characteristics. Because of that response by females, those physical characteristics contribute to the reproductive fitness of males, even though they may have no useful function in any other aspect of the animal's ecology or behavior. The colorful areas on the wings of male ducks, the red epaulets on the wings of male red-winged blackbirds (*Agelaius phoeniceus*), the red crowns on kinglets, and the elaborate tails of male peacocks are familiar examples of specialized areas of plumage that are involved in sexual behavior and display.

22.7 Oviparity

The avian mode of reproduction is limited to laying eggs. No other group of vertebrates that contains such a large number of species is exclusively oviparous. Why is this true of birds?

Constraints imposed on birds by their specializations for flight are often invoked to explain their failure to evolve viviparity. Those arguments are not particularly convincing, however, considering that bats have successfully combined flight and viviparity. Furthermore, flightlessness has evolved in at least 15 families of birds, but none of these flightless species has evolved viviparity.

Oviparity is presumed to be the ancestral reproductive mode for archosaurs, and it is retained by both crocodylians and birds. However, viviparity evolved in the marine reptiles of the Mesozoic and has evolved more than 100 times in the other major lineage of extant sauropsids—the lizards and snakes (lepidosaurs)—so the capacity for viviparity is clearly present in sauropsids.

A key element in the evolution of viviparity among lizards and snakes appears to be the retention of eggs in the oviducts of the female for some period before they are deposited. This situation occurs when the benefits of egg retention outweigh its costs. For example, the high incidence of viviparity among snakes and lizards in cold climates may be related to the ability of a female ectotherm to speed embryonic development by thermoregulation. A lizard that basks in the sun can raise the temperature of eggs retained in her body and perhaps determine the sex of her offspring, but after depositing her eggs in a nest, she has no control over their temperature and rate of development.

Birds have body temperatures of 40–42°C, whereas brooding birds regulate egg temperatures at 33–37°C. The body temperatures of adult birds may be too hot for embryonic development, and this could explain why egg retention never evolved among birds.

Egg biology

The inorganic part of eggshells contains about 98% crystalline calcite ($CaCO_3$), and the embryo obtains about 80% of its calcium from the eggshell. An organic matrix of protein and mucopolysaccharides is distributed throughout the shell and may serve as a support structure for the growth of calcite crystal (**Figure 22.29**). Eggshell formation begins in the infundibulum of the oviduct (**Figure 22.30**). Two shell membranes are secreted to enclose the yolk and albumen, and carbohydrate and water are added to the albumen.

Bird eggs are archetypal amniotic eggs (see Figure 10.13). The innermost membrane, the amnion, surrounds the embryo while the outmost membrane, the chorion, encloses all of the embryonic structures and is in contact with the inner side of the shell. The allantois has a double role: respiration and excretion. The outer potion of the allantois is highly vascularized and presses against the chorion, forming the chorioallantois—the site of embryonic gas exchange. The

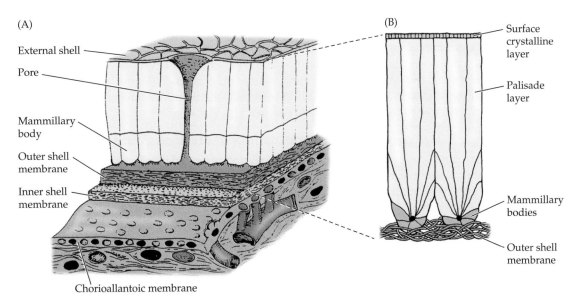

Figure 22.29 Structure of an avian eggshell. (A) Pore canals penetrate the calcified external shell region, allowing oxygen to enter and carbon dioxide and water to leave the egg. The gases are transported to and from the embryo via blood vessels in the chorioallantoic membrane. (B) Crystallization of the external shell (calcite, $CaCO_3$) begins at the mammillary bodies. Crystals grow into the outer shell membrane and upward to form the palisade layer. Changes in the chemical composition of the fluid surrounding the growing eggshell are probably responsible for the change in crystal form in the surface crystalline layer. (A, after Rahn et al. 1979; B, from Carey 1983.)

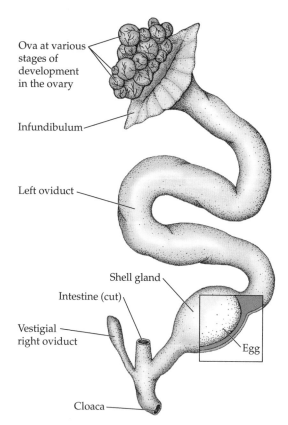

Ova at various stages of development in the ovary

Infundibulum

Left oviduct

Shell gland

Intestine (cut)

Vestigial right oviduct

Egg

Cloaca

Figure 22.30 Oviduct of a bird. Only the left ovary and oviduct are functional in most bird species. Ova released from the ovary enter the infundibulum of the oviduct. Fertilization occurs at the upper end of the oviduct, then albumen and shell membranes are secreted, and finally the egg is enclosed in calcareous shell.

interior of the allantois is the storage area for nitrogenous wastes in the form of crystals of uric acid.

Nutrition is provided by the yolk, which makes up anywhere from 20 to 70% of the volume of a newly laid egg. Lipids make up about 33% of the yolk, protein is another 20%, and water is the rest. The yolk is enclosed in the yolk sac, which is gradually absorbed into the embryo's body. Albumen, which is about 10% protein and 90% water, provides the water the embryo needs to develop.

With no means of ventilation for gas exchange and depending on a limited store of water, the permeability of the eggshell to oxygen, carbon dioxide, and water vapor is a delicate balance between retarding water loss while permitting outward diffusion of carbon dioxide and inward diffusion of oxygen. The eggshell is penetrated by an array of pores that occupy about 0.02% of the surface of an eggshell. The morphology of the pores varies in different species of birds: some pores are straight tubes, whereas others are branched, and the pore openings on the eggshell's surface may be blocked to varying degrees with organic or crystalline material. All of these factors affect the rate of evaporation of water from the egg. The eggshell grows thinner as calcium is removed from it to form the bones of the embryo, increasing the rates of diffusion of gases through the shell.

Bird eggs must lose water to hatch because the loss of water creates an air cell at the blunt end of the egg. The embryo penetrates the membranes of this air cell with its beak 1–2 days before hatching begins, and ventilation of the lungs begins to replace the chorioallantoic membrane in gas exchange. Pipping, the formation of the first cracks on the surface of the eggshell, follows about half a day after penetration of the air cell, and actual emergence begins half a day later. Shortly before hatching, the chick develops a horny projection on its upper beak. This structure is called the egg tooth, and it is used in conjunction with a hypertrophied muscle on the back of the neck (the hatching muscle) to thrust vigorously against the shell. The egg tooth and hatching muscle disappear soon after the chick hatches.

Sex determination

All birds have heterogametic sex chromosomes, as do mammals, except that in birds the female is the heterogametic sex and the male is homogametic. To avoid confusion with mammals, the sex chromosomes of female birds are designated ZW and male birds are ZZ. The presence of the W sex chromosome causes the primordial gonad to secrete estrogen, which stimulates the left gonad to develop as an ovary and the left Müllerian duct system to develop into an oviduct and shell gland. In the absence of estrogen (i.e., when the genotype is ZZ), a male develops. (This is the opposite of the sex determination process in mammals, in which a male-determining gene on the Y chromosome causes an XY individual to develop as a male.)

The evolutionary origin of genetic sex determination in birds is unclear because all crocodylians have temperature-dependent sex determination and lack heterogametic sex chromosomes. Most birds maintain relatively stable egg temperatures by incubating their eggs during embryonic development, but a lineage of galliform birds, Megapodiidae, that occurs in the Indo-Australian region is an exception. This family includes about 20 species of ground-dwelling birds that bury their eggs like crocodylians rather than incubating them as other birds do (**Figure 22.31**). The chicks are fully developed when they hatch and are not dependent on their parents.

Maternal control of sex of offspring

To add to the complications of sex determination in birds, female birds can exert some control over the sex of their offspring. Many birds lay one egg per day until all the eggs in a clutch have been produced. Eggs hatch approximately in the sequence in which they were laid, and female birds may be able to adjust the sex of their offspring in relation to the laying sequence of their eggs. A study of house finches (*Haemorhous mexicanus*) found that the eggs laid first by birds in Montana hatched into females, whereas first-laid eggs in an Alabama population hatched into males. Sex-specific growth and survival characteristics appear to be the basis for this difference, because nestlings that hatch first grow faster and reach larger adult sizes than those that hatch later. In Montana, large females survive better than

Figure 22.31 Megapodes construct nests similar to those of crocodylians. The Australian brush turkey (*Alectura lathami*) builds a mound of soil and vegetation in which the eggs are deposited. Sex determination of the offspring is affected by temperature; when brush turkey eggs are incubated at 31°C, about 75% of the hatchlings are males, but when the incubation temperature increases to 34°C, the proportion of males drops to about 28%. (Photo by Doug Becker/CC BY SA 2.0.)

smaller ones, whereas in Alabama large males are favored. The sex-biased hatching order increases chick survival by 10–20%.

22.8 Monogamy: Social and Genetic

Social vertebrates exhibit one of two broad categories of mating systems—monogamy or polygamy. **Monogamy** (Greek *mono*, "one"; *gamos*, "marriage") refers to a pair bond between a single male and a single female. Monogamy is the general social system of birds. The pairing may last for part of a breeding season, an entire season, or a lifetime.

Both parents in monogamous mating systems usually participate in caring for the young, and the universal occurrence of oviparity among birds may account for the prevalence of monogamy. Because eggs are incubated externally and the young birds consume the same food as adults, both parents can participate in incubating and caring for the young.

Monogamy does not necessarily mean fidelity to a mate, however. Genetic studies of monogamous birds have shown that extra-pair copulation (mating with a bird other than the partner) is common. Thus, some of the eggs in a nest may have been fertilized by a male other than the partner of the female that deposited them, and the male partner may have fertilized eggs of other females that are being incubated in their nests. As evidence of extra-pair copulation has accumulated, the term **social monogamy** has been introduced for species in which a male and female share responsibility for a clutch of eggs but do not demonstrate

fidelity. **Genetic monogamy** describes a social system in which a male and female share parental responsibilities and do not have extra-pair copulations.

22.9 Nests and Parental Care

Nests protect eggs not only from such physical stresses as heat, cold, and rain but also from predators. Bird nests range from shallow holes in the ground to enormous structures that represent the combined efforts of hundreds of individuals over many generations (**Figure 22.32**). The nests of passerines are usually cup-shaped structures because this shape provides support for the eggs and brooding parent. Nests can be made of woven plant materials, mud, saliva, spiderweb, and other materials. Swifts use sticky secretions from their buccal glands to cement the material together to form nests, and grebes, which are marsh-dwelling birds, build floating nests from the buoyant stems of aquatic plants.

Incubation

Some species of birds begin incubation as soon as the first egg is laid, and their eggs hatch asynchronously; other species wait to begin incubation until the clutch is complete, and their eggs hatch synchronously. In the nests of asynchronous hatchers, younger chicks often do more poorly and may even be the victims of siblicide (common in hawks and herons). During years of plentiful resources, parents may be able to care for all chicks, but otherwise smaller siblings may serve as replacements if an older sibling dies, or even facilitate the growth of an older siblings by providing additional warmth or even a meal!

Prolactin, secreted by the pituitary gland, suppresses ovulation and induces brooding behavior, at least in those species of birds that wait until a clutch is complete to begin incubation. The insulating properties of feathers that are so important in the thermoregulation of birds become a handicap during brooding, when the parent must transfer metabolic heat from its own body to the eggs. Prolactin plus estrogen or androgen stimulates the formation of brood patches in female and male birds, respectively. These brood patches are areas of bare skin on the ventral surface of a bird. The feathers are lost from the brood patch, and blood vessels proliferate in the dermis, which may double in thickness and give the skin a spongy texture. Not all birds develop brood patches, and in some species only the female has a brood patch, although the male may share in incubating the eggs. Ducks and geese create brood patches by plucking the down feathers from their breasts; they use the feathers to line their nests. Some penguins lay a single egg that they hold on top of their feet and cover with a fold of skin from the belly, thus enveloping the egg.

The temperature of eggs during brooding is usually maintained within the range of 33–37°C, although some eggs can withstand periods of cooling when the parent is off the nest. Tube-nosed seabirds are known for the long

Figure 22.32 Bird nests are enormously diverse. Some nests are no more than shallow depressions; others are elaborate structures. (A) The oystercatcher (*Charadrius vociferus*), like many shorebirds, does not construct a nest, but lays its eggs on bare ground. (B) Anna's hummingbird (*Calypte anna*) constructs a tightly woven nest lined on the inside with feathers for insulation and covered on the outside with lichen for concealment. (C) American coots (*Fulica americana*) build floating nests using the air-filled stems of aquatic plants. (D) Weaverbirds (*Philetairus socius*) of western Africa are social and build communal nests that can include hundreds of mated pairs and are occupied for generations. (A, Bridget Spencer; B, René C. Clark © Dancing Snake Nature Photography; C, Smabs Sputzer/CC BY 2.0; D, Harald Süpfle/CC BY SA 2.5; inset, Bernard Dupont/CC BY-SA 2.0.)

periods that adults spend away from the nest during foraging. Fork-tailed storm-petrels (*Oceanodroma furcata*) lay a single egg in a burrow or rock crevice. Both parents participate in incubation, but the adults forage over vast distances, and both parents may be absent from the nest for several days at a time. In Alaska, storm-petrel parents averaged 11 days of absence during an incubation period of 50 days. Eggs were exposed to ambient temperatures of 10°C while the parents were away. Experimental studies showed that storm-petrel eggs were not damaged by being cooled to 10°C every 4 days. The pattern of development of chilled eggs was like that of eggs incubated continuously at 34°C, except that each day of chilling added about a day to the total time required for the eggs to hatch.

Incubation periods are as short as 10–12 days for some species of birds and as long as 60–80 days for others. In general, large species have longer incubation periods than small species, but ecological factors also contribute to determining the length of the incubation period. A high risk of predation may favor rapid development of the eggs. Among tropical tanagers, species that build open-topped nests near the ground are probably more vulnerable to predators than are species that build similar nests farther off the ground.

The incubation periods of species that nest near the ground are short (11–13 days) compared with those of species that build nests at greater heights (14–20 days). Species of tropical tanagers with roofed-over nests have still longer incubation periods (22–24 days).

In those species that delay the start of incubation until all their eggs have been laid, an entire clutch nears hatching simultaneously. Hatching may be synchronized by clicking sounds that accompany breathing within the egg, and both acceleration and retardation of individual eggs may be involved. A low-frequency sound produced early in respiration, before the clicking phase is reached, appears to retard the start of clicking by advanced embryos. That is, the advanced embryos do not begin clicking while other embryos are still producing low-frequency sounds. Subsequently, vocalizations from advanced embryos appear to accelerate hatching by late embryos. Both effects were demonstrated in experiments with northern bobwhite quail (*Colinus virginianus*): when researchers paired two eggs, one of which had started incubation 24 hours before the other, hatching of the late-starting egg was accelerated by 14 hours and hatching of the early-starting egg by was delayed 7 hours.

(A)

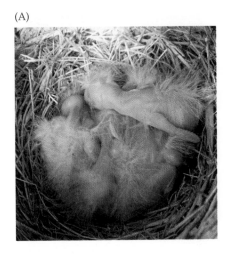

(B)

Figure 22.33 Altricial and precocial chicks. (A) Altricial chicks are nearly naked when they hatch and unable even to stand up. Their eyes are closed and they must be fed by their parents until they fledge. (B) Precocial chicks are covered with down when they hatch and can stand erect, walk, swim, and find their own food. (A, Kari Walls; B, Peter McGowan/USFWS.)

Parental care

It is a plausible inference—both from phylogenetic considerations and from fossils of coelurosaurs—that the ancestral form of reproduction in the avian lineage included depositing eggs in a well-defined nest site with the male attending the nest, eggs hatching into **precocial** young that were able to feed themselves at hatching, followed by a period of association between the young and one or both parents after hatching. All of the crocodylians that have been studied conform to this pattern, and evidence is increasing that at least some species of dinosaurs remained with their nests and young.

Extant birds follow these ancestral patterns, but not all species produce precocial young. Instead, hatchling birds show a spectrum of maturity that extends from **altricial** forms that are naked and entirely dependent on their parents for food and thermoregulation to precocial young that are feathered and self-sufficient from the moment of hatching (**Figure 22.33**).

After they hatch, altricial young are guarded and fed by one or both parents. Adults of some species carry food to nestlings in their beaks, and other species store food in the crop and later regurgitate it to feed the young. Altricial hatchlings respond to the arrival of a parent at the nest by chirping and gaping their mouths widely (**Figure 22.34**).

The sight of an open mouth appears to stimulate a parent bird to feed it, and begging by nestlings is a complicated phenomenon. The intensity with which a nestling begs does not necessarily tell a parent how hungry it is, and nestlings are most likely to give false signals when the stakes are high: when they are competing with many siblings; when their parents are likely to produce another brood, so the nestlings are competing with future siblings; or when the death or divorce of a parent means the nestlings are likely to be less related to future siblings than they are to their current siblings.

The duration of parental care is variable. The young of small passerines leave the nest about 2 weeks after hatching and are cared for by their parents for an additional 1–3 weeks. Larger species of birds, such as the tawny owl (*Strix aluco*), spend a month in the nest and receive parental care for an additional 3 months after they have fledged, and young wandering albatrosses require a year to become independent of their parents.

Brood parasitism

Some birds, including cuckoos, cowbirds, and widowbirds, lay their eggs in the nests of other species, leaving the unwilling foster parents to raise the interloper chick. Indeed, the expression "a cuckoo in the nest" is based on this

(A)

(B)

Figure 22.34 Altricial nestlings beg for food. The gaping mouths of nestlings stimulate adults to feed them. The nestling that begs most energetically is most likely to be fed, and the arrival of a parent (or almost any other disturbance) stimulates a frenzy of begging. (A) Color contrast between the flanges surrounding the mouth and the interior of the mouth provide a target for adults, as in these song thrush nestlings (*Turdus philomelos*). (B) Nestlings of some birds, such as these horned larks (*Eremophila alpestris*), have species-specific markings in the mouth. (A, NottsExMiner/CC BY SA 2.0; B, Kate Yates/BLM/CC BY 2.0.)

(A)

(B)

(C)

Cuckoo nestling　Warbler egg ejected

Figure 22.35 **Cuckoos are brood parasites.** (A) The egg of the European common cuckoo (*Cuculus canorus*) is larger than the eggs of the reed warbler (*Acrocephalus scirpaceus*) and a different color, but the host parents incubate the cuckoo egg with their own. (B) Soon after it hatches, the cuckoo nestling pushes the host eggs or hatchlings out of the nest (even when, as seen here, the host parent is watching). (C) The host parents continue to feed the cuckoo chick, even after it has grown much larger than the hosts. (A,B, blickwinkel/Alamy Stock Photo; C, Lynn Martin.)

behavior. At a minimum, the interloper diverts resources from the parents' own chicks, especially since the parasitic chick may grow larger than its foster parents before it fledges, and many parasitic chicks push the hosts eggs or hatchlings out of the nest (**Figure 22.35**).

Some species of birds appear to be oblivious to the presence of the egg of a parasitic species, even when it differs in size and color from the eggs of the host species, but other species can detect a foreign egg and either remove it or abandon the nest and start over in a new location. These species and their brood parasites may be engaged in an evolutionary race in which selection rewards the parasitic species for resembling the host's eggs more closely and rewards the host species for becoming ever better at detecting the parasitic egg.

22.10 Orientation and Navigation

For as long as people have raised and raced pigeons, it has been known that birds released in unfamiliar territory vanish from sight flying in a straight line, usually in the direction of home. The homing pigeon has become a favorite experimental animal for studies of navigation. Experiments have shown that navigation by homing pigeons (and presumably by other vertebrates as well) is complex and is based on a variety of sensory cues. On sunny days, pigeons vanish toward home and return rapidly to their lofts. On overcast days, vanishing bearings are less precise, and birds more often get lost. These observations led to the idea that pigeons use the sun as a compass.

Of course, the sun's position in the sky changes from dawn to dusk. That means a bird must know what time of day it is in order to use the sun to tell direction, and its timekeeping ability requires some sort of internal clock. If that hypothesis is correct, it should be possible to "fool" a bird by shifting its internal clock forward or backward. Such experiments have indeed supported the hypothesis. For example, if lights are turned on in the pigeon loft 6 hours before sunrise every morning for about 5 days, the birds will become accustomed to that artificial sunrise. At any time during the day, they will act as if the time is 6 hours later than it actually is. When these birds are released, they will be directed by the sun but their internal clocks will be wrong by 6 hours. That error should cause the birds to fly off on courses that are displaced by 90 degrees from the correct course for home, and that is what clock-shifted pigeons do on sunny days (**Figure 22.36**). Under cloudy skies, however, clock-shifted pigeons head straight for home despite the 6-hour error in their internal clocks. Clearly, pigeons have more than one way to navigate: when they can see the sun they use a sun compass, but when the sun is not visible they use another mechanism that is not affected by the clock-shift.

This second mechanism is probably an ability to sense Earth's magnetic field and use it as a compass. On sunny days, attaching small magnets to a pigeon's head does not affect its ability to navigate, but on cloudy days pigeons with magnets cannot return to their home lofts.

Polarized and ultraviolet light provide additional cues that pigeons probably use to determine direction, and they can also navigate by recognizing airborne odors and familiar visual landmarks. In addition, pigeons can detect extremely low-frequency sounds (infrasound), well below the frequencies that humans can hear. Those sounds are generated by ocean waves and air masses moving over mountains and can signal a general direction over thousands of kilometers, but their use as cues for navigation remains obscure.

Results of this sort are being obtained with other vertebrates as well, and lead to the general conclusion that a great deal of redundancy is built into navigation systems. Apparently there is a hierarchy of useful cues, depending on local conditions and how far a bird is from home. For example, a bird far from home relies on the sun and polarized light to navigate on clear days, but it can switch to magnetic direction sensing on overcast days. In both conditions, the bird can use local odors and recognition of landmarks as it approaches home.

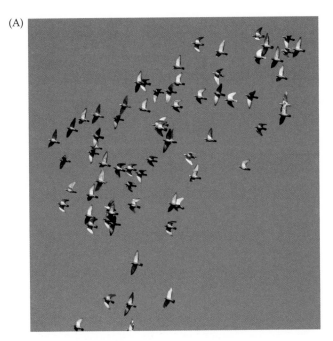

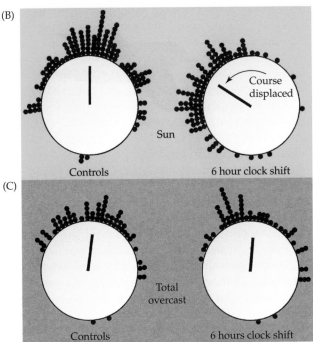

Figure 22.36 Orientation by clock-shifted pigeons.
(A) When homing pigeons are released, they depart in the direction of their home loft. (B,C) Each dot in these plots shows the direction in which a pigeon vanished from sight when it was released in the center of the large circle. The home loft is at the straight up positon, and the solid bar in each circle shows the average direction chosen by homing birds. In this experiment, pigeons were "clock-shifted" by turning on lights in the loft 6 hours before sunrise every morning for about 5 days, so that the birds became accustomed to an artificial sunrise. (B) When released at any time during a sunny day, clock-shifted pigeons acted as if the time was 6 hours later than it actually was and flew on courses that were displaced from the correct homing course. (C) Under cloudy skies, both control and clock-shifted pigeons headed directly for home along the same path. Clearly, pigeons have more than one way to navigate, and when the "sun compass" is not visible they use another mechanism that is not affected by the clock shift.
(A, Richard Elzey/CC BY 2.0; B,C, modified from Keeton 1969.)

Many birds migrate only at night. Under these conditions a magnetic sense of direction might be important. Several species of nocturnally migrating birds use star patterns for navigation. Apparently each bird fixes on the pattern of particular stars and uses their motion in the night sky to determine a compass direction. As in sun compass navigation, an internal clock is required for this sort of celestial navigation, and artificially changing the time setting of the internal clock produces predictable changes in the direction in which a bird orients.

22.11 Migration

The mobility that characterizes vertebrates is perhaps most clearly demonstrated in their movements over enormous distances. These displacements, which may cover half the globe, require both endurance and the ability to navigate. Other vertebrates migrate, some of them over enormous distances, but migration is best known among birds.

Migratory movements

Migration is a widespread phenomenon among birds. About 40% of the bird species in the Palearctic are migratory, and an estimated 5 billion birds migrate from the Palearctic every year. Migrations of Palearctic species often cross the Equator and involve movements over thousands of kilometers, especially in the case of birds nesting in northern latitudes. Short-tailed shearwaters (*Puffinus tenuirostris*), for example, make an annual migration between the North Pacific and their breeding range in southern Australia that requires a round-trip of more than 30,000 km (**Figure 22.37**). Very few Southern Hemisphere migrants cross the Equator, but movements within the Southern Hemisphere are extensive.

Many birds return each year to the same migratory stopover sites, just as they may return to the same breeding and wintering sites year after year. Migrating birds may be concentrated at high densities at certain points along their traditional migratory routes. For example, species that follow a coastal route may be funneled to small points of land, such as Cape May, New Jersey, from which they initiate long overwater flights. At these stopovers, migrating birds must find food and water to replenish their stores before venturing over the sea, and they must avoid the predators congregating at these sites. Development of coastal areas for human use has destroyed many important resting and refueling stations for migratory birds. The destruction of coastal wetlands has caused serious problems for migratory birds on a worldwide basis. Loss of migratory stopover sites may remove a critical resource from a population at a particularly stressful stage in its life cycle.

Costs and benefits of migration

The energy costs of migration must be offset by energy gained as a result of moving to a different habitat. The normal food sources for some species of birds are unavailable

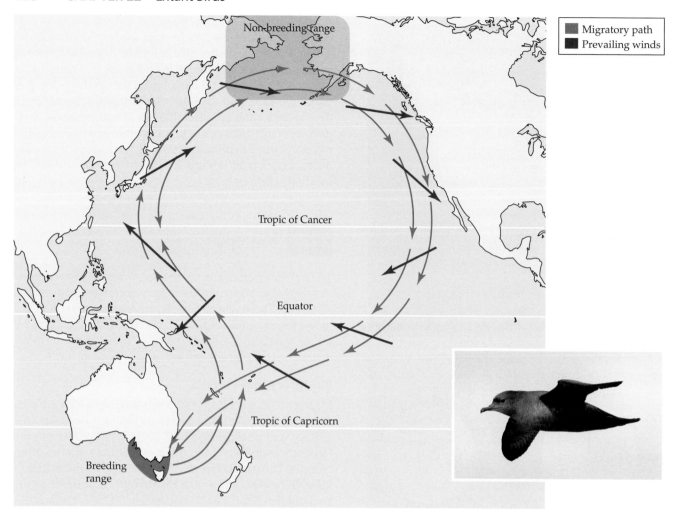

Non-breeding range

■ Migratory path
■ Prevailing winds

Tropic of Cancer

Equator

Tropic of Capricorn

Breeding range

Figure 22.37 Migratory path of the short-tailed shearwater (*Puffinus tenuirostris*).   This species takes advantage of prevailing winds in the Pacific region to reduce the energy cost of migration from its Australian breeding area to its northern non-breeding range. (After Marshall and Serventy 1956; photo by Ed Dunens/CC BY SA 2.0.)

in the winter, and the benefits of migration for those species are starkly clear. Other species may save energy mainly by avoiding the temperature stress of northern winters. In other cases, the main advantage of migration may come from breeding in high latitudes in the summer where resources are abundant and long days provide more time to forage than the birds would have if they remained closer to the Equator.

Migration of birds native to the Southern Hemisphere appears to be influenced by rainfall and aridity rather than by cold, and their movements are nomadic rather than directly between a breeding range and a nonbreeding range, as in Northern Hemisphere migrants.

22.12 Birds and Urbanization

The shift of human cultures from nomadic hunters to static agricultural societies occurred independently on every continent except Antarctica between about 12,000 and 5,000 years ago, initiating an enormous and (in evolutionary terms) exceptionally rapid change that has culminated in the intense urbanization of the modern world. In 1800, only 3% of humans lived in towns and cities, but by 1950 that fraction had increased to 30%, and in 2008 it passed the halfway mark, with 50% of humans living in urban areas. Statistics from 2016 indicated city dwellers had increased to 54.5% of the world's population, and that proportion is projected to rise to 60% by 2030.

As cities have spread, plant and animal populations have either vanished or adapted to urban conditions. Birds are often the most conspicuous non-human vertebrates in urban areas, so much so that at times they appear to have benefitted from urbanization. For a few species, that impression is probably correct. Rock pigeons (*Columba livia*) are native to northern Africa, western Asia, and southeastern Europe, but have spread to become the iconic urban bird worldwide, perhaps because cities provide nesting sites that mimic the rock ledges wild populations use. House sparrows (*Passer domesticus*) and European starlings (*Sturnus vulgaris*) have also colonized most cities.

Pigeons and starlings are the exception rather than the rule, however. Most birds found in cities are native to the region surrounding the city, but the diversity of bird species in a city is substantially lower than the diversity in surrounding natural areas, while the densities of urban populations are higher. What makes a species successful in a city, and what are the consequences for birds of living in a city?

Success in the city

Food, shelter, and nest sites are crucial elements of avian success, and cities are more hospitable to omnivorous and granivorous (grain-eating) birds than to insectivores. Architectural embellishments on buildings often provide suitable nest sites for cavity-nesting species, whereas tree-dwellers that build nests of grass and twigs face a shortage of both nest sites and nesting materials.

City birds and country birds differ even within the same species. Barbados bullfinches (*Loxigilla barbadensis*) from urban areas are bolder and better at problem-solving than their country cousins, and great tits in cities in Poland are more likely to approach an unfamiliar object than are rural birds. Some, but not all, studies have found that species that invade cities have large brains in relation to their body size, possibly indicating greater ability to adapt to novel environments. The DNA methylation patterns of rural and urban populations of the medium ground-finch (*Geospiza fortis*) on the Galápagos Islands differ substantially and may be responsible for the larger body sizes of the urban birds.

The diets of urban birds may not be as healthful as those of rural birds. The egg yolks of urban great tits have low proportions of polyunsaturated fatty acids and high proportions of saturated fatty acids. Great tits in cities have fewer and smaller offspring than great tits living in the country. Furthermore, the proportion of individuals with small brains is higher in urban areas than in rural areas, and the incidence of small brains increases with the length of time an urban population has been established.

Noise pollution

Cities are noisy; indeed, human activities in general are noisy, and anthropogenic noise pollution has changed the characteristics of animals that use vocalizations during social behaviors, including not just birds but also frogs and insects. The phrase "the peace of the countryside" describes a real phenomenon: rural habitats are quieter than cities.

Most of the noises that humans create—the roar of machinery or the hum of traffic, for example—lie at frequencies below 2.5 kHz, and these sounds can mask the vocalizations that birds use to detect prey and warn of predators. The ability of short-eared owls (*Asio flammeus*) and long-eared owls (*A. otus*) to detect the sounds made by prey was reduced by more than half at the edge of a road, and the effect of road noise on prey detection extended to 120 m from the road. Traffic noise within 20 m completely masks the alarm calls of great tits, and even at 40 m masking reduces the potential for a bird to detect an alarm call (**Figure**

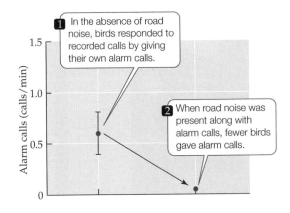

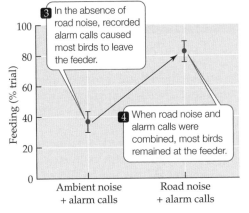

Figure 22.38 Noise pollution interferes with birds' ability to hear important sounds. Recorded alarm calls of great tits (*Parus major*) were played to wild individuals at a feeder. When the recordings contained road noise as well as alarm calls, feeding birds gave fewer alarm calls of their own and remained at the feeder instead of fleeing to safety. (After Templeton et al. 2016, CC BY 4.0; photo by Barry Marsh.)

22.38). Northern cardinals (*Cardinalis cardinalis*) eavesdrop on the alarm calls of black-capped chickadees (*Poecile atricapillus*) and tufted titmice (*Baeolophus bicolor*), and respond to those calls with predator avoidance behavior in quiet surroundings, but traffic noise abolishes that response.

Not all urban pollution is auditory. Anthropogenic electromagnetic noise disrupts magnetic compass orientation in birds, as do artificial light and polarized light reflected from shiny surfaces.

Not so sexy in the city

Most attention has been focused on the effects of anthropogenic noise on the vocalizations that birds use to advertise territorial ownership and to attract mates. Some birds change the timing of vocalization to avoid noisy periods. European robins (*Erithacus rubecula*), European blackbirds (*Turdus merula*), great tits, chaffinches (*Fringilla coelebs*), and Eurasian nuthatches (*Sitta europaea*) began singing earlier in the morning in the forest adjacent to Berlin's Tegel Airport than those species did in forest locations distant from the airport. During the day, chaffinches ceased calling while

(A)

(B)

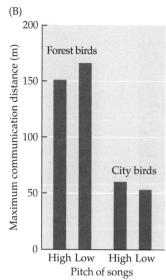

(C)

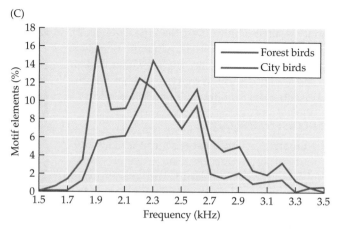

Figure 22.39 City birdsong shifts toward higher frequencies. (A) The Eurasian blackbird (*Turdus merula*) is a forest bird that has successfully colonized cities. (B) In the forest, a blackbird's low-pitched call can be heard at a greater distance than a high-pitched call. That relationship is reversed in the city, where a high-pitched call travels farther. (C) The songs of city blackbirds have more high-frequency elements (motifs) than do the songs of forest blackbirds. The peak frequency of city birdsongs lies between 2.3 and 2.4 kHz, whereas the peak frequency for forest songs is between 1.8 and 1.9 kHz. (A, hedera.baltica/CC BY SA 2.0; B, modified from Nemeth and Brum 2010; C modified from Nemeth et al. 2013.)

a plane was taking off and resumed calling in the interval before the next takeoff. In England, European robins have shifted calling from day to night in locations that are noisy during the day.

Changes in the amplitude (loudness) and frequency (pitch) have been reported for many species of birds whose songs fall within the frequency range of anthropogenic noise (**Figure 22.39**). For example, most of the acoustical energy in the calls of the Australian grey shrike-thrush (*Colluricincla harmonica*) lies between 1.5 and 4 kHz, and thus can be blocked by traffic noise, whereas the sympatric grey fantail (*Rhipidura fuliginosa*) sings at frequencies of 4–7.5 kHz, which are higher than traffic noise. Grey shrike-thrushes increase the frequency of their song in the presence of traffic noise, but grey fantails do not. A similar relationship between song frequency and traffic noise has been reported for chipping sparrows (*Spizella passerina*) in Michigan. Two call types are present, one with a minimum frequency of 3.3 kHz and a second with a minimum frequency of 4 kHz. Male chipping sparrows in the 3.3 kHz group increase the frequency of their song

in the presence of noise, but males in the 4 kHz group show no change.

Although increasing the dominant frequency of a song helps avoid masking, the increase in frequency may be a side effect of an increase in amplitude. Birds can produce high-frequency sounds at louder amplitudes than low-frequency sounds, so an increase in frequency may be an unavoidable result of singing louder. If so, that tradeoff could be a Hobson's choice[1] for birds, because high-frequency songs are less attractive to females than low frequencies. Perhaps an urban male bird can be either sexy or audible, but not both.

---

[1]"Hobson's choice" is between taking what's available or nothing at all. The expression is named for a British livery stable owner who gave his customers the choice of taking the horse in the stall closest to the door or not getting a horse.

Summary

Flight is a central characteristic of birds.

The ability to fly provides ecological and behavioral options that are not available to terrestrial vertebrates, and the aerodynamic and physiological demands of flight constrain many characters of birds, including body size and internal and external structure.

A feather consists of a calamus, which anchors the feather in the skin, and a central shaft called the rachis. In pennaceous feathers, symmetrical branches called barbs extend from the rachis, and proximal and distal barbules branching from the barbs interlock to form the vanes.

Contour feathers are the outermost feathers on a bird's body, wings, and tail. They help streamline a bird's body, making smooth junctions between the wings and head and the body that reduce turbulence.

Loss of the urinary bladder, loss of one ovary in most species, and atrophy of the female and male gonads outside the reproductive season reduce the body weight of birds, as do air-filled bones.

A bird's center of gravity lies beneath its wings, and a keel on the sternum is the origin for the major flight muscles, the pectoralis major and supracoracoideus. These muscles make up 25% of the total body weight of strong fliers, while the limb muscles may account for as little as 2%. The leg muscles make up a larger percentage of the total body weight in predatory birds that use their feet to capture prey, and a still larger percentage in swimming birds that use their feet for propulsion.

Bird flight is aerodynamically more complex than the flight of fixed-wing aircraft.

The wings of birds are both airfoils (creating lift) and propellers (creating forward thrust). The primary feathers on the outer wing provide thrust, and the secondaries on the inner wing provide lift. Small birds generate thrust only on the downstroke of the wings, but larger birds also generate thrust on the upstroke.

Long, narrow wings have high aspect ratios and provide the high lift-to-drag ratios needed by aerial foragers and oceanic birds that soar in strong winds. Short, broad wings have low aspect ratios and allow the rapid takeoff and high maneuverability needed by birds that live in forests.

Wing loading refers to the weight of a bird divided by the area of its wings. Low wing-loading values are characteristic of birds that soar in rising air currents.

Most birds have four toes on their feet.

On the ground, birds hop, run, or walk. Birds use their beak, tail, feet, and in a few cases forelimbs to climb vertical surfaces, and some species can climb in both head-up and head-down directions.

In the generalized foot structure of birds (anisodactyly), toe 1 points backward and toes 2, 3, and 4 point forward. Many climbing species (e.g., parrots and woodpeckers) and species that seize prey with their feet (hawks, owls, and others) are zygodactyl, with toes 1 and 4 pointing backward and toes 2 and 3 pointing forward. Kingfishers, kookaburras, and their relatives are syndactyl; toes 2 and 3 are fused for much of their length. Cursorial birds have reduced the number of toes; ostriches are didactyl (having only toes 3 and 4), and emus and rheas are tridactyl (having toes 2, 3, and 4). Swimming birds that use their feet for propulsion have webs between the toes (ducks and geese, toes 2–4; cormorants, toes 1–4) or lobes on toes 2–4 (grebes).

Specialization of the forelimbs as wings, combined with the loss of teeth, means that many birds rely on their beaks to capture food and on their guts to process it.

Beak shape reveals much about a bird's diet. Species that capture small items (insects or fruit) one at a time have long, pointed beaks; many seed eaters have short, thick beaks that they use to husk seeds before eating them; and carnivorous birds have hooked beaks with which they can tear pieces from their prey.

Birds have substantial skull kinesis; the upper jaw flexes upward as the mouth opens, and the lower jaw expands laterally, increasing the gape. Long-billed shorebirds that probe for prey in mud or sand can raise the upper tip of their bill to grasp prey without opening their mouth.

Specialized feeders rely on modifications of the tongue, such as the barbed tongue that a woodpecker uses to extract insect larvae from their tunnels and the hair-thin projections on the tongue tips of nectar-eating birds.

Many birds have a crop, an enlarged portion of the esophagus in which food can be stored and later swallowed or regurgitated to feed nestlings. The stomach of most birds consists of an anterior glandular portion (proventriculus) in which chemical digestion takes place, and a posterior muscular portion (gizzard) in which mechanical processing occurs.

The length of the intestine can change seasonally as a bird's diet shifts with the availability of different sorts of food. Symbiotic microorganisms ferment food in the paired ceca, which are large in herbivorous and omnivorous species of birds and small in carnivorous, insectivorous, and seed-eating species.

Bird are visually oriented, but their auditory and olfactory systems are also highly developed.

The eyes of many birds are so large that they meet in the midline of the skull. The visual acuity of most birds is approximately the same as that of humans, and even the eagle-eyed raptors are only about 2.5 times better than humans at resolving images. Passerines excel at analyzing fast-changing images; their flicker-fusion frequencies are more than twice those of humans.

The world looks different to birds than it does to us. Birds have four visual pigments in their retinal cones: red, green, and blue like humans and, for some birds, ultraviolet as well. In addition, avian cones have pigmented oil droplets that act as filters. Ultraviolet wavelengths are involved in the use of color as social signals by some birds and may help them recognize mimetic and cryptic prey.

The auditory systems of most birds have about the same sensitivity as those of humans, but are specialized to detect small differences in the frequency and temporal pattern of sound. Owls have the most sensitive hearing among birds and can capture prey in complete darkness by sound alone. Marked asymmetry of the skulls of many owls enhances their ability to pinpoint the source of a sound.

(*Continued*)

Summary (continued)

Contrary to popular belief, many birds have good olfactory sensitivity and use odors to locate and identify food, for orientation and navigation, for species recognition, and perhaps also for sex recognition.

Vision and hearing are important components of the social behavior of birds, and birds have played an important role in studies of the behavior of vertebrates.

The colors and patterns of feathers identify a bird's species, and often its sex and age, and may provide information about an individual's physical condition that potential partners can use in selecting a mate.

Carotenoid pigments are responsible for most red, orange, and yellow colors and also play a role in physiological processes. Birds obtain carotenoids from their diets, and the handicap principle proposes that the ability of a bird to allocate carotenoids to pigmentation rather than to other functions signals robustness to potential mates.

Structural colors result from scattering of light by the feather barbs. A structural color may be combined with a pigment to produce a third color.

Birds produce two kinds of sounds: vocalizations and sonations. Vocalizations are produced by passing air through the syrinx, a unique avian structure that lies at the junction of the trachea and the two bronchi. Sonations are nonvocal sounds created by feathers.

A birdsong is a long, complex vocalization, whereas a bird call is typically simple and brief. Song is controlled by the song control regions (SCRs) of the brain. A species' song is learned early in life.

Visual displays are frequently paired with songs as male birds display brightly colored patches of feathers or spread their wings or tail. Visual display structures can be elaborate, as in the tail of a male peacock.

All birds are oviparous, a character that sets them apart from most other sauropod lineages.

The first evolutionary step to development of viviparity in other sauropsids would have required an advantage to retaining the eggs for some period before depositing them in a nest. The body temperatures of birds may be too high for successful embryonic development, negating any advantage from egg retention.

Avian eggshells are formed by calcite; the shell encloses and protects the egg and is the principle source of calcium for the developing embryo. Pores in the shell permit the exchange of oxygen and carbon dioxide and a gradual loss of water.

Bird eggs are archetypal amniotic eggs. The amnion surrounds the developing embryo, the chorion is pressed against the eggshell, and a portion of the allantois is in contact with the chorion, forming the chorioallantois, which is the site of gas exchange for the embryo.

Nutrition for the developing embryo is provided by the yolk, which contains lipids and proteins. Albumen provides the water the embryo needs.

Nearly all birds have genetic sex determination; females are the heterogametic sex (ZW) and males are homogametic (ZZ). Temperature plays a role in sex determination in at least one lineage of birds, the megapodes of the Indo-Australian region.

Female birds can exert some control over the sex of their offspring.

Social monogamy is the general mating system among birds.

Avian monogamy is often social rather that genetic; that is, both parents care for a brood of young, but extra-pair copulations are common, and the chicks in a nest often have been sired by two or more males.

Oviparity may explain the high incidence of monogamy among birds, because both parents can participate in incubating the eggs and caring for nestlings.

Avian parental care commences with brooding and continues until the young have fledged (left the nest), and often for some time after that.

Bird nests can be as simple as a depression in the ground or as elaborate as a communal nest that houses dozens of pairs of birds.

Some species of birds start incubation as soon as the first egg is laid, whereas others wait until the clutch is complete. Prolactin stimulates brooding behavior, and in many species formation of a brood patch of bare skin that facilitates transfer of heat to the eggs.

Eggs hatch synchronously when incubation is delayed until the clutch is complete, and vocalizations by advanced embryos that are still in the eggs can coordinate the timing of hatching.

Hatchlings lie along a continuum of development, from extremely altricial (naked, eyes closed) to fully precocial (covered with down and able to walk or swim). Altricial hatchlings are fed by their parents, whereas precocial chicks can forage for themselves.

Some species of birds are brood parasites, laying eggs in the nests of other species to be raised by the foster parents. Hosts may detect a brood parasite's egg and either remove it or abandon the nest. Some brood parasite chicks eject the host's young from the nest.

Summary (*continued*)

The ability of some birds to orient and navigate is legendary.

Celestial navigation, using the sun by day and the stars at night, was the first orientation mechanism to be demonstrated, but birds navigate just as well under cloudy skies, so additional mechanisms were sought.

Earth's magnetic field, infrasound, odors, landmarks, and polarized and ultraviolet light are additional cues that birds and other vertebrates employ, probably in a hierarchical sequence that varies with ambient conditions.

The combination of flight and the ability to orient allows birds to undertake long-distance migrations.

About 40% of bird species in the Palearctic migrate annually, many crossing the Equator and some travelling tens of thousands of kilometers to their destinations and returning months later to their starting locations.

For some species, the benefits of migration include avoiding low temperatures and scarcity of food during the winter, whereas for other species the advantage may lie in an abundance of food and long daylight hours in northern summers. Birds in the Southern Hemisphere also move, but these movements tend to be nomadic responses to rainfall and aridity.

Most bird species have suffered as cities have spread.

Rock pigeons are the archetypal urban birds, perhaps because the ledges provided by buildings are similar to their ancestral nesting sites on cliffs. To a lesser extent, house sparrows and European starlings have colonized most cities. With those exceptions, however, city birds are a subset of the regional bird fauna.

City birds and country birds differ, even within the same species. City birds are often bolder and more exploratory than rural birds, but apparently have less healthful diets and produce fewer and smaller offspring. The incidence of birds with abnormally small brains is greater in cities than in rural areas, and the incidence of small brains increases the longer an urban population has been established.

Anthropogenic noise from machinery and automobiles is ubiquitous in cities and interferes with birds' ability to hear important sounds, such as the alarm calls of other individuals. Furthermore, anthropogenic noise lies in the lower part of the frequency range of birdsong, and interferes with the vocalizations that birds use in social behavior.

Birds respond to anthropogenic noise by changing the time of day they vocalize and by vocalizing more loudly. Increasing the amplitude of vocalization also raises its pitch, potentially making a male's song less attractive to females.

Discussion Questions

1. Imagine that you are walking in a penguin colony and you find a bird's humerus. The skin and feathers disappeared long ago, leaving only the bleached bone. You know that the bone must have come from either a penguin or a gull because those are the only two birds of appropriate size that live in the area. How can you determine which kind of bird the bone came from?

2. Many birds eat plant material, but the hoatzin (*Opisthocomus hoazin*; see Figure 22.22B) of South America is the only avian species known to employ foregut fermentation. Hoatzins are herbivorous; green leaves make up more than 80% of their diet. More than a century ago, naturalists observed that hoatzins smell like fresh cow manure, and a study of the crop and lower esophagus of hoatzins revealed that bacteria and protozoa like those found in the rumen of a cow break down plant cell walls, releasing volatile fatty acids that are absorbed in the gut. Considering the abundance of plant material that is available to birds and the widespread use of fermentative digestion by mammals, why do you think the hoatzin is the only known foregut fermenter among birds?

3. Crows are notorious nest robbers, stealing eggs from the nests of passerine birds, and they will even eat other birds if they can catch them. Adult passerines attack crows that approach their nests, darting at them and pursuing them when they fly away. What anatomical difference in the wings of passerines and crows might allow passerines to engage in such seemingly risky behavior?

4. The need to have two parents caring for hatchlings appears to be the reason that many species of birds are monogamous. But if the value of monogamy lies in rearing a brood successfully, why is extra-pair copulation so common among monogamous species of birds?

5. Some bird species have a distinctive mode of courtship called lekking. Males of lekking species assemble in groups—leks—and display. Females come to the leks and observe the displaying males. Eventually a female selects a male and mates with him. What is the fundamental difference between this situation and the formation of territories by males that call from their territories to attract females? (Hint: What does a female base her choice on in each situation?)

Additional Reading

Amo L, Dicke M, Visser ME. 2016, Are naïve birds attracted to herbivore-induce plant defences? *Behaviour* 153: 353–366.

Corfield JR and 5 others. 2015. Diversity in olfactory bulb size in birds reflects allometry, ecology, and phylogeny. *Frontiers in Neuroanatomy* 9(102): 1–16.

D'Alba L, Kieffer L, Shawkey MD. 2012. Relative contributions of pigments and biophotonic nanostructures to natural color production: A case study in budgerigar (*Melopsittacus undulatus*) feathers. *Journal of Experimental Biology* 215: 1272–1277

Jorge PE, Pinto BV. 2018. Olfactory information from the path is relevant to the homing process of adult pigeons. *Behavioral Ecology and Sociobiology* 72: 5.

Khan I and 8 others. 2015. Olfactory receptor subgenomes linked with broad ecological adaptations in Sauropsida. *Molecular Biological Evolution* 32: 2832–2843.

MacDougall-Shackleton EA, MacDougall-Shackleton SA. 2001. Cultural and genetic evolution in mountain white-crowned sparrows: Song dialects are associated with population structure. *Evolution* 55: 2568–2575

Mayr G. 2014. The origins of crown group birds: Molecules and fossils. *Frontiers in Palaeontology* 57: 231–242.

McNew SM and 6 others. 2017. Epigenetic variation between urban and rural populations of Darwin's finches. *BMC Evolutionary Biology* 17: 183.

Padget O, Dell'Ariccia G, Gagliardo A, González-Solis, Guilford T. 2017. Anosmia impair homing orientation but not foraging behaviors in free-ranging shearwaters. *Scientific Reports* 7: 9668.

Papi F, Fiore L, Fiaschi V, Benvenuti S. 1972. Olfaction and homing in pigeons. *Monitore Zoologico Italiano (N.S.)* 6: 85–95.

Prum RO and 6 others. 2015. A comprehensive phylogeny of birds (Aves) using targeted next-generation DNA sequencing. *Nature* 526: 569–573.

Salmón P, Nilsson JF, Watson H, Bensch S, Isaksson C. 2017. Selective disappearance of great tits with short telomeres in urban areas. *Proceedings of the Royal Society B* 284: 1349.

Tattersall GJ, Chaves JA, Danner RM. 2017. Thermoregulatory windows in Darwin's finches. *Functional Ecology* 2017: 1–11.

Wallraff HG. 2015. An amazing discovery: Bird navigation based on olfaction. *Journal of Experimental Biology* 218: 1464–1466.

Go to the Companion Website **oup.com/us/vertebratelife10e** for active-learning exercises, news links, references, and more.

23

Geography and Ecology of the Cenozoic Era

Throughout Earth's history, the positions of continents have affected climates and the ability of vertebrates to disperse from region to region. The geographic continuity of Pangaea in the late Paleozoic and early Mesozoic allowed for relatively easy migration, and faunas were fairly similar in composition across the globe. By the late Mesozoic, however, the continents were separating from one another, and epicontinental seas extended across the centers of North America and Eurasia. This isolation of the continental blocks resulted in the isolation of their tetrapod faunas and limited the possibilities for migration. As a result, tetrapod faunas (indeed, assemblages of all organisms) became progressively more different on different continents during the Cenozoic (**Figure 23.1**).

The radiation of the modern orders of mammals is a prominent feature of the Cenozoic era, and the Cenozoic is commonly known as the Age of Mammals. This name is misleading in two respects, however. First, the time elapsed since the start of the Cenozoic represents only about one-third of the time spanned by the total history of mammals. And second, the most diverse vertebrates of the Cenozoic, in terms of number of species, are not mammals but teleost fishes, and mammals are also much less diverse than birds.

The radiation of larger mammals may be related in part to the extinction of nonavian dinosaurs, which left the land free of large tetrapods and provided a window of opportunity for other terrestrial groups. But the radiation of mammals is also related to the radiation of angiosperm plants, which provide multilayered stands of vegetation that can sustain their own local microhabitats. Gymnosperm-dominated habitats tend to consist of tall trees with little understory or ground-level vegetation—perhaps good for tall dinosaurs, but not for mammals. Angiosperm-dominated habitats have trees with many low branches and an abundance of ground-level vegetation.

Distinct regional differences became apparent in the dinosaur faunas of the late Mesozoic, and regional differences have been a prominent feature of Cenozoic mammalian faunas, although the uniqueness of the Cenozoic mammalian faunas on different continents results in large part from early dispersal events, with subsequent

isolation. That is, the Cenozoic mammalian faunas are not the result of vicariance—original inhabitants being left on a continental block after it separated from other continents. We discuss some of the dispersals of Cenozoic mammals in Section 23.4.

23.1 Continental Geography of the Cenozoic

The breakup of Pangaea in the Jurassic began with the movement of North America that opened the ancestral Atlantic Ocean. Continental drift in the late Mesozoic and early Cenozoic moved the northern continents that had formed the old Laurasia (North America and Eurasia) from their early Mesozoic near-equatorial position into higher latitudes (**Figure 23.2**). Rifts formed in Gondwana, after which India moved northward on its separate oceanic plate, eventually to collide with Eurasia. The collision of the Indian and Eurasian plates in the middle Cenozoic (around 40 Ma) produced the Himalayas, the highest mountain range in today's world. This and other tectonic movements influenced Cenozoic climates. For example, the uplift of mountain ranges such as the Rockies, the Andes, and the Himalayas in the middle to late Cenozoic led to alterations of global and local rainfall patterns, resulting in the spread of grasslands—a new habitat type—in the higher latitudes.

South America, Antarctica, and Australia separated from Africa during the Cretaceous but maintained connections with each other into the early Cenozoic. Africa was separated from Europe and Asia by the Tethys Sea until the Miocene, and North and South America did not have a solid land connection until the Pleistocene. In the middle to late Eocene, Australia separated from Antarctica and, like India, drifted northward. Intermittent land connections between South America and Antarctica were retained until the middle Cenozoic, and Eocene Antarctic mammals were basically a subset of South American mammals. New Zealand separated from Australia in the late Cretaceous and has a diversity of endemic tetrapods, but the only native

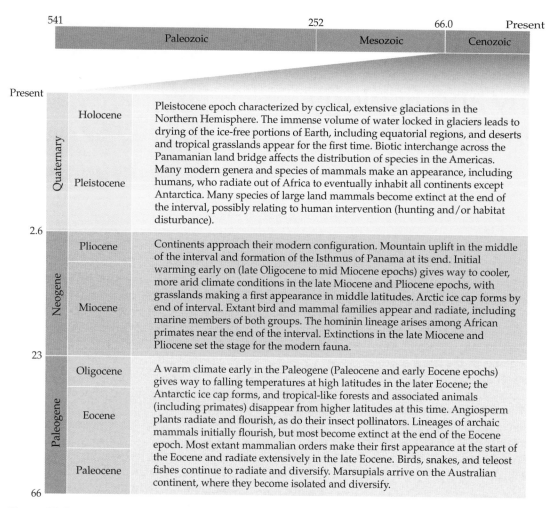

Figure 23.1 Major events of the Cenozoic era. The isolation of continental blocks into their current configurations led to the differentiation of distinctive faunas on the different continents. The rise of the hominin lineage late in the period and the subsequent dominance of modern humans also profoundly affected the vertebrates fauna.

mammals are two species of bats (which, unlike other bats, are predominantly terrestrial in their habits). Other mammals found in New Zealand today, such as possums, deer, and hedgehogs, were brought in by humans during the last few centuries.

23.2 Cenozoic Climates

The evolution of animals and plants during the Cenozoic is closely connected with climatic changes, which in turn are related to the changes in the positions of continents described above. The greater latitudinal distribution of continents, plus changes in the patterns of ocean currents that were the result of continental movements, resulted in cooling trends in high northern and southern latitudes, starting about 45 Ma when ice first formed at the South Pole. Declining levels of atmospheric carbon dioxide around

this time also contributed to global cooling. Warming in the late Oligocene and early Miocene (~25–15 Ma) reversed this cooling trend to a certain extent, but further cooling followed what is known as the Mid-Miocene Climatic Optimum and resulted in the formation of ice at the North Pole around 5 Ma. This late Cenozoic cooling led to a cycle of ice ages and interglacials that began in the Pleistocene and continues to the present. Both the fragmentation of the landmasses and the changes in climate during the Cenozoic have been important factors in the evolution of regional biotas.

A primary result of these climatic changes was a change in the ecosystems of the higher latitudes, which became cooler and drier. This change resulted in the replacement of early Cenozoic tropical-like forests with woodland and grassland starting around 45 Ma. (The term "tropical-like" indicates a forest that had a multilayered structure like that of modern tropical forests, but with a somewhat different assortment of species.)

Paleogene and Neogene climates

The world of the early Cenozoic still reflected the "hothouse" world of the Mesozoic; the tropics appear to have been hotter than those of the present day, as evidenced by the gigantic Paleocene South American snake *Titanoboa*.

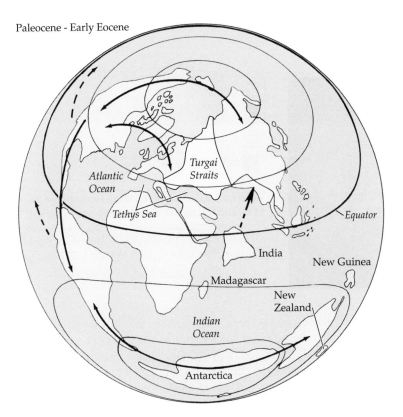

Figure 23.2 Continental positions in the late Paleocene and early Eocene. An epicontinental sea, the Turgai Strait (cross-hatching), extended between Europe and Asia. The Tethys Sea separated Europe and Asia from Africa. Dashed arrows show the direction of continental drift, and solid arrows indicate major land bridges mentioned in the text. In the Northern Hemisphere, there were two early Cenozoic routes connecting the northern continents at high but relatively ice-free latitudes. The trans-Bering Bridge (situated where the Bering Strait is today between Alaska and Siberia) connected western North American and Asia intermittently throughout the Cenozoic. A connection between eastern North America and Europe via Greenland and Scandinavia was apparent during the early Eocene.

There was an increase in the levels of atmospheric carbon dioxide at the start of the Eocene, which would have resulted in warming via a greenhouse effect. A transitory temperature spike at the Paleocene–Eocene boundary (the thermal maximum) was probably caused by the release of methane, a greenhouse gas, from shallowly buried sediments on the ocean continental shelf. Mammals responded to this temperature increase by becoming smaller (dwarfing), and a later temperature increase (the Eocene Optimum) also resulted in episodes of dwarfing. An early Eocene rise in oxygen levels to present-day amounts (see Figure 5.3) may have favored the larger placental mammals, and many of the modern orders of mammals appeared at this time.

By the middle Eocene, temperatures had started to fall, accompanied by a fall in atmospheric carbon dioxide levels (from ~1,200 parts per million to ~750 ppm), causing a "reverse greenhouse" cooling effect. The world moved into "icehouse" conditions, with the initial formation of polar ice sheets, but it did not become as cold in the higher latitudes as it is today until a few million years ago—very recently, in geological terms (**Figure 23.3**). This Eocene temperature decline was related to the Antarctic ice sheets that first formed in the late middle Eocene (~45 Ma). Cold water massed over the poles when Greenland broke away from Norway, Australia broke away from Antarctica, and the breakup of Antarctica and South America was initiated with the formation of the Drake Passage that separates those two continents today.

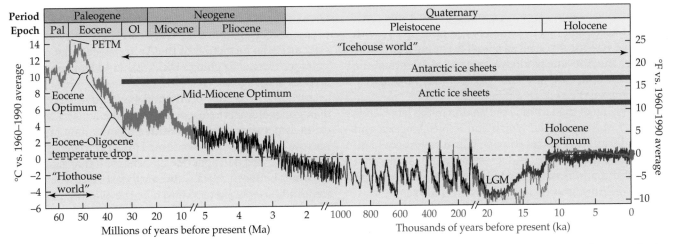

Figure 23.3 Shifting climates of the Cenozoic. Reconstructions of global air temperatures in the Cenozoic are compared with the global average air surface temperature 1960–1990 (dashed horizontal line). Note the major shifts in time scales, indicated by hatches on the x-axis and changing colors of the traces; for example, "hothouse" Earth existed for some 20 million years, half the length of the current "icehouse" conditions. The variegated traces in the Pleistocene and Holocene represent reconstructions drawn from different methodologies. Blue bars indicate the presence of Antarctic and Arctic ice sheets. The last 10,000 years have seen relatively stable climatic conditions, referred to as the Holocene Optimum. PETM, Paleocene–Eocene thermal maximum; LGM, last glacial maximum (see Figure 23.4). (Adapted from Glen Fergus/CC BY-SA 3.0.)

Pleistocene (18 ka)   Present

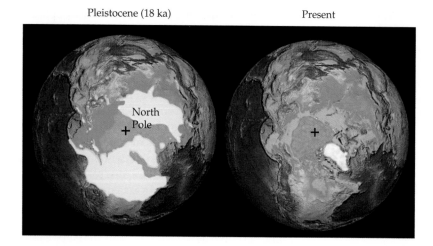

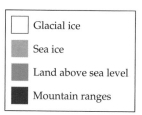

Glacial ice

Sea ice

Land above sea level

Mountain ranges

Figure 23.4 Extent of Northern Hemisphere ice 18 ka and today.   During the Pleistocene, periods of maximum glaciation saw ice sheets 3–4 km deep extending far into the temperate latitudes of the Northern Hemisphere. (Mark McCaffrey/NOAA.)

Ocean circulation carried the cold water toward the Equator, cooling the temperate latitudes. Transitory Arctic ice was present in the late middle Eocene (~46 Ma), and also in the early Oligocene (~30–34 Ma), but the current Arctic ice cap developed only about 5 Ma.

Extensive Antarctic glaciation was present during the Oligocene (from ~34 to ~27 Ma), and atmospheric CO_2 dropped to near present-day (preindustrial) levels (see Figure 5.3). The Drake Passage deepened at this time, and this may have been when the Antarctic Circumpolar Current (ACC) first formed. The Antarctic ice sheets were variable during the warm period of the late Oligocene and early Miocene; this warming has been ascribed to the isolation of the cold polar water around Antarctica in the ACC. Atmospheric levels of carbon dioxide also rose at this time, to a Neogene maximum of ~500 ppm. However, the further deepening of the Drake Passage in the middle Miocene led to a more extensive ACC and the expansion of the Antarctic ice cap about 14 Ma. Atmospheric carbon dioxide levels started their decline toward the Pleistocene low of about 180 ppm, once again bringing cooling to the higher latitudes. This cooling has persisted to the present day with occasional remissions, such as a brief warming during the Pliocene. The Miocene world was in general drier than the Paleogene world, and the combination of warmth and dryness promoted the spread of grasslands.

The closing of the Isthmus of Panama in the late Cenozoic (~2.8 Ma) connected North and South America and profoundly affected global climate. Warm waters could no longer circle the Equator, and it was during this time that the Gulf Stream formed, carrying warm water from the subtropical American regions to western Europe.

The Pleistocene ice ages

The extensive episodic continental glaciers that characterized the Pleistocene epoch represented events that had been absent from the world since the Paleozoic. These ice ages commenced about 2.5 Ma, after the formation of the Arctic ice cap, and had an important influence on vertebrate evolution in general—almost all of today's species have their origin in the Pleistocene—and on our own evolution and

even on our present civilizations. We still live in a world with abundant polar ice, but at the moment we are in a warmer (interglacial) period (see Figure 23.3), and sea levels are higher today than in glacial periods.

Four major episodes of glaciation occurred in the Pleistocene, interspersed with 20 or more minor episodes. Northern latitudes were covered with ice 3–4 km thick as far south as 38°N in North America and 50°N in Eurasia (Figure 23.4). It appears that glacial episodes get their start not from the world as a whole getting colder year-round, but from cool summers that prevent melting of winter ice.

The episodes of glaciation were caused by Milankovitch cycles, named after the astronomer who proposed them in the 1930s. There are three different cycles of variation in Earth's orbit around the sun and in the tilt of Earth's axis that affect the amount of solar radiation reaching Earth. The duration of each of the three cycles differ, from 23,000 years to 100,000 years, and normally they are out of phase with each other, like discordant keys played on a piano. Every so often, however, they align like notes making a chord, with a large effect on Earth's climate. Milankovitch cycles have existed throughout Earth's history, but it was only after the formation of the Arctic ice cap that enough polar ice existed to plunge the Northern Hemisphere into a series of ice ages.

Although the glaciers would have advanced slowly enough for animals to disperse toward the Equator, in many cases dispersal routes were limited by mountain ranges or narrow geographic connections. However, much of Alaska, Siberia, and Beringia (Pleistocene land that is now underwater between Alaska and northeast Asia) was free of ice and housed a biome known as the Mammoth Steppe or the steppe-tundra, which has since disappeared. This steppe-tundra habitat was much more productive than present-day high-latitude habitats. The steppes supported a highly diverse fauna of large mammals, including extinct forms such as mammoths and wooly rhinoceroses, extant arctic mammals such as reindeer and musk ox, and now-tropical mammals such as lions.

Drying of the ice-free portions of Earth due to the volume of water tied up in glaciers was at least as important for terrestrial ecosystems as the glaciers themselves. Many of the equatorial areas that today are covered by lowland rainforests were much drier, even arid, during glacial periods. Today's relatively mild interglacial period is apparently colder and drier than other interglacial periods in the Pleistocene. For example, during other interglacial periods hippopotamuses were found in what is now the Sahara Desert and in England.

23.3 Cenozoic Terrestrial Ecosystems

Today's world is a very cool and dry place in comparison with most of the Cenozoic (and, indeed, with most of Earth's history), and it is also quite varied in terms of habitats and climatic zones. At the start of the Cenozoic, the world was covered with tropical-like forests and their associated faunas. This angiosperm-dominated vegetation had its origins in the mid-Cretaceous, although it was not until the last 10 million years of the Cretaceous that angiosperm leaves reveal that these rainforests formed their own damp microhabitats, accompanied by dietary changes among the mammals. There are still forests of this type, now confined to equatorial regions, but there are also other types of environments at other latitudes (temperate woodland and grasslands) and also tropical grasslands.

New habitats that emerged in the Plio-Pleistocene were deserts, arctic taiga (stands of coniferous forest), and tundra (treeless shrubland). As the world became increasingly zonal from pole to Equator with respect to temperature and vegetation regimes, distinct latitudinal differences in the types of faunas became established, and the numbers of different types of animals increased enormously, corresponding with the greater diversity of habitat types. In addition, the division of the world into separate continents resulted in many faunas and floras evolving in comparative isolation, thus boosting the total global diversity.

In the warm world of the early Cenozoic, tropical-like forests were found in high latitudes—even extending into the Arctic Circle (indicating ice-free conditions year-round), with summer temperatures reaching 20°C. These arctic forests were composed of broadleaf plants, a little like the swamp-cypress forests and broadleaf floodplain forests seen today in the southeastern United States. These forests evidently could withstand 3 months of continuous light and 3 months of continuous darkness, and had species richness comparable to that of similar types of lower-latitude forests today. Vertebrates in these polar forests included reptiles such as turtles and crocodiles as well as mammals resembling—though not closely related to—the tree-dwelling primates, small tapirs, and small hippos of present-day tropical areas, whereas mammals common at lower latitudes were absent.

High-latitude tropical-like forests were established a mere million years or so into the Cenozoic, but it was not until the Eocene that more open-canopied forests capable of supporting denser undergrowth vegetation appeared, and this is when the larger mammals began to radiate into the diversity of terrestrial niches they occupy today (**Figure 23.5**). The high- and mid-latitude tropical-like forests of the Paleocene to middle Eocene partially reflected the higher temperatures of the time (see Figure 23.3), but were also related to the accompanying relatively high levels of atmospheric carbon dioxide.

Mammals radiated rapidly at the start of the Cenozoic, but many of the original lineages are now extinct. These groups are often called "archaic mammals," a condescending term stemming from our present-day perspective. Archaic mammals were mainly small to medium-size generalized forms. They included the rodentlike multituberculates, stem marsupials, and both stem and crown group placentals (such as the carnivorous creodonts and the stem ungulate condylarths). Larger specialized predators and herbivores with teeth indicating a high-fiber herbivorous diet did not appear until the late Paleocene. Members of most present-day orders did not make their first appearance until the Eocene, but there are some notable exceptions, such as Carnivora (dogs, cats, and others), whose first members were present in the Paleocene.

Modern types of birds also diversified at the start of the Cenozoic, including the passerines (songbirds), first known from the early Eocene, and large terrestrial birds, including carnivorous forms and herbivores such as the present-day ostriches. Many of the present-day groups of reptiles and amphibians had their origins in the latest Cretaceous or early Cenozoic. There was a modest radiation of terrestrial crocodyliforms in the Southern Hemisphere during the Paleocene and Eocene, but only aquatic crocodylians survived into the later Cenozoic. Among the insects, modern butterflies and moths first appeared in the middle Eocene.

In the cooler and drier world following the early Cenozoic (starting in the late middle Eocene, ~45 Ma), temperate forests and woodlands replaced the tropical-like vegetation of the higher latitudes, and tropical forests were confined to equatorial regions. Tropical-like vegetation depends not so much on high levels of temperature throughout the year, but on the absence of freezing temperatures at any time. The declining temperatures of the later Eocene also heralded seasonal climates and winter frosts in the middle and high latitudes. The tropically adapted faunas, including the higher-latitude primates and crocodylians, disappeared or retreated along with the tropical forests.

In the Neogene, both higher-latitude temperatures and levels of atmospheric carbon dioxide rose until the mid-Miocene and then declined. Extensive savannas (grasslands with scattered trees) first appeared in the early Miocene in the northern latitudes, but they did not appear in tropical areas such as East Africa until 2–3 Ma, by which time the higher-latitude grasslands had turned into treeless prairie or steppe. Fire may have been important in promoting the spread of grasslands. The change from woodlands to grasslands in the African tropics coincided with the radiation of our own genus, *Homo*, a hominid well adapted to life on the tropical savannas.

Figure 23.5 Reconstruction of a scene from the late early Eocene of Europe. The trees in this tropical-like rainforest include sequoia, pine, birch, palmetto, swamp cypress, and tree ferns. Shrub-height cycads and magnolias appear in the foreground. The birds in the foreground are ibises, with a *Gastornis* (an extinct flightless bird) in the background. The hippolike mammals are *Coryphodon*, belonging to the extinct ungulate-like order Pantodonta. An early primate, *Cantius*, climbs a liana vine. Crouching among the ferns is *Palaeonictis*, a catlike predator belonging to the extinct order Creodonta. Its potential prey (foreground) are early artiodactyls, *Diacodexis*.

The radiation of mammals in the Neogene commenced around 17 Ma and reflected these vegetational changes. Large grazing mammals such as horses, antelope, rhinoceroses, and elephants evolved along with the emerging grasslands—and with them the carnivores that preyed on them, such as large cats and dogs. Some small mammals also diversified, most notably modern types of rodents, such as rats and mice. This diversification of small mammals may explain the concurrent diversification of modern types of snakes. Late Cenozoic lizards included some very large varanids, among them the largest lizard known today, the Komodo monitor (*Varanus komodoensis*), and a now-extinct, crocodile-size terrestrial predator, *Varanus (Megalania) priscus*, from the Pleistocene of Australia. Many modern types of birds first appeared in the late Cenozoic, including birds of prey such as eagles, hawks, and vultures. The more open habitats of the later Cenozoic also favored the diversification of social insects that live in grasslands, such as ants and termites.

23.3 Biogeography of Cenozoic Mammals

The Cenozoic radiation of mammals occurred during the fractionation of the continental masses, when different stocks became isolated on different continents. This separation of ancestral stocks from one another by Earth's physical processes, rather than by their own movements, is called **vicariance**. However, only a few instances of the distribution of present-day mammals (e.g., the isolation of the monotremes in Australia and New Guinea) probably result from vicariance—that is, from animals and plants being carried passively on moving landmasses.

Other patterns of mammal distribution can be explained by **dispersal**, which reflects movements of the animals themselves, usually by the spread of populations rather than the long-distance movements of individual animals. This is the best explanation for the distribution of most of the modern groups of mammals (such as the Australian marsupials), because they were not present in their current positions before the Cenozoic.

The relatively separate evolution of different groups of mammals on different continental blocks has resulted in a duplication of mammals specialized for certain lifestyles; i.e., ecomorpohological types. For example, the role of a mammal specialized for feeding on ants and termites (myrmecophagy) is taken by different types of mammals on different continents: echidnas (monotremes) and numbats (marsupials) in Australia, anteaters and giant armadillos (placentals, Xenarthra) in South America, and pangolins (placentals, Pholidota) in Africa and tropical Eurasia. Gliding mammals have also evolved convergently, including the flying squirrels of the Northern Hemisphere (Rodentia, Sciuridae) and the marsupial flying phalangers in Australia (Diprotodonta, Petauridae and Pseudocheiridae), as well as the flying lemurs (Dermoptera, which are not true primates) of Southeast Asia, and a completely different type of flying rodent, the "scaly-tailed squirrel" of Africa (Rodentia, Anomaluridae).

The isolation of Australian mammals

Although today marsupials are considered the quintessential Australian mammals, they did not reach that continent until the early Cenozoic. Marsupials reached North America from Asia in the Late Cretaceous, where they enjoyed a modest radiation of small forms. The initial Cenozoic diversification of marsupials occurred primarily in South America, where the more basal types of marsupials persist today. Australia (including New Guinea) was populated by marsupials that probably moved from South America across Antarctica, which was warm and ice-free until about 45 Ma; the earliest known fossils are from the early Eocene (~55 Ma). Marsupial fossils are known from the Eocene of western Antarctica, and in fact, the major barrier to dispersal to Australia was probably crossing the midcontinental mountain ridge between western and eastern Antarctica. Both anatomical and molecular data show that the Australian marsupials derived from a common stock within the radiation of South American marsupials, but the details of this migration are not simple. The fossil record shows both a monotreme and possibly some early kinds of Australian marsupials in the Paleogene of Patagonia.

Once marsupials reached Australia, they enjoyed the advantages of long-term isolation, evolving to fill a variety of adaptive zones with feeding habits ranging from specialized carnivory to complete herbivory, convergently with terrestrial placentals elsewhere in the world. When humans reached Australia 65 ka, they encountered "marsupial lions" (*Thylacoleo carnifex*; **Figure 23.6A**). This remarkable predator had enormous flesh-shearing carnassial teeth, a bite force greater than any extant mammal, and a greater degree of forearm maneuverability than any extant carnivore. *Thylacoleo* is believed to have preyed on animals much larger than itself. A large retractable claw on a semi-opposable thumb could have been used for killing the prey as well as grasping it; *Thylacoleo* did not have true canines, and its canine-like incisors were blunt, indicating that they were used for holding the prey rather than stabbing like a placental lion.

The large cats are extant placental analogues of *Thylacoleo*, and jaguars (*Panthera onca*) are about the same size (**Figure 23.6B**). Extant jaguars prey on mammals that are about half their size, but this situation may reflect the extinction of large herbivorous mammals in Central and South American at the end of the Pleistocene.

Convergence is also seen among herbivores. Large macropod marsupials, such as the red kangaroo (*Macropus rufus*; **Figure 23.7A**) are primarily grazers, eating green grass, and sometimes browsing on the leaves of shrubs. Many placental herbivores, including medium-size African antelope like impala (*Aepyceros melampus*; **Figure 23.7B**) have similar dietary habits. Because vertebrates cannot digest cellulose, herbivorous mammals rely on fermentative microorganisms in the digestive system to break down plant cell walls, and these symbionts may be located in the stomach (foregut fermenters, including ruminants) or in the cecum and large intestine (hindgut fermenters). The gastric specializations of marsupials and placentals were independently derived and are anatomically different, but functionally similar. The impala is a ruminant, like cows, whereas kangaroos are described as "ruminant-like" and do not chew the cud as do ruminants.

(A)

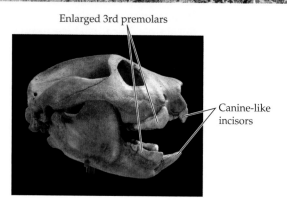

Enlarged 3rd premolars

Canine-like incisors

(B)

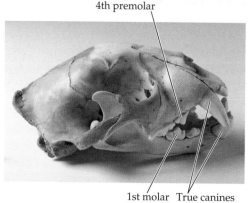

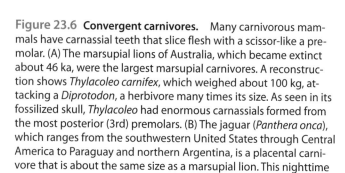

4th premolar

1st molar   True canines

Figure 23.6 **Convergent carnivores.** Many carnivorous mammals have carnassial teeth that slice flesh with a scissor-like a premolar. (A) The marsupial lions of Australia, which became extinct about 46 ka, were the largest marsupial carnivores. A reconstruction shows *Thylacoleo carnifex*, which weighed about 100 kg, attacking a *Diprotodon*, a herbivore many times its size. As seen in its fossilized skull, *Thylacoleo* had enormous carnassials formed from the most posterior (3rd) premolars. (B) The jaguar (*Panthera onca*), which ranges from the southwestern United States through Central America to Paraguay and northern Argentina, is a placental carnivore that is about the same size as a marsupial lion. This nighttime image from a camera trap shows a female jaguar with the carcass of a wild pig. The body size of jaguars has decreased since the Pleistocene, and extant jaguars prey on mammals that are about half their size, a situation that may be the result of post-Pleistocene extinction of larger prey species. The carnassials of jaguars and other carnivorans are formed by the 4th upper premolar and the 1st lower molar. (A, photo by Rom-diz; skull by Ghedoghedo/CC BY-SA 3.0; B, photo courtesy of the Northern Jaguar Project/Naturalia; skull by Wagner Souza e Silva/Museum of Veterinary Anatomy FMVZ USP/ CC BY-SA 4.0.)

Convergence also is seen among arboreal browsers. At around 10 kg, the folivorous koala (*Phascolarctos cinerus*; **Figure 23.7C**) is large for an arboreal animal. Koalas are slow-moving and lack a tail. The indri (*Indri indri*; **Figure 23.7D**), a Madagascan lemur, is a placental that is similar to koalas in diet, body form, and behavior. Unlike the grazers, both the koala and the indri are hindgut fermenters—that is, fermentation occurs in an enlarged cecum and large intestine.

Many terrestrial insectivorous mammals insert their snouts into small openings as they seek prey, and marsupial and placental insectivores have long pointed snouts (**Figure 23.8A,B**). Once the exoskeleton of an insect has been crushed, little additional oral processing is needed, and many insectivorous mammals (anteaters in particular) have a reduced dentition. Insectivorous bark-strippers provide an example of extreme convergence (**Figure 23.8C,D**). The striped possum (*Dactylopsila trivirgata*) of Indonesia,

northern Australia, and New Guinea detects insects underneath tree bark by tapping with its feet and listening with its large ears for the echoes signifying insect-containing chambers; it has chisel-like incisors to rip the bark and a specialized elongated finger to probe for insects in the wood. Its feeding behavior is almost identical to that of a lemur in Madagascar, the aye-aye (*Daubentonia madagascariensis*). The aye-aye is a little bigger than the striped possum, and morphologically more highly specialized for this behavior. The wood-boring behavior of both mammals is analogous to that of a woodpecker. Mammals with similar morphological adaptations (apatemyids) are known from North American and Europe during the Paleogene, but since the appearance of woodpeckers in the Miocene, wood-boring mammals have been confined to areas of the world where woodpeckers are absent.

There is some evidence that one type of terrestrial placental, a condylarth (a small basal ungulate), dispersed to

(A)

(B)

(C)

(D)

Figure 23.7 Convergent herbivores. (A) Red kangaroos (*Macropus rufus*) are primarily grazers, as are (B) impala (*Aepyceros melampus*). Both species are foregut fermenters. (C) The koala (*Phascolarctos cinereus*) and (D) the indri (*Indri indri*) are arboreal folivores. These species are hindgut fermenters. (A, Mike Souza/CC BY-SA 2.0; B, Chad Rosenthal/CC BY 2.0; C, Ryan Vaarsi, CC BY 2.0; D, Zigomar/CC BY-SA 3.0.)

(A)

(B)

(C)

(D)

Figure 23.8 Convergent insectivores. (A) The numbat (*Myrmecobius fasciatus*), a marsupial anteater in Australia related to carnivorous marsupials (dasyurids), and (B) the bat-eared fox (*Otocyon megalotis*), a termite-eating canid in Africa have elongated snouts and teeth that are reduced in size. Both (C) the striped possum (*Dactylopsila trivirgata*), a marsupial native to Australia and New Guinea and (D) the aye-aye (*Daubentonia madagascariensis*), a lemur from Madagascar, tap on bark to detect the presence of insects. (A, S. J. Bennett/CC BY 2.0; B, Yathin S. Krishnappa/CC BY-SA 3.0; C, Joe McKenna/CC BY-NC-ND 2.0; D, Rod Waddington/CC BY 2.0.)

Australia along with marsupials, but the lineage did not survive, and for most of the Cenozoic the only placentals in Australia were bats, which flew there from Asia. Bats are also the only mammals that reached New Zealand. Although the Australasian landmass (comprising Australia, New Zealand, and New Guinea) has been moving closer to Asia during the Cenozoic, there have been few natural migrations of Asian animals into Australia. For example, monkeys never reached Australasia, and in the absence of monkeys, tree-kangaroos have had a successful radiation in the tropical forests of northern Australia and New Guinea.

Rodents arrived in Australia by the early Pliocene, probably via dispersal along the island chain between Southeast Asia and New Guinea. Australian rodents are an interesting endemic radiation today. They are related to the mouse/rat group of Eurasian rodents but have evolved into unique Australian forms such as the small, jerboa-like hopping mice and the much larger (otter size) water rats. True mice and rats arrived later, in the Pleistocene. However, these rodents apparently had surprisingly little overall effect on the Australian marsupials. Far more serious threats came from the original invasion by humans from Asia about 65,000 years ago, the arrival of dogs (dingoes) about 4,000 years ago, and the introduction of domestic mammals such as foxes, rabbits, and cats by European colonists during the past few centuries. Today, numerous marsupial species are threatened or endangered by humans and their introduced exotic species.

The isolation of mammals on other continents

The other continents of today's world all have their own unique faunas, although none are as different from each other as the Australian fauna is from those of the rest of the world.

Mainland Africa Africa had separated from South America by the Late Cretaceous and remained isolated until it collided with Eurasia during the Oligocene and Miocene. A remarkable group of endemic mammals radiated during this period. These afrotherians include elephants and their extinct relatives, hyraxes (which are rodentlike in appearance but phylogenetically allied to elephants), aardvarks, elephant shrews, tenrecs, and golden moles. Although Africa was isolated from the rest of the world, hyraxes were larger and more diverse than today, and took the ecological roles of antelope and large pigs. Most of the mammals we think of as African (lions, hyenas, giraffes, and antelope) arrived from Eurasia in the Neogene, after Africa collided with Eurasia around 17 Ma. Today the faunas of tropical Africa and Asia are similar but different—that is, they have different genera of elephants, antelope, and large cats. Afrotheria are discussed further in Chapter 25.

Madagascar Madagascar is a large, topographically complex island off the coast of East Africa with a fauna composed almost entirely of endemic species that are descended from a small number of founding lineages. In the Middle Jurassic (170 Ma), what is now Madagascar lay in the middle of the supercontinent Gondwana. As Gondwana broke apart, the portion that included Madagascar passed through a series of splits, ultimately isolating Madagascar in the Indian Ocean about 88 Ma.

Madagascar's present-day mammalian fauna appears to have been derived from Africa in several different waves of dispersal, and includes four different lineages: lemurs (primates), tenrecs (afrotheres), carnivorans (euplerids, related to mongooses), and rodents (nesomyids, related to rats and mice). Only the rodents retain representatives in Africa. Molecular studies of the divergence of these lineages from their mainland relatives suggest that lemurs and tenrecs colonized Madagascar in the Eocene, and carnivores and rodents in the early Miocene. Unfortunately there is little to no Cenozoic fossil record in Madagascar until the latest Pleistocene. Today the entire native Madagascan fauna is severely threatened by anthropogenic habitat destruction.

The best-known endemic mammals of Madagascar are the lemurs, a basal radiation of primates known from five extant and three extinct families (Figure 23.9A–C). Their diversity reflects their long history on the island. Diverse as today's lemurs are, their diversity was much greater in the recent past. Much larger (up to gorilla size) lemurs—both terrestrial and arboreal—paralleled the radiation of the great apes among the anthropoid primates. The extinction of these giant lemurs occurred soon after the arrival of humans 2.3 ka, although anecdotal reports suggest that some species of giant lemurs may have survived until only a few hundred years ago.

In the absence of insectivores like hedgehogs, tenrecs take the role of small omnivores and insectivores of Madagascar (Figure 23.9D). Madagascan rodents are mostly mouselike, but they include the rabbit-size giant jumping rat *Hypogeomys antimena*. Madagascan carnivorans mostly resemble the African and Asian civets and mongooses, but the most remarkable of these is the fossa *Cryptoprocta ferox* (Figure 23.9E), which is a little larger than a large house cat. In the absence of true cats, the fossa has evolved to become a catlike predator, although it is more arboreal than true cats.

The Americas The North American fauna was much more distinct from that of Eurasia for most of the Cenozoic than it is today. Much of the present North American fauna (e.g., deer, bison, and rodents such as voles) arrived from Eurasia via the Bering land bridge in the Pliocene and Pleistocene, and many forms originally native to North America, such as horses and camels, are now extinct on that continent.

Some exchange between North and South American faunas occurred in the earliest Paleocene, but South America fauna was isolated from North America for much of the Cenozoic by a seaway between Panama and the northwestern corner of South America. The adaptive zone for large carnivores in South America was occupied by the now-extinct borhyaenoid metatherians (stem marsupials), and the role of large herbivores was taken by the now-extinct

(A)   (B)   (C)

(D)   (E)

Figure 23.9 Endemic mammals of Madagascar. (A–C) The radiation of lemurs on Madagascar parallels the radiation of primates in Africa and South America. Giant lemurs, such as (A) the gorilla-size *Archaeoindris fontoynonti*, were present when humans first arrived but are now extinct. Extant lemurs range in size from (B) the eastern rufous mouse lemur (*Microcebus rufus*, 30 g) to (C) the white sifaka (*Propithecus verreauxi*, 3.5 kg) and the indri (~8 kg; see Figure 23.7D). (D) The insectivorous lowland streaked tenrec (*Hemicentetes semispinosus*), one of two species of tenrecs, has converged on the body form of hedgehogs. (E) The fossa (*Cryptoprocta ferox*) is a catlike carnivoran, related to the mongooses of tropical Africa and Asia. (A, Smokeybjb/CC BY-SA 3.0; B, gailhampshire/CC BY 2.0; C, David Dennis/CC BY-SA 2.0; D, Frank Vassen/CC BY 2.0; E, Mathais Appel.)

perissodactyl-related native ungulates, such as litopterns and notoungulates. Other original Cenozoic inhabitants of South America include ameridelphian marsupials (opossums and related forms) and xenarthrans (sloths, anteaters, and armadillos). The platyrrhine primates and caviomorph rodents that are now unique to South America arrived from Africa toward the end of the Eocene; primates first appeared around 36 Ma, and rodents around 41 Ma.

The South American fauna retained its unique nature until the continent's connection with North America via the Isthmus of Panama, and commencement of an exchange of flora and fauna between the two continents known as the Great American Biotic Interchange (**Figure 23.10**). Some land connection between the two continents may have formed--at least intermittently--as long ago as the early

Miocene (a cebid primate is known from the early Miocene of Panama, 21 Ma), and a more general exchange of mammals was apparent by the late Miocene (~9 Ma) as raccoon relatives moved into South America and ground sloths into North America. However, significant faunal exchange was not apparent until the late Pliocene (~3.5 Ma), before the final closure of the seaway between the two continents at around 2.8 Ma, and the major interchange in both directions commenced at around 2.6 Ma. The Central American passageway is largely tropical forest today, but the onset of the Northern Hemisphere glaciations resulted in drying and the formation of a "savanna corridor," a habitat more favorable to large-scale migrations. Falling sea levels also would have increased available land area for migrations.

More species appear to have migrated from North to South America that in the opposite direction, and North American species appear to have been more successful than those moving in the opposite direction—about half of the mammal genera native to South America today have relatively recent North American origins. Thus, for many years the Great American Biotic Interchange was viewed as an example of the competitive superiority of Northern Hemisphere mammals. However, we now know that the newcomers and the native fauna appear to have initially coexisted on both continents, with a disparity appearing only during the Pleistocene extinctions. Although these extinctions affected the largest mammals on both continents,

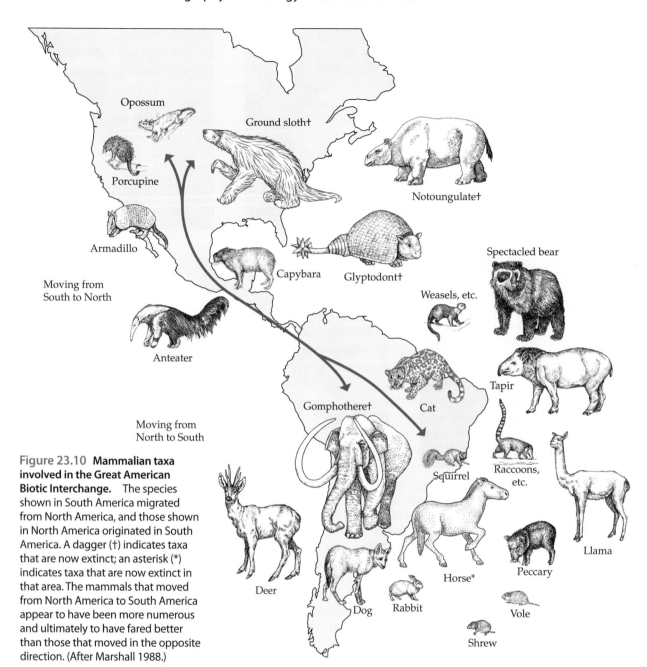

Figure 23.10 Mammalian taxa involved in the Great American Biotic Interchange. The species shown in South America migrated from North America, and those shown in North America originated in South America. A dagger (†) indicates taxa that are now extinct; an asterisk (*) indicates taxa that are now extinct in that area. The mammals that moved from North America to South America appear to have been more numerous and ultimately to have fared better than those that moved in the opposite direction. (After Marshall 1988.)

the southern immigrants in North America were affected more profoundly than were the North American forms in South America. (Although Central America is part of the North American continent, its fauna today is more like that of South America and includes monkeys, xenarthrans, and caviomorph rodents.) A key to the apparently greater success of the South American immigrants lies in an understanding of biogeography. The Equator and much of tropical America lie within South America. During times of climatic stress such as glacial periods, South America retains more equable habitats than North America does, so fewer extinctions would be expected. Our understanding of these events is also biased by the fact that most South American fossil localities are from high-latitude savanna habitats; we know very little about what was going on at that time in the tropics.

23.4 Cenozoic Extinctions

The best-known extinction of the Cenozoic occurred at the end of the Pleistocene, when about 30% of mammal genera became extinct, although other extinction events in the late Eocene and late Miocene claimed a similar number of taxa. The Pleistocene extinction (also known as the Quaternary extinction) appears dramatic because this was a selective extinction of megafauna (mammals weighing more than 20 kg), whereas earlier extinctions affected mammals of all body sizes. Other organisms, both terrestrial and marine, also experienced profound extinctions during the Eocene and Miocene, but were less affected by the Pleistocene extinction.

The late Eocene extinctions were associated with the dramatic fall in higher-latitude temperatures and the resulting environmental changes. The archaic mammals disappeared at this time, as did more modern types of mammals adapted to tropical-like forests, such as the early primates, and surviving primates became restricted to the lower latitudes. The early Cenozoic diversity of amphibians and reptiles in higher latitudes was also greatly reduced during the late Eocene.

The late Miocene extinctions were associated with global drying as well as falling higher-latitude temperatures. The major extinctions were of browsing mammals (including a diversity of large browsing horses), which suffered habitat loss as the savanna woodlands turned into open grasslands and prairie. North America was especially hard hit by the climatic events of the late Miocene because of its relatively high latitudinal position and the fact that animals could not migrate to more tropical areas in South America before the formation of the Isthmus of Panama in the Plio-Pleistocene. While the end-Pleistocene extinction represents a major event, it was preceded by some major extinctions a few million years earlier, at the end of the Pliocene; these particularly affected marine vertebrates, including cetaceans and giant sharks, and are thought to have been related to a decline in coastal productivity following changes in oceanic circulation.

Many of the megafaunal mammals that became extinct at the end of the Pleistocene are not represented by similar types today, among them glyptodonts (giant armored mammals related to armadillos) and giant ground sloths in North and South America, and diprotodontids (the largest of all marsupials, some as big as a hippopotamus) in Australia. The extinction also included large and exotic forms of more familiar mammals, such as the saber-toothed cats and mammoths of the Northern Hemisphere and Africa; the Irish elk (*Megaloceros giganteus*), cave bears (*Ursus speleas*), and woolly rhinoceros (*Coelodonta antiquitatis*) of Eurasia; and giant kangaroos (sthenurines) and giant wombats in Australia. Large terrestrial birds also suffered in these extinctions; these included herbivores such as the moas of New Zealand and the elephant birds of Madagascar, and

carnivores such as the New World phorusrhacids ("terror birds"). Although smaller mammals did not suffer as many extinctions, there was nonetheless a decline in the diversity of their communities.

There is much debate about the cause of these extinctions. The main extinctions occurred at the end of the last glacial period, which ended about 12,000 years ago, and these extinctions define the boundary between the Pleistocene and the Holocene. (Animals appear to be more vulnerable to extinction when the climate changes from glacial to interglacial rather than the other way around, probably because the effects of warming occur faster than the effects of cooling.) While climatic change is an obvious explanation for the extinctions, many scientists have noted that it is only the last glacial period—rather than any of the previous ones—that brought extinctions of such magnitude. This observation suggests that part, if not all, of the blame for megafaunal extinctions should be placed on the spread of modern humans and modern hunting techniques, which occurred at about that time. This is the overkill hypothesis. However, careful analysis shows long periods of overlap in North America between humans and megafaunal mammals—in some places exceeding 3,000 years—and this is more indicative of extinctions caused primarily by climate change, with humans perhaps delivering the final blow. We will discuss the role of humans in megafaunal extinctions further in Chapter 26.

The overkill and climate change hypotheses are not mutually exclusive, and the two forces may have acted synergistically. In addition, a growing body of geological and archeological evidence indicates that a cosmic impact event ~12.8 ka ignited widespread fires across North and South America, Europe, and Asia that could have added stress to already unstable populations.

Irrespective of the megafaunal extinctions, Pleistocene climate changes resulted in migrations and range shifts of many organisms and shaped the biogeographic distribution of present-day faunas and floras.

Summary

The Cenozoic is often known as the Age of Mammals.

While the largest land vertebrates of the Cenozoic were mammals, rather than the nonavian dinosaurs of the Mesozoic, the Cenozoic nevertheless represents only one-third of mammalian history. The most diverse vertebrates today are not mammals but teleost fishes.

Changes in the positions of the continents during the Cenozoic have affected patterns of global climate and the distribution of organisms.

As the breakup of Pangaea continued from its inception in the Mesozoic, the continents became increasingly isolated and moved into higher latitudes, with the eventual result of ice forming at the poles. Different continents had their own faunas that evolved to a greater or lesser extent in isolation, although the current distribution of mammals is largely the result of later Cenozoic migrations between continental blocks.

Continental movements affected ocean currents and created many of the present-day mountain ranges.

The mid-Cenozoic uplift of mountains affected patterns of rainfall, resulting in the spread of grasslands at ~20 Ma, and may have contributed to lowering the levels of atmospheric carbon dioxide, which would have had a "reverse greenhouse" cooling effect on global temperatures.

(Continued)

Summary *(continued)*

The Pleistocene world was cold and dry.

Episodes of extensive glaciation over the past 2.5 Ma resulted in ice sheets covering a large area of the Northern Hemisphere and in a drier world with lower sea levels, because water was tied up in the ice caps.

The world today is colder, drier, and more varied in terms of habitats and climatic zones than it has been at almost any other time in Earth's history.

Earth's ecosystems became increasingly zonal from pole to Equator during the Cenozoic, with the emergence of large areas of temperate habitat types in the higher latitudes replacing the more tropical habitats of the early Cenozoic. The tundra and taiga of the Arctic and the deserts of the lower latitudes are relatively new environments that emerged within the past 5 Ma.

Early Cenozoic terrestrial ecosystems were largely tropical.

At the start of the Cenozoic, tropical-like forests were found within the Arctic Circle, and animals resembling those of tropical forests today (e.g., primates) were also found at high latitudes. By the end of the Eocene, temperatures had fallen, tropical floras and faunas had become extinct or retreated to equatorial regions, and new radiations of faunas and floras adapted to seasonal temperatures and winter frosts were seen in Earth's mid-latitudes.

Later Cenozoic terrestrial ecosystems included extensive grasslands and temperate woodlands.

Savannas (grasslands with scattered trees) appeared at higher latitudes in the early Miocene, and in the tropics in the Plio-Pleistocene. Grassland ecosystems included radiations of large hoofed mammals and carnivores, modern types of birds and snakes, and social insects such as ants and termites.

Dispersal (rather than vicariance) has largely determined the current patterns of mammalian biogeography.

Although the mammalian faunas on different continental blocks have been largely distinct from each other in the Cenozoic, today the only really different faunas are seen in Australia and Madagascar. The isolation of mammalian faunas meant that many ecomorphs—for example, mammals specialized for eating ants and termites—evolved convergently on different continents.

The Australian marsupials evolved largely in isolation from the rest of the world.

Marsupials migrated to Australia from South America via Antarctica by 55 Ma, and Australian marsupials have a single common ancestry. There are many examples of ecomorphological convergence between Australian marsupials and placentals elsewhere in the world. Bats were the only placentals in Australia for most of the Cenozoic, until rodents migrated from Asia in the Pliocene and Pleistocene. Australian marsupials were not seriously threatened by direct competition from placentals until the past few hundred years, when humans brought domestic animals.

Other continents show their own patterns of faunal isolation.

The fauna of Africa was more distinct from that of the Northern Hemisphere until the immigration of northern mammals about 17 Ma, when Africa collided with Eurasia.

The mammals of Madagascar represent Cenozoic immigration events, but the Madagascan fauna is unique today, with its diversity of lemurs, tenrecs, and native carnivorans and rodents.

The fauna of South America was distinctively different from that of North America until about 3 Ma, when the formation of the Isthmus of Panama led to the exchange of animals between the two continents (the Great American Biotic Interchange).

Major extinction events happened in the late Eocene, in the late Miocene, and at the end of the Pleistocene.

The Eocene and Miocene extinctions affected all kinds of organisms, of all sizes, while the Pleistocene extinction affected mainly the megafauna: mammals, birds, and reptiles weighing more than 20 kg. While the Eocene and Miocene extinctions can be ascribed to climatic changes (cooling and drying), there remains much debate as to whether the Pleistocene extinction resulted from climate change or from human hunting and habitat disturbance, or from some combination of these factors.

Discussion Questions

1. In what major way is the global pattern of Earth's ecosystems different today than it was at the start of the Cenozoic?

2. Why did the faunas and floras of different continents become so different from each other during the Cenozoic?

3. Why are mammalian faunas more similar across the different continents today than they were at other times in the Cenozoic?

4. Why are the faunas and floras of Australia and Madagascar still so distinct today from those in the rest of the world?

5. Why were sea levels lower and deserts more extensive for much of the Pleistocene than they are today?

6. Why was the Southern Hemisphere less affected by the Pleistocene glaciations than the Northern Hemisphere?

7. What difference in the pattern of the Pleistocene extinction, in comparison with that of other Cenozoic extinctions, has made its probable cause so controversial?

Additional Reading

Bacon CD and 5 others. 2016. Quaternary glaciation and the Great American Biotic Interchange. *Geology* 44: 375–378.

Blois JL, Hadley EA. 2009. Mammalian response to Cenozoic climatic change. *Annual Review of Earth and Planetary Sciences* 37: 181–208.

Carrillo JD, Forasiepi A, Jaramillo C, Sánchez-Villagra MR. 2015. Neotropical mammal diversity and the Great American Biotic Interchange. *Frontiers in Genetics* 5: 451.

D'Ambrosia AR, Clyde WC, Fricke HC, Gingerich PD, Abels HA. 2017. Repetitive mammalian dwarfing during ancient greenhouse warming events. *Science Advances* 3:e1601430.

Eberle JJ, Greenwood DR. 2012. Life at the top of the greenhouse Eocene world: A review of the Eocene flora and vertebrate fauna from Canada's High Arctic. *Geological Society of America Bulletin* 124:3-23.

Firestone RB and 25 others. 2007. Evidence for an extraterrestrial impact 12,900 years ago that contributed to the megafaunal extinctions and the Younger Dryas cooling. *Proceedings of the National Academy of Sciences USA* 104: 16016-16021.

Harmon LJ. 2017. Evolution: Contingent predictability in mammalian evolution. *Current Biology* 27: R408–R430.

Head JJ and 7 others 2009. Giant boid snake from the Palaeocene Neotropics reveals hotter past equatorial temperatures. *Nature* 457: 715–718.

Lister AM. 2004. The impact of Quaternary Ice Ages on mammalian evolution. *Philosophical Transactions of the Royal Society* 359:221-241.

Meissner KJ, Bralower TJ. 2017. Volcanism caused ancient global warming. *Nature* 548: 531–532.

Meltzer DJ. 2015. Pleistocene overkill and North American mammalian extinctions. *Annual Review of Anthropology* 44:33-53.

Samonds KE and 7 others. 2012. Spatial and temporal patterns of Madagascar's vertebrate fauna explained by distance, ocean currents, and ancestor type. *Proceedings of the National Academy of Sciences USA* 109: 5352–5357.

Strömberg CAE. 2011. Evolution of grasslands and grassland ecosystems. *Annual Review of Earth and Planetary Sciences* 39: 517–544.

Wolbach WS and 26 others. 2018. Extraordinary biomass burning and impact winter triggered by the Younger Dryas cosmic impact ~12,800 years ago. 1. Ice cores and glaciers. *The Journal of Geology* 126: 185–205.

Zachos J, Pagani M, Sloan L, Thomas E, Billups K. 2001. Trends, rhythms, and abberations in the global climate 65 Ma to present. *Science* 292: 686–693.

Synapsida and the Evolution of Mammals

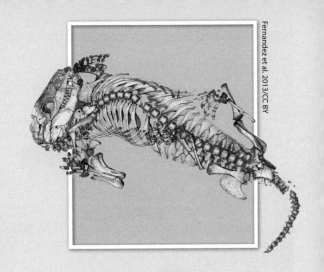

In part because we are mammals ourselves, humans tend to think of mammals as the dominant vertebrates. That perspective does not withstand examination, at least in terms of the number of species—there are barely more than half as many extant species of mammals as of birds, for example, and ray-finned fishes include as many extant species as all of the tetrapods combined—but it is pervasive. In some contexts, however, mammals are indeed exceptionally successful derived vertebrates. The size range of mammals, from shrews to whales, is impressive, as is the development of flight by bats and of echolocation by bats and cetaceans (whales and dolphins). Mammals probably have more complex social systems than any other vertebrates, and many features of their biology are related to interactions with other individuals of their species, ranging from the often-prolonged dependence of young on their mother to lifelong alliances between individuals that affect their social status and reproductive success in a group.

The anatomical and physiological characteristics of mammals—respiration with a diaphragm, hair that provides insulation, high metabolic rates, teeth with complex surfaces that process food efficiently—allow them to be successful in a wide variety of habitats. The progressive appearance of these characteristics can be traced clearly through the fossil ancestry of the mammalian lineage.

All three groups of extant mammals—monotremes, marsupials, and placentals—had appeared by the late Mesozoic, accompanied by several groups that are now extinct. We must backtrack to the end of the Paleozoic, however, to find the origin of the synapsid lineage that gave rise to the mammals.

features, distinguishes synapsids from other amniotes. Synapsids include mammals and their extinct predecessors (**Figures 24.1** and **24.2**). The latter are sometimes called "mammal-like reptiles," a term now considered incorrect, as non-mammalian synapsids (the preferred term) are not closely related to extant reptiles. Changes in the structure of the skull and skeleton of non-mammalian synapsids and their probable relation to metabolic status and the evolution of the mammalian condition are described in Section 24.3.

Synapsids were the first group of amniotes to radiate widely in terrestrial habitats. During the Late Carboniferous and throughout the Permian, synapsids were the most abundant terrestrial vertebrates, and from early in the Permian into the Early Triassic they were the top carnivores and herbivores in the food web. Most synapsids were medium-size to large animals, weighing between 10 and 200 kg (goat size to pony size), with a few weighing 500 kg or more (e.g., the cow-size *Moschops*; see Figure 24.3B). Most of the synapsid lineages disappeared by the end of the Triassic. Most surviving forms (which after the Early Cretaceous were represented only by mammals) weighed less than 1 kg (rat size or smaller).

Synapsids had their first radiation (pelycosaurs) and a significant portion of their second radiation (therapsids) in the Paleozoic—that is, before the major radiations of diapsids. The third radiation of the synapsid lineage (mammals) did not begin until the end of the Triassic and reached its peak in the Cenozoic. Nonetheless, throughout the late Paleozoic and early Mesozoic, members of the synapsid lineage were becoming increasingly mammal-like.

24.1 The Origin of Synapsids

The name *synapsid* (Greek *syn*, "together"; *apsis*, "loop, i.e., arch") refers to the single open structure (fenestra) in the temporal region of the skull (as opposed to the two openings of the diapsid skull, described in Chapters 10 and 19). This lower temporal fenestra, along with a few other skull

24.2 The Diversity of Non-Mammalian Synapsids

The two major groups of non-mammalian synapsids were pelycosaurs (Greek *pelyx*, "bowl"; *sauros*, "lizard") and therapsids (Greek *ther*, "wild beast") (**Table 24.1**). Pelycosaurs, the more basal of the two groups, were found mainly in the paleoequatorial latitudes of the Northern Hemisphere

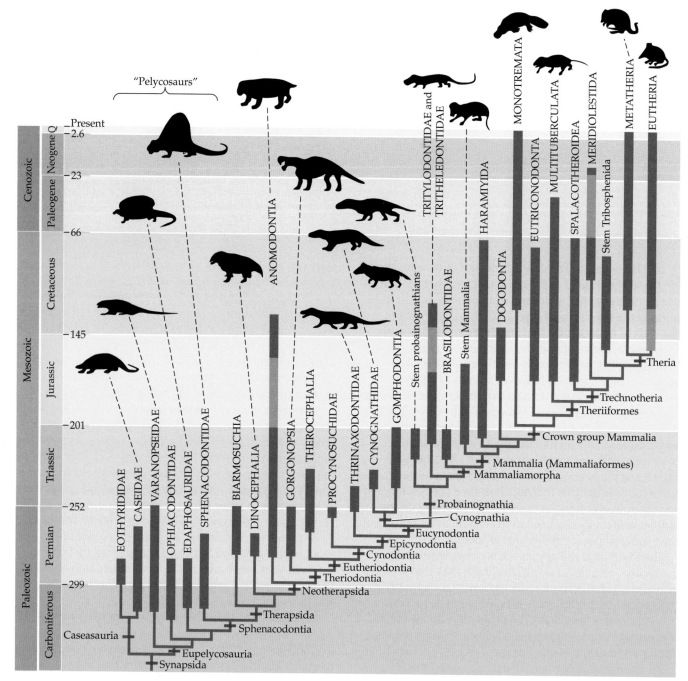

Figure 24.1 Phylogenetic relationships of Synapsida.
Probable relationships among the major groups of synapsids. Quotation marks indicate paraphyletic groups. Narrow lines show relationships only; they do not indicate times of divergence or the unrecorded presence of taxa in the fossil record. Lightly shaded bars indicate ranges of time when a taxon is believed to be present, because it is known from earlier and later times, but is not recorded in the fossil record during this interval. Only the best-corroborated relationships are shown.

(Laurasia) and were known predominantly from the Early Permian (**Figure 24.3A**). The more derived therapsids (**Figure 24.3B–E**) were found mainly in higher latitudes in the Southern Hemisphere (Gondwana). Therapsids ranged in age from the Middle Permian to the Early Cretaceous but

were known predominantly from the Late Permian to Early Triassic. Although basal synapsids were likely not endothermic, the presence of fibrolamellar bone in a diversity of fossil specimens, even in early pelycosaurs, might indicate that they had a rate of bone growth (and hence probably also a basal metabolic rate) greater than that of modern reptiles (but see Chapter 14).

Pelycosaurs: Basal non-mammalian synapsids

Pelycosaurs first appeared in the late Pennsylvanian (latest Carboniferous). They are popularly known as "sailbacks," although only a minority of them actually had sails. Pelycosaurs include the ancestors of the more derived synapsids,

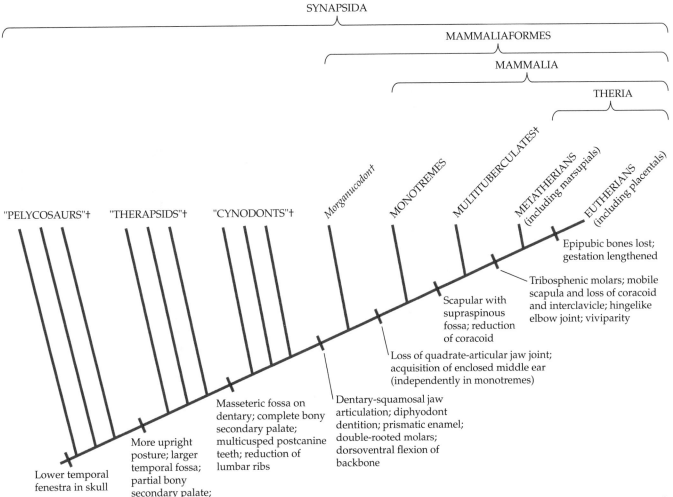

Figure 24.2 Simplified phylogenetic tree of synapsids.
Quotation marks indicate paraphyletic groups. Extinct lineages are indicated by a dagger (†).

including mammals; thus, they represent a paraphyletic assemblage. They were basically generalized amniotes, albeit with some unique specializations. There is no direct evidence about pelycosaur skin covering, but it is unlikely that they had scales like those of modern reptiles; the hard scales of modern reptiles are composed of beta-keratin, which is not produced by synapsids.

Pelycosaurs had long feet (long toes in particular) and a relatively sprawling limb posture (see Figures 24.3A and 24.4), but lateral flexion of their vertebral column was reduced. This specialization indicates that locomotion of the earliest synapsids was already somewhat more limb-based, and less reliant on axial flexion, than in the generalized amniote condition (as retained today in lizards, for example). They also retained a parietal foramen for the pineal eye, indicative of temperature regulation by behavioral means. This foramen is lost in derived cynodonts and mammals. Pelycosaurs were like other amniotes in possessing gastralia—bones in the ventral abdominal wall that help with respiration (see Chapter 14).

Many pelycosaurs were generalized carnivores, with teeth capable of piercing and ripping flesh. The herbivorous caseids (see Figure 24.3A) and edaphosaurids had blunt, peglike teeth and expanded rib cages, indicating that they had the large guts typical of herbivores. Their heads were small compared with their barrel-shaped bodies, indicating that they did not chew their food. (Extant mammalian herbivores, such as horses, have large heads with large teeth and jaw muscles, whereas the heads of herbivorous lizards, which do not chew their food, are no larger in proportion to their bodies than are the heads of carnivorous lizards.)

Sphenacodonts were the most derived pelycosaurs. They were mainly large carnivores with sharp teeth. An enlarged caninelike tooth in the maxillary bone and an arched palate (a forerunner of the later separation of the nasal passages from the mouth), plus a flange on the angular bone (the reflected lamina; see Section 24.3), are key features linking sphenacodonts to more derived synapsids.

Elongation of the neural spines of the trunk into the well-known pelycosaur sail was a remarkable feature of some edaphosaurids and sphenacodontids. The best-known pelycosaur is the sphenacodontid *Dimetrodon* (**Figure 24.4**), an animal frequently misidentified as a dinosaur. These sails evolved independently in sphenacodontids and edaphosaurids; we can infer this because sails (1) were absent

Table 24.1   Major groups of non-mammalian synapsids

Pelycosaurs

Eothyrididae: Basal pelycosaurs, small (cat size) and probably insectivorous. Early Permian of North America (e.g., *Eothyris*).

Caseidae: Large (pig size) herbivorous forms. Late Pennsylvanian (latest Carboniferous) to early Late Permian of North America and Europe (e.g., *Casea*, *Cotylorhynchus*; see Figure 24.3A).

Varanopidae: Generalized, medium-size forms. Late Pennsylvanian to Late Permian of North America, Europe, and South Africa (e.g., *Varanops*; see Figure 24.3A).

Ophiacodontidae: The earliest known pelycosaurs, although not the most basal. Medium size with long, slender heads, reflecting semi- aquatic fish-eating habits. Late Pennsylvanian to Early Permian of North America and Europe (e.g., *Archaeothyris*, *Ophiacodon*; see Figure 24.3A).

Edaphosauridae: Large herbivores, some with a sail. Late Pennsylvanian to Early Permian of North America and Europe (e.g., *Edaphosaurus*).

Sphenacodontidae: Large carnivores, some with a sail. Late Pennsylvanian to Middle Permian of North America and Europe (e.g., *Haptodus*, *Dimetrodon*; see Figure 24.4).

Noncynodont therapsids

Biarmosuchia: The most basal therapsids, medium-size (dog size) carnivores. Middle to Late Permian of Eastern Europe, South and East Africa (e.g., *Biarmosuchus*).

Dinocephalia: Medium to large (cow size) carnivores and herbivores. Some large herbivores had thickened skulls, possibly for headbutting in intraspecific combat. Middle Permian of Eastern Europe, East Asia, South Africa, and a couple of taxa from Brazil and Texas (e.g., *Titanophoneus*, *Moschops*; see Figure 24.3B).

Anomodontia: Small (rabbit size) to large herbivores, the most diverse of the therapsids. Includes the dicynodonts, which retained only the upper canines and substituted the rest of the dentition with a turtlelike horny beak. Includes burrowing and semiaquatic forms. Middle Permian to Late Triassic worldwide (including Antarctica) and Early Cretaceous of Australia (e.g., *Lystrosaurus*, *Dicynodon*; see Figure 24.6C).

Gorgonopsia: Medium to large carnivores. Middle to Late Permian of Eastern Europe and Southern Africa (e.g., *Lycaenops*, *Inostrancevia*, and *Sycosaurus*; see Figure 24.3C).

Therocephalia: Small to medium-size carnivores and insectivores. Similarities to cynodonts include a secondary palate, complex postcanine teeth, and evidence of nasal turbinate bones. Late Permian to mid-Triassic of Eastern Europe, East Asia, Antarctica, and South and East Africa (e.g., *Pristerognathus*).

Cynodont therapsids

Procynosuchidae: The most basal cynodonts, small to medium-size (rabbit size) carnivores and insectivores. Late Permian of Europe and South and East Africa (e.g., *Procynosuchus*).

Thrinaxodontidae: Medium-size (rabbit size) carnivores and insectivores. Latest Permian to Early Triassic of Eastern Europe, South Africa, South America, and Antarctica (e.g., *Thrinaxodon*; see Figure 24.3D).

Cynognathidae: Medium to large (dog size) carnivores. Early to Middle Triassic of Africa, East Asia, Antarctica, and South America (e.g., *Cynognathus*; see Figure 24.6E).

Gomphodontia: Medium-size (dog size) herbivores. Includes the diademodontids (e.g., *Diademodon*, Early Triassic of Africa, Antarctica, and Argentina), trirachodontids (e.g., *Trirachodon*, Early Triassic of Africa), and traversodontids (e.g., *Traversodon*, Middle to Late Triassic of South and East Africa, Madagascar, South America, India, eastern North American, and Europe).

Chiniquodontidae, Probainognathidae: Small to medium-sized carnivores and insectivores, forming a paraphyletic stem group to more derived cynodonts plus mammals. Middle to Late Triassic of North and South America and Madagascar (e.g., *Bonacynodon*; see Figure 24.3E).

Tritheledontidae (including Ictidosaura): Small to medium-size carnivores and insectivores. Approached mammalian condition in form of jaw joint and postcranial skeleton. Late Triassic of North and South America to Early Jurassic of South Africa (e.g., *Diarthrognathus*).

Tritylodontidae: Small, rather rodentlike omnivores and/or herbivores (e.g., *Tritylodon*, *Oligokyphus*). Previously included within the Gomphodontia, but now within the Probainognatha, close to the earliest mammals. They are known from the Late Triassic to Middle Jurassic of North America, Europe, East Asia, Antarctica, and South Africa, plus Early Cretaceous of Russia and Japan. Tritylodontids resembled the mammalian condition of the postcranial skeleton, but their skulls are rather different, and specialized for omnivory/herbivory.

Brasilodontidae: Extremely small (shrew-sized) forms from the Middle Triassic to Late Triassic of South America (e.g., *Brasilodon*; see Figure 24.6F). Almost indistinguishable from mammals, except for continued tooth replacement and lacking a well-developed ball and socket joint for the dentary-squamosal jaw articulation.

Note that large, medium, and small refer to size within a particular group. A medium-size noncynodont therapsid is much larger than a medium-size cynodont.

(A) Basal pelycosaurs

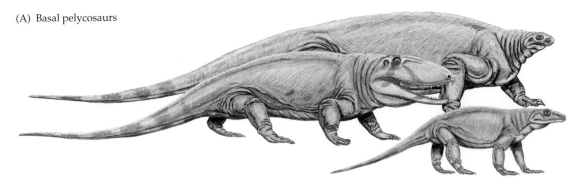

(B) Basal therapsid

(C) Derived therapsid

(D) Basal cynodont therapsid

(E) Derived cynodont therapsid (stem mammal)

Figure 24.3 Diversity of non-mammalian synapsids. (A) Three basal pelycosaur species. From back to front: the caseid *Cotylorhynchus* (herbivorous form); the ophiacodontid *Ophiacodon* (possibly a fish-eater); and the varanopsid *Varanops* (a generalized carnivore). *Cotylorhynchus* was about the size of a cow; the other two species are drawn to scale. (A more derived pelycosaur group is shown in Figure 24.4.) (B) A basal therapsid, the herbivorous dinocephalian *Moschops*. Also about the size of a cow, *Moschops* was one of the largest non-mammalian synapsids. Its thick skull has led to speculation that males engaged in head-butting. (C) A derived therapsid, the gorgonopsid *Sycosaurus*, was a carnivore about the size of a Labrador retriever. The hindlimbs were fairly upright, but the forelimbs retained a degree of the ancestral sprawling posture. (D) The basal cynodont therapsid *Thrinaxodon* was a generalized carnivore about the size of a fox. Basal cynodonts may have had some hair. (E) A derived cynodont therapsid (stem mammal), the probainognathid *Bonacynodon* was an insectivore about the size of a large rat. Various lines of evidence allow us to infer that this animal was furry. (A–C, Dmitry Bogdanov/CC BY 3.0; D, Nobu Tamura/CC BY-3.0; E, Jorge Blanco/CC BY-SA 4.0.)

from early members of both groups and (2) differed in detailed structure. Evidence of spines that had been broken and then healed suggest that a web of skin covered the spines. There is no evidence of sexual size dimorphism in pelycosaur sails, although those of males may have been more colorful than those of females.

It has long been proposed that such a large increase in the surface area of an animal would affect its heat exchange with the environment, and that pelycosaur sails were temperature-regulating devices, absorbing warmth from the sun's rays that could then be transferred to the body by blood vessels. However, detailed examination of the spines

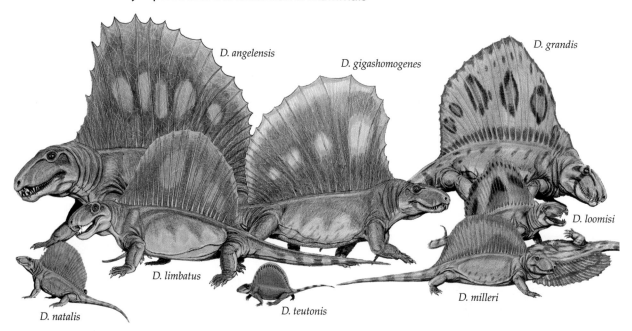

D. angelensis

D. gigashomogenes

D. grandis

D. loomisi

D. limbatus

D. natalis

D. teutonis

D. milleri

Figure 24.4 The sails of *Dimetrodon* were probably colorful. The sizes and shapes of *Dimetrodon* dorsal sails are accurately represented in this reconstruction. The colors and patterns are speculative, but it is likely that they were species-specific, as are the gular fans of extant lizards described in Chapter 17. These species of *Dimetrodon* ranged in size from 60 cm (*D. teutonis*) to 4.6 m (*D. angelensis*). (Dmitry Bogdanov/CC BY-SA 4.0.)

suggests that blood flow in the sails was not sufficient to transport heat to the body, and thus the function of the sails remains obscure.

Therapsids: More derived non-mammalian synapsids

A flourishing fauna of more derived synapsids, grouped under the general name therapsids, extended from the Middle Permian to the Early Cretaceous. Body impressions of a basal therapsid, the dinocephalian (Greek *deinos*, "terrible"; *kephale*, "head") *Estemmenosuchus* from the Middle Permian of Russia, indicate a leathery skin containing glands but with no evidence of hair. But probable hairs have been found in Late Permian coprolites (fossilized feces) from Russia and South Africa that can have come only from the more derived therapsids present at the time (although it is impossible to tell if these animals had a full furry coat). Therapsids were all fairly heavy-bodied, large-headed, stumpy-legged beasts. The image of this body form, combined with incipient mammal-like hairiness, prompted one cartoonist to declare therapsids "too ugly to survive" (**Figure 24.5**).

The earliest therapsids had various novel anatomical features that are probably related to a higher metabolic rate. They had a larger temporal fenestra, longer upper canines, and their dentition was differentiated into incisors, canines, and postcanine teeth (**Figure 24.6**). The choanae (internal nostrils) were enlarged, and a trough in the roof of the mouth (possibly covered by a soft-tissue secondary palate in life) indicates the evolution of a dedicated airway passage separate from the rest of the oral cavity, which started to be

emphasized with a partial bony separation (i.e., a secondary palate) in more derived therapsids. The entire skull was more rigid, the head was more mobile on the neck, and the neck itself was more flexible.

Therapsid limbs were held more underneath the body, resulting in a more upright posture, and the vertebrae show that little or no lateral undulation was possible; this indicates the decoupling of locomotion from respiration (see Figure 14.1), suggesting a greater level of activity than in pelycosaurs. Most therapsids lost the gastralia (seen in pelycosaurs) that are involved in respiration. This was probably correlated with their more upright posture, and with a shift to a more mammal-like mode of respiration, relying more on the ribs for lung ventilation.

Therapsids evinced a switch to a more mammal-like organization of their hindlimb muscles. Pelycosaurs retained large processes on their anterior caudal (tail) vertebrae,

Figure 24.5 Therapsids: Too ugly to survive? Cartoonist/educator Larry Gonick's classic characterization of therapsids. (From Larry Gonick, *Cartoon History of the Universe*. All rights reserved.)

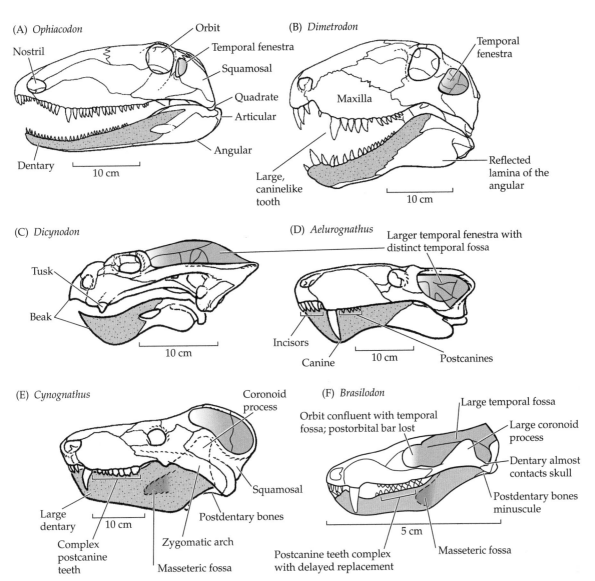

Figure 24.6 Skulls of non-mammalian synapsids. Note the increasing size of the temporal fenestra and the coronoid process of the dentary. The temporal fenestra is the opening in the skull, and the temporal fossa is a depression on the skull exposed by the temporal fenestra where the jaw muscles originate. (A) The ophiacodontid pelycosaur *Ophiacodon* may have been a semiaquatic fish eater. (B) The sphenacodontid pelycosaur *Dimetrodon* was a terrestrial predator. (C) The dicynodont therapsid *Dicynodon* was herbivorous. (D) The carnivorous gorgonopsid therapsid *Aelurognathus*. (E) The cynodont *Cynognathus*. (F) The brasilodontid cynodont *Brasilodon* was an insectivore. (A–E from Romer 1933; F from Bonaparte et al. 2005.)

reflecting retention of the caudofemoralis muscle seen in reptiles today (see Figure 24.7A). Therapsids reduced those processes but had a more expanded ilium (with an increase in the number of sacral vertebrae to three) and a greater trochanter on the femur, indicative of a switch to the predominance of the mammalian gluteal muscles (the ones that give mammals their rounded rear ends). Along with these changes in posture and hindlimb musculature, the tail of therapsids was now shorter and there was a prominent intratarsal articulation between the astragalus and calcaneum in the ankle, allowing the foot to be more turned in to effect a more upright gait (**Figure 24.7**).

Therapsid feet (especially the toes) were now shorter, reflecting the use of the feet as levers in limb-based locomotion rather than as holdfasts in axial-based locomotion; see Figure 24.7B. All tetrapods have only two phalanges in the first digit (thumb or big toe), but the primitive amniote condition (seen in pelycosaurs and retained today in lizards) is for three to five phalanges in the other digits, with digit 4 usually being the longest (i.e., a phalangeal formula of 2:3:4:5:3). Mammals have only three phalanges in digits 2–5 (we can see in ourselves the phalangeal formula of 2:3:3:3:3), and this condition was essentially reached in the earliest therapsids.

The pectoral and pelvic girdles of therapsids were less massive than those of pelycosaurs, especially in the more ventral elements (the clavicle and interclavicle in the pectoral girdle and the pubis and ischium in the pelvic girdle); the limbs were more slender; and the shoulder joint appears to have allowed more freedom in movement of the forelimbs, which would have made a longer stride possible.

(A) *Haptodus* (pelycosaur)

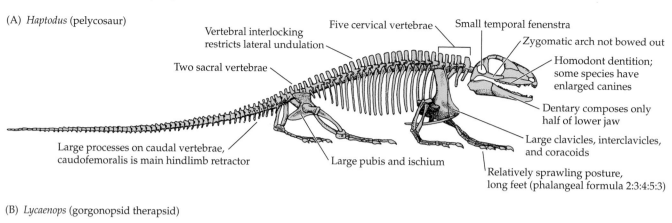

Vertebral interlocking restricts lateral undulation

Five cervical vertebrae

Small temporal fenenstra

Zygomatic arch not bowed out

Two sacral vertebrae

Homodont dentition; some species have enlarged canines

Dentary composes only half of lower jaw

Large processes on caudal vertebrae, caudofemoralis is main hindlimb retractor

Large pubis and ischium

Large clavicles, interclavicles, and coracoids

Relatively sprawling posture, long feet (phalangeal formula 2:3:4:5:3)

(B) *Lycaenops* (gorgonopsid therapsid)

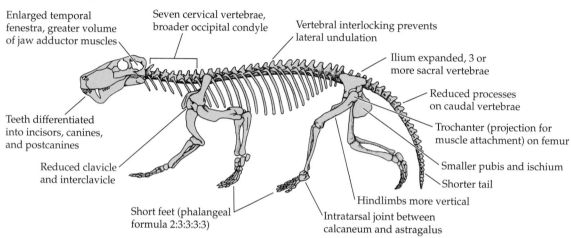

Enlarged temporal fenestra, greater volume of jaw adductor muscles

Seven cervical vertebrae, broader occipital condyle

Vertebral interlocking prevents lateral undulation

Ilium expanded, 3 or more sacral vertebrae

Reduced processes on caudal vertebrae

Trochanter (projection for muscle attachment) on femur

Teeth differentiated into incisors, canines, and postcanines

Reduced clavicle and interclavicle

Smaller pubis and ischium

Shorter tail

Short feet (phalangeal formula 2:3:3:3:3)

Hindlimbs more vertical

Intratarsal joint between calcaneum and astragalus

(C) *Thrinaxdon* (cynodont therapsid)

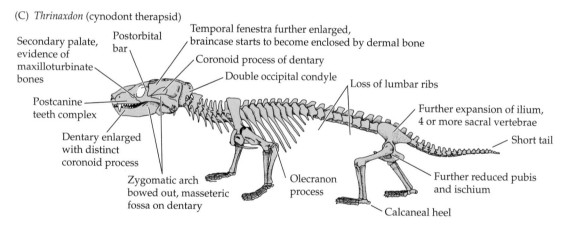

Secondary palate, evidence of maxilloturbinate bones

Postorbital bar

Temporal fenestra further enlarged, braincase starts to become enclosed by dermal bone

Coronoid process of dentary

Double occipital condyle

Loss of lumbar ribs

Further expansion of ilium, 4 or more sacral vertebrae

Short tail

Postcanine teeth complex

Dentary enlarged with distinct coronoid process

Zygomatic arch bowed out, masseteric fossa on dentary

Olecranon process

Further reduced pubis and ischium

Calcaneal heel

(D) *Megazostrodon* (early mammaliaform)

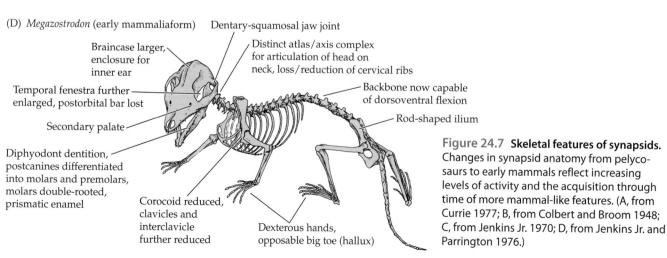

Dentary-squamosal jaw joint

Braincase larger, enclosure for inner ear

Distinct atlas/axis complex for articulation of head on neck, loss/reduction of cervical ribs

Backbone now capable of dorsoventral flexion

Temporal fenestra further enlarged, postorbital bar lost

Rod-shaped ilium

Secondary palate

Diphyodont dentition, postcanines differentiated into molars and premolars, molars double-rooted, prismatic enamel

Corocoid reduced, clavicles and interclavicle further reduced

Dexterous hands, opposable big toe (hallux)

Figure 24.7 Skeletal features of synapsids. Changes in synapsid anatomy from pelycosaurs to early mammals reflect increasing levels of activity and the acquisition through time of more mammal-like features. (A, from Currie 1977; B, from Colbert and Broom 1948; C, from Jenkins Jr. 1970; D, from Jenkins Jr. and Parrington 1976.)

Therapsid diversity

Therapsids first appeared about 267 Ma, in the Middle Permian. While pelycosaurs, with their food base in aquatic ecosystems, are known mainly from Early Permian equatorial regions that would have had large amounts of rainfall, therapsids apparently originated in a different environment. They may have evolved in drier tropical regions and adopted a food base of terrestrial plants. Pelycosaurs became extinct as sea levels rose at the end of the Early Permian and greatly reduced the area of moist tropical habitat, but therapsids were able to move into more temperate zones in the later Permian.

Like pelycosaurs, therapsids were generalized terrestrial forms that radiated into herbivorous and carnivorous forms and a variety of sizes. One of the numerous lineages of therapsids—the cynodonts—gave rise to mammals, but other therapsid lineages could be considered equally derived in their own fashions. Many of the names given to therapsid lineages reflect their mammal-like characters (**Table 24.2**).

Therapsids were the dominant large land mammals in the Late Permian, sharing center stage only with the large herbivorous parareptilian pareiasaurs. Among the more specialized forms were the herbivorous anomodonts (including the specialized dicynodonts) and the carnivorous theriodonts (gorgonopsids, therocephalians, and cynodonts); members of both lineages survived into the Triassic. One early anomodont (*Suminia*, from the Late Permian of Russia) had long fingers and toes and was probably arboreal (the first known arboreal tetrapod). Another—*Tiarajudens* from the Middle Permian of Brazil—had huge, elongated saberlike teeth, possibly used for display as in the males of some extant herbivorous mammals, such as mouse deer (*Tragulus*; see Figure 25.6G). Saberlike canines were also seen in the gorgonopsid *Inostrancevia* from the Late Permian of Russia; in this case the teeth were undoubtedly related to the animal's carnivorous lifestyle. The therocephalian *Euchambersia* had a grooved canine combined with a pocket in the maxilla (the bone in the snout housing the canine); CT scanning has supported the interpretation of these structures as being a snakelike venom-delivery system.

Dicynodonts were the most abundant large terrestrial animals in the Late Permian, and the only lineage in addition to cynodonts to survive the end-Permian extinction.

Their post-extinction survival and radiation might be attributable in part to the fact that their bones show extremely high growth rates, reducing the juvenile period when they would have been vulnerable to predation, and increasing the likelihood that they would have been able to withstand environmental perturbation.

Dicynodonts diversified into several different ecological types, including long-bodied burrowers and probable semiaquatic forms such as *Lystrosaurus*, an animal that dominated the earliest Triassic landscapes. The specializations of dicynodont skulls for herbivory included the loss of the marginal teeth (with the retention of tusklike canines; see Figure 24.6C). The toothless jaws were covered with a turtlelike horny beak, which cut vegetation via fore-and-aft motions of the lower jaw. Dicynodonts disappeared at the end of the Triassic, but a possible relict form is known from the late Early Cretaceous of Australia.

Eutheriodonts (therocephalians and cynodonts) were the major predators of the Late Permian and Early Triassic. Their jaws were characterized by the development of the coronoid process of the dentary which provided additional area for the insertion of the jaw-closing muscles, as well as a lever arm for their action (see Figure 24.6E,F). The temporal fossa in the skull was correspondingly enlarged for the origin of these muscles. Eutheriodonts also showed evidence of maxilloturbinate bones, also known as respiratory turbinates—scroll-like bones in the nasal passages that in extant mammals warm and humidify the incoming air and reduce respiratory water and heat loss. (Maxilloturbinates also apparently evolved independently in dicynodonts.)

Only the dicynodonts, cynodonts, and a few therocephalians survived the tumult of the end-Permian extinction; the first two groups diversified in the Early Triassic. But other vertebrate groups also diversified in the Triassic, most notably the diapsid reptiles that gave rise to the dinosaurs. Therapsids became an increasingly minor component of the terrestrial fauna during the Triassic and were near extinction at its end.

Cynodonts appeared in the Late Permian and had their heyday in the Triassic; non-mammalian cynodonts were mostly extinct by the end of that period. However, a few species from the tritylodont and tritheledont lineages persisted into the Jurassic and, after an apparent disappearance in the Late Jurassic, tritylodonts made a brief reappearance in the Early Cretaceous. More basal cynodonts like *Thrinaxodon* (see Figure 24.3D) flourished in the Early Triassic, but by the Late Triassic the more derived cynodonts (Eucynodontia) split into two major lineages, Cynognathia and Probainognathia. Cynognathians were the predominant forms in the Middle Triassic and had a high rate of diversification, including some large carnivorous

Table 24.2	Therapsids have descriptive names derived from Greek roots	
Group	**Meaning**	**Greek roots**
Anomodont	"abnormal tooth"	*anomo* = strange, *odous* = tooth
Cynodont, dicynodont	"dog tooth," "two dog teeth"	*kynos* = dog, *di* = two
Eucynodont	"true dog teeth"	*eu* = true, good
Gomphodont	"club teeth" (i.e., large, heavy teeth)	*gomph* = a club
Gorgonopsid	"looks like a Gorgon"	*opsis* = appearance
Theriodont	"mammal teeth"	*ther* = a wild beast
Therocephalian	"mammal head"	*kephale* = head

forms such as *Cynognathus* (Figure 24.6E), as well as a variety of medium-size to large herbivorous forms with expanded postcanine teeth with blunt cusps that are generally lumped together as "gomphodonts" (see Table 24.1). Probainognathians (including the carnivorous and insectivorous probainognathids [Figure 24.3E], chiniquodontids, and tritheledontids, the herbivorous tritylodontids, and the tiny, insectivorous brasilodontids) were smaller and less diverse than cynognathians. Probainognathians came to prominence in the Late Triassic, and it was this lineage that eventually gave rise to mammals.

Cynodonts had a variety of derived features, making them more mammal-like than other therapsids (see Figure 24.7C). The dentary was much enlarged, and the postdentary bones were reduced in size. The dentary now had a depression on its outer side, the masseteric fossa, which indicates the presence of a mammal-like **masseter muscle**, a major muscle on the outside of the jaw that in mammals operates to close the jaw (see Figures 24.6E,F and 24.8B,C). The bones that form the lower border of the synapsid temporal opening are bowed out into a **zygomatic arch** (the bony bar you can feel just below your eyes, commonly referred to as the cheekbone), which accommodates the masseter.

All cynodonts had multicusped cheek teeth (i.e., teeth with small accessory cusps anterior and posterior to the main cusp), with evidence of some occlusion (contact between teeth during biting), suggesting a more precise and more powerful bite. Multicusped teeth allow enhanced food manipulation, and they are also resistant to fracture. There was now a complete bony secondary palate.

The most derived cynodonts had an enlarged infraorbital foramen—the hole under the eye through which the sensory nerves from the snout pass back to the brain (see Figure 24.12A)—and paleoneurological studies show this to have been mammal-like in its condition. This implies a highly innervated face, perhaps indicative of a mobile, sensitive muzzle with lips and whiskers.

Cynodonts had a vertebral column differentiated into an anterior thoracic region (possessing ribs) and a posterior lumbar region (lacking ribs). The cynodont hindlimbs were now almost fully mammal-like, with a greater expansion of the iliac blade, further reduction of the pubis and ischium, development of a calcaneal heel in the ankle for the insertion of the Achilles tendon of the calf muscles, and development of a mammalian-like crurotarsal ankle joint between the proximal tarsals (primarily the astragalus) and the lower limb (primarily the tibia: see Chapter 14).

Cynodont evolution was characterized by a reduction in body size. Several Middle Triassic lineages of cynodonts were independently evolving smaller body size and, along with this, attaining various mammal-like features related to this miniaturization. Some early cynodonts were the size of large dogs, but by the Middle Triassic most were only about the size of a rabbit. The sister group to mammals, the brasilodontid cynodonts, attained the very small size of the earliest Mammaliaformes (about the size of a shrew) by the Late Triassic.

24.3 Evolutionary Trends in Synapsids

As discussed in Chapter 14, the synapsid lineage crossed a physiological boundary as the animals moved from ectothermy to endothermy, and this change was accompanied by changes in ecology and behavior. Physiology, ecology, and behavior do not fossilize directly, but some of the changes that were occurring in metabolism, ventilation, and locomotion can be traced indirectly through changes in the skull and the postcranial skeleton. Evidence from bone oxygen isotopic composition indicates that at least a degree of endothermy had evolved by the level of eucynodonts (i.e., cynodonts above the level of *Thrinaxodon*; see Figure 24.1), and bone microanatomy has been used to infer convergent degree of endothermy in derived dicynodonts (although it is possible that this higher metabolism was inherited from a common ancestor at the base of Neotherapsida). Maxilloturbinate bones, formerly considered indicative of endothermy, evolved convergently in the cynodont and dicynodont lineages. Review Figure 24.7 as we consider those changes in anatomy in the therapsid lineage that may reflect changes in metabolic rate and the evolution of endothermy:

- *Size of the temporal fenestra.* A larger fenestra (the opening in the dermal skull roof) accompanied by an increasing tendency to enclose the braincase with dermal bone indicates a greater volume of jaw musculature and implies more food eaten per day. In mammals and highly derived cynodonts, loss of the postorbital bar reflects a further enlargement in jaw muscles.

- *Condition of the lower temporal bar.* A distinct zygomatic arch bowed out from the skull in the region of the orbit, accompanied by a depression (the masseteric fossa; see Figure 24.6) on the dentary, indicates the presence of a masseter muscle originating from this bony bar and inserting on the lower jaw (**Figure 24.8**), again implying more effective food processing in cynodonts compared with other non-mammalian synapsids.

- *Lower jaw and jaw joint.* Changes here reflect an increasing compromise between food processing and hearing in synapsids, as we explain below. In the lower jaw of pelycosaurs, the tooth-bearing portion, the dentary, took up only about half of the jaw. By the level of cynodonts, the dentary had greatly expanded and the postdentary bones had been reduced in size (see Figure 24.6). In mammals, the dentary formed a new jaw joint with the skull.

- *Teeth.* Greater specialization of the dentition reflects an increased emphasis on food processing. The teeth of pelycosaurs were **homodont**, with no regionalization of size or function, except that some species had enlarged canines. In more derived synapsids, the teeth became increasingly **heterodont** (differentiated in size, form, and function; see Figure 24.6), although actual mastication with precise occlusion of the cheek teeth

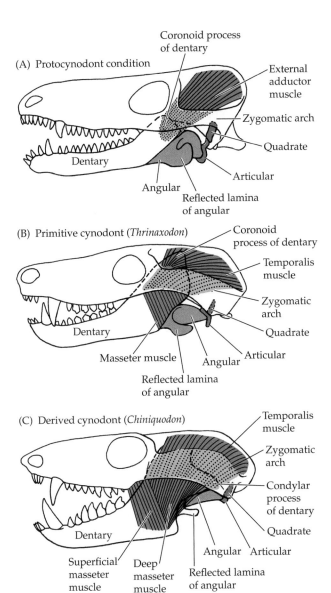

Figure 24.8 Changes in jaw muscles of cynodonts. The single external jaw adductor muscle in the protocynodont condition became differentiated into two muscles. The masseter (the new portion) originated on the zygomatic arch and inserted on the dentary, and the temporalis (in the position of the original adductor muscle) originated on the side of the cranium (the temporal fossa) and inserted on the coronoid process of the dentary. (A) The protocynodont condition with a single muscle; the coronoid process is small. (B) In a primitive cynodont, the temporalis muscle pulled the dentary backward as well as upward. A relatively small masseter was present with anteriorly oriented fibers. On contraction it would have pulled the dentary forward as well as upward, possibly balancing the forces acting on the jaw articulation. (C) In a derived cynodont (eucynodont), the deep portion of the masseter muscle pulled the jaw upward, while the superficial portion pulled it forward. (From Kemp 1982.)

is a mammalian feature. Heterodont dentition reflects a greater degree of oral food processing, indicating a higher metabolic rate. Mammals have a maximum of two sets of teeth during their lifetime, a condition called **diphyodonty**. Mammals also have double-root-ed molars and prismatic enamel capping their teeth, a feature shared with derived cynodonts.

- *Development of a bony secondary palate.* No bony secondary palate was apparent in pelycosaurs. An incipient, incomplete one was present in some noncynodont therapsids, and a complete one is present in derived cynodonts and in mammals. The bony palate helps bolster the skull against stress during food processing, and also allows the animal to eat and breathe at the same time. Both functions indicate an increase in food processing, and hence a higher metabolic rate. Most mammals have merged the originally double nasal opening into a single median one, probably reflecting an increase in the size of the nasal passages and a higher rate of ventilation.

- *Form and position of the limbs.* An upright posture indicates a higher level of activity than the sprawling posture of pelcyosaurs. All therapsids showed some degree of development of an upright posture, and the short toes of therapsids and mammals indicate that the feet are used as levers rather than holdfasts. Changes in the ankle joint in basal therapsids and cynodonts indicate increasing use of the hindfoot as a lever, while mammals have more dexterous hands and feet. Two bony lever arms for muscles associated with locomotion, a distinct olecranon process of the ulna in the forelimb and a calcaneal heel in the hindlimb, first appeared in cynodonts.

- *Shape of the limb girdles.* With the more upright posture seen in therapsids, more of the body weight was supported by the limbs, and the ventral portions of the limb girdles (pubis and ischium) could be reduced. Therapsids also showed an increase in the number of sacral vertebrae, indicating greater transference of thrust from the hindlegs. Mammals have a very reduced pubis and a rod-shaped ilium, as did the most derived cynodonts.

- *Form of the vertebral column.* The reduction of the lumbar ribs in cynodonts indicates the presence of a muscular diaphragm, evidence of a higher rate of respiration (see Chapter 14). Forward extension of the prezygapophyses of pelycosaurs and therapsids limited lateral undulation of the body, and dorsoventral flexion of the lumbar spine first appeared in mammals. Shortening of the tail, loss of the ventral processes on the caudal vertebrae, and the appearance of a trochanter on the femur show the increasing importance of the gluteal muscles for hindlimb retraction.

- *Braincase and promontorium.* A distinctive feature of mammals is the larger braincase and a distinct protrusion—the promontorium—on the petrosal bone that encloses the inner ear. The promontorium houses the cochlea.

These changes suggest that synapsids were progressively increasing locomotor speed and endurance, rates of lung ventilation, and energy intake and efficiency of oral processing of food. These changes would have been accompanied by increases in metabolic rate and, consequently, greater internal heat production. In short, synapsids were shifting from a basal amniote morphology, physiology, ecology, and behavior to progressively more mammal-like conditions.

Cynodonts have often been considered as possible endotherms based on their anatomy. Support for this notion now comes from evidence from their bone microanatomy. Cynodonts (but not other non-mammalian synapsids) resembled mammals in having small, densely packed vascular canals in their bones, suggesting that they, like mammals, had small red blood cells, probably lacking nuclei. Smaller red blood cells, which are also seen in birds (although they retain their nuclei), are associated with endothermy because they allow more rapid gas exchange.

Remember that evolution is not goal-oriented; changes toward a more mammal-like condition proceeded by a series of small steps. Presumably, each small change was advantageous for the animal at that time. All non-mammalian synapsids were well-adapted animals in their own right, not merely evolutionary stages on the road to full "mammalhood." It is only in retrospect that we can detect a pattern extending back from the modern condition to the ancestral state.

Evolution of the diaphragm

The loss of lumbar ribs in cynodonts is considered to mark the development of a muscular diaphragm. A diaphragm resolved the conflict between the roles of costal muscles in both locomotion and breathing, as we discussed in Chapter 14. The presence of a diaphragm combined with the shift to an increasingly fore-and-aft movement of limbs seen in cynodonts probably signals increased locomotor speed and endurance. Those morphological changes would have been supported by higher aerobic metabolic rates, which would have contributed to an increase in endothermal heat production.

Other mammalian functions also rely on the presence of a diaphragm. Coughing, vomiting, and defecating depend on the ability of a diaphragm to increase intra-abdominal pressure, as does parturition (giving birth), which appeared later in synapsid evolution.

Evolution of a double occipital condyle

Cynodonts developed the mammalian feature of a double occipital condyle (as opposed to the single ventral condyle seen in other amniotes; **Figure 24.9A**), allowing the head to move up-and-down on the neck (i.e., nodding), and they also developed rotational motion (shaking of the head) between the atlas and axis (the two most anterior cervical vertebrae; **Figure 24.9B**). Cynodonts probably had a range of head motions similar to those of mammals, although it was not until the most derived cynodonts that a more truly mammalian atlas/axis complex was apparent, with a projection of the axis (the dens process) that fits into the base of the atlas.

Evolution of jaws and ears

Extant mammals have a single bone in the lower jaw, the dentary, and a three-boned middle ear. The mammalian ear is more sensitive to high-frequency sounds than is the ear of other tetrapods, being able to process sound vibrations above 10 kHz. In the original synapsid condition, as in other tetrapods and in bony fishes, a tooth-bearing dentary formed the anterior half of the jaw, with a variety of bones (known collectively as postdentary bones) forming the posterior half. Jaw articulation was via the articular bone in the lower jaw and the quadrate bone (articulating with the squamosal) in the skull.

In cynodont evolution, there was a progressive enlargement in the size of the dentary and a decrease in the size of the postdentary bones and in the quadrate and the stapes (the last being the former hyomandibula, which had an articulation with the quadrate, as in the generalized gnathostome condition). In the more derived cynodonts, a condylar process of the dentary grew backward and eventually contacted the squamosal bone of the skull, and a corresponding articular depression (the glenoid) formed in the squamosal. In the cynodonts most closely related to mammals, this contact between the dentary and the squamosal formed a new jaw joint, the dentary-squamosal jaw joint. This new jaw joint coexisted for a while with the old one, but in later mammals the dentary-squamosal jaw joint was the sole one. The bones forming the old jaw joint were now the small bones (ossicles) of the middle ear: the articular became the malleus, the quadrate became the incus, and they joined with the stapes to form the distinctive three-boned mammalian middle ear. Figure 24.9A shows the difference in the position of the jaw between pelycosaurs and mammals.

The facts of this transition have been known for a century or so, but the evolutionary interpretation of them has changed over time. Almost two centuries ago, embryological studies demonstrated that the malleus and incus of the mammalian middle ear are homologous with the articular and quadrate bones forming the jaw joint of other gnathostomes. Originally it was assumed that the mammal ancestors had a lizardlike middle ear (considered to be the basal tetrapod condition), with the stapes alone forming an auditory ossicle that could transmit vibrations from the eardrum to the inner ear. There was also a tacit assumption that the mammalian condition of both ear bones and jaw articulation was inherently superior to that of other tetrapods, and that when the new jaw joint was formed, an originally single-boned middle ear was transformed into a three-boned one using some leftover jawbones.

On reconsideration, there are some problems with this traditional story. For a start, we now have good evidence that an enclosed middle ear evolved separately in modern amphibians and amniotes and probably at least three times convergently within amniotes. Genomic evidence also shows that the eardrum and middle ear cavity are not homologous between mammals and ther amniotes. Even if the mammalian type of jaw joint was indeed superior, how could researchers explain the millions of years of mammal ancestors' evolution during which the dentary was enlarging prior to contacting the skull? (Recall that evolution has no foresight.) Is the non-mammalian jaw articulation inherently inferior? An articular-quadrate joint appears to

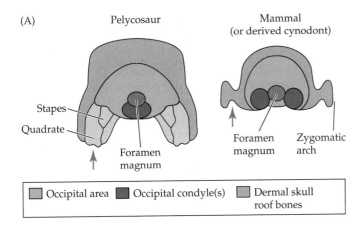

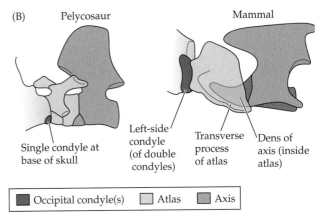

(A)

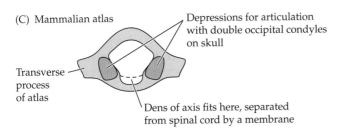

Figure 24.9 **Modifications in articulation of the synapsid skull.** The pelycosaur represents the general amniote condition, the mammal the derived synapsid condition. (A) Diagrammatic views of the occiputs of synapsid skulls. Pelycosaurs had a single occipital condyle, whereas mammals have a double occipital condyle (also the condition in cynodonts). The foramen magnum is the opening through which the spinal cord exits the skull. Red arrows shows the position of the articulation of the skull with the lower jaw. (B) Diagrammatic lateral left views of the atlas/axis complex. The pelycosaur shows a multipart atlas, with little movement between the atlas and axis. The mammal (derived cynodonts approached this condition) shows a single atlas and an axis with a process (the dens) that fits inside the base of the atlas, allowing for rotation between the two vertebrae. (C) Diagrammatic view of a mammalian axis, seen from the front.

work perfectly well in other vertebrates—no one has ever accused *Tyrannosaurus rex* of having an inferior jaw articulation! And why disrupt a perfectly functional middle ear to insert some extra bones that happened to be available? Even if a three-boned middle ear had superior acuity, there would have been a period of adjustment to a new condition that was probably less effective than the earlier condition.

As more fossils of non-mammalian synapsids were discovered, the trajectory of the evolution of the middle ear could be traced from its beginning in basal synapsids through therapsids to early mammals. In all gnathostomes, the stapes articulates with the otic capsule close to the housing of the inner ear. If this bone can vibrate, it will transmit sound waves; thus, making this bone lighter so that it vibrates will improve sound conduction. Pelycosaurs retained the basal amniote condition of a large quadrate firmly attached to the squamosal, and a large stapes that played a mechanical role in bracing at the back of the skull (see Figure 24.9A). At best, pelycosaurs may have been able to hear some low-frequency sound via bone conduction. Derived sphenacodontids had a flange of bone called the reflected lamina on the angular bone of the lower jaw (see Figures 24.6B and 24.8). This structure probably originated as the insertion for an enlarged and more elaborated pterygoideus muscle (one of the jaw-closing muscles). However, the reflected lamina became smaller and distinctly curved in derived therapsids, and eventually became the structure holding the eardrum (a homology that we can trace in mammalian development today; see Figure 24.10).

In therapsids, the stapes was reduced in size, the attachment of the quadrate to the squamosal became more mobile, and the reflected lamina became larger and thinner. This anatomy suggests that the chain of bones that make up the

middle ear of mammals was now able to have a role in sound conduction, although the relatively large size of these bones would have restricted hearing to low-frequency sound.

Using the same set of bones for two different functions—as a jaw joint and as a hearing device—might seem like a rather clumsy, makeshift arrangement, but there are many compromises in evolution; it's only possible to tinker with what is present, not to redesign structures from scratch. This trade-off between hearing and jaw function would probably have been adequate for early synapsids, because they were most likely ectotherms with low rates of food intake. But more derived synapsids with higher metabolic rates would have required greater food intake, and greater stress on the jaw joint would have interfered with the role of these bones in hearing. Thus, the evolutionary history of the jaw in synapsids has been interpreted as representing a conflict between feeding and hearing. It is also likely that, as long as the articular and quadrate bones formed the primary jaw joint, no membranous eardrum would have been present. An eardrum functions when there is an enclosed, air-filled, middle-ear cavity, which would be difficult to sustain around a working jaw joint. The reflected lamina may have acted as a type of bony eardrum or provided support for connective tissue that helped transmit low-frequency sounds to the ear ossicles and thus to the inner ear.

A depression on the outside of the dentary (the masseteric fossa) shows that the masseter muscle was first apparent in cynodonts (see Figure 24.6). In mammals, the masseter can move the lower jaw laterally and can be thought of

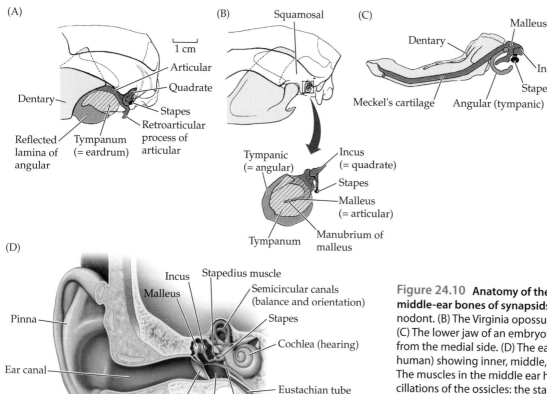

(A)

Articular
Quadrate
Dentary
Stapes
Retroarticular process of articular
Reflected lamina of angular
Tympanum (= eardrum)

1 cm

(B)

Squamosal

Tympanic (= angular)
Incus (= quadrate)
Stapes
Malleus (= articular)
Tympanum
Manubrium of malleus

(C)

Malleus
Dentary
Incus
Stapes
Meckel's cartilage
Angular (tympanic)

(D)

Incus
Malleus
Stapedius muscle
Semicircular canals (balance and orientation)
Stapes
Cochlea (hearing)
Pinna
Eustachian tube (connects middle ear cavity to pharynx)
Ear canal
Tympanum (eardrum)
Middle ear cavity
Tensor tympani muscles

Figure 24.10 Anatomy of the back of the skull and middle-ear bones of synapsids. (A) *Thrinaxodon*, a cynodont. (B) The Virginia opossum (*Didelphis virginiana*). (C) The lower jaw of an embryonic mammal, viewed from the medial side. (D) The ear of a derived mammal (a human) showing inner, middle, and external ear regions. The muscles in the middle ear help to dampen the oscillations of the ossicles: the stapedius is innervated by cranial nerve VII and the tensor tympani by cranial nerve V, showing how the middle ear bones are derived from two different embryonic sources (pharyngeal arches; see Chapter 2). (A–C, from Crompton and Jenkins Jr. 1979; D, from Sadava et al. 2017.)

as holding the lower jaw in a supportive sling. Its original function may have been relieving stresses at the jaw joint by this slinglike action (it does not appear that cynodonts could move the jaw sideways to chew, as mammals do), thereby helping resolve the conflict between jaw use and hearing.

In cynodonts the dentary bone was enlarged, accommodating all of the jaw muscle insertions. This anatomy is difficult to explain in terms of jaw mechanics, but it would have had the additional effect of reducing stresses on the postdentary bones, allowing them to become smaller and more able to vibrate to conduct sound. The cynodont jaw joint was originally interpreted as being "weak" because the postdentary bones appeared to be loose and wobbly, but a tighter articulation of these bones (as seen in other tetrapods) would have compromised their role as vibrating auditory ossicles. (Remember that every organism must be functional in its own right.)

In more derived cynodonts (Eucynodontia), the dentary became more enlarged than in more basal forms such as *Thrinaxodon*; the dentary now had a posterior condylar process, and the postdentary bones became progressively smaller. The ability to hear higher-frequency sound would now have been possible.

The transition from the non-mammalian to the mammalian condition can be visualized by comparing the back of the skull of *Thrinaxodon*, a fairly generalized cynodont (**Figure 24.10A**), with that of the Virginia opossum (*Didelphis virginiana*; **Figure 24.10B**). The lower jaw of an embryonic mammal (**Figure 29.10C**) shows that the tympanic (= angular, supporting the tympanum [eardrum]), malleus (= articular), and incus (= quadrate) develop in the same positions as seen in the adult cynodont skull, and the ancestral jaw joint persists as the articulation between the malleus and the incus. In the mammalian embryonic condition, the relative sizes of these bones in relation to the dentary are the same as in the cynodont, but in the mammal these bones do not increase in size, so in the adult they are relatively tiny. The mammalian manubrium of the malleus (see Figure 24.10B) is a new feature that also provides support for the tympanum.

A classic example of an evolutionary intermediate is provided by the tritheledontid cynodont *Diarthrognathus*. This animal gets its name from its double jaw joint (Greek *dis*, "two"; *arthron*, "joint"; *gnathos*, "jaw"); it has both the ancestral articular-quadrate joint and, next to it, the mammalian dentary-squamosal articulation (although a distinct dentary condyle, versus an incipient one, is not seen until the first true mammals). The postdentary bones are still retained in a groove in the lower jaw, and an air-filled middle-ear cavity may have been present in this highly derived cynodont. The earliest mammals retained the derived cynodont condition of the middle ear bones attached to the dentary in the lower jaw, which is where they start off in the development of modern mammals. All adult extant mammals have the middle-ear bones contained within the middle-ear cavity and completely isolated from the lower jaw (the definitive mammalian middle ear, or DMME), an arrangement that improves sensitivity to airborne sound. Developmentally, this isolation involves the loss of the embryonic Meckel's cartilage (the original gnathostome lower jawbone) and separation of the middle-ear bones from the dentary.

Many Mesozoic mammals had a condition called the partial mammalian middle ear (PMME), whereby the middle-ear bones were basically detached from the lower jaw but a connection to the dentary remained via a retained Meckel's cartilage (which was ossified in some cases). This can be determined from grooves in the dentary if the ear bones themselves are not preserved in the fossils. Phylogenetic analysis of the pattern of the middle-ear anatomy among Mesozoic mammals shows that the PMME evolved independently from the original condition twice: once in the lineage leading to monotremes and once in the lineage leading to therians. However, the final transition to the DMME occurred multiple times in mammalian lineages, including independently in monotremes, metatherians, and eutherians.

How could such a complex evolutionary pattern have come about? The evolution of the DMME is primarily a developmental issue. All mammals start their life with middle-ear bones attached to the jaws, and the genetic basis for the timing of the resorption of the Meckel's cartilage that frees the ear bones is becoming known; developmental studies show the potential for an ossified Meckel's cartilage in extant mammals. It remains to be determined to what extent direct selection for improved hearing was the major driver of the multiple evolutionary events, or whether the change from the PMME to the DMME was a by-product of other developmental changes in the cranium (although an older hypothesis that this was the result of enlargement of the brain is probably incorrect).

24.4 The First Mammals

The earliest mammals (technically mammaliaforms) are identified by some derived features of the skull that reflect enlargement of the brain and inner-ear regions, a dentary-squamosal jaw joint with a distinct ball-like projection on the dentary, teeth that occluded precisely and were replaced only once in a lifetime, prismatic enamel in the teeth, and molars with divided roots. These mammals were tiny, a few orders of magnitude smaller than most cynodonts (body weight less than 30 g—mouse size).

Figure 24.11 Two Early Jurassic mammals. *Morganucodon* and *Kuehneotherium* both were about the size of extant shrews. The teeth of the earliest mammals indicate an insectivorous diet, although even at this stage there was differentiation in the diets among different taxa. Sophisticated technology shows that *Morganucodon* (left) ate hard-bodied prey such as beetles, while *Kuehneotherium* (right) preferred softer prey such as caterpillars. Both species were probably nocturnal and solitary in their behavior, with the mother–infant bond being the only strong social structure, as is true for small insectivorous marsupials and placentals today. (Painting by John Sibbick, 2013, © Pamela Gill.)

The oldest well-known mammals from the latest Triassic and earliest Jurassic include *Morganucodon* (also known as *Eozostrodon*) and *Kuehneotherium* from Wales (**Figure 24.11**), *Megazostrodon* from South Africa (see Figure 24.7D), and *Sinoconodon* from China (see Figure 24.20). A somewhat older possible mammal is *Adelobasileus*, about 225 million years old and known only from an isolated braincase with a mammal-like promontorium from the Late Triassic of Texas. The term "Mammalia" is formally restricted to those animals bracketed by the interrelationships of surviving mammals (i.e., crown group Mammalia); the name of the crown group plus stem mammals is then Mammaliaformes, while these and the most derived cynodonts (tritylodontids, tritheledontids, and brasilodontids) are Mammaliamorpha.

How much like modern mammals were these early forms? We know about their bony anatomy, but what can we know about their soft anatomy, physiology, and behavior? Any feature shared by all extant mammals (e.g., lactation) would have been present in their common ancestor, but what can we infer about the biology of the stem mammals that predate the split between the monotremes (the platypus and echidnas) and the more derived therians (marsupials and placentals)?

We can deduce some features of the soft-tissue anatomy of the earliest mammals from the condition in monotremes. For example, monotremes lay eggs, and because egg laying is the generalized amniote condition, we can be fairly confident that the earliest mammals were egg-layers. However, monotremes are not always good examples for making deductions about the behavior and ecology of the earliest mammals. Monotremes are relatively large mammals, and they are specialized in many respects, such as in their loss of teeth and their electroreceptive beaks (see Chapter 25).

The classic defining features of mammals are hair and mammary glands. Such features rarely leave fossil evidence, although as noted there is some evidence for hair in Late Permian therapsids. However, it is possible to infer the presence of both features in the most derived cynodonts by an unexpected route—paleoneurology and knowledge of gene function. Endocasts (impressions of the endocranial space; see Figure 24.15) of the brains of probainognathids show an enlarged cerebellum, accompanied by ossification of the top of the skull, resulting in the complete loss of the pineal eye. Experiments on mice show that these features are under the control of the *Msx-2* homeogene, which also controls the development of mammary glands and a covering of body hair.

Metabolic and growth rates

We have already argued that derived cynodonts had higher metabolic rates than other therapsids, and this would have been the minimal condition for the earliest mammals. Monotremes, although endothermal, have lower metabolic rates than therians and are not good at evaporative cooling; early mammals were probably similar. Note, however, that many extant mammals exhibit facultative hypothermia—that is, they can reduce their metabolic rate for periods extending from hours to months (see Chapter 20).

The possession of maxilloturbinate bones in cynodonts has been cited as evidence for endothermy by this level of synapsid evolution. But for the soft tissue on these bones to function as an effective site for water and heat conservation, it is necessary for the nasal passage to be fully separated from the oral cavity by the muscles of the soft palate. The bony anatomy of this region suggests that the requisite soft palate and associated muscles were not present until the mammaliamorph cynodonts and the turbinates become ossified only in mammals. The maxilloturbinates may originally have increased the area of moist surfaces that contribute to evaporative cooling during panting.

Tritylodont cynodonts had **indeterminate growth**—that is, growth continued at a slow rate throughout life, as it does in most nonavian reptiles. That pattern had changed in one of the earliest stem mammals, *Morganucodon*, which had **determinate growth**—it grew rapidly as a juvenile and then ceased growing as an adult, as is the case with extant mammals. **Epiphyses** on the long bones reflect the mammalian feature of determinate growth. The epiphyses are the ends of the bones that are separated from the shaft (the diaphysis) by a zone of growth cartilage in immature mammals. At maturity, the ossification centers in the epiphyses and the shaft of the bone fuse, and the epiphyses no longer appear as distinct structures.

Skeletomuscular system

In all mammals, the dermal bones that originally formed the skull roof have grown down around the brain and

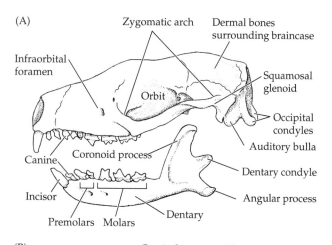

(A)

Figure 24.12 Features of the mammalian skull and skeleton. (A) Skull of a generalized mammal, a hedgehog (*Erinaceus*). The canines are reduced in this species. (B) Skeleton of the stem mammal *Agilodocodon*. (A from Lawlor 1976; B from Meng et al. 2015.)

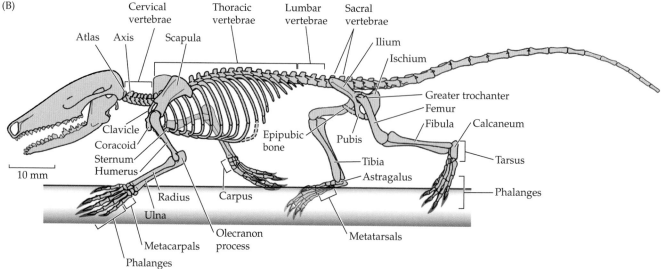

(B)

completely enclose the braincase (**Figure 24.12A**), although the details are a little different in monotremes and therians.

The earliest mammals would have retained a semi-sprawling posture (**Figure 24.12B**), as seen in generalized extant mammals such as opossums—and unlike derived modern mammals, which have more vertically positioned limbs. Mammals have a derived (crurotarsal) ankle joint, as did derived cynodonts (see Chapter 14).

Mammals share with the most derived cynodonts the rod-shaped ilium and the reduction of the pubis in the pelvic girdle, but they also have extensions on the front of the pubis (one on either side) called epipubic bones. These were first noted in marsupials, and as they are lacking in placentals it was assumed that they were used to support the pouch. However, epipubic bones are a general mammalian feature; they are important for the attachment of postural abdominal muscles, which attach directly to the pubis in placentals.

The lumbar vertebrae of cynodonts were distinct from the thoracic rib-bearing vertebrae, but lacked the zygopophyseal articulations that allow dorsoventral flexion in extant mammals. The lumbar vertebrae of mammals also have large transverse processes for attachment of the longissimus dorsi (one of the epaxial muscles) which enhances dorsoventral movement during locomotion. The ability of mammals to twist the spine both laterally and dorsoventrally may relate to their ability to lie down on their side, something that other vertebrates cannot do easily. This ability may have been important in the evolution of suckling, because the nipples are on the ventral surface of the trunk and females of many species of terrestrial mammals lie on their sides while infants suckle.

Feeding and mastication

Because hard, enamel-containing teeth are more likely to fossilize than bones, most of our information about extinct mammals comes from their teeth. Fortunately, mammalian teeth are highly informative about their owners' lifestyle.

Most vertebrates have multiply replacing sets of teeth and are hence termed **polyphyodont**. In contrast, the general mammalian condition is **diphyodonty**, and this appears to have been the condition in the earliest known mammals. Determinate growth is an essential character for diphyodonty—if the head continued to grow throughout life, teeth would have to be replaced continuously to keep up with its growth. Almost all of the earliest mammals were similar to modern mammals in that they had only two sets of teeth, and the molars were not replaced at all but instead erupted fairly late in life. (Our last molars are known as wisdom teeth, so named because they normally erupt at the age—late teens—when we supposedly have attained wisdom.)

In cynodonts, the teeth in the upper and lower jaws were the same distance apart, and chewing movements were up and down with a scissorlike closure. In early mammals (as in most extant species), the teeth in the lower jaw were closer together than the upper teeth, and chewing was only on one side of the jaw at a time, including a grinding side-to-side (lateral-to-medial) movement as well as up-and-down movement (**Figure 24.13A**). The postcanine teeth of cynodonts had the same form and were replaced continually throughout life, whereas the postcanine teeth of mammals are differentiated into premolars (replaced once) and molars (not replaced) (**Figure 24.13B**). Mammals have molars with precise occlusion, which is made possible by an interlocking

(A) Cynodont (*Thrinaxodon*)

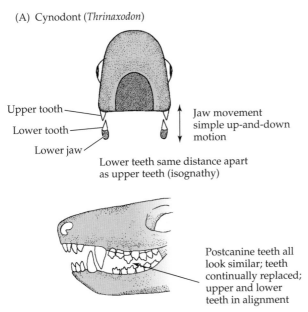

Upper tooth
Lower tooth
Lower jaw

Jaw movement simple up-and-down motion

Lower teeth same distance apart as upper teeth (isognathy)

Postcanine teeth all look similar; teeth continually replaced; upper and lower teeth in alignment

(B) Mammal (*Morganucodon*)

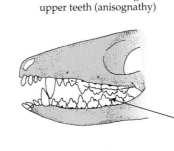

Jaw movement rotary, chewing on one side at a time

Lower teeth closer together than upper teeth (anisognathy)

Postcanine teeth divided into molars and premolars; premolars replaced once, molars not replaced; upper and lower teeth offset so they interdigitate on occlusion (e.g., one upper meshes with two lowers)

Figure 24.13 Occlusion and dentition in a cynodont and an early mammal. Cross-sectional views through the muzzles of the cynodont *Thrinaxodon* (A) and *Morganucodon* (B) show the relative positions of teeth in the upper and lower jaws (above). Lateral views of the teeth of these species are shown below.

arrangement of the upper and lower teeth. In mammals, the cusps of the upper and lower teeth interdigitate when the jaw closes, because the position of the upper teeth is offset compared with that of the lowers, so that one upper occludes with the back half of one lower and the front half of the one behind. Precise occlusion enables the cusps to cut up food very thoroughly, creating a large surface area for digestive enzymes to act on and thereby promoting rapid digestion. Only mammals **masticate** (thoroughly chew) their food in this fashion, although derived cynodonts had some degree of food processing, as evidenced by wear on their teeth.

The basal type of mammalian molar, exemplified by *Morganucodon*, had the three main cusps in a more or less straight line. In more derived early mammals, such as *Kuehneotherium*, the principal middle cusp had shifted so that the teeth assumed a triangular form in occlusal ("food's-eye") view (**Figure 24.14A**). The apex of the triangle of the upper tooth pointed inward, while that of the lower tooth pointed outward, forming an intermeshing relationship called reversed-triangle occlusion. The longer sides of these triangular teeth increased the area available for shearing action.

Therian mammals have **tribosphenic molars**, a more complex tooth that can be considered as a key adaptation

of the group (**Figure 24.14B**). These molars have multiple shearing crests, and a new cusp, the protocone, was added in the upper molars and occluded against a new basined talonid in the lower molars. The tribosphenic molar adds the function of crushing and punching to the original tooth, which acted mainly to cut and shear, and also masticates food as the teeth move out of occlusion, resulting in more food being processed per chewing cycle.

Brain, senses, and behavior

Mammals have larger brains than other vertebrates, and their brains evolved along a pathway somewhat different from that of other amniotes. Mammals are more reliant on hearing and olfaction and less reliant on vision than are most other tetrapods.

The enlarged portion of the cerebral hemispheres of mammals, the neocortex or neopallium (the area concerned with cognition), forms somewhat differently from the enlarged forebrain of derived sauropsids (see Section 14.7), although this area of the mammalian brain is not as different from that of other vertebrates as once thought. Other unique features of the mammalian brain include an infolded cerebellum (the portion of the brain concerned with neuromuscular control) and a large representation of the area for cranial nerve VII, which is associated with the facial musculature (a new feature in the mammals).

Although brains do not fossilize, their original size and shape can sometimes be determined from the inside of the skull, the endocranial cavity. If the brain fills the endocranial cavity, it leaves impressions on the cavity wall, and a virtual endocast provides details of its structure. An endocast from a derived cynodont shows a brain of a size similar to that of a modern reptile (i.e., the generalized amniote condition) but with better-developed olfactory lobes and cerebellum. The large cerebellum of cynodonts, which indicates sophisticated control of locomotion, is consistent with their mammal-like postcranial skeleton. However, the cerebral hemispheres were still small, suggesting that even derived cynodonts would not have had behavior as complex and flexible as that of most extant mammals. The portions of the brain associated with hearing and vision were also small relative to those of extant mammals. X-ray computed tomography (CT) has recently allowed researchers to discern the brain structure of early mammals (**Figure 24.15**).

Most extant small mammals are nocturnal, and this was probably the early mammalian condition. Mammals in general do not rely on vision as much as other amniotes do, and most lineages lack good color discrimination. Mammals have a retina dominated by rods (sensitive to low levels of light but not to color), and most mammals have relatively few cones (the cells that detect color and fine detail). Most mammals have only two types of opsins—visual pigments that mediate color perception—and are thus at best dichromatic (able to perceive only two basic colors. (Humans and most other anthropoids, however, are trichromatic; see Chapters 14 and 25).

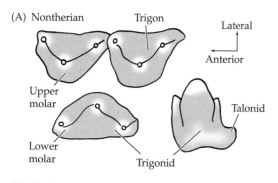

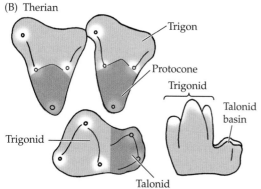

Figure 24.14 Evolution of mammalian molars.   Schematic occlusal views of the upper and lower molars of nontherian (A) and therian (B) mammals. Lower molars are also illustrated in side view. (A) Reversed triangle molars of a nontherian mammal (e.g., *Kuehneotherium*). (B) Tribosphenic molars of a therian. The new portions—the protocone in the uppers and the basined talonid in the lowers—are shaded.

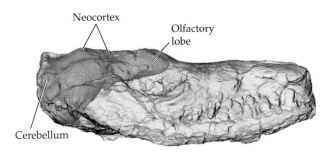

Figure 24.15 Reconstructing the brain of an early mammal. A virtual endocast of the skull of the basal mammal *Morganocudon* created using CT technology. The endocast reveals that *Morganocudon* had enlarged olfactory lobes and a larger cerebellum and neocortex than cynodonts. Crown group mammals showed further evidence of olfactory sophistication, with ossification of the olfactory (ethmoid) turbinates in the nose that provided a tenfold increase in the surface area of the olfactory epithelium. (Image by Timothy B. Rowe.)

The mammalian rod-dominated retina appears to have come about by the conversion of some cones sensitive to short-wavelength light into rods, and the cones in mammalian retinas are limited to a small area of acute vision called the fovea. Cones require a higher light intensity than rods, and diurnal mammals have a greater concentration of cones in the fovea than do nocturnal ones. Thus, most mammals have good night vision, enhanced in some species by a reflective layer called the tapetum lucidum (this is why the eyes of cats and dogs appear to shine in the dark), but relatively poor visual acuity or color vision. Other amniotes (reptiles, including birds) are mostly diurnal, so when did mammalian nocturnality evolve? Most nocturnal mammals have large eyes, and variation in orbit size among non-mammalian synapsids indicates that nocturnal behavior evolved several times within synapsids, most recently at the level of the tritylodontid cynodonts. However, genes tell a slightly different story, as we will discuss in Chapter 25, implying that full nocturnality, as seen today, took place after the monotreme–therian divergence.

Olfaction has always been an important mammalian sensory mode, probably related to mammals' primarily nocturnal behavior. Humans and other anthropoids, however, have small olfactory lobes and few functional olfactory genes, which appears to be correlated with their greater capacity for color perception than most other mammals (see Chapter 25).

The middle-ear bones of the earliest mammals were proportionally smaller and more mobile than those of their cynodont ancestors. The cochlea in the early mammalian inner ear was elongated over the cynodont condition, but not as much as in extant mammals, nor was it coiled to any extent. Thus, early mammals would have had hearing that was more acute than that of cynodonts but not as acute as that of extant mammals, especially in the range of high-pitched sounds. Most extant therians have an elongated cochlea with up to two and a half coils (although this was apparently evolved somewhat independently among

marsupials and placentals). The coils are necessary so that the structure can fit within the skull (see Figure 24.10D). The cochleas of monotremes are moderately elongated, with only half a coil. This condition appears to have been derived independently from the coiled condition of therians, and the hearing of monotremes is not as acute as that of therians. Monotremes also retain a macula sensory structure at the end of the cochlea, present in other amniotes but not in therians (this was probably also the condition in Mesozoic mammals less derived than monotremes). Monotremes lack a **pinna** (external ear; plural *pinnae*), and it is likely that the earliest mammals were also pinna-less, making reconstructions of them look rather non-mammalian. The earliest evidence of a pinna is seen in the eutriconodont *Spinolestes*, although that animal retained the transitional type of mammalian middle ear with an ossified Meckel's cartilage.

The integument: Epidermis and glands

The variety of mammalian integuments is enormous and reflects the importance of the skin in maintaining the internal body temperature in an endotherm, especially one living in a harsh climate.

Epidermis The outer layer of skin, the epidermis, varies in thickness from a few cells in some small rodents to hundreds of cell layers in elephants, rhinoceroses, and hippos (which were once grouped together as *pachyderms*, meaning "thick skins").

The tails of opossums and many rodents are covered by epidermal scales similar to those of lizards but lacking the hard beta-keratin found in birds and reptiles. Some integumentary appendages are involved in locomotion, offense, defense, and display. Claws, nails, and hooves are accumulations of keratin that protect the terminal phalanx of the digits.

Hair is composed of keratin. It grows from a deep invagination of the germinal layer of the epidermis called the **hair follicle** (**Figure 24.16**). Hair has a variety of functions in addition to insulation, including camouflage, communication, and sensation via **vibrissae** (whiskers). Vibrissae grow on the muzzle, around the eyes, and on the lower legs. Touch receptors are associated with these specialized hairs, and hair may have originated as a sensory structure. Mammalian hairs develop in association with migrating neural crest cells (see Chapter 2) that induce the formation of mechanoreceptors in the hair follicle for tactile sensation.

Pelage (fur) consists of closely placed hairs, often produced by multiple hair shafts arising from a single complex root. Its insulating effect depends on its ability to trap air, and its insulating ability is proportional to the length of the hairs and their density of spacing. The arrector pili muscles that attach midway along the hair shaft pull the hairs erect to deepen the layer of trapped air. Cold stimulates a general contraction of the arrector pili via the sympathetic nerves, as do other stressful conditions such as fear

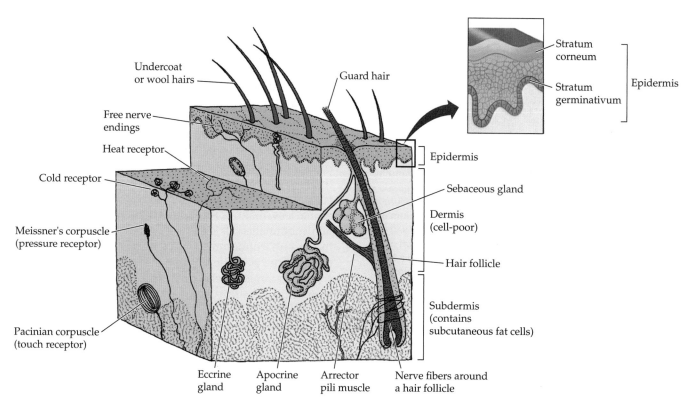

Figure 24.16 **The mammalian integument.** Most mammals have a barrier layer of flattened, tough cells (the stratum corneum) above layers of cells constantly being generated by stem cells in the stratum germinativum. Hair grows from follicles, deep invaginations of the subdermis. Mammals have glandular skin embedded with sebaceous, eccrine, and apocrine glands that secrete substances used in homeostasis (e.g., protective skin oils; sweating in humans) and chemical communication. These secretions are under hormonal and neuronal control. Receptors for various sensations, including pressure and temperature, are embedded throughout the skin. (After Ham and Cormack 1979 and Harrison and Montagna 1973.)

and anger. This autonomic reaction still occurs in hairless mammals, including humans, where it produces dimples ("goose bumps") on the skin's surface. Hair erections serve for communication as well as for thermoregulation; mammals use them to send a warning of fear or anger, as seen in the display of a puffed-up cat or the raised hackles of a dog.

Hair color depends on the quality and quantity of melanin injected into the forming hair by melanocytes at the base of the hair follicle. The color patterns of mammals are built up by the colors of individual hairs. Because exposed hair is nonliving, it wears and bleaches. Replacement occurs by growth of an individual hair or by **molting**, in which old hairs fall out and are replaced by new hairs. Most mammals have pelage that grows and rests in seasonal phases; molting usually occurs only once or twice a year.

As noted previously, hair was probably present in the most derived cynodonts. Some direct evidence of hair exists in Mesozoic mammals: the beaver-like Middle Jurassic stem mammal *Castorocauda*, a docodont (see Figure 24.20), had evidence of a dense pelage preserved as impressions in

rock, while the more derived Early Cretaceous eutriconodont *Spinolestes* has completely preserved skin structures, including, as its name suggests, small spines. The need to evolve fur as insulation may have become pressing in the latest cynodonts and earliest stem mammals, which were very small and would have had high surface/volume ratios and consequently high rates of heat loss.

Skin glands Secretory structures of the skin develop from the epidermis. Modern amphibians (but not fishes) have a glandular skin, and the condition in mammals probably represents the basal tetrapod condition, reduced in sauropsids. There are three major types of skin glands in mammals: **sebaceous**, **eccrine**, and **apocrine glands** (see Figure 24.16). Except for the eccrine glands, skin glands are associated with hair follicles, and the secretion in all of them is under neural and hormonal control. A full complement of these skin glands is found in monotremes as well as therians, and thus they may be assumed to be a basic feature of all mammals.

Sebaceous glands are found over the entire body surface. They produce an oily secretion, sebum, which lubricates and waterproofs the hair and skin. Sheep lanolin, our own greasy hair, and the grease spots that the family dog leaves on the wallpaper where it curls up in the corner are all sebaceous secretions.

In most mammals, eccrine glands are restricted to the soles of the feet, prehensile tails, and other areas that contact environmental surfaces, where they improve adhesion or enhance tactile perception. This is why your hands sweat when you are nervous—the moisture would have given your primate ancestors a better grip for escaping. In some primates (Old World monkeys and apes, including humans), eccrine glands are found over the entire body surface.

Apocrine glands have a restricted distribution in most mammals (except in ungulates, where they are used for cooling), and their secretions appear to be used in chemical communication. In humans, apocrine glands are found in the armpit and pubic regions (these are the secretions we usually try to mask with deodorant).

It is a common truism that mammals have "sweat glands," but in fact very few mammals sweat. In humans, eccrine glands produce sweat, whereas in hoofed mammals (e.g., pigs, horses) sweat is produced by apocrine glands. Other mammals induce evaporative cooling by other means; for example, dogs pant and kangaroos lick their forearms.

Many mammals have specialized scent glands that are modified sebaceous or apocrine glands. Sebaceous glands secrete a viscous substance, usually employed to mark objects, whereas apocrine glands produce volatile substances that may be released into the air as well as placed on objects. Scent glands are usually on areas of the body that can be easily applied to objects, such as the face, chin, or feet. Domestic cats often rub their face and chin to mark objects, including their owners. Many carnivorans have anal glands that deposit scents along with the feces, and apocrine anal scent glands are a well-known feature of skunks. Apocrine glands in the ear produce earwax.

Mammary glands have a more complex, branching structure than do other skin glands, but they appear to have been derived from a basal apocrine gland, because the detailed mode of secretion (exocytosis) is similar to that of apocrine glands, as are some aspects of the development of mammary glands. Note that the common notion that mammary glands are "modified sweat glands" is a serious misnomer.

Lactation and suckling

The females of all mammal species lactate, feeding their young by producing milk. Mammary glands are absent from marsupial males, but they are present and potentially functional in male monotremes and placentals (breast cancer affects human males as well as females). Males of two bat species lactate, and there are records of human males producing milk. It has long been a mystery why male placental mammals retain mammary glands but do not normally lactate.

How could lactation as a mode of nourishing infants have evolved? A small amount of proto-milk would not be sufficient, so could lactation and mammary glands have initially evolved for another reason? Obviously, the precursor to any type of feeding of the young would have to be parental care. Few extant reptiles (except crocodylians) are known to care for their young, but a fossil of an adult basal pelycosaur from the late Middle Permian of South Africa is accompanied by several juveniles, suggesting that parental care may be an ancestral synapsid trait. Similar associations

of adults and juveniles have also been reported in Early Triassic cynodonts.

Monotremes lay parchment-shelled eggs, like the eggs of most lizards and very unlike the calcareous-shelled eggs of birds. Gas exchange is reduced if the shell becomes dry, and proto-mammary glands might have prevented the eggshell from drying. Proto-mammary glands probably had the ability to secrete small quantities of organic materials, as mammalian apocrine glands do today.

Protecting eggs from microorganisms could have been another function of the proto-mammary glands. Milk contains proteins related to the lysozymes that attack bacteria—human milk has antimicrobial properties. Molecular evidence shows that mammary glands probably evolved as part of the innate immune system, and that an antimicrobial function of milk preceded its nutritional function. Once proto-mammary secretion had evolved, whether for hydration of the eggshell, antimicrobial defense, or both, selection could have led to more copious and more nutritive secretions that were ingested by the young.

Although monotremes produce milk, their mammary glands lack nipples, and nipples were probably lacking in the earliest mammals as well. However, monotreme mammary glands are associated with hairs, from which the young suck the milk.

At what point in synapsid evolution would these secretions have become essential for feeding the young? Mammalian teeth occlude precisely, unlike the teeth of reptiles. Precise occlusion is not possible with teeth that are continually being replaced, and is not apparent until the earliest mammals that essentially have a single set of teeth as an adult. Mammals are diphyodont—they have a set of milk teeth while their jaws are growing and a second set of adult teeth that are not fully in place until the jaws have reached their full size. However, an animal can be diphyodont only if it is fed milk during its early life. With a liquid diet, the jaw could grow while it had less need of teeth, and permanent teeth could erupt in a near adult-size jaw. Thus, a lineage with precise occlusion and diphyodonty must first have evolved lactation. However, we don't know if lactation originated with mammals or if it was present in some or all non-mammalian synapsids. (Note the earlier description of the paleoneurology of probainognathid cynodonts, which suggests that these may have been the first synapsids to have developed mammary glands.)

Lactation allows the production of offspring to be separated from seasonal food supply. Unlike birds, which lay eggs only when there is appropriate food for the fledglings, mammals can store food as fat and convert it into milk at a later date. Provision of food in this manner by the mother alone also means that she does not need assistance from a male to rear her young. Finally, lactation allows the young to be born at a small body size relative to the parent. The ability to suckle is a unique mammalian feature, and is related to the same changes in the oral region that allow deglutination, the unique mode of mammalian swallowing

(A) Lizard

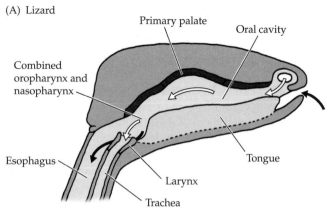

(B) Mammal

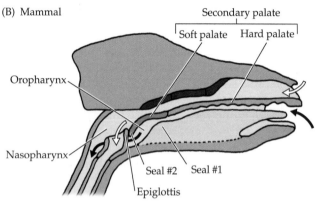

Figure 24.17 Longitudinal section of the oral and pharyngeal regions of a lizard and a mammal. (A) A lizard represents the general amniote condition, also seen in most non-mammalian synapsids. There is no secondary palate and no separation between nasal and oral passages. (B) The mammal shows the derived synapsid condition. Although a secondary palate was present in cynodonts, the seals shown here, which allow suckling and specialized swallowing, were probably uniquely mammalian (perhaps present in the most derived cynodonts). Seal #1 presses the back of the tongue against the soft palate and prevents substances from entering the pharynx. Seal #2 presses the epiglottis against the back of the soft palate, allowing air (white arrows) to enter the trachea and blocking the entrance of material from the oropharynx. Liquids (black arrows) can pass from the oropharynx, around the trachea, and into the esophagus while this seal is in place. Seal #2 is absent in postinfant humans, which is why we can "swallow the wrong way"—that is, inhale food or liquid into the trachea—and even choke to death on food.

(discussed below). Mammals can form fleshy seals against the bony hard palate with the tongue and with the epiglottis, effectively isolating the functions of breathing and swallowing (**Figure 24.17**). Mammals use these seals to suckle on the nipple while breathing through the nose. In humans, the larynx shifts ventrally during early childhood, and the posterior seal is lost. This makes us more likely to choke on our food, but also enables us to breathe through the mouth as well through the nose.

Food processing and swallowing

Mammals chew food into fine particles and swallow a discrete bolus of food, rather than swallowing large items intact with little oral preparation as do reptiles. The mammalian

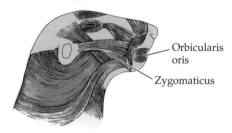

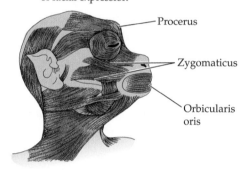

Figure 24.18 Muscles of facial expression in mammals. The orbicularis oris muscle in rodents and primates allows the lips to be pursed to form an airtight seal around a nipple during suckling. Primates have a more complex zygomaticus muscle than do rodents. Both primates and rodents can snarl, but primates can also smile, whereas rodents cannot. The procerus muscle allows primates to frown, but the procerus is virtually absent from rodents, which are unable to frown. (From Kardong 2012.)

tongue has a unique system of intrinsic muscles (innervated by cranial nerve XII) that aid in this swallowing and extensive oral food processing. To swallow, a mammal coordinates manipulation by the tongue with elevation of the soft palate and constriction of the pharynx. This process of coordinated food processing and swallowing is unique to mammals and is called **deglutination**. Mammals have a distinctive bony anatomy of the skull in this region of the posterior palate, which was approached by some derived cynodonts. The epiglottis, which blocks the opening of the trachea during swallowing, probably also appeared at this stage of mammalian evolution.

Facial musculature

Facial muscles are another unique character of extant mammals (**Figure 24.18**), although as previously noted, mobile lips, at least, may have been present in derived cynodonts. These muscles make possible our wide variety of facial expressions, but they were probably first evolved in the context of mobile lips and cheeks that enable the young to suckle. The facial muscles are thought to be homologous with the neck constrictor muscles of other amniotes (the constrictor colli, also called the sphincter colli) because both types of muscles are innervated by the facial nerve (cranial nerve VII). In reptiles, which swallow prey more or less intact, constriction of this muscle moves food down the esophagus. Because mammals chew food into small pieces, the constrictor colli could assume a different function. Elaboration of the muscles of facial expression may have occurred somewhat differently in various mammal lineages, because the facial muscles in monotremes are less complex than those in therians.

Most mammals do not have highly expressive faces. However, primates are not the only mammals capable of a diversity of facial expressions. Horses use their lips in feeding and are capable of quite a wide variety of expressions. All mammals with well-developed facial muscles display similar expressions for similar emotions; for example, the snarl of an angry human is like the snarl of an angry dog or even an angry horse.

Internal anatomy

The internal anatomy and physiology of mammals differ in numerous ways from those of other amniotes. Some of these differences relate to mammals' endothermic metabolism; similar systems have evolved convergently in the other endothermic vertebrates, birds. Other differences are uniquely mammalian and reflect mammals' evolutionary history.

Adipose tissue   Mammalian adipose tissue (white fat) is distributed as subcutaneous fat, which is important for insulation; as fat associated with various internal organs (e.g., heart, intestines, kidneys), where it may serve both a metabolic and a cushioning role; as deposits in skeletal muscles; and as cushioning for joints. Adipose tissue is not simply inert material used as an energy store or as an insulating layer, as has often been assumed. Recent studies have revealed that adipocytes (fat cells) secrete a wide variety of messenger molecules that coordinate important metabolic processes. Some small fat storage sites have specific properties that equip them to interact locally with the immune system, and possibly with other organs.

Placental mammals also have a unique type of adipose tissue known as brown fat. This tissue is specialized for thermogenesis and can break down lipids or glucose to release heat at a rate up to ten times that of the muscles. Brown fat is especially prominent in newborn mammals, where it is important for thermoregulation, and in the adults of hibernating species, where it is used to rewarm the body as the animal emerges from hibernation.

Heart   The mammalian heart has a complete ventricular septum (rendering it four-chambered) and only a single (left) systemic arch (**Figure 24.19**), although a right systemic arch is present during embryonic development and contributes to part of the circulation going to the head and arm in the adult. Mammals differ from other vertebrates in the form of their erythrocytes (red blood cells), which lack nuclei in the mature condition. Monotremes retain a small sinus venosus as a distinct chamber in the heart, so this was probably also the condition in the earliest mammals. In therians, this structure is incorporated into the wall of the left atrium as the sinoatrial node, or pacemaker.

Mammals are also unique among vertebrates in having platelets, a type of blood cell. Platelets form aggregations called thrombi that aid in blood clotting, thus preventing blood loss after injury. Unfortunately, platelets can also form thrombi within the coronary vessels, meaning that mammals are the

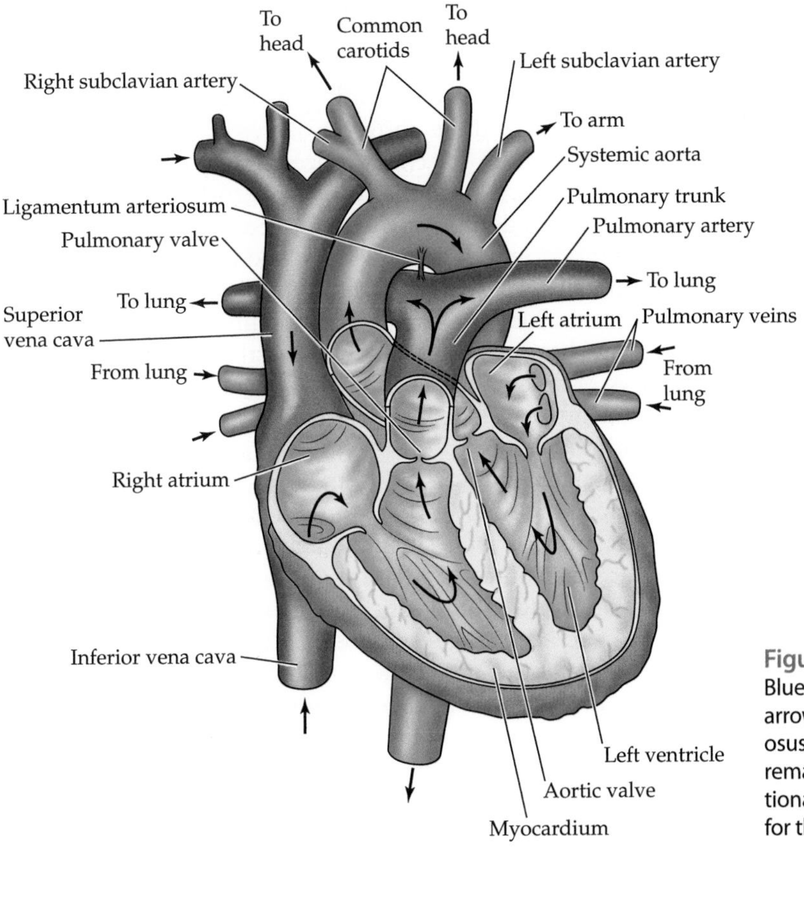

Figure 24.19 Diagrammatic view of a mammalian heart. Blue arrows indicate the passage of deoxygenated blood, red arrows the passage of oxygenated blood. The ductus arteriosus (ligamentum arteriosum in the adult condition) is the remains of the dorsal part of the pulmonary arch; it is functional in the amniote fetus, where it is used as a bypass shunt for the lungs, but closes after birth. (After Hill et al. 2016.)

only vertebrates subject to heart attacks due to platelet-induced thrombosis.

Lungs   Mammals have large, lobed lungs with a spongy appearance due to the presence of a finely branching system of bronchioles in each lung, terminating in tiny thin-walled, blind-ending chambers (the sites of gas exchange) called alveoli (see Chapter 14). The diaphragm aids the ribs in inspiration and divides the original vertebrate pleuroperitoneal cavity into a peritoneal cavity surrounding the viscera and paired pleural cavities surrounding the lungs.

Excretory system   All mammals retain the original tetrapod bladder, which was lost in many sauropsids, and excrete liquid urine. Mammals have entirely lost the renal portal system seen in other vertebrates, which supplies venous blood to the kidney in addition to the arterial blood supplied by the renal artery. Elaboration of the loop of Henle in the kidney tubules allows mammals to excrete urine that is many times more concentrated than the blood plasma (see Figures 14.10 and 14.11). Early mammals would have been like monotremes and other amniotes in having the excretory (urinary), reproductive, and alimentary (digestive) systems reach the outside via a common opening, the cloaca.

24.5   Mesozoic Mammals

The Cenozoic is often called the Age of Mammals, but the radiation of Mesozoic mammals represents two-thirds of mammalian history. Mesozoic mammals were diverse—more than 300 genera are known, the majority of them described in the past few decades. These newly described species display a range of body forms that were previously unknown, including features indicative of swimming, digging, and gliding lifestyles and a broad range of diets (**Figure 24.20**). Although most Mesozoic mammals were shrew- to mouse-size, some were as large as raccoons (~10 kg), including the carnivorous *Repenomamus*, famous for being preserved with a juvenile dinosaur in its stomach.

It was not until the Cenozoic that mammals branched out into a wider range of sizes, but it should be remembered that more than 50% of today's mammal species are in the size range of the Mesozoic forms. It has been a popular truism that the size and diversity of Mesozoic mammals were somehow suppressed by the presence of dinosaurs, and it is certainly possible that there may have been competition between mammals and juvenile dinosaurs. But recent work is indicating that patterns of plant evolution had a greater influence on Mesozoic mammal diversity than dinosaurs did (see Chapter 23).

Dual radiations of Mesozoic mammals

There were two major periods of mammalian diversification during the Mesozoic: an initial rapid radiation in the Jurassic followed by a steep decline, and a second radiation in the Early Cretaceous. At this time the northern and southern continental blocks were separated, so these two radiations of mammals occurred independently, with both groups evolving more complex types of cheek teeth. The Jurassic radiation included one clade in Gondwana (australosphenidans; Latin *australis*, "southern"; Greek *sphen*, "a wedge") that contained the origin of monotremes as well as several extinct groups, and another in Laurasia (boreosphenidans; Greek *boreas*, "north") that contained the earliest therians and multituberculates, although many of these early lineages did not survive past this period.

The rodentlike multituberculates were an important group of mammals, surviving into the Cenozoic, and will be further considered in Chapter 25. Another important radiation was of a stem mammal lineage, Haramiyida. Haramiyids are known from Europe, Asia, and Africa, and from the latest Triassic to the latest Cretaceous, although they are best known from the Jurassic. Their fossil record is sparse, but they are important because they were an early radiation of omnivorous and herbivorous mammals (with complex molars for processing food) that probably were also arboreal; some recently discovered species were specialized gliders—similar to modern gliding mammals but living in a very different floral environment—before the radiation of the angiosperm plants. Haramiyids were once thought to be related to multituberculates, which they superficially resemble. This position is still held by some researchers, but the structure of the haramiyid middle ear and shoulder girdle supports a more basal phylogenetic position for the haramiyids.

Mammal diversification was rapid during the Jurassic, and it was during this time that many of the features of later mammals emerged, including the definitive mammalian middle ear and the derived features of the dentition and shoulder girdle that characterize therians today (see Chapter 25). The earliest fossils of modern-lineage mammals are known definitively from the Early Cretaceous; these are monotremes from Australia and therians from China. A possible contender for the earliest eutherian is *Juramaia sinensis*, from the Late Jurassic of China, although there is some disagreement about whether this animal is a eutherian or a stem therian. However, note that molecular clock data implies a Late Jurassic split between metatherian and eutherian lineages.

The rate of diversification during the second radiation, which got underway near the end of the Early Cretaceous, was much slower. This radiation featured early true therians as well as a radiation of more derived multituberculates, plus some other lineages that did not survive the Cretaceous. Several evolutionary patterns started to emerge at the end of the Cretaceous in both therian and multituberculate lineages. Studies of jaws and teeth show dietary changes: multituberculates shifted from an omnivorous to a more herbivorous diet, and most therians were becoming more specialized as small insectivores and carnivores, although some marsupial lineages shifted toward

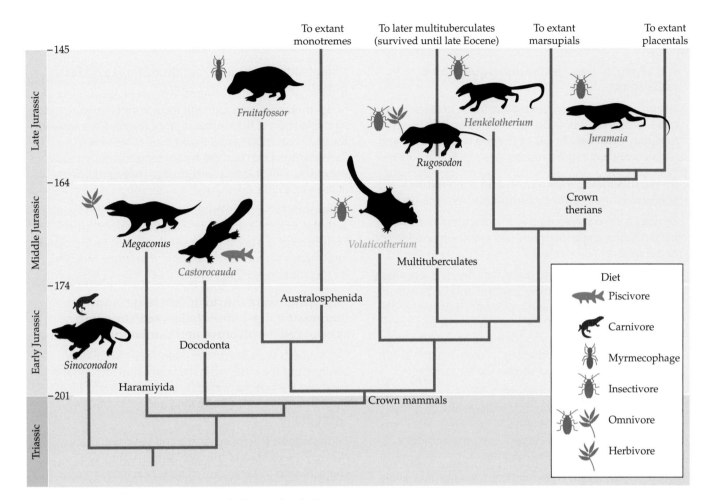

Figure 24.20 Ecological diversity and phylogenetic relationships of some newly described mammals. The mammals shown here are from the Jurassic. Image sizes are proportional to body size. Text color refers to lifestyle: terrestrial (black), arboreal (green), fossorial (brown), aquatic (blue), volant (orange). Icons in the background denote primary diet. (After Lee and Beck 2015.)

a more herbivorous diet. These changes were coincident with an increase in the diversity and dominance of the angiosperm flora, especially in the last 10 million years or so of the Cretaceous. Angiosperms are mainly insect-pollinated, and this flourishing of a new lineage of plant life may have given rise to a new diversity of insects. The coincidence of plant and mammal diversification certainly suggests that mammalian evolution was influenced more by changes in vegetation than by the presence or absence of dinosaurs.

The end-Cretaceous extinctions, famous for the demise of the nonavian dinosaurs, also affected mammals, especially marsupials and the archaic types of placentals,

although multituberculates and the founders of the modern types of placentals (as indicated by molecular data) appear to have been less affected. Mammalian species diversity rapidly recovered following the end-Cretaceous extinctions, but it was not until several million years into the Cenozoic that mammals started to radiate into a greater diversity of diets and body sizes.

There has been much debate about whether or not the radiation of extant orders of mammals was a purely Cenozoic event (as indicated by the fossil record), or whether these divergences actually had deeper roots extending millions of years back into the Cretaceous (as suggested by molecular evidence). The current consensus, based on a combination of molecular and morphological data, is that the initial diversification of extant mammal orders may have started during the last few million years of the Cretaceous (10–20 ma), but that the main diversification occurred primarily in the Cenozoic.

Summary

The synapsid lineage has undergone three radiations.

The first radiation, pelycosaurs, and much of the second radiation, therapsids, occurred in the Paleozoic. The third radiation, mammals, began at the end of the Triassic and reached its peak in the Cenozoic.

Synapsids include mammals and their extinct predecessors, the non-mammalian synapsids.

Synapsids are distinguished from other amniotes by the presence of a single lower fenestra in the temporal region of the skull.

The bone histology of synapsids indicates that, even early in their evolutionary history, they probably had a higher metabolic rate than modern reptiles do. However, endothermy was unlikely until the more derived forms, possibly convergently in the cynodonts (the lineage leading to mammals) and dicynodonts (a lineage of herbivorous forms that did not survive beyond the Mesozoic).

Synapsids were the first amniotes to radiate widely in terrestrial habitats, preceding the radiation of reptiles that occurred primarily in the Mesozoic. From the Early Permian to the Early Triassic, synapsids were the dominant terrestrial carnivores and herbivores.

Most early synapsids were medium-size to large animals (10–200 kg), with a few giants that weighed as much as 500 kg. A dramatic decrease in body size occurred during their evolution, and by the Cretaceous most synapsids weighed 1 kg or less.

Pelycosaurs and therapsids were the two major groups of non-mammalian synapsids.

Pelycosaurs, the basal group, radiated in the Early Permian, primarily in paleoequatorial regions of Laurasia. Therapsids were more derived, and are known predominantly from Late Permian to Early Triassic sites in southern Gondwana.

The sailback pelycosaurs—the sphenacodontids and edaphosaurids—were the iconic pelycosaurs, but fully developed sails were characteristic of only a few species. The function of the sails is obscure; there is no evidence that they were sexually dimorphic, seemingly ruling out a role in social behavior, and they appear to have been too poorly vascularized to have carried heat into the body.

Pelycosaurs included generalized carnivores as well as specialized herbivores and carnivores. Specialized carnivorous sphenacodonts such as *Dimetrodon* had derived dentition and an arched palate, which was a forerunner of the later separation of the mouth and nasal passages seen in mammals.

Therapsids had higher metabolic rates and faster growth rates than pelycosaurs.

Faster metabolism and growth require more energy, and differentiation of the teeth and large jaw muscles allowed therapsids to process and digest food more rapidly. Derived therapsids had the beginnings of a bony secondary palate, which allows an animal to chew and breathe at the same time.

Therapsids had a more upright stance than pelycosaurs, with the limbs held more vertically beneath the trunk. Reduction of the tail of therapsids indicates the increased importance of limb muscles, rather than trunk muscles, for locomotion.

Therapsids were the dominant large land mammals in the Late Permian, radiating into specialized herbivorous and carnivorous forms.

Dicynodonts were herbivores. They retained tusklike canines, but the other marginal teeth were lost and the toothless jaws were covered by a turtlelike horny beak. Dicynodonts included burrowing and probably semiaquatic forms.

Eutheriodonts (therocephalians and cynodonts) were the major predators of the Late Permian. They were characterized by larger jaw muscles that inserted on a new structure on the dentary bone, the coronoid process, which increased the mechanical advantage of the muscles.

Cynodonts had a variety of derived characters of the skull, jaws, and teeth that made them more mammal-like than other therapsids, including development of a zygomatic arch (cheekbone) which provided a site for the origin of a new, major, jaw-closing muscle, the masseter muscle.

Mammals are unique among tetrapods in having three bones (rather than one) in their middle ear for sound conduction.

Embryological studies dating from the early 19th century show that the two "extra" bones in the mammalian middle ear, the malleus and the incus, are homologous with the articular and the quadrate that form the jaw joint in other gnathostomes.

The earliest synapsids (pelycosaurs) had a lower jaw like other amniotes, with the dentary (the bone bearing the teeth) taking up about half the lower jaw.

In cynodonts, the dentary became progressively larger, and the postdentary bones (including the articular) and quadrate became smaller. The evolution of a masseter muscle on the outside of the lower jaw allowed the jaw to be held in a sling, taking pressure off the jaw joint and allowing the articular and quadrate to better function in hearing.

In derived cynodonts, the dentary became large enough to contact the skull, and a new jaw joint formed between the dentary and the squamosal. Some highly derived cynodonts, and many early mammals, had a double jaw joint, retaining the original one along with the new one.

The evolutionary history of the jaw in synapsids can be perceived as reflecting a conflict between hearing and eating, and both functions would have become more important in synapsids with a higher metabolic rate. Non-mammalian synapsids received sound via a chain of bones, including the quadrate and articular.

In later mammals, and convergently in different lineages, the articular and quadrate became separated from the lower jaw and enclosed in a middle-ear cavity, forming the definitive mammalian middle ear (DMME).

The earliest mammals were tiny.

Among the features that distinguished early mammals from derived cynodonts (which were also small, but not as small) were a larger brain and teeth that occluded precisely and were replaced only once.

We cannot directly know about many of the features of early mammals, such as their soft anatomy, physiology, and behavior, but much can be inferred from their bony anatomy and also by comparing extant monotremes (the platypus and echidnas) with the more derived therian mammals (marsupials and placentals).

Because monotremes lay eggs and this is the general amniote condition, we can deduce that the earliest mammals were also egg-layers.

Many extant mammals, including monotremes, have rather low metabolic rates, despite being endothermic, and are poor at evaporative cooling. The earliest mammals were probably similar.

The most derived cynodonts had indeterminate growth, meaning they grew throughout life, like most nonavian reptiles do today. The earliest mammals had determinate growth, growing rapidly as juveniles and then ceasing growth as adults, as extant mammals do.

The earliest mammals had fully enclosed the braincase with dermal bone, to form a complete bony casing.

The earliest mammals, like many today, retained a semi-sprawling stance, but their vertebrae show that they were capable of dorsoventral flexion. This is important for locomotion in therian mammals. The ability of mammals to twist the spine both laterally and dorsoventrally would have enabled early mammals to lie on their side to suckle their young.

The earliest mammals, like most extant mammals, were diphyodont. Because they are replaced only once, mammalian teeth must be durable, and teeth with prismatic enamel are a distinctive mammalian feature, as are double-rooted molars.

Mammals have interlocking upper and lower teeth. This arrangement allows precise occlusion, which in turn relates to being able to cut up food more finely and digest it more rapidly.

Extant mammals are more reliant on hearing and olfaction and less reliant on vision than most other vertebrates are.

The earliest mammals had enlarged brains with large olfactory lobes and with a larger cerebellum (concerned with balance and coordination) and larger cerebral hemispheres (concerned with cognition) than cynodonts.

Mammals have lost two of the four original vertebrate opsin genes, which code for the visual pigments in retinal cone cells. With the exception of humans and other anthropoids, most mammals have poor color vision (they have even lost some relevant genes) and relatively poor visual acuity.

Most mammalian retinas contain primarily rods (for light detection) rather than cones (for color vision and acute visual resolution). The mammalian retina probably reflects a period in evolutionary history when mammals were nocturnal and retinas evolved that emphasized vision in low light rather than color.

Mammals have hair-covered, glandular skin.

Mammals have three types of skin glands: sebaceous glands that lubricate hair, apocrine glands that produce pheromones, and eccrine glands that produce watery fluid for adhesion to the surface on the soles of feet. Most mammals do not have sweat glands; in humans eccrine glands are found all over the body and produce sweat, but in hoofed mammals it is the apocrine glands that produce sweat.

Some fossils of early mammals show the presence of hair, and paleoneurology studies show that derived cynodonts may have had at least whiskers, if not fur.

All female mammals lactate, although only therian mammals have nipples.

Mammary glands appear to have been derived from apocrine glands. Milk may originally have evolved as secretions that helped keep thin-shelled eggs moist, and that also had antimicrobial properties.

(Continued)

Summary *(continued)*

The ability to suckle is a uniquely mammalian feature, and it relates to the unique anatomy of the mammalian tongue, facial muscles, and palate.

Mammals have a unique mode of processing and swallowing food (deglutination) that involves the actions of a specialized tongue, palate, and pharynx.

These actions are also used in suckling and swallowing milk while the infant continues to breathe

The anatomy of the posterior palatal area in mammals is distinctive and related to this new mode of food processing, and it was first approached in derived cynodonts. The epiglottis, which covers the trachea during swallowing, probably also appeared at this stage of mammalian evolution.

The internal anatomy of mammals has unique features.

Adipose tissue (fat) has many important functions, including insulation, cushioning of joints and major organs, energy storage, and various metabolic functions. Mammals have a unique type of adipose tissue called brown fat, which is especially important in newborns for thermoregulation and in animals that hibernate.

The mammalian heart is four-chambered with a single (left) systemic arch. Mammals are unique in having erythrocytes that lack nuclei in the mature condition.

All mammals (unlike many sauropsids) retain the original tetrapod bladder and excrete liquid urine. An elaborated loop of Henle in the kidney tubules allows mammals to excrete urine that is many times more concentrated than the blood plasma.

The excretory, alimentary, and reproductive systems of early mammals probably emptied via a common opening, the cloaca, as they do in extant monotremes and sauropsids.

Mesozoic mammals comprise two-thirds of mammalian history; more than 300 genera are now known.

Most Mesozoic mammals were small, but they had a wide variety of lifestyles and diets. One early lineage, Haramiyida, contained specialized gliders.

An initial radiation of mammals in the Jurassic, which included one clade in Gondwana and another in Laurasia, was followed by the decline and extinction of many lineages. A second radiation, marked by the appearance of the first modern mammalian lineages, occurred in the Early Cretaceous.

The end-Cretaceous extinctions, famous for the demise of the nonavian dinosaurs, also affected mammals, especially metatherians and the archaic types of eutherians. The extant orders of mammals probably originated during the last few million years of the Mesozoic, although they are not apparent in the fossil record until the Cenozoic.

Discussion Questions

1. Why does a more upright posture (i.e., limbs more underneath the body) in therapsids indicate that they probably had a higher metabolic rate than pelycosaurs?

2. Why would the inferred appearance of a diaphragm in cynodonts make sense in terms of other attributes of these synapsids?

3. Why might a larger brain in early mammals be related (at least in part) to their process of miniaturization from the cynodont condition?

4. Why can we infer that lactation must have evolved at least by the time of the first mammal?

5. We now have a series of transitional fossils documenting the evolution of the mammalian ear. How could people be so sure, before these fossils were known, that the "new" ear ossicles in mammals (the malleus and incus) were derived from the bones that formed the original gnathostome jaw joint?

6. Why do some people think that early mammals (including most Mesozoic forms) would have lacked a pinna (external ear)?

Additional Reading

Angielczyk KD. 2009. *Dimetrodon* is not a dinosaur: Using tree thinking to understand the ancient relatives of mammals and their evolution. *Evolution, Education and Outreach* 2: 257–271.

Benoit J, Manger PR, Rubridge BS. 2016. Palaeoneurological clues to the evolution of defining mammalian soft tissue traits. *Scientific Reports* 6: 25604.

Botha-Brink J, Modesto SP. 2007. A mixed-age classed 'pelycosaur' aggregation from South Africa: Earliest evidence of parental care in amniotes? *Proceedings of the Royal Society* B 274: 2829–2834.

Brocklehurst N, Brink KS. 2017. Selection towards larger body size in both herbivorous and carnivorous synapsids during the Carboniferous. *Facets* 2: 68–84.

Gill PG and 6 others. 2014. Dietary specializations and diversity in feeding ecology of the earliest stem mammals. *Nature* 512: 303–307.

Grossnickle DM, Newham E. 2016. Therian mammals experience an ecomorphological radiation during the Late Cretaceous and selective extinction at the K-Pg boundary. *Proceedings of the Royal Society B* 283: 20160256.

Kim J-W and 8 others. 2016. Recruitment of rod photoreceptors from short-wavelength-sensitive cones during the evolution of nocturnal vision in mammals. *Developmental Cell* 37: 520–532.

Lee MSY, Beck RMD. 2015. Mammalian evolution: A Jurassic spark. *Current Biology* 25: R753–R773.

Liu L and 14 others. 2017. Genomic evidence reveals a radiation of placental mammals uninterrupted by the KPg boundary. *Proceedings of the National Academy of Sciences USA* 114: E7282–E7290. doi: 10.1073/pnas.1616744114.

Luo Z-X. 2007. Transformation and diversification in early mammal evolution. *Nature* 450: 1011–1019.

Maier W, Ruf, I. 2016. Evolution of the mammalian middle ear: A historical review. *Journal of Anatomy* 228: 270–283.

Meng Q-J and 6 others. 2017. New gliding mammaliaformes from the Jurassic. *Nature* 548: 291–296.

Meng Q-J and 5 others. 2015. An arboreal docodont from the Jurassic and mammaliaform ecological diversification. *Science* 347: 764–768.

Oftedal OT. 2011. The evolution of milk secretion and its ancient origins. *Animal* 6: 355–368.

Olivier C, Houssaye A, Jalil N-E, Cubo J. 2017. First palaeohistological inference of resting metabolic rate in an extinct synapsid, *Moghreberia nmachouensis. Biological Journal of the Linnean Society* 121: 409–419.

Rey K and 12 others. 2017. Oxygen isotopes suggest elevated thermometabolism within multiple Permo-Triassic therapsid clades. *eLife* 6:e28589.

Rodrigues PG and 6 others. 2018. Digital cranial endocast of *Riograndia guaibensis* (Late Triassic, Brazil) sheds light on the evolution of the brain in non-mammalian cynodonts. *Historical Biology,* in press.

Springer MS and 5 others. 2017. Waking the undead: Implications of a soft explosive model for the timing of placental mammal diversification. *Molecular Phylogenetics and Evolution* 106: 86–102.

Go to the Companion Website **oup.com/us/vertebratelife10e** for active-learning exercises, news links, references, and more.

© istock.com /drferry

CHAPTER
25

Extant Mammals

Cenozoic mammals are adapted to a wide variety of lifestyles and display great diversity in body size, body form, and ecology. In addition to the familiar features of lactation and hair, extant mammals have other derived characters that set them apart from other vertebrates. An understanding of mammalian diversity and specializations requires an understanding of how monotremes, marsupials, and placentals differ from each other in reproduction, some anatomical characters, and other aspects of their biology.

Although mammals are perceived as the dominant terrestrial animals of the Cenozoic, they are the smallest major vertebrate lineage in terms of species diversity, comprising about 5,500 extant species—substantially fewer than the number of amphibian species (~7,500) and just over half the number of birds or lepidosaur reptiles (~10,000 species each).

Mammals do, however, include the largest extant terrestrial and aquatic vertebrates. A large male African elephant (*Loxodonta africana*) weighs around 5,500 kg, about the same as a medium-size dinosaur like *Triceratops*. The blue whale (*Balaenoptera musculus*) weighs about 110,000 kg, which is half again the estimated weight of the largest dinosaur (the sauropod *Argentinosaurus*), making it the largest animal ever known. Mammals have great morphological diversity—no other vertebrate group has forms as different from each other as, for example, a whale and a bat. Even among strictly terrestrial mammals, there are tremendous morphological differences; consider, for example, a mole and a giraffe. It is prudent to remember, however, that mammals do not rise above other vertebrates when some measures of evolutionary success, including species diversity, are considered.

25.1 Major Lineages of Mammals

Mammalia is divided into three groups (traditionally referred to as subclasses): **Allotheria** (multituberculates, now extinct), **Prototheria** (monotremes; the platypus and echidnas), and **Theria**. Therians are further divided into two lineages:

- **Metatheria**, including marsupials and their extinct stem group relatives (**Figure 25.1**)
- **Eutheria**, including placentals and their extinct stem group relatives (**Figure 25.2**)

Although this classification does not take into account the large diversity of Mesozoic mammals (see Chapter 24), these three groups reflect basic divisions in body plans among those mammals that survived into the Cenozoic. They are all within crown group Mammalia and share several morphological characters not seen in more primitive mammals (technically called stem mammals, or non-mammalian mammaliaforms; see Figure 24.1). These characters include the reduction of the amniote pterygoid flange (see Chapter 10) to a tiny projection called the pterygoid hamulus, ossification of the maxilloturbinate bones in the nose, expansion of the occipital condyles for the articulation of the head on the neck; and the loss of various small bones from the dermal skull roof (e.g., the quadratojugal and tabular) and lower jaw (e.g., the splenial).

It should be noted that the term "placental mammal" is a bit of a misnomer, because marsupials also have a placenta, although it is less extensive than that of placentals. The term is in part a legacy of an earlier terminology that distinguished those animals that uniquely possess a marsupium—an external pouch in which the young complete development. "Marsupials" and "placentals" are commonly used to refer to the radiation of extant forms (i.e., the crown groups). The original metatherian–eutherian divergence probably occurred in Asia, because that is where the earliest members of both groups are found.

In addition to the currently accepted phylogeny of extant mammals shown in in Figures 25.1 and 25.2, the phylogeny of synapsids in Figure 24.1 reflects the current view (based on morphological and molecular data) that multituberculates (Multituberculata) are more closely related to therians than are monotremes. Extant monotremes are

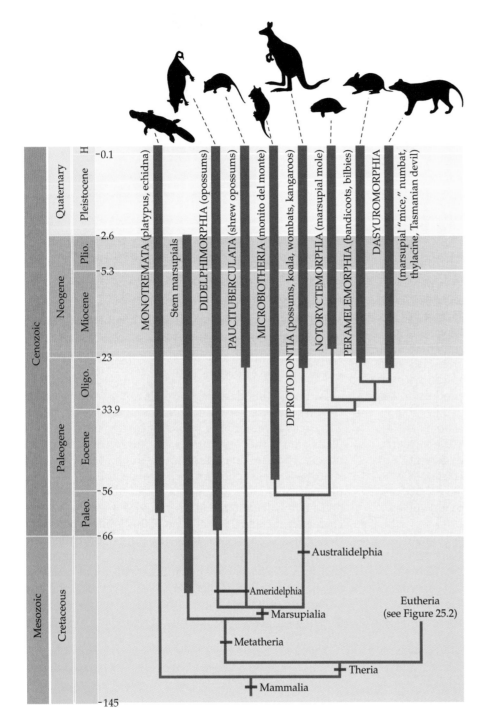

Figure 25.1 Phylogenetic relationships of extant mammals: Prototheria and Metatheria. This and Figure 25.2 (Eutheria) depict the probable relationships among the major groups of extant mammals, based on both morphological and molecular data. Narrow lines show relationships only; they do not indicate times of divergence or the unrecorded presence of taxa in the fossil record. Only the best-corroborated relationships are shown. Monotremata as shown here represents only the crown group (families Ornithodelphidae [platypus] and Tachyglossidae [echidnas]). These lineages are further described in Table 25.1.

Multituberculates

The order Multituberculata includes a few dozen families and was a very long-lived group, extending from the Late Jurassic to the middle Cenozoic. Multituberculates were mostly small mammals—rabbit size or smaller, with a few species the size of a medium-size dog. They had a narrow pelvis, leading to the inference that they were viviparous and gave birth to altricial young (**Figure 25.3**).

Multituberculates get their name from their teeth—specifically their molars, which are broad and multicusped (multituberculed), specialized for grinding rather than shearing, indicative of an omnivorous or herbivorous diet. Many multituberculates also had large, bladelike premolars, possibly used to open hard seeds. Their limbs were held in a more abducted (out to the side) position than is seen in therians. Most multituberculates were terrestrial, but a few appear to have had a prehensile tail and ankles that allowed their feet to rotate backward, and were probably arboreal. Multituberculates were probably rather like extant rodents in their mode of life. Their extinction might have been due to competition with rodents, which first appeared in the late Paleocene.

toothless as adults, so the characters of dentition on which much of mammalian phylogeny is based cannot easily be used to evaluate their relationships. However, juvenile platypuses have teeth, and teeth are known from fossil platypuses. The jaw of *Steropodon* from the Early Cretaceous of Australia shows fully formed teeth that have a triangular arrangement of cusps, unlike the less complex teeth of many Mesozoic mammals. These teeth, and other features, place monotremes within the southern radiation of mammals during the Mesozoic, Australosphenida, while the other Cenozoic mammals are contained within the northern radiation, Boreosphenida (see Chapter 24).

Monotremes

Monotremes are best known for being egg-laying mammals. They are grouped in the order Monotremata (Greek *mono*, "one"; *trema*, "hole"), referring to the single opening of the excretory and reproductive tracts (the cloaca). (An obsolete term for therians was Ditremata, referring to the separate openings of the excretory and reproductive tracts.)

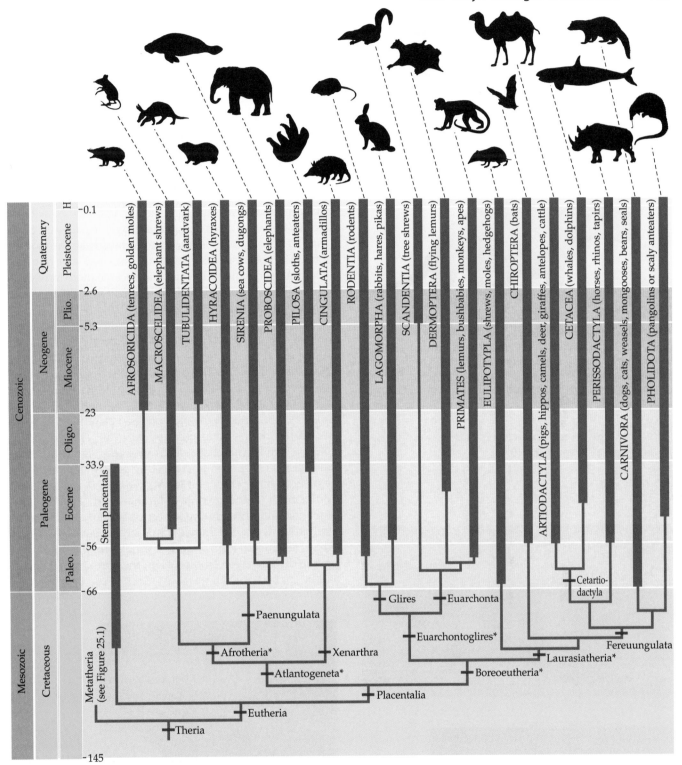

Figure 25.2 Phylogenetic relationships of extant mammals: Eutheria. Afrotheria and Xenarthra are placed here as sister taxa, although there is some debate about this; one or the other of these groups may be basal to all other placentals (that is, either Afrotheria or Xenarthra may be the sister taxon to the Boreoeutheria). Also note that some recent molecular analyses place bats (Chiroptera) as the sister group to ungulates (Cetartiodactyla + Perissodactyla). Higher taxa for which the primary (or only) evidence is molecular are indicated by an asterisk (*). These lineages are further described in Table 25.2.

There are two extant families of monotremes:

- Ornithorhynchidae (Greek *rhynchos*, "beak") contains the platypus (*Ornithorhynchus anatinus*), a semiaquatic animal that feeds on aquatic invertebrates in the streams of eastern Australia and Tasmania (**Figure 25.4A**).

- Tachyglossidae (Greek *tachy*, "swift"; *glossa*, "tongue") contains the short-beaked echidna (*Tachyglossus aculeatus*) of Australia and New Guinea, which eats mainly ants and termites; and three species of long-beaked echidnas in the genus *Zaglossus* (Greek *za*, "very") that

(A)

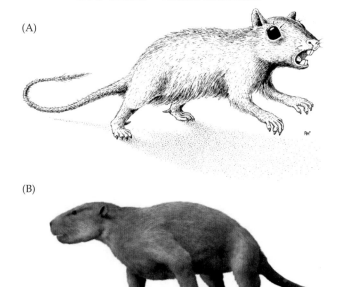

(B)

Figure 25.3 Multituberculates. (A) The Late Cretaceous Mongolian *Catopsbaatar*, a small (rat-size) scampering form. (B) The early Cenozoic North American *Taeniolabis*, a 25-kg (wombat-size) terrestrial form. (A, Zofia Kielan-Jaworowska and Jørn H. Hurum. Artwork by Bogusław Waksmundzki/CC BY 2.0; B, Nobu Tamura/CC BY-SA 4.0.)

include earthworms in their diet and are found only in New Guinea (**Figure 25.4B,C**).

(A)

(C)

Some teeth from the Cretaceous of Australia provide hints of a Mesozoic diversity of monotremes, but all these animals were fairly small, 1–2 kg at most; extant monotremes are at least twice this size. The few known Cenozoic monotreme fossils represent only echidnas or platypuses, and monotreme diversity has probably always been fairly limited. That is to say, the platypus and echidna are not the relicts of a once much larger radiation, but rather have been its mainstay. A handful of teeth and femur fragments from the Paleocene of Patagonia (the ornithorhynchid *Monotrematum sudamericanum*) provide the only evidence of monotremes outside Australia and New Guinea. Molecular data show a relatively recent (48–19 Ma) split of echidnas from the platypus. Since early Cenozoic platypuses seem to be much like the modern form, it has been suggested that echidnas may have originated from a semiaquatic lineage that became more terrestrial, a hypothesis supported by studies of their myoglobin, which resembles that of aquatic animals.

The semi-sprawling stance of monotremes (see Figure 25.4A), reminiscent of that of more generalized amniotes, may be a specialization for swimming or digging rather than a basal mammalian condition, but the elbow joint of monotremes is more like that of cynodonts than of therian mammals. Monotremes also have their own unique specializations. Both the platypus and echidnas lack teeth as adults and have a leathery bill or beak. This beak contains receptors that sense electromagnetic signals from the muscles of other animals, detecting prey underwater or in a termite nest. Thus, the leathery beaks of monotremes differ from the horny bills of birds in both structure and function. Interestingly, the Miocene platypus *Obdurodon* retained teeth but had a smaller infraorbital canal in its skull than the modern form does. This canal is for the infraorbital nerve that feeds sensation back to the brain from the front

(B)

Figure 25.4 Extant monotremes. (A) The platypus (*Ornithorhynchus anatinus*) of eastern Australia and Tasmania is the size of a large house cat. (B) The short-beaked echidna (*Tachyglossus aculeatus*) of Australia and New Guinea is the size of a corgi dog. (C) The long-beaked echidna *Zaglossus bruijni* of New Guinea is a little larger than the short-beaked echidna. (A, Brisbane City Council/CC BY 2.0; B, Steve Bittinger/CC BY 2.0; C, © Klaus Rudloff.)

Table 25.1	Extant marsupials discussed in this chapter
Order	**Lower taxa**
Ameridelphia	
Didelphimorphia	Didelphidae (opossums): ~95 species in Central and South America, 1 species in North America; semi-arboreal omnivores; 20 g to 6 kg.
Paucituberculata	Caenolestidae (shrew or rat opossums): 7 species in South America; carnivores and invertivores; 15–40 g
Microbiotheria	Microbiotheriidae (monito del monte): 1 species in South America; omnivore; 25 g
Australidelphia	
Diprotodontia	About 140 species in Australia and New Guinea; mostly herbivores; 19 g to 90 kg
	Phalangeridae (cuscuses and brushtail possums): 27 species
	Petauridae (possums, some of them gliders): 11 species
	Tarsipedidae (honey possum): 1 species
	Vombatidae (wombats): 3 species
	Phascolarctidae (koala): 1 species
	Potoroidae (bettongs, potoroos, rat kangaroos): 10 species
	Macropodidae (kangaroos and wallabies): ~66 species
Notoryctemorphia	Notoryctidae (marsupial moles): 2 species in Australia; invertivores; 50 g
Peramelemorphia	About 22 species in Australia and New Guinea, including Peramelidae (bandicoots) and Thylacomyidae (bilbies); omnivores; 100 g to 5 kg
Dasyuromorpha	Dasyuridae (marsupial "mice," quolls, Tasmanian devil): ~75 species in Australia and New Guinea; carnivores or insectivores; 12 g to 90 kg
	Myrmecobiidae (numbat): 1 species in Australia; myrmecophage, feeding primarily on termites; 350 g

of the face. The enlarged infraorbital nerve of the modern platypus not only suggests a better-developed electrosensory system but may also explain the lack of teeth, as this nerve occupies the position where the roots of the upper teeth would otherwise lie.

The platypus genome has more bird characteristics than do the genomes of therians, and monotremes are apparently similar to birds in other ways. For example, the spermatozoa of the platypus are threadlike, like those of birds, rather than having a distinct head and tail like the spermatozoa of therians. These characteristics do not represent a close link between monotremes and birds, but rather are basal amniote features that have been retained in monotremes and birds and lost in therians.

Male monotremes have a spur on the hindleg that is attached to a venom gland, which is used to poison rivals or predators. A similar spur is seen in some Mesozoic mammals, so this may not be a unique monotreme feature. Venom is a rare character in mammals, but it is present in the saliva of some shrews, moles, solenodons (rat-size insectivores), vampire bats (where it acts as an anticoagulant rather than a poison), and even primates (the slow loris).

Marsupials

Extant marsupials can be divided into seven orders (**Table 25.1**). A fundamental split divides the extant marsupials into Ameridelphia of the New World and Australidelphia of (mainly) Australia. Originally proposed on morphological grounds, this division is now supported by molecular and genetic data. Australidelphia is monophyletic, while Ameridelphia is thought to be paraphyletic, although interpretations differ as to whether Didelphimorphia or Paucituberculata is the basal group. Deltatheroidea, a now-extinct third group from the Late Cretaceous of Asia, falls outside of the grouping of extant marsupials.

The ameridelphian order Didelphimorphia includes the opossums (**Figure 25.5A**). Opossumlike forms were diverse from the Late Cretaceous to middle Cenozoic, when they were found worldwide except in Australia and Antarctica (although they were sparse outside of South America, and extremely rare in the Old World). Most of these lineages, which were mostly largely extinct by the end of the Eocene, were stem marsupials rather than true didelphimorphs. Present-day opossums (didelphids) are a diverse group of small to medium-size marsupials, mainly arboreal or semiarboreal omnivores and including animals such as the frugivorous woolly opossums and the otterlike water opossum, or yapok (*Chironectes minimus*). The other ameridelphians are Paucituberculata ("shrew" or "rat" opossums, also known as caenolestids), which are small, mainly terrestrial, shrewlike forms (**Figure 25.5B**) that are carnivores or invertivores (eating invertebrates other than insects, such as earthworms). Several extinct groups of stem marsupials lived in South America in the Cenozoic. Perhaps the most interesting of these were the borhyaenoids (order Sparassodonta, Eocene–Pliocene), which formed a large radiation of carnivorous forms, including badger-, dog-, and bearlike forms, and even a Neogene saber-toothed parallel, *Thylacosmilus*.

The remaining South American marsupial is the monito del monte (*Dromiciops gliroides*), a tiny, arboreal, mouselike animal in the montane forests of Chile and Argentina (**Figure 25.5C**). This species, which is placed in its own order (Microbiotheria) is more closely related to the Australian marsupials than to the other South American marsupials, and it may be a survivor of a lineage of originally South American marsupials that migrated across Antarctica to Australia in the early Cenozoic. Most phylogenetic analyses place this lineage as the sister taxon to the Australian marsupials, as shown in Figure 25.1. However, some analyses place Microbiotheria within the cluster of Australian marsupials, suggesting a more complex pattern of interchange

Figure 25.5 Diversity of extant and recently extinct marsupials.
(A) Didelphimorphia: The North American opossum (*Didelphis virginiana*), the size of a large cat, is the only North American marsupial. The naked tail is prehensile. (B) Paucituberculata: The mouse-size dusky shrew opossum (*Caenolestes fuliginosus*) lives in northwestern South America. (C) Microbiotheria: The mouse-size monito del monte (*Dromiciops gliroides*) lives in southwestern South America. (D) Diprotodontia: The mouse-size sugar glider (*Petaurus breviceps*) is a possum found in the forests of eastern Australia; other species live in northern Australia and New Guinea. (E) Diprotodontia: The bulldog-size common wombat (*Vombatus ursinus*) lives in eastern Australia and Tasmania. (F) Diprotodontia: The long-nosed potoroo (*Potorous tridactylus*) of eastern Australia and Tasmania is the size of a large cat. Like their kangaroo cousins, potoroos can hop, but they usually bound on all fours. (G) Peramelemorphia: The bilby, or rabbit-eared bandicoot (*Macrotis lagotis*) is the size of a large rabbit and inhabits northwestern and central Australia. (H) Dasyuromorphia: The mouse-size fat-tailed dunnart (*Sminthopsis crassicaudata*) is one of the "marsupial mice" and is found over much of southern and central Australia. (I) Dasyuromorphia: The numbat (*Myrmecobius fasciatus*), an anteater the size of a small cat. (J) Dasyuromorphia: The coyote-size thylacine (*Thylacinus cynocephalus*) of Tasmania, a generalized carnivore. The last known individual died in captivity in 1936. Despite being compared with placental canids, the thylacine was not a pack-hunting predator but a generalized carnivore. (A, Paul Hurtado/CC BY 2.0; B, courtesy of Jórge Brito M.; C, José Luis Bartheld/CC BY 2.0; D, Joe McKenna/CC BY 2.0; E, Greg Schechter/CC BY 2.0; F, Richard Thomas/CC BY-ND 2.0; G, Bernard Dupont/CC BY-SA 2.0; H, Alan Couch/CC BY 2.0; I, dilettantiquity/CC BY-SA 2.0; J, E. J. Keller, Smithsonian Institution Archives.)

between South American and Australian marsupials than a single immigration to Australia (see Chapter 23).

The Australian australidelphians fall into four orders. The largest group of marsupials is Diprotodontia (Greek *dis*, "two"; *pro*, "first"; *odous*, "tooth"). The group is named for the modified lower incisors, which project forward, rather like the incisors of rodents. This lineage includes mostly herbivorous and omnivorous forms today, although marsupial "lions," such as the Pleistocene *Thylacoleo*, appear to have evolved carnivory from a herbivorous ancestry.

The three major radiations within Diprotodontia are the phalangeroids, vombatiformes, and macropodoids. Phalangeroids (probably not a monophyletic group) represent an arboreal radiation of six families of somewhat primatelike animals, including various types of possums, three lineages of gliders, including the sugar glider (*Petaurus breviceps*; **Figure 25.5D**, which is convergent on the Northern Hemisphere "flying" squirrel); and the diminutive honey possum or noolbenger (*Tarsipes rostratus*), which is one of the few nectar-eating mammals apart from the bats. Vombatiformes include the arboreal koala (*Phascolarctos cinereus*; see Figure 23.7C) and the terrestrial burrowing wombats (**Figure 25.5E**), three species of specialist grazers with ever-growing molars. Extinct vombatiformes include the sheep- to rhinoceros-size Diprotodontidae, which looked like giant wombats and roamed the Plio-Pleistocene Australian savannas. The marsupial lions (Thylacoleonidae) also appear to fit within this grouping (see Figure 23.6A).

Macropodoids are the hopping marsupials. They include the small, omnivorous rat kangaroos (potoroos; **Figure 25.5F**) and the larger, herbivorous true kangaroos (including wallabies and the secondarily arboreal tree kangaroos). The largest kangaroos today have a body weight of about 90 kg (the size of a large man), but larger ones (up to three times that size) existed in the Pleistocene, including a radiation of one-toed, browsing sthenurine kangaroos, now all extinct. Sthenurines, with their short faces and long forearms, could be the source of legends of giant rabbits in Australia's interior. Perhaps, as desert forms, they also had large rabbitlike ears for cooling. The largest ones may have been too big to hop, and most of them appear to have been capable of walking (rather than hopping) on their hindlegs.

Notoryctemorphia includes only two species of marsupial moles, tiny beasts that, like some African placental golden moles, are sand swimmers rather than earth diggers like the true moles of the Northern Hemisphere.

Peramelemorphia includes three families: bandicoots, bilbies, and the pig-footed bandicoot (*Chaeropus ecaudatus*). The long-eared bilby (*Macrotis lagotis*; **Figure 25.5G**) looks a little like a rabbit (indeed, Australians celebrate an Easter Bilby rather than an Easter Bunny) but the bilby is omnivorous rather than herbivorous. The pig-footed bandicoot, extinct since the 1950s, was a tiny, spindly-legged, more herbivorous form that resembled the placental elephant shrews of Africa.

Dasyuromorphia is composed mainly of the carnivorous dasyurids; the numbat, or marsupial anteater; and the recently extinct thylacinids. Dasyurids include a large diversity of marsupial "mice" (**Figure 25.5H**) that would be better called marsupial shrews, since they are carnivorous and insectivorous rather than omnivorous, plus a few larger carnivores—the cat-size quolls and the badger-size Tasmanian devil (*Sarcophilus harrisii*). The numbat (*Myrmecobius fasciatus*) is a myrmecophage that eats mainly termites; it looks more like a miniature fox than like placental anteaters (**Figure 25.5I**). Like the Tasmanian devil, the thylacine

(*Thylacinus cynocephalus*; **Figure 25.5J**), occurred in recent times only on the island of Tasmania off Australia's south coast; the species was recently declared extinct, although some people hope for its continued existence.

Placentals

Extant placental mammals can be grouped into distinct lineages, but their interrelationships have been a source of controversy, which suggests that the diversification of these groups from an ancestral stock occurred very rapidly and without leaving many morphological clues. Problems remain with the resolution of the phylogeny (such as the interrelationships within Laurasiatheria). Stem placentals (i.e., eutherians falling outside the grouping of the extant orders) were thought to have been confined mainly to the Mesozoic, but the early Cenozoic order Taeniodonta (rather wombat-like animals) has now been shown to fall on the placental stem, and that may also be true of the insectivore-like Leptictida.

Over the past decade, molecular studies have resulted in a view of placental interrelationships that is very different from that of phylogenies based on morphological data alone. The molecular data support a division of placental mammals into two major groupings: a southern **Atlantogeneta** (including the African afrotheres and the South American xenarthrans) and a northern **Boreoeutheria**. Not all analyses place the afrotheres and xenarthrans together, however, and either one may be more closely related to Boreoeutheria. Although the southern lineages were isolated from the northern ones for much of the Cenozoic, molecular evidence suggests a northern origin for placentals as a whole. The current scheme of interrelationships among placentals is shown in Figure 25.1. **Table 25.2** describes the diversity of extant placentals.

Afrotheria, as the name suggests, is a grouping of endemic African mammals. A grouping of Paenungulata—the rodentlike hyraxes (**Figure 25.6B**), sirenians (manatees and the dugong, *Dugong dugon*), and proboscideans (elephants and extinct relatives)—had long been supported by morphology, but molecular data show that the aardvark (*Orycteropus afer*), elephant shrews (**Figure 25.6A**), tenrecs (Madagascan insectivores; see Figure 23.9D), and golden moles are also closely related to these larger African mammals. Afrotheria is the product of a long period of independent isolated evolution of mammals on the African continent. Other mammals that we think of as African (e.g., lions, hyenas, giraffes, and antelope) did not reach the continent until Africa collided with Eurasia around 17 Ma. Primates are the only extant long-term residents of Africa that are not afrotheres.

The grouping of sirenians, proboscideans, and the now extinct desmostylians is called Tethytheria. Both sirenians and desmostylians are aquatic forms, and some early proboscideans have anatomical features that suggest a semiaquatic existence—an interpretation supported by isotopic signatures in their teeth, and also by the myoglobin biochemistry of extant elephants. Later proboscideans may

Table 25.2	Extant placentals discussed in this chapter
Order	**Lower taxa**
Afrotheria	
Afrosoricida	About 40 species in Africa and Madagascar; includes Tenrecidae (tenrecs and otter shrews) and Chrysochloridae (golden moles); carnivores, invertivores, and insectivores; 5 g to 2 kg
Macroscelidea	Macroscelididae (elephant shrews): 19 species in Africa and Madagascar; primarily insectivores; 25–500 g
Tubulidentata	Orycteropidae: A single species, the aardvark, in Africa; myrmecophage; ~60 kg
Hyracoidea	Procaviidae: 7 species of hyraxes in Africa and the Middle East; all herbivores; ~4 kg
Sirenia	Aquatic herbivores; 140 to >1000 kg
	Dugonidae (dugong): 1 species; found in coastal regions of eastern Africa, the Red Sea, and most of coastal Asia
	Trichechidae (manatees): 3 species; found in coastal regions of Florida through the West Indies to northern South America
Proboscidea	Elephantidae (elephants): 1 species in tropical Asia, 2 species in tropical Africa; enormous terrestrial herbivores, 4500 to 7000 kg.
Xenarthra	
Pilosa	Found throughout Central and South America; 350 g to 40 kg
	Bradypodidae (three-toed sloths): 3 species; herbivores
	Megalonychidae (two-toed sloths): 2 species; herbivores
	Myrmecophagidae (anteaters): 3 species; myrmecophages
	Cyclopedidae (silky anteater): 1 species); myrmecophage
Cingulata	Dasypodidae (armadillos): 18 species; southern North America to tropical South America; most are omnivores; 120 g to 60 kg
Glires	
Rodentia	Found worldwide[a]; herbivores and omnivores; 7 g to >50 kg
	Sciuromorpha (squirrels, mountain beaver, dormice): 3 families, ~ 400 species; primarily herbivores
	Anomaluromorpha (African flying squirrels, spring hares): 2 families, 8 species; herbivores
	Castorimorpha (beavers, kangaroo rats, gophers): 3 families, ~100 species; herbivores and seed-eaters
	Myomorpha (rats, mice, voles, jerboas): 7 families, ~1600 species; omnivores and herbivores
	Hystricomorpha (African and American porcupines, capybara, guinea pig and other caviomorphs): 18 families, ~330 species; herbivores
Lagomorpha	Leporidae (rabbits and hares) and Ochotonidae (pikas): 94 species found worldwide[a]; herbivores; 180 g to 7 kg
Euarchonta	
Scandentia	Tupaiidae (tree shrews): 20 species in tropical Asia; semiarboreal omnivores; ~400 g
Dermoptera	Cynocephalidae (flying lemurs): 2 species in tropical Asia; arboreal gliding herbivores; 1–2 kg
Primates	About 450 species of terrestrial and arboreal frugivores, folivores, insectivores, and omnivores; non-human primates are found worldwide except Australasia, although they are largely tropical; 85 g to >275 kg (See Chapter 26)
Laurasiatheria	
Eulipotyphla	Found worldwide except Australasia; 2 g to 1 kg
	Erinaceidae (hedgehogs and moonrats): ~40 species; omnivores
	Talpidae (moles and desmans): ~50 species; invertivores
	Soricidae (shrews): ~385 species; carnivores and invertivores
	Solenodontidae: 2 species; burrowing insectivores of Cuba and Hispaniola
Chiroptera	Flying mammals; found worldwide (bats are the only native mammals on New Zealand); 2 g to 1.2 kg
	Megachiroptera: Pteropodidae (fruit bats or flying foxes): ~170 species; frugivores
	Microchiroptera:, 17 families, >1000 species of insectivores including Rhinolophidae (horseshoe bats) and Phyllostomidae (leaf-nosed bats). (Phyllostomidae includes the 6 species of vampire bats, obligate sangivores [blood feeders])

[a] For terrestrial species, the designation "worldwide" is understood to exclude Antarctica. In Australia, aboriginal invaders introduced dogs (dingoes) and European settlers introduced rabbits, foxes, deer, rats, and mice as well as domesticated cats, dogs, sheep, pigs, cattle, camels, and horses, although none of these groups is native to Australia.

[b] The exact number of species is difficult to determine due to a long history of domestication and interbreeding.

Table 25.2	Extant placentals discussed in this chapter
Order	**Lower taxa**
Fereuungulata	
Artiodactyla	Even-toed ungulates; found worldwide; primarily herbivores; 2–1400 kg
	Suina:
	Suidae (pigs): 18 species
	Tayassuidae (peccaries), 3 species
	Anconodonta: Hippopotamidae (hippos): 2 species
	Tylopoda: Camelidae (camels and llamas): 4–6 species[b]
	Ruminatia:
	Tragulidae (mouse deer): 10 species
	Moschidae (musk deer): 7 species
	Cervidae (deer, elk, moose, others): ~90 species
	Giraffidae (giraffes, okapi): 4 species of giraffe, 1 species of okapi
	Antilocapridae (the pronghorn "antelope"): 1 species
	Bovidae (antelope, sheep, cattle): ~140 species
Cetacea	Secondarily marine mammals; found in oceans worldwide, and in some freshwater river habitats of India and South America; 20–120,000 kg
	Mysticeti (baleen whales): suspension feeders, consume zooplankton and small crustaceans
	Balaenidae (bowhead and right whales): 3 species
	Balaenopteridae: 6 species (including the blue whale, the largest animal)
	Estrichtiidae (gray whale): 1 species
	Odontoceti (toothed whales): carnivores, feed on invertebrates and vertebrates
	Delphinidae (dolphins, orcas, pilot whales, others): 17 species
	Iniidae (South American river dolphins): 3 species
	Monodontidae (beluga and narwhal): 2 species
	Phocoenidae (porpoises): 6 species
	Physeteridae (sperm whale and pygmy sperm whale): 2 species
	Platanistidae (Ganges and Indus River dolphins): 2 species
	Ziphiidae (beaked whales): ~21 species
Perissodactyla	Odd-toed ungulates; 150–3600 kg; herbivores
	Equidae (horses, asses, zebras): 7 species; worldwide[a]
	Rhinoceratidae (rhinoceroses): 5 species in Africa, India, and southern Asia
	Tapiridae (tapirs): 5 species in the tropical Americas and Southeast Asia
Carnivora: Caniformia	Doglike forms; terrestrial caniforms found worldwide[a]; marine caniforms found along shores of all continents; mostly carnivorous, some omnivorous, 1 herbivore (giant panda); <100 g to 800 kg, (marine forms up to 2000 kg)
	Canidae (wolves, coyotes, foxes, domestic dog, others): ~30 species
	Ailuridae (red panda): 1 species
	Ursidae (bears, including the giant panda): 8 species
	Procyonidae (raccoons, coatis, kinkajous): 14 species
	Mustelidae (weasels, badgers, mink, wolverine, otters, others): ~60 species
	Mephitidae (skunks): 12 species
	Otariidae (sea lions): 14 species
	Odobenidae (walrus): 1 species
	Phocidae (seals): 19 species
Carnivora: Feliformia	Catlike forms; found worldwide[a]; almost entirely carnivorous; ~600 g to 300 kg.
	Felidae (cats, including lions, tigers, bobcats, cheetahs, domestic cat, others): ~40 species
	Nandiniidae (African palm civet): 1 species
	Prionodontidae (linsangs): 2 species
	Herpestidae (mongooses, including the meerkat): 34 species
	Hyaenidae (hyenas): 4 species, including the myrmecophageous aardwolf
	Viverridae (civets and genets): ~36 species
	Eupleridae (Malagasy carnivores): 10 species, including the fossa
Pholidota	Manidae (pangolins, also known as scaly anteaters): 8 species in tropical Africa and Asia; scale-covered myrmecophages; 2–33 kg

(A)

(B)

(C)

(D)

(E)

(F)

(G)

(H)

Canine

(I)

represent a secondary return to a more terrestrial existence from an originally semiaquatic stock. However, popular speculation that the elephant's trunk was originally a snorkel is mistaken; fossil skulls of early proboscideans show that these animals lacked trunks.

Xenarthra comprises two endemic South American orders. Pilosa includes sloths and anteaters (**Figure 25.6C**), and Cingulata includes armadillos (**Figure 25.6D**). Xenarthrans are sometimes known as edentates, meaning "without teeth." All of these animals have simplified their dental pattern, although only anteaters are completely toothless. Xenarthrans have occurred in North America only since the late Miocene (see Chapter 23). The pangolins, or scaly anteaters, of Africa and Asia (order Pholidota) were originally thought to be related to xenarthrans based

on morphological features, but these features represent convergences due to a similar lifestyle, and molecular data place Pholidota as the sister taxon to Carnivora.

All other placental orders are grouped in Boreoeutheria, which can be divided into Euarchontoglires and Laurasiatheria. Euarchontoglires can be subdivided into Glires (rodents [**Figure 25.6E**] and rabbits) and Euarchonta (primates, tree shrews, and dermopterans; the last are flying lemurs, or colugos, and are gliders rather than fliers; **Figure**

◄ **Figure 25.6 Diversity of extant placentals.** (A) Macroscelidea: The rat-size golden-rumped elephant shrew (*Rhynchocyon chrysopygus*) of central East Africa has a head like a shrew but has long legs and looks like a tiny ungulate when it runs. (B) Hyracoidea: The rabbit-size rock hyrax (*Procavia capensis*) lives in eastern and southern Africa and the Middle East. Hyraxes look somewhat like rodents but have little hooves on their feet rather than claws. Rock hyraxes bask in sunny, sheltered sites on cold winter days and in the morning after emerging from their nighttime shelters.
(C) Pilosa: The cat-size northern tamandua (*Tamandua mexicana*) from Central America and northern South America is arboreal and much smaller than its more familiar cousin, the giant anteater.
(D) Cingulata: The mouse-size pink fairy armadillo (*Clamyphorus truncatu*), also known as the pichiciego, is a subterranean dweller in arid parts of Argentina and is the smallest of the armadillos.
(E) Rodentia: The mouse-size naked mole rat (*Heterocephalus glaber*) is found in northeastern Africa. Its enormous teeth are used for chiseling earth to make underground burrows. (F) Dermoptera: The cat-size Philippine flying lemur (*Cynocephalus volans*) is known from islands in Southeast Asia. Flying lemurs, or colugos, glide rather than fly. (G) Artiodactyla: The rabbit-size lesser mouse deer (*Tragulus kanchil*) is found across Southeast Asia. Mouse deer are the smallest extant ungulates, but they are not true deer, and do not have antlers; the males have enlarged upper canines (see inset). (H) Perissodactyla: The pony-size lowland or Brazilian tapir (*Tapirus terrestris*) lives in rainforest in northern and central South America. Its feet have both hooves and a foot pad, with four toes on the forefeet and three on the hind feet. (I) Carnivora: The coyote-size aardwolf (*Proteles cristata*) is known from eastern and southern Africa. Unlike other hyenas, which are carnivorous, the aardwolf has become specialized for eating termites, and has small teeth. (Photos: A, Kim/CC BY-SA 2.0; B, F, Bernard Dupont/CC BY-SA 2.0; C, ryancandee/CC BY 2.0; D, Cliff/CC BY 2.0; E, Roman Klementschitz/CC BY-SA 3.0; G photo by Kristian Bell, G inset, Sakurai Midori; H, photo byKarelj; I, Greg Hume/CC BY-SA 3.0.)

25.6F). Bats were originally thought to be related to the euarchontans, but molecular data now place them within the Laurasiatheria, although their precise relationship to other laurasiatherian orders remains contentious.

Within Laurasiatheria, insectivores (originally considered to belong to a single order, Insectivora) were often considered to be the basal stock from which other placentals were derived. However, although modern insectivores such as shrews may superficially resemble ancestral placental mammals, they are not closely related to them. Tree shrews and elephant shrews were recognized as being morphologically distinctive (and placed into their own orders) by the mid 20th century, but more recently molecular data showed that tenrecs and golden moles were also not allied with traditional "insectivores" but instead belong with the afrotheres. The remaining insectivores (shrews, hedgehogs, and moles) are now placed in the order Eulipotyphla.

The largest grouping within Laurasiatheria is Fereuungulata (often spelled Ferungulata), mainly comprising carnivores and ungulates. The extant ungulates (hoofed mammals), include **artiodactyls** (even-toed ungulates; **Figure 25.6G**) and **perissodactyls** (odd-toed ungulates; **Figure 25.6H**). These two orders are probably sister taxa, but molecular phylogenies vary on their exactly placement with respect to Carnivora + Pholidota. Some archaic ungulates in the early Cenozoic were also omnivorous or even carnivorous, including the mesonychids, once thought to be ancestral to whales, porpoises, and dolphins (Cetacea). In South America, there were several hoofed orders that are now extinct. Two of these orders (Litopterna and Notoungulata) survived into the late Pleistocene—recently enough that collagen was available in their bones to include them in a molecular phylogeny. They were determined to be related to perissodactyls.

A surprise in the reexamination of mammalian interrelationships over the past few decades has been the realization that cetaceans are technically ungulates (even though they lack hooves), related to the artiodactyls, specifically hippos. This placement renders the traditional notion of Artiodactyla paraphyletic, and cetaceans and artiodactyls are now often grouped as the single order Cetartiodactyla (although we show them as separate orders here, as favored by some researchers). The origins and evolutionary history of cetaceans are discussed at the end of this chapter.

Carnivora (informally called carnivorans, to distinguish them from other, unrelated, carnivorous mammals) are distinguished by having a pair of specialized shearing teeth, or **carnassials**, formed from the last (fourth) upper premolar and the first lower molar. Carnivorans include not only specialized carnivores such as cats, dogs and hyenas but also the myrmecophagous aardwolf (*Proteles cristata*; **Figure 25.6I**); secondarily herbivorous forms such as the giant panda (*Ailuropoda melanoleuca*); and pinnipeds (seals, sea lions, and the walrus; see Figure 25.22), secondarily aquatic mammals that are related to bears.

Pholidota comprises the eight species of pangolins, which are scale-covered myrmecophagous mammals from Africa and southeast Asia. Pangolins include both terrestrial and arboreal species that feed only on ants and termites, tearing open their nests with the large claws on their forefeet and gathering them on a sticky tongue that is retracted by a muscle that extends back to the pangolin's pelvis.

25.2 Differences between Therians and Non-Therians

Therians are distinguished from monotremes by several derived characters, most obviously by being viviparous rather than laying eggs. (We describe modes of mammalian reproduction in Section 25.4.) In addition, therians and non-therians differ in characteristics of the skull, teeth and jaws, postcranial skeleton, and brain.

Craniodental features

Therians have completely lost the sclerotic bony rings around the eyes that are common in other amniotes, including non-mammalian synapsids. Monotremes retain sclerotic cartilages, although they do not ossify to form a bony ring. Monotremes and therians also differ in their jaw-opening muscles—the detrahens in monotremes and the digastric in therians—which reflects a legacy of their different histories of incorporating the articular bone of the lower jaw into the middle ear as the malleus. Therians also have tribosphenic molars, a distinctive new type of molar that can crush food in addition to shearing it (see Figure 24.14).

Postcranial skeletal features

Although the size of the cervical ribs is reduced in mammals relative to the condition in cynodonts, it is only in therians that the cervical ribs are further reduced and fused with the cervical vertebrae, forming the transverse processes of these vertebrae (**Figure 25.7A,B**). The therian elbow has been modified from the generalized amniote joint that allows some rotation to a hinge that limits motion to the fore-and-aft (parasaggital) plane. A more pronounced olecranon process of the ulna forms a lever arm for the triceps muscles that push the forefoot off the ground during locomotion; this arrangement is analogous to the calcaneal heel in the ankle. Despite decreased flexibility in the elbow joint, therians can rotate the radius in the lower arm around the ulna (pronation), so that the hand can be placed flat on the ground with the elbow tucked under the body, enabling their more erect stance.

The monotreme shoulder girdle is reminiscent of the reptilian condition, with a ventral midline interclavicle anterior to the sternum, and a large coracoid linking the interclavicle and sternum to the scapula (**Figure 25.7C**). Therians (with the exception of newborn marsupials; see below) have lost the interclavicle and reduced the coracoid to a small coracoid process fused with the scapula (**Figure 25.7D**). This reduction of the coracoid allows the therian glenoid fossa (the depression on the shoulder girdle for articulation with the humerus) to be oriented downward, rather than more laterally as in monotremes and other tetrapods. This change enables the more adducted posture of the therian forelimbs.

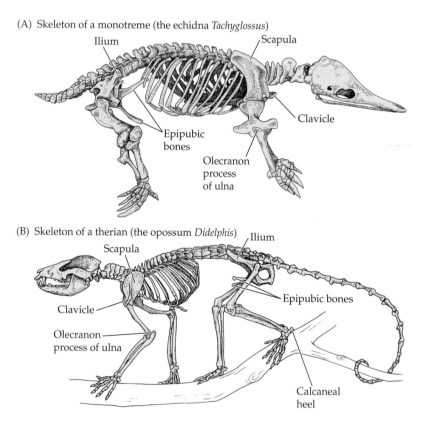

(A) Skeleton of a monotreme (the echidna *Tachyglossus*)

Ilium
Scapula
Clavicle
Epipubic bones
Olecranon process of ulna

(B) Skeleton of a therian (the opossum *Didelphis*)

Scapula
Ilium
Clavicle
Epipubic bones
Olecranon process of ulna
Calcaneal heel

Figure 25.7 Skeletal differences between monotremes and therians. Note that epipubic bones are present in monotremes and marsupials (depicted here) but have been lost in placentals. (A) The position of the limbs in monotremes, especially the forelimbs, is more abducted than in therians, reminiscent of the more general amniote condition. (B) In therians, the position of the limbs is more adducted, which enables therians to have a more derived form of locomotion, including bounding fast gaits. (C,D) Left shoulder girdles of a monotreme (C) and a therian (D). (E) The left foot and ankle bones of a therian. The astragalus is farther superimposed on the calcaneum than in the generalized mammalian condition, making the ankle action even more of a lever. (A,B, from Rogers 1986; C–E, from Walker and Liem 1994.)

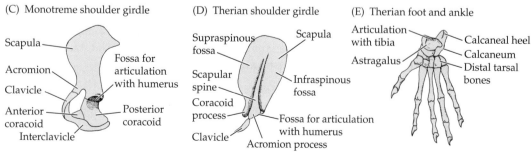

(C) Monotreme shoulder girdle

Scapula
Acromion
Clavicle
Anterior coracoid
Interclavicle
Fossa for articulation with humerus
Posterior coracoid

(D) Therian shoulder girdle

Supraspinous fossa
Scapula
Scapular spine
Infraspinous fossa
Coracoid process
Clavicle
Fossa for articulation with humerus
Acromion process

(E) Therian foot and ankle

Articulation with tibia
Astragalus
Calcaneal heel
Calcaneum
Distal tarsal bones

The shoulder joint of therians has a greater range of motion but less stability than the shoulder of monotremes, and the humerus is held in place by muscular action (the rotator cuff muscles in humans). Therians have also changed the appearance of the scapula. A scapular spine divides the blade into anterior and posterior portions (the supraspinous and infraspinous fossae). The base of the scapular spine extends out from the body of the bone into an acromion process, which is a point of muscle attachment as well as the point of articulation with the clavicle (collarbone; see Figure 25.7). The clavicle is retained in most therians but is lost in many cursorial (running) placentals (e.g., dogs and horses), and the acromion process is also reduced or lost in these mammals.

The anatomy of the therian shoulder girdle allows the scapula to move as an independent limb segment around its dorsal border, adding to the length of the stride during locomotion. Muscles rotating the base of the scapula forward originate from the cervical vertebrae. The original tetrapod humerus protractor, the supracoracoideus, has been modified into a new muscle, the supraspinatus, running from the humerus to the supraspinous fossa (**Figure 25.8A**). The supraspinatus helps stabilize the humerus, which is especially important in stabilizing the limb on the shoulder girdle while the forefoot is on the ground during locomotion. Other shoulder muscles have become modified to hold the limb girdle in a muscular scapular sling, including some new muscles derived from the postural hypaxial layer (**Figure 25.8B**). This muscle arrangement probably also increases scapular mobility during locomotion and cushions the impact of landing on the forelimbs during bounding.

While monotremes have the derived mammalian form of pelvis, with a reduced pubis, epipubic bones, and a rod-shaped ilium, therians have a modified configuration of the hip socket (acetabulum), with an inverted U-shaped articulation with the femoral head. This morphology reflects the more adducted position of the femur, associated with increased importance of the hindlimb in forward propulsion.

All mammals have a crurotarsal ankle joint between the astragalus and the tibia (see Chapter 14), but the astragalus of therians is superimposed farther on the calcaneum than in the generalized mammalian condition (**Figure 25.7E**). This configuration makes the action of the ankle joint even more of a lever, and therians also have a more pronounced calcaneal heel that increases the lever arm for the muscles retracting the foot (primarily the gastrocnemius, the calf muscle). Interestingly, this further extent of astragalar superposition apparently evolved convergently in placentals and marsupials, and their ankle bones are distinctly different.

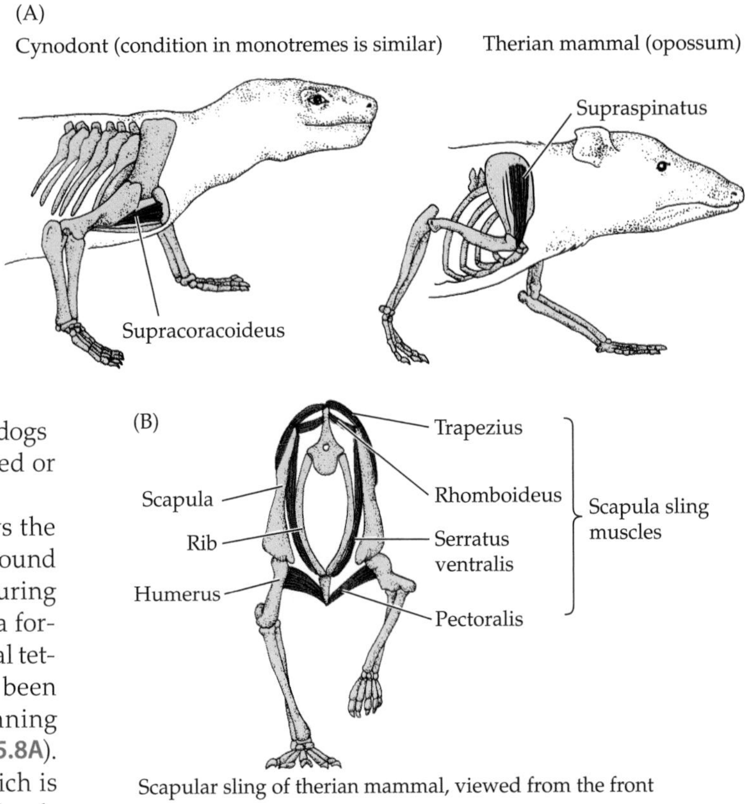

Gait and locomotion

The postcranial skeletal features discussed above can be interpreted in the context of therians adopting a new rapid gait: that of smooth, continuous bounding. Therians bound

Figure 25.8 **Specializations of shoulder girdle musculature.** (A) The supraspinatus muscle of mammals (right) is homologous to the generalized amniote supracoracoideus (illustrated here by a cynodont). The mammalian supraspinatus inserts on the scapula. (B) The rhomboideus and the serratus ventralis in the scapular sling of therians are unique muscles, derived from the hypaxial muscles. (A after Goslow Jr et al. 1989; B from Kardong 2012.)

by pushing off from the hindlimbs; dorsoventral flexion of the lumbar spine and rotation of the scapula while the forelimbs are extended increase the length of a bound. Rotation of the scapula, with the elbows tucked in under the body, allows the forelimbs to reach forward to extend the stride length. On landing, the scapular sling muscles cushion the impact, and flexion of the lumbar spine brings the back legs forward for the next bound.

This gait, at least in small mammals, is highly dependent on a flexed spine and flexed limbs that pivot from the top of the scapula/hip joints and then bend at the shoulder/knee and elbow/ankle joints in a three-part zigzag fashion (**Figure 25.9A**). The independent movement of the scapula is critical to this arrangement. Smaller mammals bound (**Figure 25.9B**), but larger mammals such as horses, representing the more derived mammalian condition, move with a stiffer back and straighter legs in a modified bounding gait called galloping (**Figure 25.9C**).

Multituberculates appear to have convergently evolved a type of bounding gait, even showing a reduction of the coracoid bone in the shoulder, which suggests they had some scapular sling musculature. But multituberculates retained a monotreme-like abducted limb posture, and

(A)

(B)

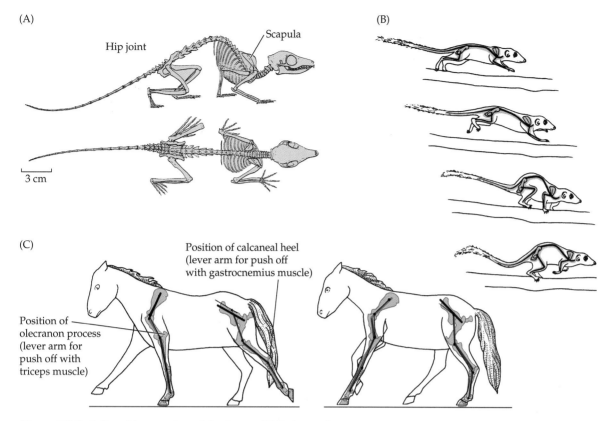

Hip joint

Scapula

3 cm

(C)

Position of calcaneal heel
(lever arm for push off
with gastrocnemius muscle)

Position of
olecranon process
(lever arm for
push off with
triceps muscle)

Figure 25.9 Gait and locomotion of therians.   (A) Skeleton of a tree shrew (*Tupaia glis*) in typical posture with flexed limbs and a quadrupedal stance. (B) Sequential phases of the bounding run in a tree shrew. Note the relatively flexed limbs and mobile back. (C) Sequential phases of the gallop in a horse; note the relatively straight limb angles and immobile back. In the general tetrapod condition, the stylopod (humerus/femur; indicated in red), zeugopod (radius + ulna/tibia + fibula; blue) and autopod (manus/pes; green) portions of the limbs are usually considered to be functional equivalents between forelimbs and hindlimbs, as shown on the horse on the left. However, with the mobility of the scapula in therians the functional equivalents change, so that the scapula is now the functional equivalent of the femur, and so on, as shown in the horse on the right in Figure 25.8C. This rearrangement allows the three-part zigzag flexion of the limbs described in the text. (A,B from Jenkins Jr. 1974; C after Biewener 1989.)

probably engaged in a rather ungainly froglike jumping, which they would not have been able to continue for more than several strides. This type of locomotion, besides being less efficient than that of therians, might have placed a limit on the body size of multituberculates.

Hearing

The evolution of the mammalian three-boned middle ear was discussed in Chapter 24, where we noted that a fully enclosed middle ear evolved convergently among the extant mammal lineages. Several other features contribute to increased auditory acuity in therians relative to the condition in monotremes. These include a long cochlea, coiled and fitting inside the skull, that is capable of better pitch discrimination (see Figure 24.10D), although the extent of coiling varies in extant marsupials and placentals, indicating a degree of independent evolution. In addition, the pinna

(external ear) helps determine sound direction. The pinna, in combination with the narrowing of the external auditory meatus (earhole) of mammals, concentrates soundwaves from the relatively large area encompassed by the external opening of the pinna to the small, thin tympanic membrane. The pinna is unique to therians; it is absent in monotremes, although it has a feathery analogue in certain owls. Most mammals can move their pinnae to detect sound, although anthropoid primates lack this ability. The auditory sensitivity of a terrestrial mammal is reduced if the pinnae are removed.

Aquatic mammals use entirely different systems to hear underwater and have reduced or lost the pinnae. Cetaceans, for example, use the lower jaw to channel soundwaves to the inner ear.

Vision

As we saw in Chapter 24, mammals have rod-dominated retinas and most have poor color vision. This is a derived condition that appears to have evolved independently in monotremes and therians. Five types of photosensitive molecules, each coded for by their own gene families, are found in vertebrates and all five were probably present in early synapsids (**Figure 25.10A**). Rods, which are effective at low light levels but do not detect colors, contain a visual pigment called RH1. The four opsins found in cones—RH2 (the so-called rod-like opsin), SWS1, SWS2 (short-wavelength-sensitive opsins), and LWS (long-wavelength-sensitive opsin)—respond to different light wavelengths and thus allow discrimination of colors (**Figure 25.10B**). Most mammals are not sensitive to ultraviolet light, but mutation

(A)

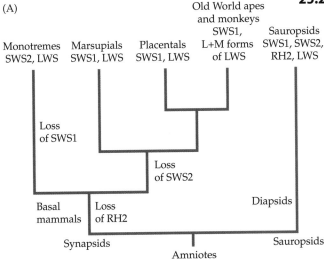

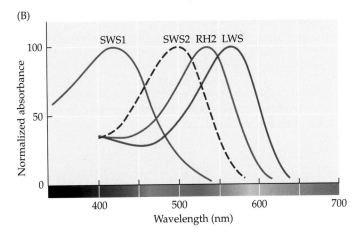

Figure 25.10 Changes in mammalian visual pigments.
(A) At least four types of opsins (pigments involved in color vision) were present in the common ancestor of synapsids and sauropsids. Sauropsids retained all of the ancestral opsins, but synapsids lost some of them. For example, monotremes lost SWS1, whereas marsupials and placentals lost SWS2. (B) The four opsins are sensitive to different wavelengths of light, as shown by their absorption curves. The photosensitivity of individuals in a given taxon depends on the combinations of opsins present. The mutations of opsins that produced trichromatic vision and UV sensitivity independently in some marsupials and placentals are not shown. (A after Jacobs 2013.)

of the SWS1 opsin gene has produced UV-sensitive opsins in a few rodents and marsupials, and UV sensitivity may be more widespread among diurnal mammals than is currently appreciated.

With multiple opsins, early amniotes would have had color vision, but loss of some opsins within the synapsid lineage led to the basic mammalian condition of dichromatism—having two opsins and limited capacity to distinguish colors (like a red-green color-blind human). All mammals retain the LWS opsin, which responds to light in the green/red range, and have lost the RH2 opsin (blue/green sensitivity). The situation with the two SWS (short-wavelength-sensitive) opsins, which respond to light at the blue/purple end of the spectrum, is more complicated.

Monotremes retain the SWS2 opsin, whereas therians retain SWS1. This distribution of opsins implies a convergent loss of color vision in monotremes and therians, suggesting that their common ancestor had both SWS1 and SWS2 opsins and was not as nocturnal as many modern mammals.

Anthropoid primates (primarily the Old World monkeys and apes) are unusual among mammals in having good color vision and a brain that is specialized for the visual sensory mode. Human trichromatic vision comes from a duplication of the LWS opsin gene into an L form with maximum sensitivity at 530 nm (more sensitive to green), and an M form with maximum sensitivity at 563 nm (more sensitive to red). This duplication may be the reason monkeys and humans are more sensitive to the color red than are most mammals. Some marsupials (e.g., the honey possum) have independently duplicated the LWS opsin gene and are also trichromatic. In contrast, many mammals (primarily nocturnal and aquatic ones, including bats and cetaceans) have lost the SWS1 opsin and its corresponding gene and are monochromatic.

Lactation

Although all mammals lactate, only therians have nipples, which allow the young to suck directly from the breast rather than from the mother's fur. Mammary hairs, probably a basal mammalian feature, are present in monotremes and marsupials, and the mammae (milk-secreting glands) develop from areola patches confined to the abdominal region. Placentals lack mammary hairs, and the mammae develop from mammary lines that form along the entire length of the ventral body surface.

The composition of milk varies little during the period of lactation in placentals. In marsupials, however, in which the mammary glands play a larger role in supplying nutrients to the offspring, milk changes markedly in composition, providing the correct balance of nutrients for each stage of development. The milk of marsupials initially is dilute and rich in protein, while later milk is more concentrated and richer in fats.

Lactation is a costly means of providing nutrients to the young. Transfer of nutrients via the milk is less efficient than transfer across the placenta, and the young also need energy to carry out their own physiological functions (e.g., respiration, excretion), which would otherwise be carried out by the mother during gestation.

Information from the genes

The hypothesis that all mammals inherited lactation from a common ancestor is supported by a comparison of their casein (milk protein) genes. Monotremes resemble other egg-laying amniotes in that the young obtain at least some nutrition from the egg yolk, but nourishment of developing monotremes is supplemented first by uterine secretions and after hatching by milk. The reduction in the contribution of the yolk to the nutrition of the developing monotreme young is reflected in the loss of genes (vitellogenin-encoding genes) that code for yolk production: there are three such genes in other amniotes, but only one functional one in monotremes, and one other is present as a pseudogene. All developing therians are nourished via a placenta, but marsupials (whose period of placenta-based nourishment is short) retain some yolk in their eggs, although placentals do not. No functional vitellogenin genes are seen in any therians, although portions of pseudogenes—nonfunctional relics of once-functional genes—remain. Comparison of the genomes of modern mammals suggests that at least some of the inactivation of vitellogenin genes took place independently in monotremes and therians, and also independently in marsupials and placentals.

Sex determination and sex chromosomes

Mammalian sex determination is genetic, and therians have heteromorphic sex chromosomes: female therians have two X chromosomes, and males have an X and a Y chromosome. The platypus, a monotreme, has multiple sex chromosomes, but the precise genetic mode of sex determination is unclear.

In therians, the male-determining gene on the Y chromosome initiates male gonadal development; in its absence, female gonads develop. Once a gonad has had its primary sex declared as female or male, one of the sex hormones—estrogen or testosterone, respectively—is produced. These hormones affect development of the secondary sexual characteristics. In humans, the genitalia, breasts, hair patterns, and differential growth patterns are secondary sexual characteristics. Horns, antlers, and color patterns are familiar sexual dimorphies in other mammals.

25.3 Differences between Marsupials and Placentals

Marsupials and placentals differ most conspicuously in their reproductive biology, and placentals differ from both marsupials and monotremes in their possession of brown fat (see Chapter 24). However, there are also major anatomical differences in the postcranial skull and skeleton of marsupials and placentals. Some of these skull differences are illustrated in **Figure 25.11**.

- The projection at the rear of the dentary, where the pterygoideus muscle inserts, bends inward in marsupials but is straight in placentals. This marsupial feature, termed the inflected (inturned) angle of the dentary, reflects differences in the muscles used during chewing.

- The nasal bones of most marsupials flare into a diamond shape posteriorly where they meet the frontal bones, whereas the nasal bones of most placentals are rectangular where they meet the frontals.

- The jugal bone of marsupials extends into the glenoid cavity where the lower jaw articulates; the jugal of most placentals ends before the glenoid cavity.

- The bones that form the base of the secondary palate of most marsupials sport two large holes (fenestrae); the palates of most placentals are solid bone.

- Many placentals have an elaboration of bone around the inner ear, the auditory bulla, which probably increases auditory acuity. Although most marsupials lack a bulla, a few have one that is formed by a different bone than the bulla of placentals.

- During ontogeny, most placentals replace all their teeth except for the molars, whereas marsupials replace only the last premolar. That is, marsupials are virtually monophyodont (only one set of adult teeth).

- The basal dentition of marsupials is five incisors on each side of the upper jaw and four on each side of the lower jaw, and one canine, three premolars, and four molars on each side of both the upper and lower jaws. The dental formula is a shorthand method of writing that information; the basal dental formula of marsupials is

$$\frac{5.1.3.4}{4.1.3.4}$$

Basal placentals have fewer incisors and molars, but more premolars, than marsupials; the basal dental formula for placentals is

$$\frac{3.1.4.3}{3.1.4.3}$$

Many species of marsupials and placentals have fewer than the basal number of teeth, but only a few species of placentals with specialized diets have more than the basal number; these include armadillos (insect specialists) and porpoises (fish-eaters). Humans have lost an incisor, two premolars, and sometimes the last molar (the wisdom tooth); the human dental formula is

$$\frac{2.1.2.2\text{--}3}{2.1.2.2\text{--}3}$$

where 2-3 indicates that the presence of a wisdom tooth is variable.

- In the brain, in addition to the anterior commissure, a structure that links the hemispheres in all amniotes, placentals have a new nerve tract linking the two cerebral hemispheres: the corpus callosum. It is often claimed that marsupials are smaller-brained than placentals, but in fact the average marsupial is not smaller-brained than the average placental. However, very large-brained mammals, such as primates and cetaceans, are found only among placentals, and it is difficult to know if that is correlated with some other aspect of placental biology.

The axial skeleton of marsupials also displays differences from that of therians:

- Almost all marsupials have epipubic bones that project forward from the pelvis (see Figure 25.7B). Once considered to be a unique character of marsupials, epipubic bones are now recognized as a basal mammalian character that was also retained in a few Cretaceous

Marsupial (opossum, *Didelphis*)

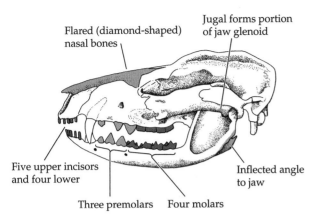

Flared (diamond-shaped) nasal bones

Jugal forms portion of jaw glenoid

Five upper incisors and four lower

Three premolars   Four molars

Inflected angle to jaw

Placental (raccoon, *Procyon*)

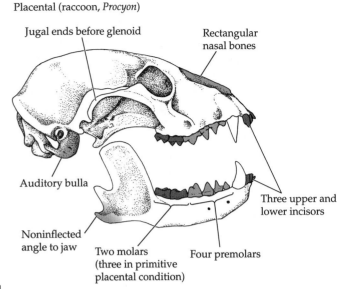

Jugal ends before glenoid

Rectangular nasal bones

Auditory bulla

Three upper and lower incisors

Noninflected angle to jaw

Two molars (three in primitive placental condition)

Four premolars

Figure 25.11 Differences in the skulls of marsupials and placentals.   The angle of inflection of the dentary, the shape of the nasal bones and jugal, number of teeth, and presence of an auditory bulla are some of the differences between marsupial and placental skulls, as described in the text. (From Lawlor 1976.)

stem placentals. Epipubic bones provide the insertion point for thigh and abdominal muscles that stiffen the torso and resist vertical bending of the trunk during locomotion, and these bones perhaps limit independent movement of the hindlimbs in gaits such as the gallop. The few quadrupedal cursorial marsupials (all extinct) had lost these bones; these included the thylacine (see Figure 25.5J) and borhyaenoids (doglike predators of South America). Two hypotheses, which are not mutually exclusive, have been proposed to explain the loss of epipubic bones by placentals: the constraints they impose on locomotion (which would also explain their loss in some marsupials), or interference that these rigid components of the abdominal wall could cause with expansion of the abdomen during pregnancy.

- Placentals all have a true patella (kneecap), which is a sesamoid bone (a bone embedded in a tendon). The patella is contained within the tendon of the quadriceps muscle, where it acts as a pulley and stabilizer for the muscles extending the knee. Almost all marsupials have a "patelloid" made of fibrocartilage rather than bone (except, strangely, for bandicoots, including the bilby, which have a bony patella).

25.4 Mammalian Reproduction

The mode of reproduction is the major and most obvious difference among the three main groups of extant mammals. All extant mammals lactate and care for their young. Monotremes, however, are unique in retaining the ancestral amniote condition of laying eggs; with the evolution of viviparity in therians, the uterine glands that add the shell and other egg components were lost, making a return to oviparity difficult or impossible.

Mammalian embryos, like those of all amniotes, have four extraembryonic membranes (see Chapter 10). In therians, some of these membranes form the placenta, which transfers nutrients, metabolic wastes, respiratory gases, and hormones between fetus and mother, and also produces hormones that signal the state of pregnancy to the mother. Fusion of the chorion with the yolk sac forms the choriovitelline membrane, and fusion of the chorion with the allantois forms the chorioallantoic membrane. The amnion is not directly involved in placentation but surrounds the embryo and provides protection.

Additionally, all mammals have a **corpus luteum** (Latin *corpus*, "body"; *lutum*, "yellowish"; plural *corpora lutea*), a hormone-secreting structure formed in the ovary from the follicular cells that remain after the egg is shed. Hormones secreted by the corpus luteum are essential for the establishment and maintenance of pregnancy. In placentals, feedback from the placenta maintains the life of the corpus luteum, suppressing ovulation and allowing the extension of gestation past a single estrus cycle.

Marsupials and placentals have profound differences in their relative lengths of gestation and lactation. Gestation never exceeds a single estrus cycle in marsupials; the young are born at an early stage and fasten onto a nipple to complete their development, with nutrition received from milk rather than via the placenta. Some, but not all, marsupials enclose these newborn offspring in a pouch. A good way to think of the reproductive differences between marsupials and placentals is that marsupials have short gestations (the length of gestation is relatively independent of body size, unlike the condition in placentals) and long periods of lactation, while placentals have long gestations and short periods of lactation (usually shorter than the period of gestation). However, large species of marsupials have longer periods of lactation than do small species.

Neonate mammals range from being highly altricial to precocial (**Figure 25.12**). Neonate monotremes and

(A) Monotreme (echidna)

(C) Altricial placental (field mouse)

(B) Marsupial (kangaroo)

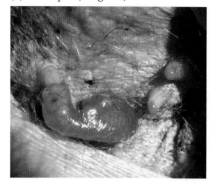

(D) Precocial placental (deer)

Figure 25.12 Mammalian neonates.
(A) A short-beaked echidna infant. Baby monotremes are called "puggles." (B) A newborn kangaroo infant suckling on a teat in its mother's pouch. (C) Altricial placentals such as these field mice are blind and helpless at birth. (D) Precocial placentals like this white-tailed deer are able to stand and follow their mothers soon after birth. (A, Stewart Nicol; B, Geoff Shaw/CC BY-ND 3.0; C, Derell Licht/CC BY-ND 2.0; D, Clay Junell/CC BY-SA 2.0.)

marsupials are both altricial, but in marsupials the head, shoulders, and forelimbs are well developed, whereas the pelvis, hindlimbs, and tail are poorly developed. This specialization of neonate anatomy is associated with the post-birth crawl to the nipple (see below). Some placentals (e.g., rodents and insectivorans) are born in a highly altricial state, whereas others (especially ungulates) are born in more developed stages, and the precocial newborn young of most ungulates can follow their mothers within a few hours of birth. All placentals, however precocial, still require a period of lactation for the transfer of essential antibodies from the mother, as well as for nutrition.

Mammalian urogenital tracts

The structures of the urinary, alimentary, and reproductive systems of vertebrates are closely associated (**Figure 25.13**). In non-amniotes and basal amniotes, all three tracts exit the body through a common opening—the cloaca—and this condition is retained in reptiles and monotremes. Only in therians do the alimentary and urogenital tracts have separate openings, with a perineum dividing them, and only in female primates (and at least some rodents) do the three tracts open independently (via the anus, vagina, and urethra). In male therians, urine and sperm both exit via the urethra; in other male amniotes, the penis is used only for sperm transmission. Therians have a modified hypaxial wall musculature for the muscles in the perineal area that control the separate urogenital and alimentary openings,

and only therians have precise control over these excretory processes among vertebrates.

The reproductive tract of monotremes retains the generalized amniote condition: the two oviducts remain separate and do not fuse in development except at the base, where they join with the urethra to form the urogenital sinus. When fertilized eggs are present, the oviducts swell to form two uteri, in which the eggs are retained for a time. Only the left oviduct is functional in the platypus.

In all therians, the ureters draining the kidney enter the base of the bladder rather than the cloaca or urogenital sinus, as in most other vertebrates, including monotremes (see Figure 25.13). In placentals, the ureters pass laterally around the developing reproductive ducts to enter the bladder. In males, the vasa deferentia (sperm ducts) loop around the ureters in their passage from the scrotum to the urethra, and in females the oviducts fuse in the midline anterior to the urogenital sinus for part of their length. All placentals have a single, midline vagina, but only a few have a single median uterus, as seen in humans. Most placentals have a uterus that is bipartite (divided lengthwise into left and right sides) for some or all of its length.

In marsupials, the ureters pass medial to the developing reproductive ducts to enter the bladder. This arrangement means that the vasa deferentia of males do not have to loop around the ureters, and in females it prevents the oviducts from fusing in the midline. The two lateral vaginas unite anteriorly, and two separate uteri diverge. The lateral vaginas are for the passage of sperm only (an arrangement complemented by the bifed penis of male marsupials). Young are born through a midline structure, the median vagina or pseudovaginal canal, which develops the first time a female gives birth.

Genitalia

Monotreme males are like non-mammalian vertebrates in having testes that are retained within the abdomen. The testes of many therians, both marsupials and placentals, descend into a scrotum and are housed outside the body. The traditional explanation for the scrotum is that it provides a cooler environment for sperm production than the

(A) **Male monotreme** (similar to general amniote condition)

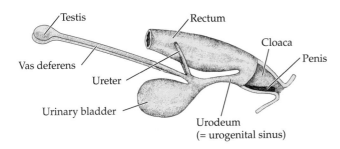

(B) **Female monotreme** (platypus)

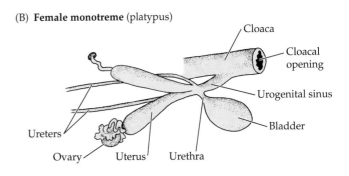

(C) **Male marsupial**

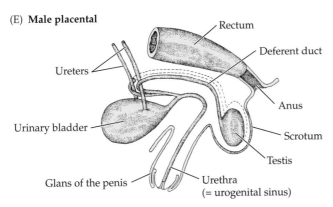

(D) **Female marsupial**

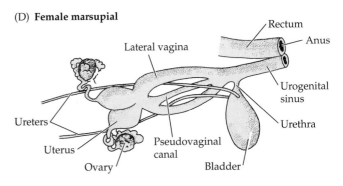

(E) **Male placental**

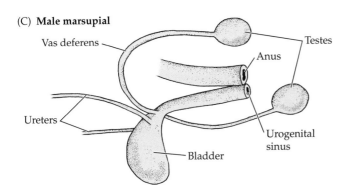

(F) **Female placental** (non-primate)

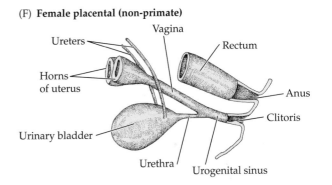

Figure 25.13 Anatomy of mammalian urogenital tracts.
The head is to the left. A, E, and F are viewed from the side. B, C, and D are viewed from the side and slightly below. (A, E, F, from Hildebrand and Goslow 1998; B–D, from Renfree 1993.)

abdomen, and some evidence supports this hypothesis. An alternative hypothesis proposes that testicular descent avoids the intra-abdominal pressure that would be exerted on the testes during bounding or galloping, which are modes of locomotion seen only in therians. The scrotum is anterior to the penis in marsupials and behind the penis in most placentals, although there are some exceptions (e.g., rabbits have a prepenile scrotum). This difference in location of the scrotum, the retention of internal testes in basal placentals (Afrotheria and Xenarthra), and a difference in physiological control of testicular descent between marsupials and placentals all lead to the conclusion that a scrotum evolved independently in the two groups of therians.

Monotremes and most marsupials have a penis with a bifid (forked) glans, whereas placentals have a single glans.

Some species of male placentals, including non-human primates, rodents, insectivores, carnivorans, and bats, have a bone in the penis, the os penis or baculum. Most male mammals extend the penis from an external sheath (normally the only visible portion), and only during urination and copulation.

The clitoris is the female homologue of the penis but is not used to pass urine. In the usual therian condition, the clitoris lies within the urogenital sinus, where it receives direct stimulation from the penis during copulation. Females of those species where the male has a baculum may have a bone in the clitoris called the os clitoridis or baubellum.

Reproductive mode of monotremes: Matrotrophic oviparity

The ovaries of monotremes are larger than those of therians, and monotremes provide the embryo with yolk, although not as much yolk as does an oviparous sauropsid of similar

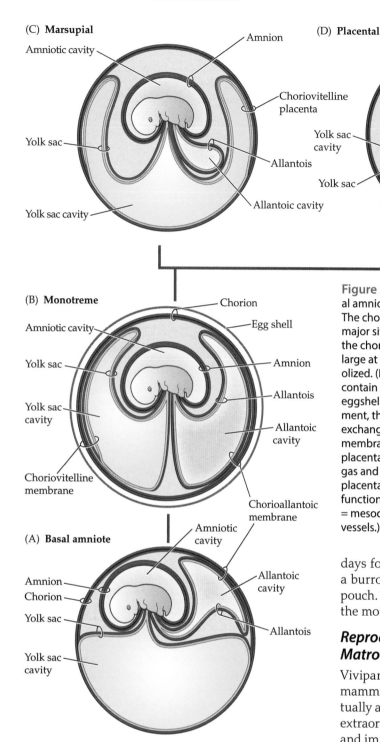

(C) Marsupial

Amniotic cavity
Amnion
Choriovitelline placenta
Yolk sac
Allantois
Yolk sac cavity
Allantoic cavity

(D) Placental

Amniotic cavity
Amnion
Chorion
Allantois
Yolk sac cavity
Yolk sac
Chorioallantoic placenta
Allantoic cavity

(B) Monotreme

Chorion
Egg shell
Amniotic cavity
Yolk sac
Amnion
Yolk sac cavity
Allantois
Allantoic cavity
Choriovitelline membrane

(A) Basal amniote

Chorioallantoic membrane
Amniotic cavity
Allantoic cavity
Amnion
Chorion
Yolk sac
Allantois
Yolk sac cavity

Figure 25.14 Types of placentation in mammals. (A) The basal amniote condition, which is retained in oviparous sauropsids. The chorion, which lies immediately inside the eggshell, is the major site of gas exchange. The allantois is apposed to a portion of the chorion and also participates in gas exchange. The yolk sac is large at the start of development and shrinks as the yolk is metabolized. (B) The eggs of monotremes and the quantity of yolk they contain are small compared with the basal amniote condition. The eggshell is highly permeable. At different times during development, the allantois, chorioallantois, and yolk sac are sites of gas exchange. (C) In marsupials, the outgrowth of the chorioallantoic membrane appears to be suppressed relative to the condition in placentals and the choriovitelline placenta is the primary site of gas and nutrient exchange. (D) In placentals, the choriovitelline placenta is usually short-lived, and a chorioallantoic placenta is functional for most of gestation. (Blue = ectoderm, red and green = mesoderm, yellow = endoderm. Only the endoderm has blood vessels.) (After Ferner and Mess 2011.)

days for echidnas). The platypus usually lays its eggs in a burrow, whereas echidnas keep their eggs in a ventral pouch. The young hatch at an extremely altricial stage, and the mother broods them for an additional 4 months.

Reproductive mode of therians: Matrotrophic viviparity

Viviparity is a common concept for us, as we are therian mammals. For animals in general, however, viviparity is actually an uncommon reproductive mode. Viviparity places extraordinary demands on the physiology of the mother, and immune systems must be modified so that the developing young are not rejected as foreign tissue. Although all therians have a placenta, there is variation in types of placentation (**Figure 25.14**), and all therian fetuses additionally receive some nutrition from uterine secretions.

A **choriovitelline placenta** ("yolk sac" placenta), formed by apposition of the yolk sac to the chorion that lines the inside of the egg membrane, is the first placental structure to appear after implantation of a fertilized egg on the wall of the uterus in all therians. A **chorioallantoic placenta**, formed from the fusion of the allantois to the chorion, is typical of placentals but not marsupials. Among marsupials, in possums, kangaroos, and phalangers the allantois

body size. As in all mammals, the eggs are fertilized in the anterior portion of the oviduct, the fallopian tube, before they enter the uterus. The yolk is not sufficient to sustain the embryo until hatching, and the eggs are retained in the uterus where they are nourished by maternal secretions and increase in size before the shell is secreted. The eggshell is parchmentlike, as in many nonavian reptiles, rather than rigid like the calcareous eggshells of birds.

Monotremes lay one or two eggs, and the young hatch soon after the egg is laid (12–14 days for the platypus, 21–28

never reaches the chorion, and in quolls the allantois is apposed to the chorion early in development but then retreats from it without forming a placental structure. In koalas and wombats, the allantois reaches the chorion, forming a rudimentary chorioallantoic placenta. Only bandicoots and the bilby form an invasive chorioallantoic placenta in addition to the choriovitelline placenta. In placentals, the choriovitelline placenta is usually transitory, although some placentals retain a choriovitelline placenta even after the chorioallantoic placenta has developed. But the choriovitelline placenta plays an important role in early development in all placentals, both for transfer of nutrients and other functions (e.g., the formation of blood cells in humans).

There is great variation in the form of the chorioallantoic placentas of placental mammals. In **hemochorial placentas** (probably the ancestral condition, and seen now in anthropoid primates and many rodents), maternal blood is in direct contact with fetal tissues. In **endotheliochorial placentas** (carnivorans and elephants), the chorionic tissue of the fetus penetrates the endometrium of the uterus as far as, but not into, the maternal blood vessels. And in **epitheliochorial placentas** (ungulates and non-anthropoid primates), the chorion is in contact with the surface of the endometrium but does not penetrate it.

The growth of the chorioallantoic membrane in developing marsupials is slower than in all other amniotes; the interpretation is that the outgrowth of this potential placental membrane has been suppressed in marsupials. In contrast to placentals, marsupials have a transient shell membrane, and placentation does not occur until the shell membrane degenerates.

Although it may appear to be an undeveloped embryo, the marsupial neonate represents a highly derived condition, with well-developed forelimbs that assist in the climb to the nipple, and a well-developed mouth for fastening onto the nipple. Guided by smell and using their front claws as holdfasts, most marsupial neonates crawl from their mother's urogenital sinus to the area of the nipples on her belly (enclosed in a pouch in many, but not all, marsupials), a distance that may be many times their own body length. The shoulder girdle of marsupial neonates retains the interclavicle and complete coracoid bones of monotremes and other amniotes (**Figure 25.15**), and this specialized anatomy enables their crawl to the nipple. This neonatal anatomy may limit the subsequent development of the shoulder girdle in adults, however, and marsupials have less variation than placentals in the form of their shoulder girdles.

In kangaroos, as in most marsupial species, the mother licks a path from the opening of the urogenital tract to the pouch but does not otherwise aid the young in its journey. A kangaroo's pouch opens anteriorly, but the pouches of most marsupials open posteriorly, shortening the distance the young must travel. A kangaroo has two nipples and can provide milk with different qualities to an attached neonate and to a joey that has left the pouch (**Figure 25.16**). Neonates receive milk with high protein content and low fat content, whereas joeys receive low-protein, high-fat milk. The composition of milk produced by the mammary glands is probably determined by how long the young spend suckling per day.

When food is plentiful, a female kangaroo can have three young at different stages of development simultaneously (see Figure 25.16). The presence of a suckling neonate stimulates the mother's pituitary gland to release prolactin, which in turn inhibits production of progesterone by the corpus luteum, holding the fertilized egg in the oviduct in **embryonic diapause**, a period of arrested development prior to implantation in the uterus. Diapause breaks and the embryo starts to develop when the neonate reduces its rate of suckling as it begins to leave the pouch. Embryonic diapause enables the mother to space successive births and to separate the time of mating and fertilization from the start of gestation. The phenomenon is particularly well developed in kangaroos and thus has often been perceived as a derived marsupial feature. However, embryonic diapause also occurs in a wide variety of placentals.

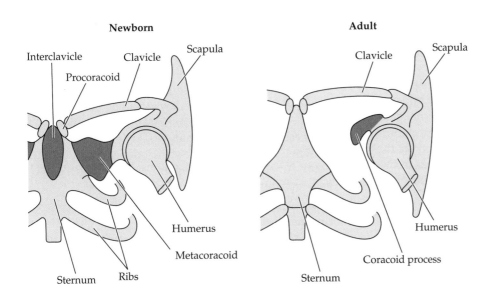

Figure 25.15 Shoulder girdle of neonate and adult marsupials. The views are from the ventral side with only the left limb shown. The neonate shoulder arch is formed by the fusion of the scapula, metacoracoid, interclavicle, and sternum. The interclavicle is lost in adults, and the metacoracoid is reduced to the coracoid process. The additional support, especially from the metacoracoid, adds rigidity to the shoulder girdle as the neonate uses its front claws as holdfasts while it wriggles to the pouch. (From Sears 2004.)

Figure 25.16 A female kangaroo can have offspring in different stages of development. The mother can suckle both a juvenile and a neonate while maintaining a fertilized embryo in diapause. (Toby Hudson/CC BY-SA 3.0.)

A neonate in the pouch (not visible) is attached to a second nipple, from which it receives a constant supply of high-protein, low-fat milk.

A joey (juvenile that has left the pouch) returning to suckle receives high-fat, low-protein milk.

The presence of a suckling neonate stimulates the mother's pituitary gland to release prolactin, initiating a hormonal cascade that maintains an early embryo in diapause.

The earliest therian condition, and the discredited notion of placental superiority

The basal therian condition probably consisted of hatching or the birth of highly altricial young. The specializations of the reproductive anatomy of marsupials and placentals cannot easily be derived from each other, and it is possible that viviparity evolved independently in the two groups. The early therians of the Cretaceous were small animals, most weighing less than 1 kg. They are likely to have had the same life-history features as small therians today— short gestation periods, several litters produced in rapid succession, and short lifespans.

One of the more bizarre facts to emerge from genomic studies is that the embryonic trophoblasts, forming the fetal side of the placenta, have an elevated expression of syncytin genes derived from endogenous retroviruses. These genes have a role in immunosuppression, and may have aided the evolution of the fusion of fetal and maternal cells in the formation of the placenta and maternal tolerance of the fetus. Such genes are diverse and widespread among placentals, and may have been responsible for the evolution of the long-term retention (i.e., past one estrus cycle) of the fetus in gestation. Some simpler versions of these genes are expressed in marsupials, and may be implicated in the initial evolution of viviparity.

Other reproductive features of therians that are probably related to viviparity include X/Y genetic sex determination; genetic imprinting, whereby maternal and paternal genes may be expressed in the offspring in a different pattern; and X-chromosome inactivation. The cells of female therians express only one of the two X chromosomes in their genome, not both at the same time; the black-and-ginger pattern of the fur of tortoiseshell and calico cats is one example of this, and cats with these coat colors are invariably female.

The marsupial mode of reproduction was once thought to be inferior to that of placentals. This opinion was based primarily on the assumption that marsupials had been unable to compete with placentals and were able to radiate only in Australia, where they apparently evolved in isolation from placental mammals. The low diversity of marsupial species (about 6% that of placentals) was also cited as evidence of their evolutionary inferiority. That analysis is faulty, however, and the low diversity of metatherians is probably in part an accident of history. The continents on which marsupials radiated had a smaller total land area during the Cenozoic than the continents that placentals inhabited. Marsupials would thus be expected to have less species diversity than placentals. South American marsupials evolved successfully alongside placentals, although they had distinct ecological roles (e.g., the large carnivores were all stem marsupials and the large herbivores were all placentals), and rodents invaded Australia about 5 Ma with little apparent effect on the local marsupials.

It seems likely that marsupials and placentals simply evolved different but equivalent reproductive strategies. Although as placentals ourselves we have historically viewed our mode of reproduction as superior, the marsupial mode has clearly been selected for in the marsupial lineage. It is not a second-best attempt, or an intermediate condition between laying eggs and having a long period of gestation. That said, the marsupial reproductive system may have placed some limitations on the ecomorphological diversity of marsupials:

- A fully aquatic marsupial could not carry altricial young in a pouch because they would be unable to breathe air, nor could a marsupial give birth underwater because the tiny neonates would be swept away by currents. There is only one semiaquatic marsupial, the South American water opossum (yapok), which seals its pouch during its short underwater forays.

- There is no marsupial group equivalent to the bats, but this may just be a matter of chance. Although gliding has evolved convergently several times among both marsupials and placentals, flight evolved only once in synapsids.

- There are no hoofed ungulates among marsupials, perhaps because marsupials cannot reduce the forefeet to nongrasping limbs because neonates would be unable to climb to the pouch. This constraint may explain why the marsupial equivalents of horses and antelope, the large kangaroos, have specialized only their hindlegs for locomotion.

- Marsupial skulls show less variability than those of placentals (even discounting extreme placental types such as cetaceans). This is probably related to the early ossification of the bones around the mouth, for fastening onto the nipple, which then constrains the developmental trajectory of the rest of the skull.

25.5 Specializations for Feeding: Teeth and Jaws

Mammals need large quantities of food to fuel their high metabolic rates, and starting to process food in the mouth rather than waiting until it reaches the stomach speeds digestion. Chewing reduces food to small particles and increases the surface area for digestive enzymes to attack.

Mammalian teeth

The mammalian complement of teeth includes (from front to rear) incisors, canines, premolars, and molars. Like jaws, teeth show extensive variation that reflects both differences in feeding and diet and in nontrophic functions (**Figure 25.17**). Canines are used to stab prey and are often lost in herbivores, but canines are also used for defense and as social signals and may be retained in herbivores in modified form. The tusks of pigs and walruses are modified canines, but the tusks of elephants are modified incisors. Upper canines are generally larger in male primates than in females (even in humans). Males of some hornless ruminants, such as mouse deer, retain large upper canines that are used in display and combat (see Figure 25.6G).

The molars of carnivores and insectivores usually have three pointed cusps (**Figure 25.17B**). Omnivorous and frugivorous mammals have reduced the cusps to rounded, flattened structures suitable for crushing and pulping (**Figure 25.17C**). They have also added a fourth cusp to the upper molars and increased the size of the talonid basin in the lower molars so that these teeth now appear square rather than triangular. These teeth are called **bunodont** (Greek *bounos*, "hill") in reference to the rounded cusps. Humans have bunodont molars, as do most primates and other omnivores such as pigs and raccoons.

In herbivores, the simple cusps of the bunodont tooth run together into ridges, or **lophs**. Lophed teeth work best when the enamel has been worn off the top of the loph to expose the underlying dentine. Each loph then consists of a pair of sharp enamel ridges lying on either side of the intervening dentine. When these teeth occlude and the jaws move laterally, food is grated between multiple sets of flat,

shearing blades. Lophed teeth have evolved convergently many times among mammals (**Figure 25.17D–F**). **Selenodont** molars (teeth with crescentic lophs running from anterior to posterior) are seen in ruminants, camelids, and the koala. **Lophodont** molars (teeth with lophs that run in a predominantly cheek-to-tongue orientation) are seen in perissodactyls such as horses and rhinos, as well as in kangaroos. Elephants, some rodents, and wombats have a more complex type of molar called **multilophed** or **lamellar**.

All herbivorous mammals face the problem of dental durability, because, as we discussed in Chapter 24, the general mammalian condition is diphyodonty, meaning one set of adult teeth must last a lifetime. Herbivorous mammals face a particular problem because they have highly abrasive diets, including fiber and sometimes silica crystals (phytoliths) in leaves and ingestion of grit or soil for species that feed close to the ground.

Grazing mammals have high-crowned, or **hypsodont** (Greek *hypsos*, "high"), cheek teeth (molars and premolars) (**Figure 25.17G**). Species with hypsodont teeth have very deep dentary bones and maxillae, and the crowns of hypsodont teeth extend deep into the bones. Cementum covers the entire tooth rather than only the root and the base of the crown as in most mammals. As the chewing surface is worn away, hypsodont teeth erupt from the base to provide a continuously renewing occlusal surface, much as the lead in a mechanical pencil is pushed up as it is worn away. Without cementum acting as filler, the individual lophs would be tall, freestanding blades and would be likely to fracture. Ultimately the entire tooth wears away and the animal can no longer eat. (Most mammals in the wild die of natural causes long before their teeth wear out, but domestic horses surviving into their late twenties and thirties often must be fed soft food because they have virtually no molars left.)

Some small mammals, such as rabbits, wombats, and some rodents, have **hypselodont** molars in which the roots do not close and the tooth is functionally ever-growing (**Figure 25.17H**). The incisors of these animals are also ever-growing, and have enamel only on the anterior surface. As the dentine wears away, it renews a sharp blade of enamel at the tip of the tooth.

Elephants do not have ever-growing cheek teeth but instead employ the novel feature of molar progression. Each molar is not only hypsodont but also greatly enlarged—the size of the entire original tooth row. Molars erupt in turn; when a molar is worn down, its remaining stub falls out of the front of the jaw, and the molar behind erupts from the back to replace it. Elephants have a total of six molars (three milk molars and three permanent molars) in each side of the upper and lower jaws.

Differences between carnivorous and herbivorous mammals

Both the teeth and jaws of carnivores differ from those of herbivores. In carnivores, the incisors seize food, the canines stab prey, the carnassials (formed by an upper

(A)

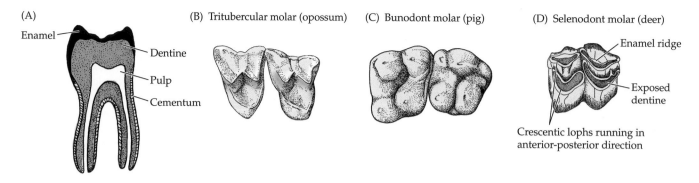

Enamel

Dentine

Pulp

Cementum

(B) Tritubercular molar (opossum)

(C) Bunodont molar (pig)

(D) Selenodont molar (deer)

Enamel ridge

Exposed dentine

Crescentic lophs running in anterior-posterior direction

(E) Lophodont molar (rhino)

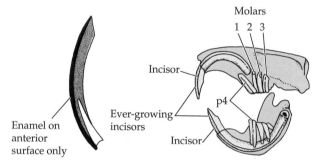

Single straight loph along outer edge

Exposed dentine

Straight lophs running in lateral-medial direction

(F) Lamellar or multilophed molar (rodent)

Enamel ridges

(G) Hypsodont teeth

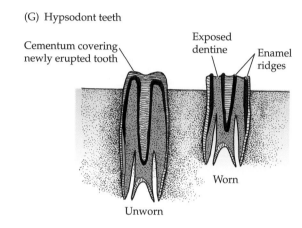

Cementum covering newly erupted tooth

Exposed dentine

Enamel ridges

Worn

Unworn

(H) Ever-growing (hypselodont) cheek teeth (3 molars plus one premolar, P4)

Molars
1 2 3

Incisor

p4

Ever-growing incisors

Incisor

Enamel on anterior surface only

(I) Carnivore carnassials (coyote) (anterior to the right)

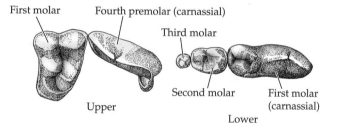

First molar

Fourth premolar (carnassial)

Third molar

Second molar

First molar (carnassial)

Upper

Lower

Figure 25.17 Structure of mammalian teeth. (A) A cross section of a molar shows the general form of mammalian teeth. The pulp cavity, which contains bone-forming cells, blood vessels, and nerves, is surrounded by dentine. Cementum forms the outer surface of the tooth below the gum line, and hard enamel covers the exposed portion. (B) Tritubercular molars (i.e., unmodified tribosphenic molars) of a generalist feeder. (C) Bunodont molars of an omnivore. (D–F) Lophed (ridged) molars of herbivores. (G) Hypsodont cheek teeth of a grazing mammal. (H) Hypselodont incisors and cheek teeth of a rodent. (I) Carnivore carnassial teeth.

premolar and a lower molar; **Figure 25.17I**)) slice through flesh, and the molars break pieces of food into fine particles. Some herbivores have lost one or both sets of incisors; ruminant artiodactyls have only lower incisors, and African rhinos lack both upper and lower incisors. Often the entire postcanine tooth row is used for mastication, and the premolars resemble the molars (i.e., they are molarized). Herbivores usually have elongated snouts, resulting in a gap between the incisors and cheek teeth called the **diastema**. The function of the diastema is uncertain. It may allow extra space for the tongue to manipulate food, or it may only reflect the elongation of the snout that allows an animal to reach into narrow spaces to eat leaves or fruit.

Jaw muscles All mammals use a combination of masseter, temporalis, and pterygoideus muscles to close the jaws, and the digastric muscle (in therians) to open the jaw. The relative sizes of the muscles and the shapes of the skulls reflect the different demands of masticating flesh and vegetation. The temporalis has its greatest mechanical advantage at moderate to large gapes, when the incisors and canines are likely to be used. Thus, a large temporalis is typical of carnivores, which use a forceful bite with their canines to kill and subdue prey (**Figure 25.18A**). The occipital region of the skull is high in carnivores, and strong muscles run to the cervical vertebrae. These muscles are probably important for helping to restrain struggling prey.

The skulls of herbivorous mammals are modified to grind tough, resistant food. The protein content of leaves and stems is usually low, and the cells are enclosed by a cell wall formed by cellulose (a complex carbohydrate). Herbivores must process large amounts of food per day. The masseter exerts crushing force at the back of the tooth row and also moves the jaw from side to side, helping the complex molars

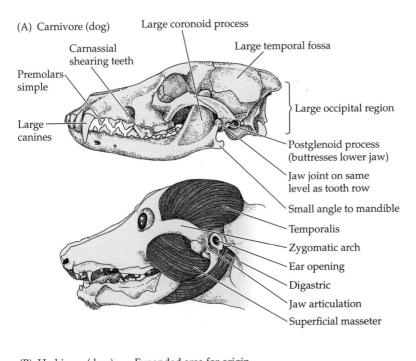

(A) Carnivore (dog)

Large coronoid process

Carnassial shearing teeth

Large temporal fossa

Premolars simple

Large occipital region

Large canines

Postglenoid process (buttresses lower jaw)

Jaw joint on same level as tooth row

Small angle to mandible

Temporalis

Zygomatic arch

Ear opening

Digastric

Jaw articulation

Superficial masseter

(B) Herbivore (deer)

Expanded area for origin of masseter muscle

Small coronoid process

Premolars molarized

Diastema

Lower canine incisiform

Small occipital region

Jaw joint above level as tooth row

Large angle to mandible

Temporalis

Ear opening

Deep part of masseter

Superficial masseter

Digastric

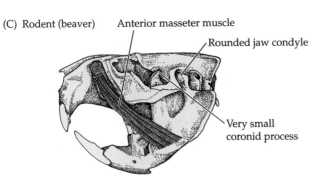

(C) Rodent (beaver)

Anterior masseter muscle

Rounded jaw condyle

Very small coronid process

Figure 25.18 Craniodental differences between carnivorous and herbivorous mammals. (A) Carnivore skull and musculature. The powerful temporalis muscle and large coronoid process power the shearing action of the carnassial teeth. (B) Herbivore skull and musculature. Herbivores chew with a grinding movement, and the masseter is more powerful than the temporalis. (C) Rodent skull showing the insertion of a portion of the masseter muscle far anterior on the skull, allowing the incisors to occlude during gnawing.

to rupture cell walls. Herbivorous mammals have a larger masseter and smaller temporalis than carnivores (**Figure 25.18B**). The posterior part of the lower jaw (where the masseter inserts) is large, and the coronoid process and temporal fossa (where the temporalis inserts and originates, respectively) are small. Herbivores usually have a fairly low occipital region because they lack the powerful muscles that allow carnivores to resist the movements of struggling prey. (Pigs, which use their snouts to overturn soil, are an exception; they have strong muscles linking the occipital region to the neck.) In many herbivorous placental mammals, the cartilaginous partition at the back of the orbit has ossified and forms a bony postorbital bar. This bar is probably important in absorbing stress from the jaws, thus protecting the braincase.

Jaw joints In carnivores, as in most mammals, the jaw joint is on the same level as the tooth row, so the teeth come into contact sequentially as the jaw closes, like the blades of a pair of scissors—a morphology well suited for cutting and shearing. The postglenoid process prevents the strong temporalis muscle from dislocating the lower jaw.

In herbivores, the jaw joint has shifted so that it is high on the skull, above the tooth row. This offset brings all the teeth in the upper and lower jaws together simultaneously, with a grinding action that shreds plant material between the lophs of the upper and lower teeth, much as the offset handle of a cooking spatula allows you to apply the entire blade of the spatula to the bottom of the frying pan while keeping your hand above the pan's rim.

Rodents: Specialized feeders

Many rodents (Latin *rodens*, "gnawing") have a highly specialized type of food processing. Their upper and lower tooth rows are the same distance apart—unlike the condition in most mammals, in which the lower tooth rows are closer together than the teeth uppers. This derived condition of rodents is combined with a rounded jaw condyle, which allows forward and backward jaw movement, and the insertion of a portion of the masseter muscle far forward on the skull so that the lower jaw can be pulled forward into occlusion (**Figure 25.18C**). This jaw apparatus allows rodents to chew on both sides of the jaw at once, which is presumably a highly efficient mode of food processing.

25.6 Specializations for Locomotion

Larger animals experience the world differently than smaller ones because of physical size and scaling effects, and large mammals are modified for specialized forms of locomotion, such as bounding.

Cursorial limb morphology

Long legs provide a long out-lever arm for the major locomotor muscles, such as the triceps in the forelimb and the gastrocnemius in the hindlimb. This arrangement favors speed of motion rather than power, and long legs allow for long strides and rapid terrestrial locomotion.

The foot posture in cursorial mammals has changed from the primitive plantigrade (foot flat on the ground) posture to a digitigrade posture (standing on tiptoe) in many carnivores, and to an unguligrade posture (like a ballerina standing en pointe) in hoofed mammals (**Figure 25.19**). Elongation is limited mostly to the distal portions of the limb: the radius and ulna in the forelimb, the tibia and fibula in the hindlimb, and the **metapodials** (a collective term used to describe the metacarpal and metatarsal bones); the humerus, femur, and phalanges are not elongated. The distal (terminal) phalanx of each foot is encased in a hoof (**Figure 25.20A**).

Muscles are limited to the proximal portion of the limb, reducing the weight of the lower limb. There is almost no muscle in a horse's leg below the wrist (the horse's so-called knee joint) or in the ankle (the hock). That anatomy makes

sense in mechanical terms because the foot is motionless at the start of each stride (as it pushes against the ground), and when it leaves the ground, it must be accelerated from zero velocity to a speed greater than the body speed as it moves forward ready for the next contact with the ground. The lighter the lower limb, the less effort is needed for these repeated accelerations.

The force of muscular contraction by the muscles in the upper limb is transmitted to the lower limb via long elastic tendons. These tendons are stretched with each stride, storing and then releasing stored elastic energy and contributing to locomotor efficiency (**Figure 25.20B**).

The leg tendons of a hopping kangaroo are an obvious example of elastic storage, as the animal bounces on landing as if using a pogo stick. However, all cursorial animals rely on energy storage in tendons for gaits faster than a walk. Even humans rely on elastic energy storage in tendons, especially in the Achilles tendon, which attaches the gastrocnemius (calf) muscle to the calcaneal heel. People who have damaged Achilles tendons—a common sports injury—find running difficult or impossible. Lengthening tendons to increase the amount of stretch and recoil may be part of the evolutionary reason for limb elongation and changes in foot posture.

Other cursorial modifications restrict the motion of the limb to a fore-and-aft plane so that most of the thrust on the ground contributes to forward movement. The clavicle is reduced or lost, the wrist and ankle bones allow motion only in a fore-and-aft plane, and the forelimb cannot be supinated. (Note how easily you can turn your hand so that your palm faces upward—a movement that actually comes from the elbow rather than the wrist. A dog can't turn its forepaw that much, and a horse has almost no ability to rotate its hoof.)

Figure 25.19 Foot posture of terrestrial mammals. (A) Plantigrade: the entire foot (phalanges to heel) is in contact with the ground. Humans are plantigrade. (B) Digitigrade: the phalanges are in contact with the ground; the metatarsals and heel are held off the ground. Dogs and cats are digitigrade. (C) Unguligrade: only the distal ends of the phalanges are in contact with the ground. Hoofed mammals like horses and cows are unguligrade. Unguligrade mammals have a reduced number of digits. (From Hildebrand and Goslow 1998.)

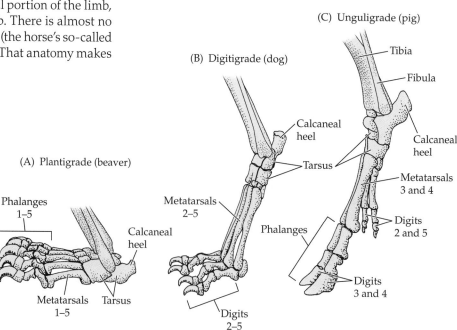

(C) Unguligrade (pig)

(B) Digitigrade (dog)

Tibia

Fibula

Calcaneal heel

Calcaneal heel

Tarsus

Metatarsals 3 and 4

(A) Plantigrade (beaver)

Metatarsals 2–5

Digits 2 and 5

Phalanges 1–5

Phalanges

Metatarsals 3 and 4

Calcaneal heel

Metatarsals 1–5   Tarsus

Digits 2–5

Digits 3 and 4

(A)

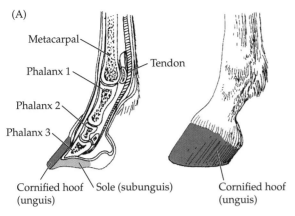

Metacarpal

Phalanx 1

Phalanx 2

Phalanx 3

Tendon

Cornified hoof (unguis)   Sole (subunguis)   Cornified hoof (unguis)

Figure 25.20 Hooves and foot tendons of cursorial ungulates. The foot of a horse illustrates specializations characteristic of cursorial ungulates, including antelope and deer. (A) *Left*: Longitudinal section of lower front foot showing the relationship of the phalanges to the hoof. *Right*: Hoof of the lower front foot. (B) Springing action of the elastic tendons in the foot. The tendon is stretched as the horse's body moves forward over the leg and shortens as the foot leaves the ground, providing additional propulsive force. (From Hildebrand and Goslow 1998.)

(B)

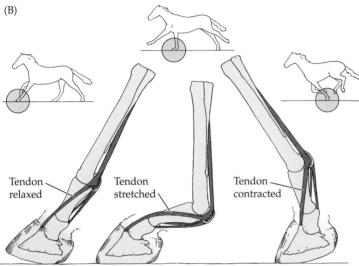

Tendon relaxed

Tendon stretched

Tendon contracted

The number of digits may also be reduced, perhaps to reduce the weight of the foot so that it can be accelerated and stopped more easily. Carnivores often reduce digit 1, but otherwise the digits are compressed together rather than reduced in number. Artiodactyls reduce or lose digits 1, 2, and 5, becoming effectively four-toed like a pig or two-toed like a deer (which is why they are called "even-toed" ungulates). Perissodactyls lose digits 1 and 5 and reduce digits 2 and 4, becoming three-toed like a rhinoceros or single-toed like a horse (which is why they are called "odd-toed" ungulates). (Cursorial flightless birds also reduce the number of toes—emus have three toes, and ostriches have only two.)

Cursorial specializations occurred convergently among many different mammalian lineages. Early Cenozoic ungulates and carnivorans were not highly cursorial, and it has long been assumed that these specializations, especially the evolution of longer legs, must have arisen in the context of predator–prey relationships. Longer legs would have given a carnivore a little more speed to pursue the herbivore, resulting in selection for ungulates with longer legs to make a faster escape.

This idea of a coevolutionary arms race between predator and prey is appealing, but the fossil record does not support it. If coevolution were the driving force, ungulates and carnivores would have evolved longer limbs at the same time, in lockstep fashion, but this is not the case. Cursorial ungulates are known from 30 Ma, whereas cursorial carnivores became apparent only within the past 5 million years. Why would ungulates evolve cursorial specializations if not to flee predators? Probably because the limb modifications that make a mammal a faster runner also make it more efficient at slower gaits, such as a walk or trot. These cursorial adaptations appeared in the middle Cenozoic, when more open habitats became more plentiful (see Chapter 23). Ungulates would have had to forage farther each day for food (as do ungulates in open habitats today), and modifications that increased endurance at slow gaits also increased speed at fast gaits. Speed became even more valuable later, when cursorial predators evolved.

Fossorial limb morphology

Fossorial mammals do several types of digging. Scratch digging at the surface (as seen in dogs) is most common. Surface scratch diggers, such as anteaters, have a stout pelvis with many sacral vertebrae involved in bracing their hindlimbs while they dig with the forelimbs. Animals that are truly fossorial (burrowing beneath the surface) have a variety of anatomical specializations. Golden moles and marsupial moles scratch dig with fore- and hindfeet, true moles use a "rotation thrust" of the forelimb, and chisel-toothed rodents such as gophers dig with their teeth.

A limb specialized for digging is almost exactly the opposite of a limb specialized for running (**Figure 25.21**). Running limbs maximize speed at the expense of power, whereas digging limbs maximize power at the expense of speed. Digging mammals achieve this mechanical advantage in the forelimb with a long olecranon process (for the triceps muscle to retract the hand) and a relatively short forearm. Digging mammals retain all five digits, tipped with stout claws for breaking the substrate. They also have large projections on their limb bones, such as the enlarged acromion process on the scapula, where strong muscles originate for retracting the limb.

(A) Left forelimb of a cursor (deer)   (B) Left forelimb of a digger (armadillo)

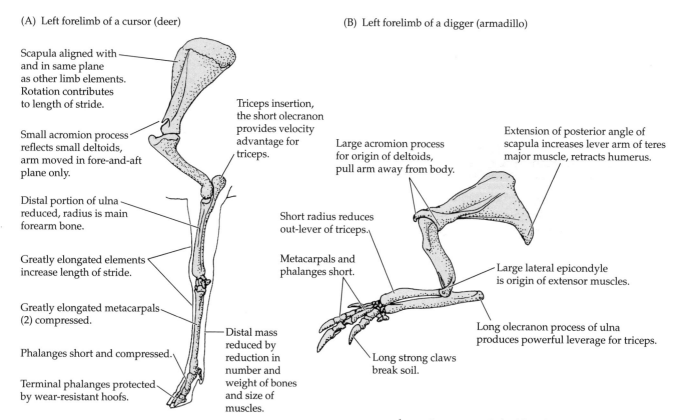

Scapula aligned with and in same plane as other limb elements. Rotation contributes to length of stride.

Triceps insertion, the short olecranon provides velocity advantage for triceps.

Small acromion process reflects small deltoids, arm moved in fore-and-aft plane only.

Large acromion process for origin of deltoids, pull arm away from body.

Extension of posterior angle of scapula increases lever arm of teres major muscle, retracts humerus.

Distal portion of ulna reduced, radius is main forearm bone.

Short radius reduces out-lever of triceps.

Greatly elongated elements increase length of stride.

Metacarpals and phalanges short.

Large lateral epicondyle is origin of extensor muscles.

Greatly elongated metacarpals (2) compressed.

Distal mass reduced by reduction in number and weight of bones and size of muscles.

Long olecranon process of ulna produces powerful leverage for triceps.

Phalanges short and compressed.

Long strong claws break soil.

Terminal phalanges protected by wear-resistant hoofs.

Figure 25.21 Contrasting forelimb structure of cursorial and fossorial mammals. (A) In the forelimb of a cursorial mammal, the radius and ulna and the metacarpals are elongated, and the feet are unguligrade. The number of phalanges is reduced. The proximal end of the ulna—the olecranon process—increases the velocity of movement of the limb. That is, the olecranon process is a short in-lever and the distal end of the ulna is a long out-lever. (B) The forelimb of a fossorial mammal is short. The long olecranon process (in-lever) and short ulna (out-lever) provide a forceful digging stroke. (From Hildebrand and Goslow 1998.)

25.7 The Evolution of Aquatic Mammals

Semiaquatic mammals are not that different from terrestrial mammals except for somewhat more paddlelike limbs and a denser fur coat, and lineages of semiaquatic mammals have evolved numerous times. Among extant mammals, we see examples in monotremes (platypus) and marsupials (water opossum, or yapok); and within placentals in the orders Eulipotyphla (water shrews, *Neomys fodiens*; and desmans, *Desmana moschata* and *Galemys pyrenaicus*); Afrosoricida (otter shrew, *Potamogale velox*, and otter tenrec, *Limnogale mergulus*); Rodentia (beavers, *Castor canadensis* and *C. fiber*; coypu, *Myocastor coypus*; muskrat, *Ondatra zibethicus*; and Australian water rat, *Hydromys chrysogaster*); Carnivora (otters [8 genera, including the sea otter *Enhydra lutris*]; mink [2 genera]; polar bear, *Ursus maritimus*; and Artiodactyla (hippopotamus, *Hippopotamus amphibius*). The fossil record adds hippopotamus-like rhinoceroses, several independent evolutions of otterlike animals (including the

recently extinct sea mink, *Neovison macrodon*), and even some semiaquatic sloths (species in the genus *Thalassocnus*).

Specialization for fully aquatic life is a different matter, however. Fully aquatic mammals—primarily marine forms that rarely or never come onto land—have evolved only three times (**Figure 25.22**): in the orders Sirenia (the dugong and manatees), Cetacea (whales, porpoises, and dolphins), and Carnivora (pinnipeds—seals, sea lions, and the walrus). Cetaceans and sirenians cannot come onto land. The pinnipeds are more amphibious, emerging onto land to court, mate, and give birth. Cetaceans and pinnipeds are carnivorous, while sirenians are herbivores. All use blubber (a thick layer of subcutaneous fat) for insulation instead of hair (or as well as hair in some pinnipeds), and show various adaptations of their physiology that allow them to stay underwater and in some cases (cetaceans and seals) to dive to great depths.

Morphological adaptations for life in water

Both feeding and locomotion underwater provide challenges for aquatic mammals. Predatory aquatic mammals use a variety of feeding mechanisms, including biting, suctioning, and filtering their prey. Locomotion may be either via the limbs (paraxial swimming) or using the undulations of the body and tail (axial swimming). Most semiaquatic mammals use the limbs to swim, as we do ourselves. Paraxial swimming is inefficient in terms of the drag forces created in the water. Fully aquatic mammals use axial swimming via dorsoventral flexion rather than lateral flexion like fishes and aquatic sauropsids. Axial swimming is a modification of the flexion of the vertebral column that is used by terrestrial mammals (**Figure 25.23**).

All fully aquatic mammals have elongated backbones, apparently by a similar modification of Hox gene expression

Figure 25.22 Diversity of marine mammals. (A) Sirenia, dugong and manatees: The West Indian manatee (*Trichechus manatus*). (B, C) Cetacea: baleen whales and toothed whales. (B) Mysticeti: The humpback whale (*Megaptera novaeangliae*). (C) Odontocei: Toothed whales include the Pacific spotted dolphin (*Stenella attenuatus*). (D–F) Pinnipedia: sea lions, the walrus, and true seals. (D) Otariidae: The California sea lion (*Zalophus californianus*). (E) Odobenidae: The walrus (*Odobenus rosmarus*). (F) Phodidae: The harbor seal (*Phoca vitulina*). (A, Keith Ramos/NFWS/CC BY 2.0; B, Christopher Michel/CC BY 2.0; C, NOAA/CC BY 2.0; D, Vasek Vinklát/CC BY 2.0; E, Capt. Budd Christman, NOAA; F, Tim Sackton/CC BY-SA 2.0.)

during development. Early cetaceans and sirenians had hindlegs, but these were lost as later forms became specialized for axial swimming. Pinnipeds, which still come out onto land at times, retain hindlegs and also retain zygapophyses in their trunk vertebrae. What appears to be a tail in pinnipeds is actually a modified pair of hindlimbs that have been turned backward. True seals are unable to change the position of the hindlimbs and are clumsy on land, but sea lions and the walrus can turn the hindlegs forward and move on land with vertical flexions of the vertebral column in an effective, although ungainly, humping fashion. Sea lions additionally have a derived type of paraxial swimming in which the forelimbs are used in synchrony

for underwater flying, creating lift in the water in concert with dorsoventral movements of the back and hindlegs.

The evolution of cetaceans

There has never been any doubt among scientists that cetaceans evolved from terrestrial mammals, but which lineage of mammals is the ancestral one was elusive for many years. Artiodactyls (even-toed ungulates) were suggested as possibly being ancestral to cetaceans as far back as the late 19th century, based on obscure anatomical details, although many researchers deemed it unlikely. But molecular and genomic research in the late 20th century showed unequivocally that cetaceans are so closely related to artiodactyls that they should be included within the order, as they form a sister group relationship with hippopotamuses. (This does not mean that cetaceans evolved from hippos, only that they shared a common ancestor some 55 Ma.)

Cetaceans have large brains for their body size, lack both hair and external ears, and have nasal openings at the top of the head (the blowhole). Some cetaceans retain a vestigial pelvis, but they have lost the other elements of the hindlimbs (although cetacean "legs" are occasionally found as developmental atavisms). Cetaceans swallow their food without chewing it; toothed whales have only a single set

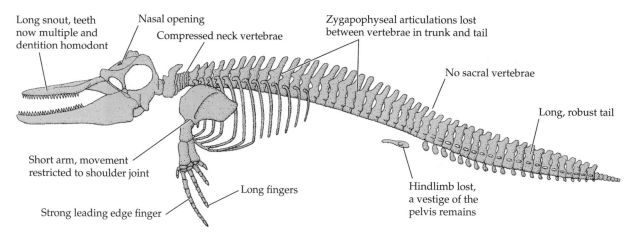

Figure 25.23 **Specializations for aquatic locomotion.** The skeleton of a whale illustrates adaptation for locomotion in water. Fully aquatic mammals have paddlelike limbs with short proximal portions and elongated phalanges. Like fish fins, these limbs are usually used for braking and steering rather than for propulsion. Unlike the lateral flexion that propels most fishes, however, fully aquatic mammals propel themselves using a dorsoventral flexion of the vertebral column that is a modification of spine flexion in terrestrial mammals. (From Macdonald 2009.)

of teeth (i.e., no tooth replacement occurs), lack incisors and canines, and have multiple simple cheek teeth.

Modern cetaceans comprise two groups: the small to medium-size toothed whales (including the large sperm whale *Physeter macrocephalus* as well as the porpoises and dolphins) and the medium-size to very large baleen whales (see Table 25.2). The toothed whales (Odontoceti, the most diverse group today) eat large prey items (fishes, squids, and other marine mammals) and have evolved a system of underwater echolocation, or sonar, to locate their prey. The baleen whales (Mysticeti, including the blue whale *Balaenoptera musculus*, the largest animal ever known) lost their teeth and evolved baleen, keratinous sheets hanging from the upper jaws, which they use to strain small items such as zooplankton from the water (suspension feeding, i.e., filtering small prey as water flows past the feeding apparatus). Mysticetes also have an exceptional ability to hear low-frequency sound, which they also can produce and use for communication. Ear regions of fossils show that this hearing evolved before mysticetes acquired baleen and large body sizes.

The fossil record   In evolving to become fully marine, cetaceans faced several challenges of both anatomy and physiology. The sequence of anatomical changes is documented by an almost perfect fossil record starting in the Eocene of Pakistan along the shores of the ancient Tethys Sea. This sequence shows a progression among early cetaceans from terrestrial forms with a full set of legs to aquatic forms with reduced and modified limbs (**Figure 25.24**).

Indohyus from the middle Eocene of India may be closely related to the ancestral lineage of cetaceans. *Indohyus* was a small (raccoon-size) artiodactyl, probably rather like a present-day mouse deer in appearance but with shorter legs and a longer tail. Its ear region shows the

characteristic cetacean derived condition (a thickened portion of the auditory bulla, the involucrum, related to hearing underwater), and its heavy (osteosclerotic) bones also indicate a semiaquatic lifestyle.

The Eocene fossil cetaceans all belong to the paraphyletic suborder Archaeoceti, which was largely extinct by the end of the epoch. The modern suborders of cetaceans, Odontoceti and Mysticeti, together grouped as Neoceti, arose from a common ancestor among the archaeocetes and first appeared in the latest Eocene, about 35 Ma.

The earliest archaeocetes were Pakicetidae, first known from the early Eocene (~52 Ma). These animals were coyote to wolf size, and although they were apparently primarily terrestrial, they had many features indicating that they were amphibious. They also had ankle joints resembling those of artiodactyls (the iconic "double-pulley" astragalus), which confirms the molecular data showing the relationship of cetaceans to these ungulates.

Pakicetids had teeth like those of modern fish eaters (e.g., sea lions), and they had orbits situated on the top of their head, like a hippopotamus, allowing them to observe their surroundings as they floated at the surface. (Later archaeocetes—protocetids—had more laterally placed eyes, indicating that they were no longer peering out above the water's surface.) Pakicetid postcranial remains show heavy bones for ballast and robust tails like those of otters, which swim with flexion of the vertebral column. The oxygen isotope ratios in the teeth of pakicetids show that they drank fresh water, not salt water, indicating that their amphibious life was conducted in rivers or at least near-shore environments. The middle ears of pakicetids were slightly, but significantly, modified for underwater hearing, including the diagnostic cetacean features of the ear region in the skull, which is how we can trace the cetacean pedigree to these terrestrial forms.

Ambulocetidae were somewhat later forms from the early middle Eocene. They were about the size of a sea lion and had hindlimbs with enormous feet, possibly a specialization for a mode of locomotion involving dorsoventral flexion and paddling with the hindfeet pointed backward like those of a seal. However, although the skeletons of ambulocetids were more modified for aquatic life than those of pakicetids, their fingers were not embedded in a webbed

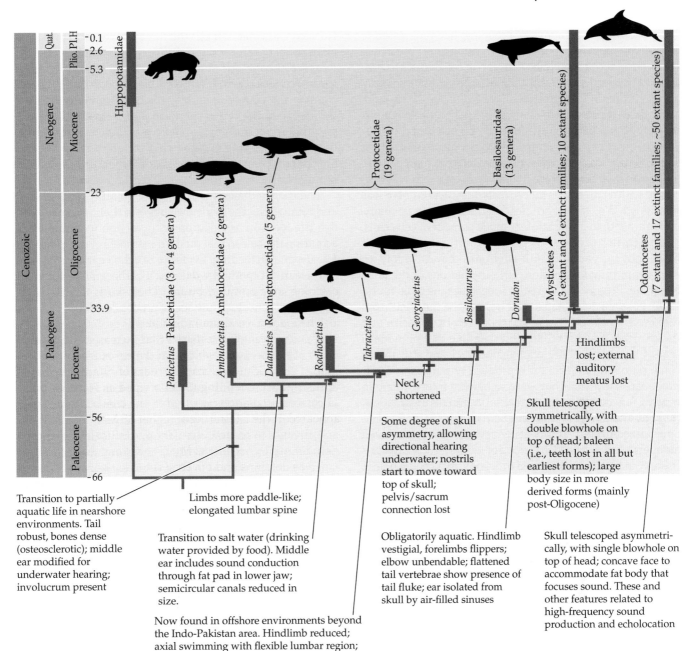

Transition to partially aquatic life in nearshore environments. Tail robust, bones dense (osteosclerotic); middle ear modified for underwater hearing; involucrum present

Limbs more paddle-like; elongated lumbar spine

Transition to salt water (drinking water provided by food). Middle ear includes sound conduction through fat pad in lower jaw; semicircular canals reduced in size.

Now found in offshore environments beyond the Indo-Pakistan area. Hindlimb reduced; axial swimming with flexible lumbar region; shorter neck; vomeronasal organ lost.

Neck shortened

Some degree of skull asymmetry, allowing directional hearing underwater; nostrils start to move toward top of skull; pelvis/sacrum connection lost

Obligatorily aquatic. Hindlimb vestigial, forelimbs flippers; elbow unbendable; flattened tail vertebrae show presence of tail fluke; ear isolated from skull by air-filled sinuses

Hindlimbs lost; external auditory meatus lost

Skull telescoped symmetrically, with double blowhole on top of head; baleen (i.e., teeth lost in all but earliest forms); large body size in more derived forms (mainly post-Oligocene)

Skull telescoped asymmetrically, with single blowhole on top of head; concave face to accommodate fat body that focuses sound. These and other features related to high-frequency sound production and echolocation

Figure 25.24 A phylogeny of whales.   This tree is based on the excellent fossil record of cetacean evolution from limbed to fully aquatic forms documented by specimens excavated in Pakistan, in regions that were once the shores of the ancient Tethys Sea. (After Zimmer 2001.)

flipper, and both fingers and toes had little hooves at the tips, betraying their ungulate origins. The fossil remains of ambulocetids are found in coastal environments, but their tooth isotopes show that they still drank fresh water. Their robust skulls and teeth suggest that they were specialized crocodile-like ambush predators feeding on large prey. Ambulocetids were a specialized side branch not directly involved in the ancestry of modern cetaceans.

Remingtonocetidae was a more derived group. Like ambulocetids, remingtonocetids were also a specialized side

branch. They were the size of seals and had relatively robust hindlimbs which, as in other early archaeocetes, were still connected to the vertebral column, so they were probably still capable of bearing their weight on land. Unlike ambulocetids, remingtonocetids had delicate skulls with long, narrow snouts and very small eyes, indicating that they fed underwater on fish. Features of the ear anatomy of both remingtonocetids and the later protocetids indicate improved underwater hearing; these include an enlarged canal in the lower jaw that in modern cetaceans houses a sound-conducting fat pad. There was also a reduction in the size of the semicircular canals, which in modern cetaceans is associated with underwater navigation. The tooth isotopes of remingtonocetids (and of all later cetaceans) show that they did not drink fresh water, but must have obtained all of their water from their food, as modern cetaceans do.

Protocetidae, a paraphyletic grouping, had progressed to a condition more like that of modern cetaceans. Protocetids had more reduced hindlimbs than earlier cetaceans. Most retained a connection of the hindlimb to the vertebral column, although they would probably have been clumsy on land. More derived protocetids retained the pelvic girdle but lost the connection between the ilium in the pelvis and the sacrum in the vertebral column. However, evidence that at least some protocetids came on land to give birth, as pinnipeds do today, can be inferred from a fossil *Maiacetus* that was preserved with a late-stage embryo in its body. The embryo was positioned so as to be born headfirst, like terrestrial mammals, rather than in the tail-first mode of modern cetaceans.

The earlier protocetids probably still relied on their hindlimbs for paddling, but later ones may have had an axial swimming locomotion, with oscillations of the lumbar spine. Protocetids were the first cetaceans to be found in offshore marine habitats and beyond the Indo-Pakistan region, as far afield as the coasts of North America and Africa. They may have had a lifestyle like that of modern seals: fully aquatic, but not obligatorily so, and still able to come onto land at least for reproductive purposes. Their inner ears show that they were somewhat intermediate in hearing function between terrestrial mammals and modern cetaceans, and were able to hear both in air and underwater. Specialized underwater hearing appears to have evolved independently in the two groups of modern cetaceans (ultrasonic in toothed whales and infrasonic in baleen whales).

Finally, Basilosauridae appeared in the later Eocene. This group was probably obligatorily aquatic. These cetaceans had lost the hindlimb connection to the vertebral column and had greatly reduced hindlimbs, although all of the skeletal elements were still present and the reduced hindlimbs might have been used as copulatory guides. (In all mammals, certain muscles run from the pelvis to the genitals, which explains why a remnant of the pelvis is retained in modern cetaceans and sirenians.) The necks of basilosaurids were short, the forelimbs were flipperlike with an immobile elbow, and the morphology of the tail vertebrae suggests that the tail had a fluke, as in modern cetaceans. The ears of basilosaurids were essentially like those of modern cetaceans, with the entire bony ear region isolated from the rest of the skull by air-filled sinuses, although they still retained an external auditory meatus (earhole). The basilosaurines, known from the Northern Hemisphere, had long bodies (up to 16 m), with greatly elongated trunk vertebrae and small heads. When first discovered, they were mistaken for sea serpents—hence the root word "saurus" in their name. The dorudontines, known from both the Northern and Southern hemispheres, were more dolphinlike in appearance and contained the ancestors of the modern cetaceans.

The initial radiation of Neoceti more or less coincided with the extinction of the archaeocetes at the Eocene–Oligocene boundary, although some archaeocetes may have survived in the seas around New Zealand until the late Oligocene. This was the time Antarctic Circumpolar Current was established, which may have led to profound changes in ocean productivity, and when higher-latitude temperatures fell dramatically. Modern cetaceans differ from archaeocetes in having a "telescoped" skull, whereby, over the course of embryonic development, the nostrils move to the top of the head and form the blowhole, and the upper and lower jaws are elongated. The mode of skull telescoping is somewhat different in mysticetes and odontocetes, indicating that this feature evolved in parallel in the two groups.

The earliest baleen whales were small (porpoise size) and retained teeth. Evidence suggests that they may have lost the teeth in association with suction feeding before acquiring the baleen that modern members of the group use for suspension feeding. Evidence of echolocation is present in the earliest toothed whales, which became adapted for catching large prey underwater. Their skulls show remarkable convergences with those of crocodiles.

The modern cetacean radiation was probably related to changes in oceanic circulation that increased the productivity of the oceans, resulting in the novel feeding strategies of echolocation-assisted predation (toothed whales) and suspension feeding (baleen whales). Another pulse of cetacean radiation occurred in the late Miocene, again concurrent with a lowering of higher-latitude temperatures and changes in oceanic circulation. At this time, many existing families became extinct, and some modern forms (such as dolphins and porpoises) made their first appearance. The gigantic baleen whales seen today did not appear until the Pliocene, along with increases in seasonal ocean upwelling providing more nutrients for their prey. Sadly, recent molecular analyses suggest that North Atlantic baleen whales were up to 20 times more abundant before the start of human commercial exploitation than they are today.

Genomic evidence   Many features that evolved as cetaceans adapted to their marine realm are not documented in their fossils. These include adaptations for diving, massive fat reserves for both metabolic and thermoregulatory functions, and changes in their senses. Land-based senses of smell and taste do not work underwater, and both vision and hearing must adapt. Study of the genomes of extant cetaceans can pinpoint possible times in cetacean history where various features were gained or lost.

The myoglobin genes of all cetaceans have improved that protein's ability to store oxygen in muscles—an essential feature for diving mammals, and one that is seen convergently in pinnipeds and beavers. Such changes in myoglobin genes can be used to establish aquatic origins for now-terrestrial mammals, as was mentioned earlier for elephants and echidnas. Further changes are seen in diving-specialist toothed whales such as the sperm whale.

Toothed whales echolocate, and in this lineage changes can be seen in genes relating to the reception of high-frequency sound, especially in the genes coding the protein prestin, which affects the hair cells in the cochlea. Echolocating bats show a similar change in their genomes.

Cetaceans have several pseudogenes associated with changes in sensory systems in water. All extant cetaceans have lost color vision, and the opsin gene for short-wavelength-light reception has become nonfunctional. This loss occurred by different mechanisms in toothed and baleen whales, showing that their common ancestor was dichromatic, like most terrestrial mammals. Cetaceans also show degeneration of the genes for olfaction and taste (fish can smell underwater, but amniotes lost the water-based sense of smell and evolved an air-based one; amphibians have both). All the olfactory genes of toothed whales have become pseudogenes; the brains of toothed whales even lack an olfactory nerve and olfactory bulb. Baleen whales, however, retain these structures, as well as some functional olfactory genes (as do pinnipeds and sirenians).

Modern cetaceans lack a vomeronasal organ (a derived olfactory organ of tetrapods; see Chapter 12) and have few or no taste buds. Fossil cetacean skulls show that the vomeronasal organ was lost at about the level of the protocetid cetaceans. Many of the genes coding for taste, especially sweet taste, have become pseudogenes. Extant baleen whales lack teeth but retain pseudogenes for the production of enamel.

All cetaceans lack hair, but they retain pseudogenes for hair production and growth. Like many hairless mammals,

cetaceans have lost their sebaceous glands, along with one of the genes involved in the glands' formation. Interestingly, hippos retain this gene despite also lacking sebaceous glands, providing molecular evidence that the common ancestor of hippos and whales was terrestrial.

25.8 Trophy Hunting

Mammals are exposed to the same detrimental human influences that we have described in discussions of other groups of vertebrates—habitat loss, harvesting for food, introduction of alien species, and spread of disease—but being killed solely to become a hunter's trophy is a risk that is largely restricted to mammals.

Trophy hunters who want a mounted head to hang on a wall target the individuals that make dramatic trophies, usually the ones with the largest horns or antlers. A genetic component accounts for 30-40% of the variance in the size of horns and antlers, and horn size should show rapid evolutionary change under strong selection pressure (**Figure 25.25**).

Horns of sheep and other members of the family Bovidae, such as cows and antelope, are formed from a bony core covered by a keratinous sheath. As in the case of other hoofed mammals with cranial appendages, these horns are used for display and fighting in male-male interactions, but unlike the antlers of deer, elk, and moose, which are shed annually, bovid horns are permanent structures that increase in size throughout the animal's life. Horn size is a key determinant of reproductive success for bighorn rams, and sexual selection favors males with large horns.

A long-term study of rams in a population of bighorn sheep (*Ovis canadensis*) on Ram Mountain in Alberta,

(A)

(B)

(C)

Figure 25.25 **Examples of selection resulting from trophy hunting.** (A) The average horn lengths of male impalas killed by trophy hunters in Africa declined by 4% between 1974 and 2008. (B) The horn lengths of sable antelope decreased by 6%. (C) Similarly, in areas of heavy trophy hunting the average antler size of Alaskan moose deceased by 15%. (A, Son of Groucho/CC BY 2.0; B, Bernard Dupont/CC BY-SA 2.0; C, U.S. National Park Service.)

Canada, has documented a decrease in average horn size that has been demonstrated to represent genetic change (**Figure 25.26**). The selection pressure placed by trophy hunters on bighorn sheep can be summarized as follows:

- To ensure that only adult rams could be shot, Alberta set the minimum horn length for "legal" rams at 80% of a full circle of curl.

- Rams with fast-growing horns reach that size in about 4 years, and 91% of these rams were killed before they reached 7 years, which is the age at which large horn size provides an advantage in mating.

- Trophy hunting thus created an unnatural selective advantage for rams with slow-growing horns that reach a smaller maximum size, and these were the individuals whose genes were passed on to succeeding generations. As a result, the average horn size of rams killed by trophy hunters decreased by 14 cm between 1974 and 2011.

Similar results have been reported for wild goats as well as for other species of wild sheep, and for moose, African antelopes, and elephants. Recovery from unnatural selection can be slow. Horn size began to recover following relaxation in hunting pressure at Ram Mountain in 1996, but had increased by only 13% as of 2010. Thus, recovery from anthropogenic selection is likely to be much slower than the original response.

Endangering the endangered: The effect of perceived rarity

The value attached to rarity is commonly used by sellers to make a product desirable, with phrases like "one of a kind" and "only five left in stock." Trophy hunters respond to the lure of rarity as readily as online shoppers, and permits to hunt in areas with a reputation for producing trophy heads can draw astonishing bids. In most jurisdictions, permits to hunt bighorn sheep are auctioned for US $20,000-40,000, but permits for Alberta regularly attract bids exceeding US $400,000.

Endangered status is also a draw for trophy hunters. In 2015 the Namibian government auctioned a permit to kill a male black rhinoceros for US $350,000. The IUCN lists this species as critically endangered, and that was part of its desirability. As species become increasingly rare, their desirability as trophies increases. Between 2004 and 2010, the fees paid by trophy hunters for species that had been moved into more endangered categories on the IUCN Red List nearly doubled, whereas permits for species that had remained in the same category or moved to a lower risk category increased by one-third.

In 2012, trophy hunting generated more than US $217 million for seven sub-Saharan African countries. Most of this revenue was derived from the big five African game species: lions, leopards, elephants, Cape buffalo, and rhinoceroses. Proponents of trophy hunting argue that the revenue generated from hunting supports a broad range of conservation activities, from establishing and maintaining game reserves to fighting poachers. This view maintains that elimination

(A)

(B)

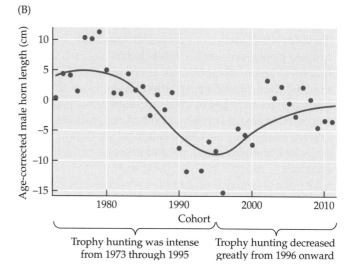

Trophy hunting was intense from 1973 through 1995 | Trophy hunting decreased greatly from 1996 onward

Figure 25.26 Changes in horn length of bighorn sheep. (A) A mature ram with horns describing a full circle of curl. Killing rams with such horns is the goal of trophy hunters. (B) The average horn size of rams in the Ram Mountain population each year from 1973 through 2010; horn sizes were standardized by correcting for the age of the individuals in the population each year. The Ram Mountain population of bighorn sheep contains approximately 500 rams. During the period of intense trophy hunting (1973–1995) a total of 52 rams were killed and the standardized horn length decreased from +5 to –9 cm as fast-growing males were removed from the population before they reached their full reproductive potential. From 1996 onward, the intensity of trophy hunting was greatly reduced, and only 4 rams were killed in 15 years. Horn size slowly increased as fast-growing rams entered the breeding population. (A, slashvee/CC BY-ND 2.0; B from Pigeon et al. 2016.)

of trophy hunting would increase the loss of biodiversity because habitat protection benefits all species in an ecological community, not just the ones that trophy hunters seek.

An opposing perspective focuses on the trophy species themselves, rather than on the community as a whole. Many species benefit from the presence of conspecifics, a phenomenon known as the Allee effect. Familiar examples include the group defenses against predators employed by many ungulates; modifications of the environment that create a suitable environmental for a species, such as construction and maintenance of dams by beavers; and cooperative breeding, such as the simultaneous birth of wildebeest calves that swamps the capacity of predators to attack them. A decline in population size thus reduces the Darwinian fitness of each individual in the population. Below a threshold size, a population may crash to extinction.

The extinction vortex

Economic theory holds that exploitation is unlikely to result in the extinction of a species: As the cost of capturing the last few individuals becomes prohibitively high, hunters should switch to species that are easier to find. This assumption may not apply to trophy hunting, however. If increasing rarity makes the remaining individuals ever more desirable to some hunters, a positive feedback loop will form that maintains or even increases hunting pressure as species numbers dwindle.

A hypothesis known as the "anthropogenic Allee effect" combines the benefits of conspecifics with the impact of rarity on trophy species. This view holds that the increased incentive provided by rarity ensures continued hunting that can reduce the size of a population until essential Allee interactions between individuals no longer occur. This situation is described as an extinction vortex (see Section 13.5).

Some life history and ecological characteristics increase the extinction risk of a species, and unfortunately several of them apply to species that are the targets of trophy hunters.

- Large herbivores often mature late and have low reproductive rates. Female African elephants, for example, are not sexually mature until they are 11 years old, and males do not reproduce until they are more than 20 years old. Females give birth to a single calf, and the interbirth interval is about 5 years. Female black rhinoceroses reach sexual maturity at about 5 years and males at 6 years; single calves are born at intervals of 3 years or more. These life history characteristics

prevent rapid repopulation when the number of individuals in a population is reduced.

- Large carnivores need a lot of space. Trophy hunting may be a factor in the decline of leopard populations in Tanzania. The home ranges of African leopards are large—an average of 229 km^2 for males and 179 km^2 for females. Only 14% of each male's home range overlaps the home ranges of females, and a reduction in the number of leopards in a population reduces the chance of encounters while a female is sexually receptive

- A high degree of sociality makes a species vulnerable to the anthropogenic Allee effect. African hunting dogs are extremely social, hunting in packs and expending substantial energy pursuing their prey. After a prey animal has been killed, spotted hyenas attempt to chase the dogs from their kill. Packs containing 6 or more dogs are able to keep hyenas away for 10–15 minutes, which is long enough for the dogs to consume the best sections of a carcass, but smaller packs of dogs lose substantial portions of their kills to hyenas. Because the dogs' hunting is so energetically costly, losing 25% of their kills to hyenas would be an unsustainable energy loss, and reduction of pack size leads to an extinction vortex.

An analysis of extinctions of 10 populations of vertebrates supported the reality of an extinction vortex. The final declines to extinction of these populations (i.e., periods without a single year-to-year increase in population size) started from an average population of 70 individuals, and three of these populations declined from more than 100 individuals. Complex interactions of multiple factors appear to have been responsible for these extinctions, including the Allee effect.

Summary

Extant mammals can be divided into three major groups: monotremes, marsupials, and placentals.

The mammalian lineages surviving into the Cenozoic were Prototheria (monotremes), Allotheria (multituberculates, now extinct), and Theria. Theria is subdivided into Metatheria (marsupials and their extinct relatives) and Eutheria (placentals and their extinct relatives). Almost all extant mammals are therians.

- Monotremes (the platypus and echidnas of Australia and New Guinea) are the only mammals that lay eggs. They appear to have been confined to Australasia for their evolutionary history, and probably were never much more diverse than they are today. They retain other generalized amniote features, such as a more sprawling stance, but they have some unique features, such as a toothless electroreceptive beak.

- Marsupials are found in South America (three orders of opossums, mostly ameridelphians) and Australasia

(four orders of australidelphians, including marsupial moles, marsupial mice, bandicoots, possums, kangaroos, wombats, and the koala). Australia has the most diversity of extant marsupials, but South American marsupials were more diverse in the past and included large carnivorous forms.

- Placentals are found worldwide (except in Antarctica) and comprise four major groups: Afrotheria (six African orders, including elephants, sirenians, and tenrecs); Xenarthra (two mostly South American orders, including sloths, anteaters, and armadillos); Euarchontoglires (five orders, including rodents and primates); and Laurasiatheria (six orders, including bats, carnivorans, hoofed mammals, and cetaceans). Afrotheria and Xenarthra are united in the Atlantogeneta (although there is some debate about this grouping). Euarchontoglires and Laurasiatheria are united in Boreoeutheria (northern mammals).

(Continued)

Summary (continued)

The extinct multituberculates (Allotheria) are more closely related to therians than are monotremes. They are known from the Late Jurassic until the middle Cenozoic. Their name comes from their multicusped molars, indicative of an omnivorous or herbivorous diet. Multituberculates were rodentlike in appearance and may have been outcompeted by true rodents, which appeared in the early Cenozoic.

Therians are distinguished by more than being viviparous.

Therians have more derived skulls and teeth than non-therians, including tribosphenic molars that can crush food as well as shearing it, and a different system of jaw-opening muscles than monotremes

Therians have a more upright posture than non-therians, including a hinge-jointed elbow that can be tucked under the body and a more hinge-like ankle joint. The coracoid and clavicle bones in the shoulder girdle have been lost, and the scapula can move as an independent limb segment, bolstered by new scapular sling muscles. All of these features contribute to the typical (and probably unique) bounding gait of therians—leaping off the hindlimbs onto the forelimbs while flexing the back.

Therians have more derived sense organs than non-therians. The cochlea in the inner ear is long and coiled, allowing for better pitch discrimination, and the pinnae (external ears), which in most species are mobile, help determine sound direction.

All extant mammals show a loss in the number of opsins (photosensitive molecules) in the cone cells of the retina, and a corresponding loss in the opsin genes, although the pattern of loss is slightly different in therians and monotremes. Most mammals are dichromatic, seeing two colors (the generalized amniote condition is to be tetrachromatic), but most anthropoid primates have duplicated one of the remaining genes, giving them trichromatic vision.

All mammals produce milk for their young (lactate), but only therians have nipples. The composition of marsupial milk varies considerably during the period of lactation.

Therians have heteromorphic sex chromosomes (XX for females, XY for males). The situation in monotremes is more complicated.

Marsupials and placentals differ in their anatomy.

The skull of marsupials differs from that of placentals in several ways, including the inturned posterior end of the dentary (the inflected angle). Marsupials have more incisors and molars than placentals, but fewer premolars.

Marsupials retain the epipubic bones in the pelvis seen in basal mammals but not in placentals.

Marsupials lack the corpus callosum that connects the two cerebral hemispheres in placentals.

Mode of reproduction is the main distinction among the three major groups of extant mammals.

Both marsupials and placentals are viviparous, in contrast to the egg-laying monotremes.

The embryos of all mammals, like those of other amniotes, have four extraembryonic membranes. In therians, some of these membranes form the placenta. All mammals have a corpus luteum, a hormone-secreting structure formed in the ovary after the egg has been shed.

Monotremes have larger ovaries than therians, which provide yolk for the developing young, and lay thin-shelled eggs that hatch soon after laying.

Marsupials and placentals differ in the relative lengths of gestation (longer in placentals) and lactation (longer in marsupials). All marsupial young complete their development fastened onto a nipple, and some marsupials house these developing young in a pouch.

Neonate monotremes and marsupials are highly altricial, but marsupial neonates have well-developed forelimbs to aid them in their climb to the nipple, and a well-developed mouth for fastening onto the nipple. The shoulder girdle of marsupial neonates retains the interclavicle and complete coracoid bones, enabling their crawl to the pouch.

Some placentals give birth to altricial young; others give birth to well-developed precocial young.

Therians have a more derived condition of the urogenital tract than monotremes.

Monotremes resemble other non-amniotes and basal amniotes in having a cloaca: a single opening for the urogenital and alimentary tracts. Therians have separate openings for the urogenital and alimentary tracts; in female primates all three tracts open independently (via the anus, vagina, and urethra). Male therians, unlike other amniotes, use their penis for both sperm transmission and urination.

In marsupials and placentals, the ureters from the kidneys enter the bladder directly, but the relationship of the ureters to the reproductive tract differs in these two groups, leading to some key differences in the internal anatomy of the reproductive system. A scrotum housing the testes outside the body is seen in most marsupials and placentals, and appears to have evolved independently in the two groups.

Therians have a more derived condition of the reproductive tract than monotremes.

All mammals have an initial choriovitelline placenta, which is transitory in placentals but forms the only placenta in most marsupials.

In all placentals and a few marsupials, the choriovitelline placenta is succeeded by the chorioallantoic placenta.

Summary (continued)

The outgrowth of the chorioallantoic membrane appears to have been suppressed in the marsupial lineage.

In placentals, the form of the placenta is highly variable as to the number of layers and the mingling of fetal and maternal tissues. The human condition of few layers and of maternal blood being in direct contact with fetal tissues is probably the ancestral condition.

Marsupials have often been considered inferior to placentals in their mode of reproduction, but it is now understood that they are simply specialized in their own fashion. The lower number of marsupial species (about 6% that of placentals) is more likely due to historical accident than to reproductive inferiority. However, the marsupial way of giving birth has limited the lineage's evolutionary flexibility. Because of the importance of specialized forelimbs and oral regions in neonates, these regions are more constrained in their adult anatomy than are those of placentals. There are no marsupial bats or hoofed forms; and the marsupial mode of birth would make a fully aquatic marsupial an impossibility.

Mammals have specializations of their skulls and teeth for different diets.

Mammals must process large quantities of food per day to fuel their high metabolic rates, and oral breakdown of food to reduce the particle size is an important starting point. Almost all mammals have a dentition consisting of incisors, canines, premolars, and molars, with the form of the molars varying in mammals with different diets. Canines, usually associated with killing prey, may be enlarged for other reasons, such as display.

Carnivorans have specialized shearing teeth called carnassials. Omnivores have low, blunt molars for crushing. Herbivores have molars with ridges (lophs) that shred fibrous food. Herbivorous diets may be abrasive, and in herbivorous mammals the cheek teeth (molars plus premolars) may be made more durable by being high crowned (hypsodont). Rodents and rabbits may have ever-growing cheek teeth. In elephants, durability of the dentition is enhanced by having molar progression.

Differences in the skulls of carnivores and herbivores reflects not only their different teeth but the relative importance of the different jaw muscles. Carnivores have a high occipital region, reflecting a large temporalis muscle best suited for exerting force at the front of the jaw. Herbivores have a deep lower jaw that reflects a large masseter muscle, which moves the jaw sideways in a grinding action; they also have a jaw joint set above the level of the tooth row, which enables them to occlude all of the cheek teeth simultaneously when the jaw moves sideways.

Rodents are specialized to move their lower jaw forward in occlusion, rather than sideways, and this enables some of them to process food on both sides of the jaw at the same time.

Mammals have specializations of their limbs for different kinds of locomotion.

The laws of physics dictate that larger mammals are more specialized for particular kinds of locomotion than are smaller mammals.

Cursorial (running) mammals have limbs modified for speed at the expense of power, and for rapid recycling at the expense of flexibility. Cursors have elongated limb segments, especially the distal elements (but not the toes); the digits may be reduced or lost, and the movement of the limbs is restricted to the fore-and-aft plane.

Fossorial (digging and burrowing) mammals have limbs modified for power at the expense of speed. They have short limb elements, with long lever arms for muscle insertion, and retain five digits with stout claws.

Aquatic mammals have a distinctive set of adaptations.

Many mammals are semiaquatic, but fully aquatic forms have evolved only three times: in cetaceans (whales, porpoises, and dolphins), sirenians (manatees and the dugong), and the more amphibious pinniped carnivorans (seals, sea lions, and the walrus). All fully aquatic mammals are streamlined for moving through water, are insulated by blubber, and have converted their limbs into flippers. Cetaceans and seals have physiological adaptations that allow them to dive to great depths.

Most semiaquatic mammals swim using their limbs. Most fully aquatic mammals swim using dorsoventral flexion of the body and tail (or modified hindlimbs in pinnipeds), and use the limbs mainly for steering and braking. Sea lions use their front flippers to "fly" through the water.

Genes and the fossil record document the evolution of cetaceans.

Scientists have always known that cetaceans evolved from terrestrial mammals. In the late 20th century, genomic evidence showed them to be most closely related to hippos among extant mammals. Modern cetaceans comprise two groups.

- The mainly smaller carnivorous toothed whales, which include dolphins and porpoises, use echolocation to find their prey.
- The plankton-feeding baleen whales can grow to enormous sizes, and filter their food through plates of baleen.

Cetacean genomes contain many pseudogenes, relicts of once-functional genes for which fully aquatic mammals have no use. These include genes for color vision, olfaction, taste, hair production, and in the case of baleen whales, tooth enamel.

Specialized genes in cetaceans include those coding for the improved ability of myoglobin to store oxygen in the muscles (also seen in some other aquatic mammals), and

Summary *(continued)*

for the reception of high-frequency sound in the toothed whales (also seen in bats).

The fossil record documents transitional forms of archaeocete cetaceans, the Eocene precursors to the modern groups, showing the progression from amphibious terrestrial forms to fully aquatic animals. The fossil record details the reduction of the hindlimbs and the conversion of the forelimbs into flippers; changes in the inner ear for underwater hearing and orientation; and changes in the oxygen isotope ratios of the teeth that show a transition from drinking fresh water to drinking sea water.

The radiation of modern cetaceans in the middle Cenozoic was probably correlated with changes in oceanic circulation that increased ocean productivity.

Trophy hunting exerts an unnatural selective pressure on some species of mammals.

Trophy hunters target healthy, mature males; these are the individuals that would normally have long reproductive lives and make the largest genetic contribution to succeeding generations.

Rarity increases the perceived value of a species to trophy hunters, resulting in increased hunting pressure as populations decline—the "anthropogentic Allee effect."

Interaction of the anthropogenic Allee effect and the ecological and life history characteristics can drive hunted species into an extinction vortex.

Discussion Questions

1. The eggs of monotremes are much smaller than those of lizards of comparable body size. What is a possible reason for this?

2. Some people have speculated that marsupials and placentals may have evolved viviparity independently. What evidence would support this?

3. Australian marsupials have been compared with similar placentals (e.g., gliding squirrels and gliding possums) as examples of convergent evolution. However, there is no equivalent among marsupials to a placental cetacean or even to a seal. What is a possible reason for this?

4. What is the information from our own retinas, and genes for color vision, that can lead to the inference that the earliest mammals were nocturnal?

5. How and why is the mobile scapula of therians essential for their form of locomotion?

6. Why do seals and sea lions retain zygapophyses in their backbones, whereas cetaceans and sirenians do not? How could we use this anatomy to determine the behavior of extinct cetaceans (and what additional evidence do we have in one fossil cetacean to back up behavioral predictions)?

Additional Reading

Abbot P, Rokas A. 2017. Mammalian pregnancy. *Current Biology* 27: R123–R128.

Asahara M, Koizumi M, Macrini TE, Hand SJ, Archer M. 2016. Comparative cranial morphology in living and extinct platypuses: Feeding behavior, electroreception, and loss of teeth. *Science Advances* 2: e1601329.

Foley NM, Springer MS, Teeling EC. 2016. Mammal madness: Is the mammal tree of life still not resolved? *Philosophical Transactions of the Royal Society B*: 371. doi.org/10.1098/rstb.2015.0140.

Freyer C, Renfree MB. 2009. The mammalian yolk sac placenta. *Journal of Experimental Biology (Mol. Dev. Evol.)* 312B: 545–554.

Gatesy J and 7 others. 2013. A phylogenetic blueprint for a modern whale. *Molecular Phylogenetics and Evolution* 66:476-506.

Haig D. 2012. Retroviruses and the placenta. *Current Biology* 22: R609–R613.

Kim J-W and 8 others. 2016. Recruitment of rod photoreceptors from short-wavelength-sensitive cones during the evolution of nocturnal vision in mammals. *Developmental Cell* 37: 520–532.

Kleisner K, Ivell R, Flegr J. 2010. Evolutionary history of testicular externalization and the origin of the scrotum. *Journal of Biosciences* 35: 27–37.

McCurry MR and 5 others. 2017. The remarkable convergence of skull shape in crocodilians and toothed whales. *Proceedings of the Royal Society B* 284: 20162348. doi.org/10.1098/rspb.2016.2348

McGowen MR, Gatesy J, Wildman DE. 2014. Molecular evolution tracks macroevolutionary transitions in Cetacea. *Trends in Ecology and Evolutionary Biology* 29: 336–345.

Mitchell KJ and 16 others. 2014. Molecular phylogeny, biogeography, and habitat preference evolution of marsupials. *Molecular Biology and Evolution* 31: 2322–2330.

Mourlam MJ, Orliac MJ. 2017. Infrasonic and ultrasonic hearing evolved after the emergence of modern whales. *Current Biology* 27: 1776–1781.

Pyenson ND. 2017. The ecological rise of whales chronicled by the fossil record. *Current Biology* 27: R558–R564.

Reidenberg JS. 2007. Anatomical adaptations of aquatic mammals. *The Anatomical Record* 290: 507–513.

Renfree MB. 2010. Marsupials: Placentals with a difference. *Placenta* 31 (suppl A): S21–S26.

Slater GJ, Goldbogen JA, Pyenson ND. 2017. Independent evolution of baleen whale gigantism linked to Plio-Pleistocene ocean dynamics. *Proceedings of the Royal Society B* 284: 20170546. http://dx.doi.org/10.1098/rspb.2017.0546

26

Primate Evolution and the Emergence of Humans

Primates have been a moderately successful group for most of the Cenozoic, although since the end of the Eocene they have been largely confined to tropical latitudes (with the exception of humans). Primates include not only the anthropoids—the group of apes and monkeys to which humans belong—but also the prosimians, animals such as bush babies and lemurs, and earlier less derived forms known only from the fossil record. Molecular techniques show that chimpanzees are the closest extant relatives of humans, and both molecular data and the fossil record indicate that the separation of humans from the African great apes occurred about 6.6 Ma. Fossils of *Australopithecus*—the sister taxon to our own genus, *Homo*—clearly show that bipedal walking arose before the appearance of a large brain. A diversity of new fossils has shown that early human evolution was much more complex and diverse than previously thought.

26.1 Primate Origins and Diversification

Humans share many biological traits with the animals variously called apes, monkeys, and prosimians. We are all members of the order Primates (Latin *primus*, "first"). The first primates were arboreal animals living in early Cenozoic forests. Humans are late-appearing primates, and complex social systems are typical of all extant primates. Table 26.1 presents a traditional classification of extant primates and their social organization.

All extant primates are by definition crown primates, formally referred to as **Euprimates** (Greek *eu*, "true"). Euprimates exclude the so-called stem primates, or **plesiadapiforms** (Greek *plesios*, "near"; Latin *ad Apis*, "toward Apis [an Egyptian bull deity]"; *formis*, "shape"). Features typical of modern primates are listed in Table 26.2. Although not unique to this group, one important characteristic of modern primates is the presence of opposable digits (thumbs and big toes) that can grasp the substrate, especially during arboreal locomotion. Opposable digits

also increase dexterity, allowing precise manipulation of objects. This precision is associated with the modification of claws into flattened and compressed nails, and the presence of a fleshy, sensitive pad on the distal ends of the fingers. This finger morphology is fundamental for behaviors such as grooming, which is necessary for social bonding.

Evolutionary trends and diversity in primates

The increase in the encephalization quotient (brain size in relation to body size) in euprimates may have been associated with shifts to a more frugivorous diet and changes in their visual system. These changes included an increase in the size of the orbits and relocation of the eyes to the front of the face. In this position, the eyes have stereoscopic vision and depth perception, and the ability to detect objects against a background. **Binocular vision** is critical for seeing cryptic fruits, and this ability means primates are a pivotal group for seed dispersal and maintenance of tropical forests. Primates are visually driven animals; olfaction is less important, and reduction in the nasal region is characteristic of primates. In addition, many of the genes responsible for odor detection in other mammals are no longer functional in some primates.

Plesiadapiforms The first primatelike mammals appeared in the early Paleocene in the Northern Hemisphere and persisted until the end of the Eocene. Plesiadapiforms were generalized arboreal primates and ranged from marmot-size to the smallest primate that has ever existed (*Toliapina*, only 6.6 g). The plesiadapiform *Purgatorius*, from the Paleocene of Montana and Saskatchewan, is the oldest and most primitive potential primate, occupying a very important position in the studies of primate origins. Different lineages had teeth indicating that they had different diets—frugivorous, gum-eating, folivorous (leaf-eating), and insectivorous (Figure 26.1).

Plesiadapiforms shared some derived features of the teeth and skeleton with euprimates, but they differed in having smaller brains and longer snouts, in lacking a postorbital bar (a bar of bone on the lateral side of the orbit), and in having large, forward-pointing incisors, making them

(A)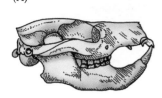

(B)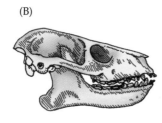

(C)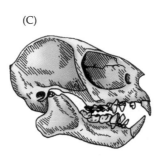

Figure 26.1 Skulls and teeth of early primates and early euprimates. (A) *Plesiadapis*, a plesiadapiform, was an arboreal frugivore/herbivore. (B) *Notharctus*, an arboreal adapoid, was a folivore. (C) *Tetonius*, an omomyoid, was an arboreal frugivore. Species are not drawn to the same scale. (After Romer 1966.)

appear superficially rodent-like. No plesiadapiform was a specialized leaper, and all of them retained claws. In contrast, all euprimates have flat nails (like our fingernails and toenails), except for the marmosets of South America, which have secondarily reverted to having claws. While claws enable small animals to cling for support on large-diameter substrates, such as tree trunks, nails allow the development of sensitive fingertips that are more useful for habitual locomotor behaviors on small-diameter supports, such as terminal branches. The appearance of a fleshy finger pad also increases manual dexterity. The first record of grasping abilities and the presence of at least one nail are found in the stem primate *Carpolestes simpsoni* (**Figure 26.2**).

Prosimians   The first true primates, or euprimates, are known from the early Eocene of North America and Eurasia,

and potentially from the late Paleocene of northern Africa. These early primates belonged to a group that has traditionally been called prosimians (Greek *pro*, "before"; Latin *simia*, "ape"). Most Eocene prosimians, including the lemurlike adapoids (three families) and tarsierlike omomyoids (two families), were larger than plesiadapiforms. They had larger brains than plesiadapiforms (though their brains were small compared with those of any modern primate), more forward-facing orbits, and more specialized arboreal features, such as relatively longer, more slender hindlimbs.

Extant prosimians include the bush babies and pottos of continental Africa, the lemurs of Madagascar, the lorises of Asia, and the tarsiers of Southeast Asia (**Figure 26.3**). Both molecular characters and a sparse fossil record support a deep split between the lorisiforms (lorises, pottos, bush babies) and lemuriforms (lemurs)—at least 40 Ma. Prosimians

Table 26.1   Classification and social organization of extant primates

Taxon	Social organization
Strepsirrhini	
Lemuriformes (aye-ayes, lemurs, indris, and sifakas)	Largely solitary or monogamous pairs
Lorisiformes (bush babies, lorises, pottos, and angwantibos)	Largely solitary
Haplorhini	
Tarsiiformes (tarsiers)	Solitary or monogamous pairs
Anthropoidea	
Platyrrhini (New World monkeys)	Monogamous pairs to large groups
Pitheciidae (sakis, uakaris, and titis)	
Cebidae (marmosets, tamarins, capuchins, squirrel monkeys, and owl monkeys)	
Atelidae (howlers, spider monkeys, woolly monkeys, and muriquis)	
Catarrhini (Old World monkeys and apes)	Small to large groups
Cercopithecoidea	
Cercopithecidae	
Cercopithecinae (vervet monkeys, guenons, mangabeys, macaques, baboons, etc.)	
Colobinae (colobus monkeys, langurs, snub-nosed monkeys, and proboscis monkeys)	
Hominoidea	
Hylobatidae (gibbons and siamangs)	Monogamous pairs to small groups
Hominidae (orangutans, gorillas, chimpanzees, and humans)	
Ponginae (orangutans)	Solitary
Homininae	
Gorillas	Single male, multifemale groups
Chimpanzees	Multimale, multifemale groups; fission–fusion
Humans	Closed social network containing several breeding males and females

Source: Modified from Fleagle 2013.

Table 26.2	Characteristics of modern primates

General retention of five functional digits on the fore- and hindlimbs; enhanced mobility of the digits, especially the thumbs and big toes, which are usually opposable to the other digits.

Specializations for leaping, such as longer hindlimbs and elongated tarsal bones.

Claws modified into flattened and compressed nails.

Sensitive tactile pads on the distal ends of the digits.

Trend toward a reduced snout and olfactory apparatus, with most of the skull lying posterior to the orbits.

Reduction in the number of teeth compared with more primitive placental mammals, but with simple bunodont (low-cusped) molars.

A complex visual apparatus with high acuity, and a trend toward development of forward-directed binocular eyes.

Presence of a postorbital bar.

Large brain relative to body mass; the cerebral cortex is particularly enlarged.

Typically only one young per pregnancy, associated with prolonged infancy and pre-adulthood.

Relatively long lifespan.

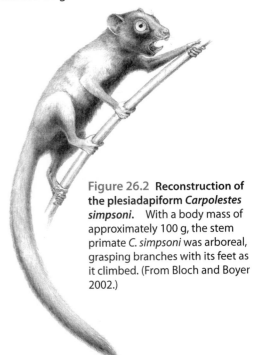

Figure 26.2 Reconstruction of the plesiadapiform *Carpolestes simpsoni*. With a body mass of approximately 100 g, the stem primate *C. simpsoni* was arboreal, grasping branches with its feet as it climbed. (From Bloch and Boyer 2002.)

(A)

(B)

(C)

(D)

(E)

(F)

Figure 26.3 Diversity of extant prosimians. (A–D) Lemuriforms. (A) Ring-tailed lemur (*Lemur catta*). (B) Indri (*Indri indri*). (C) Coquerel's sifaka (*Propithecus coquereli*). (D) Golden-brown mouse lemur (*Microcebus ravelobensis*). (E) Slender loris (*Loris tardigradus*). (F) Spectral tarsier (*Tarsius spectrum*). (A, courtesy of Amber Walker-Bolton; B, Gillian Merritt/Alamy Stock Photo; C, courtesy of Sergi López-Torres; D, courtesy of Malcolm Ramsay; E, Dr. K.A.I. Nekaris/CC BY-SA 4.0; F, S. Rohriach, CC BY 2.0.)

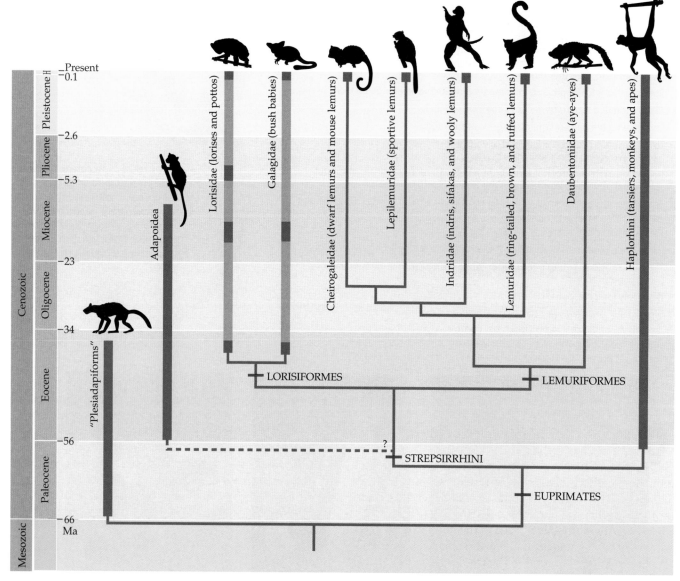

Figure 26.4 Phylogenetic relationships of strepsirrhines.
Probable relationships among strepsirrhines (Strepsirrhini) and other major primate groups. Narrow lines show interrelationships only; they do not indicate times of divergence or the unrecorded presence of taxa in the fossil record. Lightly shaded bars indicate ranges of time when the taxon is believed to be present but is not recorded in the fossil record. Question marks indicate uncertainties about relationships; quotation marks indicate paraphyletic groupings. H, Holocene. (Data from Springer et al. 2016; Ni et al. 2016; Silcox et al. 2017.)

are in general small, nocturnal, long-snouted, and relatively small-brained compared with the more derived anthropoids (Greek *anthropos*, "man") or apes and monkeys.

The group called prosimians is paraphyletic because several derived features (for example, a short snout with a dry nose rather than a wet, doglike nose) indicate that tarsiers are more closely related to the anthropoids than are the other prosimians. Molecular data support the association of tarsiers with anthropoids, and morphological characters cluster the extinct omomyoids with this lineage. An alternative division of the primates is into Strepsirrhini

(lemuriforms and lorisiforms; Greek *strepsis*, "twisted"; *rhinos*, "nose") and Haplorhini (tarsiers and anthropoids; Greek *haploos*, "simple") (see Table 26.1). Extant strepsirrhines form a monophyletic grouping (**Figure 26.4**), but there is debate about whether the extinct adapoids fit within this clade or are instead related to haplorhines or anthropoids.

In 2009, scientific attention was attracted by the discovery of a fossil adapoid, *Darwinius masillae* (**Figure 26.5**). The paleontologists who described *Darwinius* considered that new aspects of its anatomy warranted reevaluation of primate phylogeny. They placed adapoids as basal haplorhines (rather than as strepsirrhines), and later united them specifically with anthropoids. This phylogenetic rearrangement places *Darwinius* close to the origin of anthropoids, and the species was widely hailed as a new "missing link." Other paleontologists have disputed this claim, relegating *Darwinius* and the rest of the adapoids to the strepsirrhine lineage. Opinions remain divided, but whatever the systematic position of *Darwinius*, it is not

Figure 26.5 Skeleton of *Darwinius masillae*. This fossil adapoid *Darwinius masillae* is the most complete fossil primate known. Even the fur and gut contents were preserved. This remarkable specimen was found in Messel, a World Heritage Site in Germany, and is from the middle Eocene (47 Ma). (From Franzen et al. 2009, courtesy of the Natural History Museum, Oslo, Norway, CC BY.)

especially close to human origins, despite the unfortunate claims in popular media.

The diversification of early Cenozoic primates throughout the Northern Hemisphere reflects the tropical-like climates of the higher latitudes at that time. These early primates declined as climates cooled in the late Eocene, and were virtually extinct by the end of the Eocene. Even today, almost all non-human primates are restricted to the tropics, as they have been for most of the later Cenozoic, except for some excursions of apes into northern portions of Eurasia during the warming period in the early Miocene. Extant prosimians are a moderately diverse Old World tropical radiation, first known from the late middle Eocene of Asia (~45 Ma). The lemurs of the island of Madagascar have undergone an evolutionary diversification into five different families. These include some large (raccoon-size) diurnal species, such as the indri (*Indri indri*), and the nocturnal and peculiar aye-aye (*Daubentonia madagascariensis*), which uses its specialized middle finger, which is bony and narrow, to pull grubs out of tree bark.

Perhaps as recently as 2 ka lemurs were far more diverse and included giant arboreal species resembling bear-size koalas and sloths. It seems that the lemurs, in isolation from the rest of the world, evolved their own version of primate diversity, including parallels to the anthropoid apes. Unfortunately, much of this diversity is now gone, probably because of the immigration of humans to Madagascar from Asia about 2 ka. Accounts from early European explorers suggest that some of the giant lemurs may have been alive as recently as 500 years ago. The decline of lemurs is continuing, and many extant species are threatened with extinction as a result of continued destruction of forests on Madagascar.

Anthropoids Modern anthropoids are mostly larger than prosimians, with larger brains housing relatively small olfactory lobes, and they are mainly frugivorous or folivorous, although some smaller species of anthropoids get their protein supply from insects. All anthropoids except the owl monkeys are diurnal, and all have complex social systems. Anthropoids employ many types of locomotor behaviors, including terrestrial and arboreal quadrupedalism in midsize anthropoids, and vertical clinging and leaping in small South American monkeys. Large anthropoids typically employ suspensory locomotion, moving with the body hanging below the branches. A specialized form of suspensory locomotion is **brachiation**, in which the animals swing at various speeds from the underside of one branch to the underside of the next using their hands to grasp the branches. This form of locomotion is seen today in gibbons (fast brachiators) and the siamang (*Symphalangus syndactylus*; a slow brachiator), and it has evolved independently in spider monkeys.

Modern anthropoids can be distinguished from prosimians by a variety of skull features, such as a larger braincase, fusion of many paired bones, and a shorter snout (**Figure 26.6**). A bony postorbital septum is a distinguishing anthropoid feature; it prevents mechanical disturbance of the eye when the temporalis muscle contracts during chewing. Anthropoids also lack the grooming claw on the second toe that is characteristic of modern prosimians.

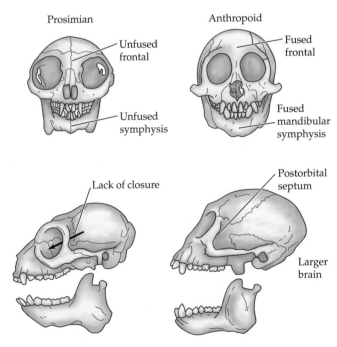

Figure 26.6 Cranial differences between prosimians and anthropoids. Anthropoids have a bony wall behind the orbit (a postorbital septum, also present in tarsiers), and the mandibular symphysis joining the two halves of the lower jaws is fused, as are the paired frontal bones. (From Fleagle 2013.)

The origin of anthropoids appears to be related to a shift from foraging at night to being diurnal, and this change in behavior was accompanied by a drastic reduction in body size. The earliest-known anthropoids were all tiny (weighing no more than a few hundred grams) and consumed a mixed diet of fruit and insects. Eosimiids have been established as stem anthropoids, predating the catarrhine–platyrrhine split.

The earliest Asian anthropoids were the Eosimiidae (Greek *eos*, "dawn"), known from the early Eocene to the early Oligocene of China, India, and Pakistan. Eosimiids include some of the smallest primates in the fossil record, weighting as little as 12 g.

Platyrrhine and catarrhine anthropoids   Modern anthropoids are Platyrrhini (Greek *platus*, "flat, broad")—the New World (broad-nosed) monkeys—and Catarrhini (Greek *catarrheo*, "downward")—the Old World (narrow-nosed) monkeys and apes (**Figure 26.7**). These two groups differ in important characters:

Figure 26.7 Diversity of extant monkeys. (A–D) Platyrrhines. (A) Pygmy marmoset (*Cebuella pygmaea*). (B) Emperor tamarin (*Saguinus imperator*). (C) Black-horned capuchin (*Sapajus nigritus*). (D) Mantled howler monkey (*Alouatta palliata*). (E–H) Catarrhines. (E) Red-tailed monkey (*Cercopithecus ascanius*). (F) L'Hoest's monkey (*Cercopithecus lhoesti*). (G) Hamadryas baboon (*Papio hamadryas*). (H) Purple-faced langur (*Semnopithecus vetulus*). (A, © iStock.com/HeitiPaves; B, courtesy of Matthew De Vries; C, courtesy of Ester Bernaldo de Quirós; D, courtesy of Montserrat Franquesa-Soler; E, F, courtesy of Lwiro Primate Rehabilitation Centre; G, Christoph Anton Mitterer/CC BY-SA 2.0; H, © Jamen Percy/123RF.)

- *Color vision* All catarrhines have trichromatic color vision, which is also seen in a few platyrrhines, such as the howler monkeys. Trichromatic color vision is produced by a duplication of the opsin gene coding for color reception in the red-green part of the spectrum. (An independent gene duplication event leading to trichromatic vision occurred among some marsupials, such as the honey possum.)

- *Olfaction* While all anthropoids have small olfactory bulbs, platyrrhine primates seem to have retained a relatively good sense of smell, and most of the genes associated with olfaction are still functional. In contrast, only 50% of the olfactory genes of catarrhines code for functional receptor proteins, and catarrhines have also lost a functional vomeronasal organ.

- *Skulls, brains, and teeth* Platyrrhines and catarrhines differ in some details of the skull, especially in the ear region. Additionally, while all modern anthropoids have relatively large brains, the fossil record shows that enlargement of the brain from a prosimian-like condition occurred independently in platyrrhines and catarrhines. Furthermore, platyrrhines retain the generalized

primate condition of having three premolars on each side of the jaw, whereas catarrhines have only two premolars.

Platyrrhines Platyrrhines, the New World monkeys, first appeared in South America in the late Eocene and are an exclusively New World radiation. They are presumed to have rafted from Africa across the South Atlantic Ocean when it was narrower than it is now. Three lineages of platyrrhines can be distinguished: pitheciids, cebids and atelids (**Figure 26.8**).

Pitheciids include the pitheciines (sakis and uakaris) and callicebines (titis). Sakis and uakaris are medium-size arboreal monkeys that share features associated with eating

Figure 26.8 Phylogenetic relationships of haplorhines. Probable phylogenetic relationships among haplorhines (Haplorhini) and other major primate groups. Narrow lines show interrelationships only; they do not indicate times of divergence or the unrecorded presence of taxa in the fossil record. Lightly shaded bars indicate ranges of time when the taxon is believed to be present but is not recorded in the fossil record. Question marks indicate uncertainties about relationships; quotation marks indicate paraphyletic groupings. H, Holocene. (Data from Beard et al. 1994; Zalmout et al. 2010; Zijlstra et al. 2013; Ni et al. 2016.)

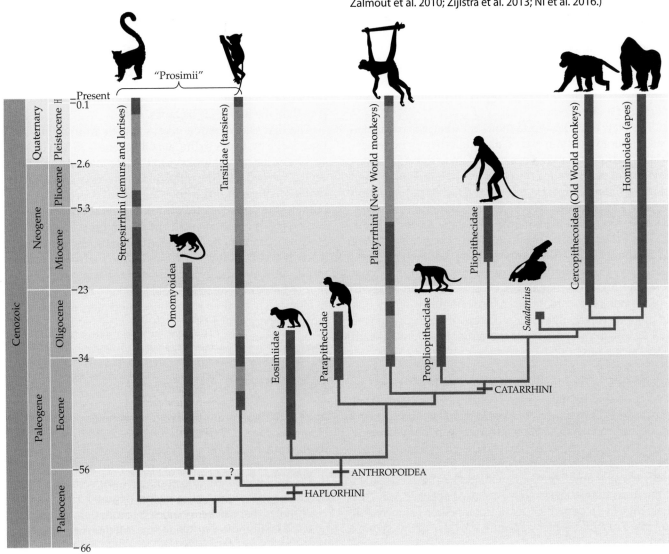

hard foods, particularly fruits and seeds with tough outer coats. Titis are unusual among New World monkeys in having a complex assortment of vocalizations, including duets sung at dawn by breeding pairs. The phylogenetic relationships of titis are controversial. Some authors think they might be the most primitive New World monkeys, but in the latest molecular analyses they appear to be closely related to pitheciines.

Cebids include the cebines (including the familiar capuchin or organ grinder monkeys and squirrel monkeys), callitrichines (marmosets and tamarins), and aotines (owl monkeys). Cebines are active arboreal quadrupeds that eat mainly fruit and insects and live in groups consisting of several males and females. Capuchin monkeys are the only cebids that retain a prehensile tail as adults. Marmosets and tamarins are small and have secondarily clawlike nails on all digits except the big toe. They have simplified molars, and most species eat gum exuded from trees. Owl monkeys are the only nocturnal anthropoids.

Atelids include the atelines (woolly monkeys, spider monkeys, and muriquis) and alouattines (howler monkeys). Atelines are distinguished by their prehensile tail and the use of brachiation aided by the tail. Howler monkeys are among the best known and most heavily studied New World monkeys. Male howlers can make very loud calls due to the enlarged hyoid bone in their throat that houses a resonating chamber. Howlers are sexually dimorphic in size and color. They are primarily folivorous and generally live in groups that include several males and females. Atelids and cebids are the only non-human primates that occur north of Panama today.

The New and Old World monkey radiations paralleled each other to a certain extent, although there is no terrestrial radiation of platyrrhines equivalent to that of baboons and macaques. Perhaps the extensive radiation of ground sloths in South America inhibited a terrestrial radiation among the primates. However, a striking parallel does exist between the spider monkeys and the gibbons—both groups are specialized brachiators with exceptionally long arms that they use to swing through the branches, and they have evolved a remarkable convergence in a wrist joint modification that allows exceptional hand rotation. Spider monkey locomotion can be distinguished from that of gibbons chiefly by spider monkeys' use of a prehensile tail as a fifth limb during locomotion. (Gibbons, like all apes, lack a tail.)

Catarrhines   Catarrhines include the Old World monkeys and apes, and humans. Stem catarrhines (i.e., forms that predate the divergence of Old World monkeys and apes) include the propliopithecids, small forms from the late Eocene and early Oligocene of Africa, and the pliopithecids, larger Miocene forms that ranged from Africa into Eurasia. The split between apes and Old World monkeys has been estimated from molecular studies to have occurred about 28 Ma. This date matches a catarrhine fossil from the late Oligocene of Arabia, *Saadanius*, which appears to be close to this split.

Catarrhines have nostrils that open forward and downward like those of humans, and they have a smaller bony nasal opening from the skull than do platyrrhines. The great apes (orangutans, gorillas, and chimpanzees) and humans are the largest extant primates, rivaled only by some of the extinct lemurs. Catarrhines never evolved prehensile tails. The group consists of two clades: the Old World monkeys (Cercopithecoidea; Greek *kerkos*, "tail"; *pithekos*, "monkey") and the apes and humans (Hominoidea; Latin *homo*, "man").

Extant Old World monkeys include two groups: colobines (Greek *kolobos*, "shortened") and cercopithecines. Colobines are found in both Africa and Asia and include colobus monkeys, langurs, proboscis monkeys, and snub-nosed monkeys. They are more folivorous than cercopithecines, and they have higher-cusped molars and a complex forestomach in which plant fiber is fermented. Colobines are primarily arboreal, with a long tail and hindlegs that are longer than their forelegs.

Cercopithecines are primarily an African radiation and include macaques, mangabeys, baboons, geladas, kipunjis, mandrills, drills, guenons, swamp monkeys, talapoins, vervet monkeys, sun-tailed monkeys, the patas monkey (*Erythrocebus patas*), and the so-called Barbary ape of Gibraltar (*Macaca sylvanus*). Cercopithecines are generally more frugivorous than colobines, and this dietary difference is reflected in their broad incisors and their flat, bunodont (low-cusped) molars. Cercopithecines are also more terrestrial than colobines, and often have a shorter tail and fore- and hindlimbs of equal length. Cercopithecines have cheek pouches in which they carry food, and hands with a longer thumb and shorter fingers than the hands of colobines.

The first Old World monkeys appear around the same time as the first apes in the late Oligocene (~25 Ma) of Africa. The radiation of monkeys in the late Miocene and Pliocene coincided with the reduction in diversity of the earlier radiation of generalized apes and apelike forms. Because humans are apes, we often think of monkeys as being the earlier, less derived members of the anthropoid radiation. However, among Old World anthropoids, the converse is actually true: apes were originally the more generalized forms, although the extant forms are specialized. The extant radiation of cercopithecoid monkeys is more derived in many respects than that of apes, and more successful in terms of species diversity.

26.2 Origin and Evolution of Hominoidea

Apes and humans are placed in Hominoidea (**Table 26.3**). Hominoids are distinguished morphologically from other recent anthropoids by a pronounced widening and dorsoventral flattening of the trunk relative to body length, so that the shoulders, thorax, and hips have become proportionately broader than in monkeys (**Figure 26.9**).

Primates that can be included in a monophyletic Hominoidea date from the late Oligocene. Modern apes (with the

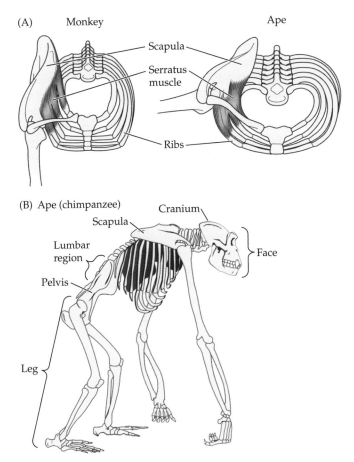

(A) Monkey   Ape

Scapula
Serratus muscle
Ribs

(B) Ape (chimpanzee)
Cranium
Scapula
Lumbar region
Face
Pelvis
Leg

Figure 26.9 Differences in the skeletons of monkeys and apes. (A) The scapulae of apes lie over a broad, flattened back, in contrast to their lateral position next to a narrow chest in monkeys and most other quadrupeds. The more curved ribs of apes place the vertebral column closer to the center of gravity, making it easier for an ape to balance in an upright position. (B) In apes, the lumbar region of the vertebral column is short, and the caudal vertebrae have become reduced to vestiges (the coccyx). These and other anatomical specializations of the trunk help apes maintain an erect posture during sitting, vertical climbing, and bipedal walking.

exception of humans) are a highly specialized radiation of large tropical animals. The apes that radiated in the Miocene were more generalized animals, and they radiated into both temperate and tropical parts of the Old World. Apes

Table 26.3	Classification of hominoids
Hominoidea (hominoids)	Gibbons, siamang, orangutans, gorillas, chimpanzees, humans
Hylobatidae (hylobatids)	Gibbons, siamang
Hominidae (hominids)	Orangutans, gorillas, chimpanzees, humans
Ponginae (pongines)	Orangutans
Homininae (hominines)	Gorillas, chimpanzees, humans
Hominini (hominins)	*Homo sapiens* and other taxa more closely related to *H. sapiens* than to chimpanzees
Homo (humans)	Species of *Homo*

and monkeys can be distinguished by their teeth, which are frequently the only remains of fossil species. Monkeys have lower molars with four cusps, whereas those of apes have five cusps. Additionally, the teeth of apes usually have lower cusps than those of monkeys, and the grooves between the posterior cusps form a distinct Y pattern.

Diversity and social behavior of extant apes

Extant apes include the Asian gibbons, siamang, and orangutans and the African chimpanzees and gorillas (**Figure 26.10**). Distinct cultures that differ in tool use and other behaviors have been described for populations of chimpanzees (see Figure 26.13) and orangutans, and chimpanzees have been shown to be capable of planning for future events.

The two species of chimpanzee are endangered, and gorillas and orangutans are critically endangered. Recent estimates show that the numbers of chimpanzees and gorillas in western equatorial Africa, a region considered to be the last refuge of these animals' tropical habitat, have declined by more than half in the past few decades, primarily as a result of commercial hunting, mechanized forest logging, and the spread of the Ebola virus. Illegal logging, fire, and conversion of lowland forest to oil palm plantations threaten orangutan populations on Borneo and Sumatra, and population sizes are declining rapidly on both islands.

Eighteen species of gibbons (*Hylobates*, *Hoolock*, and *Nomascus*) and one species of siamang (*Symphalangus*) occur in Southeast Asia, both on the mainland (from India to China) and on the islands (Borneo, Sumatra, Java, and other nearby islands). Gibbons move through the trees most frequently by brachiation and become entirely bipedal when moving on the ground, holding their arms outstretched for balance like a tightrope walker using a pole. They are the smallest apes, and they differ from other apes (and indeed the great majority of mammals) in their monogamous social systems. Mated male and female gibbons produce combined calls that are referred to as duets. Duet calls strengthen the bond between the breeding pair and act as territorial calls, sending the message that a particular area is inhabited by a pair of gibbons. Although gibbons usually live in separate populations that can be distinguished morphologically, different species of gibbons occasionally hybridize.

There are three species of orangutan: *Pongo pygmaeus* on Borneo; and *P. abelii* and *P. tapanuliensis* on Sumatra. Orangutans are extremely sexually dimorphic, with males weighing twice as much as females. Orangutans are arboreal but rarely swing by their arms, preferring slow climbing among the branches of trees. They usually move slowly using all four limbs to grasp branches, and sometimes walk bipedally on top of a branch, supporting themselves by grasping tree limbs overhead. When they move quadrupedally on the ground, orangutans close their hands into fists and support themselves on the proximal phalanges. Orangutans are generally solitary, and groups usually include only a female and her offspring. Individual

Figure 26.10 Diversity of extant apes. (A) Siamang (*Symphalangus syndactylus*). (B) Sumatran orangutan (*Pongo abelii*). (C) Eastern gorilla (*Gorilla beringei*). (D) Common chimpanzee (*Pan troglodytes*). (A, cuatrok77/CC BY-SA 2.0; B, © Anatoliy Alekseev/123RF; C, courtesy of Itsaso Vélez del Burgo Guinea; D, courtesy of Lwiro Primate Rehabilitation Centre.)

orangutans use calls to maintain social contact, however, and groups of females form in some populations when food resources permit.

Orangutans use tools for several tasks, including using leaves as drinking tools, leaf umbrellas as shelter during heavy downpours, and sticks for scratching and for getting seeds out of fruit pods. Twenty-four potential cultural variants have been proposed for orangutans, based on differential use of tools and other behaviors.

Gorillas and chimpanzees live in the tropical forests of central Africa. Both are more terrestrial than gibbons and orangutans. On the ground they move quadrupedally by knuckle-walking, a derived mode of locomotion in which they support themselves on the dorsal surface of the intermediate phalanges of digits 3 and 4, rather than closing their hands into fists, as orangutans do, or placing their weight flat on the palm of the hands, as we do when we crawl on all fours. However, gorillas and chimpanzees knuckle-walk differently: the hand and wrist of gorillas are aligned when they knuckle-walk, whereas chimpanzees knuckle-walk with a flexed wrist.

Gorillas are the largest and most terrestrial of the extant apes. Unlike orangutans, they are highly social and live in groups; but like orangutans, gorillas are sexually dimorphic in body size—males may weigh up to 200 kg, twice the mass of females—and they are the most folivorous of the apes. Two species of gorillas have been described: the western gorilla (*Gorilla gorilla*) and eastern gorilla (*G. beringei*). Gorillas use tools for activities such as food acquisition and social displays (**Figure 26.11**), but they do not have population-specific cultures like chimpanzees. A gorilla group usually consists of a single male and several females, with younger males living peripherally. Both sexes of gorillas leave their groups when they reach adulthood. In some cases, a son is permitted to stay in his father's group, assuming leadership of the group when he is fully adult. In this situation, the father remains in the group, albeit in a subordinate position to the son.

Chimpanzees are more arboreal than gorillas, climbing nimbly and moving through the canopy by walking, leaping, and with suspensory locomotion. Chimpanzees are only moderately sexually dimorphic and, like gorillas, live in groups, but with multiple males instead of a single male. The social organization of chimpanzees follows a fission-fusion pattern, meaning that small groups of chimpanzees coalesce into a larger group and then split into smaller groups. Usually the fissioning is related to the abundance and spatial distribution of food sources.

There are two species of chimpanzees (**Figure 26.12**). Although the split between the two species occurred 2.1–1.6 Ma, there was gene flow between the two lineages as recently as 200 ka. The larger and more widely distributed common chimpanzee (*Pan troglodytes*) lives primarily in central and East Africa. Whereas in the past there was a continuous biogeographic distribution for the common chimpanzee, its current distribution is patchy, with four distinct populations that correspond to four commonly recognized subspecies (**Figure 26.13**). One of the best-studied groups of common

(A)

(B)

Figure 26.11 Gorilla behaviors.   (A) A western gorilla (*Gorilla gorilla*) using a twig to extract wood-boring insects from a tree trunk. (B) Chest-thumping display by two male eastern gorillas (*G. beringei*). Both males and females slap one or both hands on the chest as an intimidation display. In males, air sacs on the sides of the throat resonate, making the thumping especially dramatic. (A, blickwinkel/Alamy Stock Photo; B, Nature Picture Library/Alamy Stock Photo.)

chimpanzees is the East African subspecies in Gombe National Park, Tanzania. The Gombe chimpanzees were initially studied by primatologist Jane Goodall as potential behavioral models for early hominins. Her study is one of the most influential works in the field of primate ethology.

A smaller species of chimpanzee, the bonobo (*Pan paniscus*), occurs in central Africa south of the Congo River. Bonobos live in more forested habitats than common chimpanzees and differ from common chimpanzees in several respects:

- Common chimpanzees use tools in the wild, whereas wild bonobos do not. A classic example of tool use by common chimpanzees is inserting twigs into anthills and termite mounds, then withdrawing the twig, sweeping off the attached insects by licking the twig, and swallowing them. Common chimpanzees also use rocks to open nuts, but do not modify rocks to use them as tools.

- Meat is a significant component of the diet of common chimpanzees, but not of bonobos. Common chimpanzees are major predators of red colobus monkeys (*Piliocolobus*). Groups of adult male chimpanzees engage in cooperative hunts for red colobus, pursuing them through the treetops while females and juveniles follow the hunt at ground level. A successful kill is shared among the males that participated in the hunt, and with females allied with the dominant males.

- Common chimpanzees are very territorial (especially against members of other groups), with a rigid male-dominance structure and frequent aggressive behaviors. In contrast, bonobos, which live in female-centered groups, do not engage in aggressive behaviors as often, and are less sexually dimorphic than common chimpanzees.

- Genital swelling signals estrus (the time of fertility and short-term sexual receptivity) in female common chimpanzees, but female bonobos (like humans) have concealed ovulation and continuous sexual receptivity.

(A)

(B)

Figure 26.12 The two species of chimpanzees.   (A) A male common chimpanzee (*Pan troglodytes*). Common chimpanzees are larger and more robust than bonobos. (B) Bonobos (*Pan paniscus*) mating face-to-face, a behavior not seen in any other non-human primate. (A, courtesy of Iulia Bădescu; B, © Sergey Uryadnikov/Shutterstock.)

(A)

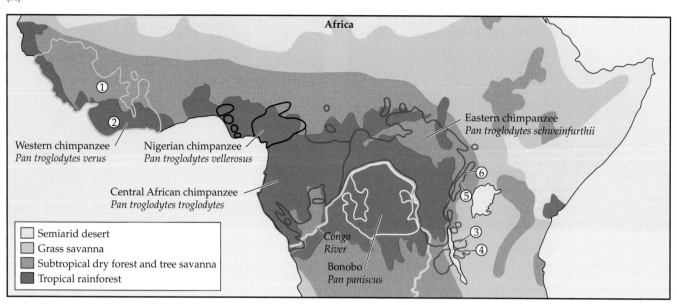

(B)

	① Bossou, Guinea	② Taï Forest, Côte d'Ivoire	③ Gombe, Tanzania	④ Mahale, Tanzania	⑤ Kibale, Uganda	⑥ Budongo, Uganda
Shake a branch to attract attention	+	+	+	+	+	+
Drag a large branch as an aggressive display	+	+	+	+	+	+
Squash a biting insect with a forefinger		+				
Open a food item by smashing it against a log or stone	+	+	+			+
Open a food item by placing it on an anvil and cracking it with a stone or stick	+	+	+			
Use a twig to extract termites from tunnels			+	+		
Use a twig to capture safari ants	+	+	+			
Use a wad of leaves as a sponge to obtain water from a puddle	+	+	+		+	+
Strike forcefully with a stick		+	+			
Throw a stick or stone at a target	+	+	+	+		
Clasp hands with another chimp during grooming		+		+	+	

Figure 26.13 Populations and cultures of common chimpanzees. (A) The four extant populations of the common chimpanzee (*Pan troglodytes*) are found in forested and savanna habitats north of the Congo River in equatorial Africa. These isolated populations are recognized as subspecies. The geographic range of the bonobo (*P. paniscus*) lies south of the Congo River. (B) Each subspecies of the common chimpanzee has many different cultures, comprising a repertoire of behaviors. Shown here are behaviors from two cultures of the western chimpanzee (*P. t. verus*) and from four cultures of the eastern chimpanzee (*P. t. schweinfurthii*). Some behaviors,

such as shaking a branch to attract attention and dragging a large branch as an aggressive display, are common to all chimpanzee cultures. A few, such as squashing a biting parasite with a forefinger, are confined to a single culture. In four of the cultures, chimpanzees open a hard food item, such as a nut, by beating it against a rock or log, but only three cultures include placing the food item on an anvil (a rock or log) and beating it with a hammer. The approximate locations of the study sites are indicated. (Data from Whiten et al. 1999.)

- The use of sex as a conciliatory behavior for bonding or conflict solving is unique to bonobos. Besides heterosexual copulation, the sexual repertoire of bonobos includes same-sex genital rubbing and penis fencing.

Populations of common chimpanzees show different combinations of behaviors. For example, western chimpanzees from Bossou, Guinea, use stones to crack nuts, while western chimpanzees from Taï National Park in Côte d'Ivoire wait for the nuts to dry out, when they are easier to open, and use a wooden hammer instead. Also, the Taï chimpanzees are known to make a slow display (called a "rain dance") when it starts raining, but this display has never been observed in Bossou.

Relationships within Hominoidea

Hominoid classification has changed considerably over the past decade. Traditionally, Hominoidea was considered to include three families: Hylobatidae (gibbons and the siamang), Pongidae (other apes, or "great apes"), and Hominidae (humans). Hylobatidae, as originally conceived, remains a valid family. However, Pongidae, a grouping of the great apes with the exclusion of humans, is paraphyletic. Molecular data show that chimpanzees are more closely related to humans than are gorillas and orangutans, and gorillas are more closely related to humans and chimpanzees than are orangutans.

Despite the fact that humans look rather different from the other great apes, and are certainly different in terms of culture and language, molecular studies show that we are very close to the other great apes genetically and that we diverged from them very recently in geological terms (chimpanzee-human split about 6.6 Ma). Thus, Hominidae now includes the great apes as well as humans. The subfamily Homininae includes humans, chimpanzees, and gorillas. (The orangutans and their fossil relatives are in the subfamily Ponginae.) Humans and their immediate ancestors (i.e., extinct taxa more closely related to *Homo sapiens* than to chimpanzees) are in the tribe Hominini. Thus, what was called a hominid in fossil studies only a few years ago (and in earlier editions of this book), meaning a fossil human, is now called a **hominin**.

Molecular data and a limited amount of fossil evidence allow us to assemble a timetable for hominoid evolution:

- Molecular data indicate that the split between apes and Old World monkeys occurred during the late Oligocene, about 28 Ma.
- Both molecular evidence and the fossil record suggest that Hylobatidae and Hominidae split about 17 Ma, and that Ponginae and Homininae separated about 13 Ma.
- Molecular evidence indicates that gorillas separated from the common ancestor of chimpanzees and humans between 10 and 8 Ma, and that gorillas separated into eastern and western populations about 3 Ma.
- Molecular evidence indicates that humans separated from their common ancestor with chimpanzees about 6.6 Ma. This date fits well with the earliest definitive hominin fossil, *Ardipithecus*, known from 5.8 Ma.
- Genetic data provide an interesting twist to this tale, suggesting that early chimpanzees and hominins interbred up until about 6.3 Ma—300 ka after the lineages separated.
- The two chimpanzee species are thought to have separated from each other about 2.1–1.6 Ma.

Diversity of fossil hominoids

The earliest fossil evidence of a hominoid comes from the late Oligocene of East Africa (about ~25 Ma) and belongs to the genus *Rukwapithecus*. Later Miocene forms include proconsulines, afropithecines, and nyanzapithecines. Early hominoids were primarily arboreal, living in forested habitats. They had bunodont molars suggesting a frugivorous diet, and they were more derived than the propliopithecid stem anthropoids of the Fayum Formation in Egypt (such as *Aegyptopithecus*). Like all apes, they lacked a tail. **Figure 26.14** presents a phylogeny of hominoids.

Proconsulines ranged from the size of a small monkey to the size of a modern human. Although they remained generalized arboreal quadrupeds, not yet showing the specialized suspensory locomotion of many later apes, their hands and feet were more capable of gripping than those of monkeys: they had longer thumbs, and the elbow joint was more stable. *Morotopithecus* had a body mass of approximately 36–54 kg and is the first hominoid to have derived features of skeletal anatomy that show a capacity for suspensory locomotion, including a highly mobile shoulder joint, a short stiff back that resisted flexion in the lumbar region, and a moderately mobile hip joint.

By the middle Miocene, more derived hominoids had diversified into a variety of ecological types. Some of these apes, such as *Afropithecus*, *Kenyapithecus*, and *Equatorius*, remained in Africa, although a specimen of *Kenyapithecus* is known from Turkey. These genera are of controversial affinities; they are probably basal hominoids (or even hominids) of some sort, and are sometimes placed within Afropithecidae (which is probably paraphyletic, and is not shown on Figure 26.14).

Other apes spread broadly into Eurasia, following the general middle Miocene warming and the connection of Africa to the Eurasian mainland. Both cercopithecine and colobine monkeys are known from Eurasia in the late Miocene and Pliocene. Later Cenozoic Eurasian apes include the dryopithecines (Greek *dryo*, "tree") and sivapithecins (after the Hindu god Shiva). Both of these groups were more derived than hylobatids, and so are included in Hominidae. Gibbons themselves were unknown in the fossil record until recently, when a possible early member of the lineage, *Yuanmoupithecus*, was described from a site in China that is dated to about 10 Ma. A small-bodied basal hominoid, *Pliobates*, from the Miocene of Spain suggests that the last common ancestor of extant hominoids might have been more gibbonlike, and less great-apelike, than previously assumed.

The dryopithecines diversified in Europe (fossils are known from France, Hungary, Greece, Italy, Moldova, and Spain) and Asia (Turkey and China), and they are now generally considered to be the sister group to extant hominids. The sivapithecins were mainly an Asian radiation and were the stem stock for modern pongines (orangutans). These Eurasian apes primarily occupied forested habitats and were somewhat modified for suspensory locomotion, although

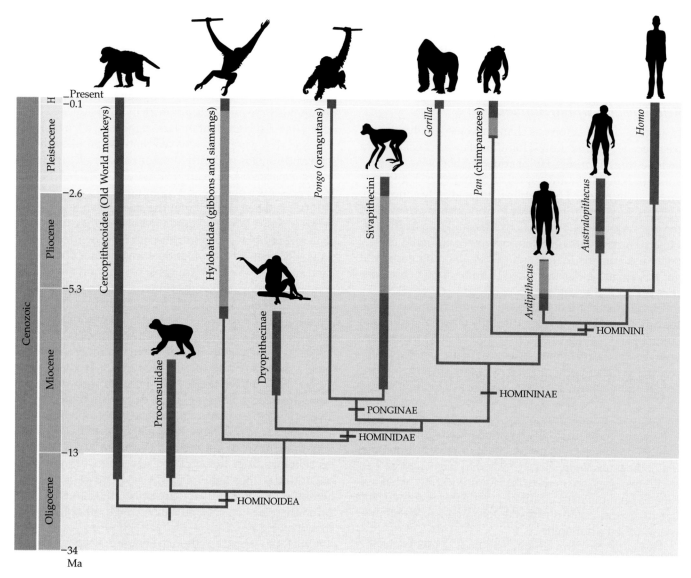

Figure 26.14 Phylogenetic relationships of hominoids.
Probable relationships among groups of hominoids (Hominoidea).
Narrow lines show interrelationships only; they do not indicate
times of divergence or the unrecorded presence of taxa in the
fossil record. Lightly shaded bars indicate ranges of time when
the taxon is believed to be present but is not recorded in the fossil
record. H, Holocene.

not to the extent seen in their modern relatives. Sivapithecins flourished from about 16 Ma (the Middle Miocene
Climatic Optimum) to 9 Ma (the start of the late Miocene)
when the climate became cooler and drier across the middle latitudes of the Northern Hemisphere (see Chapter 25).
Both the dryopithecines and sivapithecins included several
genera. A Pleistocene sivapithecin, *Gigantopithecus blacki*,
was the largest primate that has ever lived. With an estimated body mass of 300 kg, it would have been half again
the size of a male gorilla.

The various Eurasian dryopithecine genera include *Dryo-
pithecus, Hispanopithecus, Ouranopithecus, Anoiapithecus,*

Rudapithecus, Graecopithecus, Oreopithecus, Lufengpithecus,
and *Pierolapithecus* (appearing between 13 and 8 Ma),
which show features of both the skull and the postcranial
skeleton that resemble the African apes. (*Ouranopithecus*
[Greek *ourano*, "heaven"] may in fact be more closely related to Homininae). The abundance of Eurasian apes and
the absence of hominoids from Africa between 13.5 and 10
Ma have been cited as arguments for a Eurasian origin of
Homininae. However, recent finds of late Miocene African
apes that appear to be stem Homininae are challenging
this viewpoint.

Conditions for fossilization are poor in the forest habitats
of gorillas and chimpanzees, and fossils have been found
only recently. Possible basal gorillas include *Chororapithecus*
and *Nakalipithecus*, both known from Kenya between 10
and 8 Ma. However, *Nakalipithecus* predates the molecular
estimates of the split between gorillas and chimpanzees +
humans, and it may be a stem Homininae. Some teeth from
the middle Pleistocene of Kenya provide the only evidence
of fossil chimpanzees.

Although humans have evolved many of their own specializations, we have retained the ancestral characters of our clade in numerous dental and associated cranial features, while the lineages of great apes evolved their own unique derived characters. It is important to note, however, that all of the extant hominoids are derived compared with the Miocene-Pliocene ape radiation. Paleoanthropologists disagree about the appearance of the last common ancestor of humans and chimpanzees, and several models have been proposed, including African apes, orangutans, or a generalized Miocene apelike form. The fact that gibbons and orangutans are in more basal positions on the phylogeny than are humans or African apes (see Figure 26.14) does not imply that earlier apes looked like these modern forms, both of which are arboreal specialists.

26.3 Origin and Evolution of Humans

Several evolutionary trends can be identified within hominins:

- The points of articulation of the skull with the vertebral column (the occipital condyles) and the foramen magnum (the hole for the passage of the spinal cord through the skull) shifted from the ancestral position at the rear of the braincase to a position under the braincase. This change balances the skull on top of the vertebral column and signals the appearance of an upright, vertical posture.

- The braincase itself became greatly enlarged in association with an increase in forebrain size. By the end of the middle Pleistocene, a prominent vertical forehead developed in contrast to the sloping forehead of apes.

- The brow ridges and crests for muscle origins on the skull became smaller in association with the reduction in size of the muscles that attach to them. The nose became a more prominent feature of the face, with a distinct bridge and tip.

- The jaw became shorter, the canines became smaller, and the gap between the incisors and canines disappeared (**Figure 26.15**).

Early hominins

The earliest well known hominins are the australopiths (Latin *australis*, "southern"), known primarily from East and southern Africa in the Pliocene and early Pleistocene. Many new hominins have been described in the past decade, producing a confusing plethora of names and arguments about who is related to whom and when bipedalism first evolved.

The earliest potential hominins   No Miocene taxon is presumed to be a true hominin (that is, a bipedal animal on the direct human lineage), but a few candidates[1] for that designation have emerged within the past couple of decades: *Orrorin tugenensis* from Kenya, dated at 6 Ma, and *Sahelanthropus tchadensis* from the central African country of Chad, dated at 6.5 Ma. Both of these possible early hominins are found in areas that were woodland or forest.

Spectacular as these finds are, there is debate about their hominin status. The preserved fragments of *Orrorin* combine a femur that is intermediate in shape between Miocene apes and hominins with large, apelike canines and finger bones indicating climbing abilities. *Sahelanthropus* is known only from a skull, although the ventral position of the foramen magnum suggests that the head was balanced on top of the vertebral column, as in bipedal hominins. Unlike *Orrorin*, this animal had small, humanlike canines and hominin-like molars, but it also has a massive brow ridge and various cranial features that are more apelike than humanlike.

A more derived hominin, *Ardipithecus* (Amharic *ardi*, "ground floor"), comes from the Middle Awash area of Ethiopia, approximately 5.8–4.4 Ma. *Ardipithecus* (**Figure 26.16**) was probably an upright biped with a foramen magnum that opened downward. It had small canines with less sexual dimorphism than do modern African apes. Little sexual dimorphism in dentition, such as that seen in

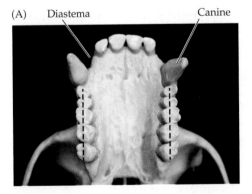 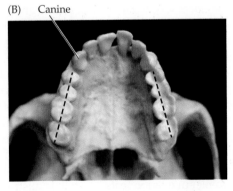

Figure 26.15 Comparison of ape and human upper jaws.   (A) Apes have long upper and lower jaws that are rectangular or U-shaped, with the molar rows parallel to each other. The canines are large and pointed, and there is a diastema (gap) between the canines and the incisors. The ape palate is flat between the parallel rows of cheek teeth. (B) Humans have short upper and lower jaws in association with the shortening of the entire muzzle. The human upper and lower jaws are V-shaped or bow-shaped, with the teeth diverging posteriorly; the canines are small and blunt; and the entire dentition is relatively uniform in size and shape without any gaps between the teeth. The human palate is prominently arched. (Courtesy of Sergi López-Torres.)

[1]*Graecopithecus freybergi*, from Greece and Bulgaria (7.2 Ma) has also been suggested as potentially the earliest hominin, but the evidence linking this species to Hominini is slim, and further research is needed to resolve the position of *Graecopithecus*.

Figure 26.16 Reconstruction of *Ardipithecus ramidus*.
Features indicating that *Ardipithecus* was bipedal include the location of the foramen magnum on the base of the skull and the structure of the pelvis and hindlimbs. Arboreal characteristics visible in this reconstruction include long arms, hands with long fingers, and a foot with long toes and a big toe that diverges widely from the remaining toes. (Copyright © Julius T. Csotonyi/Science Source.)

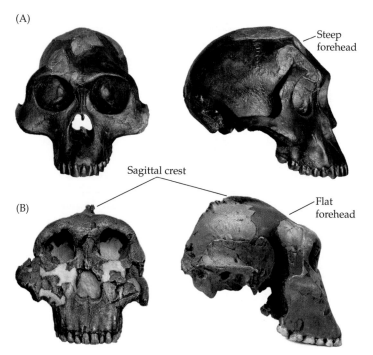

Figure 26.17 Ecomorphologic diversity in early hominins.
(A) Gracile australopiths, such as *Australopithecus afarensis*, were 1.1–1.65 m tall and lightly built. Their skull had a steep forehead. The incisors and canines were large and the molars small, suggesting a diet of items that were ripped or torn rather than crushed. (B) Robust australopiths, such as *Paranthropus boisei*, were about 1.5 m tall and powerfully built with the body proportions of a wrestler. Their skull had a flat forehead and pronounced sagittal crest that formed the origin of powerful jaw muscles. Their incisors and canines were small and their huge molars exhibited heavy wear, suggesting a diet of coarse and fibrous items that were crushed during chewing. (A, Sabena Jane Blackbird/Alamy Stock Photo; B left, PRISMA ARCHIVO/Alamy Stock Photo; B right, The Natural History Museum/Alamy Stock Photo.)

modern humans, is associated with a pair-bonding type of mating system, rather than one in which males compete for access to females, as in the African apes today.

The skeleton of *Ardipithecus* shows features that tie it both to more derived hominins and to extant great apes, and its pelvis indicates the capacity for bipedal locomotion. However, the foot had long toes and a highly divergent big toe, both features indicating that *Ardipithecus* was more arboreal than any later australopiths. The arms and hands of *Ardipithecus* were longer than those of other hominins but did not closely resemble those of the extant African apes. There is no evidence of knuckle-walking, and the wrist and finger joints suggest that the hands were used for support while walking bipedally along tree branches. All these postcranial features seen in *Ardipithecus* suggest that bipedalism might have evolved in trees.

Ardipithecus had a brain that was smaller in relation to its body size than the brain of more advanced australopiths. Its face sloped less than the face of a chimpanzee, but more than the faces of more derived hominins. In general, the anatomy of *Ardipithecus* shows that the common ancestor of chimpanzees and humans was not simply like a chimpanzee, and that chimpanzees have undergone substantial evolutionary change since their split from hominins.

There were two generalized types of australopiths: the gracile ("slender") australopiths (genus *Australopithecus*), which had less pronounced craniofacial traits and appeared earlier, and robust australopiths (genus *Paranthropus*), which had adaptations for heavy chewing and appeared more recently (**Figure 26.17**).

The gracile australopiths   Australopiths are perhaps best thought of as bipedal apes with a modified dentition that included the derived hominin features of thick enamel and small canines. Males were usually larger than females, and individuals grew and matured rapidly, a contrast to the prolonged childhood that characterizes our own species. Microwear analysis of their teeth suggests that the first gracile australopiths, at least, were primarily frugivorous, perhaps including some meat in their diet, as do common chimpanzees.

All australopiths appear to have been capable of bipedal walking. Mary Leakey and her associates found footprints with a humanlike arch at Laetoli in Tanzania, in volcanic ash beds radiometrically dated between 3.8 and 3.6 Ma. The tracks were probably made by *Australopithecus afarensis*,

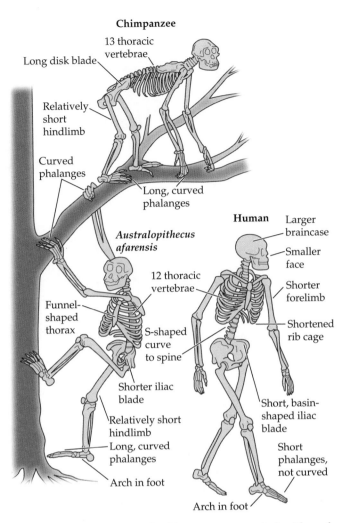

Figure 26.18 Skeletons of a chimpanzee, an australopith, and a human. Note the shorter iliac blade, less funnel-shaped rib cage, longer legs, and shorter fingers and toes of the human. The pelvis of the chimpanzee is elongate and forms part of the lever system of the hindlimbs in this quadrupedal primate. The bowl-shaped pelvis of the australopith and human supports the abdominal organs during bipedal locomotion. (From Fleagle 2013.)

which is known from fossils at the same site. Analysis of these footprints indicates that they do not differ substantially from modern human tracks made on a similar substrate, demonstrating the antiquity of bipedalism in hominin ancestry, far earlier than the appearance of an enlarged brain.

The most completely known early australopith is *Australopithecus afarensis* (see Figure 26.17A). A substantial part of the skeleton of a single young adult female, popularly known as Lucy, was found in the Afar region of Ethiopia in a deposit dated at 3.2 Ma. Lucy is the most complete pre-*Homo* hominin fossil ever found, consisting of more than 60 pieces of bone from the skull, lower jaw, arms, legs, pelvis, ribs, and vertebrae. Young but fully grown when she died, Lucy was only about 1 m tall and weighed about 30 kg. Other finds indicate that males of her species were larger, averaging 1.5 m tall and weighing about 45 kg. Lucy's teeth and lower jaw are clearly humanlike. *Australopithecus*

afarensis had a cranial volume of 380–450 cubic centimeters (cm^3), quite close to that of modern chimpanzees and gorillas.

Despite modifications for bipedalism, australopiths retained some apelike features both of limb anatomy and of the semicircular canals in the inner ear (structures responsible for orientation and balance), suggesting the retention of apelike orientation in an arboreal environment. A degree of arborealism is also reflected in the hands and feet, with the bones of the fingers and toes significantly longer and more curved than those of modern humans (**Figure 26.18**). The hands of australopiths were more humanlike than those of fully arboreal apes such as gibbons and orangutans. It appears that australopiths were able to stand and walk bipedally, but still spent much of their time in the trees and probably were not capable of sustained running. Biomechanical studies of australopiths show that they walked as we do, with an upright trunk. (The caveman image of walking with a stoop and bent hips and knees has its basis in Hollywood, not in science. It is someone's idea of an intermediate gait between that of chimpanzees and humans, and probably never existed.)

In the past decade, several new species of *Australopithecus* have been found, filling gaps in the story of human evolution. **Figure 26.19** shows one current hypothesis of how the australopith species were related to each other and to our own genus, *Homo*.

The earliest known member of the genus *Australopithecus* is now *A. anamensis*, described in 1995 from sites in Kenya and Ethiopia ranging from 4.2 to 3.9 Ma. This hominin appears to be intermediate in anatomy between *Ardipithecus* and *Australopithecus afarensis*, with an estimated body mass of about 50 kg. The fossils are associated with those of forest mammals, reinforcing the notion that early hominin evolution took place in woodlands rather than on the savanna, as was once assumed. *Australopithecus bahrelghazali*, known from a single jaw, is slightly younger than *A. anamensis* and contemporaneous with *A. afarensis*. It was described in 1995 from a site in Chad, in central Africa, demonstrating that early hominins were more widespread in Africa than had previously been supposed. A still younger species from Ethiopia, *Australopithecus garhi*, was described in 1999 from a single fragmentary skull. Dated at 2.5 Ma, the skull is only slightly older than the earliest known specimen of the genus *Homo*, and its association with some fossilized butchered animal bones has led to the speculation that it may have been the first species in the hominin lineage to eat meat and use tools.

Australopithecus africanus, which lived between 3.3 and 2.1 Ma, had heavy arm bones, suggesting it may have spent even more time in trees than *A. afarensis*. This greater degree of arborealism implies that *A. africanus* may not be in the direct evolutionary line to *Homo*. *Australopithecus africanus* apparently also differed from *A. afarensis* by including more meat in its diet. Another species, *Australopithecus sediba*, was described in 2010 from about 1.8 Ma in South Africa. This species is known from four partial skeletons and

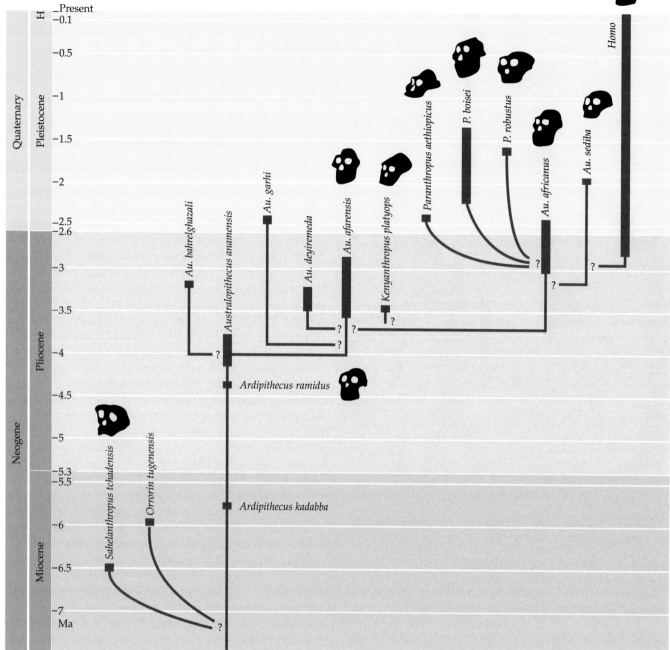

Figure 26.19 A hypothesis of the phylogenetic relationships among australopiths. Narrow lines show interrelationships only; they do not indicate times of divergence or the unrecorded presence of taxa in the fossil record. Question marks indicate uncertainties.

appears to be the closest of the australopiths to the genus *Homo* in both its cranial and postcranial features. It seems that *A. sediba* was fairly limited in its ability to eat hard foods, which might have set the stage for the emergence of *Homo*, a genus that lacks the jaw structures associated with eating tough food.

A recently discovered australopith, *Australopithecus deyiremeda*, lived between 3.5 and 3.3 Ma in present-day Ethiopia. It is known only from jaw and dental remains, and shows that features in the jaw and teeth typically associated

with more recent australopiths actually appeared earlier in the fossil record. This species confirms that there were at least two hominin species living in the same region simultaneously (*A. deyiremeda* and *A. afarensis*), and also suggests a greater diversity of australopiths than previously thought. Another early hominin, known from 3.5 Ma in Kenya, represents a new genus, *Kenyanthropus platyops*. This hominin is markedly different from its contemporary, *Australopithecus afarensis*, combining the unique features of a derived face with a less derived cranium.

The robust australopiths Ecological differences with divergent morphotypes appeared early in the evolution of hominins. The robust australopiths (see Figure 26.17B), appearing later than most gracile australopiths, were distributed in East and southern Africa from 2.5 to about 1.2 Ma. These species, apparently specialized for eating coarse, fibrous plant material, are usually placed in their own genus, *Paranthropus*—*Paranthropus robustus* from South Africa and *P. aethiopicus* and *P. boisei* from East Africa. The interrelationships of these robust australopiths are unclear, and they may not even represent a single radiation from within gracile australopiths. Robust australopiths may have arisen independently from different gracile australopith lineages.

All robust australopiths were sympatric with early *Homo* and were a highly successful radiation during the Pleistocene. It is likely that their extinction in the mid-Pleistocene was related to climatic and vegetational changes in Africa rather than to competition with early humans.

Ecological and biogeographic aspects of early hominin evolution

For many years the evolution of the hominin lineage was assumed to be related to the appearance of the African savannas, which are grasslands with widely spaced bushes and trees. The spread of African savanna environments was probably related to the formation of the Isthmus of Panama 2.5 Ma, which blocked the flow of water between the Atlantic and the Pacific Oceans at a low latitude, leading to profound global climate changes. Paleontological evidence from both the flora and fauna of East Africa suggests that the environment changed to a savanna habitat at this time. Grazing antelopes increased in abundance, and new types of carnivores appeared.

It used to be thought that the development of savanna habitats coincided with the split of the human lineage from that of the other apes. The traditional view has long been that the development of the Rift Valley, which extends from north to south in East Africa, isolated the human lineage from that of the other apes. Subsequently humans became adapted for the new open grassland habitats by adopting a bipedal gait, while the apes were relegated to the tropical forests west of the Rift Valley and remained primarily arboreal.

We now know that the emergence of broad expanses of savanna occurred only 3–2 Ma, whereas hominin bipedalism certainly extends back at least 4.4 Ma and possibly as far back as 7 Ma. Thus, it seems probable that the origins of humans and bipedalism took place in forested environments, although some isolation of humans from other hominids undoubtedly did occur.

The gracile australopiths were primarily a Pliocene radiation, preceding the expansion of savanna habitats. At the start of the Pleistocene, about 2.6 Ma and coincident with major climatic changes, the hominin lineage split into two—one lineage leading to our direct ancestors, the genus *Homo*, and the other leading to the robust australopiths. Thus, in an earlier part of the Pleistocene there were two lineages of hominins: early true humans and robust australopiths. Further climatic cooling and drying resulted in the reduced abundance of robust australopiths about 1.8

Ma, and their extinction later in the epoch. However, had the Pleistocene climatic changes been different—such as reverting to a wetter and warmer regime—it might have been our ancestors that became extinct and the robust australopiths that survived.

26.4 Derived Hominins: The Genus *Homo*

The earliest species of *Homo*, *H. habilis* (Latin *habilis*, "able"), existed in East Africa from 2.8 to 1.44 Ma. This taxon is rather poorly known and has been the subject of intense debate. Some paleoanthropologists think that *H. habilis* did not have enough derived characters to be included in the genus *Homo* and instead place it within the australopiths. Others say that *H. habilis* can be distinguished from australopiths by its larger cranial volume (500–750 cm^3 in contrast to 380–450 cm^3 for *Australopithecus afarensis*), although the braincase of *H. habilis* was small compared with that of later species of *Homo*. *Homo habilis* also differed from *Australopithecus* in having a smaller face, a smaller jaw, and dentition with smaller cheek teeth and larger front teeth. Like *Australopithecus*, *H. habilis* had a relatively small body and retained some specializations for climbing. *Homo habilis* has been found in association with stone artifacts and fossil bones with cut marks, suggesting the use of tools and hunting (or at least scavenging) behavior. *Homo habilis* co-occurred with the more derived *H. erectus* in East Africa for nearly 500 ka.

Many specialists have split the original *H. habilis* into two species: *H. rudolfensis*, known from a single skull from about 1.9 Ma; and *H. habilis* (redefined), known from 2.8 to 1.4 Ma. *Homo rudolfensis* was somewhat larger-brained than *H. habilis*, and some paleoanthropologists have concluded that it was not a member of our genus at all, but a descendant of *Kenyanthropus platyops*. **Figure 26.20** shows the phylogenetic relationships among members of the genus *Homo*.

About 1.9 Ma a new hominin appeared in the fossil record—*Homo erectus* (Latin *erectus*, "upright")—originally described in the late 19th century as *Pithecanthropus erectus* and known at that time as Java Man or Peking Man. Like modern humans, this hominin had a large body, lacked adaptations for climbing, and had relatively small teeth and jaws. There is no doubt that this taxon belongs in the genus *Homo*.

Homo erectus and Homo ergaster

Homo erectus originated in East Africa, where it coexisted for at least several hundred thousand years with two of the robust australopiths and overlapped in time with *H. habilis*. *Homo erectus* was the first intercontinentally distributed hominin. It appears to have spread to Asia at least 1.7 Ma and subsequently perhaps into Europe. This appearance of *Homo* in Asia was once hailed as a landmark in human

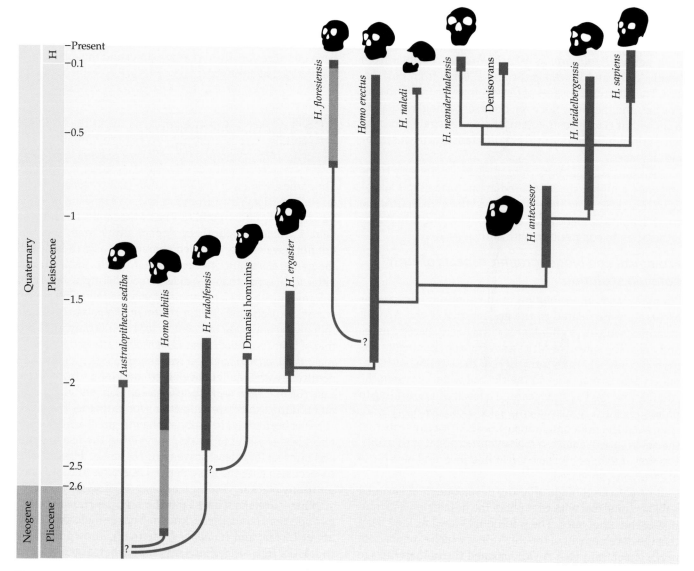

Figure 26.20 A hypothesis of the phylogenetic relationships among members of the genus *Homo*. The lines show interrelationships only; they do not indicate times of divergence or the unrecorded presence of taxa in the fossil record. Lightly shaded bars indicate ranges of time when a taxon is believed to be present, because it is known from earlier times, but is not recorded in the fossil record during this interval. Question marks indicate uncertainties about relationships.

evolution, with the notion of "out of Africa" being some kind of early human achievement similar to the first man on the moon. In fact, all kinds of other mammals were moving between Africa and Asia at this time, and the discovery of a greater variety of early hominins in Asia merely indicates that hominins, too, were migrating into Asia in the early Pleistocene.

Currently the older African form of *H. erectus* is usually called *Homo ergaster* (**Figure 26.21**), and the name *erectus* is reserved for the Asian hominin. *Homo ergaster* is thought to be more closely related to later hominins. The differences between *H. erectus* and *H. ergaster* are subtle, however, and the following description applies to both species. Several

characteristics of *Homo erectus* and *H. ergaster* represent a significant change in the evolutionary history of humans:

- They were substantially larger than earlier hominins (up to 1.85 m tall and weighing at least 65 kg—the same size as modern humans), with a major increase in female size that reduced sexual dimorphism so that males were only about 20–30% larger than females, as in our own species. The reduction in sexual dimorphism in these and later species of *Homo* implies a change from a polygynous mating system (in which males competed with each other for access to females) to monogamous pair bonding in which female choice among potential mates played a larger role.

- *Homo erectus* and *H. ergaster* had body proportions like those of anatomically modern humans. There is debate about the overall proportions of the postcranial skeleton of *H. habilis* and *H. rudolfensis* because of the fragmentary nature of the fossil material, but there is no doubt that *H. erectus* and *H. ergaster* had

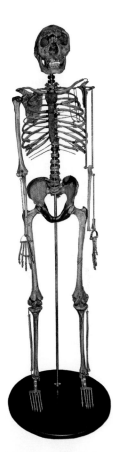

Figure 26.21 *Homo ergaster.* This skeleton of a juvenile male, known as Turkana boy, was found in 1984 near Lake Turkana, Kenya. He is thought to have been about 8 years old when he died. He was 154 cm tall at death, and his estimated adult height would have been about 163 cm. (Sabena Jane Blackbird/Alamy Stock Photo.)

the short arms, long lower legs, narrow pelvis, and barrel-shaped chest that are characteristic of modern humans.

- *Homo erectus* and *H. ergaster* also had larger brains than earlier *Homo* species, with cranial volumes ranging from 775 to 1,100 cm³ (much closer to the modern human average of 1,200 cm³). Because the brain is a metabolically demanding organ, their larger brains imply that *H. erectus* and *H. ergaster* had greater nutritional needs than earlier hominins.

- *Homo erectus* and *H. ergaster* were avid tool makers. Handaxes appeared in Africa 1.75 Ma as part of a tool industry known as the Acheulian, and were found in Europe and western Asia as well. However, no Acheulian handaxes are found in East Asia, where a more refined type of Oldowan-like tool industry is predominant.

- *Homo erectus* and *H. ergaster* were the first hominins to have a humanlike nose—broad and flat but with downward-facing nostrils. Their other facial features were less like those of modern humans: the jaw projected beyond the plane of the upper face (prognathous), the teeth were relatively large, there was no chin, the forehead was flat and sloping, and the bony eyebrow ridges were prominent.

- *Homo erectus* and *H. ergaster* were also the first hominins to have delayed tooth eruption and relatively small teeth for their body size. The smaller teeth suggest that these species cooked their food, because cooked food is easier than raw food to chew. However, the idea that *H. erectus* and *H. ergaster* controlled fire is controversial.

- Delayed tooth eruption suggests a humanlike extended childhood, which would also imply a humanlike extended lifespan and humanlike passage of learned information from one generation to the next.

The Dmanisi hominins

A group of hominins found at Dmanisi, Georgia, are dated at about 1.8 Ma. (Georgia, formerly part of the Soviet Union, is a country near Turkey.) This is the earliest date for a hominin in Eurasia. It is not clear whether this hominin (called *Homo georgicus* by some people) is an early offshoot from *H. erectus* or from an even earlier hominin species, such as *H. habilis*. A recently described skull from Dmanisi combines a small braincase of only 546 cm³ with a very large lower jaw. This cranial volume is much lower than the average for *H. erectus* (~900 cm³). Although the Dmanisi hominins resembled *H. erectus* in several ways, their smaller stature (~1.5 m), smaller brain (546–775 cm³), and the prognathous face of some individuals lead paleoanthropologists to suspect that they may have evolved from a less derived hominin. Aspects of the elbow and shoulder joints support the latter interpretation, yet the lower limbs are long—suitable for running and distance travel as seen in early *Homo*. Otherwise, the Dmanisi hominins may suggest a greater morphological variation than has ever been seen within *H. erectus* before.

Homo floresiensis

The "island rule" in evolutionary biology refers to a strange phenomenon that occurs when a lineage of animals is isolated on an island: small species evolve to larger body sizes, and large species become miniaturized. The giant land tortoises found on the Galápagos Islands and on Aldabra Island are familiar examples of island gigantism, and fossils of miniature dinosaurs have been found in deposits that formed in areas that were islands during the Mesozoic. The island of Flores in eastern Indonesia is home to the largest extant species of lizard, the Komodo monitor, and during the late Pleistocene Flores was also inhabited by a species of miniature elephant and a very small species of human, *Homo floresiensis* (**Figure 26.22**). The origin of *H. floresiensis* is still a subject of debate; the most widely accepted hypothesis is that it is derived from *H. erectus*, but some paleoanthropologists believe it might be derived from a more basal hominin than *H. erectus*.

Tools and skeletal remains of *H. floresiensis* have been found in deposits ranging from 700,000 to 50,000 years old, making the species a contemporary of modern humans during the middle and late Pleistocene. It was only about 1 m tall, much smaller than any other known species of *Homo*, and the earliest fossils of *H. floresiensis* from the research site of Mata Menge (700 ka) would have been even smaller. *Homo floresiensis* had primitive canines and premolars and advanced molars—a combination of dental traits that has not been seen in any other hominin. Because of its small size, this creature was nicknamed "the Hobbit" by the popular press. *Homo floresiensis* made a variety of stone tools: blades, scrapers, and penetrators.

Shortly after the disappearance of *H. floresiensis*, modern human remains appear, dated to ~46 ka. This raises the question of whether modern humans played a role in the extinction of the "hobbits," or even if the species interbred if there was an overlap in space and time.

Homo naledi

The most recently described species of our genus is *Homo naledi* (Sotho *naledi*, "star"). The remains of at least 15 individuals were recovered from the Rising Star cave system in South Africa. The latest studies suggest that *H. naledi* would be between 236,000 and 335,000 years old.

The only analysis of the evolutionary relationships of *H. naledi* suggests that it would group together with the other members of the *Homo* clade. *Homo naledi* would have stood about 1.5 m tall. It had humanlike shoulders and pelvis that more closely resembled those of earlier hominins, and a very small braincase. Its ribs were robust and relatively uncurved, and its vertebrae had a large spinal canal. Its foot is predominantly modern human-like, although it has more curved proximal foot phalanges. Its hand has a robust thumb and wrist morphology similar to that of modern humans, but it has longer and more curved finger bones than in most australopiths, which indicates some degree of climbing. The cranial volume of *H. naledi* was between 465 and 610 cm^3, which is within the size range of the Dmanisi hominins. Its diet probably consisted of hard and resistant food, given that it exhibits higher rates of dental chipping than other hominins. One of the enigmas surrounding *Homo naledi* is how the bones ended up deposited in the deepest chamber of the cave system (the Dinaledi Chamber), which is hardly accessible. No other animal remains were found in the chamber, and the hominin bones had not been damaged by scavengers or predators. It has been suggested that the bodies were deliberately placed deep in the cave by other individuals of its species.

Precursors of Homo sapiens

The species *Homo sapiens* (Latin *sapiens*, "wise"), as originally defined, included not only the modern types of humans but also Neandertals and some earlier forms. More recent evidence has complicated this picture, with the result that most of the forms originally included in *Homo sapiens* have been split into several different species. The name *H. sapiens* is now reserved for modern humans. The many species that are closely related to anatomically modern humans are usually referred to as **archaic *Homo sapiens***.

One of the oldest archaic *Homo sapiens* specimens known is *Homo heidelbergensis* (**Figure 26.23**), from about 600 ka. Although this specimen, from Bodo, Ethiopia, still shows

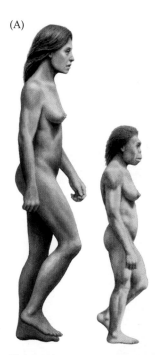

(A)

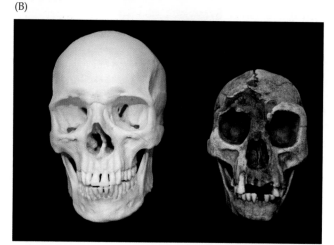

(B)

Figure 26.22 Reconstruction of *Homo floresiensis*. (A) The relative sizes of *Homo sapiens* (left) and *H. floresiensis* (right). (B) The cranial volume of *H. floresiensis* (right) was about 400 cm^3, compared with 1,200 cm^3 for extant *H. sapiens* (left). *Homo floresiensis* is considered to have been a distinct species of human with a relatively derived, though miniature, brain. (A, © Mauricio Anton/Science Source; B, Sabena Jane Blackbird/Alamy Stock Photo.)

many primitive characters (similar to those seen in *H. erectus*), its cranial volume is quite large—1,300 cm^3. A brain this size places *H. heidelbergensis* well within that of the modern human range. There is little consensus if there are members of *H. heidelbergensis* from Europe that are older than the Bodo specimen. Hominins appeared in Europe about 900 ka, but it is a matter of debate how to classify the earliest fossils. Some paleoanthropologists consider them to be *H. erectus*, others to be *H. antecessor*.

Homo antecessor is known from the site of Gran Dolina, in the Atapuerca Mountains in Spain, and is dated to 900 ka. These fossils show a mix of features of modern humans and Neandertals. This species is known for their cannibalistic practices, but they also ate hard and brittle foods because they were confronted with harsh, fluctuating environmental conditions. *Homo antecessor* was initially suggested to be a side branch of an early Pleistocene radiation of lineages in Eurasia that gave rise to Neandertals in Europe and to *Homo sapiens* in Africa.

Figure 26.23 Archaic *Homo sapiens*. *Homo heidelbergensis* is included in the broader concept of archaic forms of *Homo sapiens*. (The Natural History Museum/Alamy Stock Photo.)

The Neandertals

The first fossils of *Homo neanderthalensis* to be recognized as humans were found in the Neander Valley in western Germany in 1856. Since then, Neandertal remains have been recovered from a broad geographic range in western Eurasia. In addition to being found in Germany, Neandertals have been found in Spain, Gibraltar, France, Belgium, Italy, Croatia, Israel, Syria, Iraq, and Russia. The oldest record of a Neandertal comes from the site of Krapina, in Croatia, dated to 130 ka.

DNA analyses suggest that Neandertals were not directly ancestral to *H. sapiens*; instead the two lineages diverged from *Homo heidelbergensis* about 588 ka (see Figure 26.20). Neandertals are often popularly displayed as primitive cavemen, with the unspoken assumption that this is what non-human hominins looked like. In fact, Neandertal features represent a derived condition for hominins.

Neandertals were short and stocky compared with modern humans (**Figure 26.24**). Their stocky build has been interpreted as an adaptation for the cold conditions of Ice Age Europe, but they seem to have been more strongly affected by the Pleistocene ice age than modern humans were. Differences in resource exploitation strategies between these two hominin groups might have given anatomically modern humans an advantage over Neandertals. Neandertal brains were as large as or larger than those of modern *Homo sapiens*, but they were enlarged in a slightly different fashion. The Neandertal brain had a larger occipital area (the region at the back of the head where visual information is processed) than ours, whereas we have a larger middle temporal region (the area in which auditory information and speech and vision semantics are processed). This difference appears in the first year after birth, when the brain of a modern human infant becomes more globular; the brain of Neandertal infants, however, remained elongate.

Neandertals appear to have been much stronger than extant humans. Their front teeth typically showed very heavy wear, sometimes down to the roots. Were Neandertals processing tough, fibrous food between their front teeth or perhaps chewing hides to soften them, as do some modern aboriginal peoples? The Neandertals were stone toolmakers, producing tools known as the Mousterian tool industry, with a well-organized society and increasingly sophisticated

tools. There is also evidence of Neandertals making circular constructions of stalagmites—the oldest constructions made by humans. Whether the Neandertals had the capacity for complex speech remains controversial, but they are the first humans known to bury their dead, apparently with considerable ritual. Of special importance are burials at Shanidar Cave in Iraq that include in the grave a variety of plants recognized in modern times for their medicinal properties.

A recent molecular analysis has shown that the Neandertal diet consisted of pine cones, poplar, moss, wild mouflon, and woolly rhinoceros, although there are regional differences depending on the food sources. Neandertals probably hunted large mammals with weapons that required close contact with their prey; they did not have distance weapons (throwing spears and bows and arrows). Many skeletal remains of Neandertals show evidence of serious injury during life, and their patterns of injury resemble those of present-day rodeo bull riders. Despite the high incidence of injuries, 20% of Neandertals were more than 50 years old at the time of death; it was not until after the Middle Ages that human populations again achieved this longevity.

Modern *Homo sapiens* reached Europe and Asia between 60 and 40 ka, and European Neandertals and the populations of *Homo erectus* remaining in Asia vanished between 40 and 30 ka, with a relict population of Neandertals remaining in Gibraltar (southern Iberian Peninsula) until 28 ka. There

Figure 26.24 Reconstruction of a Neandertal. Neandertals had robust, muscular bodies with a barrel-shaped chest, large joints, short limbs, a receding forehead, a large protruding nose, prominent brow ridges, and a receding chin. (Copyright © Elisabeth Daynes/ScienceSource.)

is much debate about the role of *H. sapiens* in the disappearance of the other species of humans. Did our species gradually outcompete the others in a noncombative fashion, or was there some type of direct conflict? The climate was changing rapidly between about 45 and 30 ka when anatomically modern humans and Neandertals intermingled in Europe, and the population densities of modern humans increased tenfold during that time.

A recent study indicates that the resource exploitation strategies of modern humans gave them an advantage over Neandertals. The methods Neandertals used to obtain food appear to have remained static while the environment changed. As a result, the diets of Neandertals changed greatly over thousands of years. In contrast, anatomically modern humans changed their food-gathering methods and kept their diets relatively stable. Alternatively, another study suggests that differences in the levels of material culture of Neandertals and modern humans could have driven Neandertals to extinction. For example, Neandertals' weapons were all short-range knives and stabbing spears, whereas anatomically modern humans also had throwing spears and bows and arrows. These studies suggest that modern humans were better adapted to the new environmental conditions and that direct or indirect competition between the species contributed to the decline of Neandertals.

The Denisovans

Fragmentary remains of hominins who lived between 50 and 100 ka have been found in Denisova Cave in the Altai Mountains of southern Siberia. Very little skeletal material is known from Denisovans—only three isolated teeth and a finger bone. The Denisovan molars are distinctive: they are very large and lack traits typical of Neandertal and human molars. Recent genetic analyses of hominins from the Atapuerca site Sima de los Huesos (archaic *Homo sapiens*) in Spain show that the Sima hominins were more closely related to Neandertals than to Denisovans, indicating that the Denisovan lineage split from the Neandertal lineage about 430 ka. Unlike most cases in paleoanthropology, the main body of evidence for Denisovans being an independent hominin lineage is based on molecular studies. They were genetically distinct from Neandertals, but they were contemporaneous with both Neandertals and modern humans. Indeed, a preliminary analysis of the Denisova Cave indicated that it was occupied by Denisovans 50 ka, by Neandertals 45 ka, and by modern humans soon after that.

A recent description of two crania from Lingjing, China, could shed light on this mysterious lineage of hominins. The crania are dated to 105–125 ka, and some paleoanthropologists have raised the question of whether they actually belong to Denisovans. However, it is challenging to ascribe cranial material to a lineage of hominins that has been defined mainly genetically, with very little diagnostic fossil material.

Origins of modern humans

A single African origin of *Homo sapiens* is now supported both by the fossil record and by genetic studies of modern humans, especially the evolution of mitochondrial DNA and the Y chromosome. Few people still adhere to the older multiregional model of *H. sapiens* evolving independently in different areas, each from an already distinctive local hominin population.

Mitochondrial DNA is inherited only from the mother, because it resides in the cytoplasm of the egg, not in the nucleus, and the genome is small—only about 16,000 base pairs. Analysis of mitochondrial DNA allows one to trace the maternal lineage of an individual. A study of mitochondrial DNA from people all over the world showed that all living humans trace their mitochondria to a woman who lived in Africa about 170 ka. This hypothetical common ancestor has been called the African Eve, a phrase that obscures the biological meaning of the discovery. It does not mean there was only one woman on Earth 170 ka; instead, it means that only one woman has had an unbroken series of daughters in every generation since then.

A similar approach can be used with the Y chromosome, which is passed only from father to son. The Y chromosome has about 60 million base pairs, so it is much more difficult to study than mitochondrial DNA. An analysis of 2,600 base pairs from the Y chromosome indicates that all human males are descended from a single individual, who is estimated to have lived in Africa 59 ka. Naturally, this individual has been called the African Adam, but note that he lived considerably later than Mitochondrial Eve.

The difference between the estimates—170 ka versus 59 ka—results from uncertainty about the rates of mutation in mitochondrial DNA and the Y chromosome. What is significant is that both studies indicate that the common ancestor of modern humans lived in Africa, and this conclusion is reinforced by other genetic information. For example, there is more variation in the human genome in Africa than in the rest of the world combined, which is exactly what would be expected if humans originated in Africa. Furthermore, humans have only about one-tenth the genetic variation of chimpanzees, and that observation might indicate that human populations were once very small, passing through a genetic bottleneck.

Recent findings of human remains in Morocco indicate that the earliest modern *Homo sapiens* appeared there about 300 ka, and at that time they were probably widespread across most of the African continent. (As noted above, the earlier forms from as far back as 600 ka are known as archaic *Homo sapiens*.) From Africa, modern humans spread across the world:

- Modern humans crossed over into the Levant region of Asia (the area now called the Middle East) about 220 ka. About 75 ka, a single major dispersal of non-Africans spread across Eurasia and Australasia, reaching the islands in Southeast Asia by 73 ka and Australia by 65 ka. Humans arrived in Siberia between 48 and 44 ka, in greater central Asia by 40 ka, in eastern Asia between 39 and 36 ka, and in northeast Asia by 20 ka.

- Modern humans arrived in Europe from the Levant region about 45 ka. European populations went through a genetic bottleneck during the Last Glacial Maximum about 25 ka, with evidence of a major population influx about ~14 ka coming from the Near East.

The genetic, archaeological, and environmental evidence suggests that humans reached the Americas about 15 ka. The archaeological sites of Monte Verde in present-day southern Chile and of Schaefer and Hebior in southeastern Wisconsin provide evidence of human presence in the Americas by 14.6 ka. However, the ice-free corridor that connected Beringia with more southerly areas of North America was closed between 23 and 13.4 ka. This implies that humans had to take another route, probably along the coast.

However, the arrival to the Americas is controversial, and recent studies have proposed much older arrivals, such as 24 ka at the Bluefish Caves, Yukon, and 130 ka at the Cerutti Mastodon site, southern California, mainly based on scratch marks supposedly made by human tools. These older dates would conflict with the present hypothesis of the peopling of the Americas, and need to be further explored to assess their validity.

What happened to the humans who were already there?

Many of the areas invaded by anatomically modern humans were already home to other species of *Homo*. What happened to them? The data available a decade ago supported the "replacement hypothesis"—an interpretation of human evolution by which modern humans completely replaced the populations of other species of *Homo* they encountered, with the older populations vanishing without a trace. The replacement hypothesis has been modified by genetic studies that reveal low levels of interbreeding between anatomically modern humans and the local populations of species of *Homo* that they met as they spread from Africa.

The complete Neandertal genome was determined from DNA extracted from fossils found in Croatia. Comparison of the Neandertal genome with the genome of living humans reveals that 1–3% of the nuclear DNA of extant Eurasians came from Neandertals, not from our most recent African ancestors. A higher percentage of Neandertal DNA is found in east Asian populations. The most plausible interpretation of that observation is that modern humans interbred with Neandertals in the Middle East, where the two species coexisted between 80 and 50 ka. Sometime later a subgroup of modern *Homo sapiens* moved eastward into Asia, where they interbred with Denisovans. The descendants of this group of modern humans are found in native populations of Melanesia (the islands in the South Pacific east of Australia) and Australia, and to a lesser extent southern Asia. Melanesian and Australian populations can carry up to 0.5% of Denisovan DNA.

A 4–8% representation of other species of *Homo* (such as Neandertals or Denisovans) in the genome of non-African modern humans does not discredit the replacement hypothesis—after all, more than 90% of the genome of human populations outside Africa does represent a recent African origin—but it calls for modification of the hypothesis. Paleoanthropologists now describe the spread of modern humans as "replacement with hybridization" or "leaky replacement," to recognize the genetic contribution of other species of *Homo* to the genome of modern humans.

26.5 Evolution of Human Characteristics

Humans are distinguished from other primates by three derived features: a bipedal stance and mode of locomotion, an extremely enlarged brain, and the capacity for speech and language. Here we examine possible steps in the evolution of each of these key features and also consider the loss of body hair and the evolution of tool use.

Bipedalism

Although all modern hominoids can stand erect and walk to some degree on their hindlegs, only hominins display an erect bipedal mode of striding locomotion involving a specialized structure of the pelvis and hindlimbs, thereby freeing the forelimbs from obligatory functions of support, balance, and locomotion. The most radical changes in the hominin postcranial skeleton are associated with the assumption of a fully erect, bipedal stance in the genus *Homo* (**Figure 26.25**). Anatomical modifications include the S-shaped curvature of the vertebral column, modifications of the pelvis and position of the acetabulum (hip socket) in connection with upright bipedal locomotion, lengthening of the leg bones, and shifting of the angulation of the knee joint toward the midline of the body. Humans also differ from apes in having a longer trunk region with a more barrel-shaped (versus funnel-shaped) rib cage, resulting in a distinct waist. The humanlike waist may be a specific adaptation for bipedal walking, allowing rotation of the pelvis in striding without also involving the upper body.

The secondary curve of the spine in humans is a consequence of bipedal locomotion and forms only when an infant learns to walk. We have by no means perfected our spines for the stresses of bipedal locomotion, which are quite different from those encountered by quadrupeds. One consequence of these stresses is the high incidence of lower-back problems in modern humans.

Humans stand in a knock-kneed position, which allows us to walk with our feet placed on the midline and reduces rolling of the hips from side to side. This limb position leaves some telltale signatures at the articulation of the femur with the hip and at the knee joint. This type of bony evidence can aid paleoanthropologists in deducing whether fossil species were fully bipedal. An unfortunate consequence of this limb position is that humans, especially athletes, are prone to knee dislocations and torn knee ligaments. Because women have wider hips than men, their

(A)   (B)

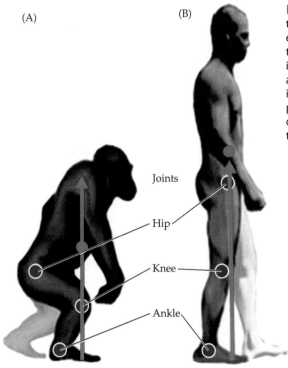

Joints

Hip

Knee

Ankle

Figure 26.25 Bipedal locomotion by chimpanzees and humans. The trunk, pelvis, and vertebral column of chimpanzees and humans have different structural characters associated with quadrupedal and bipedal locomotion. (A) When a chimpanzee walks bipedally, its center of gravity (red dot) is anterior to its hip joint and the chimpanzee must bend its legs at the hip and knee, using muscular force to hold its center of gravity over the foot that is its base of support. (B) The S-shaped curve of the human vertebral column places the center of gravity over the supporting foot and most of the weight of a human is transmitted directly through the bones of the supporting leg to the ground. (After Sockol et al. 2007.)

had some capacity to diverge from the rest of the toes in early hominins, especially in *Ardipithecus*.

Basal apes—gibbons and orangutans—tend to walk in a clumsy bipedal stance when they are on the ground. Their upright trunk, adapted for arboreal locomotion, predisposes them to do this. Wild orangutans walk bipedally in the trees, supporting themselves by grasping overhead branches with their hands. This behavior allows them to move out onto narrower supports than they could reach quadrupedally or even by hanging underneath the branch. Although orangutans are not on the direct line to humans, this behavior shows that bipedal walking could have evolved in trees instead of on the ground. This is consistent with early hominins, such as *Ardipithecus*, having a mixture of features that suggest bipedalism and also arborealism.

Large brains

The human brain increased in size threefold over a period of about 2.5 Ma. Human brains are not simply larger versions of ape brains but have several key differences, such as a relatively much larger prefrontal cortex and relatively smaller olfactory bulbs. We still do not know what selective pressures led to such large brains. Speculations include increasing ability for social interactions, conceptual complexity, tool use, dealing with rapidly changing ecological conditions, language, or a mixture of these elements.

femurs are inclined toward the knees at a more acute angle than men's femurs, and female athletes are especially prone to knee injuries.

The feet of humans show drastic modifications for bipedal, striding locomotion. The feet have become more arched, with corresponding changes in the shapes and positions of the tarsals and with close, parallel alignment of all five metatarsals and digits. In addition, the big toe is no longer opposable, as in apes and monkeys (**Figure 26.26**), and is much larger than the other toes, bearing the weight at the end of the stride. The big toe may still have

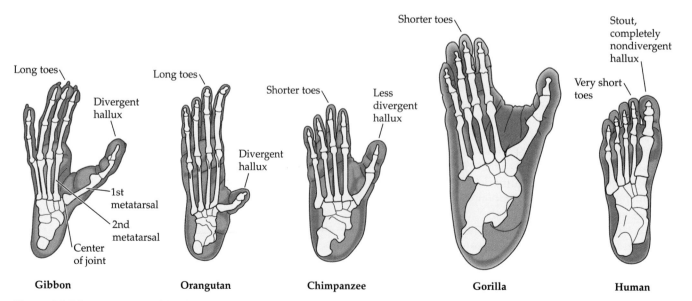

Long toes

Divergent hallux

1st metatarsal

2nd metatarsal

Center of joint

Gibbon

Long toes

Divergent hallux

Orangutan

Shorter toes

Less divergent hallux

Chimpanzee

Shorter toes

Gorilla

Stout, completely nondivergent hallux

Very short toes

Human

Figure 26.26 Comparison of the feet of extant hominoids.

Brain tissue is metabolically expensive to grow and to maintain; on a gram-for-gram basis, brain tissue has a resting metabolic rate 16 times that of muscle. Most of the growth of the brain occurs during embryonic development, but human brains continue to grow after birth and require continued energy input from the mother. Thus, selective pressures for larger brains can be satisfied only in an environment that provides sufficient energy, especially to pregnant and lactating females. The evolution of larger brains may have required increased foraging efficiency (partially achieved through larger female size and mobility) and high-quality foods in substantial quantities (partially achieved through the use of tools and fire).

Larger brains also would have required a change in life-history pattern that probably exaggerated the ancestral primate character of slow rates of pre- and postnatal development, thus lowering daily energy demands and also reducing a female's lifetime reproductive output. Our prolonged period of childhood may allow children to stay with the family long enough to acquire the knowledge necessary for survival. (Imagine the chaos if human children became sexually mature at about 3 years old, as horses do.)

Speech and language

Although other animals can produce sounds and many mammals communicate by using a specific vocabulary of sounds (as anyone who has kept domestic pets well knows), the use of a symbolic language is a uniquely human attribute. Although chimpanzees and gorillas have been taught to use some human words and form simple sentences, this is a long way from the complexity of human language. Where in human evolution did language evolve, and how can we determine this from the fossil record? The first evidence of human writing is only about 3,000 years old; obviously, language evolved before this, but how long before?

Controlled speech might not have been possible until a later stage than *Homo erectus*. In *H. erectus*, the spinal cord in the region of the thorax was much smaller than it is in modern humans. This observation suggests that *H. erectus* lacked the capacity for the complex neural control of the intercostal muscles that allows modern humans to control breathing in such a way that we can talk coherently. Furthermore, the hypoglossal canal (the exit from the skull for cranial nerve XII, which innervates the tongue muscles) is smaller in other hominins (and is also smaller in chimpanzees and gorillas) than in modern humans and Neandertals. Additionally, a specific gene involved in the production of language in humans, *FOXP2*, has the same two differences from the chimpanzee condition in both modern humans and Neandertals.

Even if more derived *Homo* species had evolved the capacity for language, they would not have been able to produce the range of vowel sounds that we can produce until a change in the anatomy of the pharynx and vocal tract had taken place. The original position of the mammalian larynx was high in the neck, right behind the base of the tongue. However, in modern humans, the larynx shifts ventrally at the age of 1–2 years, resulting in the creation of a much larger resonating chamber for vocalization. This change in the vocal tract anatomy is associated with a change in the shape of the base of the skull, so we can infer from fossil skulls when the shift in larynx position occurred. Although there are some differences of opinion, a fully modern condition of the vocal tract was probably not a feature of the genus *Homo* until modern *H. sapiens* about 50 ka. Speech also requires a change in neural capacities in order to process the rapid frequency of transmitted sounds, which are decoded at a rate much faster than other auditory signals. The processing of rapidly transmitted sounds might be related to an increase in volume around the temporal area of the brain, which is associated with speech recognition, that is not seen in Neandertals.

There is still discussion from an evolutionary point of view over the problems that come with having part of our respiratory and digestive tracts share a passageway. The descent of the larynx means that the original mammalian seal between the palate and the epiglottis has been lost, making humans especially vulnerable to choking on their food. It seems likely that a risk of choking would have been a powerful selective force acting in opposition to repositioning the larynx. In addition, the death of infants from SIDS (sudden infant death syndrome) is associated with the developmental period when the larynx is in flux. It is difficult to imagine that the ability to produce a greater range of vowel sounds could alone counteract these negative selective forces. Perhaps there was another, more immediately powerful reason for repositioning the larynx.

One advantage that repositioning the larynx affords us is the ability to voluntarily breathe through our mouths, an obviously important trait in the production of speech. However, perhaps a more important function of mouth breathing is apparent to anyone who has a bad cold. Many of us would suffocate every winter if we were unable to breathe through our mouth. It is tempting to speculate that the human capacity for well-enunciated speech owes its existence to a prior encounter of our species with the common cold virus.

Loss of body hair and development of skin pigmentation

Humans are unique among primates in their apparent loss, or at least reduction, of body hair and in their heavily pigmented skin. These features are related, as fur protects animals from deleterious effects of the sun's rays. Humans have the same number of hair follicles as other apes, but the hairs themselves are minuscule. Chimpanzees have relatively unpigmented skin, but with the loss of body hair there would have been a need to gain skin pigments. It is not clear why humans lost the majority of their body hair. Speculations include increased use of sweat glands in evaporative cooling and increased problems with skin parasites such as lice, ticks, and fleas (perhaps in association with a more sedentary way of life with groups of people living together in confined quarters).

Although we will probably never know precisely why humans lost their covering of hair, genetics enables us to figure out when this might have happened. The genes involved in human skin pigmentation appear to date from at least 1.2 Ma. This implies that hairlessness first came about with the lineage leading to *Homo sapiens*, perhaps also including the *H. erectus* lineage that branched off about 1.7 Ma. This corresponds with the time when hominins adopted a home base, which might have made them especially prone to parasitic infections and suggests that we shared hairlessness with other species, such as the Neandertals.

A study of the genetics of skin parasites provides information about when humans started to wear clothes. The human body louse is different from head lice and pubic lice in that it clings to human clothing rather than to hair. Body lice presumably evolved from head lice after humans started to wear clothes, and this separation occurred between 70 and 40 ka, broadly coincident with the emergence of modern *Homo sapiens* in Europe. Interestingly, the genetic study of skin microparasites shows that humans with different regional ancestries harbor distinct lineages of parasites, and that these associations can persist after generations spent in a completely new region. Genetic studies of skin parasites are also consistent with an out-of-Africa dispersal hypothesis. However, there are also some closely related species of parasites found in humans and gorillas with a recent date of divergence triggered by the change of hosts.

Humans from populations originating close to the Equator have dark skin, whereas people from populations farther from the Equator have lighter skin. A balance between protection against the damaging effects of ultraviolet (UV) light and the need for vitamin D synthesis appears to offer the best general explanation of this phenomenon. Folic acid is essential for normal embryonic development, and UV light breaks down folic acid in the blood. Melanin in the skin blocks penetration by UV light, thereby protecting folic acid. Too much melanin in the skin creates a different problem, however, because vitamin D is converted from an inactive precursor into its active form by UV light in blood capillaries in the skin. Thus, human skin color probably represents a compromise—enough melanin to protect folic acid while still permitting enough UV light penetration for vitamin D synthesis.

Human technology and culture

Tool use is estimated to occur in less than 1% of non-human animal species. Although tool use has been recorded in four lineages of vertebrates (fishes, crocodylians, birds, and mammals), only passerine birds and primates use tools for a variety of tasks.

Many primate lineages use tools. Capuchin monkeys crack nuts with rocks, as do common chimpanzees, and some macaques use rocks to open oysters. Squirrel monkeys have been observed in the wild flinging down sticks in agonistic behaviors, and capuchins use leaves as cups to carry water. Orangutans use sticks to reach fruit hanging out of reach and branches to check the depth of water in a pond. Orangutans and common chimpanzees were once thought to be the only apes to use tools for food acquisition, but this behavior has recently been observed in gorillas as well. Bonobos have not been observed using tools in the wild, although captive bonobos and common chimpanzees are equally diverse tool users in most contexts. For example, bonobos have been observed to use fallen branches as ladders and to use branches as weapons.

Manipulation and preparation of stone tools are traits limited to hominins. The earliest simple stone tools that have been recognized (known as Lomekwian tools) were found in West Turkana, Kenya. They are irregular in shape, and chipped to have a sharp point. Lomekwian tools appeared about 3.3 Ma and predate the appearance of any species of the genus *Homo*. The better-known Oldowan tools, which are generally circular and chipped on one side to produce a sharp edge, appeared about 2.7 Ma.

Oldowan tools remained the standard for a million years until they were replaced by Acheulean tools, which appeared in East Africa about 1.76 Ma. These tools, which differ from Oldowan tools in having a distinct long axis and being chipped on both sides, include cleavers and so-called hand axes. They were apparently made by *Homo erectus* and changed very little in style for the next 1.5 Ma. The lack of any dramatic advances in tool manufacture over these immense periods of time is surprising, especially in light of the spread of these tools to southwest Asia and western Europe.

It is often assumed that much of the evolution of human tool use and culture developed in the context of humans hunting other animals. The phrase "Man the Hunter" was coined in the 1960s. Anthropologists have pointed out that much of our perception of human evolution as being an upward and onward quest has more to do with Western cultural myths of the Hero's Tale than with anthropological data. The archaeological evidence for hunting by early hominins, such as stone tools and bones with cut marks, suggests that early humans were scavengers rather than hunters, and occasionally engaged in cannibalistic practices. Furthermore, the corollary phrase—"Woman the Gatherer"—has been discredited. Studies of the few remaining hunter–gatherer societies have shown that tasks are shared by both sexes: men do substantial amounts of gathering, and men and women both contribute to the heavy tasks of butchering large prey, cutting it into manageable pieces, and carrying it back to the home base.

26.6 Why Is *Homo sapiens* the Only Surviving Hominin Species?

We live in an unusual time because only in the past 27 ka has there been just one species of hominin. Throughout hominin history (possibly with the exception of the first 500 ka), several species of hominins have coexisted. As recently as 30 ka our species, *Homo sapiens*, shared the planet with at least three other species—*H. neanderthalensis*, the Denisovans, and *H. floresiensis*—and *H. erectus* might have been present as well.

Homo erectus was long believed to have disappeared between 300 and 200 ka, but recent discoveries have shown that it survived at least until 150 ka and possibly until 40 ka. Remains of *H. erectus* from Java have been dated between 53,000 and 27,000 years old by two methods and between 60,000 and 40,000 years old by a third method. The sediments in which these fossils are found are complex and hard to decipher, making these young dates controversial, but if they are correct, late-surviving *H. erectus* were contemporaries of Neandertals, Denisovans, modern *H. sapiens*, and potentially *H. floresiensis*. It is hard to imagine how it would feel to live with other species of humans to whom we were as closely related as dogs are to wolves and coyotes.

Hybridization among species of Homo

The occurrence of more than one species of *Homo* in a particular location led to interbreeding in the last 100 ka: early modern humans and modern humans interbred with Neandertals, modern humans interbred with Denisovans, Neandertals interbred with Denisovans, and Denisovans interbred with an unknown population of hominins whose skeletal remains are yet to be discovered (**Figure 26.27**). This "mystery hominin" could potentially be an offshoot of Asian *H. erectus*.

Human populations outside Africa carry approximately 1–3% of genetic material that is derived from Neandertals, but an anatomically modern human from Peştera cu Oase, Romania, dated to approximately 40 ka, had 6–9%,

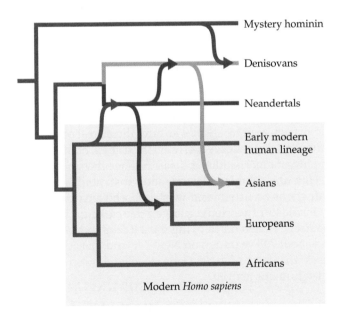

Figure 26.27 Genetic introgression between different lineages of hominins. Studies of ancient DNA show that modern Eurasian populations carry Neandertal DNA, reflecting an interbreeding event before European and Asian populations separated, but after the Eurasian lineage separated from the African lineage. Some modern Asian populations carry Denisovan DNA as well. Neandertals also interbred with Denisovans, and there is evidence of DNA from a "mystery hominin" lineage present in Denisovan populations. (After Callaway 2016.)

suggesting that this individual had a Neandertal ancestor as recently as four to six generations back.

Interbreeding has influenced the genetic variability that we see in present-day human populations. While much of the genetic inheritance that modern Eurasians have received from Neandertals includes archaic genes that boost immune response and define cranial and brain morphology, some Neandertal genes are linked to modern diseases, such as depression, skin lesions, blood clots, and urinary tract disorders.

26.7 Humans and Other Vertebrates

A review of the biology of vertebrates must include consideration of the effect of the current dominance of *Homo sapiens* over other members of the vertebrate clade. Never before in Earth's history has a single species so profoundly affected the abundance and prospects for survival of so many other species.

The current status of primates is a microcosm of the plight of all vertebrates. At the World Conservation Congress in September 2016, the IUCN (International Union for the Conservation of Nature) announced that four of the seven species of great apes are Critically Endangered—one step away from extinction. (The recently described orangutan species *Pongo tapanuliensis* has not yet been evaluated by the IUCN, but will probably also be listed as Critically Endangered, given that it is the rarest of all great apes.) The situation is almost as dire for the other species of primates: 60% are threatened with extinction and 75% have declining populations. Illegal hunting, which poses the greatest threat to many primates, has led to a decline of more than 70% in the number of eastern gorillas (*Gorilla beringei*) in the past 20 years and to significant losses among the other species.

Humans as superpredators and environmental disruptors

Humans are qualitatively and quantitatively different from all other predators. Most apex predators prey on young and aged individuals in a population, because those are the easiest ones to capture and kill. In contrast, weapons (from spears to automatic rifles) allow humans to kill the largest and fittest adults in a population. And that's what humans do, killing adult individuals 14 times more frequently than other predators do.

In addition to their direct impact as predators, humans have influenced other vertebrates by changing their environments. Tens of thousands of years ago, humans were modifying the environment with fire. Agriculture, which originated independently on every continent except Antarctica between 12 and 5 ka, and its accompanying transition from nomadic bands to settled populations increased the pace and scope of anthropogenic environmental change. Urbanization and accelerating climate change are the most recent manifestations of indirect human impacts on other vertebrates. Even the presence of humans can affect other

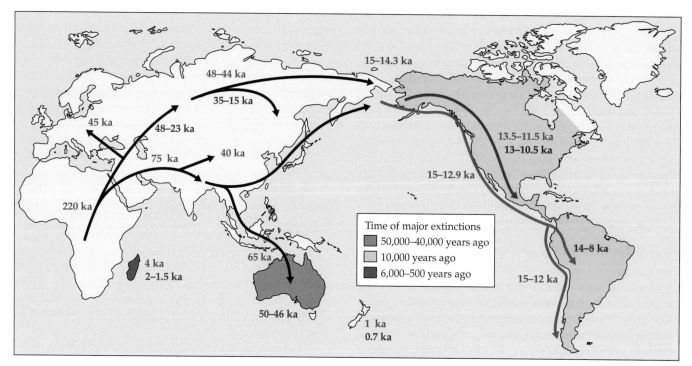

Figure 26.28 Correspondence between the spread of human populations and the extinction of native vertebrates. Modern humans began to leave Africa about 220 ka and spread eastward across Eurasia and into the Americas. The dates of extinctions of Pleistocene megafauna correspond with estimates of the times humans arrived in Australia, the Americas, and major islands such as New Zealand and Madagascar. Dates in red indicate human arrivals, dates in blue indicate megafaunal extinctions.

predators; playbacks of sounds associated with humans caused mountain lions to flee from feeding sites and reduced the total time they spent feeding by more than half.

Megafaunal extinctions

The diversity of vertebrates increased slowly throughout the Paleozoic and early Mesozoic, and then more rapidly during the past 100 Ma. This overall increase has been interrupted by periods of extinction. Extinction is as normal a part of evolution as species formation, and the duration of most species in the fossil record appears to be from 1 to 10 Ma. But the pattern of long-lived species changed at about the time that humans became the dominant species in many parts of the world. We are in the midst of the "sixth extinction," which is progressing at a rate more than 10,000 times the rate of the end-Cretaceous extinction.

Extinctions begin soon after humans arrive The first humans to reach Australia were met by a megafauna that included representatives of four groups: marsupials, flightless birds, tortoises, and echidnas. The largest Australian land animals in the Pleistocene were several species of herbivorous mammals in the genus *Diprotodon*, the largest of which probably weighed about 2,000 kg. The largest kangaroos weighed 250 kg, and echidnas reached weights of 20–30 kg and were waist-high to an adult human. The flightless bird *Genyornis newtoni* was twice the height of a human, and the horned turtles in the family Meiolaniidae were nearly as large as a Volkswagen Beetle car. Perhaps the most dramatic species in the Australian megafauna was a monitor lizard (*Varanus*

[*Megalania*] *priscus*) 5.5 m long. The horned turtles, the monitor lizard, *Genyornis*, and all marsupial species weighing more than 100 kg became extinct after humans arrived.

The most striking evidence of the impact of humans on other vertebrates is the apparent correspondence in the times of the arrival of humans on continents and islands and the extinction of the megafauna (**Figure 26.28**). Humans reached northern Australia around 65 ka, occupied the dry center of the continent by 49 ka, and had moved south to Tasmania by 39 ka. On a continent-wide basis, the Australian megafauna had disappeared by about 9 ka after the arrival of humans. Similar correspondences between the dates of human arrival and megafaunal extinctions exist for the other continents.

Humans colonized islands later than continents, and extinctions occurred later on islands: between 10 and 4 ka on islands in the Mediterranean Sea, 4 ka on islands in the Arctic Ocean north of Russia, 2–1.5 ka on Madagascar, and about 700 years ago on New Zealand. In each of these cases, the dates of extinctions closely followed the dates when humans arrived.

Were humans solely responsible? The multiple coincidences between the arrival of humans and extinctions of native species shown in Figure 26.29 provide strong circumstantial evidence of a role for humans in megafaunal extinctions, but not proof that humans were solely responsible. Other factors were undoubtedly involved, including climate change, changes in vegetation, the arrival of alien species in addition to humans, and perhaps new pathogens and parasites introduced by these alien species.

Few authorities argue that humans were the sole cause of megafaunal extinctions; rather, a synergistic interaction of events stressed populations of large mammals and birds to the point where human hunting tipped them into an extinction vortex. Fossil sites provide persuasive evidence that humans killed enormous numbers of large animals:

- Sites in Eurasia dated between 40 and 15 ka contain the remains of as many as 166 woolly mammoths (*Mammuthus primigenius*) per site.

- When the Maori arrived in New Zealand, they found nine species of moas (flightless birds, some species as large as 250 kg). Genetic analysis of fossils indicates that moa populations had been large and stable for thousands of years, but all nine species were extinct within 100 years after humans arrived. The role of Maori hunters in the extinction of moas is amply documented. A butchering site on the sand dunes at Kaupokonui on the North Island of New Zealand contains the remains of hundreds of individuals of three species of moas. Uncooked moa heads and necks had been left in piles to rot, while the legs were roasted. At Wairau Bar on the South Island, the ground is littered with the bones of moas, an estimated 9,000 individuals plus 2,400 eggs, and at Waitaki Mouth are the remains of an estimated 30,000–90,000 moas.

- Hundreds of eggshells of the 200-kg flightless bird *Genyornis newtoni* that were cooked in campfires have been found at 2,000 sites in Australia, all dated between 53.9 and 43.4 ka. After that, *Genyornis* was extinct and campfire sites contain only eggshells from emus (*Dromaius*).

These species, and the other species of megafauna driven to extinction by humans, were diverse in terms of their phylogeny, ecology, and geography. What they had in common was large body size; extinctions were overwhelmingly concentrated among species with adult weights greater than 44 kg. Life-history characteristics of large species of vertebrates, especially mammals, include late maturation and late first reproduction, single offspring at intervals from 1 to 3 years, long reproductive lifespans, and low population densities. Large species cannot rebuild populations rapidly, and as a result they are vulnerable to overkill.

Is this the Anthropocene?

The overwhelming global impact of humans has led to a proposal that we are now in a new epoch of the Quaternary, the Anthropocene (Greek *anthropos*, "human"; *kainos*, "specific period of time"). The suggestion has popular appeal, but reaction from the scientific community has been mixed.

Human activities unquestionably influence global processes, in some cases exceeding the magnitude of natural processes. Increasing levels of carbon dioxide resulting from burning fossil fuels have initiated climate change on a global basis, as well as raising sea levels and acidifying the oceans. Extreme weather events are becoming more frequent and more severe, and extremes of rainfall and drought are increasing.

Worldwide impacts of humans can be traced back for hundreds or thousands of years, and include megafaunal extinctions, phenotypic changes in species, effects of agriculture on current plant communities, and reduction in genetic diversity of animal populations. However, geological time periods are based on stratigraphic markers—evidence in geological deposits and ice cores—and many stratigraphers, the geologists who study these markers, are not convinced that we are truly in a new epoch. Potential stratigraphic markers have been proposed: more than 200 minerals and mineral-like materials are the result of human activities. Portland cement is the most familiar of these, and others include new minerals produced by alteration of existing minerals in the walls of mine tunnels and in mine dumps. Plastic has been proposed as a stratigraphic marker of the Anthropocene. Scraps of plastic are widespread in sedimentary deposits, and microplastics (particles smaller than 1 mm in diameter that are present in health and beauty products, including skin lotions and toothpaste) are ubiquitous in ocean sediments (Figure 26.29).

The Anthropocene Working Group, composed of geologists, archeologists, and climate and atmospheric scientists, is currently preparing a proposal for the International Commission of Stratigraphy and the International Union of Geological Sciences. When those groups have voted on the proposal, we will know whether humans' enormous environmental footprint will receive formal stratigraphic recognition as a new geological epoch.

Figure 26.29 Plastic makes up nearly 70% of all ocean litter. Large pieces of plastic are broken down by ultraviolet light and ocean turbulence into microplastic debris that is deposited in marine sediments. (Copyright © iStock.com/jacus.)

Summary

The first primatelike mammals are known from the early Paleocene, but the earliest true primates are not known until the early Eocene (or perhaps the late Paleocene).

Early euprimates were initially present in North America and the Old World. All were arboreal, and some had larger brains relative to their body size than other mammals of that time had.

By the middle Eocene, some of the earliest anthropoids had populated Asia and soon spread to Africa. After climates cooled in the late Eocene, primates were confined to tropical regions until the early Miocene.

By the Oligocene, two distinct groups of extant anthropoids had evolved: the platyrrhine monkeys of the New World tropics and the catarrhine monkeys and apes of the Old World.

The apes and humanlike species, including *Homo sapiens*, are grouped in Hominoidea.

Many morphological features distinguish hominoids from other catarrhines, and enlargement of the brain has been a major evolutionary force molding the shape of the hominoid skull, especially in the later part of human evolution.

The first known hominoids occurred in the late Oligocene, about 25 Ma. By the late Miocene, hominoids had diversified and spread throughout Africa, Europe, and Asia.

The genetic closeness of humans and great apes (orangutans, gorillas, and chimpanzees) has led to a regrouping of the traditional Hominidae to include these apes as well as humans.

Humans and their fossil relatives are assigned to a tribe, Hominini, within Hominidae.

A variety of hominin fossils have been found in late Pliocene and early Pleistocene deposits of Africa.

The earliest well known hominin was *Ardipithecus*, although other, less complete forms have been described.

The australopiths, known from 5.8 to 1.2 Ma, diversified into gracile and robust forms.

Australopiths were bipedal but retained many characters associated with arborealism and had a relatively small, apelike cranial volume.

The earliest member of the genus, *Homo habilis*, dates from about 2.8 Ma, appearing sometime after the earliest stone tools found in East Africa (~3.3 Ma).

Homo erectus lived in Africa and Eurasia from about 1.8 Ma to at least as recently as 150 ka and possibly up to 40 ka. *Homo erectus* had a cranial volume approaching the lower range of *Homo sapiens*, made stone tools, and might have used fire.

Fossils from Dmanisi, Georgia, dated to 1.8 Ma, seem to be closely related to *Homo erectus*, but their cranial volume is significantly smaller.

Homo naledi is a recently discovered species from the Rising Cave system in South Africa that lived between 236 and 335 ka. The cranial volume of *H. naledi* was within the range of that of the Dmanisi hominins.

Homo floresiensis, or "the hobbit," was a 1-m-tall hominin from the island of Flores in Indonesia. It lived between 700 and 50 ka and is thought to be derived from some form of *Homo erectus*.

Archaic *Homo sapiens* included forms closely related to anatomically modern humans, such as *Homo heidelbergensis* and *Homo antecessor*.

Neandertals were a group of Eurasian hominins that coexisted with modern humans. They appeared 130 ka in Europe. The last remaining populations of Neandertals are known from Gibraltar, dated as recently as 28 ka.

The Denisovans were an enigmatic group of hominins from the Altai Mountains of Siberia. Their skeletal remains are fragmentary; they are thought to have lived between 50 and 30 ka.

Homo sapiens, the only surviving species of the tribe Hominini, originated in Africa about 300 ka.

Bipedalism appeared in the earliest stages of hominin evolution.

Ardipithecus already showed evidence of bipedal posture, including an anteriorly positioned foramen magnum and a short and broad pelvis with laterally oriented iliac blades.

Other anatomical modifications for bipedal locomotion include an S-shaped curvature of the vertebral column and the angulation of the knee joint toward the midline of the body.

Summary (continued)

A cranial volume comparable to that of modern humans was not seen until the evolution of *Homo heidelbergensis* and other archaic forms.

The increase in brain volume in anatomically modern humans might have been caused by social and conceptual complexity, tool use, rapid climate change, language, or the combined action of all these elements.

The loss of hair in our lineage might have been related to increased evaporative cooling of the body or to problems with parasites.

Changes in skin pigmentation could have been associated with striking a balance between UV light–induced destruction of folic acid and the need for UV light for synthesis of vitamin D.

For much of human evolutionary history, several species of *Homo* have coexisted.

At least three archaic species of *Homo* were in existence 30 ka, and two of them interbred with modern humans.

Modern human populations from Eurasia carry a small percentage (1–3%) of Neandertal DNA, and native populations from Australia, Melanesia, and southern Asia can carry up to 0.5% of Denisovan DNA.

The present-day situation, with *Homo sapiens* as the sole existing hominin, is a phenomenon of the past 27 ka or less.

The dates when humans reached continents and islands correspond closely with the dates of extinction of large vertebrates.

Humans are superpredators, able to kill large animals that have no natural predators as adults.

Megafauna species have low population densities, late maturation, low birth rates, and small litter sizes; consequently, they are vulnerable to overkill by human predation.

Humans changed the environment, initially with fire and later with agriculture and urbanization, and generally to the detriment of other species of vertebrates. Global climate change is the most recent and farthest-reaching example of the environmental impact of humans.

The magnitude of human impact has led to a proposal that we are in a new epoch, the Anthropocene.

The effects of humans are undeniable and can be traced back for thousands of years, but geological periods are based on evidence in geological strata, and scientists are currently divided about whether or not there are stratigraphic markers for the Anthropocene.

Microplastic debris—a major form of pollution—may be stable in sedimentary deposits. Plastic has been proposed as a stratigraphic marker of the Anthropocene.

Discussion Questions

1. There are no (nonhuman) primates today in North America. What elements of the Earth's climate in the Eocene allowed primates to live in North America, and what climatic changes accompanied their disappearance from this continent?

2. During the Miocene, when temperatures became warm at higher latitudes, monkeys and apes radiated from Africa into Eurasia. What prevented primates from reappearing in North America during this time?

3. What is the functional significance of the S-shaped bend in the vertebral column of *Homo sapiens*?

4. The South American monkeys (platyrrhines) first appear in the fossil record of that continent in the Oligocene, and we deduce that they must have arrived there at least by the late Eocene. This reasoning involves some rather complex arguments about crossing the Atlantic Ocean. Why don't we think that the platyrrhines might have reached South America a lot earlier when the Atlantic Ocean was narrower and traversing it would have been easier?

5. Humans originated in Africa, and their closest relatives, the chimps and gorillas, are known only from Africa. So why is there speculation that this lineage (the Homininae) had its origins in Eurasia?

6. Suppose that Bigfoot (or Sasquatch, the legendary primate that is imagined to live in the Pacific Northwest) or the Yeti (the equally imaginary primate that is supposed to live in the Himalayas) were found to be real and derived from *Homo erectus*. What moral responsibility would we (*Homo sapiens*) have to a species that was as closely related to us as a wolf is to a dog? Would we feel less responsible if it was more closely related to another great ape?

Additional Reading

Begun D. 2010. Miocene hominids and the origins of the African apes and humans. *Annual Review of Anthropology* 39: 67–84.

Finlayson C. 2005. Biogeography and evolution of *Homo*. *Trends in Ecology and Evolutionary Biology* 20: 457–463.

Fuss J, Spassov N, Begun DR, Böhme M. 2017. Potential hominin affinities of *Graecopithecus* from the Late Miocene of Europe. *PLoS ONE* 12: e0177127.

Gross M. 2016. Chimpanzees, our cultured cousins. *Current Biology* 26: 83–101.

Jablonski N, Chaplin G. 2003. Skin deep. *Scientific American* 13(2): 72–79.

Kay R. 2015. New World monkey origins. *Science* 347: 1068–1069.

McNulty K. 2010. Apes and tricksters: The evolution and diversification of human's closest relatives. *Evolution and Education Outreach* 3: 322–332.

Posth C and 10 others. 2017. Deeply divergent archaic mitochondrial genome provides lower time boundary for African gene flow in Neanderthals. *Nature Communications* 8: 16046.

Reed DL, Light JE, Allen JM, Kirchman JJ. 2007 Pair of lice lost or parasites regained? The evolutionary history of anthropoid primate lice. *BMC Biology* 5: 7.

Richter D and 11 others. 2017. The age of the hominin fossils from Jebel Irhoud, Morocco, and the origins of the Middle Stone Age. *Nature* 546: 293–296.

Rightmire G. 1998. Human evolution in the middle Pleistocene: The role of *Homo heidelbergensis*. *Evolutionary Anthropology* 6: 218–227.

Silcox MT, López-Torres S. 2017. Major questions in the study of primate origins. *Annual Review of Earth and Planetary Sciences* 45: 113–137.

Shoenemann R. 2006. Evolution of the size and functional areas of the human brain. *Annual Review of Anthropology* 35: 379–406.

Strait D. 2010. Evolutionary history of the australopithecines. *Evolution and Education Outreach* 3: 341–352.

Tattersall I, Schwartz J. 2009. Evolution of the genus *Homo*. *Annual Review of Earth and Planetary Sciences* 37: 67–92.

Go to the Companion Website **oup.com/us/vertebratelife10e** for active-learning exercises, news links, references, and more.

Etymology of Some Biological Names and Terms

a, prefix meaning "without"; Greek

ad Apis, "toward Apis" (an Egyptian bull deity)

akantha, "spine"; Greek

aktis, "ray"; Greek

allos, "different"; Greek

amphis, "double"; Greek

an, prefix meaning "without"; Greek

anas, "duck"; Latin

aner (andros), "male"; Greek

ankylos, "stiffening"; Greek

anomos, "strange"; Greek

anthropos, "man"; Greek

apo-, prefix meaning "away from"; Greek

apsis, "loop"; Greek

archaios, "ancient"; Greek

archon, "ruler"; Greek

ardi, "ground floor"; Amharic (an Ethiopian language)

arthron; "joint"; Greek

atrium, "hall" or "entry room"; Latin

australis, "southern"; Latin

auto, "prefix meaning self"; Greek

avis, "bird"; Latin

azhdarkho, a dragon, Uzbek

baino, "to walk"; Greek

bios, "life"; Greek

blastos, "bud"; Greek

bole, "dart"; Greek

boreas, "north"; Greek

bounos, "hill"; Greek

bous "ox"; Greek

branchia, "gills"; Greek

caecilia, "slow worm"; Latin

canis, "dog"; Latin

catarrheo, "downward"; Greek

chondros, "gristle"; Greek

chorde, "string"; Greek

chronos, "time"; Greek

cleistos, "closed"; Greek

cornu, "horn"; Latin

corpus, "body"; Latin

cranium, "skull"; Latin

crocodilus, "crocodile"; Latin

cupula, "small tub"; Latin

daktylos, "finger" or "toe"; Greek

deinos, "terrible"; Greek

deire, "neck"; Greek

derma, "skin"; Greek

di-, dis-, prefixes meaning "two"; Greek

dia-, prefix meaning "across"; Greek

domesticus, "belonging to the household"; Latin

dorsum, "back"; Latin

dromas, "running"; Greek

dryos, "tree"; Greek

durus, "hard"; Latin

dys-, prefix meaning "bad"; Greek

elasma, "plate"; Greek

elopos, "marine fish"; Greek

enantios, "opposite"; Greek

endon, "within"; Greek

epi-, prefix meaning "on" or "above"; Greek

eu-, ev- prefixes meaning "true"; Greek

eurys, "wide"; Greek

familiaris, "domestic"; Latin

fenestra. "window"; Latin

forma, "shape"; Latin

fossor, "digger"; Latin

frons, "brow"; Latin

gamos, "marriage"; Greek

gastros, "stomach"; Greek

genesis, "origin"; Greek and Latin

germen, "bud"; Latin

glossa, "tongue"; Greek

glyphe, "carving"; Greek

gnathos, "jaw"; Greek

gomph, "club"; Greek

gymnos, "naked"; Greek

gyne, "female"; Greek

habilitatem, "able"; Latin

hadros, "bulky"; Greek

hals, "the sea", hence "salt"; Greek

haploos, "simple"; Greek

heteros, "different"; Greek

histion, "sail"; Greek

holos, "whole, entire"

homo, "man"; Latin

homoios, "alike"; Greek

hyoeides, "shaped like the letter upsilon"; Greek

hypsos. "high"; Greek

ichnos, "track"; Greek

ichthys, "fish"; Greek

imperator, "ruler"; Latin

ischion, "hip joint"; Greek

isos, "similar"; Greek

kainos "recent"; Greek

kephale, "head"; Greek

keras, "horn"; Greek

kerkos, "tail"; Greek

klados, "branch"; Greek

kleis, "clavicle"; Greek

kleptes, "thief"; Greek

koilos, "hollow"; Greek

koinos, "shared"; Greek

kolobos, "shortened"; Greek

krainion, "skull"; Greek

kryptos, "hidden"; Greek

kyklos, "circle"; Greek

kynos, "dog"; Greek

kytos, "cell"; Greek

laganon, "pancake"; Greek

lekithos, "yolk of an egg"; Greek

lepisma "scale"; Greek

leptos, "thin, small"; Greek

lissos, "smooth"; Greek

lithos, "stone"; Greek

lutum, "yellowish"; Latin

magnus, "great"; Latin

maia, "good mother"; Greek

mamma, "teat"; Greek

manus, "hand"; Latin

margo, "border"; Latin

mater, "mother"; Latin

mesos, "middle"; Greek

metron, "measure"; Greek

mono-, prefix meaning "one" or "single"; Greek

morphe, "form"; Greek

naledi, "star"; Sotho (an African language)

neos, "new"; Greek

nomos, "order"; Greek

notos, "the back"; Greek

odous, "tooth"; Greek

onychos, "claw"; Greek

oon, "an egg"; Greek

ophis, "snake"; Greek

opisthen, "behind"; Greek

opsis, "appearance"; Greek

ornis, "bird"; Greek

osteon, "bone"; Greek

ostrakon, "shell"; Greek

oura, "tail"; Greek

oxys, "sharp"; Greek

pachys, "thick"; Greek

pais, "child"; Greek

palaios, "ancient"; Greek

-para-, prefix meaning "beside" or "near"; Greek

passer, "sparrow"; Latin

pelyx, "bowl"; Greek

phago, "to eat"; Greek

phaneros, "visible"; Greek

phoros, "bearer"; Greek

photos, "light"; Greek

phylon, "tribe"; Greek

physeter, "blower"; Greek

picis, "fish"; Latin

pithekos, "monkey"; Greek

plast, "formed"; Greek

platos, "flat, broad"; Greek

plax, "plate"; Greek

plesios, "near"; Greek

pleura, "the side"; Greek

pneuma, "breathing"; Greek

poly-, prefix meaning "many"; Greek

poros, "small opening"; Greek

pous, "foot"; Greek

pro-, prefix meaning "earlier" or "before"; Greek

pseudes, "false"; Greek

pteron, "wing" or "fin"; Greek

raptor, "robber"; Latin

rhinos, "nose"; Greek

rhynchos, "beak"; Greek

rodens, "gnawing"; Latin

sapiens, "wise"; Latin

sarkos, "flesh"; Greek

sauros, "lizard"; Greek

selachos, "cartilaginous fish"; Latin

simia, "ape"; Latin

simos, "flat-nosed"; Greek

solen, "pipe"; Greek

souchos, "crocodile"; Greek

sphen, "wedge"; Greek

splanchnon, "viscera"; Greek

stegos, "roof"; Greek

stenos, "narrow"; Greek

stoma, "mouth"; Greek

streptos. "twisted"; Greek

stylos, "pillar"; Greek

sym-, syn-, prefixes meaning "together"; Greek

tachy, "swift"; Greek

taxo, "to arrange"; Greek

teleos, "entire"; Greek

terra, "earth"; Latin

testudo "tortoise"; Latin

tetra, "four"; Greek

thallasa, "the sea"; Greek

thanatos, "death"; Greek

theca (Latin) or *thekos* (Greek), "a case to put something in"

ther, "wild beast"; Greek

thermos, "heat"; Greek

thyreos, "shield"; Greek

tomos, "cut"; Greek

trema, "hole"; Greek

trophos, "one who feeds"; Greek

venter, "belly"; Latin

xiphos, "sword"; Greek

za, "very"; Greek

zoon, "an animal"; Greek

zygos, "joining"; Greek

Glossary

A

activity temperature range The range of body temperatures an animal (especially an ectotherm) maintains during the part of the day when it is thermoregulating.

advertisement calls Vocalizations used by individuals to announce their presence, such as mating calls and territorial calls.

aerobic scope model A hypothesis proposing that endothermy is a consequence of selection for high rates of aerobic metabolism that support high levels of physical activity. (See **parental care model, thermogenic opportunity model, warmer is better model**.)

Afrotheria The grouping of endemic African mammals, including orders Proboscidea (elephants), Sirenia (dugong and manatees), Hyracoidea (hyraxes), Tubulidentata (aardvark), Macroscelidea (elephant shrews) and Afrosoricida (tenrecs and golden moles).

Agnatha The basal paraphyletic grouping of vertebrates that lack jaws. Includes the extant cyclostomes (lampreys and hagfishes), and the extinct conodonts and ostracoderms.

air capillaries Interconnected small tubes that radiate from the parabronchi; gas exchange occurs in faveoli in the walls of the air capillaries.

air sacs Spaces in the respiratory passages of sauropsids with faveolar lungs in which air is briefly stored during through-flow lung ventilation.

allantois One of the extraembryonic membranes of amniotes; stores embryonic wastes and fuses with the chorion to form the chorioallantoic membrane, which functions in gas exchange.

allometry The condition in which the relative sizes of parts of the body change during ontogenetic growth. (See **isometry**.)

Allotheria One of the traditional three major subgroups of mammals—includes the extinct multituberculates. (See **Prototheria** and **Theria**.)

altricial Of neonates, helpless at birth or hatching. (See **precocial**.)

alula The feather-covered second digit of a bird's wing.

alveolar lung A lung in which gas exchange occurs in spherical closed-ended chambers at the ends of a series of treelike dichotomous branches of the airways. Airflow in alveolar lungs is tidal. (See **faveolar lung**.)

ammonotely Excretion of nitrogenous wastes primarily as ammonia. (See **ureotely, urcotely**.)

amnion One of the extraembryonic membranes of amniotes; the inner membrane surrounding the embryo.

amniotes Vertebrates with the derived character of three fetal membranes: amnion, chorion, and allantois.

amphioxus Small, superficially fishlike marine animals, the cephalochordates or lancets.

amphistylic A type of jaw suspension in fishes in which the anterior end of the upper jaw is attached to the cranium and the posterior end is supported by the hyomandibula.

ampullae of Lorenzini Electroreceptors in the skin on the snout of elasmobranchs.

anadromous Of fishes, migrating from seawater to freshwater to reproduce. (See **catadromous**.)

anapsid Describes a skull with no temporal fenestrations.

ancestral character See **plesiomorphy**.

angiosperms Flowering, seed-producing, vascular plants. Most plants today, including trees and grasses. (See **gymnosperms**.)

annuli (singular **annulus**) Folds in the skin that encircle or partly encircle the body.

antidiuretic hormone (ADH) A hypothalamic hormone released in response to increased blood osmolality or reduced blood volume that promotes recovery of water by the kidney. Also known as **vasopressin**.

apex predator A predator species that is at the top of its food web.

apnea "Without breath"; holding the breath as in diving.

apocrine gland Skin gland found in restricted areas (under armpits and in the pubic region of humans); produces an oily secretion used in chemical communication. In hoofed mammals, apocrine glands are found over the body surface and produce sweat for evaporative cooling.

apomorphy A character that has changed from its ancestral state—i.e., a derived character.

aposematic Having a character, such as color, sound, odor, or behavior, that advertises an organism's noxious properties.

apteria Regions of skin without feathers.

aquaporins Tubular proteins that form water channels through plasma membranes.

Artiodactyla An order of extant hoofed mammals, the even-toed ungulates, having either four toes (pigs, peccaries and hippos) or two main toes (camels and llamas, mouse deer, musk deer, true deer, giraffe, pronghorn, and bovids [antelope, sheep and cattle]).

aspect ratio The ratio of the length of a wing to its width.

Atlantogeneta Name given to the grouping of southern mammals, groups Afrotheria and Xenarthra.

atrium (plural **atria**) 1. A chamber of the heart of vertebrates. 2. A chamber surrounding the gill slits of urochordates and cephalochordates. The atrium exits to the exterior through a pore. 3. In a general sense, an empty space within a structure.

autodiastylic A type of jaw suspension in fishes in which projections from the upper jaw attach it rigidly to the cranium.

autotomy Sacrificing a body part (e.g., tail, skin, limb) to escape from a predator.

B

barbs The primary branches from the rachis of a feather that compose the vane.

barbules Extensions from the barbs of a feather that hook to barbules from an adjacent barb, providing the rigidity that makes a vane an aerodynamic surface.

Bilateria Animals that at some point in their lives have a body plan with sides that are mirror images of each other.

binocular vision Stereoscopic vision by two eyes that have overlapping fields of view.

binomial nomenclature The Linnean system of identifying a species with a generic and specific name—e.g., *Homo sapiens*.

blastopore The original opening of the central cavity of an embryo.

Boreoeutheria Name given to the grouping of northern mammals, groups Euarchontoglires and Laurasiatheria.

brachiation A specialized form of arboreal locomotion in which animals swing from the underside of one branch to another.

brachiopatagium The primary wing of pterosaurs.

branchiostegal rays A fanlike series of dermal bones on the underside of the skull, forming the floor of the gill chamber

bristles Specialized feathers with a stiff rachis and without barbs, or with barbs on only a portion of the rachis.

bryophytes Land plants that lack internal channels for movement of water, today represented by mosses, liverworts, and hornworts. (See **tracheophytes**.)

buccopharyngeal pumping Drawing air or water into the mouth and pharynx and expelling it with movements of the floor of the mouth.

bunodont Molar teeth that are square in shape, with four low rounded cusps that pulp the food. Typical of omnivorous mammals, including humans.

C

calamus The short, tubular base of the rachis of a feather that implants the feather in its follicle.

carapace The dorsal portion of the shell of a turtle. (See **plastron**.)

carnassials The specialized pair of shearing teeth of mammals in the order Carnivora, comprising the last (4th) upper premolar and the first lower molar.

catadromous Of fishes, migrating from freshwater to seawater to reproduce. (See **anadromous**.)

centrum (plural **centra**) The portion of a vertebra that surrounds the notochord.

Cephalochordata One of two groups of extant non-vertebrate chordates, the lancets. (See **Urochordata**.)

ceratotrichia The fin rays of chondrichthyan fishes; unjointed and formed from keratin.

cerebral vesicle Any of the three divisions of the early embryonic brain.

choanae Internal openings of the nostrils in the oral cavity. Seen in lungfishes and tetrapods today (possibly convergently evolved), in contrast to the condition in most fishes where U-shaped nostrils open and exit to the outside.

chondrocranium A structure that surrounds the brain. Initially formed by cartilage that is replaced by endochondral bone in most bony fishes and tetrapods.

Chordata The phylum that includes vertebrates.

chorioallantoic placenta The late-forming placenta of therian mammals, formed from the fusion of the allantois to the chorion. In placental mammals this is the major placenta, while only a few marsupials (bandicoots, bilbies, koala, wombat) form a transitory chorioallantoic placenta towards the end of gestation. (See **choriovitelline placenta, hemochorial placenta, endotheliochorial placenta, epitheliochorial placenta**.)

chorion One of the extraembryonic membranes of amniotes; the outer membrane surrounding the contents of the egg.

choriovitelline placenta The early-forming placenta of all therian mammals, produced by apposition of the yolk sac to the chorion. In marsupial mammals this is the major placenta, in placentals it is transitory. (See **chorioallantoic placenta, hemochorial placenta, endotheliochorial placenta, epitheliochorial placenta**.)

cloaca The common opening of the reproductive and excretory tracts.

coelom A body cavity lined with mesodermal tissue.

cone cells Photoreceptor cells in the retina that are sensitive to a narrow range of wavelengths (different types of cones are sensitive to different wavelengths); used to perceive colors and for visual acuity.

connective tissues Tissues that provide structural support, protection, and strength: mineralized tissues (bone and cartilage), adipose tissue (fat), blood cells, and flexible tendons and ligaments.

conodonts Extinct Paleozoic jawless vertebrates known almost entirely from tooth microfossils. Their relationship to other vertebrates is uncertain.

continental drift Movement of continents across Earth's surface.

contour feathers Feathers that streamline a bird, reducing turbulence by smoothing transitions, as between the neck and trunk.

conus arteriosus An elastic chamber anterior to the ventricle of some gnathostomes.

coprolites Fossilized feces.

corpus luteum A hormone-secreting structure found in mammalian ovaries, formed from the follicular cells that remain after the egg is shed. Produces hormones essential for establishment and maintenance of pregnancy.

Cretaceous Terrestrial Revolution (KTR) The diversification of terrestrial organisms in the mid-Cretaceous, marking the point at which diversity on land outstripped that in the oceans.

crocodyloid joint A type of mesotarsal ankle joint in which the main articulation is a peg on the astragalus fitting into a socket on the calcaneum.

crop An enlarged portion of the esophagus of birds that is specialized for temporary storage of food.

cross-current exchange system An anatomical arrangement of air passages and blood vessels in which air and blood pass in opposite directions, but do not follow parallel pathways as they do in a counter-current exchange system.

crown group Members of a crown group have all of the derived characters of a lineage. (See **stem group**.)

crurotarsal joint A joint in which the axis of bending in the foot runs between the astragalus and calcaneus; characteristic of pseudosuchians. (See **mesotarsal joint**.)

cryptodires Turtles that bend the neck vertically to retract the head. (See **pleurodires**.)

cupula A cup-shaped gelatinous secretion of a neuromast organ in which the kinocilium and stereocilia are embedded.

cursorial Specialized for running.

D

Darwinian fitness The genetic contribution of an individual to succeeding generations relative to the contributions of other members of its population.

deglutination A physiological suite of coordinated food-processing movements of the tongue, soft palate, and pharynx, unique to mammals.

dentary The tooth-bearing dermal bones forming the anterior part of the lower jaw in osteichthyans.

derived character See **apomorphy**.

dermal bone Dermal bone (sometimes called membrane bone) is formed in the skin and lacks a cartilaginous precursor.

dermatocranium Dermal bones that cover a portion of the skull.

determinate growth Ontogenetic growth that ceases when an individual becomes an adult. (See **indeterminate growth**.)

Deuterostomia Animals in which the blastopore becomes the anus. (See **Protostomia**.)

diapause A period of arrested embryonic development.

diapsid Describes a skull with two temporal fenestrations (upper and lower) on each side. (See **anapsid**, **synapsid**.)

diastema A gap in the row of teeth, usually seen in herbivores between the incisors (plus canines if present) and the cheek teeth (molars and premolars).

diphyodont The condition of having only two sets of teeth per lifetime: in humans, the milk teeth and the permanent teeth.

diplorhiny The condition of having two nostrils and nasal sacs, formed from paired nasal placodes.

dispersal The movement of organisms over geological time resulting from the movement of the organisms themselves. (See **vicariance**.)

distal convoluted tubule The portion of a mammalian nephron adjacent to the collecting duct that changes the concentration of the ultrafiltrate by actively transporting Na⁺.

down feathers Feathers with a short rachis that provide insulation and, in some species of birds, waterproofing.

drag Backward force opposed to forward motion.

durophagous Eating hard-shelled food items.

E

eccrine gland Skin gland found on the surfaces that contact the substrate in most mammals, the soles of the hands and feet and underside of the tail (if prehensile); produce a watery secretion that aids adhesion. In humans and some other primates, eccrine glands are found over the body surface and produce sweat for evaporative cooling. (See **apocrine gland**.)

ectoderm The outermost embryonic germ layer.

ectothermy The end of the continuum of thermoregulatory modes at which most of the energy used to raise body temperature comes from external sources (e.g., solar radiation, warm substrate).

embryonic diapause The situation in which a fertilized embryo is retained in a state of arrested development.

embryophytes Land plants, so called because they retain the embryo within the maternal tissue during early development.

endochondral bone Endochondral bone is made up of osteocytes formed in a cartilaginous precursor deep within the body.

endoderm The innermost embryonic germ layer.

endostyle A ciliated glandular groove on the floor of the pharynx of non-vertebrate chordates that is homologous with the thyroid gland of vertebrates.

endotheliochorial placenta A chorioallantoic placenta in which the tissues of the fetus penetrate the lining of the uterus, but not the lining of the maternal blood vessels. Seen in carnivores and elephants. (See **epitheliochorial placenta**.)

endothermy The end of the continuum of thermoregulatory modes at which most of the energy used to raise body temperature comes from internal sources (i.e., metabolism).

epicontinental sea A shallow sea that extends into the interior of a continent.

epigenetic effect Modification of gene expression during development by non-genetic factors, such as temperature.

epiphyses The ends of the long bones, separated by a zone of growth cartilage from the shaft of the bone of mammals during growth; growth ceases when epiphyses fuse with the shaft.

epithelial tissues Tightly connected cells that form the boundaries between the inside and outside of the body and between compartments of the body.

epitheliochorial placenta Chorioallantoic placenta in which the tissues of the fetus contact the lining of the uterus, but do not penetrate it. Seen in ungulates and non-anthropoid primates. (See endotheliochorial placenta.)

euprimates Crown primates.

Eutheria Placental mammals and their extinct stem group relatives. (See **Methatheria, Theria**.)

euryhaline Capable of living in a wide range of salinities. (See **stenohaline**.)

explosive breeding A mating system in which the breeding season is very short. (See **prolonged breeding**.)

extant phylogenetic bracketing Using the character states of two extant lineages to infer the character states of extinct taxa that lie between the extant taxa in a phylogeny.

extinction vortex A situation in which feedback among multiple processes drives a population into an irreversible spiral to extinction.

F

faveolar lung A lung in which gas exchange occurs in cuplike chambers (faveoli) lining the walls of the airways.

Air flows in one direction in faveolar lungs—from posterior to anterior (See **alveolar lung**.)

fetal membranes Membranes derived from the fetus: amnion, chorion, allantois.

filoplumes Hairlike feathers with few barbs that sense the position of contour feathers.

flow-through ventilation The one-way flow of air in a faveolar lung. (See **tidal ventilation**.)

forebrain The anteriormost portion of the developing brain; contains olfactory and visual sensory structures.

fossorial Specialized for burrowing.

furcula The avian wishbone, formed by fusion of the two clavicles at their central ends.

G

gastralia Bones in the ventral body wall of some amniotes, also known as ventral ribs, important in lung ventilation of some taxa.

gastroliths Stones in a muscular gizzard that pulverize food.

genetic monogamy A breeding system in which a male and female mate only with each other. (See **social monogamy**.)

genetic sex determination (GSD) The situation in which the sex of an individual is determined by heteromorphic sex chromosomes; e.g., XX/XY for mammals, ZZ/ZW for birds.

gigantism A life history pattern characterized by growing to an extremely large body size.

gigantothermy The ability of a very large organism to maintain a stable body temperature warmer than its environment due to its low surface area/volume ratio.

gill slits See **pharyngeal slits**.

gizzard The posterior muscular stomach of birds and other archosaurs. (See **proventriculus**.)

glomerulus A capillary tuft associated with a kidney nephron that produces an ultrafiltrate of the blood.

gnathostomes Vertebrates that possess jaws: among extant vertebrates all except lampreys and hagfishes.

Gondwana The southern supercontinent of the Paleozoic and early Mesozoic, formed from South America, Antarctica, and Australia.

gymnosperms Plants that reproduce by seeds (rather than by spores) but do not produce flowers. Include the extant conifers and cycads. (See **angiosperms**.)

gynogenesis A mode of reproduction of all-female species in which an egg must be activated by sperm from a male of a bisexual species before it begins development.

H

hair cells Sensory receptors found in the inner ear of vertebrates and in the lateral lines of fishes and aquatic amphibians.

hair follicle The invagination of the germinal layer of the epidermis that contains individual hairs.

Haversian system Cylindrical units of bone formed by concentric layers of mineralized tissue around a capillary and a venule.

hemochorial placenta Chorioallantoic placenta in which the maternal blood is in direct contact with the tissues of the fetus. Seen in anthropoid primates (including humans) and rodents, may be the basal mammalian condition.

hermaphroditic The condition of an individual animal having functioning gonads of both sexes.

heterochrony Changes in the timing of gene expression during development.

heterodont Description of a dentition where the teeth are regionalized into different forms. (See **homodont**.)

heterometry A change in the intensity of a gene's expression during development.

heterothermy 1. Regional heterothermy—maintaining different temperatures in different parts of the body. 2. Temporal heterothermy—changing the set-point of temperature regulation, as in hibernation.

heterotopy A change in the location of a gene's expression during development.

hindbrain The posterior portion of the developing brain; will control motor activities, respiration, and circulation.

holostylic (holostyly) A type of jaw suspension in fishes in which the upper jaw is fused to the cranium. (See **hyostylic**.)

hominid (Hominidae) Orangutans, gorillas, chimpanzees, and humans, and extinct forms more closely related to these taxa than to gibbons and siamangs.

hominin (Hominini) Humans and extinct forms more closely related to humans than to chimpanzees.

hominine (Homininae) Humans, chimpanzees, gorillas, and extinct forms more closely related to these taxa than to orangutans.

homocercal A form of the tail fin seen in teleost fishes in which the upper and lower lobes are symmetrical. (See **heterocercal**.)

homodont Description of a dentition in which all the teeth are of the same type. (See heterodont.)

Hox genes Found throughout the animal kingdom, the genes of the Hox complex specify the regional identity of cells along the anterior-posterior axis of the developing embryo (i.e., the specification of head versus thorax versus caudal structures), Duplications of the Hox complex in vertebrate taxa are believed to have led to the evolution of distinctive structures.

hybridogenesis 1. A mode of reproduction of all-female species in which DNA from a male of a bisexual species enters the egg but is eliminated at each generation. 2. Formation of a new species by hybridization of two parental species.

hyostylic (hyostyly) The condition of jaw suspension in which the upper jaw is attached to the skull only by the hyoid apparatus. (See **holostylic**.)

hyperdactyly Having more than five digits in the forefeet or hindfeet.

hyperosmolal Of a solution with higher solute concentration, hence lower water potential, than the comparison solution.

hyperphalangy An increase in the number of bones (phalanges) in the digits.

hyposmolal Of a solution with lower solute concentration, hence higher water potential, than the comparison solution.

hypothalamus A structure in the floor of the midbrain that is involved in neural-hormonal integration.

hypselodont Hypsodont teeth that have become ever-growing by delaying the formation of the root and continuing growth of the tooth throughout life. Seen in rabbits, some rodents, and wombats.

hypsodont Lophed molar teeth that have tall crowns (but retain the tooth root); on tooth eruption much of the crown is contained within the jaw bones, but emerges throughout life as the exposed crown wears down. Enables mammalian teeth to resist wear; typical of grazing mammals such as horses and many bovids.

I

indeterminate growth Growth that continues throughout life. (See **determinate growth**.)

ingroup The evolutionary group under study. (See **outgroup**.)

integumentary sensory organs Pressure receptors on the surface of the skin of crocodylians.

isometry The condition in which the relative sizes of parts of the body do not change during ontogenetic growth. (See **allometry**.)

isosmolal Of solutions with the same water potential.

K

kinocilium A sensory cell in the neuromast organ.

L

lateral line system The sensory system on the body surface of fishes and aquatic amphibians that detects water movement.

lateral plate mesoderm The ventral part of the mesoderm, surrounding the gut.

lateral somitic frontier The boundary in gnathostome embryos between two developmental domains or modules: the primaxial domain (formed from the somite alone—the backbone and associated muscles derive from this domain) and the abaxial domain (formed from contributions from both the somite and the lateral plate mesoderm—the limbs derive from this domain).

Laurasia The northern supercontinent of the later Paleozoic, formed from Laurentia and Siberia.

Laurentia The northern supercontinent of the early Paleozoic, formed from North America, Greenland, Scotland, and part of Asia.

lecithotrophy A mode of fetal nutrition in which a developing embryo receives its nourishment from the egg yolk. (See **matrotrophy**.)

lepidotrichia The fin rays of osteichthyan fishes: jointed and formed from bone.

lophodont Lophed molar teeth with straight lophs that run primarily in the cheek-to-tongue direction along the tooth; seen in perissodactyls and kangaroos.

lophs Molar teeth in which the cusps seen in bunodont teeth are run together into ridges, which grate the food rather than pulping it. Seen in herbivorous mammals.

M

marsupials Mammals that give birth to young after a brief period of intrauterine development. Most marsupials have a pouch in which the young complete development. (See **Metatheria**.)

masseter muscle Jaw-closing muscle of cynodonts and mammals.

masticate To chew food with the back teeth.

matrotrophy A mode of fetal nutrition in which the reproductive tract of the mother supplies the energy for a developing embryo. (See **lecithotrophy**.)

maxillae The posterior tooth-bearing dermal bones in the upper jaw of osteichthyans.

mesoderm The central embryonic germ layer.

mesotarsal joint A distinctive joint in the middle of the tarsus (ankle), between proximal and distal rows of tarsal bones; the axis of bending in the foot is a simple hinge. (See **crurotarsal joint**.)

metapodials The collective term given to the bones of the palm of the hand (the metacarpals) and the sole of the foot (the metatarsals).

Metatheria Marsupials and their extinct stem group relatives. (See **Eutheria**, **Theria**.)

midbrain The central portion of the developing brain; becomes the pathway for incoming sensory information and outgoing motor signals.

Modern Synthesis The blending of genetics, natural selection, and population biology that occurred in the 1930s and 1940s.

molecular scaffolding Using molecular data to determine the branching pattern of phylogenies and morphological data to date the branching.

molting Loss of old hairs and replacement by new ones.

monogamy A mating system in which a male and female are paired during breeding. (See **genetic monogamy**, **social monogamy**.)

monophyletic lineage An evolutionary lineage that includes an ancestor and all of its descendants. (See **paraphyetic lineage**.)

monorhiny The condition of having a single nostril and nasal sac, formed from a single midline nasal placode.

monotremes A monophyletic lineage of extant mammals that retain the ancestral character of oviparity: echidnas and the platypus.

muscle tissues Tissues that contain the filamentous proteins actin and myosin, which together cause muscle cells to contract and exert force.

myomeres Blocks of striated muscle fiber arranged along both sides of the body; most conspicuous in fishes.

N

neocortex The new portion of the cerebrum in the forebrain of mammals, with a distinctive six-layered structure; site of higher-order brain functions such as cognition.

nephron The basic functional unit of the kidney.

neural crest An embryonic tissue that forms many structures unique to vertebrates.

neural tissues Tissues that include neurons (cells that transmit information via electric and chemical signals) and glial cells that form an insulative barrier between neurons and surrounding cells,

neuromast organs Clusters of sensory hair cells and associated structures usually enclosed in the lateral lines of fishes and aquatic amphibians.

notochord A dorsal stiffening rod that gives the phylum Chordata its name.

O

olfactores The phylogenetic grouping that contains urochordates (tunicates) and vertebrates.

opisthoglyphous Having enlarged teeth (fangs) in the rear of the upper jaw. (See **proteroglyphous**, **solenoglyphous**.)

orobranchial chamber The internal cavity formed by the mouth and gill regions.

osmosis Movement of water across a semipermeable membrane from a solution of higher water potential to a solution of lower water potential.

osteoderm A bone embedded in the skin.

ostracoderms Extinct Paleozoic jawless vertebrates with an external covering of dermal bone, more closely related to gnathostomes than the extant jawless lampreys and hagfishes.

outgroup A reference group that is less closely related to the group under study (which is known as the ingroup) than the members of the ingroup are related to each other.

oviparity A mode of reproduction in which a female deposits eggs that develop outside her body.

P

paedomorphosis Retention of larval or embryonic characters into adult life.

Pangaea The supercontinent that formed in the Carboniferous and persisted until the Middle Jurassic.

parabronchi The third level (smallest) air passages in faveolar lungs.

paraphyletic lineage An evolutionary lineage that does not include an ancestor and all of its descendants. (See **monophyletic lineage**.)

parental care model A hypothesis proposing that endothermy is a consequence of selection acting on the effect of high maternal body temperature on embryonic development and parental care of young. (See **aerobic scope model, thermogenic opportunity model, warmer is better model**.)

parsimony In phylogenetic systematics, the branching sequence that requires the fewest changes to go from the ancestral condition to the derived condition.

parthenogenesis Reproduction by females without fertilization of the ova by males.

pelage A coat of fur, made up of closely spaced individual hairs.

pelvic patch A vascularized area in the pelvic region of anurans where most of the uptake of water occurs.

pennaceous feather A feather having a central shaft with a vane on each side.

perichondral bone Bone that forms in the perichondral membrane around cartilage or bone.

Perissodactyla An order of extant hoofed mammals, the odd-toed ungulates, having either three toes (rhinos and tapirs) or one toe (horses, asses, and zebras).

Phanerozoic eon The last 541 million years of Earth's 4.6-billion-year history, including the Paleozoic, Mesozoic, and Cenozoic eras. The period during which multicellular life burgeoned and radiated. (See **Precambrian**.)

pharyngeal arches The gill supports between the pharyngeal gill slits.

pharyngeal jaws Dermal tooth plates in the back of the pharynx in neopterygian fishes that are now fused to mobile gill arch elements and act as an additional pair of jaws for food manipulation and processing.

pharyngeal pouches Outpocketings between the pharyngeal arches.

pharyngeal slits Openings in the pharynx that were ancestrally used in filter feeding.

pharynx The throat region.

phenotypic plasticity The ability of a genotype to produce different phenotypes as a result of epigenetic modification of developmental processes.

photophore A cell specialized to produce light.

phylogenetic systematics The grouping of taxa according to the evolutionary relationships.

phylogeny Evolutionary relationships based on the branching sequence of lineages.

physoclistous Of fishes, lacking a connection between the air bladder and the gut.

physostomous Of fishes, having a connection between the air bladder and the gut.

pineal body A light-sensitive structure in the dorsal part of the forebrain, sometimes known as the "third eye"; also known as the pineal gland or epiphysis. Produces the hormone melatonin.

pineal organ A light-sensitive organ and/or endocrine gland in the brain.

pinna The external ear, seen in therian mammals.

pituitary gland An endocrine organ formed in part from a ventral outgrowth of the midbrain that is involved in neural-hormonal integration.

placentals A monophyletic lineage of extant mammals that retain the fetus in the uterus until it has reached an advanced stage of development. (See **Eutheria**.)

placoid scales The scales of chondrichthyans, which are formed from an outer layer of enameloid and an inner layer of dentine. Also known as dermal denticles, because they resemble teeth.

plastron The ventral portion of the shell of a turtle. (See **carapace**.)

plate tectonics The mechanism by which continental drift occurs by the movement of the underlying tectonic plates.

plesiadapiforms Stem primates.

plesiomorphy A character that is unchanged from its ancestral condition; i.e., an ancestral character.

pleurodires Turtles that bend the neck horizontally to retract the head. (See cryptodires.)

polyphyodont The condition of having multiple sets of replacement teeth per lifetime. (See **diphyodont**.)

postcranial pneumatization Air spaces in the bones of the axial skeleton.

Precambrian The more than 4 billion years preceding the Phanerozoic. Covering more than 85% of Earth's planetary history, the Precambrian encompasses the lifeless Hadean and the advent of the earliest life in the Archean and Proterozoic. (see **Phanerozoic**.)

precocial Of neonates, well developed and capable of locomotion soon after birth or hatching. (See **altricial**.)

premaxilla The anterior tooth-bearing dermal bone in the upper jaw of osteichthyans.

primaries The feathers on the outmost portion of the wing.

proavians Derived nonavian theropods ("almost birds").

prolonged breeding A mating system in which the breeding season can last for months. (See **explosive breeding**.)

proprioception The neural mechanism that senses the positions of the limbs in space.

protandry A pattern of life history in which an individual begins life as a male and later transforms to a female. (See **protogyny**.)

proteroglyphous Of snakes, having enlarged teeth (fangs) in the front of the upper jaw. (See **opisthoglyphous, solenoglyphous**.)

protogyny A pattern of life history in which an individual begins life as a female and later transforms to a male. (See **protandry**.)

Protostomia Animals in which the blastopore becomes the mouth. (See **Deuterostomia**.)

Prototheria One of the traditional three major subgroups of mammals—includes the monotremes. (See **Allotheria, Theria**.)

protrusible jaws Jaws that can be extended forward from the head, in teleost fishes aided by the loose attachment to the skull of the dermal elements of the upper jaw (premaxilla and maxilla).

proventriculus The anterior glandular stomach of birds and other archosaurs. (See **gizzard**.)

proximal convoluted tubule The portion of a mammalian nephron adjacent to the glomerulus that changes the concentration of the ultrafiltrate by actively transporting Na+.

pterylae Tracts of follicles from which feathers grow.

pycnofibers Fine, hairlike projections on the skin of some pterosaurs.

pygostyle The fused caudal vertebrae of a bird.

R

rachis The central shaft of a feather.

ram ventilation Creation of a respiratory current across the gills created by a fish swimming with its mouth open.

rectrices (singular **retrix**) Tailfeathers.

regional heterothermy Maintaining different tissue temperatures in different parts of the body.

remiges (singular **remix**) Wing feathers.

rete mirabile "Wonderful net"; intertwined capillaries that exchange heat or dissolved substances between countercurrent flows.

rod cells Photoreceptors in the retina sensitive to a wide range of wavelengths; used to perceive low levels of light but not for high visual acuity.

S

Sauropsida The lineage of amniotes represented by extant nonavian reptiles and birds and their extinct relatives. (See **Synapsida**.)

sebaceous gland Skin gland found over the surface of the body; sebaceous glands produce an oily secretion that lubricates hairs.

secondaries The inner wing feathers. (See **primaries**.)

secondary lamellae Microscopic projections from the gill filaments within which gas exchange occurs.

selenodont Lophed molar teeth with crescentic lophs that run in an anterior-posterior direction along the tooth; seen in camelids, ruminant artiodactyls, and the koala.

semiplume A feather that combines a large rachis with an entirely plumulaceous vane.

shared derived character See **synapomorphy**.

simultaneous hermaphroditism A pattern of life history in which an individual has functional ovaries and testes and can function as a male or female in reproduction.

sinus venosus The posteriormost chamber of the heart of nonamniotes; receives blood from the systemic veins.

sister group The monophyletic lineage that is most closely related to the monophyletic lineage being discussed.

social monogamy A breeding system in which a male and female mate and cooperate to raise their young, but either or both of them may have extra-pair matings as well. (See **genetic monogamy**.)

solenoglyphous Of snakes, having enlarged teeth (fangs) in the front of the upper jaw that rotate when the mouth is open. (See **opisthoglyphous**, **proteroglyphous**.)

solute A substance dissolved in a liquid.

somite A member of a series of paired segments of the embryonic dorsal mesoderm of vertebrates.

sonations Nonvocal sounds used for communication.

spermatophore A packet of sperm transferred from male to female during mating of most salamanders.

splanchnocranium The pharyngeal skeleton associated with the gills.

stem group Extinct forms in a lineage that lack some of the derived characters that define the **crown group**.

stenohaline Capable of living only within a narrow range of salinities. (See **euryhaline**.)

stereocilia Hair cells within the cupula of a neuromast organ.

symplesiomorphy An ancestral character that is shared by two or more taxa.

synapomorphy A derived character shared by two or more taxa and postulated to have been inherited from their common ancestor.

synapsid Describes a skull with a single lower temporal fenestration on each side.

Synapsida The lineage of amniotes represented by extant mammals and their extinct relatives. (See **sauropsids**.)

syrinx The site of vocalizations by birds; a unique structure that lies at the junction of the trachea with the two bronchi.

systematics The evolutionary classification of organisms.

T

tapetum lucidum Shiny crystals of guanine behind the retina that reflect light back through the retina.

tarsometatarsus A bone formed by fusion of the distal tarsal elements with the metatarsals in birds and some nonavian dinosaurs. (See **tibiotarsus**.)

taxon (plural **taxa**) A group of organisms of any rank—e.g., species, genera, families, etc.

taxonomy The discipline that assigns names to organisms.

temperature-dependent sex determination (TSD) A non-genetic sex-determining mode in which the temperature of an embryo during development determines the sex of the individual.

temporal fenestrations Openings in the skull of amniote tetrapods, through which muscles pass from the skull roof to the lower jaw.

temporal heterothermy Maintaining different body temperatures at different times, as in hibernation.

Theria One of the traditional three major subgroups of mammals; includes the Metatheria (marsupials and their extinct stem group relatives) and the Eutheria (placentals and their extinct stem group relatives). (See **Allotheria** and **Prototheria**.)

thermogenic opportunity model A hypothesis proposing that the adaptive value of endothermy lies in permitting nocturnal activity. (See **aerobic scope model**, **parental care model**, **warmer is better model**.)

thermoregulation Control of body temperature.

tibiotarsus A bone formed by fusion of the tibia and proximal tarsal elements in birds and some nonavian dinosaurs. (See **tarsometatarsus**.)

tidal ventilation The passage of water or air in through the mouth or nose, and then exiting back out again through the mouth or nose. (See **flow-through ventilation**.)

tooth whorl A structure in the jaws of extant cartilaginous fishes and some extinct forms that comprises rows of replacing teeth that are not embedded into the jaw itself but form in the skin.

tracheophytes Vascular plants; land plants that have internal channels for the conduction of water. (See **Bryophytes**.)

tribophenic molars Three-cusped molars well adapted to cutting and shearing; they are characteristic of therian mammals.

turbinates Scroll-like bones in the nasal passages covered by moist tissue that warms and humidifies air on inspiration and recovers water and heat on expiration.

U

uncinate processes Bony projections of the ribs of birds on which respiratory muscles insert.

urea cycle The enzymatic pathway by which urea is synthesized from ammonia

ureotely Excretion of nitrogenous wastes primarily as urea. (See **ammonotely, uricotely**.)

uricotely Excretion of nitrogenous wastes primarily as salts of uric acid. (See **ammonotely, ureotely**.)

Urochordata One of two groups of extant non-vertebrate chordates, the tunicates. (See **Cephalochordata**.)

uropatagium The posterior wing present in some pterosaurs.

urostyle A solid rod of bone formed by fused posterior vertebrae.

V

vane The aerodynamic surface of a feather formed by interlocking barbules on the barbs that extend from the rachis.

ventricle A chamber; the ventricle of the heart is the portion that applies force to eject blood.

vestibular apparatus The sensory structures within the inner ear that monitor and control balance and orientation.

vibrissae Sensory hairs around the snout, also known as whiskers.

vicariance The movement of organisms on drifting continental blocks.

viviparity The reproductive mode in which the mother gives birth to fully formed young.

vomeronasal organ (also knowns as **Jacobson's organ**) A chemosensory organ in the roof of the mouth of tetrapods.

W

warmer is better model A hypothesis proposing that the adaptive value of endothermy lies in the faster rates of biochemical and physiological processes at high temperature. (See **aerobic scope model, parental care model, thermogenic opportunity model**.)

Weberian apparatus In otophysian fishes, small bones that connect the gas bladder to the inner ear.

wing loading The weight of a bird divided by the wing area.

X

Xenarthra The grouping of endemic South American mammals, including the orders Pilosa (sloths and anteaters) and Cingulata (armadillos).

Y

yolk sac An extraembryonic membrane present in all vertebrates that encloses the yolk.

Z

zygapophyses Articular processes formed by the neural arch of a vertebra.

zygodactylous A type of foot in which the toes are arranged in two opposable groups.

zygomatic arch The bowed-out lower border of the synapsid temporal opening seen in cynodonts and mammals. In humans this is the cheek bone.

Illustration Credits

Chapter 1 **1.5A,B** Fondon JW, Garner HR. 2004. Molecular origins of rapid and continuous morphological evolution. *Proceedings of the National Academy of Sciences USA* 101: 18058–18063. Copyright © 2004 National Academy of Sciences USA. **1.8B,C** Abzhanov A, Protas M, Grant BR, Grant PR, Tabin CJ. 2004. Bmp4 and morphological variation of beaks in Darwin's finches. *Science* 305: 1462–1465.

Chapter 2 **2.3A** Patel NH. 2004. Evolutionary biology: Time, space and genomes. *Nature* 431: 28–29. Reprinted by permission of Springer Nature © 2004. **2.5A** Gilbert SG, Barresi MJF. 2016. *Developmental Biology*, 11th Ed. Sinauer Associates, Sunderland, MA. **2.6** Kingsley JS (1854–1929). 1912. *Comparative Anatomy of Vertebrates.* Philadelphia, P. Blakiston's Son & Co. **2.8** Gilbert SG, Barresi MJF. 2016. *Developmental Biology*, 11th Ed. Sinauer Associates, Sunderland, MA **2.9A** Lankester ER, Ridewood WG. 1908. *Guide to the Gallery of Fishes.* British Museum (Natural History), Department of Zoology. **2.9B** Lankester ER, Ridewood WG. 1908. *Guide to the Gallery of Fishes.* British Museum (Natural History), Department of Zoology; *and* Maisey JG. 1996. *Discovering Fossil Fishes.* Basic Books, New York. **2.10** Bone Q, Marshall MA. 1982. *Biology of Fishes.* Blackie, Glasgow. **2.14** Kardong KV. 2012. *Vertebrates: Comparative Anatomy, Function, Evolution*, 6th Ed. McGraw Hill, New York.

Chapter 3 **3.1B** Elliott DK. 1987. A reassessment of *Astraspis desiderata*, the oldest North American vertebrate. *Science* 237: 190–192. **3.2A** Hildebrand AM, Goslow G. 1998. *Analysis of Vertebrate Structures*, 5th Ed. John Wiley & Sons, New York. **3.2B,C,D** Smith HM 1960. *Evolution of Chordate Structure: An Introduction to Comparative Anatomy.* Brooks/Cole, a division of Cengage Learning, Boston. **3.3A** Jarochowska E, Munnecke A. 2016. Late Wenlock carbon isotope excursions and associated conodont fauna in the Podlasie Depression, eastern Poland: A not-so-big crisis? *Geological Journal* 51: 683–703. **3.3B** Purnell MA. 1994. Skeletal ontogeny and feeding mechanisms in conodonts. *Lethaia* 27(2): 129–139; *and* Aldridge RJ. 1987. *Palaeobiology of Conodonts.* Ellis Horwood Ltd. for the British Micropalaeontological Society, London. **3.6** McCoy VE and 15 others. 2016. The "Tully monster" is a vertebrate. *Nature* 532: 496–499. **3.7A,C** Martini FH. 1998. Secrets of the slime hag. *Scientific American* 279(10): 70–75. **3.9A-D** Moy-Thomas JA, Miles RS. 1971. *Palaeozoic Fishes.* Chapman & Hall, London; 3.9D after Stensio EA. 1932 The cephalaspids of Great Britain. London: British Museum (Natural History), https://archive.org/details/cephalaspidsofgr00sten. **3.11A** Miyashita T. 2016. Fishing for jaws in early vertebrate evolution: A new hypothesis of mandibular confinement. *Biological Reviews* 91: 611–657. **3.12** Gillis JA, Tidswell O. 2017. The origin of vertebrate gills. *Current Biology* 27: 729–732. **3.13** Gai Z, Donoghue PCJ, Zhu M, Janvier P, Stampanoni M. 2011. Fossil jawless fish from China foreshadows early jawed vertebrate anatomy. *Nature* 476: 324–327. Reprinted by permission of Macmillan Publishers Ltd. **3.14A,B** Kuratani S. 2004. Evolution of the vertebrate jaw: Comparative embryology and molecular developmental biology reveal the factors behind evolutionary novelty. *Journal of Anatomy* 205: 335--347, Figure 5. **3.16A,B** Shearman RM, Burke AC. 2009. The laterial somitic frontier in ontogeny and phylogeny. *Journal of Experimental Biology B, Molecular Development and Evolution* 312: 603–612. **3.17** Kusakabe R, Kuratani S. 2005. Evolution and developmental patterning of the vertebrate skeletal muscles: Perspectives from the lamprey. Developmental Dynamics 234: 824–834. **3.18** Friedman M, Sallan LC. 2012. Five hundred million years of extinction and recovery: A Phanerozoic survey of large-scale diversity patterns in fishes. *Palaeontology* 55: 707–742. **3.19A, 3.20A** Moy-Thomas JA, Miles RS. 1971. *Palaeozoic Fishes.* Chapman & Hall, London. **3.20B** Woodward AS. 1891. Catalogue of fossil fishes in the British Museum (Natural History), Pt. II. London: British Museum (Natural History).

Chapter 4 **4.1A, 4.2A** Townsend DW. 2012. *Oceanography and Marine Biology.* Sinauer Associates, Sunderland, MA. **4.4C,D** Flock A. 1967. Ultrastructure and function in the lateral line organs. In Cahn PH (ed.), *Lateral Line Detectors: Proceedings.* Indiana University Press, Bloomington. **4.5** Schwartz E. 1974. *Handbook of Sensory Physiology*, Volume 3, Part 3. Fessard A (ed.). Springer Verlag, Berlin & New York. **4.6** Moller P. 1995. *Electric Fishes: History and Behavior.* London: Chapman & Hall. **4.6** Fish from Bennett MVL. 1970 Comparative physiology: Electric organs. *Annual Review of Physiology* 32: 471–528. **4.7A,B** Catania KC. 2015. Electric eels concentrate their electric fields to induce involuntary fatigue in struggling prey. *Current Biology* 25: 2889–2898. **4.7C** Catania KC. 2016. Leaping eels electrify threats, supporting Humboldt's account of a battle with horses. Proceedings of the National Academy of Sciences USA 113: 6979–6984. **4.8B,C** Josberger EE and 6 others. 2016. Proton conductivity in ampullae of Lorenzini jelly. *Science Advances* 13 May 2016: E1600112. © The Authors, some rights reserved; exclusive licensee American Association for the Advancement of Science. CC BY-NC 4.0. Permission granted by AAAS. **4.9A–E** Kalmijn AJ. 1971. The electric sense of sharks and rays. *Journal of Experimental Biology* 55: 371–383.

Chapter 5 **5.4** Scotese CR, McKerrow WS. 1990. Revised World Maps and Introduction. *The Geological Society Memoir 12, 1-21: Palaeozoic Palaeogeography and Biogeography.* London. **5.6** Benton M. 2014. *Vertebrate Palaeontology*, 4th Edition. Wiley-Blackwell. ISBN: 978-1-118-40764-6. **5.7** Smith RMH, Botha-Brink J. 2014. Anatomy of a mass extinction: Sedimentological and taphonomic evidence for drought-induced die-offs at the Permo-Triassic boundary in the main Karoo Basin, South Africa. *Palaeogeography, Palaeoclimatology, Palaeoecology* 396: 99–118.

Chapter 6 **6.4A,B** Moy-Thomas JA, Miles RA. 1971. *Paleozoic Fishes.* Springer, New York. **6.4B** (fish), Moy-Thomas JA. 1971.

Subclass Chondrichthyes. Infraclass Elasmobranchii. In: *Palaeozoic Fishes*, Springer, New York. **6.4C** Benton MJ. 2005. *Vertebrate Paleontology*, 3rd ed. Blackwell Science Ltd.; Zangerl R. 2004. Chondrichthyes I: Paleozoic Elasmobranchi. *Handbook of Paleoichthyology*. Verlag Dr. Friedrich Pfeil, Munich. **6.4D** Lund R. 1986. On *Damocles serratus*, nov. ge. et sp. (Elasmobranchii: Cladodontida) from the Upper Mississippian Bear Gulch Limestone of Montana. *Journal of Vertebrate Paleontology* 6: 12–19, ©1986 The Society for Vertebrate Paleontology. **6.5A,B** Ramsay JB and 6 others. 2015. Eating with a saw for a jaw: Functional morphology of the jaw and tooth-whorl in *Helicopron davisii. Journal of Morphology* 276: 47–64. **6.6** Moy-Thomas JA, Miles RA. 1971. *Paleozoic Fishes*. Springer Verlag, New York.

Chapter 7   7.1 Vélez-Zuazo X, Agnarsson I. 2011. Shark tales: A molecular species-level phylogeny of sharks (Selachimorpha, Chondrichthyes). *Molecular Phylogenetics and Evolution* 58: 207–217. Aschliman NC and 5 others. 2012. Body plan convergence in the evolution of skates and rays (Chondrichthyes: Batoidea) *Molecular Phylogenetics and Evolution* 63: 28–42. **7.2A,B,D,E** Wilga CD. 2002. A functional analysis of jaw suspension in elasmobranchs. *Biological Journal of the Linnean Society* 75: 483–502. Copyright © 2002 by The Linnean Society of London. **7.2C** Goodrich ES. 1909. *A Treatise on Zoology*. Part IX: Vertebrata Craniata (First Fascicle: Cyclostomes and Fishes). Sir Ray Lankester (ed). Adam and Charles Black Publishers. London. Agent: The Macmillan Company, New York. **7.4** Townsend DW. 2012. *Oceanography and Marine Biology*. Sinauer Associates, Sunderland, MA. **7.6** Gardiner JM, Atema J, Hueter RE, Motta PJ. 2014. Multisensory integration and behavioral plasticity in sharks from different ecological niches. *PLoS One* 9(4): e93036; CC-BY 4.0. **7.7** Carey FG, Casey JG, Pratt HL, Urquhart D, McCosker JE. 1985. Temperature, heat production and heat exchange in lamnid sharks. *Memoirs of the Southern California Academy of Sciences* 9: 92–108. **7.8** Whitnack LB, Simkins DJ Jr, Motta PJ. 2011. Biology meets engineering: The structural mechanics of fossil and extant shark teeth. *Journal of Morphology* 272: 169–179; Whitenack LB, Motta PJ. 2010. Performance of shark teeth during puncture and draw: Implications for the mechanics of cutting. *Biological Journal of the Linnean Society* 100: 271–286, © The Linnean Society of London, with permission of John Wiley & Sons Ltd. **7.13** Didier DA.1995. Phylogenetic systematics of extant chimaeroid fishes (Holocephali, Chimaeroidea). *American Museum Novitates* 3119, 89 pp. **7.14** Klimley P. 1999. Sharks beware. *American Scientist* 87: 488–491 (Nov/Dec 1999).

Chapter 8   8.5 Gill EL. 1923. Permian fishes of the genus *Acentrophorus. Proceedings of the Zoological Society of London 1923*, pp. 19–40 **8.6A** Gunther ACLG.1880. *Introduction to the Study of Fishes*. Adam and Charles

Black, Edinburgh. **8.7** Bellwood DR, Goatley CHR, Bellwood O, Delbarre DJ, Friedman M. 2015. The rise of jaw protrusion in spiny-rayed fishes closes the gap on elusive prey. *Current Biology* 25: 2696–2700.

Chapter 9   9.3C Heiligenberg, W. 1977. *Principles of Electrolocation and Jamming Avoidance in Electric Fish: A Neurological Approach*. Springer-Verlag, Berlin. **9.6** Bradbury, JW, Vehrencamp, SL. 2011. *Principles of Animal Communication*, 2nd ed. Sinauer Associates, Sunderland, MA. **9.8D** Okada N, Takagi Y, Tanaka M, Tagawa M. 2003. Fine structure of soft and hard tissues involved in eye migration in metamorphosing Japanese flounder (*Paralichthys olivaceus*). *Anatomical Record* 273A: 663–668. **9.10** Lindsey CC. 1978. Form, function and locomotory habits in fish. *Fish Physiology* 7: 1–100 **9.18C** Schnell NK, Johnson GD. 2017. Evolution of a functional head joint in deep-sea fishes (Stomiidae). *PLoS ONE* 12(2): e0170224. CC-BY 4.0. **9.19** Pietsch T. 2005. Dimorphism, parasitism, and sex revisited: Modes of reproduction among deep-sea ceratioid anglerfishes (Teleostei: Lophiiformes). Ichthyological Research 52: 207–236. **9.21** Carey FG, Teal JM. 1966. Heat conservation in tuna fish muscle. *Proceedings of the National Academy of Sciences USA* 56: 1464–1469. **9.22B, 9.23** Wegner, N, Snodgrass OE, Dewar H, Hyde JR. 2015. Whole-body endothermy in a mesopelagic fish, the opah, *Lampris guttatus. Science* 348: 786–789. Data on tuna from Childers JC, Snyder S, Kohin S. 2011. Migration and behavior of juvenile North Pacific albacore (*Thunnus alalunga*). US Department of Commerce, NOAA Fisheries. Publications of the University of Nebraska, Lincoln.

Chapter 10   10.3A (1) Shubin NH, Daeschler EB, Jenkins FA Jr. 2013. Pelvic girdle and fin of *Tiktaalik roseae. Proceedings of the National Academy of Sciences USA* 11: 893–899. (2) Daeschler EB, Shubin NH, Jenkins FA Jr. 2006. Devonian tetrapod-like fish and the evolution of the tetrapod body plan. *Nature* 440: 757–763. **10.3B** Coates MI. 1996. The Devonian tetrapod *Acanthostega gunnari* Jarvik: Postcranial anatomy, basal tetrapod interrelationships, and patterns of skeletal evolution. *Transactions of the Royal Society of Edinburgh: Earth Sciences* 87: 363–421. Copyright © 1996 Royal Society of Edinburgh. **10.3C** Ahlberg PE, Clack JA, Blom H. 2005. The axial skeleton of the Devonian tetrapod *Ichthyostega. Nature* 437: 137–140. **10.4** Long JA. 2011. *The Rise of Fishes: 500 Million Years of Evolution*, 2nd Ed. Johns Hopkins University Press, Baltimore, MD. **10.6A** Saxena A, Cooper KL. 2016. Evolutionary biology: Fin to limb within our grasp. *Nature* 537: 176–177. **10.6B** Left: Ahlberg PE, Milner AR. 1994. The origin and early development of tetrapods. *Nature* 368: 507–514. Right: Clack JA, Coates MI. 1990. Polydactyly in the earliest known tetrapod limbs. *Nature* 347: 66–99. **10.7A** Hohn-Schultz B, Preuschoft H, Witzel U, Distler-Hoffman C. 2013. Biomechanical

and functional preconditions for terrestrial lifestyle in basal tetrapods, with special consideration of *Tiktaalik roseae. Historical Biology, An International Journal of Paleobiology* 25: 167–181. Courtesy of Taylor & Francis Publishers, www.informaworld. com. **10.07B** Kawano SM, Blob RW. 2013. Propulsive forces of mudskipper fins and salamander limbs during terrestrial locomotion: Implications for the invasion of land. *Integrative and Comparative Biology* 53: 283–294, by permission of Oxford University Press. **10.08A–C** Schoch RR. 2014. *Amphibian Evolution The Life of Early Land Vertebrates*. Blackwell, West Sussex, UK. Figure 5.2. **10.15** (1) Kemp TS. 1982. *Mammal-Like Reptiles and the Origin of Mammals*. Academic Press, New York. (2) Romer AS, Price LW. 1940. *Review of the Pelycosauria*. Geological Society of America Special Papers 28: 1–538.

Chapter 11   11.3 Deban SM, Wake DB. Roth G. 1997. Salamander with a ballistic tongue. *Nature* 389: 27–28. **11.12C** Ryan MJ. 1992. *Tungara Frog: a Study in Sexual Selection and Communication*. University of Chicago Press. **11.13B,E, 11.14D,E** Crump ML. 2015. Anuran reproductive modes: Evolving perspectives. i 49: 1–16. **11.17B,C** Taylor EH. 1968. *The Caecelians of the World: A Taxonomic Review*. University of Kansas Press, Lawrence, KS. **11.18** Gilbert SF, Barresi MJF. 2016. *Developmental Biology*, 11th Ed. Sinauer Associates, Sunderland, MA. **11.22A** Taboada C and 9 others. 2017. Naturally occurring fluorescence in frogs. *Proceedings of the National Academy of Sciences USA* 114: 4. **11.22B** Guayasamin JM and 5 others. 2017. A marvelous new glassfrog (*Centrolenidae hyalinobatrachium*) from Amazonian Ecuador. *ZooKeys* 673: 1-20. CC BY 4.0. **11.23** von Byern J and 6 others. 2015. Morphological characterization of the glandular system in the salamander *Plethodon shermani* (Caudata: Plethodontidae). *Zoology* 118: 334–347. **11.25C** Brodie ED Jr, Nussbaum RA, DiGiovanni M. 1984. Antipredator adaptations of Asian salamanders (Salamandridae). *Herpetologica* 40: 56–68. Allen Press Publishing Services. **11.25D–F** Jared C and 6 others. 2015. Venomous frogs use heads as weapons. *Current Biology* 25: 2166–2170. **11.26** IUCN Red List 2016-3.

Chapter 12   12.2A Rockwell H, Evans FG, Pheasant HC. 1938. Comparative morphology of the vertebrate spinal column: Its form as related to function. *Journal of Morphology* 63: 87–117. **12.2B** Cunningham DJ. 1903. *Textbook of Anatomy*. William Wood and Co., New York. **12.4A–D** Schoch RR. 2014. *Amphibian Evolution: The Life of Early Vertebrates*. John Wiley & Sons, Chichester, UK. **12.5A** Carroll RL, Baird DB. 1972. Carboniferous stem-reptiles of the family Romeriidae. *Bulletin of the Museum of Comparative Zoology* 143: 321–363. **12.8** McMahon TA, Bonner JT. 1983. *On Size and Life*. Scientific American Books, New York. **12.12D** Sivak JG. 1976. Optics of the eye of the "four-eyed fish"(*Anableps anableps*). *Vision Research* 16: 531–534, IN6. **12.14A** Hillenius WJ. 1992.

The evolution of nasal turbinates and mammalian endothermy. *Paleobiology* 18: 17–29. **12.17A** Warnecke L, Turner JM, Geiser F. 2008. Torpor and basking in a small arid zone marsupial. *Naturwissenschaften* 95: 73–78. **12.17B** Van Mierop LHS and Barnard S. 1978. Further observations on thermoregulation in the brooding female *Python molurus bivattatus* (Serpentes: Boidae). *Copeia* 1978: 615–621.

Chapter 13   13.4 Estes R, Berberian P. 1970. Paleoecology of a Late Cretaceous vertebrate community from Montana. *Breviora* 343: 1–35. **13.04.14,15** (silhouettes) Csiki-Sava Z. 2016. East Side Story: The Transylvanian latest Cretaceous continental vertebrate record and its implications for understanding Cretaceous–Paleogene boundary events. *Cretaceous Research* 57: 662–698. **13.04.17** (silhouette) Luo ZX. 2007. Transformation and diversification in early mammal evolution. *Nature* 450: 1011–1019. \

Chapter 14   14.1 Carrier DR. 1987. The evolution of locomotor stamina in tetrapods: Circumventing a mechanical constraint. *Paleobiology* 13: 326–341. **14.2** Kemp TS. 2007. Origin of higher taxa: Macroevolutionary process and the case of the mammals. *Acta Zoologica* (Stockholm) 88: 3–22. **14.5A** Overton F, MD. 1897. *Applied Physiology.* American Book Co., New York, Cincinnati, Chicago. **14.6A,B** Sadava DS, Hillis DM, Heller HC, Hacker SD. 2016. *Life: The Science of Biology*, 11th Ed. Sinauer Associates & Macmillan Publishing, New York. **14.9A,B** Tattersall GJ. 2016. Reptile thermogenesis and the origins of endothermy. *Zoology* 119: 403-405. **14.10A,B, 14.11** Sadava DS, Hillis DM, Heller HC, Hacker SD. 2016. *Life: The Science of Biology*, 11th Ed. Sinauer Associates & Macmillan Publishing, New York. **14.12** Hill RW, Wyse GA, Anderson M. 2016. *Animal Physiology*, 4th Edition. Sinauer Associates, Sunderland, MA. **14.13** Davis LE, Schmidt-Nielsen B, Stolte H, Bookman LM. 1976. Anatomy and ultrastructure of the excretory system of the lizard, *Sceloporus cyanogenys*. *Journal of Morphology* 149: 279–326. **14.15** Jerison HJ. 1970. Gross brain indices and the analysis of fossil endocasts. In Noback C, Montagna W (eds.), *The Primate Brain. Advances in Primatology*, Vol. 1. Appleton-Century-Crofts, New York.

Chapter 15   15.2 Nagy KA, Medica PA. 1986. Physiological ecology of desert tortoises in southern Nevada. *Herpetologica* 42(1): 73–92. **15.3C** Nagy KA. 1973. Behavior, diet, and reproduction on a desert lizard, *Sauromalus obesus*. *Copeia* March 5, 1973, 93–102. **15.5** Scholander PF, van Dam L, Kanwisher JW, Hammel HT. 1957. Supercooling and osmoregulation in Arctic fish. *Journal of Cellular Physiology* 49: 5–24. **15.8, 15.9, 15.10** Pough FH. 1980. The advantage of ectothermy for tetrapods. *American Naturalist* 115: 92–112. Copyright © 1980 The University of Chicago Press. **15.10** Pough FH. 1980. The advantage of ectothermy

for tetrapods. *American Naturalist* 115: 92–112. Copyright © 1980 The University of Chicago; Smith FA and 8 others. 2003. Body mass of late Quaternary mammals. *Ecology* 84: 3403; Dunning JB Jr. 2008 *CRC Handbook of Avian Body Masses*, 2nd. Ed. CRC Press, Boca Raton, FL.

Chapter 16   16.1 Zangerl R. 1969. The turtle shell. In Gans C (ed.), *Biology of the Reptilia, Vol. 1: Morphology A*, pp. 311–320. Academic Press, London and New York. **16.2A,C** Werneburg I, Wilson LAB, Parr WCH, Joyce WG. 2015. Evolution of neck vertebral shape and neck retraction at the transition to modern turtles: An integrated geometric morphometric approach. *Systematic Biology* 64: 187–204. **16.5** Pough FH and 5 others. 2016. *Herpetology*, 4th ed. Sinauer Associates, Sunderland, MA. **16.6A** Bull JJ. 1980. Sex determination in reptiles. *Quarterly Review of Biology* 55: 3–21. **16.6B** Yntema CL. 1976. Effects of incubation temperature on sexual differentiation in turtles. *Journal of Morphology* 150: 453–462. **16.6C** Bull JJ, Vogt RC. 1979. Temperature-dependent sex determination in turtles. *Science* 206: 1186–1188. **16.9** Luschi P, Hays GC, Del Seppia C, Marsh R, Papi F. 1998. The navigational feats of green sea turtles migrating from Ascension Island investigated by satellite telemetry. *Proceedings of the Royal Society B* 265: 2279–2284. **16.10** Lohmann KJ, Cain SD, Dodge SA, Lohmann CMF. 2001. Reginal magnetic fields and navigational markers for sea turtles. *Science*: 294: 364–366; Fuxjager MJ, Eastwood BS, Lohmann KJ. 2011. Orientation of hatchling loggerhead sea turtles along a transoceanic magratory pathway. *Journal of Experimental Biology* 214: 2504–2508.

Chapter 17   17.4A, C Gans C. 1974. *Biomechanics: An Approach to Vertebrate Biology.* Williams & Wilkins, Philadelphia. **17.6B** Astley, HC, Jayne, BC. 2009. Arboreal habitat structure affects the performance and modes of locomotion of corn snakes (*Elaphe guttata*). *Journal of Experimental Zoology A: Ecological Genetics and Physiology* 311: 207–216. **17.10** Gans C. 1961. The feeding mechanism of snakes and its possible evolution. *American Zoologist* 1: 217–227. **17.17** Echelle, AA, Echelle AF, Fitch HS. 1971. A comparative analysis of aggressive display in 9 species of Costa Rican *Anolis*. *Herpetologica* 27: 271–288. **17.22** Stevenson RD, Peterson CR, Tsuji, J. 1985. The thermal dependence of locomotion, tongue-flicking, digestion, and oxygen consumption in the wandering garter snake. *Physiological and Biochemical Zoology* 565: 48–57. Copyright © 1985 The University of Chicago Press. **17.23** Sinervo B and 26 others. 2010. Erosion of lizard diversity by climate change and altered thermal niches. *Science* 328: 894–899.

Chapter 18   18.1 Wermuth H, Fuchs K. 1978. *Bestimmen von Krokodelin und iher Haute.* Gustav Fischer Verlag, Stuttgart. **18.2D** Grigg G. 2015. *Biology and Evolution of Crocodylians.* Illustrations by David Kirshner. Cornell University Press,

Ithaca, NY. **18.4D** Krause DW and 5 others. 2010. Overview of the discovery, distribution, and geological context of *Simosuchus clarki* (Crocodyliformes: Notosuchia) from the Late Cretaceous of Madagascar. *Journal of Vertebrate Paleontology* 30/sp1: 4–12; sculpture by Boban Filipovic, photo by Luci Betti-Nash. **18.6A–D** Walmsley CW and 8 others. 2013. Why the long face? The mechanics of mandibular symphysis proportions in crocodiles. *PLoS ONE* 8(1): e53873. CC BY 4.0. **18.7A** Grigg G. 2015. *Biology and Evolution of Crocodylians.* Illustrations by David Kirshner. Cornell University Press, Ithaca, NY. **18.8C** Dinets V, Brueggen JC, Brueggen JD. 2015. Crocodilians use tools for hunting. *Ecology & Evolution* 27: 01/02/15. Reprinted by permission of Taylor & Francis Ltd. http://www.tandfonline.com.

Chapter 19   19.3 From https://archosaurmusings.wordpress.com/2010/10/01/interview-with-jeff-martz/. Reprinted by permission of Jeffrey Martz. **19.9** Zhou CF and 5 others. 2017. Earliest filter-feeding pterosaur from the Jurassic of China and ecology evolution of Pterodactyloidea. *Royal Society Open Science* 4: 160672. CC BY 4.0 **19.12** Benton MJ, Forth J, Langer MC. 2014. Models for the rise of dinosaurs. *Current Biology* 24: R87–R95. CC BY 3.0. **19.14, 19.15, 19.16, 19.19** Fastovsky DE, Weishampel DB. 2016. *Dinosaurs: A Concise Natural History*, 3rd ed. Cambridge University Press, Cambridge **19.20A–C** Peterson JE, Dischler C, Longrich NR. 2013. Distributions of cranial pathologies provide evidence for head-butting in dome-headed dinosaurs (Pachycephalosauridae). *PLoS ONE* 8(7): e68620. **19.21A–E** Charig A. 1979. *A New Look at the Dinosaurs.* Mayflower Books, New York. **19.22A,B** Fastovsky DE, Weishampel DB. 2005. The Evolution and Extinction of the Dinosaurs, 2nd Ed. Illustrated by John Sibbick. Copyright © Cambridge University Press 1996, 2005. **19.23A,B** Osborn HF, Mook CC. 1921. *Camarasaurus, Amphicoelias*, and other sauropods of Cope. *Memoirs of the American Museum of Natural History*, new series, vol. 3, pt. 3. **19.23C,D** Hatcher JB. 1901. *Diplodocus* (Marsh): Its osteology, taxonomy, and probable habits, with a restoration of the skeleton. *Memoirs of the Carnegie Museum* 1: 1–63. **19.24A,B** Salisbury SW and 5 others. 2016. The dinosaurian ichnofauna of the Lower Cretaceous (Valanginian–Barremian) Broome Sandstone of the Walmadany area (James Price Point), Dampier Peninsula, Western Australia. *Journal of Vertebrate Paleontology* 36 , Iss. Sup1. CC BY-NC-ND 4.0 with permission of Steven W. Salisbury. **19.25** Fastovsky DE, Weishampel DB. 2016. *Dinosaurs: A Concise Natural History*, 3rd ed. Cambridge University Press, Cambridge.

Chapter 20   20.2 Humphries MM, Careau V. 2011. *Heat for Nothing or Activity for Free? Evidence and Implications of Activity-Thermoregulatory Heat Substitution.* Oxford University Press, Oxford. **20.3A** Scholander

PF, Hock V, Walters V, Johnson F, Irving L. 1950. Heat regulation in some arctic and tropical mammals and birds. *Biological Bulletin* 99: 237–258. **20.3B** West, GC. 1972. Seasonal differences in resting metabolic rate of Alaska ptramigan. *Comparative Biochemistry and Physiology* 42A: 867–876. **20.7B,C** Wang LCH, Hudson JW. 1978. *Strategies in Cold: Natural Torpidity and Thermogeneisis.* Academic Press. By permission of L. C. H. Wang. **20.8** Beuchat C.A. 1990. Body size, medullary thickness, and urineconcentrating ability in mammals. *American Journal of Physiology and Regulatory Integrative Comparative Physiology* 258: R298-R308. **20.11B** Schmidt-Nielsen K, Schmidt-Nielsen B, Jarnum SA, Houpt TR. 1956. Body temperature of the camel and its relation to water economy. *American Journal of Physiology* 188: 103–112.

Chapter 21   21.1 Silhouettes (all CC BY Unported): 1, Dann Pigdon. Public Domain Dedication 1.0; 2, Matt Martyniuk. CC BY NC-SA 3.0; 3, Matt Martyniuk. CC BY 3.0; 4, Nobu Tamura, vectorized by T. Michael Keesey. CC BY 3.0; HuttyMcphoo, vectorized by T. Michael Keesey. CC BY-SA 3.0. **21.2B** Vogt C. 1880. *Achaeopteryx macrura,* an intermediate form between birds and reptiles. *Ibis* 4: 434–456. archosourmusings.wordpress.com/2008/10/26/the-changing-legs-of-archaeopteryx/. **21.2C** Carney RM, Vinther J, Shawkey MD, D'Alba L, Ackermann J. 2012. New evidence on the colour and nature of the isolated *Archaeopteryx* feather. *Nature Communications* 3 no. 637. **21.3A–C** Chen, P, Dong Z, Zhen S. 1998. An exceptionally well-preserved theropod dinosaur from the Yixian formation of China. *Nature* 391: 147-152. **21.4** Tree: Witmer LM. 2009. Fuzzy origins for feathers. *Nature* 458: 293–295; Drawings: Xu X, Zheng X, Yu H. 2010. Exceptional dinosaur fossils show ontogenetic development of early feathers. *Nature* 464: 1338–1341. **21.6A,B** Wiemann J and 7 others. 2017. Dinosaur origin of egg color: Oviraptors laid blue-green eggs. *PeerJ* 5: e3706, CC BY 4.0. **21.7** Graph: Lee MSY, Cau A, Naish D, Dke GJ. 2014. Sustained miniaturization and anatomical innovation in the dinosaurian ancestors of birds. *Science* 345: 562–566. Silhouettes (all CC BY Unported): 1,2 Scott Hartman, CC BY NC-SA 3.0; 3, Ian Reid, CC BY 3.0; 4,5, Matt Martyniuk, CC BY NC-SA 3.0; 6, Gareth Monger, CC BY 3.0. **21.8** Padian K. 1996. Early bird in slow motion. *Nature* 382: 400-401. **21.10** Dial KP. 2003. Wing-assisted incline running and the evolution of flight. *Science* 299: 402–404. **21.11A** Sereno PC, Chenggang R. 1992. Early evolution of avian flight and perching: New evidence from the Lower Cretaceous of China. *Science* 255: 845–848. **21.12** Feduccia A. 1995. Explosive evolution in Tertiary birds and mammals. *Science* 267: 637–638.

Chapter 22   22.1 Tree: Mayr G. 2014. The origins of crown group birds: Molecules and fossils. *Frontiers in Palaeontology* 57: 231–242; Prum RO and 6 others. 2015. A comprehensive phylogeny of birds (Aves) using targeted next-generation DNA sequencing. *Nature* 526: 569–573. Silhouettes: 1, 9, 11, 12, 14 Ferran Sayol, PDD (Public Domain Dedication) 1.0; 2, John Gould, vectorized by T. Michael Keesey, PDD 1.0; 3,Catherine Yasuda, PDD 1.0; 4,5,7, 20, 22 Steven Traver, PDD 1.0; 6, Annalee Blysse, PDD 1.0; 8, Rebecca Groom, CC BY 3.0 Unported; 10, Public Domain Mark 1.0; 13, 15, Michael Scroggie, PDD 1.0; 16, Estelle Bourdon, CC BY-SA 3.0 Unported; 17, Gorden E. Robertson, CC BY-SA 3.0 Unported; 18, Lip Kee Yap (modified), CC BY-SA 3.0 Unported; 19, Matt Martyniuk, CC BY 3.0 Unported; 21, Darren Naish, vectorized by T. Michael Keesey, CC BY 3.0 Unported; 23, Matt Martyniuk, vectorized by T. Michael Keesey, CC BY-SA 3.0 Unported **22.2A,B** Lucas AM, Stettenheim PR. 1972. *Avian Anatomy and Integument. Agricultural Handbook 362.* US Department of Agriculture, Washington, DC. **22.6A** Gill F. 2007. *Ornithology,* 3rd Ed. WH Freeman, New York. **22.9B** Lee S, Kim J, Park H, Jablonski PG, Choi H. 2015. The function of the alula in avian flight. *Scientific Reports* 5, No. 9914. **22.10A** Storer TI. 1943. *General Zoology.* McGraw-Hill, New York. **22.10B** Feduccia A. 1980. *The Age of Birds.* Harvard University Press, Cambridge, MA. **22.11, 13** Peterson RT. 1978. *The Birds,* 2nd Ed. Life Nature Library, Time Life Books. **22.21A,B** Leiber A. 1907. Vergleichende Anatomie der Spetzunge. *Zoologica* 20: 1–79. **22.21C** Eckstorm FH. 1901. The Woodpeckers. Houghton Mifflin, Boston. **22.24B** Norberg RA. 1978. Skull asymmetry, ear structure and function, and auditory localization in Tengmalm's owl, *Aegolius funereus* (Linne). *Philosophical Transactions of the Royal Society of London B* 282: 325–410. **22.25** Corfield JR and 5 others. 2015. Diversity in olfactory bulb size in birds reflects allometry, ecology, and phylogeny. *Frontiers in Neuroanatomy* 9: 102. **22.27B,C** D'Alba L, Kieffer L, Shawkey MD. 2012. Relative contributions of pigements and biophotonic nanostructures to natural color production: A case study in budgerigar (*Melopsittacus undulatus*) feathers. *Journal of Experimental Biology* 215: 1272–1277. **22.28** MacDougall-Shackleton EA, MacDougall-Shackleton SA. 2001. Cultural and genetic evolution in mountain white-crowned sparrows. Evolution 55: 2568–2575. **22.29A** Rahn et al. 1979. How bird eggs breathe. *Scientific American* 240(2): 46–55. **22.29B** Carey C. 1983. Structure and function of avian eggs. In Johnston RF (ed) *Current Ornithology*, Vol 1. *Current Ornithology* 351: 69–103. Springer, Boston, MA. **22.36B,C** Keeton WT. 1969. Orientation by pigeons: Is the sun necessary? *Science* 165: 922–928. **22.37** Marshall AJ, Serventy DL. 1956. The breeding cycle of the Short-Tailed Shearwater, *Puffinus tinuirostris,* in relation to trans-equatorial migration and its environment. *Proceedings of the Zoological Society of London* 127: 489–510. Copyright © 1956 by the Zoological Society of London, by permission of John Wiley & Sons. **22.38A,B** Templeton CN, Zollinger SA, Brumm H. 2016. Traffic noise drowns out great tit alarm calls. *Current Biology* 26: R1167–R1176. **22.39B** Nemeth E, Brumm H. 2010. Birds and anthropogenic noise: Are urban songs adaptive? *American Naturalist* 176: 465–475. **22.39C** Nemeth E and 6 others. Bird song and anthropogenic noise: Vocal constraints may explain why birds sing higher-frequency songs in cities. *Proceedings of the Royal Society B:* 280: 21022798.

Chapter 22   23.10 Marshall LG. 1988. Land mammals and the Great American Interchange. *American Scientist* 76, July/August: 380–3833

Chapter 24   Opener Fernandez V and 6 others. 2013. Synchrotron reveals Early Triassic odd couple: Injured amphibian and aestivating therapsid share burrow. Butler RJ, ed. *PLoS ONE* 8(6): e64978. **24.6A–E** Romer AS. 1933. *Vertebrate Paleontology.* University of Chicago Press, Chicago. **24.6F** Bonaparte JF, Martinelli AG, Shultz CL. 2005. On the sister-group of mammals *Brasilodon* and *Brasilitherium* (Cynodontia, Probainognathia) from the Late Triassic of southern Brazil. *Revista Brasileira de Paleontologia* 8: 25–46. **24.7A** Currie PJ. 1977. A new haptodontine sphenacodont (Reptilia: Pelycosauria) from the Upper Pennsylvanian of North America. *Journal of Paleontology* 51: 927–942. **24.7B** Colbert EH, Broom R. 1948. The mammal-like reptile *Lycaenops.* *Bulletin of the American Museum of Natural History* 89: 353–404. **24.7C** Jenkins FA Jr. 1970. Cynodont postcranial anatomy and the "prototherian" level of mammalian organization. *Evolution* 24: 230–252. **24.7D** Jenkins FA Jr., Parrington FR. 1976. Postcranial skeletons of the Triassic mammals *Eozostrodon, Megazostrodon* and *Erythrotherium.* *Philosophical Transactions of the Royal Society of London B* 273: 387–431. **24.08** Kemp TS. 1982. *Mammal-Like Reptiles and the Origin of Mammals.* Elsevier, New York and Amsterdam. **24.10A-C** Crompton AW, Jenkins FA Jr. 1979. Origin of mammals. In Lillegraven et al. (eds) *Mesozoic Mammals: The First Two-Thirds of Mammalian History.* University of California Press, Berkeley. **24.10D** Sadava D, Hillis DM, Heller HC, Hacker SD. 2017. *Life: The Science of Biology,* 11th Ed. Sinauer Associates/Macmillan, New York. **24.12A** Lawlor TE. 1979. *Handbook to the Orders and Families of Living Mammals.* By permission of Barbara Lawlor. **24.12B** Meng QJ and 5 others. 2015. An arboreal docodont from the Jurassic and mammaliaform ecological diversification. *Science* 347: 764–768. **24.16** Ham AW, Cormack DW. 1979. *Histology,* 8th Ed. J.B. Lippincott, Philadelphia; Harrison RJ, Montagna EW. 1973. *Man,* 2nd Ed. Prentice-Hall, Englewood Cliffs, NJ. **24.18** Kardong, KV. 2012. Vertebrates: *Comparative Anatomy,* 6th Ed. McGraw-Hill, New York. **24.19** Hill RW, Wyse GA, Anderson M. 2016. *Animal Physiology,* 4th Ed. Sinauer Associates, Sunderland, MA. **24.20** Lee

MSY, Beck RMD. 2015. Mammalian evolution: A Jurassic spark. *Current Biology* 25: R753-R773.

Chapter 25   25.7A,B Rogers E. 1986. *Looking at Vertebrates.* Pearson Education, Ltd. **25.7C–E** Walker WF Jr, Liem K. 1994. *Functional Anatomy of Vertebrates*, 2nd Ed. Brooks/Cole, Cengage Learning. **25.8A** Goslow GE Jr, Dial KP, Jenkins FA Jr. 1989. The avian shoulder: An experimental approach. *American Zoologist* 29: 287–301. **25.8B** Kardong K. 2012. *Vertebrates: Comparative Anatomy, Function, Evolution*, 6th Ed. McGraw-Hill, New York. **25.9A,B** Jenkins FA Jr. 1974. *Primate Locomotion.* Elsevier, New York and Amsterdam. **25.9C** Biewener AA. 1989. Mammalian terrestrial locomotion and size. *BioScience* 39: 786–783. **25.10A** Jacobs, G. 2013. Losses of functional opsin genes, short-wavelength cone photopigments, and color vision: A significant trend in the evolution of mammalian vision. *Visual Neuroscience* 30: 39–53. **25.11** Lawlor TE. 1979. *Handbook to the Orders and Families of Living Mammals.* By permission of Barbara Lawlor. **25.13A,E,F** Hildebrand M, Goslow G. 1998. *Analysis of Vertebrate Structures*, 5th Ed. John Wiley & Sons. **25.13B,C,D** Renfree MB. 1993. Ontogeny, genetic control, and phylogeny of female reproduction in monotreme and therian mammals. In Szalay FS, Novacek MJ, McKenna MC (eds), *Mammal Phylogeny.* Springer, New York. **25.14** Ferner K, Mess A. 2011. Evolution and development of fetal membranes and placentation in amniote vertebrates. *Respiratory Physiology & Neurobiology* 178: 39–50. **25.15** Sears KE. 2004. Constraints on the morphological evolution of marsupial shoulder girdles. *Evolution* 58: 2353–2370. **25.19, 20, 21** Hildebrand M and Goslow G. 1998. *Analysis of Vertebrate Structures*, 5th Ed. John Wiley & Sons, New York. **25.23** Macdonald DW. 2009. *Encyclopedia of Mammals.* Oxford University Press, Oxford. ISBN: 9780199567997. **25.24** Zimmer C. 2001. *Evolution: The Triumph of an Idea.* Harper Collins, New York. **25.26B** Pigeon G, Festa-Bianchet M, Coltman DW, Pelletier F. 2016. Intense selective hunting leads to artificial evolution in horn size. *Evolutionary Applications* 9: 521–530.

Chapter 26   26.1 Romer AS. 1933. *Vertebrate Paleontology.* University of Chicago Press, Chicago. **26.2** Bloch JI, Boyer DM. 2002. Grasping primate origins. *Science* 298:1606–1610. **26.4** Springer MS and 11 others. 2012. Macroevolutionary dynamics and historical biogeography of primate diversification inferred from a species supermatrix. *PLoS ONE* 7(11): e49521; Ni X, Li Q, Li L, Beard KC. 2016. Oligocene primates from China reveal divergence between African and Asian primate evolution. *Science* 352: 673–677; Silcox MT, Bloch JI, Boyer DM, Chester SGB, López-Torres S. 2017. The evolutionary radiation of plesiadapiforms. *Evolutionary Anthropology* 26: 74–94. **26.5** Franzen JL and 5 others. 2009. Complete primate skeleton from the Middle Eocene of Messel in Germany: Morphology and paleobiology. *PLoS ONE* 4(5): e5723. **26.6** Fleagle JG. 2013. *Primate Adaptation and Evolution*, 3rd edition. Elsevier. **26.8** Beard KC, Qi T, Dawson MR, Wang B, Li C.1994. A diverse new primate fauna from middle Eocene fissure-fillings in southeastern China. *Nature* 368: 604–609; Zalmout IS and 12 others. 2010. New Oligocene primate from Saudi Arabia and the divergence of apes and Old World monkeys. *Nature* 466: 360–364; Zijlstra JS, Flynn LJ, Wessels W. 2013. The westernmost tarsier: A new genus and species from the Miocene of Pakistan. *Journal of Human Evolution* 65: 544–550; Ni X, Li Q, Li L, Beard KC. 2016. Oligocene primates from China reveal divergence between African and Asian primate evolution. *Science* 352: 673–677; **26.13** Whiten A. and 8 others. 1999. Cultures in chimpanzees. *Nature* 399: 682–685 and online supplemental information. **26.18** Fleagle JG. 2013. *Primate Adaptation and Evolution*, 3rd Ed. Elsevier. **26.25** Sockol MD, Raichlen DA, Pontzer H. 2007. Chimpanzee locomotor energetics and the origin of human bipedalism. *Proceedings of the National Academy of Sciences USA* 104: 12265–12269. Copyright © 2007 National Academy of Sciences USA. **26.27** Callaway E. 2016. Evidence mounts for interbreeding bonanza in ancient human species. *Nature News* 17 February 2016.

Index

Numbers in *italic* refer to information in an illustration or table.

ABOUT THE BOOK

Editor: Andrew D. Sinauer

Production Editors: Carol J. Wigg and Laura Green

Copyeditor: Elizabeth Pierson

Photo Editor: Mark Siddall

Rights and Permissions: Michele Beckta

Indexer: Grant Hackett

Production Manager: Christopher Small

Book Design: Joan Gemme

Cover Design: Joan Gemme

Book Production: Joan Gemme

Illustration Program: Elizabeth Morales

Book and Cover Manufacturer: Sheridan